AF382854

Progress in Mathematics

Volume 162

Series Editors

H. Bass
J. Oesterlé
A. Weinstein

Singularities

The Brieskorn Anniversary Volume

V.I. Arnold

G.-M. Greuel

J.H.M. Steenbrink

Editors

Springer Basel AG

Editors:

V.I. Arnold
Department of Geometry and Topology
Steklov Mathematical Institute
8, Gubkina Stree
117966 Moscow GSP-1
Russia

G.-M. Greuel
Fachbereich Mathematik
Universität Kaiserslautern
D-67653 Kaiserslautern
Germany

and

CEREMADE
Université Paris-Dauphine
Place du Maréchal de Lattre de Tassigny
Postfach 3049
F-75775 Paris Cedex 16e
France

J.H.M. Steenbrink
Subfaculteit Wiskunde
Katholieke Universiteit Nijmegen
Toernooiveld
NL-6525 ED Nijmegen
The Netherlands

1991 Mathematics Subject Classification 14B05, 32SXX, 58C27

A CIP catalogue record for this book is available from the Library of Congress, Washington D.C., USA

Deutsche Bibliothek Cataloging-in-Publication Data

Singularities : the Brieskorn anniversary volume / V. I. Arnold ... ed.
  (Progress in mathematics ; Vol. 162)
  ISBN 978-3-0348-9767-9   ISBN 978-3-0348-8770-0 (eBook)
  DOI 10.1007/978-3-0348-8770-0

© 1998 Springer Basel AG
Originally published by Birkhäuser Verlag, Basel, Switzerland in 1998
Printed on acid-free paper produced of chlorine-free pulp. TCF ∞

ISBN 978-3-0348-9767-9

9 8 7 6 5 4 3 2 1

*Dedicated to
Egbert Brieskorn
on the Occasion of
His 60th Birthday*

Prof. Dr. Egbert Brieskorn

Contents

Chapter 1: Classification and Invariants

Chapter 2: Deformation Theory

Chapter 3: Resolution

Contentsxi

Chapter 4: Applications

Preface

In July 1996, a conference in honour of the 60th birthday of Egbert Brieskorn was held at the Mathematische Forschungsinstitut Oberwolfach. It was organised by Gert-Martin Greuel, Vladimir Arnol'd and Joseph Steenbrink. Most people invited to the conference had been influenced in one way or another by Brieskorn's work in singularity theory. We were particularly happy to meet so many people from the Arnol'd school at the conference; it was the first time we were able to enjoy their contribution to this extent.

This volume contains papers on singularity theory and its applications, and almost all of them have been written by participants of the conference. In many cases, they are extended versions of the talks given at the conference. The diversity of subjects of the contributions reflects the positioning of singularity theory as a discipline between topology, analysis and geometry, combining ideas and techniques from all of these fields, as well as the broadness of interest of Brieskorn himself.

We have arranged the papers according to four main aspects of singularity theory: deformations, classification and invariants, resolution and applications. The latter category contains papers on such diverse topics as affine hypersurfaces, trigonometric functions, Fermion loops and knot theory.

Apart from the regular schedule of talks on singularity theory, one day of the conference consisted of lectures devoted to several aspects of the jubilant's mathematical career, and finished with a musical serenade. This part of the program has not been incorporated in this volume, the only exception being the first paper, "Some aspects of Brieskorn's mathematical work" by G.-M. Greuel.

All papers published here are original contributions, which will not be published elsewhere. All of them have been carefully refereed.

Some Aspects of Brieskorn's Mathematical Work

Gert-Martin Greuel
Universität Kaiserslautern
Fachbereich Mathematik
Erwin-Schrödinger-Str.
67663 Kaiserslautern
GERMANY
e-mail: greuel@mathematik.uni-kl.de

Lieber Egbert, dear colleagues and friends!

I am very happy that this conference on singularities at this wonderful Mathematisches Forschungsinstitut Oberwolfach can take place on the occasion of Egbert Brieskorn's 60th birthday, which was a little bit more than a week ago. There have been several conferences on singularities in Oberwolfach – but this is certainly a special one.

In this conference we have a special day, the "Brieskorn-day" today, and I am especially happy that this day became possible.

As you can clearly see from the programme, the today's speakers are Brieskorn's teacher Prof. Hirzebruch, four of Brieskorn's students and, of course, Prof. Brieskorn himself.

I am particularly grateful to Prof. Hirzebruch that when I asked him whether he could give a talk at this occasion, he did not hesitate but immediately said yes. He will give us many personal and exciting details of the wonderful discovery of the relation between exotic spheres and singularities.

I should also like to thank very much Heidrun Brieskorn, Matthias Kreck and Joseph Steenbrink for preparing a music programme for tonight. They immediately started to exercise when they arrived (even before!). I am sure that we shall have some wonderful music tonight. Thank you very much!

Last but not least I want to thank one person especially for his participation at this conference: it is Brieskorn himself. Actually, you can believe me that it was not a trivial task to convince him to come. As many of you know, Brieskorn does not like ceremonies like this, in particular if they concern his own person. Probably he already thinks I should stop talking now. In some sense I would like to agree. On the other hand, I am convinced that Egbert Brieskorn

deserves this special day in honour of the person and of his mathematical work
which was important

- for his students

- for the unfolding of singularity theory, and

- for mathematics as a whole.

Before I start to talk about some aspects of Brieskorn's mathematical work, let
me mention that Brieskorn is a person whose interest, knowledge and activities
reach far beyond mathematics.

He loves music and, by the way, knows a lot about the theory of music.
He is definitely a very political person with strong opinions. He was actively
engaged in the peace movement and is still engaged in projects for saving the
environment. As you know, during the past years, he has become a semipro-
fessional historian in connection with the life and the work of Felix Hausdorff.
Actually, he is the editor of the book "Felix Hausdorff zum Gedächtnis I".
And – who is surprised – there will soon be a second volume with Brieskorn's
biography of Felix Hausdorff.

The later work of Brieskorn, with important historical and philosophical
contributions, as well as his textbooks is, however, not subject of this short
overview.

Now let me start with a very short review of part of Brieskorn's mathematical work.

As you will see, the talks of today are all related to some mathematical theory
which nowadays is a grown-up theory, but where, sometime at the beginning,
there was a discovery of Brieskorn or a development of a germ of a theory by
Brieskorn.

Moreover, as I shall try to explain, in all of Brieskorn's work you see the
idea of unity of mathematics. Brieskorn's work is led by the idea to combine
different mathematical structures, different mathematical categories.

Historically speaking the following different structures are involved. (I hope
this will not be too schematic but casts some light on his work):

differential	—	*analytic*	(exotic spheres)
resolution	—	*deformation*	(simultaneous resolutions of ADE sin-gularities)
Lie groups	—	*equations*	(construction of singularities from the corresponding simple Lie groups)
transcendental	—	*algebraic*	(construction of the local Gauß-Ma-nin-connection)
continuous	—	*discrete*	(generalized Braid groups, Milnor lat-tices and Dynkin diagrams)

It is quite interesting to notice that perhaps in almost all cases these different structures correspond to the two parts of our brain, as Arnol'd explained in his talk yesterday.

Already in his first paper, which, as far as I know, emerged from his dissertation and which has the title

"Ein Satz über die komplexen Quadriken", Math. Annalen 155 (1964),

he proves:

Let X be a complex n-dimensional Kähler manifold diffeomorphic to Q_n (n-dimensional projective quadric), then

(i) n odd $\Rightarrow$ X is biholomorphic to Q_n

(ii) n even, $n \neq 2$ $\Rightarrow$ $c_1(X) = \pm ng$ ($H^2(X,\mathbb{Z}) \cong \mathbb{Z}g$, g positive) and if $c_1(X) = ng$, then X is biholomorphic to Q_n.

(If $n = 2$ then $X = \mathbb{P}^1 \times \mathbb{P}^1$ has infinitely many different analytic structures, the Σ_{2m} of Hirzebruch)

This was an exact analogue of a previous theorem of Hirzebruch and Kodaira about the complex projective space. (An earlier announcement of the result appeared 1961 in the Notices of the AMS.)

The next paper

"Über holomorphe $\mathbb{P}_n$-Bündel über $\mathbb{P}_1$", Math. Ann. 157 (1965)

treats the same question and gives a complete answer (including the Hirzebruch Σ-surfaces).

His next paper was

"Examples of singular normal complex spaces which are topological manifolds", Proc. Nat. Acad. Sci. 55 (1966).

This paper contains already the Brieskorn singularity

$$X : \quad z_0^3 + z_1^2 + \cdots + z_n^2 = 0 \,.$$

He proves: *if $n \geq 4$, n odd, then X is a topological manifold.*

The result at that time was a big surprise, since in 1961 Mumford had published in his well-known paper "The topology of normal singularities of an algebraic surface and a criterion for simplicity", Publ. Math. IHES 8 (1961), that such phenomena are not possible for surfaces.

Now, I should like to switch to the paper

"Beispiele zur Differentialtopologie von Singularitäten", Inventiones Math. 2 (1966).

The results of this paper were a sensation in the mathematical world.

Brieskorn showed that the just discovered exotic spheres (by Kervaire and Milnor) appear as neighbourhood boundaries of singularities and, therefore, can be described by real algebraic equations! As an example, I should like to mention that

$$\{x_1^{2k-1} + x_2^3 + x_3^2 + x_4^2 + x_5^2 = 0\} \cap S^9\,, \quad k = 1,\ldots,28,$$

represent all 28 different differentiable structures on the topological 7-sphere.

As far as I understood, the story of discovery of this result was also very exciting and Hirzebruch, who was himself involved in this discovery, will tell us very interesting details.

The next two papers

> **"Über die Auflösung gewisser Singularitäten von holomorphen Abbildungen"**, Math. Ann. 166 (1966),

and

> **"Die Auflösung der rationalen Singularitäten holomorpher Abbildungen"**, Math. Ann. 178 (1968),

contain a proof of the fact that the rational double points admit a simultaneous resolution (after base change), i.e., let X, S be smooth, $\dim X = 3$, $\dim S = 1$ and $f : X \to S$ be a morphism such that $\mathrm{Sing}\,(f) = \{x\}$ and $(X_{f(x)}, x)$ is a RDP, then there exists a diagram

$$
\begin{array}{ccc}
X' & \xrightarrow{\ \psi\ } & X \\
{\scriptstyle f'}\downarrow & & \downarrow{\scriptstyle f} \\
T & \xrightarrow[\ \varphi\ ]{} & S
\end{array}
$$

where X', T are smooth, $\mathrm{Sing}\,(f') = \emptyset$, φ is a smooth covering of S with $\varphi^{-1}(f(x)) = \{t\}$, ψ is proper, surjective and $\psi|_{X_t'}$ is a resolution of the singularities of $X_{f(x)}$.

Moreover, he describes all simultaneous resolutions in terms of invariants of the group G, defining the quotient singularity of type A_k, D_k or E_6, E_7, E_8. These two papers and the next one had an enormous influence on the deformation theory of 2-dimensional singularities as well as on the minimal model programme for 3-folds.

The paper

> **"Rationale Singularitäten komplexer Flächen"**, Inv. Math. 4 (1968),

contains a description of the resolution of quotient singularities and a proof that

$$\mathbb{C}\{x,y,z\}/(x^2 + y^3 + z^5)$$

is the only 2-dimensional factorial analytic ring which is not regular. The rational surface singularities and, in particular, Brieskorn's work about these play an important role in the subsequent deformation theory of surface singularities. I just mention Riemenschneider, Wahl and later, in connection with the minimal model programme of Mori, Kollar, Reid and others.

One of Brieskorn's shortest papers is certainly one of his most important ones:

"Singular elements of semi-simple Algebraic Groups", Intern. Congress Math. (1970).

In this famous paper Brieskorn shows how to construct the singularity of type ADE directly from the simple complex Lie group of the same type. Moreover, he constructs the whole semi-universal deformation.

At the end of that paper Brieskorn says:

"Thus we see that there is a relation between exotic spheres, the icosahedron and E_8."

which expresses explicitly Brieskorn's idea of the unity of mathematics. But he continues:

"But I still do not understand why the regular polyhedra come in."

I think that even today there is some mystery in these connections of such different parts of mathematics.

Peter Slodowy, who himself developed the theory of singularities and algebraic groups further, will give us a talk about this fascinating subject.

In 1970 Brieskorn published

"Die Monodromie der isolierten Singularitäten von Hyperflächen", Manuscripta Math. 2 (1970).

In this paper he constructed the local Gauß-Manin connection of an isolated hypersurface singularity.

This construction gave an algebraic method to compute the characteristic polynomial of the monodromy, and in this way combined topological and algebraic structures. In his paper Brieskorn proves that the eigenvalues of monodromy are roots of unity by a really very beautiful argument using the solution to Hilbert's 7th problem.

This was the time when I was a student in Göttingen, and it was my task to generalize his paper to complete intersections in my Diplomarbeit and later in my dissertation. Much later the work of Brieskorn was taken up by Scherk and Steenbrink and especially by Morihiko Saito who made a tremendous machinery out of it. Also Claus Herling continued Brieskorn's work and applied it to obtain theorems of Torelli type for singularities. He will explain to us the magic Brieskorn lattice H_0''.

Now, let me mention Brieskorn's work about "continuous versus discrete structures" which concerns (generalized) braid groups and actions of these.

In his paper

> **"Die Fundamentalgruppe des Raumes der regulären Orbits einer endlichen komplexen Spiegelungsgruppe"**, Inventiones Math. 12 (1971),

Brieskorn shows that the fundamental group of E_{reg}/W of regular orbits of a complex reflection group W has a presentation with generators g_s, $s \in S$, and relations

$$g_s g_t g_s \cdots = g_t g_s g_t \cdots$$

where both sides have m_{st} factors and where (m_{st}) is a Coxeter matrix. These groups are generalized braid groups and were called Artin-groups by Brieskorn and Saito in

> **"Artin-Gruppen und Coxeter-Gruppen"**, Inventiones Math. 17 (1972).

In that paper these groups were studied from a combinatorial point of view and the authors solved the word problem and the conjugation problem.

The connection to singularity theory comes from the fact that for W of type A_n, D_n, E_6, E_7, E_8 the space E_{reg}/W is the complement of the discriminant of the semi-universal deformation of a simple singularity of the same type. This follows from Brieskorn's work "Singular elements of semi-simple algebraic groups" at the International Congress in Nice.

Now, the classical braid group of n strings B_n acts on the set of "distinguished bases" of the Milnor lattice. In his later work Brieskorn and several of his students worked on this subject and Brieskorn expresses at several places that the understanding of this action should be essential for understanding the geometry of the versal unfolding.

The first step is to understand the deformation relations between singularities within a fixed modality class.

The classification of isolated hypersurface singularities with respect to their modality by V.I. Arnol'd is certainly one of the most important achievements of singularity theory. The adjacencies (deformation relations) between these singularities are important as well and still the subject of research articles.

In the paper

> **"Die Hierarchie der 1-modularen Singularitäten"**, Manuscripta Math. 27 (1979),

Brieskorn gives all possible deformation relations among Arnold's list of 1-modular (unimodal) singularities. The knowledge of all deformation relations is important by itself but is particularly interesting because of the different other characterizations of this class of singularities.

The deformation relations among the unimodular singularities were related by Brieskorn to a theory which seems to be really far away from singularity theory, the theory of partial compactifications of bounded symmetric domains. This is done in the survey article

"The unfolding of exceptional singularities", Nova Acta Leopoldina 52 (1981).

In the introduction, Brieskorn describes the fascinating relation between these apparently unconnected theories:

> *"On the one hand we have the deformation theory of the singularities in the boundary layer. It has three strata – corresponding to the simply elliptic singularities, the cusps and the exceptional singularities. And it has three stems, corresponding to the tetrahedron, the octahedron, and the icosahedron. On the other hand, we have the three quadratic forms $E_k \perp U \perp U$ obtained from the three exceptional forms E_6, E_7, E_8 by adding two hyperbolic planes. These three forms correspond to the three stems. To each of the forms is associated a bounded symmetric domain D of type IV and two unbounded realizations belonging to the 0- and 1-dimensional boundary components F_0 and F_1 of D. Corresponding to these there are canonically defined arithmetic quotients $D/\Gamma, D/Z_\Gamma(F_0)$ and $D/Z_\Gamma(F_1)$ and their partial Baily-Borel compactifications identify with the deformation spaces associated to the singularities in the three strata: exceptional singularities, cusps, and simply elliptic singularities."*

Still investigating the unimodal singularities which constitute, after the zeromodal (or simple or ADE) singularities, the next class in Arnold's hierarchy of singularities, Brieskorn gives a very fine and detailed study of the Milnor lattice of the 14 exceptional unimodal singularities in

"Die Milnorgitter der exzeptionellen unimodularen Singularitäten", Bonner Math. Schriften 150 (1983).

The Milnor lattice is the integral middle homology together with the quadratic intersection form with respect to a distinguished basis. It is an arithmetic coding of (part of) the geometry of the versal unfolding and it is a great challenge to see to what extent it reflects the essential features of this geometry. This problem which is embedded in a whole programme is again considered in the paper

"Milnor lattices and Dynkin diagrams", Proc. of Symp. in Pure Math. 40 (1983).

Brieskorn poses the question

> *"To which extent is this subtle geometry (the geometry of the unfold-ing of a singularity) reflected in the invariants associated to these singularities?"*

In the words of Arnol'd, this programme is the attempt to build a bridge between the two parts of our brain.

By the work of Gusein-Zade and Ebeling we understand a lot more, but I guess we are still far away from a complete understanding of the relation between the continuous and the discrete structure of a singularity.

Wolfgang Ebeling will talk on these themes.

The last paper I should like to mention is

"Automorphic Sets and Braids and Singularities", Contemporary Mathematics 78 (1988).

In this paper Brieskorn gives a survey on the action of the braid group on the set of distinguished bases of an isolated singularity. Moreover, he introduces the general concept of an automorphic set Δ, which unifies many investigations about the action of the braid group.

His statement in the introduction of that paper

> *"The beauty of braids is that they make ties between so many dif-ferent parts of mathematics, combinatorial theory, number theory, group theory, algebra, topology, geometry and analysis, and, last but not least, singularities."*

shows very clearly Brieskorn's strong belief in the unity of mathematics.

Now I come to the end of my talk. I hope I could explain some aspects of Brieskorn's mathematical work and point out that the idea or the wish to show the unity of mathematics, apart from its different realizations, was perhaps one of the leading principles of Brieskorn's work.

I should like to finish with a citation. In January 1992 there was a special colloquium in honour of Felix Hausdorff in Bonn. Brieskorn started his talk by citing Hausdorff. Hausdorff had spoken the following words at the grave of the mathematician Eduard Study and had cited Friedrich Nietsche from "Zaratustra" with the words.

> *"Trachte ich denn nach dem Glück?*
> *Ich trachte nach meinem Werke."*

(Do I aim for happiness? I do aim for my work.)

Brieskorn said, that this was certainly Hausdorff's leading principle and I should like to add, Brieskorn's too.

Publication List

Prof. Dr. Egbert Brieskorn

1. *Ein Satz über die komplexen Quadriken.* Math. Ann. **155**, 184–193 (1964).

2. *Über holomorphe $\mathbb{P}_n$-Bündel über $\mathbb{P}_1$.* Math. Ann. **157**, 343–357 (1965).

3. *Über die Auflösung gewisser Singularitäten von holomorphen Abbildungen.* Math. Ann. **166**, 76–102 (1966).

4. *Examples of singular normal complex spaces which are topological manifolds.* Proc. Nat. Acad. Sci. **55**, 1395–1397 (1966).

5. *Beispiele zur Differentialtopologie von Singularitäten.* Invent. Math. **2**, 1–14 (1966).

6. *Rationale Singularitäten komplexer Flächen.* Invent. Math. **4**, 336–358 (1968).

7. *Die Auflösung der rationalen Singularitäten holomorpher Abbildungen.* Math. Ann. **178**, 255–270 (1968).

8. *Some complex structures on products of homotopy spheres.* (With A. Van de Ven). Topology **7**, 389–393 (1968).

9. *Die Monodromie der isolierten Singularitäten von Hyperflächen.* Manuscripta Math. **2**, 103–161 (1970).

10. *Die Fundamentalgruppe des Raumes der regulären Orbits einer endlichen komplexen Spiegelungsgruppe.* Invent. Math. **12**, 57–61 (1971).

11. *Die Monodromie der isolierten Singularitäten von Hyperflächen.* Mathematika, Moskva **15**, No. 4, 130–160 (1971).

12. *Singular elements of semi-simple algebraic groups.* Actes Congr. intern. Math., Nice 1970, 2, 279–284 (1971).

13. *Sur les groupes de tresses, d'après V.I. Arnold.* Séminaire Bourbaki 1971/72, Exposé No. 401. Lecture Notes Math. 317, 21–44 (1973).

14. *Artin-Gruppen und Coxeter-Gruppen.* (With Kyoji Saito). Invent. Math. **17**, 245–271 (1972).

15. *Vue d'ensemble sur les problèmes de monodromie.* In: Singularités à Cargèse, Astérisque 7 et 8, 195–212 (1973).

16. *Artin-Gruppen und Coxeter-Gruppen.* (With Kyoji Saito). Matematika, Moskva **18**, No. 6, 56–79 (1974).

17. *Sur les groupes de tresses.* Matematika, Moskva **18**, No. 3, 46–59 (1974).

18. *Singularities of complete intersections.* (With G.-M. Greuel). In: Manifolds, Proc. Intern. Conf. Manifolds relat. Top. Topol., Tokyo 1973, 123–129 (1975).

19. *Die Fundamentalgruppe des Raumes der regulären Orbits einer endlichen komplexen Spiegelungsgruppe.* Uspehi mat. Nauk **30**, No. 6 (186), 147–151 (1975).

20. *Special singularities – resolution, deformation and monodromy.* A series of survey lectures given at the twenty-first Summer Research Institute of the AMS July 29 – August 16, 1974, Humboldt State Univ., Arcata/California. Duplicated typoscript, 96 pages.

21. *Über die Dialektik in der Mathematik.* In: Mathematiker über Mathematik. (Hrsg. M. Otte). Springer-Verlag, Berlin etc., 221–286, (1974).

22. *O dia lektice v matematic I–III.* (Czech translation). Pokropky Mat. Fyz. & Astr. **24**, 33–43, 89–103, 163–173 (1979).

23. *Singularitäten.* Jber. DMV **78**, 93–112 (1976).

24. *Die Hierarchie der 1-modularen Singularitäten.* Manuscripta Math. **27**, 183–219 (1979).

25. *The unfolding of exceptional singularities.* Nova acta Leopoldina, NF **52**, No. 240, 65–93 (1981).

26. *Ebene algebraische Kurven.* (With H. Knörrer). Birkhäuser, Basel, Boston, Stuttgart, XI, 964 pages (1981).

27. *Plane algebraic curves.* (English translation by John Stillwell). Birkhäuser, Basel, Boston, Stuttgart, VI, 721 pages (1986).

28. *Milnor lattices and Dynkin diagrams.* Singularities, Summer Inst., Arcata/Calif. 1981, Proc. Symp. Pure Math. **40**, Part 1, 153–165 (1983).

29. *Die Milnorgitter der exzeptionellen unimodularen Singularitäten.* Bonner Math. Schr. **150**, 225 pages (1983).

30. *Lineare Algebra und analytische Geometrie I.* Noten zu einer Vorlesung mit historischen Anmerkungen von Erhard Scholz. Vieweg & Sohn, Braunschweig, Wiesbaden, VIII, 636 pages (1983).

31. *Lineare Algebra und analytische Geometrie II.* Noten zu einer Vorlesung mit historischen Anmerkungen von Erhard Scholz. Vieweg & Sohn, Braunschweig, Wiesbaden, XIV, 534 pages (1985).

32. *Automorphic sets and braids and singularities.* Braids, AMS-IMS-SIAM Jt. Summer Res. Conf., Santa Cruz/Calif. 1986, Contemp. Math. **78**, 45–115 (1988).

33. Brieskorn, Egbert (ed.), *Felix Hausdorff zum Gedächtnis. Band I: Aspekte seines Werkes.* (In memoriam Felix Hausdorff. Vol. I: Aspects of his work). Vieweg, Wiesbaden, 286 pages (1986).

34. *Gustav Landauer und der Mathematiker Felix Hausdorff.* In: Gustav Landauer im Gespräch. Symposium zum 125. Geburtstag. Hrsg. Hanna Delf, Gert Mattenklott. Conditio Judaica, Band 18. Max Niemeyer Verlag, Tübingen, 105–128 (1997).

35. *Gibt es eine Wiedergeburt der Qualität in der Mathematik?* To appear in: E. Neuenschwander (ed.): Wissenschaft zwischen Qualitas und Quantitas. Birkhäuser, Basel, 138 pages (ca. 1998).

36. *Felix Hausdorff zum Gedächtnis. Band II: Elemente einer Biographie.* (In memoriam Felix Hausdorff. Vol. II: Elements of a biography.) To appear in Vieweg, Wiesbaden, ca. 550 pages (ca. 1998).

Chapter 1

Classification and Invariants

Progress in Mathematics, Vol. 162, © 1998 Birkhäuser Verlag Basel/Switzerland

On Schappert's Characterization of Strictly Unimodal Plane Curve Singularities

Yuri A. Drozd *
Faculty of Mechanics and Mathematics
Kyiv Taras Shevchenko University
252033 Kyiv
UKRAINE

Gert-Martin Greuel
Universität Kaiserslautern
Fachbereich Mathematik
Erwin-Schrödinger-Straße
67663 Kaiserslautern
GERMANY

Dedicated to Egbert Brieskorn on the occasion of his 60th birthday

Introduction

The representation theory of curve singularities (more precisely, of their local rings) has turned out to be closely related to their deformation properties. Namely, as was shown in [6],[9],[7], such a ring R is of *finite type*, that is has only finitely many torsion-free indecomposable modules (up to isomorphism), if and only if it dominates one of the so called *simple plane curve singularities* in the sense of [1]. In [4] the authors have shown that R is of *tame type*, that is it has essentially only 1-parameter families of indecomposable torsion-free modules, if and only if it dominates one of the *unimodal plane curve singularities* of type T_{pq} (T_{pq2} in the classification of [1]).

These singularities form the "serial" part of the list of all unimodal plane curve singularities. There are also 14 "exceptional" ones, which happen to be wild, that is they possess n-parameter families of (non-isomorphic) indecomposable modules for arbitrary large n. The *bimodal* plane curve singularities in

---

*Supported by DFG and International Science Foundation, grant RKJ000.

the sense of [1] are also wild. Nevertheless, in [11] was shown that all uni- and bimodal plane curve singularities possess only 1-parameter families of *ideals*. In [13] these singularities are called *strictly unimodal* and we prefer to use this terminology.

The aim of our paper is to show that the strictly unimodal plane curve singularities are in some sense "universal" among those having not more than 1-parameter families of ideals. Namely, we prove that a curve singularity has this property if and only if it dominates one of the strictly unimodal plane curve singularities. Moreover, we prove this result for curve singularities over an algebraically closed field of arbitrary characteristic. For this we use, instead of the definition of such singularities by the corresponding equations, their characterization via *parametrization* given in [11]. Note that it follows from [6],[9],[7] that a curve singularity has only finitely many non-isomorphic ideals if and only if it is of finite type (in contrast with the case of 1-parameter families). We use the parametric characterization as a *definition* in positive characteristic (with special care in characteristic 2) and call them *ideal-unimodal* in view of the main theorem of this paper.

The proof of this theorem follows the same scheme as that of the main result on tameness from [4]. Namely, we first introduce some "overring conditions" for the ring R and show that whenever they do not hold, R possess 2-parameter families of non-isomorphic ideals. Then we show that these conditions imply that R dominates a strictly unimodal plane curve singularity. To accomplish the proof, we need also to show that any strictly unimodal plane curve singularity has not more than 1-parameter families of ideals. But indeed, one can calculate all ideals of these rings. This has already been done in [11] and [12]. Although Schappert used the "definition via equations", one can verify (and we do it here for three most complicated examples) that his calculations depend only on the parametrization of these rings. This calculation of ideals shows that all strictly unimodal plane curve singularities really have only 1-parameter families of ideals. Moreover, using the parametrization, we can extend this result to curve singularities over algebraically closed fields of positive characteristic, that is, to ideal-unimodal singularities.

1   Preliminaries

Notation 1.1 Throughout this article we use the following notations:

- R denotes a complete local noetherian ring without nilpotent elements.

- Q its full ring of fractions.

- $\mathfrak{m} = \operatorname{rad} R$ its unique maximal ideal.

- $k = R/\mathfrak{m}$, the residue field of R.

- R_0 its *normalization*, i.e. its integral closure in Q.

- $R_i = \mathfrak{m}^i R_0 + R$ (a local ring for $i > 0$).

- $\mathfrak{m}_i = \mathfrak{m}^i R_0 + \mathfrak{m}$ (the maximal ideal of R_i for $i > 0$).

- $d(M) = \dim_{\mathbf{k}}(M/\mathfrak{m}M)$, the minimal number of generators of an R-module M.

- $d_i = d(R_i)$.

Later on we suppose $\mathbf{k}$ to be algebraically closed.

Definition 1.2 R is said to be a *curve singularity* provided it satisfies the following conditions:

1. R is a $\mathbf{k}$-algebra and $R/\mathfrak{m} = \mathbf{k}$.

2. R is of Krull dimension 1.

Such rings are just the completions of the local rings of points of reduced algebraic curves over the field $\mathbf{k}$.

It is known that, in this case, d_0 is finite and, moreover, $d(I) \leq d_0$ for each R-ideal I (cf. [3]).

Recall the definitions related to families of R-modules (cf. [5],[10]). We shall consider here only *full* R-*ideals*, i.e. ideals I, such that $QI = Q$ (later we omit the epithet "full").

Definition 1.3 Let X be an algebraic variety over $\mathbf{k}$ and $\mathcal{I}$ an $R \otimes \mathcal{O}_X$-ideal-sheaf, such that $Q\mathcal{I} = Q \otimes \mathcal{O}_X$ (the tensor product is over $\mathbf{k}$). Call $\mathcal{I}$ a *family of ideals with base* X if it is flat over $\mathcal{O}_X$ and if $\mathcal{I}/r\mathcal{I}$ is $\mathcal{O}_X$-flat for each non-zero divisor $r \in R$.

A series of such families, which are in some sense *universal*, can be constructed as follows. Consider the subvariety $\mathsf{B}(d)$ of the Grassmannian $\mathsf{Gr}(d, R_0/R)$, consisting of those subspaces, of R_0/R of codimension d, which are R-submodules in R_0/R. The pre-image $\mathcal{I}(d)$ in $R_0 \otimes \mathcal{O}_{\mathsf{B}(d)}$ of the canonical locally free sheaf of corank d on $\mathsf{B}(d)$ is then a family of R-ideals and any other family can be "glued" from the inverse images of the families $\mathcal{I}(d)$ (cf. [5], Proposition 3.5 and Corollary 3.6). Hence, we are able to define, following [10], the *number of parameters for* R-*ideals*, $\mathsf{par}(1, R)$.

Definition 1.4 Let $\mathsf{B}(d, i)$ the subset of $\mathsf{B}(d)$ consisting of points x such that the set $\{\, y \in \mathsf{B}(d) \,|\, \mathcal{I}(d)(y) \simeq \mathcal{I}(d)(x)\,\}$ (which is locally closed) has dimension i and define

$$\mathsf{par}(1, R) = \max_{d,i}\{\, \dim \mathsf{B}(d, i) - i \,\}.$$

Note that $\mathsf{B}(d, i)$ is also locally closed in $\mathsf{B}(d)$ and that, intuitively, $\mathsf{par}(1, R)$ is the maximal number of independent parameters of isomorphism classes of R-ideals (= torsion free R-modules of rank 1). Hence, $\mathsf{par}(1, R)$ may be considered as the maximal dimension of a component of a moduli space parametrizing non-isomorphic R-ideals.

We say that a ring R' *dominates* R if $R \subseteq R' \subseteq R_0$. In this case, evidently, $\mathrm{par}\,(1, R') \leq \mathrm{par}\,(1, R)$. It follows from [7], [3] that $\mathrm{par}\,(1, R) = 0$ if and only if R dominates one of the so-called *simple (or 0-modal) plane curve singularities* in the sense of [1] (cf. also [13], [8]). We are going to prove an analogous fact concerning the *strictly unimodal* plane curve singularities (cf. [1], [13]) [1] and extend this to arbitrary characteristic.

As $\mathbf{k}$ is algebraically closed, we may suppose that $R_0 = \prod_{i=1}^{s} D_i$, where $D_i = \mathbf{k}[[t_i]]$ (formal power series rings). The number s is called *the number of branches* of R. Let $t = (t_1, t_2, \ldots, t_s)$ and v_i be the standard valuation in D_i. For any element $r = (r_1, r_2, \ldots, r_s) \in R_0$ define its *(multi-)valuation* as the vector $\mathbf{v}(r) = (v_1(r_1), v_2(r_2), \ldots, v_s(r_s))$. In Table 1 we prefer to present the plane curve singularities in a parametrized form, that is given by their generators x, y as *complete* subalgebra of R_0. Such a presentation has the advantage that it is almost independent of the characteristic — only char $\mathbf{k} = 2$ needs extra conditions. In the table the valuations $\mathbf{v}(x)$ and $\mathbf{v}(y)$ are given.

In view of Theorem 2.1 we propose the following definition:

Definition 1.5 A plane curve singularity with complete local ring $R \subset R_0$ is called *ideal-unimodal* (IUS) if its maximal ideal admits generators x, y whose (multi-)valuation satisfies the conditions of Table 1.

According to Theorem 2.1, ideal-unimodal (IUS) is the same as strictly unimodal for plane curve singularities of characteristic 0. For char $\mathbf{k} > 0$ we wish to reserve the name strictly unimodal (SUS) to singularities defined by their deformation properties (as in [1], [13]). As, at the time of this writing, the strictly unimodal singualrities have not been classified in positive characteristic, we have to distinguish between IUS and SUS for char $\mathbf{k} > 0$.

We also use the following definition and notation.

Definition 1.6 Let $\{a_1, \ldots, a_d\}$, $a_i \in \mathfrak{m}$, represent a basis of $\mathfrak{m}/(\mathfrak{m} \cap t\mathfrak{m}R_0)$, $\mathbf{v}_j = \mathbf{v}(a_j)$. The set $\{\mathbf{v}_1, \ldots, \mathbf{v}_d\}$ will be called *a valuation type* of R and denoted by $\mathrm{val}\,(R)$.

2   Main theorem

2.1   Formulation

We pass now to the main theorem. In addition to the notations 1.1, let $\tilde{I} = t^2\mathfrak{m}R_0 + \mathfrak{m}$, $\tilde{R} = \mathrm{End}\,\tilde{I}$ and A_0 the 4-dimensional $\mathbf{k}$-algebra having a basis

---

[1]In [1] these singularities are called "uni-" and "bimodal" (with respect to right-equivalence), while in [13] they are called "strictly unimodal" (with respect to contact equivalence). We use the latter terminology, which is more adequate in our situation.

Table 1

Type	s	$\mathbf{v}(x)$	$\mathbf{v}(y)$	Condition	Name
E	1	(3)	(l)	$l = 7, 8, 10, 11$	$E_{12}, E_{14}, E_{18}, E_{20}$
	2	$(1, 2)$	(∞, l)	$l = 4, 5, 6, 7$	$E_{2,2p-1}, E_{13}, E_{3,2p-1}$ $(p \geq 1), E_{19}$
	3	$(1, 1, 1)$	(∞, k, l)	$l = 2, 3,\ k \geq l$	$E_{l,2(k-l)}$
T	2	$(2, k)$	$(l, 2)$	k, l odd, $lk > 4$	$T_{k+2,l+2,2}$
	3	$(1, 1, k)$	$(\infty, l, 2)$	k odd, $lk \geq 4$	$T_{k+2,2(l+1),2}$
	4	$(1, \infty, 1, k)$	$(\infty, 1, l, 1)$	$lk \geq 1$	$T_{2(k+1),2(l+1),2}$
W	1	(4)	(l)	$l = 5, 6, 7\ (*)$	$W_{12}, W^{\#}_{1,2p-1}$ $(p \geq 1), W_{18}$
	2	$(1, 3)$	(∞, l)	$l = 4, 5$	W_{13}, W_{17}
	2	$(2, 2)$	$(3, l)$	$l = 3 \qquad (*)$	$W_{1,0}, W^{\#}_{1,2p}\ (p \geq 1)$
				$l \geq 5, \text{odd}$	$W_{1,l-3}$
	3	$(1, 1, 2)$	$(\infty, 1, 3)$	$l \geq 2$	$W_{1,2l-3}$
Z	2	$(1, l)$	$(\infty, 3)$	$l = 4, 5, 7, 8$	$Z_{11}, Z_{13}, Z_{17}, Z_{19}$
	3	$(1, \infty, 2)$	$(\infty, 1, l)$	$l = 2, 3, 4, 5$	$Z_{0,p}, Z_{12}, Z_{1,2p-1}$ $(p \geq 1), Z_{18}$
	4	$(1, \infty, 1, 1)$	$(\infty, 1, l, k)$	$l = 1, 2,\ k \geq l$	$Z_{l-1,2(k-l)}$

$(*)$ If char $\mathbf{k} = 2$, extra conditions in case W are required: if $s = 1$ then $l = 5$, if $s = 2$, $v(x) = (2, 2)$ and $l = 3$ then $x^2 - y^3 \notin t^7 R_0$.
(This excludes W_{18} $(s = 1)$ and all $W^{\#}_{1,p}$ $(s = 1$ or $s = 2)$.)

$\{1, a, b, ab\}$ with $a^2 = b^2 = 0$ (these notations will be used only in the case char $\mathbf{k} = 2$).

Theorem 2.1 *Let* R *be a curve singularity. The following conditions are equivalent:*

1. $\mathrm{par}\,(1, \mathrm{R}) \leq 1$.

2. R *dominates a simple or ideal-unimodal plane curve singularity.*

3. *(a)* $\mathsf{d}(\mathsf{R}_0) \leq 4$;

   (b) $\mathsf{d}(\mathsf{R}_1) \leq 3$;

   (c) $\mathsf{d}(\mathsf{R}_2 + e\mathsf{R}) \leq 3$ *for any idempotent* $e \in \mathsf{R}_0$, *such that* $\mathsf{d}(e\mathsf{R}_0) = 1$ *(provided it exists);*

   (d) if $\mathsf{d}(\mathsf{R}_0) = 3$, *then* $\mathsf{d}(\mathsf{R}_3) \leq 2$;

   (e) if $char\,\mathbf{k} = 2$, *then* $\tilde{R}/\tilde{I} \not\simeq \mathsf{A}_0$.

The proof of Theorem 2.1 (3. $\Rightarrow$ 2.) implies i) of the following corollary:

Corollary 2.2

   i) *Let* R *be a plane curve singularity. Then* R *is simple or ideal-unimodal if and only if the equivalent conditions of theorem 2.1 are satisfied.*

   ii) *If char* $\mathbf{k} = 0$, *then* R *dominates a simple or strictly unimodal plane curve singularity if and only if the equivalent conditions of Theorem 2.1 are satisfied.*

Part ii) follows from the following parametric classification of strictly unimodal curve singularities.

Proposition 2.3 *Let char* $\mathbf{k} = 0$ *and* R *be a plane curve singularity. Then* R *is ideal-unimodal if and only if it is strictly unimodal.*

Remark 2.4 One can see that the condition 3(c) of the theorem means that either $\mathsf{d}(\mathsf{R}_2) \leq 2$ or $em \not\subseteq \mathfrak{m} + \mathfrak{m}^3 \mathsf{R}_0$.

It is perhaps worth giving a more geometric interpretation of these overring conditions.

- $\mathsf{R}_0 = \prod_{i=1}^{s} \mathbf{k}[[t_i]]$ is the normalization of R and $\mathsf{d}_0 = \mathsf{d}(\mathsf{R}_0)$ the usual multiplicity of the local ring R.

- $\mathsf{R}_1 = \mathbf{k} + \prod_{i=1}^{s} t_2^{m_i} \mathbf{k}[[t_i]]$ is the maximal local overring of R, having the same multiplicity vector $(m_1, \ldots, m_s)$ as R, where $m_i = $ multiplicity of the i-th branch; $\mathsf{d}_1 = \mathsf{d}(\mathsf{R}_1) = \dim_{\mathbf{k}}(\mathsf{R}_0/\mathsf{R}_1) + 1$.

- In general, we have $R_0 \supset R_1 \supset R_2 \supset \cdots \supset R$, $R_{i+1} = \mathbf{k} + \mathfrak{m}R_i$, hence $d_i = d(R_i) = \dim_{\mathbf{k}}(R_{i+1}/R_i) + 1$. R_i is the maximal local overring of R such that $R_i/\mathfrak{m}^i R_0 = R/\mathfrak{m}^i R_0$.

- If $e = (e_1, \ldots, e_s)$, $e_i \in \{0,1\}$, is an idempotent such as $d(eR_0) = 1$, then $e_i = 1$ for some i, $e_j = 0$ for $j \neq i$ and the i-th branch of R is nonsingular. Hence, $eR = \mathbf{k}[[t_i]]$ and $d(R_2 + eR) \leq 3$ is a condition on the remaining branches of R.

2.2 Proofs

Proof of Theorem 2.1. $1. \implies 3$. Suppose first that $d = d_0 \geq 5$. Consider the factoralgebra $A = R_0/\mathfrak{m}R_0$. If V is any subspace in A, then its pre-image $M(V)$ in R_0 is an R-submodule. Moreover, if V and U are two subspaces in A such that $AV = AU = A$, then, evidently, $M(V) \simeq M(U)$ if and only if $U = aV$ for some invertible element $a \in A$. Consider now the subset $\mathrm{Gr}_0(m, A)$ of the Grassmannian $\mathrm{Gr}\,(m, A)$, consisting of all such subspaces V that $AV = A$. Obviously, it is an open subset, hence, an algebraic variety over $\mathbf{k}$ of dimension $m(d - m)$. The algebraic group $G = A^*/k^*$ of dimension $d - 1$ is acting on this variety, and different orbits of this action correspond to non-isomorphic R-ideals. In particular, there are families of non-isomorphic ideals of dimension $\geq \dim \mathrm{Gr}_0(m, A) - \dim G = (m-1)(d-m-1)$ for $1 \leq m \leq d-1$. But, as $d > 4$, $\dim G \leq \dim \mathrm{Gr}_0(2, A) - 2$. In view of [5] (Corollary 3.9), this implies $\mathsf{par}\,(1, R) \geq 2$.

Let now $d(R_1) \geq 4$. Note that $\mathrm{rad}\,R_1 = \mathfrak{m}R_0$ and $R_1/\mathrm{rad}\,R_1 = \mathbf{k}$. Thus, the algebra $A' = R_1/\mathfrak{m}R_1$ is local with radical $J = \mathfrak{m}R_0/\mathfrak{m}R_1$. Moreover, $\mathfrak{m}R_1 \supseteq (\mathfrak{m}R_0)^2$, whence $J^2 = 0$. Then, for any subspace $W \subseteq J$, the subspace $V = \mathbf{k} + W$ is a subalgebra in A and its pre-image $M(V)$ is a subring of R_0. Hence, taking different subspaces $W \subseteq J$, we get non-isomorphic R-modules. As $\dim J = d_1 - 1$, $\dim \mathrm{Gr}\,(2, J) = m(d_1 - m - 1)$ for $1 \leq m \leq d_1$ and, hence, $\mathsf{par}\,(1, R) \geq 2$, if $d_1 \geq 4$.

Remark 2.5 The same observations show that $\mathsf{par}\,(1, R) \geq (\frac{d_0-2}{2})^2$ for d_0 even respectively $\geq (\frac{d_0-1}{2})(\frac{d_0-3}{2})$ for d_0 odd and if $d_0 = d_1$ $\mathsf{par}\,(1, R) \geq \frac{d_1}{2}(\frac{d_1-2}{2})$ for d_1 even respectively $\geq (\frac{d_1-1}{2})^2$ for d_1 odd.

To complete the proof, we need two simple lemmas.

Lemma 2.6 *Let $I, J \subset R$ be full ideals. Then*

$$dim\,(I/JI) = dim\,R_0/JR_0 - dim\,(JI/rI)$$

for any generator r of the principal R_0-ideal JR_0.

Proof. Since J is full, r is a nonzero divisor of R_0 and the snake lemma of

$$
\begin{array}{ccccccccc}
0 & \longrightarrow & I & \longrightarrow & R_0 & \longrightarrow & R_0/I & \longrightarrow & 0 \\
 & & \downarrow r & & \downarrow r & & \downarrow r & & \\
0 & \longrightarrow & I & \longrightarrow & R_0 & \longrightarrow & R_0/I & \longrightarrow & 0
\end{array}
$$

shows $\dim R_0/rR_0 = \dim I/rI$ which is equal to $\dim I/JI + \dim JI/rI$. $\square$

Lemma 2.7 *Suppose that* $d_0 = d_1 = \cdots = d_k$, $k \geq 1$. *Let* $r \in \mathfrak{m}$, *such that* $rR_0 = \mathfrak{m}R_0$. *Then*

1. $r, r^2, \ldots, r^k$ *form a basis of* $\mathfrak{m}_{k+1}/\mathfrak{m}^{k+1}R_0$.

2. $r\mathfrak{m}_i = \mathfrak{m}\mathfrak{m}_i$ *and* $\dim \mathfrak{m}_i/\mathfrak{m}\mathfrak{m}_i = d_0$ *for* $i = 0, \ldots, k$.

Proof. To prove the first assertion, consider the dimensions

$$
c_j = \dim \left(\mathfrak{m}_j / \mathfrak{m}^j R_0 \right)
$$

and note that for $j > 0$

$$
d_j = 1 + \dim \left(\mathfrak{m}_j / \mathfrak{m}_{j+1} \right) = 1 + c_j + \dim \left(\mathfrak{m}^j R_0 / \mathfrak{m}^{j+1} R_0 \right) - c_{j+1}.
$$

Evidently, $\dim \left(\mathfrak{m}^j R_0 / \mathfrak{m}^{j+1} R_0 \right) = d_0$ for all j. So, we have $c_{j+1} = c_j + 1$ for $1 \leq j \leq k$, whence, $c_j = j - 1$, $0 \leq j \leq k + 1$. In particular, $\dim \left(\mathfrak{m}_{k+1} / \mathfrak{m}^{k+1} R_0 \right) = k$. As, of course, the elements $r, r^2, \ldots, r^k$ are linear independent modulo $\mathfrak{m}^{k+1} R_0$, they form a basis of this vector space.

Now note that $r\mathfrak{m}_k \subseteq \mathfrak{m}\mathfrak{m}_k$ and $\dim \left(\mathfrak{m}_k / r\mathfrak{m}_k \right) = \dim \left(R_0 / \mathfrak{m}R_0 \right) = d_0$ in view of Lemma 2.6. But the result just obtained implies that

$$
\begin{aligned}
\dim \left(\mathfrak{m}_k / \mathfrak{m}\mathfrak{m}_k \right) &= \dim \left(\mathfrak{m}_k / \mathfrak{m}_{k+1} \right) + \dim \left(\mathfrak{m}_{k+1} / \mathfrak{m}\mathfrak{m}_k \right) \\
&= d_0 - 1 + \dim \left(\mathfrak{m}_{k+1} / (\mathfrak{m}^2 + \mathfrak{m}^{k+1} R_0) \right) = d_0.
\end{aligned}
$$

Therefore, $r\mathfrak{m}_k = \mathfrak{m}\mathfrak{m}_k$ $\square$

If $r^{j+1} b_i \in r^{j+1} R_0$, $i = 1, \ldots, l$, are a basis of $\mathfrak{m}_{j+1}/\mathfrak{m}_{j+2} = (\mathfrak{m} + r^{i+1} R_0)/(\mathfrak{m} + r^{j+1} R_0)$, then $r^j b_i$ are linear independent in $\mathfrak{m}_j/\mathfrak{m}_{j+1}$. Hence

Lemma 2.8 $d_{j+1} \leq d_j$ *for every* $j \geq 0$.

Now suppose that $d_0 = d_3 = 3$. Consider the factoralgebra $F = R_2/\mathfrak{m}^2\mathfrak{m}_2$. Choosing $r \in \mathfrak{m}$ as in Lemma 2.7, we see that $\mathfrak{m}^i\mathfrak{m}_2 = r^i\mathfrak{m}_2$ for all i, $\dim \left(\mathfrak{m}_2/r\mathfrak{m}_2 \right) = 3$ and $\dim F = 7$. Of course, $r \notin r\mathfrak{m}_2$, so we can choose $r, u, v \in \mathfrak{m}_2$ linear independent modulo $r\mathfrak{m}_2$. Then $\{1, r, u, v, r^2, ru, rv\}$ is a basis of F. Now F contains a 2-parameter family of subalgebras, containing the image of R (i.e. 1 and r), namely, the subalgebras $A(\lambda, \mu)$ with bases $\{1, r, u + \lambda v + \mu r v, r^2, r u + \lambda r v\}$. Then their pre-images in R_2 form a 2-parameter family of overrings of R, hence, of pairwise non-isomorphic R-ideals.

Remark 2.9 It follows from [3] that in this case we can only obtain *families of overrings*, as there are at most two non-isomorphic ideals with a fixed endomorphism ring.

At last, suppose that $\mathsf{d}(\mathsf{R}_2 + e\mathsf{R}) = 4$ for some idempotent $e \in \mathsf{R}_0$, such that $\mathsf{d}(e\mathsf{R}_0) = 1$. As we have noted, this means: $\mathsf{d}(\mathsf{R}_2) = 3$ and $e\mathfrak{m} \subseteq \mathfrak{m} + \mathfrak{m}^3\mathsf{R}_0$. Of course, the idempotent e is primitive and $e\mathsf{R} = e\mathsf{R}_0$. Denote $\mathsf{R}' = (1 - e)\mathsf{R}$; $\mathsf{R}'_k = (1 - e)\mathsf{R}_k$; $\mathsf{d}'_k = \mathsf{d}(\mathsf{R}'_k)$. Then $\mathsf{d}'_0 = \mathsf{d}'_1 = \mathsf{d}'_2 = 3$. Hence, we can apply Lemma 2.7 and choose an element $r \in \mathfrak{m}$, such that $\{r, r^2\}$ form a basis of $\mathfrak{m}\mathsf{R}'_2/\mathfrak{m}^3\mathsf{R}'_0$. Consider the factoralgebra $F = (e\mathsf{R} + \mathsf{R}_1)/(\mathfrak{m}^2 + \mathfrak{m}^3\mathsf{R}_0)$. If $\{r, u, v\}$ is a basis of $\mathfrak{m}\mathsf{R}'_0/\mathfrak{m}^2\mathsf{R}'_0$, then a basis of F can be chosen in the form: $\{1, e, r, u, v, r u, r v\}$. The subspaces $V(\lambda) = V(\lambda_0, \lambda_1, \lambda_2)$ with the bases $\{1, e + \lambda_0 u + \lambda_1 v + \lambda_2 r v, r, \lambda_0 r u + \lambda_1 r v\}$, where $\lambda_0 \neq 0$, form a 3-dimensional family of R-submodules in F. Thus, they define a 3-dimensional family $M(\lambda)$ of R-ideals. But it follows from [5] (proof of Proposition 3.6), that the ideals, isomorphic to some fixed $M(\lambda)$, form a subvariety of dimension $\dim(V(\lambda)/S(\lambda))$, where $S(\lambda) = \{a \in F \mid aV(\lambda) \subseteq V(\lambda)\}$. As $\dim V(\lambda) = 4$ and $S(\lambda)\{1, r, \lambda_0 r u + \lambda_1 r v\}$, this dimension is at most (indeed, equals) 1. Hence, again $\mathsf{par}(1, \mathsf{R}) \geq 2$.

If $\mathrm{char}\,\mathbf{k} = 2$ and $\tilde{\mathsf{R}}/\tilde{I} \simeq \mathsf{A}_0$, consider the subspaces $A(\lambda, \mu) \subset \mathsf{A}_0$ with bases $\{1, a + \lambda b + \mu a b\}$. They are subalgebras in A_0 (as $\mathrm{char}\,\mathbf{k} = 2$), hence, their pre-images in $\tilde{\mathsf{R}}$ form a 2-parameter family of overrings of R, hence, of non-isomorphic R-ideals $\qquad\qquad\square$

$3 \implies 2$. Take any ring R satisfying the conditions 3(a–e). It is known (cf. [3],[7])that if R has only finitely many non-isomorphic ideals then it dominates one of the simple pane curve singularities. So, we may supposethat R has infinitely many non-isomorphic ideals, i.e. $\mathsf{d}_0 \geq 3$ and if $\mathsf{d}_0 = 3$, then also $\mathsf{d}_1 = 3$ (cf. ibid.). Suppose first that $s = 1$, where s is the number of branches. Then the condition (a) implies that $\mathrm{val}(\mathsf{R}) = \{3\}$ or $\{4\}$. In the first case the condition (d) easily implies that R contains also an element of the valuation $l \in \{7, 8, 10, 11\}$. But then it dominates a SUS of type E (cf. the list). If $\mathrm{val}(\mathsf{R}) = \{4\}$, then the condition (b) impliesthat R contains an element of the valuation $5 \leq l \leq 7$, hence, dominates a SUS of type W. If $\mathrm{char}\,\mathbf{k} = 2$, the condition (e) implies also that R contains an element of the valuation 5: otherwise $\tilde{I} = t^6\mathsf{R}_0 + \mathbf{k} + \mathbf{k}t^4$, hence $\tilde{\mathsf{R}} = t^2\mathsf{R}_0 + \mathbf{k}$ and $\tilde{\mathsf{R}}/\tilde{I} \simeq \mathsf{A}_0$.

Let now $s = 2$. If $\mathsf{d}_0 = 3$ then $\mathrm{val}(\mathsf{R}) = \{(1, 2)\}$ (up to a numbering of the branches; later we omit this notice). Again the condition (d) implies that R contains an element of the valuation (∞, l) with $4 \leq l \leq 7$, hence, dominates a SUS of type E. Suppose that $\mathsf{d}_0 = 4$. Then the following cases can occur:

- $\mathrm{val}\,(\mathsf{R}) \supset \{(1,3)\}$. Then the condition (b) implies that R contains an element with the valuation either $(\infty,4)$ or $(\infty,5)$, hence, dominate a SUS of type W.

- $\mathrm{val}\,(\mathsf{R}) \supset \{(2,2)\}$. Again (b) implies that R contains an element with the valuation $(3,l)$, i.e. dominates a SUS of type W. Again, if char $\mathbf{k}=2$ and $x^3 - y^2 \in t^7 \mathsf{R}_0$, we get that $\tilde{\mathsf{R}}/\tilde{I} \simeq \mathsf{A}_0$.

- $\mathrm{val}\,(\mathsf{R}) = \{(2,k),(l,2)\}$. Then R dominates a SUS of type T.

- $\mathrm{val}\,(\mathsf{R}) = \{(1,l),(\infty,3)\}$. Now the condition (c) obviously implies that $l \le 8$ (and not 6), hence, R dominates a SUS of type Z.

If $s = \mathsf{d}_0 = 3$, then $\mathrm{val}\,(\mathsf{R}) = \{(1,1,1)\}$ and the condition (d) implies that R contains am element of the valuation $(\infty,3,l)$, i.e. dominates a SUS of type E. Let $s = 3$, $\mathsf{d}_0 = 4$. If $\mathrm{val}\,(\mathsf{R})$ consists of only one vector, then it is $(1,1,2)$ and the condition (b) implies immediately that R contains also an element of the valuation $(\infty,3,l)$, hence, dominate a SUS of type W. If $\mathrm{val}\,(\mathsf{R})$ consists of 2 vectors, then there are the following possibilities:

- $\mathrm{val}\,(\mathsf{R}) = \{(1,1,k),(\infty,l,2)\}$. Then R dominates a SUS of type T.

- $\mathrm{val}\,(\mathsf{R}) = \{(1,\infty,l),(\infty,1,2)\}$. Then the condition (c) implies that $l \le 5$, hence, R dominates a SUS of type Z.

If $\mathrm{val}\,(\mathsf{R})$ consists of 3 vectors, they may be chosen as

$$\{(1,\infty,k),(\infty,1,l),\ (\infty,\infty,2)\}$$

and R dominates a SUS of type T.

At last, let $s = 4$. The condition (b) implies that $\mathrm{val}\,(\mathsf{R})$ has at least 2 vectors. If there are really only 2 of them, then either $\mathrm{val}\,(\mathsf{R}) = \{(1,\infty,k,1),(\infty,1,1,l)\}$ or $\mathrm{val}\,(\mathsf{R}) = \{(1,\infty,1,1),(\infty,1,k,l)\}$. In the first case R dominates one of the SUS of type T, while in the latter case the condition (c) implies that $k \le 2$, thus R dominates a SUS of type Z. Finally, if $\mathrm{val}\,(\mathsf{R})$ contains 3 vectors, one can easily see that R dominates a SUS of type T $\square$

Proof of Proposition 2.3. The parametric form of an SUS, as computed by Schappert ([11]), satisfies the conditions of Table 1, hence it is an IUS.

Let R be an IUS. In the following subsection we classify curve singularities of multiplicity 3 and 4 given in parametric form. This implies the sixth column of Table 1 (notations of [1]), hence ideal-unimodal singularities are strictly unimodal.

2.3 Classification in characteristic 0

Let char $\mathbf{k} = 0$. Any local ring of a plane curve singularity of multiplicity 3 or 4 admits generators x, y with $v(x)$ and $v(y)$ as in columns 3 and 4 of Table 1, together with a few more cases. Starting from these valuations, we deduce a relation between x and y and determine its position in Arnold's list of singularities [1], classifying in this way all parametrizations of plane curve singularities of multiplicity 3 and 4.

We proceed by increasing s, the number of branches.

One branch $(s = 1)$:

- (Multiplicity 3) $v(x) = 3, v(y) = l$:
  after a change of the parametrization, we may assume (since char $\mathbf{k} = 0$) that

$$x = t^3, \ y = t^l + t^m p(t^3),$$

  $l \not\equiv 0(3)$, $m + l \equiv 0(3)$, $l \geq 4$, $p(0) \neq 0$. These satisfy the relation

$$y^3 - 3yx^{(l+m)/3}p(x) - x^l - x^m p^3(x).$$

  Following Arnold's determinator ([1], Para. 16), we obtain

$$\begin{aligned} l &= 3k+1 &&\Rightarrow \mathsf{E}_{6k}, \\ l &= 3k+2 &&\Rightarrow \mathsf{E}_{6k+2}. \end{aligned}$$

  (Only $k = 2, 3$ give strictly unimodal singularities.)

- (Multiplicity 4) $v(x) = 4, v(y) = l$:
  we may assume

$$x = t^4, \ y = t^{4k+1}p(t) + t^{4n+2}q(t) + t^{4m+3}r(t),$$

  $k, n, m \geq 1$, $p, q, r \in \mathbf{k}[[t^4]]$, $l = \min\{4k+1, 4n+2, 4m+3\}$.
  We obtain the relation $(p, q, r \in \mathbf{k}[[x]])$,

$$\begin{aligned} y^4 &- y^2(2q^2x^{2n+1} + 4prx^{k+m+1}) \\ &- y(4p^2qx^{2k+n+1} + 4qr^2x^{2m+n+2}) \\ &- p^4x^{4k+1} + q^4x^{4n+2} - 4pq^2rx^{k+2n+m+2} + 2p^2r^2x^{2k+2m+2} - r^4x^{4m+3}. \end{aligned}$$

  The determinator gives

$$\begin{aligned} l &= 4k+1 &&\Rightarrow \mathsf{W}_{12k}, \\ l &= 4n+2 &&\Rightarrow \mathsf{W}^{\#}_{n,2i-1}, \ i = 2(k-n) \text{ if } k \leq m, \\ & && \qquad\qquad\quad\, i = 2(m-n)+1 \text{ if } k > m, \\ l &= 4m+3 &&\Rightarrow \mathsf{W}_{12m+6}. \end{aligned}$$

  (Only $k = n = m = 1$ provide strictly unimodal singularities.)

Two branches $(s = 2)$:

- (Multiplicity 3) $v(x) = (1, 2)$, $v(y) = (\infty, l)$:

$$x = (t_1, t_2^2), \ y = (0, t_2^l + t_2^m p(t^2)),$$

$l \geq 3, l + m \equiv 1(2), p(0) \neq 0$, with relation $(p \in K[[x]])$

$$y(y^2 + x^{2k} - x^{2k+2j-1}p^2 - 2x^k y) \text{ if } l = 2k, m = 2k + 2j - 1$$

and

$$y(y^2 + x^{2k+1} - x^{2(k+j)}p^2 - 2x^{k+j}yp) \text{ if } l = 2k + 1, m = 2(k + j).$$

$l = 2k, m = 2k + 2j - 1 \Rightarrow \mathsf{E}_{k,2j-1}, j \geq 1 \ (\mathsf{J}_{k,2j-1} \text{ in } [1])$
$l = 2k + 1 \Rightarrow \mathsf{E}_{6k+1}.$
(Strictly unimodal for $k = 2, 3, D_4$ for $l = 2$.)

- (Multiplicity 4, 2 singular branches, 2 tangents) $v(x) = (2, k)$, $v(y) = (l, 2)$:

$$x = (t_1^2, t_2^{2m+1}p(t_2^2) + t_2^{2m+2}q(t_2^2)), \ y = (t_1^{2n+1}, t_2^2)$$

$k = 2m + 1, \ l = 2n + 1, m \geq 1, n \geq 1$ with relation $(p, q \in \mathbf{k}[[y]], p(0) \neq 0, q(0) \neq 0)$

$$(y^2 - x^{2n+1})\,(x^2 - y^{2m+1}p^2 + y^{m+2}y^2 - 2xy^{m+1}q),$$

$\Rightarrow \mathsf{T}_{k+2,l+2,2}$
(even for $k = 1$ or $l = 1$ if $kl > 4$. Note that $\mathsf{T}_{3,p+6,2} = \mathsf{E}_{2,p}$).

- (Multiplicity 4, 1 singular branch, 1 tangent) $v(x) = (1, 3)$, $v(y) = (\infty, l)$:

$$x = (t_1, t_2^3), \ y = (0, t_2^{3k+1}p(t_2^3) + t_2^{3n+2}q(t_2^3) + t_2^{3m+3}r(t_2^3))$$

$k, n, m \geq 1, \ l = \min\{3k + 1, 3n + 2, 3m + 3\}$ with relation

$$y(y^3 - x^{3k+1}p^3 - 3x^{k+n+1}ypq - x^{3n+2}q^3 - 3x^{m+1}y^2 r \\ +3x^{k+n+m+2}pqr + 3x^{2m+2}yr^2 - x^{3m+3}r^3).$$

$l = 3k + 1 \Rightarrow \mathsf{W}_{12k+1},$
$l = 3n + 2 \Rightarrow \mathsf{W}_{12n+5},$
$l = 3m + 3, n \geq k > m \Rightarrow \mathsf{Z}^{m+1}_{6(m+k)-5}.$

- (Multiplicity 4, 2 singular branches, 1 tangent) $v(x) = (2, 2)$, $v(y) = (2k + 1, l)$:

$$x = (t_1^2, t_2^2), \ y = (t_1^{2k+1}, t_2^{2n+1}p(t_2^2) + t_2^{2m}q(t_2^2)),$$

$k \geq 1$, $l = \min\{2n+1, 2m\}$ with relation $(p, q \in K[[x]],\ p(0) \neq 0,\ q(0) \neq 0)$

$$(y^2 - x^{2k+1})\,(y^2 - x^{2n+1}p^2 - 2x^m y^q + x^{2m}q^2).$$

$$
\begin{aligned}
k = n < m \quad &\Rightarrow \quad \mathsf{W}_{k,0}\ (p(0) \neq 1),\ \mathsf{W}^{\#}_{k,2j}\ (p(0) = 1) \\
&\qquad \text{with } j = ord(y^2(t_2) - t_2^{2k+1}) - (2k+1), \\
k < n, m \quad &\Rightarrow \quad \mathsf{W}_{k,2(n-k)}, \\
k \geq m \geq 2 \quad &\Rightarrow \quad \mathsf{Y}^m_{r,s}\ (1 \leq s \leq r).
\end{aligned}
$$

- (Multiplicity 4, 1 singular branch, 2 tangents) $v(x) = (1, l)$, $v(y) = (\infty, 3)$:

$$x = (t_1, t_2^l + t_2^k p(t_2^3)),\ y = (0, t_2^3),$$

$k > l \geq 4$, $k + l \equiv 0(3)$, with relation $(p \in \mathbf{k}[[y]], p(0) \neq 0)$

$$y(x^3 - y^l - y^k p^3 - 3xy^{(k+l)/3}p).$$

$l = 3p + 1 \Rightarrow \mathsf{Z}_{6p+5}$,
$l = 3p + 2 \Rightarrow \mathsf{Z}_{6p+7}$,
$(l = 3 : \mathsf{E}_{6j}$ (if $k = 3j + 1$), E_{6j+2} (if $k = 3j + 2$)).

Three branches $(s = 3)$:

- (Multiplicity 3, 1 tangent) $v(x) = (1, 1, 1)$, $v(y) = (\infty, l, k)$:

$$x = (t_1, t_2, t_3),\ y = (0, t_2^l, t_3^{l+n}p(t_3)),$$

$l \geq 2, n \geq 0$, with relation $(p \in K[[x]], p(0) \neq 0)$

$$y(x^l - y)(x^{l+n}p - y).$$

$n = 0 \Rightarrow \mathsf{E}_{l,0}\ (p(0) \neq 1)$,
$n > 0 \Rightarrow \mathsf{E}_{l,2n}\ (\mathsf{J}_{l,2n}$ in [1].)

- (Multiplicity 4, 2 smooth branches with same tangent, 2 tangents) $v(x) = (1, 1, k)$, $v(y) = (\infty, l, 2)$:

$$x = (t_1, t_2, t_3^k q(t_3^2)),\ y = (0, t_2^l p(t_2), t_3^2),$$

k odd, $kl \geq 2$; relation $(p(0) \neq 0, q(0) \neq 0)$

$$y(y - x^l p(x))(x^2 - y^k q^2(y)).$$

$\Rightarrow \mathsf{T}_{k+2,2(l+1),2}.$

- (Multiplicity 4, 2 smooth branches, 1 tangent) $v(x) = (1, 1, 2)$, $v(y) = (\infty, l, 2n + 1)$,

$$x = (t_1, t_2, t_3^2), \ y = (0, t_2^l p(t_2), t_3^{2n+1} q(t_3^2)),$$

$l \geq 2, n \geq 1$; relation $(p(0) \neq 0, q(0) \neq 0)$

$$y(y - x^l p(x))(y^2 - x^{2n+1} q^2(x)),$$

$l \geq n + 1 \Rightarrow W_{n,2(l-n)-1}$,
$l \leq n \Rightarrow Z_{6(n-l)}^{12l}$.

- (Multiplicity 4, 2 smooth branches with different tangents, 2 tangents) $v(x) = (1, \infty, 2)$, $v(y) = (\infty, 1, 2n + 1)$:

$$x = (t_1, 0, t_3^2), \ y = (0, t_2, t_3^l + t_3^k q(t_3^2)),$$

$k > l \geq 2$, relation $(q(0) \neq 0)$ for $l = 2j + 1$, k even:

$$xy(y^2 - x^l - x^k q^2 - 2yx^{k/2}q + 2x^k q^2).$$

$\Rightarrow Z_{6j+6}$,
for $l = 2j$ we have the relation

$$xy(y^2 + x^l - x^k q^2 - 2x^j y)$$

$\Rightarrow Z_{j-1,k-l} \ (Z_{0,p} = X_{1,p} = T_{4,4+p,2})$.

Four branches $(s = 4)$:

- (4 tangents, 3 tangents, or 2 tangents, each belonging to 2 branches) $v(x) = (1, \infty, 1, k)$, $v(y) = (\infty, 1, l, 1)$:

$$x = (t_1, 0, t_3, t_4^k p(t_4)) \ y = (0, t_2, t_3^l, t_4),$$

$k, l \geq 1$; relation

$$yx(x^l - y)(x - y^k p).$$

$\Rightarrow T_{2(k+1),2(l+1),2}$.

- (2 tangents, one belonging to 3 branches) $v(x) = (1, \infty, 1, 1)$, $v(y) = (\infty, 1, l, k)$:

$$x = (t, 0, t, t), \ y = (0, t, t^l q(t), t^k p(t)),$$

$k \geq l \geq 1$; relation $(p(0) \neq 0, q(0) \neq 0)$

$$yx(x^l q - y)(x^k p - y).$$

$\Rightarrow Z_{l-1,2(k-l)}$.

- (1 tangent) $v(x) = (1,1,1,1), v(y) = (\infty, k, l, m)$:

$$x = (t_1, t_2, t_3, t_4), \quad y = (0, t^k, t^l p(t), t^m q(t)),$$

$1 < k \leq l \leq m$; relation $(p(0) \neq 0, q(0) \neq 0)$

$$y(x^k - y)(x^l p - y)(x^m q - y).$$

$$k < l \leq m \quad \Rightarrow \quad \mathsf{Z}^k_{l-k-1, 2(m-l)},$$
$$k = l < m \quad \Rightarrow \quad \mathsf{X}_{k,j}(j \geq 0) \text{ or } \mathsf{Y}^k_{r,s}(1 \leq s \leq r).$$
(None is strictly unimodal.)

3   Ideals of ideal-unimodal plane curve singularities

Now we have to prove the implication $2 \Longrightarrow 1$, that is to show that any IUS from Table 1 has only 1-parameter families of ideals. Indeed, this was done in Schappert's paper [12]. Though Schappert supposed that $\operatorname{char} \mathbf{k} = 0$ and used another definition of IUS, one can check that his calculations are valid for our list too, independent of the characteristics. To demonstrate it and because Schappert's thesis has not been published, we show below examples of such calculations (in somewhat different form). Moreover, we have chosen the most complicated cases.

Our calculations are based on the following simple observation (cf. [3]). Let R be a curve singularity, $\mathfrak{m}$ its maximal ideal and $\mathsf{S} = \operatorname{End}_\mathsf{R}(\mathfrak{m}) = \{a \in \mathsf{Q} \mid a\mathfrak{m} \subseteq \mathfrak{m}\}$. We keep these notations through the whole section and also put $\mathfrak{n} = \operatorname{rad} \mathsf{S}$, $\mathsf{S}' = \operatorname{End}_\mathsf{S}(\mathfrak{n})$. For any R-ideal I, $\mathsf{S}I$ is an S-ideal and $\mathfrak{m}I = \mathfrak{m}\mathsf{S}I \subset I \subseteq \mathsf{S}I$. Consider the vector space $V = V(I) = I/\mathfrak{m}I$. It is a *generating subspace* in $W = \mathsf{S}I/\mathfrak{m}\mathsf{S}I$, i.e. such that $FV = W$, where $F = \mathsf{S}/\mathfrak{m}$, and one can easily check that $I \simeq I'$ if and only if $\bar\gamma V(I) = V(I')$ for some map $\bar\gamma : F \to F$ induced by an automorphisms γ of $\mathsf{S}I$. Moreover, $\mathsf{S} \neq \mathsf{R}$, whenever R is not a discrete valuation ring. Therefore, we can calculate the ideals "inductively", ascending by overrings. Note also that any plane curve singularity R is *Gorenstein* [2], i.e. $\operatorname{inj.dim}_\mathsf{R}\mathsf{R} = 1$. Hence, R has as only minimal overring R' and any R-ideal is either principal or an R'-ideal. In particular, $\operatorname{par}(1, \mathsf{R}) = \operatorname{par}(1, \mathsf{R}')$.

Note that all IUS of type T are known to be *tame*, i.e. have at most 1-parameter families of indecomposable torsion-free modules of any rank [4]. So we have to consider only the IUS of types W, Z and E.

3.1   Ideals of singularities of type W

Here we consider the case, when $s = 1$ and R contains elements x, y with $v(x) = 4$, $v(y) = 7$ ("type W_{18}" in Arnold's classification). It is convenient to suppose here that $t = x^2/y$. Of course, we suppose also that $\operatorname{char} \mathbf{k} \neq 2$. It is easy to verify that then $R \supset t^{18}R_0$ and, if R is a plane curve singularity, its minimal overring contains even $t^{14}R_0$. Hence, we may restrict ourselves by the case, when R is the smallest subalgebra of R_0 containing $t^{14}R_0$ and generated modulo $t^{14}R_0$ by x and y. Thus,

$$R = \langle 1, x, y, x^2, xy, x^3 \rangle + t^{14}R_0 \,,$$

where, as usual, we denote by $\langle a_1, \ldots, a_m \rangle$ the k-subspace generated by $a_1, \ldots, a_m$. An obvious calculation shows that in this case

$$S = \langle 1, x, y, x^2 \rangle + t^{10}R_0$$

and

$$S' = \langle 1, z, x \rangle + t^6 R_0 \,,$$

where $z = y/x$. As $v(z) = 3$, it follows from [6] or [9] that S' has only finitely many non-isomorphic ideals (it is the simple plane curve singularity of type E_6). Moreover, in these articles the precise list of such ideals is given. Namely, they are, except R_0 and S' itself:

$$A = S'\langle 1, t^5 \rangle, \quad A^* = S'\langle 1, t \rangle, \quad A' = S'\langle 1, t^2 \rangle \,.$$

Here A and A' are overrings of S' and A^* the module dual to A with respect to the duality described e.g. in [3].

Now, for each of these ideals, say M, we have to calculate $M/\mathfrak{n}M$. Here is their list (we write "r" for the image of an element $r \in M$ in $M/\mathfrak{n}M$ too):

$$S' = \langle 1, z, z^2, z^3 \rangle, \quad A = \langle 1, z, t^5, t^6 \rangle, \quad A^* = \langle 1, t, z, t^6 \rangle,$$
$$A' = \langle 1, t^2, z, t^5 \rangle, \quad R_0 =: \langle 1, t, t^2, t^3 \rangle \,.$$

Now one can easily write down all generating subspaces V from each $W = M/\mathfrak{n}M$ (up to automorphisms of M). Consider the case $M = A^*$ (the most complicated). First, as V is generating, it has to contain at least two elements of the form: $a_1 = 1 + \lambda_1 z + \lambda_2 t^5$ and $a_2 = t^2 + \mu_1 z + \mu_2 t^5$. Dividing by a_1 (which is the image of an invertible element from A'), we may suppose that $a_1 = 1$. Suppose that $V = \langle a_1, a_2 \rangle$. If $\mu_1 \neq 0$, one can replace V by $(1 - (\mu_1 + \mu_2)(2\mu_1)^{-1})V$, thus obtaining a subspace of the same form but with $\mu_2 = 0$. Hence, we get two 1-parameter families of S-ideals:

$$F_1(\mu) = S\langle 1, t^2 + \mu z \rangle, \quad F_2(\mu) = S\langle 1, t^2 + \mu t^5 \rangle \,.$$

Suppose now that V contains another element, which can be chosen in the form $b = \eta_1 z + \eta^2 t^5$. If $\eta_1 \neq 0$, we may suppose that $\mu_1 = 0$ and then, multiplying by $1 - \eta_2/\eta_1 z$, also $\eta_2 = 0$, i.e. $b = z$. But then, multiplying by $1 - \mu_2 t^3$, we get $\mu_2 = 0$, which gives only one more ideal:

$$I_1 = S\langle 1, t^2, z \rangle \,.$$

If $\eta_1 = 0$, we may suppose $\mu_2 = 0$, obtaining 1-parameter family:

$$F_3(\mu) = \mathsf{S}\langle 1, t^2 + \mu z, t^5 \rangle .$$

Of course, we have to add to those families also the ideal A' itself (correspond-ing to $V = W$). But we can remark that $\mathsf{d}(\mathsf{A}') \leq 4$, hence $\dim \mathsf{A}'/\mathsf{m}\mathsf{A}' \leq 4$ and $\mathsf{m}\mathsf{A}' = \mathsf{n}\mathsf{A}'$. Thus, any generating subspace in $\mathsf{A}'/\mathsf{m}\mathsf{A}'$ (with respect to S) must coincide with $\mathsf{A}'/\mathsf{m}\mathsf{A}'$ itself. Therefore, we need not consider the case $\mathsf{S}I = \mathsf{A}'$ when calculating the R-ideals. The same argument is valid, of course, in each case, when we have an S-ideal L such that $\mathsf{n}L = \mathsf{m}L$ (it is always the case, if $\dim(L/\mathsf{n}\mathsf{S}) = \mathsf{d}_0$): we may exclude them while calculating the R-ideals. In particular, here we may exclude all S'-ideals.

Quite analogous observations give us the following list of S-ideals (which are not S'-ideals):

$\mathsf{S}'L = \mathsf{S}'$:

$$F_4(\mu) = \mathsf{S}\langle 1, z + \mu z^3 \rangle, \quad F_5(\mu) = \mathsf{S}\langle 1, z^2 + \mu z^3 \rangle, \quad \mathsf{S} = \mathsf{S}\langle 1 \rangle,$$
$$I_2 = \mathsf{S}\langle 1, z^3 \rangle, \quad I_3 = \mathsf{S}\langle 1, z, z^2 \rangle, \quad I_4 = \mathsf{S}\langle 1, z, z^3 \rangle, \quad I_5 = \mathsf{S}\langle 1, z^2, z^3 \rangle .$$

$\mathsf{S}'L = \mathsf{A}$:

$$F_6(\mu) = \mathsf{S}\langle 1, z + \mu t^5 \rangle, \ \mu \neq 0, \quad F_7(\mu) = \mathsf{S}\langle 1, z + \mu t^5, t^6 \rangle, \ \mu \neq 0,$$
$$F_8(\mu) = \mathsf{S}\langle 1, t^5 + \mu t^6 \rangle, \quad F_9(\mu) = \mathsf{S}\langle 1, z, t^5 + \mu t^6 \rangle, \quad I_6 = \mathsf{S}\langle 1, t^5, t^6 \rangle .$$

$\mathsf{S}'L = \mathsf{A}^*$:

$$F_{10}(\mu) = \mathsf{S}\langle 1, t + \mu z \rangle, \quad F_{11}(\mu) = \mathsf{S}\langle 1, t + \mu z, t^6 \rangle, \quad I_7 = \mathsf{S}\langle 1, t, z \rangle .$$

$\mathsf{S}'L = \mathsf{R}_0$:

$$I_8 = \mathsf{S}\langle 1, t, t^2 \rangle .$$

Now we can pass to R-ideals I. Put $M = \mathsf{S}I$. It would be one of the ideals $F_{1-11}(\mu)$, I_{1-8} or S. One can easily check that in the following cases $\mathsf{m}M = \mathsf{n}M$, so we need not to consider them:

$$
\begin{array}{ll}
F_i(\mu) & \text{for } i \neq 4, 8, 10 , \\
F_i(\mu) & \text{for } i = 4, 10 \text{ and } \mu \neq 0 , \\
I_i & \text{for } i \neq 2 .
\end{array}
$$

In the case $M = F_4(0)$ we have: $M/\mathsf{m}M = \langle 1, z, t^{13} \rangle$. Hence, the only proper generating subspace is $\langle 1, z \rangle$, which gives one new ideal:

$$I_9 = \mathsf{R}\langle 1, z \rangle .$$

Analogously, in the case $M = F_{10}(0)$ we obtain also one new ideal:

$$I_{10} = \mathsf{R}\langle 1, t \rangle \, .$$

In the case $M = F_8$ we have $M/\mathfrak{m}M = \langle 1, t^5 + \mu t^6, t^{10} \rangle$. Hence, there is again only one proper generating subspace, namely, $\langle 1, t^5 + \mu t^6 \rangle$ and we obtain a new 1-parameter family:

$$F_{12}(\mu) = \mathsf{R}\langle 1, u(\mu) \rangle \, , \quad \text{where } u(\mu) = t^5 + \mu t^6 \, .$$

In the case $M = \mathsf{S}$ we have: $M/\mathfrak{m}M = \langle 1, t^{10}, t^{13} \rangle$. As there is no element with valuation 3 in S, we get here a new family parametrized by the projective line:

$$F_{13}(\lambda_0 : \lambda_1) = \mathsf{R}\langle 1, \lambda_0 t^{10} + \lambda_1 t^{13} \rangle \, .$$

At last, the case $M = I_2$ also gives a new 1-parameter family:

$$F_{14}(\mu) = \mathsf{R}\langle 1, z^3 + \mu t^{10}, t^{13} \rangle \, .$$

Thus, we have described all R-ideals and proved that $\mathsf{par}\,(1, \mathsf{R}) = 1$. Quite similar (mainly easier) calculations show that $\mathsf{par}\,(1, \mathsf{R}) = 1$ all other singularities of type W.

Remark. In the list of Schappert [11] the ideals $F_2(\mu)$ (which are indeed overrings of R) and I_3 are missing.

3.2   Ideals of singularities of type Z

Now consider the singularities of type Z. Here we suppose that $s = 2$ and R contains elements x, y with $\mathbf{v}(x) = (1, 8)$, $\mathbf{v}(y) = (\infty, 3)$ (IUS of type Z_{19}). One can check that in this case $\mathsf{R} \supset (t_1^3, t_2^{17})\mathsf{R}_0$ and, moreover, its minimal overring contains (t_1^2, t_2^{14}). Hence, we may suppose that

$$\mathsf{R} = \langle 1, x, y, y^2, y^3, xy, y^4 \rangle + (t_1^2, t_2^{14})\mathsf{R}_0 \, .$$

It is convenient to take for t_1 the first component of x and choose t_2 in such way that $y^3 = (0, t_2)x$. Now

$$\mathsf{S} = \langle 1, x, y, y^2, y^3 \rangle + (t_1, t_2^{11})\mathsf{R}_0 \quad \text{and} \quad \mathsf{S}' = \mathsf{D}_1 \oplus \mathsf{S}_2 \, ,$$

where $\mathsf{D}_1 = \mathbf{k}[[t_1]]$ and $\mathsf{S}_2 = \mathbf{k}[[t_2^3, t_2^5]]$ (the simple plane curve singularity of type E_8). Here is the list of all S_2-ideals (cf. [6],[9]) given by their generators over S_2 and over S:

Ideal	S_2-generators	S-generators
S_2	1	$1,\ t_2^5,\ t_2^{10}$
A	$1,\ t_2^7$	$1,\ t_2^5,\ t_2^7$
A*	$1,\ t_2^2$	$1,\ t_2^2,\ t_2^7$
B	$1,\ t_2^4$	$1,\ t_2^4,\ t_2^5$
B*	$1,\ t_2$	$1,\ t_2,\ t_2^5$
B'	$1,\ t_2^2,\ t_2^4$	$1,\ t_2^2,\ t_2^4$
D_2	$1,\ t_2,\ t_2^2$	$1,\ t_2,\ t_2^2$

Any (full) S'-ideal is of the form $D_1 \oplus N$, where N is an S_2-ideal. Consider first the S-ideals I such that $S'I = D_1 \oplus A$. Then $\mathfrak{n}I = \mathfrak{n} + \langle (0, t_2^{10}) \rangle$,

$$S'I/\mathfrak{n}I = W = \langle (1,0), (0,1), (0, t_2^5), (0, t_2^7) \rangle$$

and any generating subspace $V \subseteq W$ contains an element of the form $(1, a)$ and also elements of the form $(\alpha_1, 1 + \mu_1 t_2^5), (\alpha_2, t_2^7 + \mu_2 t_2^5)$. Multiplying by $(1, 1 - \mu_1 t_2^5)$, we may suppose that V contains $(\alpha_1, 1)$. Then, if $\dim V = 2$, there are only two possibilities:

$$V = \langle (1,1), (0, \mu t_2^5 + t_2^7) \rangle \quad \text{and} \quad V = \langle (0,1), (1, \mu t_2^5 + t_2^7) \rangle .$$

If $\dim V = 3$, we can add also an element of the form (β, t_2^5) or $(1, 0)$ It gives three more possibilities:

$$V = \langle (1,1), (0, t_2^5), (0, t_2^7) \rangle, \quad V = \langle (0,1), (\lambda_0, t_2^5), (\lambda_1, t_2^7) \rangle,$$
$$V = \langle (1,0), (0,1), (\mu t_2^5 + t_2^7) \rangle,$$

where $(\lambda_0 : \lambda_1)$ is a point of the projective line. Hence, here is the list of the corresponding S-ideals (which are not S'-ideals):

$$
\begin{aligned}
F_1(\mu) &= S\langle (1,1), (0, \mu t_2^5 + t_2^7) \rangle, \\
F_2(\mu) &= S\langle (0,1), (1, \mu t_2^5 + t_2^7) \rangle, \\
F_3(\lambda_0 : \lambda_1) &= S\langle (0,1), (\lambda_0, t_2^5), (\lambda_1, t_2^7) \rangle, \\
F_4(\mu) &= S\langle (1,0), (0,1), (\mu t_2^5 + t_2^7) \rangle, \\
I_1 &= S\langle (1,1), (0, t_2^5), (0, t_2^7) \rangle .
\end{aligned}
$$

Note that the ideals $F_4(\mu)$ are decomposable (as modules), while all other ideals of this list are indecomposable.

Analogous calculations give the following list of all S-ideals I, which are not S'-ideals:

$$S'I = S':$$

$$
\begin{aligned}
F_5(\lambda_0 : \lambda_1) &= S\langle(\lambda_0, 1), (\lambda_1, t_2^5)\rangle, \\
S &= S\langle(1, 1)\rangle, \\
I_2 &= S\langle(1, 1), (0, t_2^{10})\rangle, \\
I_3 &= S\langle(1, 1), (0, t_2^5), (0, t_2^{10})\rangle, \\
I_4 &= S\langle(0, 1), (1, t_2^5), (0, t_2^{10})\rangle, \\
I_5 &= S\langle(0, 1), (1, t_2^{10})\rangle, \\
I_6 &= S\langle(0, 1), (0, t_2^5), (1, t_2^{10})\rangle, \\
I_7 &= S\langle(1, 0), (0, 1)\rangle, \\
I_8 &= S\langle(1, 0), (0, 1), (0, t_2^5)\rangle, \\
I_9 &= S\langle(1, 0), (0, 1), (0, t_2^{10})\rangle.
\end{aligned}
$$

$$S'I = D_1 \oplus A^*, \quad \mathfrak{n}I = \mathfrak{n} + \langle(0, t_2^5), (0, t_2^{10})\rangle:$$

$$
\begin{aligned}
F_6(\lambda_0 : \lambda_1) &= S\langle(\lambda_0, 1), (\lambda_1, t_2^2)\rangle, \\
F_7(\lambda_0 : \lambda_1) &= S\langle(\lambda_0, 1), (\lambda_1, t_2^2), (0, t_2^7)\rangle, \\
I_{10} &= S\langle(0, 1), (0, t_2^2), (1, t_2^7)\rangle, \\
I_{11} &= S\langle(1, 0), (0, 1), (0, t_2^2)\rangle.
\end{aligned}
$$

$$S'I = D_1 \oplus B, \quad \mathfrak{n}I = \mathfrak{n} + \langle(0, t_2^7), (0, t_2^{10})\rangle:$$

$$
\begin{aligned}
F_8(\mu) &= S\langle(1, 1), (0, t_2^4 + \mu t_2^5)\rangle, \\
F_9(\mu) &= S\langle(0, 1), (1, t_2^4 + \mu t_2^5)\rangle, \\
F_{10}(\lambda_0 : \lambda_1) &= S\langle(0, 1), (\lambda_0, t_2^4), (\lambda_1, t_2^5)\rangle, \\
F_{11}(\mu) &= S\langle(1, 0), (0, 1), (0, t_2^4 + \mu t_2^5)\rangle, \\
I_{12} &= S\langle(1, 1), (0, t_2^4), (0, t_2^5)\rangle.
\end{aligned}
$$

$$S'I = D_1 \oplus B^*, \quad \mathfrak{n}I = \mathfrak{n} + \langle(0, t_2^4), (0, t_2^7), (0, t_2^{10})\rangle:$$

$$
\begin{aligned}
F_{12}(\lambda_0 : \lambda_1) &= S\langle(\lambda_0, 1), (\lambda_1, t_2)\rangle, \\
F_{13}(\lambda_0 : \lambda_1) &= S\langle(\lambda_0, 1), (\lambda_1, t_2), (0, t_2^5)\rangle, \\
I_{13} &= S\langle(0, 1), (0, t_2), (1, t_2^5)\rangle, \\
I_{14} &= S\langle(1, 0), (0, 1), (0, t_2)\rangle.
\end{aligned}
$$

$$S'I = D_1 \oplus B', \quad \mathfrak{n}I = \mathfrak{n} + \langle(0, t_2^5), (0, t_2^7), (0, t_2^{10})\rangle:$$

$$
\begin{aligned}
I_{15} &= S\langle(1, 1), (0, t_2^2), (0, t_2^4)\rangle \\
I_{16} &= S\langle(0, 1), (1, t_2^2), (0, t_2^4)\rangle \\
I_{17} &= S\langle(0, 1), (0, t_2^2), (1, t_2^4)\rangle.
\end{aligned}
$$

$S'I = \mathsf{R}_0$:

$$
\begin{aligned}
I_{18} &= \mathsf{S}\langle (1,1),(0,t_2),(0,t_2^2)\rangle \\
I_{19} &= \mathsf{S}\langle (0,1),(1,t_2),(0,t_2^2)\rangle \\
I_{20} &= \mathsf{S}\langle (0,1),(0,t_2),(1,t_2^2)\rangle .
\end{aligned}
$$

Here μ denotes an element of $\mathbf{k}$ and $(\lambda_0 : \lambda_1)$ a point of the projective line over $\mathbf{k}$.

Now pass to the calculation of R-ideals, which are not S-ideals. Again one can verify that for the following S-ideals M we have $\mathfrak{m}M = \mathfrak{n}M$, so we do not need to consider the case, when $\mathsf{S}I = M$:

$$
\begin{aligned}
F_i(\mu) & \qquad \text{for } i \neq 1,8 , \\
F_i(\lambda_0 : \lambda_1) & \qquad \text{for all } i , \\
F_8(\mu) & \qquad \text{for } \mu \neq 0 , \\
I_i & \qquad \text{for } i \neq 2,7 .
\end{aligned}
$$

For $M = F_1(\mu)$, $M/\mathfrak{m}M = \langle (1,1),(0,\mu t_2^5 + t_2^7),(0,t_2^8)\rangle$. If $\mu \neq 0$, the multiplication by itself maps the second of these elements to the third one, which is congruent modulo $\mathfrak{m}M$ to $\mu^{-1}t_2^{10}$. Hence, here we obtain only the following new 1-parameter family of ideals:

$$
F_{14}(\mu) = \mathsf{R}\langle (1,1),(0,\mu t_2^5 + t_2^7)\rangle , \quad \mu \neq 0 .
$$

If $\mu = 0$, we obtain the following new family:

$$
F_{15}(\mu) = \mathsf{R}\langle (1,1),(t_2^7 + \mu t_2^8)\rangle .
$$

In the case $M = F_8(0)$, $M/\mathfrak{m}M = \langle (1,1),(0,t_2^4 + \mu t_2^5),(0,t_2^8)\rangle$. Hence, we also obtain only one new family of ideals:

$$
F_{16}(\mu) = \mathsf{R}\langle (1,1),(t_2^4 + \mu t_2^5)\rangle .
$$

The remaining cases: S and I_i ($i = 2,7$), give the ring R itself and the following ideals:

$$
\begin{aligned}
F_{17}(\lambda_0 : \lambda_1) &= \mathsf{R}\langle (1,1),(\lambda_0 t_1, \lambda_1 t_2^{13})\rangle , \\
F_{18}(\mu) &= \mathsf{R}\langle (1,1),(\mu t_1, t_2^{10})\rangle , \\
I_{21} &= \mathsf{R}\langle (1,0),(0,1)\rangle , \\
I_{22} &= \mathsf{R}\langle (1,t_2^{13}),(0,1)\rangle .
\end{aligned}
$$

Therefore, we have proved that in this case also $\mathsf{par}\,(1,\mathsf{R}) = 1$. Quite similar calculations prove the same for all other singularities of type Z.

Remark: Here the family F_{13} is missing in Schappert's list of ideals.

3.3   Ideals of singularities of type E

Now consider the case of singularities E. In this case $\mathsf{d}_0 = 3$, so it follows from [3] that each R-ideal is isomorphic either to an overring of R or to its dual module. Therefore, we only need to find all overrings. But if I is an overring of R then $\mathsf{S}I$ is also an overring of S. Hence, at each stage of our inductive process we may restrict ourselves to overrings. To be complete, we always mark, which of these overrings are Gorenstein (i.e. self-dual).

Here we suppose that $s = 3$ and R contains elements x, y with $\mathbf{v}(x) = (1,1,1)$ and $\mathbf{v}(y) = (\infty, k, 4)$ with $k > 3$ (IUS of type $\mathsf{E}_{2,p}$). Of course, we suppose here that $x = t = (t_1, t_2, t_3)$. Again, passing to the minimal overring, we may suppose that R contains $(t_1^{k+2}, t_2^{k+2}, t_3^5)\mathsf{R}_0$ and is generated by x and y modulo this ideal. Then S contains $(t_1^{k+1}, t_2^{k+1}, t_3^4)\mathsf{R}_0$ and is generated modulo this ideal by x and z, where $y = xz$ and $\mathbf{v}(z) = (\infty, k-1, 2)$ (it is an IUS of type $\mathsf{E}_{1,p}$). The ring S' contains $(t_1^k, t_2^k, t_3^3)\mathsf{R}_0$ and is generated modulo this ideal by x and z. Put now $\mathfrak{n}' = \operatorname{rad}\mathsf{S}'$, $\mathsf{S}'' = \operatorname{End}(\mathfrak{n}')$. Then S'' is generated by x and z', where $\mathbf{v}(z') = (\infty, k-2, 1)$ and $z = xz'$. Hence, S'' has only finitely many ideals up to isomorphism (it is a simple plane curve singularity of type D), cf. [6],[9]. Namely, here is the list of the overrings of S'' (except S'' itself): .

$$
\begin{aligned}
\mathsf{A}_i &= \mathsf{S}''\langle 1, (0, t_2^i, 0)\rangle \quad (1 \le i \le k-2), \\
\mathsf{B}_i &= \mathsf{S}''\langle 1, (0,0,1), (0, t_2^i, 0)\rangle \quad (0 \le i \le k-2), \\
\mathsf{B}_{01} &= \mathsf{S}''\langle 1, (1,0,0)\rangle, \\
\mathsf{B}_{02} &= \mathsf{S}''\langle 1, (0,1,0)\rangle, \\
\mathsf{R}_0 &= \mathsf{S}''\langle (1,0,0), (0,1,0), (0,0,1)\rangle.
\end{aligned}
$$

Among them, only A_i are non-Gorenstein. Moreover, $M\mathfrak{n}'' = M\mathfrak{n}'$ for $M = \mathsf{R}_0$ or $M = \mathsf{B}_i$, $i < k-2$, where $\mathfrak{n}'' = \operatorname{rad}\mathsf{S}''$, so we do not need further to consider these overrings. Note also that in the case of B_{k-2} the generator $(0, t_2^i, 0)$ above is superfluous.

As S is Gorenstein, its overrings, except S itself, are those of S'. Here is the list of factoralgebas $M/\mathfrak{n}'M$ for the overrings M of S'':

$$
\begin{aligned}
\mathsf{S}''/\mathfrak{n}' &= \langle 1, z', (0,0,t_3^2)\rangle, \\
\mathsf{A}_i/\mathfrak{n}'\mathsf{A}_i &= \langle 1, (0,0,t_3), (0, t_2^i, 0)\rangle, \\
\mathsf{B}_{k-2}\mathfrak{n}'\mathsf{B}_{k-2} &= \langle 1, (0,0,1), (0, t_2^{k-2}, 0)\rangle, \\
\mathsf{B}_{01}/\mathfrak{n}'\mathsf{B}_{01} &= \langle 1, (1,0,0), (0, t_2, 0)\rangle, \\
\mathsf{B}_{02}/\mathfrak{n}'\mathsf{B}_{02} &= \langle 1, (0,1,0), (0, t_2, 0)\rangle.
\end{aligned}
$$

It is now easy to find all proper subalgebras of these algebras and, hence, the overrings of S (except S itself), which are not overrings of S'':

$$
\begin{aligned}
S' &= S\langle 1, (0, 0, t_3^3)\rangle, \\
A_{k-1} &= S\langle 1, (0, 0, t_3^2)\rangle, \\
F_i(\mu) &= S\langle 1, (0, t_2^i, \mu t_3)\rangle \quad (1 \le i \le k-2), \\
B_{k-1} &= S\langle 1, (0, 0, 1)\rangle, \\
B_{01}' &= S\langle 1, (1, 0, 0)\rangle, \\
B_{02}' &= S\langle 1, (0, 1, 0)\rangle.
\end{aligned}
$$

Among them only S', A_{k-1} and $F_i(0)$ are non-Gorenstein. Note also that $M\mathfrak{m} = M\mathfrak{n}'$ for all overrings of S'', so we do not need to consider them further. To find all overrings of R, which are not overrings of S, calculate the factoralgebras $M/\mathfrak{m}M$ for the rings of the preceding list and S:

$$
\begin{aligned}
S/\mathfrak{m} &= \langle 1, z, (0, 0, t_3^4)\rangle, \\
S'/\mathfrak{m}S' &= \langle 1, z, (0, 0, t_3^3)\rangle, \\
A_{k-1}/\mathfrak{m}A_{k-1} &= \langle 1, (0, 0, t_3^2), (0, 0, t_2^{k-1})\rangle, \\
F_i(\mu)/\mathfrak{m}F_i(\mu) &= \langle 1, (0, t_2^i, \mu t_3), (0, 0, t_3^2)\rangle, \\
B_{k-1}/\mathfrak{m}B_{k-1} &= \langle 1, (0, 0, 1), (0, t_2^{k-1})\rangle, \\
B_{01}'/\mathfrak{m}B_{01}' &= \langle 1, (1, 0, 0), (0, t_2^2, 0)\rangle, \\
B_{02}'/\mathfrak{m}B_{02}' &= \langle 1, (0, 1, 0), (t_1^2, 0, 0)\rangle.
\end{aligned}
$$

It gives us the following list of overrings of R, which are not overrings of S (except R itself):

$$
\begin{aligned}
R' &= R\langle 1, (0, 0, t_3^4)\rangle, \\
F_{k-1}'(\mu) &= R\langle 1, \mu z + (0, 0, t_3^3)\rangle, \\
F_{k-1}(\mu) &= R\langle 1, (0, \mu t_2^{k-1}, t_3^2)\rangle \quad (\mu \ne 1), \\
F_i'(\mu) &= R\langle 1, (0, t_2^i, t_3^2)\rangle \quad (1 \le i \le k-2), \\
B_k &= R\langle 1, (0, 0, 1)\rangle, \\
B_{01}'' &= R\langle 1, (1, 0, 0)\rangle, \\
B_{02}'' &= R\langle 1, (0, 1, 0)\rangle.
\end{aligned}
$$

Here $F_{k-1}(\mu)$, B_k, B_{01}'' and B_{02}'' are Gorenstein.

Thus, we have proved that $\mathrm{par}\,(1, R) = 1$. Analogous calculations show the same for all other IUS of type E, which accomplishes the proof of Theorem 2.1.

References

[1] Arnold V.I., Varchenko A.N., Gusein-Zade S.M., Singularities of Differentiable Maps, Vol. 1, Birkhäuser, Boston-Basel-Stuttgart, 1985.

[2] Bass H., On the ubiquity of Gorenstein rings, Math. Z. **82** (1963), 8–28.

[3] Drozd Yu.A., Ideals of commutative rings, Mat. Sbornik, 101 (1976) 334–348.

[4] Drozd Yu.A., Greuel G.-M., Cohen-Macaulay module type, Compositio Math. 89 (1993) 315–338.

[5] Drozd Yu.A., Greuel G.-M., Semicontinuity for representations of Cohen-Macaulay rings, Preprint Nr. 247, Fachbereich Math. Univ. Kaiserslautern, 1993. To appear in Math. Ann. 1996

[6] Drozd Yu.A., Roiter A. V., Commutative rings with a finite number of indecomposable integral representations Izv. Aad. Nauk SSSR. Ser. Mat. 31 (1967) 783–798.

[7] Greuel G.-M., Knörrer H., Einfache Kurvensingularitäten und torsionfreie Moduln, Math. Ann. 270 (1985) 417–425.

[8] Greuel G.-M.; Kröning H., Simple singularities in positive characteristic, Math. Z. **203**, (1990) 339–354.

[9] Jacobinski H., Anneaux commutatifs avec un nombre fini de réseaux indécomposable, Acta Math. **118** (1967) 1–31.

[10] Knörrer H., Torsionfreie Moduln bei Deformation von Kurvensingularitäten, In: Greuel G.-M., Trautmann G. (ed.) Singularities, Representations of Algebras and Vector Bundles, Lambrecht 1985. Lecture Notes in Math., Vol. 1273, Springer, Berlin-Heidelberg-New York (1987) 150–155.

[11] Schappert A., A characterization of strict unmodal plane cure singularities, In: Greuel G.-M., Trautmann G. (ed.) Singularities, Representations of Algebras and Vector Bundles, Lambrecht 1985. Lecture Notes in Math., Vol. 1273, Springer, Berlin-Heidelberg-New York (1987) 168–177.

[12] Schappert A., Kurvensingularitäten und Isomorphieklassen von Moduln, Dissertation, Universität Kaiserslautern, 1990.

[13] Wall C.T.C., Classification of unimodal isolated singularities of complete intersections, In: Orlik P. (ed.) Singularities, Arcata 1981. Proc. Sympos. Pure Math. 40(2) (1983) 625–640.

Progress in Mathematics, Vol. 162, © 1998 Birkhäuser Verlag Basel/Switzerland

Geometric Quotients of Unipotent Group Actions II

Gert-Martin Greuel
Universität Kaiserslautern
Fachbereich Mathematik
Erwin-Schrödinger-Strasse
67663 Kaiserslautern
GERMANY

Gerhard Pfister
Universität Kaiserslautern
Fachbereich Mathematik
Erwin-Schrödinger-Strasse
67663 Kaiserslautern
GERMANY

Dedicated to Egbert Brieskorn on the occasion of his 60th birthday

Introduction

Let G be a unipotent algebraic group over K (a field of characteristic 0) which acts rationally on an affine scheme $X = \operatorname{Spec} A$ over K, where A is a commutative K-algebra. The problem of finding sufficient and manageable conditions to guarantee that the geometric quotient X/G exists is of fundamental interest in the theory of moduli spaces for local objects such as isolated singularities or (Cohen-Macaulay) modules over the local ring of a singularity (cf. [L-P], [G-P2], [G-H-P], [H]).

In [G-P1] we derived such conditions which are complemented in this paper. These conditions are even useful when the geometric quotient does not exist globally. Namely, they allow the construction of a stratification of X into locally closed G-stable subschemes on which the geometric quotient exists. If the action of G is sufficiently explicitly given, say in terms of coordinates of X and generators of G, then the stratification can be described explicitly in terms of these data. Note that for unipotent groups, in contrast to reductive groups, the existence of a geometric quotient depends in general not only on X and G but also on the action, that is, knowledge about the action is necessary. The purpose of [G-P1] was to prove existence criteria which were as general and as explicit as possible. In all applications so far, the explicit description of the strata was the key point to being able to describe the strata in terms of invariants of the singularities or modules.

On the other hand, the explicit formulation in terms of coordinates and generators made the statements of the theorems in [G-P1] somewhat technical, even in the case of a free action, which is an important cornerstorne for the general theory (cf. Theorem 3.10 in [G-P1]). One of the equivalent conditions of that theorem (loc. cit.) was the vanishing of $H^1(G, A)$ (the usual algebraic group cohomology), in particular we showed that $H^1(G, A) = 0$ is equivalent to $\operatorname{Spec} A \longrightarrow \operatorname{Spec} A^G$ being a trivial geometric quotient (which implies in particular that A^G is of finite type over K if A is of finite type over K). Moreover, we proved that $H^1(G, A) = 0$ implies that $\operatorname{Spec} A \longrightarrow \operatorname{Spec} A^G$ is a principal fibre bundle with group G, that is, $\operatorname{Spec} A \longrightarrow \operatorname{Spec} A^G$ is faithfully flat and the canonical map $A \otimes_{A^G} A \longrightarrow A \otimes_K K[G]$ is an isomorphism (cf. [M-F], Def. 0.10).

In this note, which is intended to be a supplement of [G-P1], we prove the converse of the last statement, providing the following conceptual, necessary and sufficient condition for $H^1(G, A) = 0$.

Theorem: *Let A be a commutative K-algebra and G a unipotent algebraic group over K acting rationally on $\operatorname{Spec} A$. Then the following are equivalent:*

(i) $H^1(G, A) = 0$;

(ii) $\operatorname{Spec} A \longrightarrow \operatorname{Spec} A^G$ is faithfully flat and the canonical map

$$(*) \qquad A \otimes_{A^G} A \longrightarrow A \otimes_K K[G] \text{ is an isomorphism.}$$

Moreover, if A is reduced, then (i) and (ii) are equivalent to

(ii') $\operatorname{Spec} A \longrightarrow \operatorname{Spec} A^G$ is faithfully flat and the canonical map

$$(**) \qquad A \otimes_K A \longrightarrow A \otimes_K K[G] \text{ is surjective.}$$

Condition $(**)$ means that $X \times_K G \longrightarrow X \times_K X$ is a closed immersion, that is, the action is free in the sense of Mumford (cf.[M-F], Def. 0.8). We ignore whether we can drop the assumption of A being reduced in (ii'). Note that $(*)$ implies $(**)$ but that $(*)$ does not imply the flatness of A over A^G, cf. the example in [D-F], examined at the end of this paper.

That (i) implies (ii) follows from [G-P1], Theorem 3.10 and Remark 3.11; (ii') is a trivial consequence of (ii). The remaining implications are proved in this paper.

The equivalence of (i) and (ii) was already mentioned in [K-M-T], but some arguments in the proof seemed to be insufficient. More recently, in [D-F-G], a result was proved which states (in our terms) the implication (ii) $\Rightarrow$ (i) for G the additive group of $K = \mathbb{C}$ and A the polynomial ring over $\mathbb{C}$. In any case, here we give an elementary proof of the following slightly more general fact.

If (*) holds, and if the canonical map $\operatorname{Spec} A \longrightarrow \operatorname{Spec} A^G$ is flat, then $\operatorname{Spec} A$ is mapped onto an open set $U \subset \operatorname{Spec} A^G$ such that $\operatorname{Spec} A \twoheadrightarrow U$ is a geometric quotient and a principal fibre bundle with group G.

This article was inspired by discussions with C. Hertling, when we tried to extend the results of [G-H-P], in order to construct moduli spaces for semiquasi-homogeneous hypersurface singularities without fixing the principal part. We could not prove the existence of a geometric quotient as an algebraic $\mathbb{C}$-scheme. From the examples of Deveney and Finston we learned that, additionally, at least the flatness of A as an A^G-module is necessary. Condition (ii') shows that it is also sufficient if A is reduced. Although we could not prove the existence of a geometric quotient as an algebraic $\mathbb{C}$-variety under the assumption "$\operatorname{Spec} A \longrightarrow \operatorname{Spec} A^G$ surjective and (∗∗) holds", in our application Hertling was able to prove the existence of a geometric quotient as a complex space.

The following conjecture points in the same direction (G and A as above):

Conjecture. Assume that G acts freely on $\operatorname{Spec} A$ (in the sense of Mumford). Then there exists an étale covering $\{\operatorname{Spec} B_i\}$ of $\operatorname{Spec} A$ and a lifting of the action of G to B_i such that $H^1(G, B_i) = 0$.

Notice that under our assumption the quotient exists in the category of algebraic spaces (cf. [P, Theorem 3.7]). Our conjecture says that this quotient is locally trivial. We prove this under a slightly different assumption. We should like to emphasize that passing to an étale covering is necessary, as we show at the end of this paper.

As in [G-P1] we prefer to work with the Lie algebra L of G. Since G is unipotent and char $K = 0$ this is equivalent. Also the Lie algebra cohomology ([C-E]) coincides in this case with the group cohomology.

1   Special representations

Let L be an n-dimensional nilpotent K-Lie algebra. We deduce the vanishing of $H^1(L, K[X_1, \ldots, X_n])$ for certain special representations of L in $\operatorname{der}_K A[X_1, \ldots, X_n]$, in particular for the representation of L on the coordinate algebra $K[G(L)] \cong K[L] = K[X_1, \ldots, X_n]$ of its associated unipotent group $G(L)$ derived from the left regular representation of $G(L)$ on $K[G(L)]$. This result is perhaps known to the specialists but we could not find a reference. In any case, it is an immediate consequence of Theorem 3.10 in [G-P1]. In order to apply that theorem we need a description of the left regular action in terms of coordinates.

Let $X, Y \in L$ be two elements and $H(X, Y) = \sum_{i>0} H^i(X, Y)$ the series of Campbell-Hausdorff (cf. [G]), where

$$
\begin{aligned}
H^1(X, Y) &= X + Y, \\
H^2(X, Y) &= \tfrac{1}{2}[X, Y], \\
H^3(X, Y) &= \tfrac{1}{12}([X, [X, Y]] + [Y, [Y, X]]), \ \dots
\end{aligned}
$$

and $H^i(X, Y) = 0$ for large i since L is nilpotent.

Consider L as an affine K-variety. Then the multiplication $H : L \times L \longrightarrow L$ gives L the structure of a unipotent algebraic group which we call $G(L)$, with Lie-algebra isomorphic to L (cf. [D-G]).

Let $\{\delta_1, \dots, \delta_n\}$ be a basis of L and $[\delta_i, \delta_j] = \sum_k C_k^{ij} \delta_k$. The choice of a basis defines an isomorphism $G(L) \cong \operatorname{Spec} K[X_1, \dots, X_n]$ and then the co-multiplication

$$
m : K[X_1, \dots, X_n] \longrightarrow K[X_1, \dots, X_n] \otimes_K K[X_1, \dots, X_n],
$$

is given in terms of the chosen coordinates by

$$
m(X_k) = X_k \otimes 1 + 1 \otimes X_k + \tfrac{1}{2} \sum_{ij} C_k^{ij} X_i \otimes X_j + \dots .
$$

Via $G(L) \cong \operatorname{Spec} K[X_1, \dots, X_n]$ the Lie-algebra L is represented as a subalgebra of $\operatorname{Der}_K K[X_1, \dots, X_n]$ with basis $\{\delta_j\}$ and $\delta_j(X_k) = \delta_{jk} + \sum_i C_k^{ij} X_i + h_{jk}$ (δ_{jk} the Kronecker symbol, $h_{jk} \in (X_1, \dots, X_n)^2$), the derived left regular representation of L on $K[G(L)] \cong K[L] = K[X_1, \dots, X_n]$.

Since L is nilpotent, we can choose the basis $\{\delta_1, \dots, \delta_n\}$ such that $C_k^{ij} = 0$ if $k \leq \max\{i, j\}$. This implies that L acts on $K[L]$ via

1. $\delta_j(X_j) = 1$,

2. $\delta_j(X_k) = 0$ if $j > k$,

3. $\delta_j(X_k) \in K[X_1, \dots, X_{k-1}]$ if $j < k$

and, in particular,

3.' $\delta_i \delta_j(X_k) = 0$ if $i \geq k$.

Using Theorem 3.10 in [G-P1], we obtain the following

PROPOSITION 1.1 *The derived left regular representation of L on $K[L]$ satisfies* $H^1(L, K[L]) = 0$.

COROLLARY 1.2 *Let L be an n-dimensional nilpotent K-Lie-algebra. There exists a faithful representation $\rho : L \longrightarrow \operatorname{Der}_K K[X_1, \dots, X_n]$ such that*

$$
H^1(L, K[X_1, \dots, X_n]) = 0.
$$

COROLLARY 1.3 *Let A be a commutative K-algebra (with unit 1), L an n-dimensional nilpotent K-Lie-algebra and $\varphi : L \longrightarrow \mathrm{Der}_K A$ a representation such that the elements of $\varphi(L)$ are locally nilpotent. Let $\rho : L \longrightarrow \mathrm{Der}_K K[X_1, \ldots, X_n]$ be any representation satisfying*

$$H^1(L, K[X_1, \ldots, X_n]) = 0.$$

Then for the tensor-product representation $\varphi \otimes \rho : L \longrightarrow \mathrm{Der}_K A[X_1, \ldots, X_n]$ we have

(i) $H^1(L, A[X_1, \ldots, X_n]) = 0$.

(ii) Let $\{\delta_1, \ldots, \delta_n\}$ be a basis of L such that $[\delta_i, \delta_j] \in \sum_{\ell > \max\{i,j\}} K\delta_\ell$ (such a basis does always exist) then $(\exp(-X_1\delta_1) \circ \cdots \circ \exp(-X_n\delta_n))(A) = A[X_1, \ldots, X_n]^L$.

(iii) Consider a basis $\{\delta_1, \ldots, \delta_n\}$ of L as in (ii) and extend it trivially to $A[X_1, \ldots, X_n]$ (that is $\delta_j(X_i) = 0$), then

$$\delta_i \circ \exp(-X_1\delta_1) \circ \cdots \circ \exp(-X_n\delta_n)$$
$$= \exp(-X_1\delta_1) \circ \cdots \circ \exp(-X_n\delta_n) \circ \left(\delta_i + \sum_{k>i} \xi_{i_k}\delta_k\right)$$

for suitable $\xi_{i_k} \in K[X_1, \ldots, X_n]$.

Proof. The first claim follows because $H^1(L, K[X_1, \ldots, X_n]) = 0$. To prove (ii) let $\hat{\delta}_i = \rho \otimes \varphi(\delta_i)$. Denote by

$$s : A[X_1, \ldots, X_n] \quad \rightarrow A[X_1, \ldots, X_n]^L$$

the section (cf. [G-P1]) defined by

$$s(h) = (\exp T_1\hat{\delta}_1 \circ \cdots \circ \exp T_n\hat{\delta}_n)(h)(T_1 = -X_1, \ldots, T_n = -X_n),$$

where $(T_1 = -X_1, \ldots, T_n = -X_n)$ means evaluation at $T_i = -X_i$.

If $a \in A$ then $\hat{\delta}_i(a) = \delta_i(a)$. This implies $s(a) = (\exp(-X_1\delta_1) \circ \cdots \circ \exp(-X_n\delta_n))(a)$, that is

$$\exp(-X_1\delta_1) \circ \cdots \circ \exp(-X_n\delta_n)(A) \subseteq A[X_1, \ldots, X_n]^L.$$

If $h = a + \sum X_i h_i$, $a \in A$, then $s(h) = s(a)$ because $s(X_i) = 0$. This proves (ii). Since $[\delta_i, \delta_j] \in \sum_{\ell > \max\{i,j\}} K\delta_\ell$ (iii) holds. $\qquad\square$

The following corollary points towards the conjecture in the introduction. It is an improvement of Remark 3.12 of [G-P1], where we assumed that L is abelian. Note that $\det(\delta_i(a_j)) \in A^*$ implies that the action is set theoretically free.

COROLLARY 1.4 *Let A be a commutative K-algebra with 1 and $L \subset \mathrm{der}_K(A)$ an n-dimensional nilpotent K-Lie algebra. Assume that there exist $\delta_1, \ldots, \delta_n \in L$ and $a_1, \ldots, a_n \in A$ such that $\det(\delta_i(a_j))$ is a unit in A. Let $X_1, \ldots, X_n$ be indeterminates and define $F_i := \exp(-X_1\delta_1) \circ \cdots \circ exp(-X_n\delta_n)(a_i)$. Then*

$$B := A[X_1, \ldots, X_n]/(F_1, \ldots, F_n)$$

is étale over A, the action of L on A lifts to B and $H^1(L, B) = 0$.

Proof. By Corollary 1.3 there exists a faithful representation of L such that $H^1(L, K[X_1, \ldots, X_n]) = 0$. Using any such representation, we define the tensor product representation of L on $A[X_1, \ldots, X_n]$ as in 1.3. Then the F_i are invariant under L by 1.3 (ii). The vanishing of $H^1(L, B)$ is now a consequence of 1.3(i) and Theorem 3.10 in [G-P1].

To prove that $A \to B$ is étale, we may assume that $\delta_1, \ldots, \delta_n$ are chosen as in (ii) of 1.3 and $\hat{\delta}_1, \ldots, \hat{\delta}_n \in \mathrm{Der}_K A[X_1, \ldots, X_n]$ are extensions of $\delta_1, \ldots, \delta_n$ such that $\hat{\delta}_i(F_j) = 0$ for all i, j. Let $\bar{\delta}_i := \delta_i - \hat{\delta}_i$, $i = 1 \ldots n$ (the δ_i are trivially extended to $A[X_1, \ldots, X_n]$); then $\bar{\delta}_i \in \mathrm{Der}_A A[X_1, \ldots, X_n]$.

Furthermore, by 1.3 (iii),

$$
\begin{aligned}
\bar{\delta}_i(F_j) &= \delta_i(F_j) \\
&= \delta_i \circ \exp(-X_1\delta_1) \circ \cdots \circ \exp(-X_n\delta_n)(a_j) \\
&= \exp(-X_1\delta_1) \circ \cdots \circ \exp(-X_n\delta_n)(\delta_i(a_j) + \textstyle\sum_{k>i} \xi_{i_k}\delta_k(a_j)).
\end{aligned}
$$

Then

$$
\begin{aligned}
\det(\bar{\delta}_i(F_j)) &= \exp(-X_1\delta_1) \circ \cdots \circ \exp(-X_n\delta_n)\left(\det(\delta_i(a_j) + \textstyle\sum_{k>i} \xi_{i_k}\delta_k(a_j))\right) \\
&= \exp(-X_1\delta_1) \circ \cdots \circ \exp(-X_n\delta_n)\left(\det(\delta_i(a_j))\right) \\
&= \det(\delta_i(a_j)).
\end{aligned}
$$

This implies that $\det(\bar{\delta}_i(F_j))$ is a unit and, therefore,

$$A \to A[X_1, \ldots, X_n]/(F_1, \ldots, F_n)$$

is étale. $\square$

2   Free actions

We are now going to prove the main theorem which was explained in the introduction.

THEOREM 2.1 *Let A be a commutative K-algebra. Let $L = \sum_{i=1}^{n} K\delta_i \subseteq \mathrm{Der}_K(A)$ be an n-dimensional nilpotent Lie-algebra and assume that the δ_i are locally nilpotent. Assume, moreover, that*

1. *Spec $A \longrightarrow$ Spec A^L is faithfully flat,*

2. *the canonical map $A \otimes_{A^L} A \longrightarrow A[Z_1, \ldots, Z_n]$ defined by L,*

$$a \otimes b \rightsquigarrow a \cdot (\exp Z_1 \delta_1 \circ \dot{s} \circ \exp Z_n \delta_n)(b)$$

is an isomorphism.

Then $H^1(L, A) = 0$.

Proof. We prove the theorem by induction on $n = \dim L$. In the case $n = 1$ let $L = K\delta$ and $\int A^L = \{a \in A \mid \delta(a) \in A^L\}$.

The sequence

$$0 \longrightarrow \int A^L \longrightarrow A \xrightarrow{\delta^2} A$$

is exact. Let L act on $A[Z]$ by $\delta(Z) := -1$, then $B := \exp(-Z\delta)(A) \subseteq A[Z]^L$, and B is via $A^L \subseteq A \xrightarrow{\sim} B$ an A^L-algebra. Since B is flat over A^L,

$$0 \longrightarrow \int A^L \otimes_{A^L} B \longrightarrow A \otimes_{A^L} B \xrightarrow{\delta^2 \otimes 1_B} A \otimes_{A^L} B$$

is exact.

By assumption, $A \otimes_{A^L} A \longrightarrow A[Z]$ is an isomorphism, that is, we may identify $A \otimes_{A^L} B$ with $A[Z]$ and obtain the commutative diagram:

$$
\begin{array}{ccccccc}
0 & \longrightarrow & \int A^L \otimes_{A^L} B & \longrightarrow & A \otimes_{A^L} B & \xrightarrow{\delta^2 \otimes 1_B} & A \otimes_{A^L} B \\
 & & \downarrow & & \| & & \| \\
0 & \longrightarrow & \int B & \longrightarrow & A[Z] & \xrightarrow{\delta^2} & A[Z].
\end{array}
$$

Also by assumption $Z = \sum \xi_i h_i$, $\xi_i \in A$, $h_i \in B$. On the other hand, $\delta(Z) = -1$ whence $\delta^2(Z) = 0$. This implies that we can choose the ξ_i to be in $\int A^L$, by the above diagram.

Then $-1 = \delta(Z) = \sum \delta(\xi_i) h_i$ and, in particular,

$$-1 = \sum \delta(\xi_i) h_i(Z = 0).$$

Now $\delta(\xi_i) \in A^L$ and $h_i(Z = 0) \in A$. If $\mathfrak{a} = \langle \{\delta(\xi_i)\}_i \rangle$ denotes the A^L-ideal generated by the $\delta(\xi_i)$ in A^L, then $\mathfrak{a}A = A$. By faithful flatness we have $\mathfrak{a} = A^L$, that is there are $\eta_i \in A^L$ such that $1 = \sum \delta(\xi_i)\eta_i = \delta(\sum \xi_i \eta_i)$.

If we define $x := \sum \xi_i \eta_i \in A$ then $\delta(x) = 1$ and this implies $H^1(L, A) = 0$ ([G-P1], 3.10).

Now assume the theorem for $(n-1)$-dimensional Lie-algebras. Let $L = \sum_{i=1}^{n} K\delta_i$ and assume $\delta_n \in Z(L)$, where $Z(L)$ denotes the centre of L and let $L_0 := K\delta_n$. We shall prove that $H^1(L_0, A) = 0$. By Corollary 1.2 we can extend the action of L to $A[Z_1, \ldots, Z_n]$ with the properties (i), (ii) of Corollary 1.3.

As before, we consider the exact sequence

$$0 \longrightarrow \int A^L \longrightarrow A \longrightarrow A^{n^2},$$
$$a \rightsquigarrow (\delta_i \delta_j(a))$$

and put $B := \exp(-Z_1\delta_1) \circ \cdots \circ \exp(-Z_n\delta_n)(A)$, $\int A^L = \{a \in A \mid \delta(a) \in A^L$ for all $\delta \in L\}$. Then the following commutative diagram has exact rows:

$$
\begin{array}{ccccc}
\int A^L \otimes_{A^L} B & \longrightarrow & A \otimes_{A^L} B & \longrightarrow & A^{n^2} \otimes_{A^L} B \\
\| & & \| & & \| \\
\int B & \longrightarrow & A[Z_1,\ldots,Z_n] & \longrightarrow & A[Z_1,\ldots,Z_n]^{n^2} \\
& & h & \rightsquigarrow & \delta_i\delta_j(h)
\end{array}
$$

By assumption we have $Z_n = \sum \xi_i h_i$, $\xi_i \in A$, $h_i \in B$. As in the case $n = 1$ (since $\delta_i\delta_n(Z_n) = \delta_n\delta_i(Z_n) = 0$) we obtain a presentation of Z_n with $\xi_i \in \int A^L$ and deduce $H^1(L_0, A) = 0$ and $A^{L_0}[x] = A$ for a suitable $x \in A$.

Now $\bar{L} = L/L_0$ acts on A^{L_0}. In order to proceed by induction, we have to verify that

1. Spec $A^{L_0} \longrightarrow$ Spec A^L is faithfully flat,

2. $A^{L_0} \otimes_{A^L} A^{L_0} \longrightarrow A^{L_0}[Z_1,\ldots,Z_{n-1}]$ is an isomorphism.

The first property is clear because $A^{L_0} \subset A = A^{L_0}[x]$ is faithfully flat, and Spec $A \longrightarrow$ Spec A^L is surjective.

Consider the following commutative diagram:

$$
\begin{array}{ccc}
A^{L_0} \otimes_{A^L} A^{L_0} & \xrightarrow{\;\;m_0\;\;} & A^{L_0}[Z_1,\ldots,Z_{n-1}] \\
\Big\downarrow i \;\Big\uparrow \pi & & \Big\downarrow j \;\Big\uparrow \psi \\
A \otimes_{A^L} A & \xrightarrow[\;\;m_1\;\;]{} & A[Z_1,\ldots,Z_n] \\
\| & & \| \\
A^{L_0}[x] \otimes_{A^L} A^{L_0}[x] & & A^{L_0}[x][Z_1,\ldots,Z_n]
\end{array}
$$

defined by

- $i(a \otimes b) = a \otimes b$,
- $j(h(Z_1,\ldots,Z_{n-1})) = h(Z_1,\ldots,Z_{n-1})$,
- $\pi(a(x) \otimes b(x)) = a(0) \otimes b(0)$

- $\psi(h(x, Z_1,..., Z_n)) = h(0, Z_1,..., Z_{n-1}, -\exp Z_1\delta_1 \circ \cdots \circ \exp Z_{n-1}\delta_{n-1}(x))$,

- $m_0(a \otimes b) = a \cdot \exp Z_1\delta_1 \circ \cdots \circ \exp Z_{n-1}\delta_{n-1}(b)$,

- $m_1(a(x) \otimes b(x)) = a(x) \cdot \exp Z_1\delta_1 \circ \cdots \circ \exp Z_{n-1}\delta_{n-1} \circ \exp Z_n\delta_n(b(x))$.

m_1 is an isomorphism by assumption, ψ is obviously surjective and i is injective. This implies that m_0 is an isomorphism and 2. is proved.

By induction hypothesis we obtain $H^1(L/L_0, A^{L_0}) = 0$. Together with $A^{L_0}[x] = A$ this implies $H^1(L, A) = 0$. $\qquad\square$

COROLLARY 2.2 *Let A be a commutative K-algebra, which we assume to be reduced. Let $L = \sum_{i=1}^n K\delta_i \subseteq \mathrm{Der}_K(A)$ be a nilpotent Lie-algebra and assume that the δ_i are locally nilpotent. Assume, moreover, that*

1. *A is a faithfully flat A^L-algebra.*

2. *The map $A \otimes_K A \longrightarrow A[Z_1, \ldots, Z_n]$ defined by L is surjective.*

Then $H^1(L, A) = 0$.

Proof. We have to prove that the canonical map $A \otimes_{A^L} A \longrightarrow A[Z_1, \ldots, Z_n]$ is injective.

In [G-P1], proof of Proposition 1.6, we proved that there is a dense open subset $\cup D(f_i) \subseteq \mathrm{Spec}\, A$, $f_i \in A^L$ such that $\mathrm{Spec}\, A_{f_i} \longrightarrow \mathrm{Spec}\, A^L_{f_i}$ is a trivial quotient. This implies that

$$(A \otimes_{A^L} A)_{f_i} = A_{f_i} \otimes_{A^L_{f_i}} A_{f_i} \longrightarrow A_{f_i}[Z_1, \ldots, Z_n]$$

is an isomorphism.

Since $\cup D(f_i)$ is dense, we obtain that $A \otimes_{A^L} A \longrightarrow A[Z_1, \ldots, Z_n]$ is injective, which proves the corollary. $\qquad\square$

COROLLARY 2.3 *Let A and L be as in Theorem 2.1 (respectively, moreover, that A is reduced). Assume that the map $A \otimes_{A^L} A \longrightarrow A[Z_1, \ldots, Z_n]$ defined by L is an isomorphism (respectively the map $A \otimes_K A \longrightarrow A[Z_1, \ldots, Z_n]$ defined by L is surjective) and A is a flat A^L-algebra. Then there is an open subset $U \subseteq \mathrm{Spec}\, A^L$ such that $\mathrm{Spec}\, A \longrightarrow U$ is a locally trivial geometric quotient.*

Example. (cf. [D-F])

$$A = K[x_1, x_2, y_1, y_2, z], \quad \delta = x_1\tfrac{\partial}{\partial x_2} + y_1\tfrac{\partial}{\partial y_2} + (1 + x_1 y_2^2)\tfrac{\partial}{\partial z}.$$

We obtain $A^\delta = K[u_1, u_2, u_3, u_4, u_5]$ with $u_1 = x_1$, $y = y_1$, $y_3 = x_1 y_2 - x_2 y_1$, $u_4 = 3y_1 z - x_1 y_2^3 - 3y_2$, $u_5 = 3x_1^3 z - 3x_1^2 x_2 y_2^2 + 3x_1 x_2^2 y_1 y_2 - x_2^3 y_1 - 3x_1 x_2$, with relation $u_2 u_5 - u_1^2 u_4 - u_3^3 - 3u_1 u_3 = 0$.

We define $F(t) := (\exp t\delta)(z) = \frac{1}{3}x_1 y_1^2 t^3 + x_1 y_1 y_2 t^2 + (1 + x_1 y_2^2)t + z$. Then $d := 9 \cdot \mathrm{disc}(F) = x_1(x_1^2 y_2^6 - 6x_1 y_1 y_2^3 z + 6x_1 y_2^4 + 9y_1^2 z^2 - 18y_1 y_2 z + 9y_2^2) + 4 = u_1 u_4^2 + 4 \in A^\delta$ and we obtain that $B_1 := A_d[t]/F$ is étale over A_d. δ extends to B_1 by $\delta(t) = -1$ and then we have $B_1^\delta[t] = B_1$.

Altogether we see: $\mathrm{Spec}\, A = D(x_1) \cup D(d)$, on the open set $D(x_1)$ a trivial quotient exists since $A_{x_1}^\delta[x_2] = A_{x_1} =: B_2$, but on $D(d)$ the quotient does not exist since over $D(d)$ we have fibres of different dimensions (here flatness fails, although the action is free in the sense of Mumford, cf. [D-F]). On the other hand, we have constructed an explicit étale covering $\{\mathrm{Spec}\, B_1, \mathrm{Spec}\, B_2\}$ of $\mathrm{Spec}\, A$ with $H^1(K\delta, B_i) = 0$, as it should be, according to the conjecture of the introduction.

References

[C-E]   Cartan, H; Eilenberg: Homological algebra. Princeton University Press, 1956.

[D-F]   Deveney, J.K.; Finston, D.R.: A proper G_a Action on $\mathbb{C}^5$ which is not locally trivial. Proc. AMS **123** (3), 651–655 (1995).

[D-F-G]   Deveney, J.K.; Finston, D.R.; Gehrke, M.: G_a Actions on $\mathbb{C}^n$. Comm. Algebra, **22**(12), 4977–4988 (1994).

[D-G]   Demazure, M.; Gabriel, P.: Groupes Algébriques. North-Holland, Amsterdam 1970.

[G]   Godement, R.: Introduction à la Théorie des Groupes de Lie. Publ. Math. de l'Université Paris VII, 1982.

[G-P1]   Greuel, G.-M.; Pfister, G.: Geometric quotients of unipotent group actions. Proc. London Math. Soc. (3) **67** 75–105, 1993.

[G-P2]   Greuel, G.-M.; Pfister, G.: Moduli spaces for torsion free modules on curve singularities. I. J. Algebraic Geometry **2**, 81–135 (1993).

[G-H-P]   Greuel, G.-M.; Hertling, C.; Pfister, G.: Moduli spaces of semiquasihomogeneous singularities with fixed principal part. Preprint 264, University of Kaiserslautern, 1995. To be published in J. Algebraic Geometry.

[H]   Hadan, C.: Modulräume für ebene Kurvensingularitäten. Dissertation, Berlin 1996.

[K-M-T]   Kambayashi, T.; Miyanishi, M.; Takeuchi, M.: Unipotent algebraic groups. Lecture Notes in Mathematics 414, Springer, Berlin 1974.

[L-P]   Laudal, O.; Pfister, G.: Local moduli and singularities. Lecture Notes in Mathematics 1310 (Springer, Berlin, 1988).

[M-F]   Mumford, D.; Fogarty, J.: Geometric invariant theory. 2nd enlarged edn, Ergebnisse der Mathematik und ihrer Grenzgebiete 34 (Springer, Berlin, 1982).

[P]   Popp, H.: Moduli Theory and Classification Theory. Lecture Notes in Mathematics 620 (Springer, Berlin 1977).

Progress in Mathematics, Vol. 162, © 1998 Birkhäuser Verlag Basel/Switzerland

Hodge Numbers for Isolated Singularities of Non-degenerate Complete Intersections

Helmut A. Hamm
Mathematisches Institut der WWU
Einsteinstr. 62
48161 Münster
GERMANY

Introduction

In the early time of singularity theory one of the main objects of interest has been the link of an isolated singularity. This is an odd-dimensional differentiable manifold; an interesting question was under which condition it is an exotic sphere. The first case which has been studied in detail was that of a hypersurface singularity. In this case the cohomology of the link could be determined using the monodromy map on the Milnor fibre. In the case of a complete intersection the relation between link and Milnor fibre is more complicated.

In order to calculate the Betti numbers of the link it is useful to study finer invariants. For arbitrary isolated singularities Steenbrink [16] has introduced a mixed Hodge structure on the cohomology of the link as well as on the vanishing cohomology (i.e. the reduced cohomology of the Milnor fibre). The main object of this paper is to show how one can calculate the corresponding Hodge numbers in the case of a complete intersection which satisfies a certain nondegeneracy condition (in the framework of Newton diagrams). It will turn out that it is only necessary to look at the link *or* at the vanishing cohomology. In the hypersurface case, the vanishing cohomology has been studied by Danilov [3], see also [18]. Here we prefer to work with the link because this is perhaps a bit simpler, and extend the investigation to the case of a complete intersection.

Furthermore we admit V-manifolds because this has numerical advantages. In order to handle toric V-manifolds easily we represent them as quotients of quasi-affine varieties by some torus action, see [5] or [1]. We will see that finally everything can be reduced to the calculation of Hodge numbers of complete intersections in a torus; here we may use the results of Danilov and Khovanskii [6], see also [11].

The author would like to thank the referee for his remarks and the suggestion to add examples, as well as his teacher by whom he was introduced more than twenty-five years ago into singularity theory[1].

1   Mixed Hodge numbers for the link and the vanishing cohomology

This chapter is based on results of J. Steenbrink [16], see also [17], there are, however, some modifications of the approach.

1.1. First let us look at the link. Let us suppose that X is a complex analytic subset of $\mathbf{C}^N$ which is purely of dimension $n > 0$ and has an isolated singularity at 0. Let $0 < \epsilon < \epsilon' \ll 1$, and let $K := S_\epsilon \cap X$ be the link of X. It is convenient to replace X by $B_{\epsilon'} \cap X$, where $B_{\epsilon'}$ is the open ball of radius ϵ' around 0.

The cohomology groups of K carry a mixed Hodge structure. In fact, let $\pi\colon \tilde{X} \to X$ be a V-resolution of $(X, 0)$, which means that $\tilde{X}$ is a V-manifold, π is proper and holomorphic, $D := \pi^{-1}(\{0\})$ (the exceptional set) is a divisor with V-normal crossings, and $\pi|(\tilde{X} \setminus D) \to X \setminus \{0\}$ is biholomorphic, see [15]. We assume moreover that π is projective. There is a natural way to define a mixed Hodge structure on the cohomology groups $H^k(D)$ (see [4], [15]) and $H^k(\tilde{X}, \tilde{X} \setminus D) = H_D^k(\tilde{X})$ (see [7], [16]). We will always take cohomology with complex coefficients. Furthermore, if $\tilde{X}$ is not smooth the role of the complex of germs of holomorphic forms is taken over by the complex which is denoted by $\tilde{\Omega}_{\tilde{X}}^{\cdot}$ in [15](1.7), for simplicity it will be denoted here by $\Omega_{\tilde{X}}^{\cdot}$.

LEMMA 1.1.1 *[16] We have an exact sequence:*

$$\ldots \longrightarrow H^k(\tilde{X}, \tilde{X} \setminus D) \longrightarrow H^k(D) \longrightarrow H^k(K) \longrightarrow \ldots$$

Proof. This follows from the exact cohomology sequence

$$\ldots \longrightarrow H^k(\tilde{X}, \tilde{X} \setminus D) \longrightarrow H^k(\tilde{X}) \longrightarrow H^k(\tilde{X} \setminus D) \longrightarrow \ldots$$

because D is a deformation retract of $\tilde{X}$ and K is a deformation retract of $X \setminus \{0\}$ which is isomorphic to $\tilde{X} \setminus D$. $\square$

Now we may introduce in a unique way a mixed Hodge structure on the cohomology groups of K such that we get an exact sequence of mixed Hodge structures. In fact, the exact sequence above splits into short exact sequences. Here, the following result which is due to M. Goresky and R. MacPherson is essential:

---

[1] "like ... Eggeberth he would caligulate by multiplicables the alltitude and malltitude ..." (J.Joyce, *Finnegans Wake*)

THEOREM 1.1.2 *(cf. [8], [16]) The mapping $H^k(\tilde{X}) \to H^k(\tilde{X} \setminus D)$ is surjective for $k < n$ and the zero map for $k \geq n$.*

Proof. The proof is based on the decomposition theorem which is due to Bernstein-Beilinson-Deligne-Gabber [2] in the algebraic and to M. Saito [19] in the analytic context. The proof of Theorem 1.1.2 (in the algebraic case) is given in [G-M] in the language of homology. We will proceed in a slightly different way.

The decomposition theorem of M. Saito ([19] Cor. 5.4.8) implies that $R\pi_* \mathbf{C}_{\tilde{X}} \simeq \mathcal{IC}_X \oplus \mathcal{M}$ where $\mathcal{M}$ is some complex of sheaves concentrated on 0 given by a complex M of $\mathbf{C}$-vector spaces, which implies $H^k(\tilde{X}) \simeq IH^k(X) \oplus H^k(M)$. Now it is well-known that $IH^k(X) \simeq H^k(X \setminus \{0\})\}$ if $k < n$, $IH^k(X) \simeq im\{H^k(X) \to H^k(X \setminus \{0\})\}$ if $k = n$ and $IH^k(X) \simeq H^k(X)$ if $k > n$, because X has an isolated singularity (see e.g. [N] p. 283). This implies that $H^k(\tilde{X}) \to H^k(\tilde{X} \setminus D) \simeq H^k(X \setminus \{0\})$ is surjective for $k < n$. Furthermore, $\mathcal{M}$ is concentrated on 0, so the map $H^k(M) \to H^k(X \setminus \{0\})$ is zero. Now let $k \geq n$. Then $H^k(X) = 0$, so $IH^k(X) = 0$. This implies that the map $H^k(\tilde{X}) \to H^k(\tilde{X} \setminus D) \simeq H^k(X \setminus \{0\})$ is zero. $\qquad \square$

Obviously this implies

THEOREM 1.1.3 *(cf. [16] (1.12)) One has short exact sequences:*

$$0 \longrightarrow H^k(\tilde{X}, \tilde{X} \setminus D) \longrightarrow H^k(D) \longrightarrow H^k(K) \longrightarrow 0 \ \ if \ k < n,$$

$$0 \longrightarrow H^n(\tilde{X}, \tilde{X} \setminus D) \longrightarrow H^n(D) \longrightarrow 0,$$

$$0 \longrightarrow H^{k-1}(K) \longrightarrow H^k(\tilde{X}, \tilde{X} \setminus D) \longrightarrow H^k(D) \longrightarrow 0 \ \ if \ k > n.$$

Since $H^k(\tilde{X}, \tilde{X} \setminus D) \to H^k(D)$ is a mapping of mixed Hodge structures we define a mixed Hodge structure on $H^k(K)$ uniquely in such a way that the exact sequences in Theorem 1.1.3 are sequences of mixed Hodge structures.

In fact, this mixed Hodge structure is independent of the choice of the projective V-resolution $\pi \colon \tilde{X} \to X$: Let $\pi_j \colon \tilde{X}_j \to X$, $j = 1, 2$, be two such V-resolutions with exceptional divisors D_1 and D_2, and let $\psi \colon \tilde{X}_3 \to \tilde{X}_1 \times_X \tilde{X}_2$ be a V-resolution of $(\tilde{X}_1 \times_X \tilde{X}_2, D_1 \times D_2)$, $D_3 := \psi^{-1}(D_1 \times D_2)$. This yields a V-resolution $\pi_3 \colon \tilde{X}_3 \to X$ of $(X, 0)$. For $j = 1, 2$ we have a square of mixed Hodge structures

$$\begin{array}{ccc} H^k(\tilde{X}_j, \tilde{X}_j \setminus D_j) & \longrightarrow & H^k(D_j) \\ \downarrow & & \downarrow \\ H^k(\tilde{X}_3, \tilde{X}_3 \setminus D_3) & \longrightarrow & H^k(D_3) \end{array}$$

Because of Theorem 1.1.3 we get that the induced mixed Hodge structure on $H^k(K)$ is the same in all three cases.

Now let us introduce the following notation: If the cohomology groups $H^k(A, B)$ are equipped with a mixed Hodge structure, let

$$h^{kpq}(A, B) := \dim Gr_F^p Gr_{p+q}^W H^k(A, B) \text{ and}$$

$$e^{pq}(A, B) := \sum (-1)^k h^{kpq}(A, B)$$

The notation $e^{pq}(A, B)$ has been introduced by [6], but with respect to cohomology with compact supports instead of ordinary cohomology.

It is known that $h^{kpq}(D) = 0$ unless $p + q \leq k$, since D is a compact algebraic variety.

Furthermore it is known that the cup product gives a non-degenerate pairing of Hodge structures: $H^k(D) \otimes H^{2n-k}(\tilde{X}, \tilde{X} \setminus D) \longrightarrow H^{2n}(\tilde{X}, \tilde{X} \setminus D) \simeq \mathbf{C}(-n)$. Therefore we have: $h^{kpq}(\tilde{X}, \tilde{X} \setminus D) = h^{2n-k,n-p,n-q}(D)$. This implies that $h^{kpq}(\tilde{X}, \tilde{X} \setminus D) = 0$ unless $p + q \geq k$. Since we know now that the exact sequence in Lemma 1.1.1 is compatible with mixed Hodge structures we can deduce

LEMMA 1.1.4 *(cf. [16](1.15))*: $h^{kpq}(K) = h^{2n-k-1,n-p,n-q}(K)$.
This means that it is sufficient to look at the Hodge numbers $h^{kpq}(K)$ for $k < n$.

Similarly because of Theorem 1.1.3 we obtain

THEOREM 1.1.5 *The Hodge numbers of D determine the Hodge numbers of K. In particular, if $k < n$:*

$$\begin{cases} h^{kpq}(K) = h^{kpq}(D), & \text{if } p + q < k, \\ h^{kpq}(K) = h^{kpq}(D) - h^{2n-k,n-p,n-q}(D), & \text{if } p + q = k, \\ h^{kpq}(K) = 0, & \text{if } p + q > k. \end{cases}$$

REMARK 1.1.6 If we replace X by an isomorphic space germ the Hodge numbers of K are not changed.

1.2. It is well-known that the germ of $(X, 0)$ of X at 0 is isomorphic to a germ defined by polynomials because we have an isolated singularity. So we may assume because of Remark 1.1.6 that there is a purely n-dimensional algebraic variety V which is smooth outside 0 and represents the germ $(X, 0)$, i.e. X is an open subset of V. Then there is a more direct way to define a mixed Hodge structure on the cohomology of the link because we have a canonical mixed Hodge structure on the local cohomology groups $H^k(V, V \setminus \{0\})$, and $H^k(V, V \setminus \{0\}) \simeq \tilde{H}^{k-1}(K)$. We will not discuss the exact relationship with the mixed Hodge structure introduced in Section 1.1 but show that the corresponding Hodge numbers are the same. From the numerical point of view the advantage of the approach in this section is that we need only an algebraic (not projective) V-resolution. We start with the following consequence of the decomposition theorem:

THEOREM 1.2.1 *Suppose that V is compact.*

a) *$H^k(V \setminus \{0\})$ and $H^k(V)$ have a pure Hodge structure if $k < n$ resp. $k > n$,*

b) *The map $Gr_k^W H^k(V) \to Gr_k^W H^k(V \setminus \{0\})$ is injective for $k \leq n$ and surjective for $k \geq n$.*

Proof. Let $\tilde{\pi} \colon \tilde{V} \to V$ be an algebraic V-resolution of $(V, 0)$ and $D := \tilde{\pi}^{-1}(\{0\})$. Since $Gr_k^W H^{k-1}(D)$ vanishes the mapping $\tilde{\pi}$ induces an injective mapping $Gr_k^W H^k(V, \{0\}) \simeq Gr_k^W H^k(\tilde{V}, D) \to Gr_k^W H^k(\tilde{V}) \simeq H^k(\tilde{V})$. Furthermore, the decomposition theorem in [2] (Th. 6.2.5) implies that $R\tilde{\pi}_* \mathbf{C}_{\tilde{V}} \simeq \mathcal{IC}_V \oplus \mathcal{M}$ where $\mathcal{M}$ is some complex of sheaves concentrated on 0 given by a complex M of $\mathbf{C}$-vector spaces, which implies $H^k(\tilde{V}) \simeq IH^k(V) \oplus H^k(M)$. (As the referee has remarked, one can show that this decomposition is compatible with Hodge structures, reducing to the case where $\tilde{\pi}$ is projective and applying Saito's decomposition theorem, which shows that $IH^k(V)$ is equipped with a pure Hodge structure of weight k.) Therefore we obtain induced mappings $Gr_k^W H^k(V, \{0\}) \to IH^k(V)$ and $Gr_k^W H^k(V, \{0\}) \to H^k(M)$. The latter is the zero map since $\mathcal{M}$ is concentrated on 0, so the former is injective. By duality we obtain that $IH^k(V) \to Gr_k^W H^k(V \setminus \{0\})$ is surjective. In total, we obtain that the map $Gr_k^W H^k(V) \to Gr_k^W H^k(V \setminus \{0\})$ splits into the composition $Gr_k^W H^k(V) \to IH^k(V) \to Gr_k^W H^k(V \setminus \{0\})$ where the first map is injective (we were allowed to pass from $H^k(V, \{0\})$ to $H^k(V)$) and the second one is surjective.

As mentioned earlier, it is well-known that $IH^k(V) \simeq H^k(V \setminus \{0\})\}$ if $k < n$, $IH^k(V) \simeq im\{H^k(V) \to H^k(V \setminus \{0\})\}$ if $k = n$ and $IH^k(V) \simeq H^k(V)$ if $k > n$, because V has an isolated singularity. The rest is easy. $\qquad\square$

Now let us look at the exact cohomology sequence of mixed Hodge structures

$$\ldots \to H^k(V, V \setminus \{0\}) \to H^k(V) \to H^k(V \setminus \{0\}) \to \ldots$$

Of course, we know that $Gr_r^W H^k(V) = 0$ for $r > k$, $Gr_r^W H^k(V \setminus \{0\}) = 0$ for $r < k$. Because of Theorem 1.2.1 we get:

$$Gr_r^W H^k(V, V \setminus \{0\}) = 0 \text{ if } k \leq n, r \geq k \text{ or } k > n, r < k.$$

Furthermore, we obtain exact sequences:

$$0 \to Gr_k^W H^k(V) \to H^k(V \setminus \{0\}) \to Gr_k^W H^{k+1}(V, V \setminus \{0\}) \to 0 \text{ if } k < n,$$

$$0 \to Gr_k^W H^k(V, V \setminus \{0\}) \to H^k(V) \to Gr_k^W H^k(V \setminus \{0\}) \to 0 \text{ if } k > n,$$

and isomorphisms $Gr_r^W H^{k+1}(V, V \setminus \{0\}) \simeq Gr_r^W H^{k+1}(V)$ if $k < n, r < k$ as well as $Gr_r^W H^k(V, V \setminus \{0\}) \simeq Gr_r^W H^{k-1}(V \setminus \{0\})$ if $k > n, r > k$.

In this section we consider on $H^k(K)$ the mixed Hodge structure coming from the isomorphism $\tilde{H}^k(K) = H^{k+1}(V, V \setminus \{0\})$. As above, let $\tilde{\pi} \colon \tilde{V} \to V$ be an algebraic (not necessarily projective!) V-resolution of $(V, 0)$ and $D := \tilde{\pi}^{-1}(\{0\})$. Then we have for the corresponding Hodge numbers:

THEOREM 1.2.2

a) $h^{kpq}(K) = h^{2n-k-1,n-p,n-q}(K)$

b) *If $k < n$:*

$$\begin{cases} h^{kpq}(K) = h^{kpq}(D)\,, & \text{if } p+q < k, \\ h^{kpq}(K) = h^{kpq}(D) - h^{2n-k,n-p,n-q}(D)\,, & \text{if } p+q = k, \\ h^{kpq}(K) = 0\,, & \text{if } p+q > k. \end{cases}$$

c) $e^{pq}(K) = e^{pq}(D) - e^{n-p,n-q}(D).$

Proof. Without loss of generality we may assume that V is compact.

a) We have $\tilde{H}^k(V) = H^k(V, \{0\}) \simeq H^k(\tilde{V}, D)$, $H^k(V \setminus \{0\}) \simeq H^k(\tilde{V} \setminus D) \simeq H^{2n-k}(\tilde{V}, D)$. Therefore we get the duality statement from the preceding considerations.

b) This follows using the exact cohomology sequence of the pair $(\tilde{V}, D)$ and the preceding considerations.

c) The exact sequence of mixed Hodge structures

$$\ldots \to H^k(V) \to H^k(V \setminus \{0\}) \to H^{k+1}(V, V \setminus \{0\}) \to \ldots$$

leads to a corresponding sequence of mixed Hodge structures

$$\ldots \to H^k(V, \{0\}) \to H^k(V \setminus \{0\}) \to H^k(X \setminus \{0\}) \to \ldots \text{ or}$$

$$\ldots \to H^k(\tilde{V}, D) \to H^k(\tilde{V} \setminus D) \to H^k(\tilde{X} \setminus D) \to \ldots$$

where $H^k(\tilde{X} \setminus D) \simeq H^k(K)$, so $e^{pq}(K) = e^{pq}(\tilde{V} \setminus D) - e^{pq}(\tilde{V}, D) = e^{n-p,n-q}(\tilde{V}, D) - e^{pq}(\tilde{V}, D) = e^{n-p,n-q}(\tilde{V}) - e^{n-p,n-q}(D) - e^{pq}(\tilde{V}) + e^{pq}(D)$, and $e^{n-p,n-q}(\tilde{V}) = e^{pq}(\tilde{V})$. $\qquad\square$

COROLLARY 1.2.3 *The Hodge numbers for $H^k(K)$ with respect to the mixed Hodge structures introduced in Section 1.1 and 1.2 coincide.*

Proof. This follows from Lemma 1.1.4, Theorem 1.1.5 and 1.2.2, taking $\tilde{\pi}$ to be projective. $\qquad\square$

1.3. Now let us consider the situation of Section 1.1 or 1.2, and let $\tilde{\pi}$ and D be accordingly chosen. If we are in the situation of Section 1.1 let us assume that we have an automorphism $\gamma\colon X \to X$ of finite order which admits a lifting $\tilde{\gamma}\colon \tilde{X} \to \tilde{X}$; in the situation of Section 1.2 we assume the same with V and $\tilde{V}$ instead of X and $\tilde{X}$. We have $\gamma(K) \subset K$ and $\tilde{\gamma}(D) \subset D$. Of course, $\gamma^*\colon H^k(K) \to H^k(K)$ and $\tilde{\gamma}^*\colon H^k(D) \to H^k(D)$ are automorphisms of mixed Hodge structures of finite order. So we get a splitting of $H^k(K)$ into subspaces where γ^* acts as multiplication by a root of unity λ. Therefore we get corresponding Hodge numbers $h_\lambda^{kpq}(K)$. In the context of duality it is important to note that $h_\lambda^{kpq}(K) = h_{\lambda^{-1}}^{kpq}(K)$ because $\lambda^{-1} = \bar{\lambda}$. It will be convenient to define $\underline{h}^{kpq}(K) := \sum_\lambda h_\lambda^{kpq}(K) < \lambda > \in \mathbf{Z}[\mathbf{C}^*]$. We can proceed similarly in the case of D, using $\tilde{\gamma}$ instead of γ. Then we have the obvious analogues of Lemma 1.1.4, Theorem 1.1.5 and Theorem 1.2.2:

THEOREM 1.3.1

a) $h_\lambda^{kpq}(K) = h_\lambda^{2n-k-1,n-p,n-q}(K)$.

b) *If* $k < n$:

$$\begin{cases} h_\lambda^{kpq}(K) = h_\lambda^{kpq}(D), & \text{if } p+q < k, \\ h_\lambda^{kpq}(K) = h_\lambda^{kpq}(D) - h_\lambda^{2n-k,n-p,n-q}(D), & \text{if } p+q = k, \\ h_\lambda^{kpq}(K) = 0, & \text{if } p+q > k. \end{cases}$$

c) $e_\lambda^{pq}(K) = e_\lambda^{pq}(D) - e_\lambda^{n-p,n-q}(D)$.

1.4. Again let us return to the situation of Section 1.1 or 1.2 but in addition let us assume now that X is a complete intersection. Let us recall the following consequence of the local Lefschetz theorem [9]:

LEMMA 1.4.1

$$H^k(K) \simeq \mathbf{C} \text{ if } k = 0, 2n - 1 \text{ and } n > 1,$$

$$H^k(K) = 0 \text{ if } k \neq 0, n - 1, n, 2n - 1.$$

Of course, $h^{000}(K) = 1$ for $n > 1$. Therefore it is sufficient to look at the Hodge numbers $h^{n-1,p,q}(K)$ which are determined by the corresponding numbers $e^{pq}(K)$ because of Theorem 1.1.5. For simplicity let $\tilde{e}^{pq} := e^{pq} - 1$ if $p = q = 0$ and $\tilde{e}^{pq} := e^{pq}$ otherwise. So:

LEMMA 1.4.2 *If* $n > 1$ *we have* $h^{n-1,p,q}(K) = (-1)^{n-1}\tilde{e}^{pq}(K)$ *if* $p+q \leq n-1$, $h^{n-1,p,q}(K) = 0$ *otherwise.*

For $n = 1$ the same holds with e^{pq} instead of $\tilde{e}^{pq}$. In the situation of Section 1.3 we may add in Lemma 1.4.2 everywhere a subscript λ.

Furthermore we can deduce from Theorem 1.1.3 and Lemma 1.4.1 immediately:

LEMMA 1.4.3

   a) The mixed Hodge structure on $H^k(D)$ is pure of weight k if $k \neq n-1$.

   b) If X is smooth the mixed Hodge structure on $H^k(D)$ is pure of weight k for any k.

1.5. In general, in order to compute the Hodge numbers of K it is useful to consider the following situation. Similarly as above, let $(Y, 0) \subset (\mathbf{C}^N, 0)$ be the germ of a complex analytic subset Y which is purely of dimension $n+1$ and has 0 as isolated singularity. Let $f: Y \to \mathbf{C}$ be a holomorphic function such that $f|Y \setminus \{0\}$ is nonsingular. Then $X := f^{-1}(\{0\})$ is a complex analytic subset of $\mathbf{C}^N$ which is purely of dimension n and has 0 as isolated singularity. Let $0 < \epsilon < \epsilon' \ll 1$, and let $K := S_\epsilon \cap Y$ and $K_0 := S_\epsilon \cap X$ the corresponding links of Y and X. Again it is convenient to replace the sets Y and X by their intersections by $B_{\epsilon'}$.

Similarly as before, let $\pi: \tilde{Y} \to Y$ be a V-resolution of $(Y, 0)$, i.e. $\tilde{Y}$ is a V-manifold, π is proper and holomorphic, $D := \pi^{-1}(\{0\})$ is a divisor with V-normal crossings, and $\pi|(\tilde{Y} \setminus D) \to Y \setminus \{0\}$ is biholomorphic. Furthermore, we ask that π is projective and that $\pi^{-1}(X)$ is a divisor with V-normal crossings; let D_0 be the intersection of the strict transform $\tilde{X}$ of X by D. Then the restriction of π to $\tilde{X}$ defines a V-resolution of $(X, 0)$ with D_0 as exceptional set.

There is a natural way to endow the cohomology groups of (K, K_0) with a mixed Hodge structure: the mapping $H^k(K) \to H^k(K_0)$ is induced by an epimorphism of the complexes used in [16] to define the mixed Hodge structures on $H^k(K)$ and $H^k(K_0)$, so we may take the kernel. Furthermore let us recall the local Lefschetz theorem, see [9]:

THEOREM 1.5.1 $H^k(K, K_0) = 0$ *for* $k < n$.

Together with the results on the mixed Hodge structure before we obtain:

COROLLARY 1.5.2

   a) $H^k(K_0) \simeq H^k(K)$ if $k < n-1$.

   b) There is an exact sequence of mixed Hodge structures:

$$0 \to H^{n-1}(K) \to H^{n-1}(K_0) \to H^n(K, K_0) \to H^n(K) \to 0$$

Proof. By Theorem 1.1.5 and Lemma 1.1.4 we know that $h^{npq}(K) = 0$ if $p + q > n$ and $h^{npq}(K_0) = 0$ if $p + q \leq n$, so $H^n(K) \to H^n(K_0)$ is the zero map. The rest follows from Theorem 1.5.1. $\square$

COROLLARY 1.5.3 *The Hodge numbers for K_0 can be calculated from the Hodge numbers for K and the numbers $e^{pq}(K, K_0) = e^{pq}(K) - e^{pq}(K_0)$.*

Proof. Because of Corollary 1.5.2 and Lemma 1.1.4 we need only look at $h^{n-1,p,q}(K_0)$. For $p + q \geq n$ we have $h^{n-1,p,q}(K_0) = 0$. Therefore let $p + q < n$. If $k > n$ we know that $h^{k-1,p,q}(K_0) = h^{kpq}(K) = 0$, so $h^{kpq}(K, K_0) = 0$. Because of Theorem 1.5.1 we obtain $h^{npq}(K, K_0) = (-1)^n e^{pq}(K, K_0)$, so we can calculate $h^{n-1,p,q}(K_0)$ by Corollary 1.5.2b). $\qquad\square$

1.6. Now let us turn to the vanishing cohomology. We keep the notations of the preceding section.

It is well-known that for $0 < \alpha \ll \epsilon$ the map $f \colon f^{-1}(S_\alpha) \to S_\alpha$ defines a differentiable fibre bundle, where $S_\alpha := \{ t \in \mathbf{C} \mid 0 < |t| < \alpha \}$. Let $F = F_t$ be the corresponding fibre – the Milnor fibre – and $h \colon F \to F$ the corresponding monodromy. According to [16] there is a canonical mixed Hodge structure on the cohomology groups $H^k(F)$ and $H^k_c(F)$.

Now it is known that, for any k, $h^* \colon H^k(F) \to H^k(F)$ is quasi-unipotent. This means that the eigenvalues are roots of unity. Therefore we may choose an integer $e > 0$ such that, for any k, $(h^*)^e \colon H^k(F) \to H^k(F)$ is unipotent, which means that the original eigenvalues are e-th roots of unity. Let $\hat{Y} := Y \times_{\mathbf{C}} \mathbf{C}$, where the fibre product is taken with respect to f and $\mathbf{C} \to \mathbf{C} : t \mapsto t^e$. Note that $\hat{Y}$ has an isolated singularity at 0. Let $\hat{K}$ be the link of $\hat{Y}$. We have an automorphism γ of $\hat{Y}$ of finite order: $(y, t) \mapsto (y, \zeta t)$, where $\zeta := \mathrm{e}^{2\pi i/e}$, which induces an automorphism of $\hat{K}$. Furthermore, we have a mapping $\hat{f} \colon \hat{Y} \to \mathbf{C}$, given by the projection of $Y \times_{\mathbf{C}} \mathbf{C}$ onto the second factor. Then the Milnor fibre $\hat{F}_t$ of $\hat{f}$ is isomorphic to F_{t^e}. Now γ maps $\hat{F}_t$ onto $\hat{F}_{\zeta t}$, which corresponds to the action of the semisimple part of h on F_{t^e}. The monodromy of $\hat{f}$ corresponds to the unipotent part of the monodromy of f.

THEOREM 1.6.1 *There are exact sequences of mixed Hodge structures:*

$$\ldots \to H^{k-1}(K_0) \to H^k_c(F) \xrightarrow{j} H^k(F) \to H^k(K_0) \to \ldots$$

$$\ldots \to H^k(\hat{K}) \to H^k(F) \xrightarrow{V} H^k_c(F)(-1) \xrightarrow{\partial} H^{k+1}(\hat{K}) \to \ldots$$

Proof. This is shown in [16] (2.3),(2.6b) under the additional hypothesis that we obtain a semistable smoothing of X by resolution of the singularity of $\hat{Y}$. In general this requires a further base change, which can be accomplished replacing e by some multiple ed. Let $\tilde{K}$ be the new link; $\hat{K}$ is the quotient of $\tilde{K}$ by a cyclic group of order d which acts trivially on the cohomology of $\tilde{K}$ because of the choice of e. Therefore it does not matter whether we look at the cohomology of $\hat{K}$ or $\tilde{K}$. $\qquad\square$

COROLLARY 1.6.2 *There are isomorphisms of mixed Hodge structures:*

a) $H^k(\hat{K}) \simeq H^k(F) \simeq H^k(K_0)$, $k < n - 1$,

b) $H^{n-1}(\hat{K}) \simeq H^{n-1}(F)$,

c) $0 \to H^{n-1}(F) \to H^{n-1}(K_0) \to H^n_c(F) \overset{j}{\to} H^n(F) \to H^n(K_0) \to H^{n+1}_c(F) \to 0$,

d) $0 \to H^n(\hat{K}) \to H^n(F) \overset{V}{\to} H^n_c(F)(-1) \overset{\partial}{\to} H^{n+1}(\hat{K}) \to 0$,

e) $H^{n+1}_c(F) \simeq H^{n+2}(\hat{K})(1)$,

f) $H^{k-1}(K_0) \simeq H^k_c(F) \simeq H^{k+1}(\hat{K})(1)$, $k > n + 1$.

Proof. This follows from the fact that $H^k(F) \simeq H^{2n-k}_c(F) = 0$, $k > n$, because F is Stein. $\qquad\square$

In a straightforward way we get the following notation: If the cohomology groups $H^k_c(A, B)$ are equipped with a mixed Hodge structure, let

$$h^{kpq}_{(c)}(A, B) := \dim Gr^p_F Gr^W_{p+q} H^k_c(A, B) \text{ and}$$

$$e^{pq}_{(c)}(A, B) := \sum (-1)^k h^{kpq}_{(c)}(A, B)$$

COROLLARY 1.6.3 $h^{kpq}_{(c)}(F) = h^{2n-k,n-p,n-q}(F)$.

Proof. This follows from Corollary 1.6.2 or directly from [16] (2.6e). $\qquad\square$

COROLLARY 1.6.4 *The Hodge numbers of F can be calculated from the Hodge numbers of $\hat{K}$ and the numbers $e^{pq}(F)$.*

THEOREM 1.6.5 *It is equivalent to calculate*

a) *the Hodge numbers of $\hat{K}$ and K_0,*

b) *the Hodge numbers of F.*

Proof. First let us assume that the Hodge numbers of $\hat{K}$ and K_0 are known. Because of Corollary 1.6.2a), b) we have $h^{kpq}(F) = h^{kpq}(\hat{K})$ for $k < n$; since $h^{kpq}(F) = 0$ for $k > n$ it remains to calculate $h^{npq}(F)$.

First assume that $p + q < n$. Then we know that $h^{npq}(K_0) = 0$, so we get from 1.6.2b),c),d):

(i) $h^{n-1,p,q}(\hat{K}) - h^{n-1,p,q}(K_0) + h^{npq}_{(c)}(F) - h^{npq}(F) = 0$

(ii) $h^{n,p+1,q+1}(\hat{K}) - h^{n,p+1,q+1}(F) + h^{npq}_{(c)}(F) - h^{n+1,p+1,q+1}(\hat{K}) = 0$

By subtraction we obtain:

$$h^{n,p+1,q+1}(F) - h^{npq}(F) \;=\; h^{n-1,p,q}(K_0) - h^{n-1,p,q}(\hat{K})$$
$$+ h^{n,p+1,q+1}(\hat{K}) - h^{n+1,p+1,q+1}(\hat{K})$$

so we can proceed inductively.

Similarly, if $p + q \geq n$, we have $h^{n-1,p,q}(K_0) = 0$, so we obtain instead of (i), using 1.6.2e):

(iii) $h^{npq}_{(c)}(F) - h^{npq}(F) + h^{npq}(K_0) - h^{n+2,p+1,q+1}(\hat{K}) = 0.$

Subtracting (ii) from (iii) we obtain in this case:

$$h^{n,p+1,q+1}(F) - h^{npq}(F) \;=\; h^{n,p+1,q+1}(\hat{K}) - h^{npq}(K_0)$$
$$- h^{n+1,p+1,q+1}(\hat{K}) + h^{n+2,p+1,q+1}(\hat{K})$$

so we may proceed by descending induction.

Now let us suppose that the Hodge numbers of F are known. By Corollary 1.6.3 we also know the numbers $h^{kpq}_{(c)}(F)$. From 1.6.2a),b) we obtain the Hodge numbers $h^{kpq}(\hat{K})$ for $k < n$.

If $p + q \leq n + 1$ we have $h^{n+1,p,q}(\hat{K}) = 0$, so 1.6.2d) gives

$$h^{npq}(\hat{K}) = h^{npq}(F) - h^{n,p-1,q-1}_{(c)}(F)$$

If $p + q > n + 1$, we have $h^{npq}(\hat{K}) = 0$. So we know the Hodge numbers $h^{kpq}(\hat{K})$ for $k \leq n$, by duality we obtain those for $k > n$.

Finally, $h^{kpq}(K_0) = h^{kpq}(F)$ for $k < n - 1$. If $p + q \leq n$ we obtain from (i):

$$h^{n-1,p,q}(K_0) = h^{n-1,p,q}(F) + h^{npq}_{(c)}(F) - h^{npq}(F)$$

If $p + q \geq n$ we have $h^{n-1,p,q}(K_0) = 0$. So we know $h^{kpq}(K_0)$ for $k \leq n - 1$, by duality we obtain those for $k > n - 1$. $\qquad\square$

In fact we may consider the finer invariants h^{kpq}_{λ}, where λ is an e-th root of unity, see Section 1.3, because of the automorphism γ. Let $\pi \colon \tilde{Y} \to Y$ be a V-resolution of $(Y, 0)$ such that $\pi^{-1}(X)$ is a divisor with V-normal crossings, see the preceding section. Let $\overline{Y}$ be the normalization of the fibre product of π and $t \mapsto t^e$. Then the induced map $\hat{\pi} \colon \overline{Y} \to \hat{Y}$ is a V-resolution of the singularity of $(\hat{Y}, 0)$, and the action on $\hat{Y}$ extends to an action on $\overline{Y}$.

REMARK 1.6.6

a) $h^{kpq}_1(\hat{K}) = h^{kpq}(K)$

b) $h^{kpq}_{\lambda}(K_0) = 0$ for $\lambda \neq 1.$

Finally let us observe that a much simpler analogue to Theorem 1.6.5 can be obtained when looking at the numbers e^{pq} instead of the Hodge numbers:

THEOREM 1.6.7

$$
\begin{aligned}
e^{pq}(K_0) &= e^{pq}(F) - e^{n-p,n-q}(F), \\
e^{pq}(\hat{K}) &= e^{pq}(F) - e^{n-p+1,n-q+1}(F), \\
e^{pq}(F) &= e^{p+1,q+1}(F) + e^{pq}(K_0) - e^{p+1,q+1}(\hat{K}),
\end{aligned}
$$

so it is equivalent to calculate

   a) the numbers $e^{pq}(K_0)$ and $e^{pq}(\hat{K})$,

   b) the numbers $e^{pq}(F)$.

The same holds if e^{pq} is replaced by e_λ^{pq}.

Proof. The first two formulas follow directly from Theorem 1.6.1 and Corollary 1.6.3. If we replace in the second one (p, q) by $(p + 1, q + 1)$ and subtract the result from the first one we obtain the third formula. $\qquad\square$

1.7. In addition let us assume now that $(Y, 0)$ is a complete intersection. Of course the same holds for X. Let us recall:

LEMMA 1.7.1 $H^k(F) = 0$ *unless* $k = 0, n$, $H^0(F) \simeq \mathbf{C}$ *if* $n > 0$.

Of course, $h_1^{000}(F) = 1$ for $n > 0$. Therefore it is sufficient to look at the Hodge numbers $h^{npq}(F)$. Obviously we have

LEMMA 1.7.2   *If* $n > 0$ *we have* $h^{npq}(F) = (-1)^n \tilde{e}^{pq}(F)$
and $h_\lambda^{npq}(F) = (-1)^n \tilde{e}_\lambda^{pq}(F)$.

For $n = 0$ this lemma holds with e^{pq}, e_λ^{pq} instead of $\tilde{e}^{pq}, \tilde{e}_\lambda^{pq}$.

Because of this lemma and Theorem 1.6.7 we will concentrate on the calculation of the numbers e^{pq} resp. e_λ^{pq} for the link in this case.

2   Nondegenerate complete intersections

2.1. First let us recall that to each fan $\mathcal{F}$ in $\mathbf{R}^m$ there corresponds an m-dimensional toric variety $T_{\mathcal{F}}$, see [12] or [14]. If $\mathcal{F}$ is simplicial, i.e. the cones in $\mathcal{F}$ are spanned by linearly independent vectors, $T_{\mathcal{F}}$ is a V-manifold, i.e. locally the quotient of a manifold by a finite group. We will prove this in a special case in Lemma 2.1.2 below.

In the sequel we will suppose that $\mathcal{F}$ is a fan in $\mathbf{R}^m$ with the following properties:

a) $\mathcal{F}$ is a subdivision of the cone $\mathbf{R}_+^m$,

b) the proper faces of $\mathbf{R}_+^m$ are cones of $\mathcal{F}$,

c) $\mathcal{F}$ is simplicial,

d) $\mathcal{F}$ contains at least one edge different from the coordinate axes.

Because of a) we have a canonical proper algebraic map $\pi\colon T_{\mathcal{F}} \to \mathbf{C}^m$, and b) implies that $\pi|T_{\mathcal{F}} \setminus D \to \mathbf{C}^m \setminus \{0\}$ is biholomorphic. Here $D := \pi^{-1}(\{0\})$. As we shall see, $T_{\mathcal{F}}$ is a V-manifold because of c), and D a divisor with V-normal crossings because of d). Altogether we get that π defines an algebraic V-resolution of $(\mathbf{C}^m, 0)$. On the cost of heavier calculations we could achieve that π is even projective but because of Section 1.2 this is not necessary.

If in addition $\mathcal{F}$ is simple, i.e. if the generators of the intersections of the edges of each cone by $\mathbf{N}^m$ can be extended to a basis of $\mathbf{Z}^m$, one knows that $T_{\mathcal{F}}$ is even smooth.

We recall the description of a toric V-manifold as a quotient of some quasi-affine variety by some torus action, see [5] or [1]; this approach is very natural in view of the example of the ordinary projective space. We will use here a variant which takes care of the fact that our fan $\mathcal{F}$ constitutes a subdivision of $\mathbf{R}_+^m$:

Let $p_1, \ldots, p_r$ be the generators of the semigroups obtained intersecting the edges of $\mathcal{F}$ by $\mathbf{N}^m$ which are different from the vectors of the canonical basis. Note that $p_{ij} > 0$ for all i, j and that $r > 0$ because of d). Then we consider the following fan $\mathcal{F}'$ in $\mathbf{R}^{r+m}$: instead of p_j we take the canonical basis vector e_j of $\mathbf{R}^{r+m}$, and the canonical basis of $\mathbf{R}^m$ is replaced by the last vectors $e_{r+1}, \ldots, e_{r+m}$ of the canonical basis of $\mathbf{R}^{r+m}$. The cones in $\mathcal{F}'$ are by definition those whose edges correspond to the edges of a cone of $\mathcal{F}$. We have a canonical map $\kappa\colon T_{\mathcal{F}'} \to T_{\mathcal{F}}$.

Note that $\mathcal{F}'$ is a subfan of the fan generated by the positive octant, so $T_{\mathcal{F}'}$ is an open subset of $\mathbf{C}^{r+m}$; in fact it is a union of orbits with respect to the canonical action of the torus $(\mathbf{C}^*)^{r+m}$ on $\mathbf{C}^{r+m}$. Now we have an action of $(\mathbf{C}^*)^r$ on $\mathbf{C}^{r+m}$: $c \circ (\zeta_1, \ldots, \zeta_r, z_1, \ldots, z_m) := (c_1\zeta_1, \ldots, c_r\zeta_r, c^{-p^1} z_1, \ldots, c^{-p^m} z_m)$, where $p^i := (p_{1i}, \ldots, p_{ri})$. Then we have:

THEOREM 2.1.1 *The map κ induces an isomorphism $T_{\mathcal{F}'}/(\mathbf{C}^*)^r \simeq T_{\mathcal{F}}$.*

Proof. This follows from the fact that the monomials $\zeta_1^{\langle p_1, q \rangle} \cdot \ldots \cdot \zeta_r^{\langle p_r, q \rangle} z^q$, $q \in \mathbf{Z}^m$, are a basis of the vector space of the $(\mathbf{C}^*)^r$-invariant Laurent polynomials. $\square$

As indicated at the beginning of this section we may use this in order to show that $T_{\mathcal{F}}$ is a V-manifold. We will need the details of the proof later on.

LEMMA 2.1.2 *$T_{\mathcal{F}}$ is a V-manifold.*

Proof. Let σ be an m-dimensional cone of $\mathcal{F}$ and σ' the corresponding cone of $\mathcal{F}'$. Without loss of generality we may assume that σ is spanned by $p_1, \dots, p_l, e_{s+1}, \dots, e_m$. Note that these vectors are linearly independent since σ is simplicial, so $l = s$. Then $T_{\sigma'} = \mathbf{C}^s \times (\mathbf{C}^*)^r \times \mathbf{C}^{m-s}$. Let $S_{\sigma'} = \mathbf{C}^s \times \{(1, \dots, 1)\} \times \mathbf{C}^{m-s} \subset T_{\sigma'}$. Then $T_{\sigma'}$ is the saturation of $S_{\sigma'}$ with respect to the action of $(\mathbf{C}^*)^r$. The stabilizer $G_{\sigma'}$ of this subset is a finite group. This implies that $T_\sigma \simeq T_{\sigma'}/(\mathbf{C}^*)^r \simeq S_{\sigma'}/G_{\sigma'}$ is a V-manifold. $\qquad\square$

By composition of the canonical map $\kappa \colon T_{\mathcal{F}'} \to T_{\mathcal{F}}$ with π we get a map $\pi' \colon T_{F'} \to \mathbf{C}^m$. Let $D' := \kappa^{-1}(D) = {\pi'}^{-1}(\{0\})$. Obviously we have

LEMMA 2.1.3

    a) $\pi'(\zeta_1, \dots, \zeta_r, z_1, \dots, z_m) = (\zeta^{p^1} z_1, \dots, \zeta^{p^m} z_m)$,

    b) $D' = T_{\mathcal{F}'} \cap \{\zeta_1 \cdot \ldots \cdot \zeta_r = 0\}$.

Of course, the irreducible components of D' are $D_1', \dots, D_r'$, with $D_j' := T_{F'} \cap \{\zeta_j = 0\}$. Using the techniques of the proof of Lemma 2.1.2 we obtain, with $D_j := \kappa(D_j')$:

LEMMA 2.1.4 *$D = \kappa(D')$ is a divisor with V-normal crossings. The irreducible components are $D_1, \dots, D_r$.*

2.2. Now let us turn to complete intersections defined by holomorphic functions $f_1, \dots, f_k$. The question is how to find an appropriate fan $\mathcal{F}$ as above.

Let $f = \sum a_q z^q$ be a convergent power series in m variables with complex coefficients, $f(0) = 0$. Then let $\operatorname{supp} f := \{q \mid a_q \neq 0\} \subset \mathbf{N}^m$ be the support of f. If the support of f intersects each coordinate axis f is called convenient.

Let M be a subset of $\mathbf{N}^m \setminus \{0\}$. M is called convenient if M intersects each coordinate axis. Let $\mathbf{C}[M]$ resp. $\mathbf{C}\langle M \rangle$ be the vector space of polynomials resp. convergent power series f such that $\operatorname{supp} f \subset M$. Of course, $\mathbf{C}[M] = \mathbf{C}\langle M \rangle$ if M is finite. Note that we have a natural map $\rho_M \colon \mathbf{C}\langle \mathbf{N}^m \rangle \to \mathbf{C}\langle M \rangle$ which is defined by omitting all monomials which do not correspond to a point of M.

Now let $M_1, \dots, M_k$ be non-empty subsets of $\mathbf{N}^m \setminus \{0\}$, $M := M_1 + \dots + M_k$. If M is convenient, i.e. $M_1, \dots, M_k$ are convenient, let Δ_M^+ be the convex hull of $M + \mathbf{N}^m$.

In general the definition of Δ_M^+ is more complicated. First, let Δ_M' be the convex hull of $M + \mathbf{N}^m$. Let Δ_M^+ be the intersection of $\mathbf{R}^m$ and all closed halfspaces H such that ∂H contains an $m-1$-dimensional compact face of Δ_M' and 0 is not contained in H. Indeed we have $\Delta_M^+ = \Delta_M'$ if M intersects each coordinate axis.

We can associate to Δ_M^+ a dual fan and consequently a toric variety. We subdivide the dual fan in such a way that the new fan $\mathcal{F}$ is simplicial. Furthermore we ask that it contains at least one edge which does not lie on an coordinate axis and that the proper faces of $\mathbf{R}_+^m$ are not subdivided. One could even achieve that $\mathcal{F}$ is simple, on the cost of massy calculations.

A fan $\mathcal{F}$ which is obtained in this way will be called adapted to M. It is easy to see that it has the properties which were demanded in Section 2.1.

From now on we suppose that $\mathcal{F}$ is chosen as in (2.1). For $j = 1, \dots, k$ let $f_j = \sum a_{jq} z^q$ be a convergent power series with *supp* $f_j \subset M_j$. Let $d_{ij} := \min\{\langle p_i, q\rangle \mid q \in M_j\}$, $F_j := \sum a_{jq}\zeta_1^{\langle p_1, q\rangle - d_{1j}} \cdot \ldots \cdot \zeta_r^{\langle p_r, q\rangle - d_{rj}} z^q$. Note that $d_{ij} > 0$, and we may define an action of $(\mathbf{C}^*)^r$ on $\mathbf{C}^k$ by $c \circ t := c^{-d^j} t$, where $d^j := (d_{1j}, \dots, d_{rj})$. Then $(F_1, \dots, F_k)$ is $(\mathbf{C}^*)^r$-equivariant, i.e. $F_j(c \circ (\zeta, z)) = c_1^{-d_{1j}} \cdot \ldots \cdot c_r^{-d_{rj}} F_j(\zeta, z)$, $j = 1, \dots, k$.

Let us assume that $f_1, \dots, f_k$ are holomorphic on an open ball $B := B_{\epsilon'}$. Let $X := B \cap \{f_1 = \dots = f_k = 0\}$. Then $F_1, \dots, F_k$ are holomorphic on $B' := {\pi'}^{-1}(B) \subset T_{\mathcal{F}'}$; let $X' := B' \cap \{F_1 = \dots = F_k = 0\}$.

LEMMA 2.2.1 ${\pi'}^{-1}(X) = D' \cup X'$.

Proof. $f_j \circ \pi' = \zeta^{d^j} F_j$. $\qquad\qquad\square$

Now let $\tilde{X} := \kappa(X')$. We will now introduce a condition under which $\tilde{X}$ is a V-manifold and intersects D (i.e. the components of D as well as their intersections) V-transversally.

Let σ be a cone in $\mathcal{F}$. Without loss of generality we may assume that it is spanned by $p_1, \dots, p_l, e_{s+1}, \dots, e_m$. Since σ is simplicial these vectors are linearly independent, so we have $l \leq s$. Furthermore we assume $l > 0$, in this case let us call σ non-trivial. For $j = 1, \dots, k$ let M_j^σ be the set of points $q \in M_j$ such that $\langle p_i, q\rangle = d_{ij}$, $i = 1, \dots, l$. Note that this set is finite. Let $f_j^\sigma := \rho_{M_j^\sigma}(f_j) = \sum a_{jq} z^q$, where the sum extends only over those q with $\langle p_i, q\rangle = d_{ij}$, $i = 1, \dots, l$. Let us consider the following conditions:

($\mathbf{C}_\sigma$) The mapping $(f_1^\sigma, \dots, f_k^\sigma)\colon \mathbf{C}^m \to \mathbf{C}^k$ has no critical point z such that $f_1^\sigma(z) = \dots = f_k^\sigma(z) = 0$, $z_1 \cdot \ldots \cdot z_s \neq 0$, and $z_{s+1} = \dots = z_m = 0$.

($\mathbf{N}_\sigma$) The mapping $(f_1^\sigma, \dots, f_k^\sigma, z_{s+1}, \dots, z_m)\colon \mathbf{C}^m \to \mathbf{C}^k$ has no critical point z such that $f_1^\sigma(z) = \dots = f_k^\sigma(z) = 0$, $z_1 \cdot \ldots \cdot z_s \neq 0$, and $z_{s+1} = \dots = z_m = 0$.

(C) The condition ($\mathbf{C}_\sigma$) holds for any non-trivial cone σ of $\mathcal{F}$.

(N) The condition ($\mathbf{N}_\sigma$) holds for any non-trivial cone σ of $\mathcal{F}$. In this case we call $(f_1, \dots, f_k)$ non-degenerate with respect to $\mathcal{F}$.

Obviously we have

LEMMA 2.2.2 *The condition* $(\mathbf{N}_\sigma)$ *implies* $(\mathbf{C}_\sigma)$.

As for condition $(\mathbf{C})$ see [10].

Let M_j^+ be the union of all M_j^σ such that σ is a non-trivial cone of $\mathcal{F}$.

LEMMA 2.2.3 *If $M_1,\dots,M_k$ are convenient and $\mathcal{F}$ is adapted to M there is a Zariski-open dense subset U of the finite-dimensional vector space $\mathbf{C}[M_1^+] \times \dots \times \mathbf{C}[M_k^+]$ such that* $(\mathbf{N})$ *holds as soon as* $(\rho_{M_1^+}(f_1),\dots,(\rho_{M_k^+}(f_k))) \in U$.

Proof. Let σ be a non-trivial cone in $\mathcal{F}$. As above, we may assume that it is spanned by $p_1,\dots,p_l,e_{s+1},\dots,e_m$ with $0 < l \leq s$. Let $V_\sigma := \mathbf{C}[M_1^\sigma] \times \dots \times \mathbf{C}[M_k^\sigma]$. It is easy to see that the equations $f_1(z) = \dots = f_k(z) = 0$ define a submanifold of codimension k in the manifold $W_\sigma := \{(z, f_1,\dots,f_k) \in \mathbf{C}^m \times V_\sigma \mid z_1 \cdot \dots \cdot z_s \neq 0, z_{s+1} = \dots = z_m = 0\}$. According to the theorem of Bertini-Sard, the regular values of the canonical map $W_\sigma \to V_\sigma$ form a Zariski-open dense subset U_σ of V_σ. Obviously, for any point of U_σ condition $(\mathbf{N}_\sigma)$ holds. The rest is clear. $\qquad\square$

For $p \in (\mathbf{N} \setminus \{0\})^m$ let $f_j^p := \sum a_{jq} z^q$, where the sum extends only over those q with $\langle p, q \rangle = \min_{q' \in supp\, f_j} \langle p, q' \rangle$ (we suppose here $f_j \neq 0$.) In the convenient case condition $(\mathbf{N})$ can be reformulated as follows:

REMARK 2.2.4 If M is convenient, $supp\, f_j = M_j$ for $j = 1,\dots,k$ and if $\mathcal{F}$ is adapted to M condition $(\mathbf{N})$ means that $(f_1,\dots,f_k)$ is non-degenerate at 0, i.e. that for any $p \in (\mathbf{N} \setminus \{0\})^m$ the mapping $(f_1^p,\dots,f_k^p)\colon (\mathbf{C}^*)^m \to \mathbf{C}^k$ has no critical point z such that $f_1^p(z) = \dots = f_k^p(z) = 0$.

Now let us apply condition $(\mathbf{C})$.

THEOREM 2.2.5 *Assume that* $(\mathbf{C})$ *holds and ϵ' is small enough. Then*

   a) X' is smooth and intersects D' transversally,

   b) X is a complete intersection of dimension $n = m - k$ which is smooth except at 0,

   c) $\tilde{X}$ is a V-manifold which intersects D V-transversally.

Proof.

   a) It is sufficient to show that X' is smooth along D' and intersects D' transversally. Let σ be an m-dimensional non-trivial cone of $\mathcal{F}$, spanned by $p_1,\dots,p_t,e_{t+1},\dots,e_m$ with $t > 0$. Let us look at a point $(\zeta',z') \in X' \cap T_{\sigma'} \cap D'$. We may assume that $\zeta'_1 = \dots = \zeta'_l = 0, \zeta'_{l+1} \cdot \dots \cdot \zeta'_r \neq 0, z'_1 \cdot \dots \cdot z'_s \neq 0, z'_{s+1} = \dots = z'_m = 0$, where $0 < l \leq t \leq s$. We want to show that X' is smooth and transversal to $D'_1 \cap \ldots \cap D'_l = \{\zeta_1 = \ldots = \zeta_l = 0\}$ at (ζ', z'). After renumbering of the coordinates $z_1, \ldots, z_t$ if necessary we may assume that $p_1, \ldots, p_l, e_{l+1}, \ldots, e_m$ are linearly independent. Then (ζ', z') belongs to the saturation of the subset S of $T_{\sigma'}$ defined by $\zeta_{l+1} = \ldots = \zeta_r = 1$, so we may assume $(\zeta', z') \in S$. Let σ_0 be the subcone of σ spanned by $p_1, \ldots, p_l, e_{s+1}, \ldots, e_m$. Then F_j coincides with $f_j^{\sigma_0}$ along $S \cap \{\zeta_1 = \ldots = \zeta_l = 0\}$, so our claim follows from the condition $(\mathbf{C}_{\sigma_0})$.

b),c) Let us begin as in the proof of a). Note that the functions $F_1, \ldots, F_k$ are by definition weighted homogeneous, in particular we have the following identities:

$$\zeta_i \partial F_j / \partial \zeta_i - p_{i1} z_1 \partial F_j / \partial z_1 - \ldots - p_{im} z_m \partial F_j / \partial z_m = -d_{ij} F_j \, , \; i = 1, \ldots, r$$

so at the point $(\zeta', z') \in X' \cap S \cap D'$ we get $p_{i1} z'_1 \partial f_j^{\sigma_0} / \partial z_1 (\zeta', z) + \ldots + p_{im} z'_m \partial f_j^{\sigma_0} / \partial z_m = 0 \, , \; i = 1, \ldots, l$. Since the matrix $(p_{ij})_{i,j=1,\ldots,l}$ is invertible we obtain from condition $(\mathbf{C}_{\sigma_0})$ that (ζ', z) is not a critical point of $(F_1, \ldots, F_k)|S$. $\qquad\square$

COROLLARY 2.2.6 *Under the assumption of Theorem 2.2.5 $\pi|\tilde{X}: \tilde{X} \to X$ is a V-resolution of $(X, 0)$ with exceptional divisor $D_0 := D \cap \tilde{X}$.*

2.3. From now on let us assume that $(\mathbf{C})$ holds and that ϵ' is chosen small enough. Then we are in the situation of Section 1.4. Let K be the corresponding link.

Using Lemma 1.4.2 and Theorem 1.2.2 we get:

THEOREM 2.3.1 *If $n > 1$ we have $h^{n-1,p,q}(K) = (-1)^{n-1}(\tilde{e}^{pq}(D_0) - e^{n-p,n-q}(D_0))$ if $p + q \leq n - 1$, $h^{n-1,p,q}(K) = 0$ otherwise.*

Notice that for the proof of Theorem 2.3.1 we may pass to the case where $f_1, \ldots, f_k$ are polynomials without affecting D_0 and the isomorphism class of the germ of X, cf. Remark 1.1.6.

So we must calculate the numbers $e^{pq}(D_0)$.

Recall that each toric variety decomposes into orbits, each orbit O_σ corresponding to a cone σ of the defining fan (see below). Here, D is the union of those orbits which correspond to nontrivial cones of $\mathcal{F}$. Therefore we get by the additivity of the numbers $e^{pq}_{(c)}$:

PROPOSITION 2.3.2 $e^{pq}(D_0) = \sum e^{pq}_{(c)}(O_\sigma \cap \tilde{X})$, *where the sum extends over all non-trivial cones σ.*

So we must calculate $e^{pq}_{(c)}(O_\sigma \cap \tilde{X})$. First let us describe O_σ, where σ is a non-trivial cone of F, spanned by $p_1, \dots, p_l, e_{s+1}, \dots, e_m$ where $0 < l \leq s$. Note that $O_{\sigma'} = \{(\zeta, z) \mid \zeta_1 = \dots = \zeta_l = z_{s+1} = \dots = z_m = 0, \zeta_{l+1} \cdot \dots \cdot \zeta_r \cdot z_1 \cdot \dots \cdot z_s \neq 0\}$. Let $S^*_\sigma := \{(\zeta, z) \in O_{\sigma'} \mid \zeta_{l+1} = \dots = \zeta_r = 1\}$. Obviously, S^*_σ may be identified with $(\mathbf{C}^*)^s$. Then $O_\sigma = \kappa(S^*_\sigma)$, and the stabilizer of S^*_σ is $(\mathbf{C}^*)^l \times \{(1, \dots, 1)\}$. So O_σ may be indentified with the algebraic quotient of $(\mathbf{C}^*)^s$ with respect to the induced $(\mathbf{C}^*)^l$ action. This yields the well-known description of O_σ: Let $W_\sigma := \{q \in \mathbf{Z}^m \mid\ <p_i, q> = 0, i = 1, \dots, l, q_{s+1} = \dots = q_m = 0\}$. Then $O_\sigma \simeq Spec\,\mathbf{C}[W_\sigma]$.

Let $M^{\sigma|}_j$ be the set of all $q \in M^\sigma_j$ with $q_\nu = 0$ for all $\nu > s$. Let K_σ be the set of all j such that $M^{\sigma|}_j \neq \emptyset$, $k_\sigma := |K_\sigma|$. If $M_1, \dots, M_k$ are convenient we have $K_\sigma = \{1, \dots, k\}$. Note that $F_j|S^*_\sigma$ corresponds to the polynomial $f^{\sigma|}_j = \sum_q a_{jq} z^q$, where the sum extends over all $q \in M^{\sigma|}_j$. Obviously $f^{\sigma|}_j = 0$ if $j \notin K_\sigma$.

Now we have the following description of $O_\sigma \cap \tilde{X}$. For any $j \in K_\sigma$ choose a point $q^*_j \in \mathbf{Z}^m$ such that $\langle p_i, q^*_j \rangle = d_{ij}, i = 1, \dots, l, q^*_{s+1} = \dots = q^*_m = 0$. Let $q'_1, \dots, q'_{s-l}$ be a basis of W_σ. The choice of this basis gives an isomorphism of $(\mathbf{C}^*)^{s-l} = Spec\,\mathbf{C}[\mathbf{Z}^{s-l}]$ onto $Spec\,\mathbf{C}[W_\sigma] \simeq O_\sigma$. For $j \in K_\sigma$ we can write $f^{\sigma|}_j$ in the form $z^{q^*_j} g_j(z^{q'_1}, \dots, z^{q'_{s-l}})$. Then we have:

LEMMA 2.3.3

a) $O_\sigma \cap \tilde{X}$ *is isomorphic to* $\{w \in (\mathbf{C}^*)^{s-l} \mid g_j(w) = 0, j \in K_\sigma\}$.

b) *If* $(f^{\sigma|}_j)_{j \in K_\sigma}$ *is non-degenerate the same holds for* $(g_j)_{j \in K_\sigma}$.

Here recall the following definitions. Let $g_1, \dots, g_k$ be arbitrary nontrivial Laurent polynomials in m variables: $g_j = \sum_q b_{jq} z^q$, where $q \in \mathbf{Z}^m$ and the sum is finite. For $p \in \mathbf{Z}^m$ let $g^p_j = \sum_q b_{jq} z^q$, where the sum extends only over those q with $\langle p, q \rangle = \min_{q' \in supp\, g_j} \langle p, q' \rangle$. Remember that $(g_1, \dots, g_k)$ is called non-degenerate if for any $p \in \mathbf{Z}^m$ the mapping $(g^p_1, \dots, g^p_k) : (\mathbf{C}^*)^m \to \mathbf{C}^k$ has no critical point such that $g^p_1(z) = \dots = g^p_k(z) = 0$. In this case the complete intersection in the torus $(\mathbf{C}^*)^m$ defined by $g_1 = \dots = g_k = 0$ is called non-degenerate, too.

LEMMA 2.3.4 *There is a Zariski open dense subset U of* $\mathbf{C}[M^+_1] \times \dots \times \mathbf{C}[M^+_k]$ *such that for any non-trivial cone $\sigma \in \mathcal{F}$ the mapping $(f^{\sigma|}_j)_{j \in K_\sigma}$ is non-degenerate as soon as* $(\rho_{M^+_1}(f_1), \dots, (\rho_{M^+_k}(f_k)) \in U$.

Proof. Similarly to the proof of Lemma 2.2.3 which refers to the special case where $M_1, \dots, M_k$ are convenient. $\square$

REMARK 2.3.5 Suppose that M is convenient, $supp\, f_j = M_j$ for $j = 1, \ldots, k$, $\mathcal{F}$ is adapted to M, and $(f_1, \ldots, f_k)$ is non-degenerate at 0. Then $(f_1^{\sigma|}, \ldots, f_k^{\sigma|})$ is non-degenerate for any non-trivial cone $\sigma \in \mathcal{F}$ (note that $K_\sigma = \{1, \ldots, k\}$ in this case).

So it remains finally to calculate the numbers $e_{(c)}^{pq}$ for a non-degenerate complete intersection in a torus. Here we refer to the algorithm given by [6] (see also [11] and the following section). In the following examples, however, we can argue more directly.

EXAMPLE 2.3.6 Let $m = 3$, $k = 2$, $M_1 := \{(4,0,0),(1,2,0)\}$, $M_2 := \{(3,2,0),(0,5,0),(0,0,60)\}$, $f_1 := z_1^4 + z_1 z_2^2$, $f_2 := z_1^3 z_2^2 + z_2^5 - z_3^{60}$. Then $M := M_1 + M_2 = \{(7,2,0),(4,5,0),(4,0,60),(4,4,0),(1,7,0),(1,2,60)\}$. The 2-dimensional compact faces of the convex hull Δ_M' of $M + \mathbf{N}^2$ lie in the hyperplanes given by the equations $10q_1 + 15q_2 + q_3 = 100$, $12q_1 + 12q_2 + q_3 = 96$. So $\Delta_M^+ = \{(q_1, q_2, q_3) \in \mathbf{R}^3 \mid q_1 \geq 0, q_2 \geq 0, q_3 \geq 0, 10q_1 + 15q_2 + q_3 \geq 100, 12q_1 + 12q_2 + q_3 \geq 96\}$. The dual fan $\mathcal{F}$ has edges spanned by e_1, e_2, e_3, p_1, p_2, where $p_1 := (10, 15, 1)$, $p_2 := (12, 12, 1)$. It is obtained by joining the edges spanned by e_1, e_2, e_3 to the edge spanned by p_1, then joining the edge spanned by p_2, which lies inside the cone spanned by p_1, e_1, e_3, to the edges of this cone. Altogether the following sets span a cone in $\mathcal{F}$:
$\{e_1, e_2, p_1\}, \{e_2, e_3, p_1\}, \{e_1, e_3, p_2\}, \{e_1, p_1, p_2\}, \{e_3, p_1, p_2\}$, as well as all subsets of these.
Obviously, $\mathcal{F}$ is already simplicial but not simple (because $\det(e_2, e_3, p_1) = 10 \neq \pm 1$).
As above, let $d_{ij} := \min\{< p_i, q > \mid q \in M_j\}$. Then $d_{11} = 40$, $d_{21} = 36$, $d_{12} = d_{22} = 60$. Let σ be a cone in $\mathcal{F}$.
If σ contains p_1 but not p_2 we have $f_1^\sigma = f_1$, $f_2^\sigma = z_1^3 z_2^2 - z_3^{60}$.
If σ contains p_2 but not p_1 we have $f_1^\sigma = z_1 z_2^2$, $f_2^\sigma = f_2$.
If σ contains p_1 and p_2 we have $f_1^\sigma = z_1 z_2^2$, $f_2^\sigma = z_1^3 z_2^2 - z_3^{60}$.
Now it is easy to verify condition (**C**); however (f_1, f_2) is not non-degenerate with respect to $\mathcal{F}$ since $(\mathbf{N}_\sigma)$ is violated for the cone σ spanned by $\{e_1, p_2\}$.
Now let us look at $O_\sigma \cap \tilde{X}$, where σ is a non-trivial cone of $\mathcal{F}$. This set is non-empty only if σ is spanned by $\{p_1\}$ or by $\{e_1, p_2\}$; in the first case we get a set B_1 of 12 points, in the second case a set B_2 of 5 points. In fact, in the first case we have $K_\sigma = \{1, 2\}$, and we may choose $q_1^* = (0, 0, 40)$, $q_2^* = (0, 0, 60)$, $q_1' = (1, 0, -10)$, $q_2' = (0, 1, -15)$. Then $f_j^{\sigma|} = z^{q_j^*} g_j(z^{q_1'}, z^{q_2'})$, where $g_1(w) = w_1^4 + w_1 w_2^2$, $g_2(w) = w_1^3 w_2^2 - 1$, and $O_\sigma \cap \tilde{X}$ is isomorphic to $\{w \in (\mathbf{C}^*)^2 \mid g_1 = g_2 = 0\}$ which consists of 12 points. In the second case we have $K_\sigma = \{2\}$, and using $q_2^* = (0, 0, 60)$, $q_1' = (0, 1, -12)$ we see that $O_\sigma \cap \tilde{X}$ is isomorphic to $\{w \in \mathbf{C}^* \mid w^5 - 1 = 0\}$ which consists of 5 points. Altogether, we obtain the following Hodge numbers for the link K:
$h^{000}(K) = h^{111}(K) = 12 + 5 = 17$, $h^{kpq} = 0$ otherwise.

EXAMPLE 2.3.7 Take $m = 4$, $k = 2$, $f_1 := z_1^3 + z_2^2 - z_3^2$, $f_2 := z_1^2 + z_2^2 + z_3^2 - z_4^4$. Then we get a fan $\mathcal{F}$ whose edges are spanned by $e_1, e_2, e_3, e_4, p_1, p_2$, where $p_1 := (2, 3, 3, 1)$, $p_2 := (2, 2, 2, 1)$. The following sets generate a cone in $\mathcal{F}$: $\{e_1, e_2, e_3, p_1\}$, $\{e_2, e_3, e_4, p_1\}$, $\{e_1, e_2, e_4, p_2\}$, $\{e_1, e_3, e_4, p_2\}$, $\{e_1, e_2, p_1, p_2\}$, $\{e_1, e_3, p_1, p_2\}$, $\{e_2, e_4, p_1, p_2\}$, $\{e_3, e_4, p_1, p_2\}$, as well as all subsets of these. Then condition **C** is satisfied.

Let us look at $O_\sigma \cap \tilde{X}$, where σ is a non-trivial cone of $\mathcal{F}$. This set is non-empty only if σ is spanned by $\{p_1\}$, $\{p_2\}$, $\{p_1, e_2\}$, $\{p_1, e_3\}$, $\{p_2, e_1\}$, $\{p_2, e_4\}$, or $\{p_1, p_2\}$. In order to determine the Hodge numbers of the link K it is sufficient to calculate the numbers $e_c^{pq}(O_\sigma \cap \tilde{X})$ for these seven cones. For the first two cones we get a one-dimensional set, for the others a zero-dimensional one. So we concentrate upon the first two cones.

If σ is spanned by $\{p_1\}$, $B_1 := O_\sigma \cap \tilde{X}$ is isomorphic to $\{w \in (\mathbf{C}^*)^3 \,|\, w_1^3 + w_2^2 - w_3^2 = 0, w_1^2 = 1\}$, i.e to $\{(w_2, w_3) \in (\mathbf{C}^*)^2 \,|\, w_2^2 - w_3^2 = \pm 1\}$. So we have two connected components, either of them is isomorphic to the complements of four points in $\mathbf{C}^*$, i.e to the complement of six points in $\hat{\mathbf{C}}$. This implies $e_{(c)}^{11}(B_1) = 2$, $e_{(c)}^{00}(B_1) = 2 - 12 = -10$, $e_{(c)}^{pq}(B_1) = 0$ otherwise.

If σ is spanned by $\{p_2\}$, $B_2 := O_\sigma \cap \tilde{X}$ is isomorphic to $\{w \in (\mathbf{C}^*)^3 \,|\, w_2^2 - w_3^2 = 0, w_1^2 + w_2^2 + w_3^2 = 1\}$, i.e to $\{w \in (\mathbf{C}^*)^3 \,|\, w_1^2 + 2w_2^2 = 1, w_3 = \pm w_2\}$ and therefore to B_1.

In each of the other cases we get a set of four points.

Altogether we obtain that $e^{11}(D_0) = 4$, $e^{00}(D_0) = 20 - 20 = 0$, $e^{pq}(D_0) = 0$ otherwise, so $e^{pq}(K) = 0$ for all p, q by Theorem 1.2.2. By Lemma 1.4.1 this implies $h^{000}(K) = h^{100}(K) = 1$, $h^{0pq}(K) = h^{1pq}(K) = 0$ otherwise (see also Theorem 2.3.1).

2.4. In addition let us now suppose that we are given a linear automorphism γ of $\mathbf{C}^m$ of finite order. For simplicity we assume that γ is given by some diagonal matrix. We assume that $f_1, \dots, f_k$ are chosen such that $f_j(\gamma(z)) = c_j f_j(z)$ for some $c_j \neq 0$. Then X is γ-invariant. We may extend the action of γ to $T_{\mathcal{F}'}$ by $\gamma(\zeta, z) := (\zeta, \gamma(z))$. Then $F_j(\gamma(\zeta, z)) = c_j F_j(\zeta, z)$, and γ induces an automorphism $\tilde{\gamma}$ of $\tilde{X}$. Therefore we may apply Section 1.3 and study the Hodge numbers $h_\lambda^{kpq}(K)$ which can be calculated by a straightforward generalization of the theory developped before. In Section 1.3 we had introduced invariants $\underline{h}^{kpq}$, in an analogue way we can define $\underline{e}^{pq}$ and $\underline{e}_{(c)}^{pq}$. Then the only question is how to calculate the numbers $\underline{e}_{(c)}^{pq}$ for a non-degenerate complete intersection in a torus (in the example below we can argue directly).

Let us assume that the latter is of the form $\{z \in (\mathbf{C}^*)^m \,|\, g_1 = \dots = g_k = 0\}$, and that $\gamma(z) = (c_1 z_1, \dots, c_m z_m)$, $g_j(\gamma(z)) = \lambda_j g_j(z)$, $j = 1, \dots, k$. For any λ let $W_\lambda = \{q \in \mathbf{Z}^m \,|\, c^q = \lambda\}$. Then we may proceed as in [6], with the following modifications:

A trivial case is $k = 0$, since γ acts trivially on the cohomology of the torus, because the action of γ extends to an action of the connected group $(\mathbf{C}^*)^m$.

So we have $\underline{e}^{pq}_{(c)}((\mathbf{C}^*)^m) = e^{pq}_{(c)}((\mathbf{C}^*)^m)$, where $\mathbf{Z}$ is canonically embedded into $\mathbf{Z}[\mathbf{C}^*]$.

As in [6] we now first look at the case of a hypersurface, i.e. $k = 1$. The only necessary modification is that we must replace the number $|M|$ of points of a subset M of $\mathbf{Z}^m$ by $\sum_{q \in M} \langle c^q \rangle = \sum_\lambda |M \cap W_\lambda| \langle \lambda \rangle$.

As in [D-K], the general case can be reduced to the hypersurface case, by looking at the hypersurface in $\mathbf{C}^k \times (\mathbf{C}^*)^m$ given by $h(t, z) := t_1 g_1(z) + \ldots + t_k g_k(z)$. Here we take the action of γ on $\mathbf{C}^k \times (\mathbf{C}^*)^m$ given by $(t, z) \mapsto (t_1 \lambda_1^{-1}, \ldots, t_k \lambda_k^{-1}, \gamma(z))$. The rest is straightforward. For an alternative approach see [11].

EXAMPLE 2.4.1 In Example 2.3.6 let γ be the automorphism of $\mathbf{C}^3$ given by $(z_1, z_2, z_3) \mapsto (z_1, z_2, e^{2\pi i/60} z_3)$. Then f_1 and f_2 are γ-invariant. Each of the sets B_1 and B_2 is permuted cyclically.

As in Section 1.3, we put $\underline{h}^{kpq}(K) := \sum_\lambda h^{kpq}_\lambda(K) < \lambda > \in \mathbf{Z}[\mathbf{C}]$ and $div(X - c_1) \cdot \ldots \cdot (X - c_r) := < c_1 > + \ldots + < c_r >$. Then we get:
$\underline{h}^{000}(K) = \underline{h}^{111}(K) = div(X^{12} - 1)(X^5 - 1)$, $\underline{h}^{kpq}(K) = 0$ otherwise.

EXAMPLE 2.4.2 *rm In Example 2.3.7 let γ be the automorphism of $\mathbf{C}^4$ given by $(z_1, z_2, z_3, z_4) \mapsto (z_1, z_2, z_3, i z_4)$. Then f_1 and f_2 are γ-invariant. Remember that B_1 has two connected components; these are permuted by γ. Each component is the complement of four points in a torus, and γ^2 acts as the identity on the cohomology of the torus but not on the removed points. On the other hand, the components of B_2 are fixed by γ, each component is isomorphic to the complements of four points in a torus, the cohomology of the torus is fixed by γ but not the removed points. This leads to the following result:*
$\underline{e}^{11}_{(c)}(B_1) = div(X^2 - 1)$, $\underline{e}^{00}_{(c)}(B_1) = div(X^4 - 1)^{-2}(X^2 - 1)^{-1}$, $\underline{e}^{11}_{(c)}(B_2) = div(X - 1)^2$, $\underline{e}^{00}_{(c)}(B_2) = div(X^2 - 1)^{-4}(X - 1)^{-2}$.
After some calculation which is similar to the preceding example we get $\underline{e}^{11}(D_0) = div(X^2 - 1)(X - 1)^2$, $\underline{e}^{00}(D_0) = div(X^2 - 1)^{-1}(X - 1)^2$, *so* $\underline{e}^{00}(K) = -\underline{e}^{22}(K) = div(X^2 - 1)^{-1}(X - 1)^2$, $\underline{e}^{pq}(K) = 0$ *otherwise. This implies:*
$\underline{h}^{000}(K) = div(X - 1)$, $\underline{h}^{100}(K) = div(X + 1)$, $\underline{h}^{0pq}(K) = \underline{h}^{1pq}(K) = 0$ *otherwise, or:* $h^{000}_1(K) = h^{100}_{-1}(K) = 1$, $h^{0pq}_\lambda(K) = h^{1pq}_\lambda(K) = 0$ *otherwise.*

2.5. Up to now it has been explained how to calculate the Hodge numbers of the link of a non-degenerate complete intersection. Now let us turn to the vanishing cohomology. As explained in the first chapter one may restrict here to the consideration of a link, too, but the starting point is a bit different now.

Let $k > 0$, let $M_1, \ldots, M_k$ be non-empty subsets of $\mathbf{N}^m \setminus \{0\}$. For $j = 1, \ldots, k$ let $f_j = \sum a_{jq} z^q$ be a convergent power series with $supp f_j \subset M_j$. We may assume that $f_1, \ldots, f_k$ are holomorphic on an open ball $B := B_{\epsilon'}$. Let $Y := B \cap \{f_1 = \ldots = f_{k-1} = 0\}$. Let us assume that Y is a complete intersection which is smooth outside 0, and that $f_k|Y$ is non-singular outside 0. This implies that $X := Y \cap \{f_k = 0\} = B \cap \{f_1 = \ldots = f_k = 0\}$ is a complete intersection which is smooth outside 0, too. So we are in the situation of Section 1.5, with f_k instead of f.

Because of the base change made in Section 1.6 it is useful to replace Y by the graph of $f_k|Y$: $G := \{(z_1, \ldots, z_{m+1}) \in B \times \mathbf{C} \,|\, g_1 = \ldots = g_k = 0\}$, where $g_j(z_1, \ldots, z_{m+1}) := f_j(z_1, \ldots, z_m)$ for $j = 1, \ldots, k-1$ and $g_k(z_1, \ldots, z_{m+1}) := f_j(z_1, \ldots, z_m) - z_{m+1}$. Note that G is also a complete intersection with an isolated singularity. Under the identification of Y with G, the map f_k corresponds to the map g_{k+1}: $g_{k+1}(z_1, \ldots, z_{m+1}) := z_{m+1}$.

The hypothesis $supp\, f_j \subset M_j$, $j = 1, \ldots, k$ corresponds to the hypothesis that $supp\, g_j \subset M_j' := M_j \times \{0\}$, $j = 1, \ldots, k-1$ and $supp\, g_k \subset M_k' := \{e_{m+1}\} \cup M_k \times \{0\}$. Note that $M_1', \ldots, M_{k-1}'$ are not convenient even if $M_1, \ldots, M_{k-1}$ are convenient; this is why we did not restrict to the convenient case in the preceding sections.

Because of these considerations we may restrict to the case where the last function is given by a coordinate, i.e. we look at the following situation:

Let $k \geq 0$, and let $M_1, \ldots, M_k$ be non-empty subsets of $\mathbf{N}^{m+1} \setminus \{0\}$. For $j = 1, \ldots, k$ let $f_j = \sum a_{jq} z^q$ be a convergent power series with $supp\, f_j \subset M_j$. We may assume that $f_1, \ldots, f_k$ are holomorphic on an open ball $B := B_{\epsilon'}$. Let $Y := B \cap \{f_1 = \ldots = f_k = 0\}$. Let us assume that Y is a complete intersection which is smooth outside 0, and that the restriction of f_{k+1} to Y is non-singular outside 0, where $f_{k+1}(z) := z_{m+1}$. This implies that $X := B \cap \{f_1 = \ldots = f_{k+1} = 0\}$ is a complete intersection which is smooth outside 0, too.

Now choose a fan $\mathcal{F}$ in $\mathbf{R}^{m+1}$ as in Section 2.1, e.g. by the procedure indicated in Section 2.2. Let $p_1, \ldots, p_r$ be accordingly chosen. Obviously we have by Lemma 2.1.3:

LEMMA 2.5.1 $f_{k+1}(\pi'(\zeta, z)) = \zeta_1^{p_{1,m+1}} \ldots \zeta_r^{p_{r,m+1}} z_{m+1}$.

Note that π' defines a V-resolution of $f_{k+1}|Y$. It is semi-stable if $f_{k+1} \circ \pi'$ vanishes of first order along D', i.e. if $p_{j,m+1} = 1$ for all j; in this case the corresponding monodromy is unipotent. This can be achieved by a base change: Let $e > 0$ be a common multiple of $p_{1,m+1}, \ldots, p_{r,m+1}$. Let $\hat{Y}$ be the fibre product of $f_{k+1}|Y$ and $\mathbf{C} \to \mathbf{C}$: $t \mapsto t^e$. This means that we substitute in $f_1, \ldots, f_k$ the variable z_{m+1} by z_{m+1}^e. The sets M_j are changed by replacing $q \in M_j$ by $(q_1, \ldots, q_m, eq_{m+1})$. The fan $\mathcal{F}$ is accordingly changed in a dual way: p is replaced by $(ep_1, \ldots, ep_m, p_{m+1})$. Now it is clear that each p_j is changed in such a way that the last component becomes 1 since we have to normalize such that all components are relatively prime. In total we get:

LEMMA 2.5.2 *The number e fulfills the condition of Section 1.6.*

This means that we can apply the results of Section 2.3 in order to calculate the Hodge numbers of the vanishing cohomology, and even the numbers $h_\lambda^{npq}(F)$ (see Section 2.4). Note that when applying Section 2.4 we put
$$\gamma(z_1,\ldots,z_{m+1}) := (z_1,\ldots,z_m, e^{2\pi i/e} z_{m+1}).$$

EXAMPLE 2.5.3 Let $m = 2$, $k = 2$, $M_1 := \{(4,0),(1,2)\}$, $M_2 := \{(3,2),(0,5)\}$, $f_1 := z_1^4 + z_1 z_2^2$, $f_2 := z_1^3 z_2^2 + z_2^5$. As above, we put $g_1(z_1,z_2,z_3) := f_1(z_1,z_2)$, $g_2(z_1,z_2,z_3) := f_2(z_1,z_2) - z_3$, $M_1' := M_1 \times \{0\}$, $M_2' := \{e_3\} \cup M_2 \times \{0\}$. Let us associate to M_1' and M_2' a fan $\mathcal{F}$ as in 2.2. Then $\mathcal{F}$ looks as follows:
Let $p_1 := (2,3,12)$, $p_2 := (1,1,5)$. The following sets generate a cone in $\mathcal{F}$: $\{e_1, e_2, p_1\}, \{e_2, e_3, p_1\}, \{e_1, e_3, p_2\}, \{e_1, p_1, p_2\}, \{e_3, p_1, p_2\}$, as well as all subsets of these.
Now $e = 60$ is the least common multiple of p_{13} and p_{23}. Therefore we substitute the variable z_3 by z_3^{60}. This transforms g_1 and g_2 into the two polynomials studied in Example 2.3.6 (and 2.4.1). The substitution changes p_1 and p_2 into $(120, 180, 12)$ and $(60, 60, 5)$ which in fact after normalization gives the vectors p_1 and p_2 of Example 2.3.6. Furthermore we must look at the automorphism studied in Example 2.4.1.
Since K_0 is empty we obtain from Theorem 1.6.7: $h_\lambda^{000}(F) = e_\lambda^{00}(F) = -e_\lambda^{11}(\hat{K}) = h_\lambda^{111}(\hat{K})$. Therefore we get from Example 2.4.1 the following result: $\underline{h}^{000}(F) = div(X^{12} - 1)(X^5 - 1)$, $\underline{h}^{kpq}(F) = 0$ otherwise.

EXAMPLE 2.5.4 Let $m = 3$, $k = 2$, $f_1 := z_1^3 + z_2^2 - z_3^2$, $f_2 := z_1^2 + z_2^2 + z_3^2$. Again, we put $g_1(z_1, z_2, z_3, z_4) := f_1(z_1, z_2, z_3)$, $g_2(z_1, z_2, z_3, z_4) := f_1(z_1, z_2, z_3) - z_4$. Let us associate to $supp\, g_1$, $supp\, g_2$ a fan $\mathcal{F}$ as in 2.2. Then $\mathcal{F}$ looks like in Example 2.3.7, however, with $p_1 := (2,3,3,4)$, $p_2 := (1,1,1,2)$. Then $e = 4$ is the least common multiple of p_{14} and p_{24}, so we substitute z_4 by z_4^4 and obtain Example 2.3.7, together with the automorphism γ studied in Example 2.4.2. This means that $e_\lambda^{pq}(\hat{K})$ has just been calculated in Example 2.4.2. The numbers $e^{pq}(K_0)$ are easily calculated if we consider in Example 2.3.7 only the cones which contain e_4: $e^{00}(K_0) = -e^{11}(K_0) = 4$ if $p = q = 0$ and $e^{pq}(K_0) = 0$ otherwise. In total, according to Theorem 1.6.7 and Lemma 1.7.2: $\underline{h}^{000}(F) = div(X - 1)$, $\underline{h}^{100}(F) = div(X + 1)$, $\underline{h}^{111}(F) = div(X^2 - 1)(X - 1)^2$, $\underline{h}^{lpq}(F) = 0$ otherwise. This means $h_1^{000}(F) = h_{-1}^{100}(F) = h_{-1}^{111}(F) = 1$, $h_1^{111}(F) = 3$, $h_\lambda^{lpq}(F) = 0$ otherwise.

References

[1] M. Audin: *The topology of torus actions on symplectic manifolds.* Birkhäuser: Basel (1991).

[2] A.A. Beilinson; J. Bernstein; P. Deligne: *Faisceaux pervers.* Astérisque **100**, 1–172, (1982).

[3] V.I. Danilov, *Newton polyhedra and vanishing cohomology*. Funct. Anal. Appl. **13**, 103–115 (1979).

[4] P. Deligne: *Théorie de Hodge III*. Publ. Math. I. H. E. S. **44**, 5–77, (1975).

[5] T. Delzant: *Hamiltoniens périodiques et image convexe de l'application moment*. Bull. Soc. Math. France **116**, 315–339, (1988).

[6] V.I. Danilov; A.G. Khovanskii: *Newton polyhedra and an algorithm for computing Hodge-Deligne numbers*. Izv. Akad. Nauk SSSR Ser. Mat. **50**, no. 5 , 925–945, (1986)= Math. USSR Izv. **29**, 279–298, (1987).

[7] A. Fujiki: *Duality of mixed Hodge structures of algebraic varieties*. Publ. Res. Inst. Math. Sci. **16**, 635–667, (1980).

[8] M. Goresky; R. MacPherson: *On the topology of complex algebraic maps*. In: Proc. Int. Conf. on Alg. Geometry (La Rábida, 1981), Springer Lecture Notes **961**, 119–129, (1982).

[9] H.A. Hamm: *Lokale topologische Eigenschaften komplexer Räume*, Math. Ann. **191**, 235–252, (1971).

[10] H.A. Hamm: *Hodge numbers of affine complete intersections*. Schr. FSP Komplexe Mannigfaltigkeiten, Heft 140, Münster (1992).

[11] H.A. Hamm: *Hodgezahlen vollständiger Durchschnitte in Tori*. In preparation.

[12] G. Kempf; F. Knudsen; D. Mumford; B. Saint-Donat: *Toroidal embeddings I*. Springer Lecture Notes **339**, (1973).

[13] V. Navarro Aznar: *Sur la théorie de Hodge des variétés algébriques à singularités isolées*. Astérisque **130**, 272–307, (1985).

[14] T. Oda: *Convex bodies and algebraic geometry*. Springer, Berlin (1988).

[15] J.H.M. Steenbrink: *Mixed Hodge structure on the vanishing cohomology*. Proc. Real and Complex Singularities, Oslo, pp. 525–562, (1976).

[16] J.H.M. Steenbrink: *Mixed Hodge structures associated with isolated singularities*. In Proc. Summer Inst. on Singularities, Arcata 1981, part 2, pp. 513–536. Proc Symp. Pure Math. **40**, (1983).

[17] J.H.M. Steenbrink: *Monodromy and weight filtration for smoothings of isolated singularities*. Compos. Math. **97**, 285–293, (1995).

[18] J.H.M. Steenbrink; M.G.M. van Doorn: *A supplement to the monodromy theorem*. Abh. Math. Sem. Univ. Hamburg **59**, 225–233, (1989).

[19] M. Saito: *Modules de Hodge polarisables*. Publ. RIMS Kyoto Univ. **24**, 849–995, (1988).

Progress in Mathematics, Vol. 162, © 1998 Birkhäuser Verlag Basel/Switzerland

Differential Invariants of Embeddings of Manifolds in Complex Spaces

Weiming Huang
Department of Mathematics
Beijing Normal University
Beijing
CHINA

Joseph Lipman *
Department of Mathematics
Purdue University
W. Lafayette IN 47907
USA

Abstract

Let V be a reduced complex space, W a complex submanifold, and let (V', W') be another such pair. Let $f : V \to V'$ be a homeomorphism with $f(W) \subset W'$, such that f and f^{-1} are both continuously (real-) differentiable. Then f induces a component – (with multiplicity) – preserving homeomorphism $\mathbf{f}_0$ from the normal cone $C(V, W)$ to $C(V', W')$, respecting the natural $\mathbb{R}^*$ actions on these cones. Moreover, though $\mathbf{f}_0$ need not respect the $\mathbb{C}^*$ actions nevertheless the induced map on Borel-Moore homology $f_* : H_*(W) \to H_*(W')$ takes the Segre classes of the components of $C(V, W)$ to $\pm$those of the corresponding components of $C(V', W')$. In particular we recover the differential invariance of the multiplicity of W in V.

Introduction

In studying singularities one is interested in invariants, analytic (biholomorphic) or topological. And it can be an occasion for celebration when an analytic invariant turns out to be topological. For example, a famous open problem of Zariski is to determine whether the multiplicity of a hypersurface germ in $\mathbb{C}^n$ is invariant under ambient homeomorphisms.

In between the analytic and topological domains, there is a large and relatively unexplored territory populated by *differential* invariants, i.e, data which are associated to complex spaces and which are always the same for two C^s-homeomorphic spaces ($s > 0$). The multiplicity of a reduced complex space germ is such a differential invariant, for $s = 1$ [GL], but not a topological one, even for ambient homeomorphisms of curves in $\mathbb{C}^3$.

---

*Partially supported by the National Security Agency

In this paper we consider a reduced complex space V with an r-dimensional connected submanifold $i: W \hookrightarrow V$. Assume for simplicity that all the irreducible components of V have the same dimension, say d, and that they all properly contain W. Let $\mathcal{I}$ be the kernel of the natural surjection $\mathcal{O}_V \to i_* \mathcal{O}_W$, let $\mathcal{G}$ be the graded $\mathcal{O}_W$-algebra $\oplus_{m \geq 0} i^*(\mathcal{I}^m/\mathcal{I}^{m+1})$, and let $C(V,W) := \mathrm{Specan}(\mathcal{G})$ be the normal cone of W in V (see §1), with (reduced, irreducible) components $(C_j)_{j \in J}$. The components P_j of the projectivized normal cone $P = P(V,W) := \mathrm{Projan}(\mathcal{G}) \xrightarrow{\wp} W$ correspond naturally to those of $C(V,W)$. For each j let $[P_j] \in H_{2d-2}(P)$ (Borel-Moore homology) be the natural image of the fundamental class of P_j. P carries a canonical invertible sheaf $\mathcal{O}(1)$, with Chern class, say, $c \in H^2(P, \mathbb{Z})$. The *Segre class* $s_i(C_j) \in H_{2r-2i}(W)$ is defined by $s_i(C_j) := \wp_*\big([P_j] \cap c^{d-1-r+i}\big)$. [1]

Our motivating result is that *these Segre classes are, up to sign, C^1 invariants of the pair (V,W)*. (For a precise statement see Theorem (6.3).)

We first prove that the normal cone $C(V,W)$ is a differential invariant, even "as a cycle": given a second pair $V' \supset W'$, then any C^1 homeomorphism $f: V \to V'$ with f^{-1} also C^1 and $f(W) = W'$ induces a homeomorphism $\mathbf{f}_0$ from $C(V,W)$ onto $C(V',W')$ such that $\mathbf{f}_0$ maps each irreducible component of $C(V,W)$ onto one of $C(V',W')$ having the same multiplicity. (See Theorem (4.3.1); the case where W is a point was an important part of [GL].) This is shown via the standard deformation (see §2) of V to $C(V,W)$, *restricted however to real parameters t*. (So we have the trivial family $V_t \cong V$ for $0 \neq t \in \mathbb{R}$, together with $V_0 \cong C(V,W)$.) Of course the trivial part of this deformation, away from $t = 0$, behaves functorially; and one needs to show that the functoriality "extends continuously" to the entire deformation. This is done in Theorem (3.3), via the derivative of f. In §4 we prove the differential invariance of the multiplicities of the components by interpreting these numbers as intersection multiplicities along the components of V_0, and noting that such intersection numbers are known to be topological invariants.

Now in order to get at the Segre classes we must pass from $C(V,W)$ to $P(V,W)$, and so we have to quotient out the natural $\mathbb{C}^*$ action. The problem is that we used the derivative of f to establish functoriality of $C(V,W)$, and that derivative is only *real*-linear. Thus the $\mathbb{C}^*$ action may not be functorial.

To deal with this problem, we construct in §5 the *relative complexification* of $C := C(V,W)$ (in fact, of any cone over W), an analytic subset $\widetilde{C} \subset C \times_W C$ whose fibers are real-analytically isomorphic to the complexifications of the fibres of C, at least almost everywhere over W. This $\widetilde{C}$, together with a natural real-analytic $\mathbb{C}^*$ action, is indeed C^1-functorial (Theorem (5.3.1)). But we have not been able to extract any Segre classes directly from $\widetilde{C}$. Instead we use the $\mathbb{C}^*$-stable, analytic subset $\Lambda(C) \subset \widetilde{C}$ consisting of pairs (c_1, c_2) of points of C

---

[1] When V and W are algebraic varieties, this definition connects to the algebraic one in [Fn, Chap.4] via the cycle map of *ibid.*, §19.1.

such that one of them lies in the orbit of the other with respect to the natural $\mathbb{C}^1$ action (reviewed in §1). Using the functoriality of $\widetilde{C}$, we find that $\Lambda(C)$ *is* $\mathbb{C}^1$-*functorial*. Furthermore, off its vertex section, $\Lambda(C)$ together with its induced $\mathbb{C}^*$ action is topologically isomorphic to the rank two bundle $\mathcal{O}(1) \oplus \mathcal{O}(-1)$ (minus its 0-section) over $P(V, W)$. It follows that the Segre classes of the components of this rank two bundle become differential invariants, up to sign, when pushed down from $P(V, W)$ to W. These pushed-down classes are easily seen to be the Segre classes $s_i(C_j)$ such that $i - 1 - $(codimension of W in V) is even. The remaining Segre classes can be obtained similarly, just by changing (V, W) to $(V \times \mathbb{C}^1, W \times \{0\})$ (which doesn't affect the total Segre class, but changes the codimension by one). For details see §§5–6.

Incidentally, with $e_j :=$ multiplicity of the component C_j of $C := C(V, W)$, the Segre classes $s_i(C)$ can be defined by $s_i(C) := \sum_j e_j s(C_j)$ (cf. [Fn, p. 74, Lemma 4.2]; the sums here are "locally finite" with respect to decomposition into irreducible components [BH, p. 465, 1.7]). As above, the e_j are differential invariants; but because of the sign ambiguity in Theorem (6.3), $s_i(C)$ may not be a differential invariant – though its image in $H_*(W, \mathbb{Z}_2)$ is.

In particular, $s_0(C) = m(V, W)[W]$, where $m(V, W)$ is the multiplicity of W in V [Fn, §4.3]. Hence Theorem (6.3) implies that $m(V, W)$ *is a differential invariant*. (That is the main result of [GL], where a more straightforward proof is given.)

1 Normal cones

We begin with a brief review of some facts about normal cones, facts which are "well-known" but not, as a whole, easily accessible in the literature.

(1.1) Let $(V, \mathcal{O}_V)$ be a reduced complex analytic space, and let $(W, \mathcal{O}_W)$ be a (not necessarily reduced) complex subspace of V . Let $\mathcal{I}$ be the kernel of the surjection $\mathcal{O}_V \twoheadrightarrow i_*\mathcal{O}_W$ corresponding to the inclusion $i : W \hookrightarrow V$. The graded $\mathcal{O}_W - algebra$-$\mathrm{gr}_W(V) := \oplus_{m \geq 0} i^*(\mathcal{I}^m/\mathcal{I}^{m+1})$ is finitely presentable, since $i^*(\mathcal{I}^m/\mathcal{I}^{m+1})$ is coherent for all m [MT, p 2, Prop. 1.4]. So one can define the *normal cone $C(V, W)$ of V along W* to be

$$C(V, W) := \mathrm{Specan}(\mathrm{gr}_W(V)) .$$

(For the definition of Specan, see [Ho, p. 19–02].) This cone is naturally equipped with a map

$$p = p(V, W) : C(V, W) \to W,$$

together with a "vertex" section

$$\sigma = \sigma(V, W) : W \to C(V, W)$$

$(p \circ \sigma = \text{identity})$, corresponding, via functoriality of Specan, to the obvious maps $\mathcal{O}_W \leftrightarrows \mathrm{gr}_W(V)$. Moreover, with $\mathbb{C}^1$ the affine line there is the map

$$\mu : \mathbb{C}^1 \times C(V, W) \to C(V, W)$$

corresponding to the map of $\mathcal{O}_W$-algebras $\mathrm{gr}_W(V) \to \mathrm{gr}_W(V)[T]$ (T an indeterminate) whose restriction to $\mathcal{I}^m/\mathcal{I}^{m+1}$ is multiplication by T^m ($m \geq 0$).

One checks via the corresponding $\mathcal{O}_W$-algebra maps that there are commutative diagrams (with "id" standing for "identity" and "mpn" for "multiplication"):

$$
\begin{array}{ccc}
\mathbb{C}^1 \times C(V, W) & \xrightarrow{\;\mu\;} & C(V, W) \\
{\scriptstyle pr_2}\big\downarrow & & \big\downarrow {\scriptstyle p} \\
C(V, W) & \xrightarrow[\;p\;]{} & W
\end{array}
$$

$$
\begin{array}{ccccc}
\mathbb{C}^1 \times (\mathbb{C}^1 \times C(V, W)) & \xrightarrow{\;id\times\mu\;} & \mathbb{C}^1 \times C(V, W) & \xrightarrow{\;\mu\;} & C(V, W) \\
\big\| & & & & \big\| \\
(\mathbb{C}^1 \times \mathbb{C}^1) \times C(V, W) & \xrightarrow[\;mpn\times id\;]{} & \mathbb{C}^1 \times C(V, W) & \xrightarrow[\;\mu\;]{} & C(V, W)
\end{array}
$$

$$
\begin{array}{ccc}
\mathbb{C}^0 \times C(V, W) & \xrightarrow{\;\sim\;} & C(V, W) \\
{\scriptstyle 1\times id}\big\downarrow & & \big\| \\
\mathbb{C}^1 \times C(V, W) & \xrightarrow[\;\mu\;]{} & C(V, W)
\end{array}
\qquad
\begin{array}{ccc}
\mathbb{C}^0 \times C(V, W) & \xrightarrow{\;\sim\;} & C(V, W) \\
{\scriptstyle 0\times id}\big\downarrow & & \big\downarrow {\scriptstyle \sigma\circ p} \\
\mathbb{C}^1 \times C(V, W) & \xrightarrow[\;\mu\;]{} & C(V, W)
\end{array}
$$

Restricting attention to underlying point sets, if for $a \in \mathbb{C}$ and $x \in C(V, W)$ we set $ax := \mu(a, x)$, then

$$p(ax) = p(x), \quad a_1(a_2 x) = (a_1 a_2)x, \quad 1x = x, \quad 0x = \sigma p(x).$$

Remark (1.1.1) The foregoing holds with $\mathrm{gr}_W(V)$ replaced by any finitely presented graded $\mathcal{O}_W$-algebra $\mathcal{G} = \oplus_{m \geq 0} \mathcal{G}_m$ ($\mathcal{G}_0 = \mathcal{O}_W$, and every $\mathcal{G}_m$ is a coherent $\mathcal{O}_W$-module).

(1.2) To get a picture of $p : C(V, W) \to W$ near a point $w \in W$, we embed the triple (V, W, w) locally into some $\mathbb{C}^n$, as follows. In the local ring $\mathcal{O}_{V,w}$ let $(\tau_1, \tau_2, \ldots, \tau_s)$ generate the ideal corresponding to the germ of W. Denoting convergent power series rings by $\mathbb{C}\langle \cdots \rangle$, pick a surjective $\mathbb{C}$-algebra homomorphism

$$\alpha : \mathbb{C}\langle T_1, T_2, \ldots, T_{r+s} \rangle \twoheadrightarrow \mathcal{O}_{V,w} \qquad (T_i \text{ indeterminates})$$

such that $\alpha(T_{r+i}) = \tau_i$ $(1 \leq i \leq s)$. Correspondingly, with $n := r + s$, there is an open neighborhood V^* of w in V, an open neighborhood U of 0 in $\mathbb{C}^n$, a holomorphic map $\theta : V^* \to U$, and holomorphic functions $\varphi_i : U \to \mathbb{C}$ $(i = 1, 2, \ldots, e)$ such that

(i) θ induces an isomorphism of V^* onto the reduced analytic subspace V' of U consisting of the common zeros of the φ_i:

$$V' := \{\, z \in U \mid \varphi_1(z) = \varphi_2(z) = \cdots = \varphi_e(z) = 0 \,\}.$$

(ii) θ maps $W^* := W \cap V^*$ isomorphically onto the analytic space

$$W' := L \cap V' = L \times_{\mathbb{C}^n} V' \subset V'$$

where L is the reduced r-dimensional space

$$L := \{\, (z_1, \ldots, z_n) \in U \mid z_{r+1} = z_{r+2} = \cdots = z_n = 0 \,\}.$$

(iii) $\theta(w) = 0$.

The embedding θ induces an isomorphism

$$C(V, W) \times_W W^* = C(V^*, W^*) \xrightarrow{\sim} C(V', W')$$

compatible with the canonical maps p, σ, and μ. So let us simply consider the case where $V = V'$ and $W = W'$. Then $\mathcal{I} = \mathcal{J}\mathcal{O}_V$, where $\mathcal{J}$ is the $\mathcal{O}_U$-ideal generated by the coordinate functions $\xi_{r+1}, \ldots, \xi_n$ (i.e., $\xi_h(z_1, \ldots, z_n) = z_h$).

With $j : L \hookrightarrow U$ the inclusion, there is an isomorphism of graded $\mathcal{O}_L$-algebras

$$\mathrm{gr}_L(U) := \oplus_{m \geq 0} j^*(\mathcal{J}^m/\mathcal{J}^{m+1}) \xrightarrow{\sim} \mathcal{O}_L[T_1, \ldots, T_s]$$

whose inverse takes T_h to the section of $j^*(\mathcal{J}/\mathcal{J}^2)$ given by ξ_{r+h} $(1 \leq h \leq s)$; and so we have an isomorphism

$$C(U, L) \xrightarrow{\sim} (L \times \mathbb{C}^s) \subset (\mathbb{C}^r \times \mathbb{C}^s) = \mathbb{C}^n.$$

This isomorphism identifies $p(U, L)$ with the projection $\mathrm{pr}_1 : L \times \mathbb{C}^s \to L$, and $\sigma(U, L)$ with the map $\mathrm{id} \times 0 : L \xrightarrow{\sim} L \times \{0\} \hookrightarrow L \times \mathbb{C}^s$. Furthermore, we have the closed immersion

$$C(V, W) \hookrightarrow C(U, L) \tag{1.2.1}$$

corresponding to the natural surjection $\mathrm{gr}_L(U) \twoheadrightarrow \mathrm{gr}_W(V)$. There results a commutative diagram, whose horizontal arrows represent embeddings:

$$\begin{array}{ccc} C(V, W) & \longrightarrow & L \times \mathbb{C}^s \subset \mathbb{C}^n \\ {\scriptstyle p}\big\downarrow\big\uparrow{\scriptstyle \sigma} & & {\scriptstyle \mathrm{pr}_1}\big\downarrow\big\uparrow{\scriptstyle \mathrm{id}\times 0} \\ W & \longrightarrow & L \end{array} \tag{1.2.2}$$

The action of $\mathbb{C}^1$ on $C(V, W)$ (via μ) is induced by the action on $C(U, L) \cong L \times \mathbb{C}^s$, easily checked to be given on underlying point sets by

$$a(x, z) = (x, az) \qquad (a \in \mathbb{C}, \ x \in L, \ z \in \mathbb{C}^s). \qquad (1.2.3)$$

In particular, the analytic group $\mathbb{C}^* = \mathbb{C}^1 - \{0\}$ acts freely on $C(V, W) - \sigma(W)$. The points of $C(V, W)$ – identified via (1.2.2) with a subvariety of $W \times \mathbb{C}^s$ – can be specified by equations as follows. Let $w \in W \subset L$. For any open neighborhood N of w in L, for $x \in N$, and for any polynomial

$$F(T_1, \ldots, T_s) \in \Gamma(N, \mathcal{O}_L)[T_1, \ldots, T_s],$$

let $F_x \in \mathbb{C}[T_1, \ldots, T_s]$ be the polynomial obtained from F by evaluating coefficients at x, and define the function $\widetilde{F} : N \times \mathbb{C}^s \to \mathbb{C}$ by

$$\widetilde{F}(x, y) = F_x(y_1, \ldots, y_s) \qquad (x \in N, \ y \in \mathbb{C}^s).$$

Set $V_N := V \cap (N \times \mathbb{C}^s)$. (Recall that $V \subset \mathbb{C}^n = \mathbb{C}^r \times \mathbb{C}^s$.) Then:

(1.2.4) The point $(w, z) \in W \times \mathbb{C}^s$ is in $C(V, W) \Leftrightarrow$ for every $m \geq 0$ and for every N and F as above with F homogeneous of degree m, if the function $\widetilde{F}|V_N$ is in $\Gamma(V_N, \mathcal{I}^{m+1})$ then $\widetilde{F}(w, z) = 0$.

The *proof*, an exercise on the definition of Specan, is left to the reader.

Remark. The following "initial form" characterization (1.2.5), suggested by [Hi2, p. 18, Remark 3.2], is readily seen to be equivalent to the one in (1.2.4). For s-tuples $\nu = (\nu_1, \ldots, \nu_s)$ of non-negative integers, we set $|\nu| := \nu_1 + \cdots + \nu_s$; and for $z = (z_1, \ldots, z_s) \in \mathbb{C}^s$, we set $z^\nu := z_1^{\nu_1} z_2^{\nu_2} \ldots z_s^{\nu_s}$.

(1.2.5) The point $(w, z) \in W \times \mathbb{C}^s$ is in $C(V, W) \Leftrightarrow$ for all open neighborhoods N_1 of w in $\mathbb{C}^r$ and N_2 of 0 in $\mathbb{C}^s$, and for all $m \geq 0$, if the holomorphic functions $f_\nu : N_1 \times N_2 \to \mathbb{C}$ are such that $\sum_{|\nu|=m} f_\nu(x, y) y^\nu = 0$ for all $(x, y) \in V \cap (N_1 \times N_2)$, then $\sum_{|\nu|=m} f_\nu(w, 0) z^\nu = 0$.

(Equivalently: for all holomorphic functions $f : N_1 \times N_2 \to \mathbb{C}$ vanishing on $V \cap (N_1 \times N_2)$ and such that $\lim_{t \to 0} t^{-m} f(x, ty) < \infty$ for all $(x, y) \in N_1 \times N_2$, we have $\lim_{t \to 0} t^{-m} f(w, tz) = 0$.)

(1.3) Now here is a geometric description of $C(V, W)$. As in (1.2), we identify (V, W) with $(V', W') \subset (U, W') \subset (\mathbb{C}^r \times \mathbb{C}^s, \mathbb{C}^r)$. We denote by π_f the projection $\mathbb{C}^r \times \mathbb{C}^s \to \mathbb{C}^s$ ("f" stands for "fiber").

Proposition. *The point $(w, z) \in W \times \mathbb{C}^s = C(U, W)$ is in $C(V, W)$ iff there exist sequences $v_i \in V$, $a_i \in \mathbb{C}$ $(0 < i \in \mathbb{Z})$ such that $v_i \to w$ and $a_i \pi_f\, v_i \to z$. Moreover, for any $(w, z) \in C(V, W)$, there exist such a_i, v_i with all the a_i real and positive.*

Proof. Suppose that there are sequences $v_i \in V$, $a_i \in \mathbb{C}$, such that $v_i \to w$ and $a_i \pi_f\, v_i \to z$. Set $v_i = (x_i, y_i)$, so that $x_i \to w$, $y_i \to 0$, and $a_i y_i = a_i \pi_f\, v_i \to z$.

With notation as in (1.2.5), we have then (assuming, as we may, that $v_i \in N_1 \times N_2$):

$$\sum_{|\nu|=m} f_\nu(w,0)z^\nu = \lim_i \sum_{|\nu|=m} f_\nu(x_i,y_i)(a_i y_i)^\nu = \lim_i a_i^m \sum_{|\nu|=m} f_\nu(x_i,y_i)(y_i)^\nu = 0.$$

Thus $(w,z) \in C(V,W)$.

For the converse, we have the following stronger statement, due to Hironaka [Hi, p. 131, Remark (2.3)].

Lemma (1.3.1). *If $(w,z) \in C(V,W)$ and $z \neq 0$, then there exists a real analytic map $\varphi : (-1,1) \to V$ with $\varphi(0) = w$, $\varphi(t) \notin W$ if $t \neq 0$, and such that*

$$z/|z| = \lim_{t\to 0^+} \pi_f\, \varphi(t)/|\pi_f\, \varphi(t)|\,.$$

A variant of Hironaka's proof will be given below, in (2.3).

2 Specialization to the normal cone

With $i : W \hookrightarrow V$ and $\mathcal{I}$ as in (1.1), consider the graded $\mathcal{O}_V$-algebra

$$\mathcal{R} = \mathcal{R}_{\mathcal{I}} := \oplus_{n \in \mathbb{Z}} \mathcal{I}^n T^{-n} \subset \mathcal{O}_V[T, T^{-1}]$$

where T is an indeterminate and $\mathcal{I}^n$ is defined to be $\mathcal{O}_V$ for all $n \leq 0$. By [MT, p. 2, Prop. 1.4], $\mathcal{R}$ is finitely presentable, so we can set

$$\mathbf{V} = \mathbf{V}_W := \operatorname{Specan}(\mathcal{R}_{\mathcal{I}})\,.$$

$\mathbf{V}$ is called the *specialization of (V,W) to $C(V,W)$*, see [LT, pp. 556–557].

The terminology is explained as follows. We have natural maps

$$W \times \mathbb{C}^1 \xrightarrow{\alpha} \mathbf{V} \xrightarrow{\beta} V \times \mathbb{C}^1$$

where α is the closed immersion corresponding to the obvious $\mathcal{O}_V$-algebra homomorphism

$$\mathcal{R} \twoheadrightarrow \mathcal{R}/\mathcal{I}T^{-1}\mathcal{R} \xrightarrow{\sim} \oplus_{n \geq 0} (\mathcal{O}_V/\mathcal{I})T^n = i_*\mathcal{O}_W[T]\,,$$

and β corresponds to the $\mathcal{O}_V$-algebra inclusion $\mathcal{O}_V[T] \hookrightarrow \mathcal{R}$. Note that $\beta \circ \alpha$ is the closed immersion $i \times \mathrm{id} : W \times \mathbb{C}^1 \hookrightarrow V \times \mathbb{C}^1$. Let $\mathbf{t}$ be the composition

$$\mathbf{t} : \mathbf{V} \xrightarrow{\beta} V \times \mathbb{C}^1 \xrightarrow{\mathrm{pr}} \mathbb{C}^1\,.$$

Denote the fiber $\mathbf{t}^{-1}(0)$ by $\mathbf{V}_0$.

Proposition (2.1)

(i) *The map $\mathbf{t}$ is flat.*

(ii) *β induces an isomorphism of $\mathbf{V} - \mathbf{V}_0$ onto $V \times (\mathbb{C}^1 - \{0\})$.*

(iii) *There is a natural commutative diagram*

$$
\begin{array}{ccccc}
W & \xrightarrow{\ \sigma\ } & C(V,W) & \xrightarrow{\ p\ } & W \\
{\scriptstyle (id,0)}\downarrow{\scriptstyle \simeq} & & {\scriptstyle \simeq}\downarrow{\scriptstyle \rho} & & \downarrow{\scriptstyle (i,0)} \\
W \times \{0\} & \xrightarrow{\ \alpha\ } & \mathbf{V}_0 & \xrightarrow{\ \beta\ } & V \times \mathbb{C}^1
\end{array}
$$

with σ and p as in (1.1), and ρ an isomorphism.

Thus $\mathfrak{t}$ gives us a flat family of closed immersions, isomorphic to $i : W \hookrightarrow V$ wherever $\mathfrak{t} \neq 0$ and to $\sigma : W \hookrightarrow C(V,W)$ where $\mathfrak{t} = 0$.

Proof. We have $\mathrm{pr}^{-1}(0) = V \times \{0\} = \mathrm{Specan}(\mathcal{O}_V[T]/T\mathcal{O}_V[T])$, and it follows that $\mathbf{V}_0 = \mathrm{Specan}(\mathcal{R}/T\mathcal{R}) \subset \mathrm{Specan}(\mathcal{R})$. But there is an obvious isomorphism $\mathcal{R}/T\mathcal{R} \xrightarrow{\sim} \oplus_{n\geq 0} \mathcal{I}^n/\mathcal{I}^{n+1}$, whence an isomorphism $\rho : C(V,W) \xrightarrow{\sim} \mathbf{V}_0$.

The surjection $\mathcal{R}/T\mathcal{R} \twoheadrightarrow \mathcal{R}/(T\mathcal{R} + \mathcal{I}T^{-1}\mathcal{R})$ is naturally identifiable with the obvious surjection of $\oplus_{n\geq 0} \mathcal{I}^n/\mathcal{I}^{n+1}$ onto its degree 0 component $i_*\mathcal{O}_W$; thus the restriction of α to $\mathbf{V}_0$ gets identified with $\sigma : W \hookrightarrow C(V,W)$, and so the left square in (iii) commutes. The right square commutes because it is obtained by applying the functor Specan to a (clearly) commutative diagram of graded $\mathcal{O}_V$-algebras.

A morphism of analytic spaces $f : X \to \mathbf{V}$ factors through $\mathbf{V} - \mathfrak{t}^{-1}(0)$ iff the corresponding map $\Gamma(\mathbf{V}, \mathcal{R}) \to \Gamma(X, \mathcal{O}_X)$ sends T to a unit, i.e., $\mathcal{R} \to f_*\mathcal{O}_X$ factors through $\mathcal{R}[T^{-1}]$. Consequently

$$
\mathbf{V} - \mathbf{V}_0 = \mathrm{Specan}(\mathcal{R}[T^{-1}]) = \mathrm{Specan}(\mathcal{O}_V[T, T^{-1}]),
$$

and (ii) follows.

In particular, off $\mathbf{V}_0$ the map $\mathfrak{t}$ coincides with the projection pr, which is flat. Since T is not a zero-divisor in $\mathcal{R}$, therefore the germ of $\mathfrak{t}$ in the local ring of any point on $\mathbf{V}_0$ is not a zero-divisor (see e.g., [Ho, p. 19-07, Corollaire]), and so $\mathfrak{t}$ is flat everywhere along $\mathbf{V}_0$ too. This proves (i). $\qquad\square$

(2.2) Now let us see how the above specialization looks locally.

Assume as in (1.2) that $(V, W) \subset (\mathbb{C}^{r+s}, \mathbb{C}^r)$. Let $\xi_1, \ldots, \xi_{r+s}$ be the coordinate functions on $\mathbb{C}^{r+s}$, and for $i = 1, 2, \ldots, s$, set $\eta_i := \xi_{r+i}|V$. We embed $\mathbf{V}$ into $\mathbb{C}^{r+s+1}$ as follows. There is a surjective $\mathcal{O}_V$-algebra homomorphism

$$
\psi : \mathcal{O}_V[T_1', \ldots, T_s', T] \twoheadrightarrow \mathcal{R}
$$

$$
\text{with} \quad \psi(T_i') = \eta_i T^{-1} \quad (1 \leq i \leq s), \qquad \psi(T) = T.
$$

Correspondingly, there is an embedding $\mathbf{V} \hookrightarrow V \times \mathbb{C}^{s+1} \hookrightarrow \mathbb{C}^r \times \mathbb{C}^s \times \mathbb{C}^{s+1}$. But for each i, $\eta_i - T_i'T$ is a global section of the kernel of ψ; therefore the embedding factors through the subspace of $\mathbb{C}^r \times \mathbb{C}^s \times \mathbb{C}^{s+1}$ where these functions vanish, i.e., the reduced subspace whose points are of the form $(x_1,\ldots,x_r,ay_1,\ldots,ay_s,y_1,\ldots,y_s,a)$, a subspace isomorphic to $\mathbb{C}^{r+s+1}$.

With $\mathbf{V}$ so regarded as a subspace of $\mathbb{C}^{r+s+1}$, the maps $\alpha : W \times \mathbb{C}^1 \to \mathbf{V}$ and $\beta : \mathbf{V} \to V \times \mathbb{C}^1$ are given on underlying point sets by

$$\begin{aligned}
\alpha(x_1,\ldots,x_r,a) &= (x_1,\ldots,x_r,0,\ldots,0,a), \\
\beta(x_1,\ldots,x_r,y_1\ldots,y_s,a) &= (x_1,\ldots,x_r,ay_1,\ldots,ay_s,a).
\end{aligned}$$

The map t is induced by projection to the last coordinate. For $a \neq 0$, β maps the fiber $\mathbf{V}_a := \mathsf{t}^{-1}(a)$ isomorphically onto $V \times \{a\}$, i.e.,

$$\mathbf{V}_a = \{\,(x_1,\ldots,x_r,y_1\ldots,y_s,a) \mid (x_1,\ldots,x_r,ay_1,\ldots,ay_s) \in V \,\}. \qquad (2.2.1)$$

The embedding of $C(V,W) = \mathbf{V}_0$ in $\mathbf{V} \subset \mathbb{C}^r \times \mathbb{C}^s \times \mathbb{C}^{s+1}$ arises from the surjection $\overline{\psi}$ obtained from ψ by modding out T. This $\overline{\psi}$ factors as

$$\mathcal{O}_V[T_1',\ldots,T_s'] \twoheadrightarrow \mathcal{O}_W[T_1',\ldots,T_s'] \twoheadrightarrow \mathcal{R}/T\mathcal{R}.$$

Comparing this embedding to (1.2.1), we find that the underlying point set of $\mathbf{V}_0$ consists of all $(w,0,z,0) \in \mathbb{C}^r \times \mathbb{C}^s \times \mathbb{C}^s \times \mathbb{C}^1$ with $(w,z) \in \mathbb{C}(V,W)$, where $C(V,W)$ is regarded as being embedded into $\mathbb{C}^r \times \mathbb{C}^s$ as in (1.2); and then passing as above from $\mathbb{C}^r \times \mathbb{C}^s \times \mathbb{C}^s \times \mathbb{C}^1$ to $\mathbb{C}^{r+s+1}$, we can write

$$\mathsf{t}^{-1}(0) = \mathbf{V}_0 = \{\,(w,z,0) \in \mathbb{C}^r \times \mathbb{C}^s \times \mathbb{C}^1 \mid (w,z) \in C(V,W) \,\}. \qquad (2.2.2)$$

(2.3). *To prove (1.3.1)*, we first note that since t is flat, therefore $\mathbf{V}_0$ is nowhere dense in $\mathbf{V}$, so that for any point $(w,z,0) \in \mathbf{V}_0$, there exists an analytic map

$$\phi : \mathbb{D} \to \mathbf{V} \qquad (\mathbb{D} := \text{unit disc in } \mathbb{C}^1)$$

$$\text{such that} \quad \phi(\mathbb{D} - \{0\}) \subset \mathbf{V} - \mathbf{V}_0 \quad \text{and} \quad \phi(0) = (w,z,0).$$

(This follows, e.g., from the Nullstellensatz and from the algebraic fact that in a noetherian local ring A – like the stalk at $(w,z,0)$ of $\mathcal{O}_\mathbf{V}$ – any prime ideal is the intersection of all prime ideals $\wp$ containing it and such that $\dim(A/\wp) = 1$.) Set

$$\phi(\xi) = (\lambda(\xi),\mu(\xi),\tau(\xi)) \in \mathbb{C}^r \times \mathbb{C}^s \times \mathbb{C}^1 \qquad (\xi \in \mathbb{D}).$$

For ξ sufficiently small, $\tau(\xi)$ is given by a convergent power series

$$\tau(\xi) = a\xi^q + a_1\xi^{q+1} + \ldots \qquad (a \neq 0,\ q > 0).$$

With $b \in \mathbb{C}$ such that ab^q is real and positive, we have then, for real $t > 0$:

$$\lim_{t \to 0^+} \tau(bt)/|\tau(bt)| = 1. \qquad (2.3.1)$$

Assuming, as we may, that $|b| = 1$, consider the real analytic map $\varphi : (-1,1) \to V$ given by

$$\varphi(t) = \big(\lambda(bt),\, \mu(bt)\tau(bt)\big) \in V \qquad (t \in (-1,1)),$$

see (2.2.1). Then $\pi_f\,\varphi(t) = \mu(bt)\tau(bt)$; and since $\mu(0) = z$ and $\tau(0) = 0$, (1.3.1) results from (2.3.1). $\qquad\qquad \square$

3   Differential functoriality of the specialization over $\mathbb{R}$

In this section we look at maps of analytic spaces primarily in terms of underlying topological spaces.

(3.1). Let $\mathfrak{t} : \mathbf{V} \to \mathbb{C}$ be the specialization of (V, W) to $C(V, W)$, see §2. The *specialization over* $\mathbb{R}$ (or $\mathbb{R}$*-specialization*) of (V, W) to $C(V, W)$ is the real analytic space

$$_{\mathbb{R}}\mathbf{V} := \mathbb{R} \times_{\mathbb{C}} \mathbf{V} = \mathfrak{t}^{-1}(\mathbb{R}).$$

As in §2, we have natural maps

$$W \times \mathbb{R} \xrightarrow{\ \alpha\ } {_{\mathbb{R}}\mathbf{V}} \xrightarrow{\ \beta\ } V \times \mathbb{R}.$$

The fibers $\mathbf{V}_a := \mathfrak{t}^{-1}(a)$ $(a \in \mathbb{R})$ of $\mathfrak{t} : {_{\mathbb{R}}\mathbf{V}} \to \mathbb{R}$ are all real-isomorphic, via β, to V, except for $\mathbf{V}_0 \cong C(V, W)$.

(3.2) Let (V, W) be as in (1.1), and let (V', W') be another such pair. Define $\mathfrak{t}' : {_{\mathbb{R}}\mathbf{V}'} \to \mathbb{R}$ as above (with respect to $W' \subset V'$). Let $f : V \to V'$ be a C^1 (continuously differentiable) map such that $f(W) \subset W'$.

We recall the definition of C^1 map. A map $g : V \to \mathbb{R}^n$ is C^1 at $v \in V$ if for some analytic germ-embedding $(V, v) \hookrightarrow (\mathbb{C}^N, 0)$, there is an open neighborhood U of 0 in $\mathbb{C}^N$ and a C^1 map $U \to \mathbb{R}^n$ whose restriction to $V \cap U$ coincides with that of g. A germ-map $\gamma : (V, v) \to (V', v')$ is C^1 if its composition with some embedding $(V', v') \hookrightarrow (\mathbb{C}^M, 0)$ is C^1 at v. (If this property of γ holds for one choice of embeddings then it holds for any choice.) Finally, the above map f is C^1 if its germ at each $v \in V$ is C^1.

Define the C^1 map $\mathbf{f} : {_{\mathbb{R}}\mathbf{V}} - \mathbf{V}_0 \longrightarrow {_{\mathbb{R}}\mathbf{V}'} - \mathbf{V}_0'$ to be the composition

$$_{\mathbb{R}}\mathbf{V} - \mathbf{V}_0 \xrightarrow[\ \beta\ ]{\sim} V \times (\mathbb{R}^1 - \{0\}) \xrightarrow[f \times id]{} V' \times (\mathbb{R}^1 - \{0\}) \xrightarrow[\ \beta'^{-1}\ ]{\sim} {_{\mathbb{R}}\mathbf{V}'} - \mathbf{V}_0'.$$

Theorem (3.3). *With preceding notation, assume further that W is a complex submanifold of the analytic space V. Then the map $\mathbf{f}$ has a unique extension to a continuous map (still denoted $\mathbf{f}$) : ${_{\mathbb{R}}\mathbf{V}} \to {_{\mathbb{R}}\mathbf{V}'}$; and the following diagram commutes:*

$$
\begin{array}{ccccc}
W \times \mathbb{R} & \xrightarrow{\ \alpha\ } & {_{\mathbb{R}}\mathbf{V}} & \xrightarrow{\ \beta\ } & V \times \mathbb{R} \\
{\scriptstyle f \times id}\big\downarrow & & {\scriptstyle f}\big\downarrow & & \big\downarrow{\scriptstyle f \times id} \\
W' \times \mathbb{R} & \xrightarrow[\ \alpha'\ ]{} & {_{\mathbb{R}}\mathbf{V}'} & \xrightarrow[\ \beta'\ ]{} & V' \times \mathbb{R}
\end{array}
\qquad (3.3.1)
$$

In particular, $\mathfrak{t}' \circ \mathbf{f} = \mathfrak{t}$. The restriction $\mathbf{f}_0$ of $\mathbf{f}$ to $\mathbf{V}_0 = C(V, W)$ is a continuous map from $C(V, W)$ to $C(V', W')$, fitting into a commutative diagram

$$
\begin{array}{ccc}
\mathbb{R} \times C(V,W) & \xrightarrow{\ id \times \mathbf{f}_0\ } & \mathbb{R} \times C(V',W') \\[2pt]
\ \downarrow{\scriptstyle \mu} & & \ \downarrow{\scriptstyle \mu} \\[6pt]
C(V,W) & \xrightarrow{\ \mathbf{f}_0\ } & C(V',W') \\[2pt]
{\scriptstyle p}\uparrow\downarrow{\scriptstyle \sigma} & & {\scriptstyle p'}\uparrow\downarrow{\scriptstyle \sigma'} \\[6pt]
W & \xrightarrow{\ f\ } & W'
\end{array}
\qquad (3.3.2)
$$

see (1.1), and for each $w \in W$, the restriction of $\mathbf{f}_0$ to $p^{-1}(w)$ is real-analytic.[2]

Proof. The assertions need only be verified near an arbitrary point $\nu \in \mathbf{V}_0$, so we can introduce coordinates as in (2.2). To be more precise, let $\pi : \mathbf{V} \to V$ be the canonical map, corresponding to the inclusion $\mathcal{O}_V \hookrightarrow \mathcal{R}$; and define $\pi' : \mathbf{V}' \to V'$ similarly. Let $w := \pi(\nu) = pp^{-1}(\nu) \in W$ (see (2.1), noting that π is β followed by the projection $V \times \mathbb{C}^1 \to V$), and let $w' := f(w) \in W'$. Choose neighborhoods V^* of w in V and V'^* of w' in V' such that $f(V^*) \subset V'^*$ and such that $(V^*, W \cap V^*, w)$ and $(V'^*, W' \cap V'^*, w')$ can be embedded into $(\mathbb{C}^r \times \mathbb{C}^s, \mathbb{C}^r, 0)$ and $(\mathbb{C}^{r'} \times \mathbb{C}^{s'}, \mathbb{C}^{r'}, 0)$ respectively, as in (1.2). Then $\mathbf{V}^* := \pi^{-1}(V^*)$ is the specialization of V^* to $C(V^*, W \cap V^*)$. From the definition of $\mathbf{f}$ and the relation between β and π, we see that $f\pi = \pi'\mathbf{f}$, so that $\mathbf{f}$ maps ${}_{\mathbb{R}}\mathbf{V}^* - \mathbf{V}_0$ into $\mathbf{V}'^* := \pi'^{-1}(V'^*)$. Hence we may – and do – assume that $(V, V', \mathbf{V}, \mathbf{V}') = (V^*, V'^*, \mathbf{V}^*, \mathbf{V}'^*)$, coordinatized as in (1.2) and (2.2). We may assume further, because W is a *submanifold* of V, that W is actually identical with the flat space L in (1.2).

Uniqueness of the extension holds because ${}_{\mathbb{R}}\mathbf{V} - \mathbf{V}_0$ is dense in ${}_{\mathbb{R}}\mathbf{V}$, as follows via (2.2.1) and (2.2.2) from Proposition (1.3): setting $v_i = (x_i, y_i)$ there, and with a_i real and positive, the sequence (x_i, a_iy_i, a_i^{-1}) in ${}_{\mathbb{R}}\mathbf{V} - \mathbf{V}_0$ has limit $(w, z, 0)$. (Since $y_i \to 0$, therefore $a_i \to \infty$ if $z \neq 0$; and if $z = 0$ then we can take $y_i = 0$ and $a_i = i$ for all i.)

Commutativity of the right half of (3.3.1) can be checked on the dense set ${}_{\mathbb{R}}\mathbf{V} - \mathbf{V}_0$, where it holds by the definition of $\mathbf{f}$. The left half can be also be checked outside of $\mathbf{V}_0$ (since $W \times (\mathbb{R} - \{0\})$ is dense in $W \times \mathbb{R}$), and there it is obvious because β' is bijective and $\beta \circ \alpha = i \times \mathrm{id}$, etc., see §2.

Now let us show that the asserted extension of $\mathbf{f}$ exists. The question comes down to the existence, for each $\nu \in \mathbf{V}_0$, of a point $\nu' \in \mathbf{V}'$ such that every sequence $(\nu_i)_{i>0}$ in ${}_{\mathbb{R}}\mathbf{V} - \mathbf{V}_0$ with $\nu_i \to \nu$ satisfies $\lim \mathbf{f}(\nu_i) = \nu'$. After embedding ${}_{\mathbb{R}}\mathbf{V}$ into $\mathbb{C}^r \times \mathbb{C}^s \times \mathbb{R}^1$ as above, we have $\nu = (0, z, 0)$ for some $z \in \mathbb{C}^s$, and $\nu_i = (x_i, y_i, a_i)$. The description of β preceding (2.2.1) gives an expression for $\mathbf{f}$ in coordinates:

$$
\mathbf{f}(x, y, a) = (\xi, a^{-1}\eta, a), \quad \text{where } (\xi, \eta) := f(x, ay).
$$

---

[2] See also Remark (5.4.3) below.

The question thus becomes whether the sequence $\mathbf{f}(x_i, y_i, a_i) = (\xi_i, a_i^{-1}\eta_i, a_i)$ has a limit depending only on z. Since $x_i \to 0$, $y_i \to z$, and $a_i \to 0$, and since f is continuous, therefore $(\xi_i, \eta_i) \to f(0,0) = (0,0)$, so that $\xi_i \to 0$. It remains to investigate $\lim a_i^{-1}\eta_i$.

By the definition of C^1 map, there exists a neighborhood U^* of $(0,0)$ in $\mathbb{C}^r \times \mathbb{C}^s$ and a C^1 map $F : U^* \to \mathbb{C}^{r'} \times \mathbb{C}^{s'}$ agreeing with f on $V \cap U^*$. To simplify, we multiply F by a C^∞ function $\psi : \mathbb{C}^r \times \mathbb{C}^s \to \mathbb{R}$ which takes the value 1 on a small neighborhood U_1 of $(0,0)$ and vanishes outside a compact subset $\bar{U}$ of U^*; then after replacing V by $V \cap U_1$, and F by the extension of ψF which takes the value $(0,0)$ outside $\bar{U}$, we may assume that $U^* = \mathbb{C}^r \times \mathbb{C}^s$. We may also assume that $F(\mathbb{C}^r \times \{0\}) \subset \mathbb{C}^{r'} \times \{0\}$ (take $U_1 \subset U$ where U is as in (1.2), recall that $L = W$, see above, and that $f(W) \subset W'$).

Denote the derivative of F at (x, y) – a real-linear map from $\mathbb{C}^r \times \mathbb{C}^s$ to $\mathbb{C}^{r'} \times \mathbb{C}^{s'}$ – by $DF_{(x,y)}$. Set $F(x_i, 0) =: (x_i', 0)$. Let $\mathrm{pr}_2 : \mathbb{C}^{r'} \times \mathbb{C}^{s'} \to \mathbb{C}^{s'}$ be the projection, let q^j $(1 \le j \le 2s')$ be the real coordinate functions on $\mathbb{C}^{s'}$, and set $F^j := q^j \circ \mathrm{pr}_2 \circ F$. We are concerned with the limits (as $i \to \infty$):

$$\begin{aligned}
\lim_i q^j(a_i^{-1}\eta_i) &= \lim_i a_i^{-1} q^j \mathrm{pr}_2\big((\xi_i, \eta_i) - (x_i', 0)\big) \\
&= \lim_i a_i^{-1}\big(F^j(x_i, a_i y_i) - F^j(x_i, 0)\big).
\end{aligned}$$

But a_i being *real*, the Mean Value Theorem gives

$$\begin{aligned}
\lim_i a_i^{-1}\big(F^j(x_i, a_i y_i) - F^j(x_i, 0)\big) &= \lim_i DF^j_{(x_i, b_{ij}a_i y_i)}(0, y_i) \quad (0 < b_{ij} < 1) \\
&= DF^j_{(0,0)}(0, z),
\end{aligned}$$

the last equality by continuity of DF (needed only at points of W). Thus, the extended $\mathbf{f}$ exists.

It is clear that $\mathbf{f}$ maps $\mathbf{V}_0$ into $\mathbf{V}_0'$. Commutativity of (3.3.2) follows, via (2.2.2), (1.2.2), and (1.2.3), from the description of $\mathbf{f}_0$ entailed by the foregoing, viz.

$$\mathbf{f}_0(0, z, 0) = \big(0, \mathrm{pr}_2 DF_{(0,0)}(z), 0\big). \tag{3.3.3}$$

This description also shows that the restriction of $\mathbf{f}_0$ to $p^{-1}(w)$ is real-analytic (even *real-linear* in these coordinates). $\qquad\square$

For any subvariety (i.e., reduced analytic subspace) V_1 of V, set $W_1 := W \times_V V_1$, so that the deformation of V_1 to $C(V_1, W_1)$ is canonically embedded in $\mathbf{V}$. If in the preceding proof we have $(0, z, 0) \in C(V_1, W_1)$, then by (1.3), we can choose $(x_i, y_i, a_i) \to (0, z, 0)$ such that $(x_i, a_i y_i) \in V_1$, and consequently:

Corollary (3.4). *If V_1 and V_1' are subvarieties of V and V' respectively, and if $f(V_1) \subset V_1'$, then $\mathbf{f}_0$ maps $C(V_1, W_1)$ continuously into $C(V_1', W_1')$.*

Remark (3.5). The same proof as in (3.3) shows that the C^1 map $\mathbf{F}$ defined by

$$\mathbf{F}(x,y,a) := (\xi, a^{-1}\eta, a) \quad ((\xi,\eta) := F(x,ay)) \quad (x \in \mathbb{C}^r,\ y \in \mathbb{C}^s,\ 0 \neq a \in \mathbb{R})$$

extends continuously to a map (still denoted $\mathbf{F}$) from $\mathbb{C}^r \times \mathbb{C}^s \times \mathbb{R}$ to $\mathbb{C}^{r'} \times \mathbb{C}^{s'} \times \mathbb{R}$ such that $\mathsf{t}' \circ \mathbf{F} = \mathsf{t}$, where now t and t' denote the respective projections to $\mathbb{R}$.

4 Multiplicities of components of C(V, W)

By *component* of a complex analytic space Z (not necessarily reduced) is meant an irreducible component of the reduced space Z_{red}. Let Y be such a component, with inclusion map $j : Y \hookrightarrow Z$, and let $\mathcal{P}$ be the defining $\mathcal{O}_Z$-ideal of Y, i.e., the kernel of the natural map $\mathcal{O}_Z \to j_*\mathcal{O}_Y$. Let y be any point of Y. Then the stalk $\mathcal{P}_y$ is an intersection of finitely many minimal prime ideals P_i in $\mathcal{O}_{Z,y}$, corresponding to the local components of Y at y.

Proposition-Definition (4.1) *The length e of the local artin ring $(\mathcal{O}_{Z,y})_{P_i}$ depends only on Y, and not on y or P_i. This integer is called the* multiplicity *of Y in Z, and denoted $e_{Y,Z}$.*

Proof. For each $n \geq 0$, set $\mathcal{G}_n := j^*(\mathcal{P}^n/\mathcal{P}^{n+1})$, a coherent $\mathcal{O}_Y$-module. Let k be the residue field of $(\mathcal{O}_{Z,y})_{P_i}$, and set

$$e_n := \dim_k(\mathcal{G}_{n,y}) \otimes_{\mathcal{O}_{Y,y}} k,$$

so that $e_n = 0$ for $n \gg 0$ and $e = \sum_{n=0}^{\infty} e_n$. Then by [Ho, p. 20-10, Prop. 6] the module $\mathcal{G}_n$ is locally free of rank e_n outside a nowhere dense analytic subspace of Y. Thus e_n (for given n), and hence e, depends only on Y. $\qquad\qquad\square$

Remark (4.1.1). If U is an open subset of Z, and $Y_1, \ldots, Y_r$ are the components of $Y \cap U$, then clearly $e_{Y_i,U} = e_{Y,Z}$ for all i.

(4.2). Let (V, W) be as in (1.1), and let $(V_\lambda)_{\lambda \in \Lambda}$ be the family of all components of V. For each λ, let $W_\lambda := (W \times_V V_\lambda) \subset V_\lambda$, and let $\mathcal{I}_\lambda$ be the defining $\mathcal{O}_{V_\lambda}$-ideal of W_λ. Set

$$\mathcal{R}_\lambda := \oplus_{n \in \mathbb{Z}} \mathcal{I}_\lambda^n T^{-n} \subset \mathcal{O}_{V_\lambda}[T, T^{-1}]$$

and

$$\mathbf{V}_\lambda := \operatorname{Specan}(\mathcal{R}_\lambda),$$

the specialization of (V_λ, W_λ) to $C(V_\lambda, W_\lambda)$. With notation as in §2, there is an obvious commutative diagram, whose vertical arrows are closed immersions:

$$
\begin{array}{ccccc}
W_\lambda \times \mathbb{C}^1 & \xrightarrow{\ \alpha_\lambda\ } & \mathbf{V}_\lambda & \xrightarrow{\ \beta_\lambda\ } & V_\lambda \times \mathbb{C}^1 \\
\downarrow & & \downarrow & & \downarrow \\
W \times \mathbb{C}^1 & \xrightarrow[\ \alpha\ ]{} & \mathbf{V} & \xrightarrow[\ \beta\ ]{} & V \times \mathbb{C}^1
\end{array}
\qquad (4.2.1)
$$

The $\mathbf{V}_\lambda$ are all the components of $\mathbf{V}$: this need only be verified outside the nowhere dense analytic subset $\mathbf{V}_0$, where it follows from (2.1)(ii). With t_λ the restriction of t to $\mathbf{V}_\lambda$ we have

$$C(V,W) = t^{-1}(0) = \bigcup_\lambda t_\lambda^{-1}(0) = \bigcup_\lambda C_\lambda(V_\lambda, W_\lambda).$$

Now W is covered by open subsets $U \subset V$ meeting only finitely many V_λ, and for such a U, $p^{-1}(U) \subset C(V,W)$ meets $C(V_\lambda, W_\lambda)$ only for those same λ; so the family $C(V_\lambda, W_\lambda)$ is locally finite in $C(V,W)$. Hence every component of $C(V,W)$ is a component of $C(V_\lambda, W_\lambda)$ for at least one and at most finitely many λ. Conversely, if $\dim V_\lambda = \dim V$ then every component of $C(V_\lambda, W_\lambda)$ is a component of $C(V,W)$ (since $\dim C(V,W) = \dim V$, by (2.1)).

Proposition (4.2.2). *Assume that V is equidimensional, i.e., all the components V_λ of V have the same dimension. Let C_* be a component of $C(V,W)$. Then*

$$e_{C_*, C(V,W)} = {\sum_\lambda}^* e_{C_*, C(V_\lambda, W_\lambda)}$$

the sum being over all λ such that C_* is a component of $C(V_\lambda, W_\lambda)$.

Proof. Note that after fixing $y \in C_*$ we can replace V by any open subset V^* containing $p(y)$ ($p : C(V,W) \to W$ the canonical map): first, by (4.1.1), the component of $C_* \cap p^{-1}(W \cap V^*)$ containing y has multiplicity $e_{C_*, C(V,W)}$ in $p^{-1}(W \cap V^*) = C(V^*, W \cap V^*)$, and similarly for $C(V_\lambda \cap V^*, W_\lambda \cap V^*)$; and second, though $V_\lambda \cap V^*$ may no longer be irreducible, that doesn't matter because (4.2.2) is clearly equivalent to a similar statement in which we assume only that $V = \cup V_\lambda$ where each V_λ is a union of components of V (all having the same dimension as V) and no two V_λ have a common component. So pick V^* as in (1.2), and embed $\mathbf{V}$ in $\mathbb{C}^{r+s+1}$ as in (2.2).

Now let B_* be a local component of C_* at y, and let P be the prime ideal in $\mathcal{O}_{\mathbf{V},y}$ consisting of germs of functions vanishing on B_*. Let $t \in \mathcal{O}_{\mathbf{V},y}$ be the germ of the function $t : \mathbf{V} \to \mathbb{C}$, so that $\mathcal{O}_{\mathbf{V},y}/(t) = \mathcal{O}_{C(V,W),y}$, see (2.1). Then $e_{C_*, C(V,W)}$ is, by definition, the length of the artin local ring $\left(\mathcal{O}_{\mathbf{V},y}/(t)\right)_P$, i.e., (since t is flat and hence t is not a zero-divisor in $\mathcal{O}_{\mathbf{V},y}$) the multiplicity of the ideal $t(\mathcal{O}_{\mathbf{V},y})_P$. But by the equality of algebraic and topological intersection numbers (see e.g., [GL, p. 184, Fact]), that multiplicity is the intersection number $i\left((\mathbb{C}^{r+s} \times \{0\}) \cdot \mathbf{V}, C_*\right)$ defined in [BH, p. 482, 4.4]. (The intersection takes place in $\mathbb{C}^{r+s+1}$.) Similarly, with $\mathbf{V}_\lambda \subset \mathbf{V}$ as in (4.2.1), we have $e_{C_*, C(V_\lambda, W_\lambda)} = i\left((\mathbb{C}^{r+s} \times \{0\}) \cdot \mathbf{V}_\lambda, C_*\right)$. So the conclusion results from the equality

$$i\left((\mathbb{C}^{r+s} \times \{0\}) \cdot \mathbf{V}, C_*\right) = {\sum_\lambda}^* i\left((\mathbb{C}^{r+s} \times \{0\}) \cdot \mathbf{V}_\lambda, C_*\right)$$

given in [BH, p. 483]. $\square$

(4.3) Suppose next that we have two equidimensional reduced analytic spaces V and V', along with complex submanifolds $W \subset V$ and $W' \subset V'$. We consider a situation as in §3, where there is a C^1 map $f : (V, W) \to (V', W')$; and we assume that f is *invertible,* i.e., that there is a C^1 map $g : (V', W') \to (V, W)$ such that $f \circ g$ and $g \circ f$ are both identity maps. Then by Theorem (3.3), f and g naturally induce inverse homeomorphisms $\mathbf{f}$ and $\mathbf{g}$ between $\mathbf{V}$ and $\mathbf{V}'$, restricting to homeomorphisms $\mathbf{f}_0$ and $\mathbf{g}_0$ between the respective subspaces $C(V, W)$ and $C(V', W')$. [3]

Theorem (4.3.1). *Under the preceding circumstances, the homeomorphism $\mathbf{f}_0$ gives a one-one multiplicity-preserving correspondence between the components of $C(V, W)$ and those of $C(V', W')$.*

Proof. The one-one correspondence obtains because any homeomorphism of analytic spaces maps each component of the source onto a component of the target, [GL, p. 172, (A8)]. We need to show that corresponding components C_* and C'_* have the same multiplicity (in $C(V, W)$, $C(V', W')$ respectively). The proof which follows is essentially the same as that in [GL, §D], to which we refer for more details.

Let $y \in C_* \subset \mathbf{V}_0$, and, $\pi : \mathbf{V} \to V$ being the canonical map, let $v := \pi(y)$. Using (4.1.1), and arguing as in the beginning of the proof of (3.3), we find that we may replace $\mathbf{V}$ by $\pi^{-1}(V^*)$ where V^* is an "embeddable" neighborhood of v (i.e., V^* is as in (1.2)) such that $V'^* := f(V^*)$ is also embeddable; and we may replace $\mathbf{V}'$ by $\pi^{-1}(V'^*)$. Thus we reduce to where V and V' are embedded in some $\mathbb{C}^n$ and $\mathbb{C}^{n'}$ respectively, with $W = L$, see (1.2), and similarly for W'. Then as in the proof of (3.3) we can assume, after replacing V by a smaller neighborhood of v if necessary, that there is a C^1 map $F_n : \mathbb{C}^n \to \mathbb{C}^{n'}$ agreeing with f on V; and similarly assume that there is a C^1 map $G_{n'} : \mathbb{C}^{n'} \to \mathbb{C}^n$ agreeing with g on V'. We then define inverse C^1 maps

$$\mathbb{C}^n \times \mathbb{C}^{n'} \underset{G}{\overset{F}{\rightleftarrows}} \mathbb{C}^{n'} \times \mathbb{C}^n$$

by
$$\begin{aligned} F(x, y) &:= \bigl(y + F_n(x),\ x - G_{n'}(y + F_n(x))\bigr), \\ G(x', y') &:= \bigl(y' + G_{n'}(x'),\ x' - F_n(y' + G_{n'}(x'))\bigr), \end{aligned}$$

and verify that for $x \in V$ (resp. $x' \in V'$) we have

$$F(x, 0) = \bigl(f(x), 0\bigr) \quad \text{resp.} \quad G(x', 0) = \bigl(g(x'), 0\bigr).$$

Hence, if we embed V and V' in $\mathbb{C}^{n+n'}$ by

$$V \xrightarrow{\sim} V \times \{0\} \hookrightarrow \mathbb{C}^n \times \mathbb{C}^{n'} \quad \text{resp.} \quad V' \xrightarrow{\sim} V' \times \{0\} \hookrightarrow \mathbb{C}^{n'} \times \mathbb{C}^n,$$

---

[3] Continuity of the derivative of f (resp. g) need only hold at points of W (resp. W'), see proof of (3.3).

and correspondingly embed $_\mathbb{R}V$ and $_\mathbb{R}V'$ in $\mathbb{C}^{n+n'} \times \mathbb{R}$, see (2.2), then (3.5) gives us *inverse homeomorphisms*

$$\mathbb{C}^{n+n'} \times \mathbb{R} \underset{\mathbf{G}}{\overset{\mathbf{F}}{\rightleftarrows}} \mathbb{C}^{n+n'} \times \mathbb{R}$$

with $\mathbf{F}(_\mathbb{R}V) \subset {}_\mathbb{R}V'$ and $\mathbf{G}(_\mathbb{R}V') \subset {}_\mathbb{R}V$. And finally, in view of (4.2.2) and (3.4) we can replace V by a component V_λ, i.e., we may assume V to be irreducible, of dimension, say, d.

Now the underlying idea is that, as we have just seen, the multiplicity of a component is an intersection multiplicity, and as such should be invariant under the homeomorphism $\mathbf{F}$. Technical complications arise from working with $_\mathbb{R}V$ rather than with $\mathbf{V}$ (which has been necessitated by the real derivatives of f and g being not necessarily complex-linear).

Setting $N := n + n'$, we first deduce from [BH, p. 475, 2.15], applied to the inclusion of $\mathbb{C}^N \times \mathbb{R}^1$ (with fixed orientation) into $\mathbb{C}^N \times \mathbb{C}^1$, and to the smooth locus U of $\mathbf{V} - \mathbf{V}_0$ (which has real codimension ≥ 2 in $Y := \mathbf{V}$), that $_\mathbb{R}V$ has a fundamental class ρ in the Borel-Moore homology $H_{2d+1}(_\mathbb{R}V)$. Using the projection formula, we see further that $\pm\rho$ is the intersection of the fundamental classes of $\mathbb{C}^N \times \mathbb{R}$ and of $\mathbf{V}$. (Strictly speaking, the intersection class lies in $H^\Phi_{2d+1}(\mathbb{C}^N \times \mathbb{C}^1)$ where Φ is the family of closed subsets of $_\mathbb{R}V$; but that group is canonically isomorphic to $H_{2d+1}(_\mathbb{R}V)$.) [4] Then associativity of the intersection product and the relation

$$\begin{aligned}\mathbf{V}_0 &= (\mathbb{C}^N \times \{0\}) \cap {}_\mathbb{R}V = (\mathbb{C}^N \times i\mathbb{R}^1) \cap {}_\mathbb{R}V \\ &= (\mathbb{C}^N \times i\mathbb{R}^1) \cap (\mathbb{C}^N \times \mathbb{R}^1) \cap \mathbf{V} \quad (i = \sqrt{-1})\end{aligned}$$

show that $e_{C_*, C(V,W)}$ is the intersection number $i\big((\mathbb{C}^N \times \{0\}) \cdot {}_\mathbb{R}V, C_*\big)$ (in $\mathbb{C}^N \times \mathbb{R}^1$), see [GL, p. 176, (B.5.2)]. Given the topological invariance (up to sign) of intersection numbers, the principal remaining problem is to show that the map $\mathbf{F}_* : H_{2d+1}(_\mathbb{R}V) \to H_{2d+1}(_\mathbb{R}V')$ induced by $\mathbf{F}$ takes ρ to $\pm$ the fundamental class ρ' of $_\mathbb{R}V'$. (The corresponding statement for $\mathbb{C}^N \times \{0\}$ is straightforward.) One can proceed as in [GL, §(D.4)]. Another way, since $\mathbf{F}_*$ is an isomorphism, is to show that

$$H_{2d+1}(_\mathbb{R}V) \cong \mathbb{Z} \cong H_{2d+1}(_\mathbb{R}V'),$$

generated, necessarily, by ρ and ρ' respectively. This we now do.

Recall that for any locally compact space X, there are canonical isomorphisms

$$H_i(X \times \mathbb{R}^1) \xrightarrow{\sim} H_{i-1}(X) \qquad (i \in \mathbb{Z}).$$

These arise, upon identification of $\mathbb{R}^1$ with the open unit interval $(0, 1)$, from the following exact sequence associated to the inclusion of the pair of points $\{0, 1\}$

---

[4] Cf. [GL, p. 175, (B.3.5)], where the second $\overline{S}$ ($= {}_\mathbb{R}V$) should be S ($= \mathbf{V}$).

into the closed unit interval $I := [0,1]$, see [BH, p. 465, 1.6]:

$$\cdots \longrightarrow H_i(X) \oplus H_i(X) \xrightarrow{\alpha} H_i(X \times I) \longrightarrow H_i(X \times \mathbb{R}^1)$$
$$\xrightarrow{\beta} H_{i-1}(X) \oplus H_{i-1}(X) \xrightarrow{\gamma} H_{i-1}(X \times I) \longrightarrow \cdots$$

The point is that the (proper) projection $X \times I \to X$, being a homotopy equivalence, induces for every i an isomorphism $H_i(X \times I) \xrightarrow{\sim} H_i(X)$, whose inverse is given by $H_i(X) \xrightarrow{\sim} H_i(X \times \{a\}) \longrightarrow H_i(X \times I)$ for any $a \in I$ [BH, p. 465, 1.5]; hence α is surjective, and β maps $H_i(X \times \mathbb{R}^1)$ isomorphically onto the kernel of γ, which is isomorphic to $H_{i-1}(X)$ (diagonally embedded in $\oplus$). As a corollary, we note that for any integers $i \neq j$, with $j \geq 0$, we have

$$H_i(\mathbb{R}^j) \cong H_{i-j}(\mathbb{R}^0) = 0, \tag{4.3.2}$$

the last equality by [BH, p. 464, 1.3]. (Similarly, $H_j(\mathbb{R}^j) = \mathbb{Z}$.) Now consider the exact sequence

$$0 = H_{2d+1}(\mathbf{V}_0) \longrightarrow H_{2d+1}(_\mathbb{R}\mathbf{V}) \longrightarrow H_{2d+1}(_\mathbb{R}\mathbf{V} - \mathbf{V}_0) \xrightarrow{\delta} H_{2d}(\mathbf{V}_0) \xrightarrow{\epsilon} H_{2d}(_\mathbb{R}\mathbf{V})$$

see [BH, p. 465, 1.6]. Note that $\mathbf{V}_0$ has complex dimension d, by (2.1)(i), hence cohomological dimension $2d$ [BH, p. 475, 3.1], whence the vanishing of $H_{2d+1}(\mathbf{V}_0)$ see [BH, p. 467, (1)]. By (2.1)(ii), $_\mathbb{R}\mathbf{V} - \mathbf{V}_0$ is homeomorphic to the disjoint union of two copies of $V \times \mathbb{R}^1$. Since V is, by assumption, irreducible, we have

$$H_{2d+1}(V \times \mathbb{R}^1) \cong H_{2d}(V) \cong \mathbb{Z},$$

the first isomorphism as above, the second by [BH, p. 476, 3.3]. Thus $H_{2d+1}(_\mathbb{R}\mathbf{V})$ is free, of rank 1 or 2. (The rank is > 0 because $\rho \neq 0$, since as above, ρ gives rise via intersection to $e_{C_*, C(V,W)} > 0$.) Moreover, $H_{2d}(\mathbf{V}_0)$ is torsion-free [BH, p. 482, 4.3]. It will therefore suffice to show that δ is not the zero map. We do this by noting, with $[C_\mu]$ the fundamental class of the component C_μ of $C(V, W) = \mathbf{V}_0$, that

$$\epsilon\left(\sum_\mu \pm e_{C_\mu, C(V,W)}[C_\mu]\right) = 0. \tag{4.3.3}$$

Indeed, with the right choice of $\pm$, the left side is the image under ϵ of the intersection class $(\mathbb{C}^N \times \{0\}) \cdot {}_\mathbb{R}\mathbf{V}$ (see above). But by compatibility of intersections with "enlargement of families of supports" [BH, p. 468, 1.12], we have a commutative diagram, where $H^Z(-)$ stands for the Borel-Moore homology of $\mathbb{C}^N \times \mathbb{R}^1$ with supports in closed subsets of Z:

$$
\begin{array}{ccc}
H_{2N}^{\mathbb{C}^N \times \{0\}}(-) \times H_{2d+1}^{_\mathbb{R}\mathbf{V}}(-) & \xrightarrow{\text{intersect}} & H_{2d}^{\mathbf{V}_0}(-) = H_{2d}(\mathbf{V}_0) \\
{\scriptstyle\text{natural}} \downarrow & & \downarrow {\scriptstyle \epsilon} \\
H_{2N}^{\mathbb{C}^N \times \mathbb{R}^1}(-) \times H_{2d+1}^{_\mathbb{R}\mathbf{V}}(-) & \xrightarrow{\text{intersect}} & H_{2d}^{_\mathbb{R}\mathbf{V}}(-) = H_{2d}(_\mathbb{R}\mathbf{V})
\end{array}
$$

in which the lower left corner vanishes, by (4.3.2); and (4.3.3) results. $\qquad\square$

5   Relative complexification of the normal cone

We now construct the *relative complexification* of a cone C, and for $C = C(V, W)$ establish C^1 functorial properties of this complexification (Theorem (5.3.1)).

Let C be a cone over a complex space W, i.e., $C = \mathrm{Specan}(\mathcal{G})$ for some finitely presented graded $\mathcal{O}_W$-algebra $\mathcal{G}$, see (1.1.1). Assume that all the irreducible components of C have the same dimension, and that all the fibers of the canonical map $C \to W$ have positive dimension. For example, if $V \supset W$ is as in (1.1), with V equidimensional and W nowhere dense in V, then (2.1) implies that $C(V, W)$ is equidimensional, of dimension $\dim C = \dim V > \dim W$, and hence the fibers of $p : C(V, W) \to W$ are all positive-dimensional.

Recall that a subset of a complex space X is *Zariski-open* if its complement is an analytic subset of X. (Analytic subsets of X are understood to be *closed*, defined locally by the vanishing of sections of $\mathcal{O}_X$.)

Lemma (5.1). *There exists a unique analytic subset $\widetilde{C}$ of $C \times_W C$ such that with $\tilde{p} : \widetilde{C} \to W$ the natural composition $\widetilde{C} \hookrightarrow C \times_W C \to W$,*

(i) *for any open dense $U \subset W$, $\tilde{p}^{-1}(U)$ is dense in $\widetilde{C}$; and*

(ii) *there is a dense Zariski-open subset W_0 of W such that for every $w \in W_0$, the reduced fiber $\widetilde{C}_w := \tilde{p}^{-1}(w)_{\mathrm{red}}$ is*

$$\widetilde{C}_w = \bigcup_{i \in I_w} C_w^i \times C_w^i \subset C \times_W C,$$

$(C_w^i)_{i \in I_w}$ being the family of irreducible components of the cone $C_w := p^{-1}(w)$.

In fact $\widetilde{C}$ is a union of irreducible components of $C \times_W C$, and so is stable under the natural $\mathbb{C}^1 \times \mathbb{C}^1$ action (given by μ in (1.1.1)).

We will call $\tilde{p} : \widetilde{C} \to W$ the *relative complexification* of $p : C \to W$. That's because for almost all $w \in W$ (e.g., $w \in W_0$), $\widetilde{C}_w$ is real-analytically isomorphic to a reduced complexification of $(C_w)_{\mathrm{red}}$, see (5.3.0). For example, if $\mathcal{G}$ is the symmetric algebra of a finite-rank locally free $\mathcal{O}_W$-module, i.e., C is a complex vector bundle over W, then $\widetilde{C} = C \times_W C$ together with the natural addition on the fibers and the $\mathbb{C}^1$ action specified immediately before Thm. (5.3.1) below, is just the usual complexification of the real vector bundle underlying C.

Proof. Uniqueness is immediate: if $(\widetilde{C}', W_0')$ and $(\widetilde{C}'', W_0'')$ are two pairs satisfying the conditions of (5.1), then $W_0' \cap W_0''$ is open and dense in W, and so $\widetilde{C}'$ and $\widetilde{C}''$ are both equal to the closure in $C \times_W C$ of

$$\bigcup_{w \in W_0' \cap W_0''} \left(\bigcup_{i \in I_w} C_w^i \times C_w^i \right).$$

As for existence, with $\sigma : W \to C$ as in (1.1.1) let C^* be the reduced space $C_{\mathrm{red}} - \sigma(W)$, on which $\mathbb{C}^*$ acts freely, preserving fibers of p; and set

$$P := C^* / \mathbb{C}^* = \mathrm{Projan}(\mathcal{G})_{\mathrm{red}}.$$

(Projan is constructed, in analogy with Specan, by pasting together subspaces of relative projective spaces $W_\alpha \times \mathbb{P}^{N_\alpha}$, with (W_α) a suitable open cover of W.) Let $\overline{P}$ be the normalization of P, and let

$$\phi : \overline{P} \to W, \qquad \Phi : \overline{P} \times_W \overline{P} \to W$$

be the natural maps, both of which are proper. Consider the commutative diagram

$$
\begin{array}{ccc}
\overline{P} & \xrightarrow{\ \Delta\ } & \overline{P} \times_W \overline{P} \\[2pt]
\phi' \downarrow & & \downarrow \Phi' \\[6pt]
\mathrm{Specan}(\phi_* \mathcal{O}_{\overline{P}}) =: S & \xrightarrow{\ \delta\ } & T := \mathrm{Specan}(\Phi_* \mathcal{O}_{\overline{P} \times_W \overline{P}}) \\[6pt]
\downarrow & & \downarrow \\[6pt]
W & = & W
\end{array}
\qquad (5.1.1)
$$

whose sides are the Stein factorizations of ϕ and Φ respectively [Fi, p. 71], where Δ is the diagonal map, and where δ corresponds to the natural map of $\mathcal{O}_W$-algebras

$$\Phi_* \mathcal{O}_{\overline{P} \times_W \overline{P}} \to \Phi_* \Delta_* \mathcal{O}_{\overline{P}} = \phi_* \mathcal{O}_{\overline{P}}.$$

Since S is proper over W, the map δ is proper (in fact, finite), and so $\delta(S)$ is an analytic subset of T. Let $\overline{Z} := \Phi'^{-1} \delta(S)$, an analytic subset of $\overline{P} \times_W \overline{P}$. [5] Let $\widetilde{Z}$ be the image of $\overline{Z}$ under the natural finite map $\overline{P} \times_W \overline{P} \to P \times_W P$, so that $\widetilde{Z}$ is an analytic subset of $P \times_W P$. Let $\widetilde{C}^* \subset C^* \times_W C^*$ be the inverse image of $\widetilde{Z}$ under the quotient map $C^* \times_W C^* \to P \times_W P$.

For any $w \in W$, the fiber P_w is non-empty (since C_w has positive dimension); the points in S_w correspond to the connected components of $\overline{P}_w$ (since S_w is finite and the fibers of $\phi' : \overline{P} \to S$ are non-empty and connected); the points of T_w correspond to the connected components of $\overline{P}_w \times \overline{P}_w$; and from commutativity of (5.1.1) it follows for any $s \in S_w$ that if $\phi'^{-1}(s) = \overline{D}$ (a connected component of $\overline{P}_w$), then $\Phi'^{-1}(\delta s) = \overline{D} \times \overline{D}$ (the connected component

---

[5] $\overline{Z}$ can be defined without reference to Stein factorization as being the support of the cokernel of the natural map $\Phi^* \Phi_* \mathcal{J} \to \mathcal{O}_{\overline{P} \times_W \overline{P}}$, where $\mathcal{J}$ is the kernel of $\mathcal{O}_{\overline{P} \times_W \overline{P}} \to \Delta_* \mathcal{O}_{\overline{P}}$. We do need Stein factorization to derive (5.1.2) below; but there might well be a more elementary argument.

of $\bar{P}_w \times \bar{P}_w$ containing $\Delta \bar{D}$). Thus

$$\bar{Z}_w = \bigcup_{i=1}^{m_w} \bar{D}_w^i \times \bar{D}_w^i \tag{5.1.2}$$

where m_w is the cardinality of S_w, and $\bar{D}_w^1, \ldots, \bar{D}_w^{m_w}$ are the connected components of $\bar{P}_w$.

Now, there exists a dense Zariski-open subspace W_0 of W such that:

(a) W_0 is locally irreducible (as holds, e.g., at any smooth point of W_{red}).

(b) The natural (proper) map $\varphi : \tilde{Z}_{\text{red}} \to W_{\text{red}}$ is flat everywhere on $\varphi^{-1}(W_0)$ (Frisch's generic flatness theorem [BF, (1.17)(2), (2.4), (2.5)(2), (2.7)(1)]).

(c) For each $w \in W_0$, the fiber C_w is equidimensional, of dimension equal to the codimension c_w of $\sigma(W)$ in C at $\sigma(w)$, see (1.1.1). (Apply generic flatness of the proper map $P \to W_{\text{red}}$, keeping in mind that C – and hence P – is equidimensional.)

(d) For each $w \in W_0$, the fiber P_w is reduced and the natural map $\pi_w : \bar{P}_w \to P_w$ is a normalization of P_w. (Generic simultaneous normalization for the map $P \to W_{\text{red}}$, see [BF, Theorem (2.13)].)

In view of (d), for $w \in W_0$, if $D_w^1, \ldots, D_w^{n_w}$ are the irreducible components of P_w, then $n_w = m_w$, see above, and after relabeling we have $\pi_w^{-1}(D_w^i) = \bar{D}_w^i$ for all i. Hence the decomposition of $\tilde{Z}_w$ into irreducible components is

$$\tilde{Z}_w = \bigcup_{i=1}^{m_w} D_w^i \times D_w^i \qquad (w \in W_0). \tag{5.1.3}$$

Since the fibers of $C^* \times_W C^* \to P \times_W P$ (resp. $C^* \to P$) are all isomorphic to the manifold $\mathbb{C}^* \times \mathbb{C}^*$ (resp. $\mathbb{C}^*$), it follows, for $w \in W_0$, that the irreducible components of $\tilde{C}_w^*$ are the reduced spaces $C_w^{*i} \times C_w^{*i}$, where the C_w^{*i} are the irreducible components of $C_w^* := C_w - \sigma(w)$. So we are approaching our goal.

Any irreducible component Γ^* of $\tilde{C}^* \cap q^{-1}(W_0)$ ($q : C \times_W C \to W$ the natural map) is contained in a component Γ of $C \times_W C$. By (c) and (5.1.3), the fibers $\tilde{Z}_w$ are equidimensional, each component having dimension $2 \dim P_w = 2(c_w - 1)$. It follows then from (a) and (b) that for any irreducible component Z^* of $\tilde{Z} \cap \varphi^{-1}(W_0)$ and any $z \in Z^*$,

$$\dim_z Z^* = \dim_w W + 2(c_w - 1) \qquad (w = \varphi(z));$$

and therefore for any $x \in \Gamma^*$,

$$\dim \Gamma^* = \dim_w W + 2c_w \geq \dim \Gamma \qquad (w = q(x)). \tag{5.1.4}$$

Hence $\dim \Gamma^* = \dim \Gamma$ and

$$\Gamma^* = \Gamma \cap (C^* \times_W C^*) \cap q^{-1}(W_0), \tag{5.1.5}$$

so that Γ is Zariski open in Γ^*.

Finally, let $\widetilde{C}$ be the union of all those components Γ of $C \times_W C$ which contain a component, say Γ^*, of $\widetilde{C}^* \cap q^{-1}(W_0)$. Every such Γ – and hence $\widetilde{C}$ – is mapped into itself under the $\mathbb{C}^1 \times \mathbb{C}^1$ action, since the image of the multiplication map $\mathbb{C}^1 \times \mathbb{C}^1 \times \Gamma \to C \times_W C$ is irreducible and contains Γ.

Let r be the restriction of $\tilde{p} := q|_{\widetilde{C}}$ to the Zariski open subset $\widetilde{C}^* \cap q^{-1}(W_0)$ of $\widetilde{C}$. In view of (5.14), in which

$$2c_w = 2(\dim P_w + 1) = 2 \dim C_w = \dim_x q^{-1}q(x) \geq \dim_x r^{-1}r(x),$$

a theorem of Remmert [Fi, p. 142, 3.9], guarantees that r is an open map. So for any open dense $U \subset W$, $r^{-1}(U \cap W_0)$ is dense in $\widetilde{C}^* \cap q^{-1}(W_0)$, which is in turn dense in $\widetilde{C}$. Thus (5.1)(i) holds.

To finish, observe for $w \in W_0$ that the components C_w^i of C_w are given by $C_w^i = C_w^{*i} \cup \{\sigma(w)\}$ $(1 \leq i \leq m_w)$, and that by (5.1.5),

$$
\begin{aligned}
\widetilde{C}_w^* \subset \widetilde{C}_w \;\; &\subset \;\; \widetilde{C}_w^* \cup \left(\{\sigma(w)\} \times C_w\right) \cup \left(C_w \times \{\sigma(w)\}\right) \\
&= \;\; \bigcup_{i=1}^{m_w} \left[(C_w^{*i} \times C_w^{*i}) \cup (\{\sigma(w)\} \times C_w^i) \cup (C_w^i \times \{\sigma(w)\}) \right] \\
&= \;\; \bigcup_{i=1}^{m_w} (C_w^i \times C_w^i).
\end{aligned}
$$

Since $\widetilde{C}_w^* = \cup_i (C_w^{*i} \times C_w^{*i})$ is dense in $\cup_i (C_w^i \times C_w^i)$, and $\widetilde{C}_w$ is closed, therefore (5.1)(ii) results. $\qquad\square$

Example (5.2). Again let $\mathcal{G} = \oplus_{m \geq 0} \mathcal{G}_m$ $(\mathcal{G}_0 = \mathcal{O}_W)$ be a finitely-presentable $\mathcal{O}_W$-algebra, set $C := \mathrm{Specan}(\mathcal{G})$, $P := \mathrm{Projan}(\mathcal{G})$, and let $p : C \to W$, $\wp : P \to W$ be the canonical maps. Points $x \in P$ correspond to $\mathbb{C}^1$-orbits of points in $C \setminus \sigma(W)$: the "line" L_x corresponding to x lies in the fiber $C_{\wp(x)}$.

Assume that $\mathcal{G}_m = \mathcal{G}_1^m$ for all $m \gg 0$. Let $\mathcal{L} \xrightarrow{\pi} C$ be the proper map obtained by blowing up $\sigma(W)$ where $\sigma : W \to C$ is the vertex section (see (1.1.1)). Then $\mathcal{L} \cong \mathrm{Specan}\big(\mathrm{Sym}\,\mathcal{O}_P(1)\big)$ is the canonical line bundle on P, and $\pi^{-1}\sigma(W) = \epsilon(P)$ where $\epsilon : P \to \mathcal{L}$ is the zero-section (cf. [GD, (8.7.8)]).

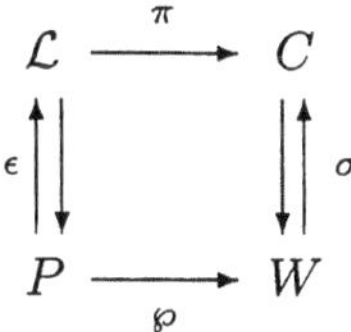

For any $x \in P$, π maps the fiber $\mathcal{L}_x$ bijectively onto the line $L_x \subset C_{\wp(x)}$. The map π is compatible with the multiplication maps $\mu_{\mathcal{L}}$, μ_C of (1.1), i.e., the following diagram commutes:

$$
\begin{array}{ccc}
\mathbb{C}^1 \times_P \mathcal{L} & \xrightarrow{\ \mu_{\mathcal{L}}\ } & \mathcal{L} \\
{\scriptstyle id \times \pi} \downarrow & & \downarrow {\scriptstyle \pi} \\
\mathbb{C}^1 \times_W C & \xrightarrow[\ \mu_C\ ]{} & C
\end{array}
$$

as can be checked, e.g., via commutativity of the diagrams

$$
\begin{array}{ccc}
\big(\oplus \mathcal{O}_P(n)\big)[T] & \xleftarrow{\ \text{mpn by } T^n\ } & \mathcal{O}_P(n) \\
{\scriptstyle \text{natural}} \uparrow & & \uparrow {\scriptstyle \text{natural}} \\
\big(\oplus \wp^* \mathcal{G}_n\big)[T] & \xleftarrow[\ \text{mpn by } T^n\ ]{} & \wp^* \mathcal{G}_n
\end{array}
\qquad (n \geq 0).
$$

Now consider the proper map

$$
\pi \times \pi : \mathcal{L} \times_P \mathcal{L} \hookrightarrow \mathcal{L} \times_W \mathcal{L} \to C \times_W C.
$$

The restriction of this map to the complement of $\epsilon(P) \times_P \epsilon(P)$ ($\cong P$) is clearly injective. From (5.1)(ii) it follows that for any $x \in \wp^{-1}(W_0)$ the image of the induced map $\mathcal{L}_x \times \mathcal{L}_x \to C_{\wp(x)} \times C_{\wp(x)}$ lies in $\widetilde{C}$. Hence, $\widetilde{C}$ being closed in $C \times_W C$, if $\wp^{-1}(W_0)$ is dense in P (i.e., every component of P meets $\wp^{-1}(W_0)$) then the entire image of $\pi \times \pi$ lies in $\widetilde{C}$. Furthermore, if the fibers C_w all have dimension 1, then the maps $\wp$ and $\pi \times \pi$ are both finite, the image of $\pi \times \pi$ is $\widetilde{C}$ itself, and $\pi \times \pi$ induces a *homeomorphism*

$$
(\mathcal{L} \times_P \mathcal{L}) \setminus P \xrightarrow{\ \sim\ } \widetilde{C} \setminus W
$$

where we have identified P (resp. W) with $\epsilon(P) \times_P \epsilon(P)$ (resp. $\sigma(W) \times_W \sigma(W)$).

All this happens, e.g., for $C := C(V, W)$ when W is a codimension-one submanifold of a reduced complex space V and V is equimultiple along W, because of Schickhoff's theorem [Li, p. 121, (2.6)].

(5.3). Every complex space $(V, \mathcal{O}_V)$ has a *conjugate space* $\overline{V}$, equal to $(V, \mathcal{O}_V)$ as a topological space with a sheaf of rings, but with $\mathcal{O}_{\overline{V}} = \mathcal{O}_V$ considered to be a $\mathbb{C}$-algebra via the composition

$$
\mathbb{C} \xrightarrow{\ \text{conjugation}\ } \mathbb{C} \xrightarrow{\ \text{natural}\ } \mathcal{O}_V,
$$

see [Hi, Definition (1.10)].

The identity map $V \to \overline{V}$ is a real-analytic isomorphism. Complex conjugation ρ_n in $\mathbb{C}^n$, along with the sheaf-isomorphism $\mathcal{O}_{\overline{\mathbb{C}^n}} \xrightarrow{\sim} \rho_{n*}\mathcal{O}_{\mathbb{C}^n}$ taking a holomorphic function $f(z)$ on an open set U to the holomorphic function $\rho_1 f(\rho_n z)$ on $\rho_n^{-1}(U)$, is a complex-analytic isomorphism of $\mathbb{C}^n$ onto $\overline{\mathbb{C}^n}$. Hence for any analytic subset V of $\mathbb{C}^n$, $\rho_n(V)$ can be regarded as an analytic subset of $\overline{\mathbb{C}^n}$ isomorphic to V, or as an analytic subset of $\mathbb{C}^n$ isomorphic to $\overline{V}$. With $(V_i)_{i \in I}$ the family of (reduced) irreducible components of V, we set

$$V^{\mathrm{c}} := \bigcup_{i \in I} (V_i \times \overline{V_i}) \subset (V \times \overline{V}). \tag{5.3.0}$$

We call V^{c} the *reduced complexification* of V, or simply the *complexification* of V when V itself is reduced. The reduced space V_{red} can be identified via the diagonal map with a real-analytic subvariety of V^{c}.

For example, with reference to (5.1), for each $w \in W$, there are natural inclusions

$$(C_w)^{\mathrm{c}} \underset{j_w}{\rightarrow\hookrightarrow} C_w \times \overline{C_w} \underset{l_w}{\rightarrow\hookrightarrow} C \times_W C,$$

where l_w is a real-(but not necessarily complex-)analytic embedding. For $w \in W_0$, we have $l_w j_w((C_w)^{\mathrm{c}}) = \widetilde{C}_w$. *From now on,* when we regard a reduced fiber $\widetilde{C}_w$ ($w \in W_0$) as a complex space, we mean it to be identical as such with $(C_w)^{\mathrm{c}}$. (Thus we do not mean it to be a *complex* subspace of $C \times_W C$.) And when we refer to a $\mathbb{C}^1$ action on $C \times_W C$ or on one of its analytic subsets (for instance, $\widetilde{C}$) we mean the one given on point sets by

$$a(x, x') = (ax, \overline{a}x') \qquad (\text{see } (1.1.1))$$

where now $\overline{a}$ is the complex conjugate of $a \in \mathbb{C}$. This action is real-analytic, being obtained from the natural $\mathbb{C}^1 \times \mathbb{C}^1$ action on $C \times_W C$ via the (real-analytic) map $a \mapsto (a, \overline{a})$ from $\mathbb{C}^1$ to $\mathbb{C}^1 \times \mathbb{C}^1$.

The next result allows us to regard $\widetilde{C}$ as a "differential functor."

Consider a C^1 map $f : (V, W) \to (V', W')$ where now V and V' are reduced equidimensional complex analytic spaces and W (resp. W') is a nowhere-dense submanifold of V (resp. V'). Let

$$\mathbf{f}_0 : C := C(V, W) \to C(V', W') =: C'$$

be the continuous map in Theorem (3.3).

Let C_g ("g" for "generic") be the union of those components of C whose image in W is not nowhere-dense. [6] Note that the (locally finite) union of the images of all the remaining components of C is nowhere dense in W, so that

---

[6] In fact the image is the same as that of the corresponding component of the projectivized normal cone $P(V, W)$, and so is an analytic subset of W. Thus any component of C_g maps onto a component of W.

the fibre $(C_g)w$ is the same as C_w for all w in some dense open subset of W. Identify C with the diagonal in $C \times_W C$. Lemma (5.1) implies that the non-empty (hence dense) Zariski open subset $\tilde{p}^{-1}W_0 \cap C_g$ of C_g is contained in $\widetilde{C}$; and hence $C_g \subset \widetilde{C}$.

Assume further that the open subset of W on which the induced map $W \to W'$ is a submersion is dense in W, so that, submersions being open maps, the inverse image of any nowhere-dense subset of W' is nowhere dense in W. It follows that $\mathbf{f}_0(C_g) \subset C'_g$; and we set $\mathbf{f}_{0g} := \mathbf{f}_0|_{C_g}$.

Theorem (5.3.1). *In the preceding situation, let U be a dense open subset of W_0 such that $(C_g)_w = C_w$ for all $w \in U$. Then $\mathbf{f}_{0g}$ extends uniquely to a continuous map $\tilde{f} : \widetilde{C} \to \widetilde{C}'$, not depending on the choice of W_0 or of U, such that the following diagram commutes,*

$$
\begin{array}{ccc}
\widetilde{C} & \xrightarrow{\ \tilde{f}\ } & \widetilde{C}' \\
\tilde{p} \downarrow & & \downarrow \tilde{p}' \\
W & \xrightarrow{\ f\ } & W'
\end{array}
$$

and such that for each $w \in U$ the resulting map of (reduced) fibers $\widetilde{C}_w \to \widetilde{C}'_{f(w)}$ is complex-analytic. Moreover, $\tilde{f}$ commutes with the $\mathbb{C}^1$ actions on $\widetilde{C}$ and $\widetilde{C}'$.

Remark (5.3.2). Let $(V, W) \xrightarrow{f} (V', W') \xrightarrow{g} (V'', W'')$ be two maps satisfying the hypotheses of (5.3.1). Then gf also satisfies these hypotheses, because a composition of submersions is a submersion, and because submersions being open maps, the inverse image under f of a dense open subset of W' is dense and open in W. Since, clearly, $(\mathbf{gF})_0 = \mathbf{g}_0\mathbf{f}_0$, we conclude from uniqueness in (5.3.1) and the denseness of $f^{-1}(W'_0) \cap W_0$ in W_0 that $\widetilde{gf} = \tilde{g}\tilde{f}$.

To begin the *proof* of Theorem (5.3.1), we recall some simple facts.

Lemma (5.3.3) *Let $V^c \subset V \times \overline{V}$ be the complexification of a reduced complex space V, so that V^c contains the diagonal $\Delta_V \subset V \times V = V \times \overline{V}$ as a real-analytic subspace. Then the only complex-analytic subset Z of V^c containing Δ_V is V^c itself.*

Proof. Let V_0 be the (open, dense) smooth locus of V. Then $\overline{V_0}$ is the smooth locus of $\overline{V}$, for example because smoothness at a point $v \in V$ means that the local ring $\mathcal{O}_{V,v} = \mathcal{O}_{\overline{V},v}$ is regular. Then $(V_0)^c$ is a dense open subset of V^c, and the closed set $Z \cap (V_0)^c$ contains Δ_{V_0}; hence we can replace V by V_0, i.e., we may assume that V is a manifold.

If $Z \neq V^c$ then for some i, Z intersects the connected open and closed subspace $(V_i \times \overline{V_i})$ of V^c nowhere densely; so there exists for some $u \in \Delta_V$ a neighborhood U together with an isomorphism $\theta : (U, u) \xrightarrow{\sim} (B, 0)$ where B is an open ball in some $\mathbb{C}^n$, and a non-zero holomorphic function $h : U \times \overline{U} \to \mathbb{C}$ vanishing on $Z \cap (U \times \overline{U})$, hence on $\Delta_U \cap (U \times \overline{U})$. There is a holomorphic open immersion $\Theta : U \times \overline{U} \to \mathbb{C}^n \times \mathbb{C}^n$ given, with ρ = complex conjugation, by

$$\Theta(v, w) = \left(\frac{\theta(v) + \rho\theta(w)}{2}, \frac{\theta(v) - \rho\theta(w)}{2\sqrt{-1}} \right),$$

taking Δ_U onto an open subset of $\mathbb{R}^n \times \mathbb{R}^n \subset \mathbb{C}^n \times \mathbb{C}^n$. All the derivatives of the holomorphic function $h \circ \Theta^{-1} : \Theta(U \times \overline{U}) \to \mathbb{C}$ vanish everywhere on $\Theta(\Delta_U)$, and hence h vanishes everywhere in a neighborhood of Δ_U, contradicting the assumption that h is non-zero (since $U \times \overline{U}$ is connected). $\qquad\square$

Corollary (5.3.3.1). *If a holomorphic map $\varphi : V^c \to Y$ maps Δ_V into an analytic subset W of Y, then φ maps all of V^c into W.*

Corollary (5.3.3.2). *If two complex-analytic maps from V^c to a complex space X agree on Δ_V, then they must be identical.*

Proofs. For (5.3.3.1), let Z in (5.3.3) be $\varphi^{-1}(W)$. For (5.3.3.2), let $\varphi : V^c \to X \times X$ in (5.3.3.1) be the map whose cordinates are the two maps in question, and let W be the diagonal of $X \times X$. $\qquad\square$

Uniqueness in Theorem (5.3.1) follows, in view of (5.1)(i), from (5.3.3.2) applied to each of the fibers $\widetilde{C}_w$ ($w \in U$). (For independence from W_0 and U, note that the intersection of two dense open subsets of W is again a dense open subset ...) This uniqueness guarantees that it is enough to prove existence with W replaced by an arbitrary member of an open covering (W_α) of W. (The global $\tilde{f}$ over all of W can then be obtained from the local maps $\tilde{f}_\alpha : \widetilde{C} \times_W W_\alpha \to \widetilde{C}'$ by pasting.) Thus, as in §(1.2), we can identify W with an open neighborhood of the origin in $\mathbb{C}^r$, embed C in $W \times \mathbb{C}^s$ ($p : C \to W$ being induced by projection to the first factor), and hence embed $C \times_W C$ in $\mathbb{C}^r \times \mathbb{C}^s \times \mathbb{C}^s$; and similarly for $p' : C' \to W'$, ...

Let $\lambda : \mathbb{C}^s \times \mathbb{C}^s \to \mathbb{C}^s \times \mathbb{C}^s$ be the real-linear automorphism taking (y, z) to (u, v), where with $i = \sqrt{-1}$ and $\bar{z}$ the complex conjugate of z,

$$u = \frac{y + \bar{z}}{2}, \qquad v = \frac{y - \bar{z}}{2i}.$$

The inverse automorphism is given by

$$y = u + iv, \qquad z = \bar{u} + i\bar{v}.$$

Then $y = z$ if and only if u and v are both real, i.e., λ maps the diagonal of $\mathbb{C}^s \times \mathbb{C}^s$ onto $\mathbb{R}^s \times \mathbb{R}^s \subset \mathbb{C}^s \times \mathbb{C}^s$.

Now recall from (3.3.3) that we can represent $\mathbf{f}_0$ locally by

$$\mathbf{f}_0(w, z) = \big(f(w), L_w(z) \big)$$

where

$$L_w : \mathbb{C}^s = \mathbb{R}^{2s} \to \mathbb{R}^{2s'} = \mathbb{C}^{s'} \tag{5.3.4}$$

is a real-linear map which depends continuously on w. In view of the relation $\lambda(y,y) = \big(\mathrm{re}(y), \mathrm{im}(y)\big)$, we see that the preceding identification of $\mathbb{C}^s$ (diagonally embedded in $\mathbb{C}^s \times \mathbb{C}^s$) with $\mathbb{R}^{2s}$ is given by λ. Hence, if $L_w^{\mathbb{c}}$ is the $\mathbb{C}$-linear map

$$L_w^{\mathbb{c}} := L_w \otimes_{\mathbb{R}} \mathbb{C} : \mathbb{C}^{2s} \to \mathbb{C}^{2s'},$$

then the continuous map $\tilde{f} : W \times \mathbb{C}^{2s} \to W' \times \mathbb{C}^{2s'}$ defined by

$$\tilde{f}(w,x) = \big(f(w),\, \lambda^{-1} L_w^{\mathbb{c}} \lambda(x)\big) \tag{5.3.5}$$

is an extension of $\mathbf{f}_0$ such that $q'\tilde{f} = fq$ (with q, q' the respective projections to W and W').

As before, complex conjugation $\rho_s : \mathbb{C}^s \to \mathbb{C}^s$ induces a complex-analytic isomorphism $\overline{B} \xrightarrow{\sim} \rho_s(B)$ for any analytic subset $B \subset \mathbb{C}^s$. The composition $\overline{\lambda}$ of λ with the real-linear map $\mathbb{C}^s \times \mathbb{C}^s \to \mathbb{C}^s \times \mathbb{C}^s$ taking $(y, \overline{z})$ to (y, z) is complex-linear. Thus if A and B are analytic subsets of $\mathbb{C}^s$, then λ induces a complex-analytic isomorphism of $A \times \overline{B}$ onto the analytic subset $\lambda(A \times B) = \overline{\lambda}(A \times \overline{B})$ of $\mathbb{C}^s$. In particular, λ maps $\mathbb{C}^s \times \overline{\mathbb{C}^s}$ isomorphically onto $\mathbb{C}^s \times \mathbb{C}^s$. Hence for every $w \in W$, $\tilde{f}$ induces a complex-analytic map

$$(C_w)^{\mathbb{c}} \subset C_w \times \overline{C_w} \subset \mathbb{C}^s \times \overline{\mathbb{C}^s} \to \mathbb{C}^{s'} \times \overline{\mathbb{C}^{s'}}.$$

Moreover, one checks that $\tilde{f}$ commutes with the $\mathbb{C}^1$ action on $W \times \mathbb{C}^s \times \overline{\mathbb{C}^s}$ (resp. $W \times \mathbb{C}^{s'} \times \overline{\mathbb{C}^{s'}}$) given by $c(w, x_1, x_2) = (w, cx_1, \overline{c}x_2)$.

We need only show now that $\tilde{f}(\widetilde{C}) \subset \widetilde{C}'$. Set $U := f^{-1}(W_0') \cap W_0$. Because of (5.3.3.1), it suffices, since $\tilde{p}^{-1}(U)$ is dense in $\widetilde{C}$, see (5.1)(i) and (5.3.2), that $\tilde{f}(C_w) \subset \widetilde{C}'$ for each $w \in U$; and that's so since

$$\tilde{f}(C_w) = \mathbf{f}_0(C_w) \subset C'_{f(w)} \subset (C'_{f(w)})^{\mathbb{c}} = \widetilde{C}'_{f(w)}.$$

(5.4). A certain subvariety $\Lambda(C) \subset \widetilde{C}$ will play an important role in the subsequent discussion of Segre classes.

Recall from Example (5.2) the canonical line bundle $\mathcal{L} \to P := \mathrm{Projan}(\mathcal{G})$. Assume that every component of P meets $\wp^{-1}(W_0)$, where $\wp : P \to W$ is the canonical map; or equivalently, that every component of C meets $p^{-1}(W_0)$, where $p : C \to W$ is the canonical map. (This assumption holds, e.g., for $C :=$ $C(V, W)$ when W is a submanifold of a reduced complex space V and V is equimultiple along W, by a theorem of Schickhoff [Li, p. 121, (2.6)].) Then (5.2) gives a natural map $\mathcal{L} \times_P \mathcal{L} \to \widetilde{C}$. Also, since $\widetilde{C}_w$ contains the diagonal of $C_w \times C_w$ for all $w \in W_0$ (see Lemma (5.1)(ii)), therefore $\widetilde{C}$ contains the dense subset $\{\,(x, x) \mid p(x) \in W_0\,\}$ of the diagonal of $C \times_W C$, and so $\widetilde{C}$ contains the entire diagonal of $C \times_W C$.

Let $\mathcal{L}^*$ be the real-analytic complex line bundle conjugate to $\mathcal{L}$, got by replacing every local trivialization $\varphi_U : U \times \mathbb{C}^1 \xrightarrow{\sim} \mathcal{L}_{|U}$ (U open in P) by its composition with $U \times \mathbb{C}^1 \xrightarrow{\mathrm{id}_U \times \rho} U \times \mathbb{C}^1$ ($\rho :=$ complex conjugation). Via the family $\{\mathrm{id}_U \times \rho\}$ we get an isomorphism of real-analytic spaces $\rho_{\mathcal{L}} : \mathcal{L} \xrightarrow{\sim} \mathcal{L}^*$. This preserves fibers over P, and addition on the fibers, but is not a line-bundle isomorphism since

$$\rho_{\mathcal{L}}(ax) = \bar{a}x \qquad (a \in \mathbb{C},\ x \in \mathcal{L}).$$

Indeed, in the topological category $\mathcal{L}$ is isomorphic to a unitary bundle [Hz, p. 51, III], and so $\mathcal{L}^*$ is isomorphic to the dual bundle $\mathcal{L}^{-1} := \mathrm{Specan}\big(\mathrm{Sym}\,\mathcal{O}_P(-1)\big)$.

We identify the rank-two complex vector bundle $\mathcal{L} \oplus \mathcal{L}^*$ over P with $\mathcal{L} \times_P \mathcal{L}^*$. The composition

$$\gamma :\ \mathcal{L} \times_P \mathcal{L}^* \xrightarrow{\mathrm{id} \times \rho_{\mathcal{L}}}\ \mathcal{L} \times_P \mathcal{L} \xrightarrow{(5.2)}\ \widetilde{C}$$

commutes with the respective $\mathbb{C}^1$ actions. The image $\Lambda = \Lambda(C)$ of γ is an analytic subset of $\widetilde{C}$, being the image of the proper map $\pi \times \pi$ of §5.2. It consists of all points $(x, x') \in C \times_W C$ such that $x = ax'$ ($a \in \mathbb{C}$) or $x' = a'x$ ($a' \in \mathbb{C}$). In other words (verification left to reader):

$$\Lambda(C) = \{\, (b'cx, b\bar{c}x) \mid b', b \in \mathbb{R},\ c \in \mathbb{C},\ x \in C \,\}. \tag{5.4.1}$$

Recalling that λ identifies the diagonal of $\mathbb{C}^s \times \mathbb{C}^s$ with $\mathbb{R}^{2s}$, we see from (5.3.4) etcthat the map $\tilde{f}$ of Theorem (5.3.1) takes the diagonal of $C \times_W C$ into the diagonal of $C' \times_{W'} C'$. Moreover $\tilde{f}$ commutes with the $\mathbb{C}^1$ action on $\widetilde{C}$, as well as with the natural $\mathbb{R} \times \mathbb{R}$ action (since the maps L_w^c and λ used to construct $\tilde{f}$ both commute with the $\mathbb{R} \times \mathbb{R}$ action on $\mathbb{C}^s \times \mathbb{C}^s$). Hence:

Corollary (5.4.2). *Under the assumptions of Theorem* (5.3.1), *$\tilde{f}(\Lambda(C)) \subset \Lambda(C')$.*

6 Segre classes

As in §5, $C := \mathrm{Specan}(\mathcal{G})$ is a cone, with $\mathbb{C}^1$ action on $C \times_W C$ given on point sets by

$$a(x, x') = (ax, \bar{a}x').$$

Assume that all the irreducible components of the complex space W have the same dimension, say r. We identify W with its image under $\Delta \circ \sigma$ where $\sigma : W \to C$ is the vertex section and $\Delta : C \to C \times_W C$ is the diagonal map.

(6.1) For any closed analytic $\mathbb{C}^1$-stable subset Υ of $C \times_W C$,[7] all of whose irreducible components have the same complex dimension, we define the *topological Segre classes*

$$s_i(\Upsilon) \in H_{2(r-i)}(W) := H_{2(r-i)}(W, \mathbb{Z}) \qquad \text{(Borel-Moore homology)}$$

---

[7]From (5.3.3.2) it follows that Υ is actually $\mathbb{C}^1 \times \mathbb{C}^1$-stable.

as follows: Let Q be the topological quotient of $\Upsilon \setminus W$ under the induced (free) $\mathbb{C}^*$ action. The action preserves fibers over W, so the canonical map $(\Upsilon \setminus W) \to W$ induces a map $\nu : Q \to W$, which is *proper*. To see this, since Q is closed in the $\mathbb{C}^*$-quotient of $(C \times_W C) \setminus W$, we may assume $\Upsilon = C \times_W C$, and then, since the question is local over W, the definition of Specan allows us to assume that C is a closed subset of $W \times \mathbb{C}^s$ for some s (the zero-set of finitely many homogeneous polynomials in s variables, with coefficients which are analytic functions on W – see (1.2)). Then $C \times_W C$ is closed in $W \times \mathbb{C}^s \times \mathbb{C}^s$, so we may assume $\Upsilon = W \times \mathbb{C}^s \times \mathbb{C}^s$ (and $\Delta\sigma(W) = W \times \{0\} \times \{0\}$), with $\mathbb{C}^*$ action given by

$$a(w, z, z') = (w, az, \bar{a}z'). \tag{6.1.1}$$

For this action, every point in $(W \times \mathbb{C}^s \times \mathbb{C}^s) \setminus (W \times \{0\} \times \{0\})$ is equivalent to a point in $W \times S^{4s-1}$ where S^{4s-1} is the unit sphere in $\mathbb{C}^{2s}$; so there is a surjection $W \times S^{4s-1} \twoheadrightarrow Q$ whose composition with ν is the (proper) projection $W \times S^{4s-1} \to W$, whence ν itself is proper.

Next, *the quotient map* $q : \Upsilon \setminus W \to Q$ *is a principal real-analytic* $\mathbb{C}^*$-*bundle.* One can verify this via an open covering (U_ι) of Q together with commutative diagrams

$$
\begin{array}{ccc}
U_\iota \times \mathbb{C}^* & \xrightarrow{\;\phi_\iota\;} & q^{-1}U_\iota \\
{\scriptstyle \text{proj'n}} \downarrow & & \downarrow {\scriptstyle q} \\
U_\iota & = \!\!= \!\!= & U_\iota
\end{array}
$$

where each ϕ_ι is a real-analytic homeomorphism commuting with the respective $\mathbb{C}^*$ actions (the action on $U_\iota \times \mathbb{C}^*$ being given by multiplication in $\mathbb{C}^*$). As before we reduce to consideration of the action (6.1.1) on $W \times \mathbb{C}^s \times \mathbb{C}^s$. The real-analytic homeomorphism $(w, z, z') \mapsto (w, z, \overline{z'})$ transforms the action into the relative diagonal one of $W \times (\mathbb{C}^{2m} \setminus \{0\})$ over W, the quotient of which is $W \times \mathbb{CP}^{2m-1}$, and here everything becomes straightforward.

Lemma (6.1.2). *The quotient map* q *takes the non-singular locus* V *of* $\Upsilon \setminus W$ *onto an open subset* $U \subset Q$ *which is naturally a $2n$-dimensional real-analytic oriented manifold, and such that* $Q \setminus U$ *has topological dimension* $\leq 2n - 2$.

The proof is given below.

Lemma (6.1.2) guarantees that Q has a fundamental class $[Q] \in H_{2n}(Q)$ [BH, p. 469, Prop. 2.3]. Now let $c \in H^2(Q, \mathbb{Z})$ be the first Chern class of the principal C^*-bundle $q : \Upsilon \setminus W \to Q$, and, with $r := \dim W$, set

$$s_i(\Upsilon) := \nu_*([Q] \cap c^{i+n-r}) \in H_{2r-2i}(W),$$

where $\cap$ denotes "cap product" [BH, p. 505, Thm. 7.2], and

$$\nu_* : H_{2r-2i}(Q) \to H_{2r-2i}(W)$$

is defined because ν is proper [BH, p. 465, 1.5].

Example (6.1.3). If C is a vector bundle over W, with conjugate C^* (cf. (5.4)), then $C \times_W C$ with its $\mathbb{C}^*$ action (cf. (6.1.1)) can be identified with the bundle $C \oplus C^*$ with its standard (diagonal) $\mathbb{C}^*$ action; and the total Segre class

$$s(C \oplus C^*) := \sum_{i \geq 0} s_i(C \oplus C^*) \in \oplus_{i \geq 0} H_{2r-2i}(W)$$

is the cap product of the fundamental class $[W]$ with the multiplicative inverse (in the graded cohomology ring $\oplus_j H^j(W)$) of the total Chern class $\mathrm{ch}(C \oplus C^*)$ (cf. [Fn, p. 71, Prop. 4.1], where everything is algebraic, but corresponds to topological constructs as in *ibid.* Chap. 19). And since C^* is topologically isomorphic to the dual bundle of C [Hz, p. 51, III], we have, with $c_j \in H^{2j}(W)$ the j-th Chern class of C,

$$\mathrm{ch}(C \oplus C^*) = (1 + c_1 + c_2 + c_3 + \dots)(1 - c_1 + c_2 - c_3 + \dots),$$

the total Pontrjagin class of C [Hz, p. 65, Thm. 4.5.1].

In particular, this applies to $C(V, W)$ when V is a complex manifold and W is a submanifold (so that $C(V, W)$ is the normal bundle).

Proof of (6.1.2). The singular locus $S := \mathrm{Sing}(\Upsilon)$ is a closed analytic C^1-stable subset of Υ, of complex dimension $\leq n$. As above, $S \setminus W$ is a principal $\mathbb{C}^*$-bundle over $q(S \setminus W)$, and so $q(S \setminus W) = Q \setminus q(V)$ has topological dimension $\leq 2n - 2$. So we can prove Lemma (6.1.2) by choosing for each $z \in V$ an open neighborhood $V_z \subset V$ in such a way that the sets $q(V_z)$ (which are open, since q is an open map) carry charts for a $2n$-dimensional canonically orientable real-analytic manifold structure on $q(V)$.

Let $N \subset \Upsilon$ be any neighborhood of the point $z_0 := \lim_{a \to 0} az$ ($a \in \mathbb{C}^*$). Replacing z by az for suitable a, we may assume that $z \in N$. Thus we may assume that $\Upsilon \subset W \times \mathbb{C}^s \times \mathbb{C}^s$, W being identified with $W_0 \times \{0\} \times \{0\}$ where W_0 is an open neighborhood of the origin in some $\mathbb{C}^r$, that $z_0 = (0, 0, 0)$, and that Υ is given in a polydisk neighborhood N_0 of z_0 by the vanishing of finitely many convergent power series

$$f_i(w, x, y) = \sum_{\alpha, \beta} c_{i\alpha\beta}(w) x^\alpha y^\beta$$

where $x^\alpha = x_1^{\alpha_1} \dots x_n^{\alpha_n}$, etc. Moreover, the $\mathbb{C}^*$ action is as in (6.1.1). For any $(w_0, x_0, y_0) \in N_0$ and $t \in (0, 1)$, then (since Υ is C^1-stable),

$$\sum_{\alpha, \beta} c_{i\alpha\beta}(w_0) x_0^\alpha y_0^\beta = 0 \implies \sum_{\alpha, \beta} t^{|\alpha|+|\beta|} c_{i\alpha\beta}(w_0) x_0^\alpha y_0^\beta = 0 \qquad (|\alpha| := \alpha_1 + \dots + \alpha_s, \dots);$$

and it follows easily that for each $m \geq 0$,

$$\sum_{|\alpha|+|\beta|=m} c_{i\alpha\beta}(w_0) x_0^\alpha y_0^\beta = 0.$$

So we may assume that the f_i are homogeneous polynomials in x and y.

Furthermore (see (6.1.1)), for each $\theta \in \mathbb{R}$,

$$\sum_{|\alpha|+|\beta|=m} e^{\sqrt{-1}\theta(|\alpha|-|\beta|)} c_{i\alpha\beta}(w_0) x_0^\alpha y_0^\beta = 0,$$

and it follows, for fixed i, that $|\alpha| - |\beta|$ has the same value for all α, β such that $|\alpha| + |\beta| = m$ and $c_{i\alpha\beta} \neq 0$; in other words, f_i is a bihomogeneous polynomial in the two sets of variables x, y.

Now, on the open set O_1 where the coordinate x_1 does not vanish, q is induced by the map $\widetilde{q}: W \times \mathbb{C}^s \times \mathbb{C}^s \to W \times \mathbb{C}^{s-1} \times \mathbb{C}^s$ given by

$$\widetilde{q}(w, x_1, \ldots, x_s, y_1, \ldots, y_s) = (w, \frac{x_2}{x_1}, \ldots, \frac{x_s}{x_1}, \frac{y_1}{\overline{x_1}}, \ldots, \frac{y_s}{\overline{x_1}})$$

where "$-$" denotes "complex conjugate." And since $f_i(w, x, y) = \sum_{\alpha, \beta} c_{i\alpha\beta}(w) x^\alpha y^\beta$ is bihomogeneous,

$$f_i(w, x, y) = x_1^{|\alpha|} \overline{x_1}^{|\beta|} f_i(w, 1, \frac{x_2}{x_1}, \ldots, \frac{x_s}{x_1}, \frac{y_1}{\overline{x_1}}, \ldots, \frac{y_s}{\overline{x_1}}).$$

Hence $q(\Upsilon \cap O_1)$ is homeomorphic to the complex-analytic variety U_1 defined by the vanishing of the power series $f_i(w, 1, \xi_2, \ldots, \xi_s, \eta_1, \ldots, \eta_s)$. Standard arguments show that for any $x \in V \cap O_1$, U_1 is a manifold in a neighborhodd of $q(x)$.

Had we used a different embedding of Υ into $W \times \mathbb{C}^s \times \mathbb{C}^s$ (but with the same projection to W, and the same $\mathbb{C}^*$ action (6.1.1)), then the resulting chart would be complex-analytically equivalent to the one just described – that is a special case of the fact that if two graded $\mathcal{O}_W$-algebras have isomorphic Specans, then they are isomorphic and so have isomorphic Projans.

Similarly, working in O_i, where x_i doesn't vanish, we get another manifold chart; but on the overlap $O_i \cap O_1$ the two charts differ by a *real-analytic* coordinate transformation, of the form

$$(\ldots, \xi_i, \ldots, \xi_j, \ldots, \eta_k, \ldots) \mapsto (\ldots, \frac{1}{\xi_i}, \ldots, \frac{\xi_j}{\xi_i}, \ldots, \frac{\eta_k}{\overline{\xi_i}}, \ldots).$$

Similar remarks apply to the open sets O'_k where y_k doesn't vanish, and to the overlaps $O_i \cap O'_k$.

It should now be more or less apparent how $U := q(V)$ can be made into a real-analytic $2n$-dimensional manifold. The manifold U is *canonically orientable* because, q being a $\mathbb{C}^*$-bundle map, for each $u \in U$ there is an open set $O \subset \mathbb{C}^n$ together with a real-analytic homeomorphism ψ from O onto an open neighborhood U_u of u in U fitting into a commutative diagram

$$
\begin{array}{ccc}
O \times \mathbb{C}^* & \xrightarrow{\ \phi\ } & q^{-1}U_u \\
{\scriptstyle \text{proj'n}}\downarrow & & \downarrow{\scriptstyle q} \\
O & \xrightarrow[\ \psi\ ]{} & U_u
\end{array}
$$

where ϕ is an *orientation-preserving* homeomorphism commuting with the respective $\mathbb{C}^*$ actions; and one checks that the charts ψ provide an orientation for U.

(6.2) We return to the situation in (5.4), assuming as we did there that every irreducible component of the cone C meets $p^{-1}(W_0)$. We assume further that both C (hence P) and W are equidimensional, with $\dim W < \dim C$. We are going to relate the Segre classes of components of $\Lambda(C)$ with the Segre classes of components of C, as described, algebraically, in [Fn, Chap. 4].

More specifically, the Segre classes $s_i(C_j) \in H_{2\dim W - 2i}(W)$ of the irreducible components C_j of C can be defined topologically as above (and more easily, because we need only deal with the *complex*-analytic $\mathbb{C}^1$-action given by μ in (1.1.1)), cf. [Fn, Chap. 19]): viz., if P_j is the component of P corresponding to C_j (P_j is topologically the $\mathbb{C}^*$-quotient of $C_j \setminus \sigma(W)$), and $\iota_j : P_j \hookrightarrow P$ is the inclusion; if $\mathcal{L}_j := \iota_j^* \mathcal{L}$ and c_j is its first Chern class; and if $\wp : P \to W$ is, as before, the canonical map, then

$$s_i(C_j) := \wp_* \iota_{j*}\left([P_j] \cap c_j^{\dim C - 1 - \dim W + i}\right) = \wp_*\left(\iota_{j*}[P_j] \cap c^{\dim C - 1 - \dim W + i}\right).$$

where, with c the first Chern class of $\mathcal{L}$ – so that $c_j = \iota_j^* c$ – the equality is given by the projection formula [BH, p. 507, 7.5]. (Note that C_j is a cone, P_j is its projectivization, and $\mathcal{L}_j$ is the canonical line bundle on P_j.)

Now the construction of Segre classes in §6.1 applies in particular when $W = P$ and $C = \mathcal{L}$, in which case $C \times_W C$ with its $\mathbb{C}^*$ action can be identified with the rank two bundle $\mathcal{L} \oplus \mathcal{L}^*$ with its standard (diagonal) $\mathbb{C}^*$ action. As noted in (5.4), $\mathcal{L}^*$ is topologically isomorphic to $\mathcal{L}^{-1}$, so the total Chern class $\mathrm{ch}(\mathcal{L}_j \oplus \mathcal{L}_j^*)$ is $1 - c_j^2$ where c_j is the first Chern class of $\mathcal{L}_j$. Hence (see Example (6.1.3))

$$s(\mathcal{L}_j \oplus \mathcal{L}_j^*) = [P_j] \cap (1 + c_j^2 + c_j^4 + \ldots).$$

As noted in (5.4), the proper map $\gamma : (\mathcal{L} \oplus \mathcal{L}^*) \setminus P \to \Lambda(C) \setminus W$ is bijective, hence is a homeomorphism, and it commutes with the respective $\mathbb{C}^*$ actions, but since it involves one complex conjugation it reverses the natural orientations. Since homeomorphisms of analytic spaces take components to components [GL, p. 172, (A8)], it follows that any irreducible component of $\Lambda(C)$ is $\mathbb{C}^1$-stable, so that its total Segre class is defined, and indeed can be obtained by applying $-\wp_*$ to the Segre class of the corresponding component of $\mathcal{L} \oplus \mathcal{L}^*$. The components in question correspond to those of P, and so to those of C. Hence, for the component of $\Lambda(C)$ corresponding to the component C_j the total Segre class is

$$-\wp_* \iota_{j*} s(\mathcal{L}_j \oplus \mathcal{L}_j^*) = -\sum_{i \geq 0} s_{2i - \dim C + 1 + \dim W}(C_j) \in \oplus_{i \geq 0} H_{2(\dim C - 2i - 1)}(W).$$

We can thus recover from $\Lambda(C)$ about half of the total Segre class $s(C_j)$. To recover the rest, proceed likewise with $\Lambda(C \times \mathbb{C}^1)$, noting that

$$s(C \times \mathbb{C}^1) = s(C) \qquad \text{(cf. [Fn, p. 71, 4.1.1)],}$$

$$\text{and of course} \qquad \dim(C \times \mathbb{C}^1) = \dim C + 1.$$

Here $C \times \mathbb{C}^1$ is viewed as the cone corresponding to the grading of $\mathcal{G}[T]$ (T an indeterminate) with degree n piece $\oplus_{i=0}^n \mathcal{G}_i T^{n-i}$, a cone whose components are naturally in one-one correspondence with those of C.

Theorem (6.3). *Let V and V' be reduced equidimensional complex spaces, and let $W \subset V$ and $W' \subset V'$ be nowhere dense equidimensional complex submanifolds. Let $f : V \to V'$ be a C^1 homeomorphism such that f^{-1} is C^1 and $f(W) = W'$. Let C_j be an irreducible component of $C := C(V, W)$ and let C_j' be the corresponding component of $C' := C(V', W')$ (see Theorem (4.3.1)). Then*

$$f_* s(C_j) = \pm s(C_j').$$

Proof. As in Corollary (5.4.2), f induces a homeomorphism $\tilde{f} : \Lambda(C) \to \Lambda(C')$ which takes W to W' (see (5.3.5)), and which commutes with the $\mathbb{C}^*$ actions. Similarly, the map $f \times 1 : V \times \mathbb{C}^1 \to V' \times \mathbb{C}^1$ induces a homeomorphism

$$\Lambda\big(C(V \times \mathbb{C}^1, W \times \{0\})\big) = \Lambda(C \times \mathbb{C}^1) \to \Lambda(C' \times \mathbb{C}^1) = \Lambda\big(C(V' \times \mathbb{C}^1, W' \times \{0\})\big).$$

These homeomorphisms respect irreducible components [GL, p. 172, (A8)], and so the induced homology maps take fundamental classes of components to fundamental classes of components, up to multiplication by ± 1.

The theorem results easily now from the foregoing procedure to recover the Segre classes of components of C from those of the corresponding components of $\Lambda(C)$ and $\Lambda(C \times \mathbb{C}^1)$. $\qquad\square$

References

[BF] J. Bingener; H. Flenner: *On the fibers of analytic mappings.* In: Complex Analysis and Geometry, p. 45–101 Univ. Ser. Math. Plenum, New York, 1993.

[BH] A. Borel, A. Haefliger: *La classe d'homologie fondamentale d'un espace analytique.* Bull. Soc. math. France **89**, 461–513, (1961).

[Fi] G. Fischer: Complex Analytic Geometry. Lecture Notes in Math. 538, Springer Verlag, New York, 1976.

[Fn] W. Fulton: Intersection Theory. Springer Verlag, New York, 1984.

[GD] A. Grothendieck, J. Dieudonné: *Éléments de Géometrie Algébrique* II. Publ. Math. Inst. Hautes Études Scientifiques **8**, (1961).

[GL] Y.-N Gau, J. Lipman: *Differential invariance of multiplicity on analytic varieties.* Invent. math. **73**, 165–188, (1984).

[Hi] H. Hironaka: *Normal cones in analytic Whitney stratifications.* Publ. Math. Inst. Hautes Études Scientifiques, **36**, 127–138, (1969).

[Hi2] H. Hironaka: Introduction to the Theory of Infinitely Near Singular Points. Instituto "Jorge Juan" de Matematica, Madrid, 1974. (Memorias de Matematica del Instituto "Jorge Juan," 28)

[Ho] C. Houzel: *Géométrie analytique locale.* In: Familles d'Espaces Complexes et Fondements de la Géométrie Analytique. (Séminaire Henri Cartan, 1960/61, fascicule 2.) Institut Henri Poincaré, Paris, 1962.

[Hz] F. Hirzebruch: Topological Methods in Algebraic Geometry. (Third enlarged edition) Springer Verlag, New York, 1966.

[Li] J. Lipman: *Equimultiplicity, reduction, and blowing up.* In: Commutative Algebra, 111–147. Lecture Notes Pure ApplMath. **68**, (Editor RDraper); Marcel Dekker, New York, 1982.

[LT] D. T. Lê, B. Teissier: *Limites d'espaces tangents en géométrie analytique.* Comment. Math. Helvetici, **63**, 540–578, (1988).

[MT] M. Lejeune, B. Teissier: Contributions à l'Étude des Singularités. Thèse d'État, Centre de Mathématiques, École Polytechnique, Paris, (1973).

Progress in Mathematics, Vol. 162, © 1998 Birkhäuser Verlag Basel/Switzerland

On the Spectrum of Curve Singularities

András Némethi
The Ohio State University
Columbus, OH 43210, USA

Dedicated to Egbert Brieskorn on the occasion of his 60th birthday

1 Introduction

There is the following crucial problem in the singularity theory: how can we recognize that a singular germ is hypersurface, or complete intersection singularity. Or: is there any criterion which distinguishes the isolated hypersurface singularities among the isolated singularities? These kind of questions arise in many areas of the algebraic geometry and singularity theory, we mention here only one.

A. Durfee [3] conjectured that the Milnor fiber of an isolated complete intersection singularity of dimension two is negative. Although J. Wahl [23] found a smoothing of a (non-hypersurface) singularity with positive signature, there is still a strong belief that the conjecture is true for hypersurfaces. But, if the conjecture is true, then what is special in hypersurfaces, which makes their signature negative?

In this paper we try to understand this question for hypersurfaces of type $f(x,y) + z^N$. In [12, 10], we give a series of properties of the resolution graphs of germs $f : (X,x) \to (\mathbf{C},0)$, defined on a normal surface singularity (X,x), which distinguish the plane curve singularities. Here, we present some properties ($\{\mathcal{P}(N)\}_{2 \leq N \leq \infty}$) of the spectrum, which codifies a part of the information of the mixed Hodge structure on the vanishing cohomology of the germs. The property $\mathcal{P}(N)$, for plane curve singularities is equivalent to the negativity of the equivariant signature $\sigma_{\neq 1}(f + z^N)$ of the germ $f(x,y) + z^N$. We prove that these properties are valid for plane curve singularities (in particular, we prove Durfee's conjecture for hypersurfaces of type $f(x,y) + z^N$) (cf. 5.2), and we give germs $f : (X,x) \to (\mathbf{C},0)$ (with (X,x) non-smooth) which do not satisfy any of the properties $\mathcal{P}(N)$.

In the paper we emphasize the arithmetical flavour of the spectrum and its connections with number theory, namely with generalized Dedekind sums.

2   The properties $\mathcal{P}(N)$ $(2 \leq N \leq \infty)$.

We start this section with some definitions (their motivation will come later, cf. 2.7).

2.1. Definitions. Let $\mathbf{N}^{\mathbf{Q}}$ be the free abelian monoid generated by $\mathbf{Q}$: its elements are finite sums of the form $\sum_{r \in \mathbf{Q}} n_r(r)$, where $n_r \in \mathbf{N} = \{0, 1, \dots\}$. We will use the notation S_N for $\sum_{1 \leq k < N}(k/N)$ $(N \geq 2)$. If $S = \sum_{r \in (0,1]} n_r(r)$, we define the symmetric element with respect $r = 1$ by $\tilde{S} = n_1(1) + \sum_{r \in (0,1)}\big(n_r(r) + n_r(2-r)\big)$. The degree of $S = \sum n_r(r)$ is given by $\mu(S) = \sum n_r \in \mathbf{N}$, the "signature of S" by $\sigma(S) = \sum_r n_r \mathrm{sign}\,(\sin \pi r) = \sum_{r \notin \mathbf{Z}}(-1)^{[r]}n_r$ (where $[\cdot]$ denotes the integer part), and

$$m(S) = \sum_r n_r \sum_{k \in \mathbf{Z}}(-1)^k m\Big([r, 1+r] \cap [k, k+1]\Big),$$

where, for an interval $I \subset \mathbf{R}$, $m(I)$ denotes its length.

The multiplicative structure of $\mathbf{N}^{\mathbf{Q}}$ is generated by $(r) \otimes (s) = (r+s)$. Then it is easy to see that

$$m(S) = \lim_{N \to \infty} \frac{\sigma(S \otimes S_N)}{N}.$$

2.2. Definition. a) Fix an element $S = \sum_{r \in (0,1]} n_r(r)$ and an integer $N \geq 2$. We say that S satisfies the property $\mathcal{P}(N)$ if $\sigma(\tilde{S} \otimes S_N) < 0$.
b) An element $S = \sum_{r \in (0,1]} n_r(r)$ satisfies the property $\mathcal{P}(\infty)$ if $m(\tilde{S}) < 0$.

2.3. Remark. If $S = \sum_{r \in (0,1)} n_r(r)$, then $\sigma(\tilde{S} \otimes S_N) = 2\sigma(S \otimes S_N)$ and similarly $m(\tilde{S}) = 2m(S)$. Therefore, $\mathcal{P}(N)$ (resp. $\mathcal{P}(\infty)$) is equivalent to $\sigma(S \otimes S_N) < 0$ (resp. $m(S) < 0$).

We are really interested in elements S provided by the spectrum of an isolated curve singularity.

2.4. Example. Let (X, x) be a normal surface singularity, with the property that its link K_X is rational homology sphere (i.e. $H_1(K_X, \mathbf{Q}) = 0$). (This condition is not really necessary, but it will simplify our presentation. For the general situation, see for example, [14].) Let $f : (X, x) \to (\mathbf{C}, 0)$ be a germ of an analytic function which defines a one-dimensional isolated singularity. We consider an embedded resolution $\phi : (\mathcal{Y}, D) \to (X, f^{-1}(0))$ of $(f^{-1}(0), x) \subset (X, x)$ (here $D = \phi^{-1}(f^{-1}(0))$). Let $E = \phi^{-1}(x)$ be the exceptional divisor and $E = \cup_{w \in \mathcal{W}} E_w$ be its decomposition in irreducible divisors. All these irreducible divisors are rational. Let $\cup_{a \in \mathcal{A}} S_a$ be the irreducible decomposition of the strict transform S of $f^{-1}(0)$. Then $D = E \cup S$. Let G_f be the resolution graph of f, i.e. its vertices $\mathcal{V} = \mathcal{W} \coprod \mathcal{A}$ consist of the nonarrowhead vertices $\mathcal{W}$ (corresponding to the irreducible exceptional divisors), and arrowhead vertices

$\mathcal{A}$ (correponding to the strict transform divisors of D). We will assume that no irreducible exceptional divisor has an self intersection and $\mathcal{W} \neq \emptyset$. If two irreducible divisors corresponding to $v_1, v_2 \in \mathcal{V}$ have an intersection point then (v_1, v_2) $(= (v_2, v_1))$ is an edge of G_f. The set of edges is denoted by $\mathcal{E}$.

For any $w \in \mathcal{W}$, we denote by $\mathcal{V}_w$ the set of vertices $v \in \mathcal{V}$ adjacent to w. The graph G_f is decorated by the self intersection (or Euler-) numbers $e_w := E_w \cdot E_w$ for any $w \in \mathcal{W}$.

For any $v \in \mathcal{V}$ let m_v be the multiplicity of $f \circ \phi$ along the irreducible divisor corresponding to v. In particular, for any $a \in \mathcal{A}$ one has $m_a = 1$. The multiplicities satisfy the following relations. For any $w \in \mathcal{W}$ one has:

$$(2.5) \qquad\qquad e_w m_w + \sum_{v \in \mathcal{V}_w} m_w = 0.$$

These relations determine the multiplicities $\{m_w\}_{w \in \mathcal{W}}$ in terms of the autointersection numbers $\{e_w\}_w$. For an edge $e = (v_1, v_2)$ define $m_e = \text{g.c.d.}(m_{v_1}, m_{v_2})$.

The vanishing cohomology $H^1(F, \mathbf{C})$ of the Milnor fiber F of f has a natural mixed Hodge structure. The associated spectrum $Sp(f)$ is in the interval $(-1, 1)$, and it is symmetric with respect to $r = 0$ (for details, see for example, [21, 20, 14]). We define $S(f) = \sum(r + 1)$, where the sum is over the spectral elements r of $Sp(f)$ which are in the interval $(-1, 0]$. Obviously $\tilde{S}(f)$ is exactly $Sp(f)$ shifted one unit to the right. In particular $\mu(\tilde{S}(f)) = \dim H^1(F, \mathbf{C})$.

The element $S(f)$ can be computed from the graph G_f as follows:

2.6. Proposition. [20, 14] *Define* $R_w^s := \sum_{v \in \mathcal{V}_w} \{sm_v/m_w\}$, *where* $\{\cdot\}$ *denotes the fractional part. Then*

$$S(f) = (\#\mathcal{A} - 1)(1) + \sum_{e \in \mathcal{E}} \sum_{0 < s < m_e} \left(\frac{s}{m_e}\right) + \sum_{w \in \mathcal{W}} \sum_{0 < s < m_w} (-1 + R_w^s)\left(\frac{s}{m_w}\right).$$

2.7. Remark. The relation between the above example and the properties $\mathcal{P}(N)$ is the following. If $g : (\mathbf{C}^3, 0) \to (\mathbf{C}, 0)$ is an isolated singularity, then its equivariant signature σ_λ for $\lambda \neq 1$ can be computed from its spectrum $Sp(g)$ only (for the exact formula, see e.g. [9, 11]). In general σ_1 cannot be computed from $Sp(g)$ alone (cf. 6.10. [11]). If $g(x, y, z) = f(x, y) + z^N$, then $Sp(g)$, shifted by one unit, is exactly $\tilde{S}(f) \otimes S_N$ (cf. [18, 22]); and $\sigma_{\neq 1}(f + z^N) = \sigma(\tilde{S}(f) \otimes S_N)$. The equivariant signature $\sigma_1(f + z^N)$ is zero if the monodromy $h_1(g)$ corresponding to the eigenvalue one has no Jordan blocks of size two. In particular, (cf. [6]) if f is irreducible, then $\sigma_1(g) = 0$. In general $\sigma_1(g) = h_1^{22}(g) \geq 0$ (cf. 6.7. [11]), hence $\sigma(\tilde{S}(f) \otimes S_N) \leq \sigma(f + z^N)$.

Now, the conjecture of Durfee [3], mentioned in the introduction, says that the signature of any hypersurface singularity $g : (\mathbf{C}^3, 0) \to (\mathbf{C}, 0)$ is negative. Property $\mathcal{P}(N)$ $(2 \leq N < \infty)$ for $S(f)$ corresponds exactly to this conjecture. The function $N \mapsto \sigma(\tilde{S}(f) \otimes S_N)$ is a sum of a periodic and a linear function (say $P(N)$ and $c \cdot N$); (cf. 4.2). Hence $\mathcal{P}(\infty)$ is equivalent to $c < 0$.

We expect that the above results (valid in the hypersurface situation) have the corresponding analogues in the general case, (i.e. when f is not plane curve singularity). On the other hand, the properties $\mathcal{P}(N)$ make sense for arbitrary germ $f : (X,0) \to (\mathbf{C},0)$. The goal of the present paper is to show that the properties $\mathcal{P}(N)$ distinguish the plane curve singularities among the general germs $f : (X,0) \to (\mathbf{C},0)$.

We end this section with the following remark: for any fixed $2 \leq N_1 < N_2 < \infty$, the properties $\mathcal{P}(N_1)$, $\mathcal{P}(N_2)$ and $\mathcal{P}(\infty)$ are independent.

2.8. Example. Consider a sufficiently small $1 >> r > 0$.

a) $S = 2(1/2 - r) + (1 - r)$ satisfies $\mathcal{P}(\infty)$ and $\mathcal{P}(3)$, but not $\mathcal{P}(2)$;

b) $S = (r) + 2(1/2 + r)$ satisfies $\mathcal{P}(2)$, but not $\mathcal{P}(3)$ and $\mathcal{P}(\infty)$.

3 Positive results. The case of plane curve singularities

In this section we study the set $S(f)$, where $f : (\mathbf{C}^2,0) \to (\mathbf{C},0)$ is an isolated singularity.

I. Combinatorial-arithmetical approach

Since $S(f)$ is constant under the μ-constant deformations, $S(f)$ depends only on the topological type of the embedding $(f^{-1}(0),0) \subset (\mathbf{C}^2,0)$ (this fact can be read also from 2.6). Moreover, in the irreducible case, the topological invariants of this embedding behave rather additively with respect to the splice decomposition of the of $f^{-1}(0)$ (or of the graph G_f) (see, for example, [4]). Hence, one can expect that for irreducible f the invariant $S(f)$ have a nice representation in terms of the Newton pairs of f. In the Brieskorn case $f(x,y) = x^p + y^q$, $((p,q) = 1)$ one has $S(f) = \sum(l/p + k/q)$, where the sum is over $0 < l < p$, $0 < k < q$, $l/p + k/q < 1$. It is convenient to use the notation $S(p,q)$ for this element of $\mathbf{N}^{\mathbf{Q}}$. One can expect that for general irreducible f, $S(f)$ can be obtained by "iterating" the Brieskorn case. The following theorem provides such a formula (cf. 2.2 in [20], and an unpublished paper of M. Saito).

3.1. Theorem. *Assume that* $f : (\mathbf{C}^2,0) \to (\mathbf{C},0)$ *is irreducible with Newton pairs* $(p_i,q_i)_{i=1}^r$. *Define the integers* $\{a_i\}_{i=1}^r$ *by*

$$a_1 = q_1 \text{ and } a_{i+1} = q_{i+1} + p_{i+1}p_i a_i \text{ if } i \geq 1.$$

(These numbers are the decorations of the splice diagram of f, *cf. [4].) Then*

$$S(f) = \sum_{i=1}^r S_i, \text{ where } S_i = \sum \left(\frac{k/a_i + l/p_i + t}{p_{i+1}p_{i+2} \cdots p_r} \right),$$

where the second sum is over $0 < k < a_i$, $0 < l < p_i$, $k/a_i + l/p_i < 1$ *and* $0 \leq t \leq p_{i+1} \cdots p_r - 1$ *(if* $i = r$, *then* $S_r = S(a_r, p_r)$).

Proof. If $\#\mathcal{V}_w > 2$, then w is called "rupture point". Consider the relation given by (2.6). If $\#\mathcal{V}_w = 1$, then (by 2.5) $R_w^s = 0$, if $\#\mathcal{V}_w = 2$, then $R_w^s = 1$ iff $m_w | sm_v$ ($v \in \mathcal{V}_w$). By an elementary computation one can show that $S(f)$ is concentrated only in the rupture points $\{w_1, \ldots, w_r\}$ of G_f, namely:

$$(3.2) \qquad S(f) = \sum_{i=1}^{r} \sum_{\substack{0 < s < m_{w_i} \\ m_{w_i} \nmid sd_i}} \left(-1 + \sum_{v \in \mathcal{V}_{w_i}} \{sm_v/m_{w_i}\} \right) \left(\frac{s}{m_{w_i}} \right),$$

where $d_i := g.c.d.\left((m_v)_{v \in \mathcal{V}_{w_i}}, m_{w_i} \right)$ (cf. [20] and [19]).

Now, $m_{w_i} = a_i p_i p_{i+1} \cdots p_r$ (for details, see e.g. [4]). Consider the integers b_i and c_i such that $b_i p_i + c_i a_i = 1$. Then the multiplicities $\{m_v\}$ ($v \in \mathcal{V}_{w_i}$), modulo m_{w_i}, are: $(-b_i p_i p_{i+1} \cdots p_r, -c_i a_i p_{i+1} \cdots p_r, p_{i+1} \cdots p_r)$ (cf. [15, 19]). Hence: $d_i = p_{i+1} \cdots p_r$. Therefore $R_{w_i}^s = \{-b_i s/a_i\} + \{-c_i s/p_i\} + \{s/a_i p_i\}$. Now, write $s = kp_i + la_i + ta_i p_i$, where $0 \leq k < a_i$, $0 \leq l < p_i$, $0 \leq t < p_{i+1} \cdots p_r$. Then $m_{w_i} | sd_i$ iff $(k,l) = (0,0)$, and $R_{w_i}^s = \{-k/a_i\} + \{-l/p_i\} + \{k/a_i + l/p_i\}$, hence the result follows. $\qquad \square$

By (3.1) the Milnor number $\mu(f)$ of f is $2\mu(S(f)) = 2\sum_i \mu(S_i)$, where $2\mu(S_i) = (a_i - 1)(p_i - 1)p_{i+1} \cdots p_r$. The next theorem gives a similar additivity formula for the signature $\sigma(f + z^N)$ of $f(x,y) + z^N$.

Define $((x))$ by $\{x\} - 1/2$ if $x \notin \mathbf{Z}$, and $((x)) = 0$ otherwise.

3.3. Theorem. *Let f be as in (3.1), and $D_i := g.c.d.(N, p_{i+1} \cdots p_r)$ for $1 \leq i < r$, and $D_r = 1$. Then*

$$\sigma(S(f) \otimes S_N) = \sum_{i=1}^{r} D_i \cdot \sigma(S(a_i, p_i) \otimes S_{N/D_i}).$$

Proof. First we consider two lemmas:

3.4. Lemma. *Let $\alpha \in (0,1)$, and $N \geq 2$ an integer. Then:*

$$\sigma((\alpha) \otimes S_N) = 2((N\alpha)) - 2N((\alpha)).$$

The proof is easy and it is left to the reader (write $i/N \leq \alpha < (i+1)/N$).

3.5. Lemma.

$$\sum_{t=0}^{P-1} \left(\left(N \cdot \frac{x+t}{P} \right) \right) = (N, P) \cdot \left(\left(\frac{N}{(N,P)} \cdot x \right) \right).$$

Proof. Use the Fourier sine expansion:

$$((x)) = -\frac{1}{\pi} \sum_{k=1}^{\infty} \frac{1}{k} \sin 2\pi kx,$$

(or an elementary, longer computation). $\qquad \square$

Now (3.4) and (3.5) applied for $x = k/a_i + l/p_i$ gives: $\sigma(S_i \otimes S_N) = D_i \cdot \sigma(S(a_i, p_i) \otimes S_{N/D_i})$, which ends the proof. $\qquad \square$

The discussion (2.7) gives:

3.6. Corollary. *Let f be as in (3.1), and D_i as in (3.3). Then:*

$$\sigma(f + z^N) = \sum_{i=1}^{r} D_i \cdot \sigma(x^{a_i} + y^{p_i} + z^{N/D_i}).$$

Now, for the Brieskorn singularities it is well known that $\sigma(x^a + y^p + z^N) < 0$, and actually $\sigma(x^a + y^p + z^N) < (a-1)(p-1)(N-1)/3$, therefore $\lim_{N \to \infty} \sigma(x^a + y^p + z^N)/N < 0$ as well (cf. 4.3). This gives:

3.7. Corollary. *For any irreducible plane curve singularity f, the element $S(f)$ satisfies the properties $\mathcal{P}(N)$, $(2 \leq N \leq \infty)$.*

Unfortunately, the above arithmetical results (i.e. 3.1 and 3.6) have no analogs in the reducible case (at least known by the author). In the next subsection, we will prove the analog of (3.7) for reducible plane curves.

II. Geometrical approach

We start with the following theorem, which supports the belief that Durfee's conjecture ([3]) is true for hypersurface singularities.

3.8. Theorem. *Assume that the germ $f : (\mathbf{C}^2, 0) \to (\mathbf{C}, 0)$ defines an isolated singularity and it has r irreducible components. Then the signature of $g(x, y, z) = f(x, y) + z^N$ satisfies:*

$$\sigma(f + z^N) \leq -(N - 1)(r - 1) \leq 0.$$

Proof. We start with the following remark. Assume that $g : (\mathbf{C}^3, 0) \to (\mathbf{C}, 0)$ defines an isolated singularity with Milnor lattice $L(g)$. If g_t is a deformation of g such that $g_0 = g$ and g_t, for $t \neq 0$ small, has k singular points with Milnor lattices $L_1, \ldots, L_k$, then there is a lattice embedding $\oplus_{i=1}^{k} L_i \hookrightarrow L(g)$. If c is the codimension of the embedding, then: $(*)$ $\sigma(g) \leq c + \sum_i \sigma(L_i)$. Now consider a deformation $\{f_t\}_t$ of f such that $f_0 = f$ and f_t, for $t \neq 0$ small, has exactly $\delta(f)$ nodes, where $\mu(f) = 2\delta(f) - r + 1$ (cf. [7]). The deformation $g_t = f_t + z^N$ gives an embedding of $\delta(f)$ copies of $L(x^2 + y^2 + z^N)$ in $L(f + z^N)$ with codimension $c = (N - 1)(\mu - \delta)$, hence $(*)$ gives the wanted inequality. $\qquad\square$

Since $\sigma(\tilde{S}(f) \otimes S_N) \leq \sigma(f + z^N)$ (cf. 2.7), (3.7) and (3.8) gives:

3.9. Corollary. *For any plane curve singularity f, the element $S(f)$ satisfies the properties $\mathcal{P}(N)$, $(2 \leq N \leq \infty)$.*

4   The arithmetical approach revisited. Dedekind sums

We will consider again a germ $f : (X, x) \to (\mathbf{C}, 0)$ as in (2.4). For simplicity, we will assume that the graph G_f has only one rupture point w.

Since in (2.6) the contribution of the non-rupture points is cancelled with the contribution of the edges, one has:

$$(4.1) \qquad S(f) \; = \; (\#\mathcal{A} - 1)(1) + \sum_{0<s<m_w} \Big(-1 + \sum_{v\in V_w} \{sm_v/m_w\}\Big)\Big(\frac{s}{m_w}\Big).$$

Using (3.4) and $\sum_{0<s<b}((as/b)) = 0$ (apply 3.5 for $x=0$), one has:

$$\sigma(\tilde{S}(f) \otimes S_N) \; = \; -(\#\mathcal{A} - 1)(N - 1)$$
$$+ 4 \cdot \sum_{0<s<m_w} \sum_{v\in V_w} \Big(\Big(\frac{sm_v}{m_w}\Big)\Big)\Big(\Big(\Big(\frac{Ns}{m_w}\Big)\Big) - N\Big(\Big(\frac{s}{m_w}\Big)\Big)\Big).$$

We recall that the generalized Dedekind sum (cf. [17, 24]) is defined by:

$$s(b,c;a) \; = \; \sum_{0<s<a} \Big(\Big(\frac{sb}{a}\Big)\Big)\cdot\Big(\Big(\frac{sc}{a}\Big)\Big) \quad \text{(where } a>0\text{)}.$$

Notice that $s(b,1;a) = s(1,b;a)$ is exactly the classical Dedekind sum.

4.2. Corollary. *Let (X,x) and f be as above. Then*

$$\sigma(\tilde{S}(f) \otimes S_N) \; = \; -(\#\mathcal{A} - 1)(N - 1)$$
$$+ 4 \cdot \sum_{v\in V_w} \Big(s(m_v, N; m_w) - N \cdot s(m_v, 1; m_w)\Big),$$

and $\qquad m(\tilde{S}(f)) \; = \; -(\#\mathcal{A} - 1) - 4 \sum_{v\in V_w} s(m_v, 1; m_w).$

In particular, the properties $\mathcal{P}(N)$ are related to the arithmetical properties of the Dedekind sums (associated with the multiplicity structure of G_f).

4.3. Example. Consider again $f : (\mathbf{C}^3, 0) \to (\mathbf{C}, 0)$, $f(x,y) = x^p + y^a$, $(a,p) = 1$, $a > 1$, $p > 1$. Fix b and c such that $bp + ca = 1$. Then $m_w = ap$ and $\{m_v\} = \{-bp, -ca, 1\}$ (cf. the proof of 3.1). We recall the following properties of the Dedekind sums:

$$s(b,c;a) \; = \; (a,b,c)\cdot s\Big(\frac{b}{(a,b)}, \frac{c}{(a,b,c)}; \frac{a}{(a,b)}\Big),$$
$$\text{and:} \quad s(b,c;a) \; = \; s(kb, kc; a) \text{ if } (k,a) = 1.$$

Using these identities one has: $m(\tilde{S}(f))/4 = s(p,1;a) + s(a,1;p) - s(1,1;ap)$. Finally, the famous Dedekind's Reciprocity Law says that:

$$s(b,1;a) + s(a,1;b) = -\frac{1}{4} + \frac{a^2 + b^2 + (a,b)^2}{12ab}.$$

Therefore: $m(\tilde{S}(f)) = -(a^2 - 1)(p^2 - 1)/(3ap) < 0$, in particular:

$$\sigma(x^a + y^p + z^N) = -N\frac{(a^2 - 1)(p^2 - 1)}{3ap} + 4 \sum_{v\in V_w} s(m_v, N; ap).$$

Therefore our Corollary (4.2) gives, in this very particular case, (via Brieskorn formula of the signature [2]), the number of lattice points in the tetrahedron $(0,0,0),(0,0,a),(0,p,0),(N,0,0)$ in terms of Dedekind sums. This problem is well-known in number theory, it was solved by Mordell [8] (in the case when a,p,N are relatively prime numbers), and recently by Pommersheim [16] (in the general case). We invite the reader to complete our example for the case $(a,p) \neq 1$ (in order to provide a new proof of Pommersheim's formula, cf. [12]).

5   Germs without property $\mathcal{P}(N)$.

5.1. Fix an integer $k \geq 1$. Consider the graph $G(k)$:

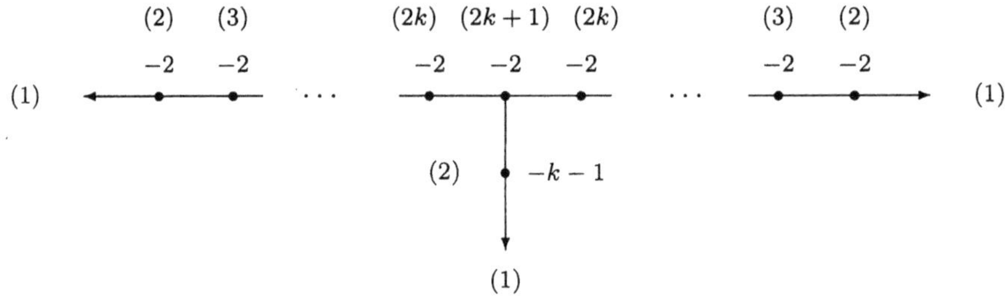

$$
\begin{array}{c}
\underset{-2}{\overset{(2)}{\bullet}}\;\;\underset{-2}{\overset{(3)}{\bullet}}\;\cdots\;\underset{-2}{\overset{(2k)}{\bullet}}\;\underset{-2}{\overset{(2k+1)}{\bullet}}\;\underset{-2}{\overset{(2k)}{\bullet}}\;\cdots\;\underset{-2}{\overset{(3)}{\bullet}}\;\underset{-2}{\overset{(2)}{\bullet}}
\end{array}
$$

with a vertical branch from the central vertex $(2k+1)$: a vertex (2) with weight $-k-1$, and a free end (1). The two horizontal arrows both carry (1).

One can verify that the intersection matrix of the graph is negative definite (here it is useful the method given in [4], page 154). Then by a theorem of Grauert [5], there exists a germ $f_k : (X_k,x) \to (\mathbf{C},0)$ as in (2.4), such that $G_{f_k} = G(k)$. Then, (4.1) gives:

$$
S(f) = 2(1) + \sum_{1 \leq s \leq k} \left(\frac{s}{2k+1} \right),
$$

therefore: $\quad m(\tilde{S}(f_k)) = -2 + 2 \sum_{1 \leq s \leq k} \left(1 - \frac{2s}{2k+1} \right) = -2 + \frac{2k^2}{2k+1}.$

This shows that there is no upper bound for $m(\tilde{S}(f))$, in general. By (4.2):

$$
\sigma_N^{(k)} := \sigma(\tilde{S}(f_k) \otimes S_N) = 2 + m(\tilde{S}(f_k)) \cdot N + P^{(k)}(N),
$$

where $P^{(k)}(N)$ is an odd periodic function in N, of period $2k+1$.

If $k=2$, then $\sigma_N^{(2)} = 2 - 2N/5 + P^{(2)}(N)$, where for $N = 0,;\pm 1;\pm 2$ one has: $P^{(2)}(N) = 0; \mp 8/5; \pm 4/5$. In particular, $\sigma_N^{(2)} \geq 0$ iff $N = 2,3,4,5,7,9$.

If $k=3$, then $\sigma_N^{(3)} = 2 + 4N/7 + P^{(3)}(N)$, where for $N = 0; \pm 1; \pm 2; \pm 3$ one has: $P^{(3)}(N) = 0; \mp 18/7; \pm 6/7; \pm 2/7$. In particular, $\sigma_N^{(3)} > 0$ for any $N \geq 2$. The case $N=2$ also can be computed explicitly. By the Dedekind's Reciprocity Law one has:

$$
s(1,1;2k+1) = s(2,2;2k+1) = \frac{k(2k-1)}{6(2k+1)}, \quad s(2,1;2k+1) = \frac{k(k-2)}{6(2k+1)}.
$$

Therefore, (4.2) gives: $\sigma_2^{(k)} = -2 + 2k \geq 0$.

Now, assume that $k \geq 4$ and $N \geq 3$. Then by the inequality:

$$|s(b, c; a)| \leq \frac{(a-1)(a-2)}{12a}; \quad (a > 0),$$

(cf. [12, 13]), and by (4.2):

$$\sigma_N^{(k)} \geq 2 + N(-2 + \frac{2k^2}{2k+1}) - \frac{2k(2k-1)}{2k+1} > 0.$$

Therefore, for $k \geq 3$, $S(f_k)$ does not satisfy the property $\mathcal{P}(N)$ for any $2 \leq N \leq \infty$, and for $k = 2$, $\mathcal{P}(N)$ is satisfied only if $N \notin \{2, 3, 4, 5, 7, 9\}$.

5.2. Final Remarks. a). Actually, for any $c \in \mathbf{R}$, we can define similar properties as above: we say that $S = \sum_{r \in (0,1]} n_r(r)$ satisfies $\mathcal{P}(N, c)$ if $\sigma(\tilde{S} \otimes S_N) < c \cdot \mu(\tilde{S} \otimes S_N)$. The case $c = -1/3$ is optimal: if $c < -1/3$, then for sufficiently large d, $\mathcal{P}(d, c)$ is not true for $S(x^d + y^d)$. By another conjecture of Durfee ([3]), $\mathcal{P}(N, -1/3)$ is true for any plane curve singularity f.

b). For a plane curve singularity f, the negativity of the signature of $f + z^N$ was firstly proved by Ashikaga [1] in a very long paper with a rather sophisticated method. Actually, he proved that the order of $\sigma(f + z^N)$ is less than $-N \cdot m(f)^2/3$, where $m(f)$ is the multiplicity of f. In [12, 13], by a more conceptual approach, we reprove these facts, together with some stronger results. For example, in [12] we prove that $\mathcal{P}(N, -1/3)$ is true for any N and any irreducible plane curve singularity f. In [13] we consider the case when f is not irreducible, and we prove that $\mathcal{P}(N, -1/3)$ is true with the following restrictions: we assume that if N is smaller than a bound $B(f)$ then $\lambda - 1$ is not an eigenvalue of the monodromy of $f + z^N$.

References

[1] Ashikaga, T.: The Signature of the Milnor fiber of Complex Surface Singularities on Cyclic Coverings, preprint, (1995).

[2] Brieskorn, E.: Beispiele zur Differentialtopologie von Singularitäten, *Invent. Math.*, **2.**, 1–14 (1966).

[3] Durfee, A.: The Signature of Smoothings of Complex Surface Singularities, *Math. Ann.*, **232**, 85–98 (1978).

[4] Eisenbud, D. and Neumann, W.: *Three-Dimensional Link Theory and Invariants of Plane Curve Singularities*, Ann. of Math. Studies **110**, Princeton University Press, 1985.

[5] Grauert, H.: Über Modifikationen und exzeptionelle analytische Mengen, *Math. Ann.*, **146**, 331–368 (1962).

[6] Lê Dũng Tráng: Sur les noeuds algébriques, *Compositio Math.*, **25**, 281–321 (1972).

[7] Milnor, J.: *Singular Points of Complex Hypersurfaces*, Annals of Math. Studies, **Vol. 61**, Princeton University Press, 1868.

[8] Mordell, L.J.: Lattice points in a tetrahedron and generalized Dedekind sums, *J. Indian Math.*, **15**, 41–46 (1951).

[9] Némethi, A.: The equivariant signature of hypersurface singularities and eta-invariant, *Topology*, **34**, 243–259 (1995).

[10] Némethi, A.: The eta-invariant of variation structures I, *Topology and its Applications*, **67**, 95–111 (1995).

[11] Némethi, A.: The real Seifert form and the spectral pairs of isolated hypersurface singularities, *Compositio Math.*, **98**, 23–41 (1995).

[12] Némethi, A.: Dedekind sums and the signature of $f(x,y) + z^N$, submitted.

[13] Némethi, A.: Dedekind sums and the signature of $f(x,y) + z^N$, II, submitted.

[14] Némethi, A. and Steenbrink, J.: Spectral pairs, mixed Hodge modules and series of plane curve singularities, *New York Journal of Math.*, August 16 (1995) (http://nyjm.albany.edu:8000/j/v1/Nemethi-Steenbrink.html)

[15] Neumann, W.: Splicing Algebraic Links, *Advanced Studies in Pure Math.*, **8**, 349–361 (1986). (Proc U.S.-Japan Seminar on Singularities 1984)

[16] Pommersheim, J.E.: Toric varieties, lattice points and Dedekind sums, *Math. Ann.*, **295**, 1–24 (1993).

[17] Rademacher, H.: Generalization of the Reciprocity formula for Dedekind sums, *Duke Math. Journal*, **21**, 391–397 (1954).

[18] Scherk, J. and Steenbrink, J.H.M.: On the Mixed Hodge Structure on the Cohomology of the Milnor Fiber, *Math. Ann.*, **271**, 641–665 (1985).

[19] Schrauwen, R.: Topological Series of Isolated Plane Curve Singularities, *Eiseignement Mathématique*, **36**, 115–141 (1991).

[20] Schrauwen, R., Steenbrink, J. and Stevens, J.: Spectral Pairs and Topology of Curve Singularities, *Proc. Sympos. Pure Math.*, **53**, 305–328 (1991).

[21] Steenbrink, J.H.M.: The spectrum of hypersurface singularities, *Asterisque*, **179–180**, 211–223 (1989.

[22] Varchenko, A.N.: Asymptotic Hodge structure in the vanishing cohomology, *Math. USSR Izv.*, **18**, 469–512 (1982).

[23] Wahl, J.: Smoothings of normal surface singularities, *Topology*, **20**, 219–246 (1981).

[24] Zagier, D.: Higher dimensional Dedekind sums, *Math. Ann.*, **202**, 149–172 (1973).

Progress in Mathematics, Vol. 162, © 1998 Birkhäuser Verlag Basel/Switzerland

Embedding Nonisolated Singularities into Isolated Singularities

Mihai Tibăr*
Université d'Angers
Département de Mathématiques
2, bd Lavoisier
49045 Angers Cedex 01, FRANCE

Dedicated to Professor Egbert Brieskorn

1 Introduction

Let $f : (\mathbb{C}^n, 0) \to (\mathbb{C}, 0)$ be a complex analytic function germ with arbitrary singular locus. By adding to f a Brieskorn-Pham polynomial $x_1^{N_1} + \cdots + x_n^{N_n}$ in generic coordinates and with high enough powers N_i, one obtains a function germ g with isolated singularity. If we put coefficients in the Brieskorn-Pham polynomial, then f can be viewed as a very singular deformation of an isolated singularity g. In case $\dim(\mathrm{Sing}\, f) = 1$ one would have, for $k \gg 0$, a series of functions with isolated singularities $g_k = f + \varepsilon y^k$, sometimes called Iomdin series. The most familiar one is maybe the A_d-series: $x_1^2 + \cdots + x_{n-1}^2 + \varepsilon x_n^{d+1}$ with limit A_∞: $x_1^2 + \cdots + x_{n-1}^2$. For Iomdin series, one has formulas which relate a certain (topological) invariant of f to the same invariant of g_k, for instance the Euler characteristic of the Milnor fiber [Io], [Lê-1], the zeta function of the monodromy [Si], the spectrum [St], [Sa-1].

What one can see from these formulas is that the "limit" f of the series contains information on the isolated singularities g_k. For instance, under certain conditions, some of the eigenvalues of the monodromy of g_k, or some of the roots of the Bernstein polynomial, occur in the limit too. Reciprocally, roughly speaking, the limit seems to be "part" of each term of the series.

This behavior motivates the following question, raised by B. Malgrange:

Is there a way to embed the Milnor fibre of f into the Milnor fibre of some isolated singularity g? More generally: *is there a natural mapping between the corresponding sheaves of vanishing (or neighbouring) cycles?*

---

*I thank Claude Sabbah for discussions related to this paper and the referee for some remarks which improved the presentation

Let us fix our notations and degree of generality which we keep throughout this paper: $(\mathbf{X}, x)$ will be a complex analytic space germ of dimension n embedded into $(\mathbb{C}^m, 0)$, for some sufficiently large $m \in \mathbb{N}$. Let then $f : (\mathbf{X}, x) \to (\mathbb{C}, 0)$ be an analytic function germ, $f \not\equiv 0$ and let $\mathcal{S} := \{\mathcal{S}_j\}_{j \in \Lambda}$ be some Whitney stratification of $\mathbf{X}$. We call *singular locus of f* with respect to $\mathcal{S}$ the germ at x of the closed analytic set $\mathrm{Sing}_{\mathcal{S}} f := \bigcup_{j \in \Lambda} \mathrm{Sing}(f_{|\mathcal{S}_j})$ (it is closed since $\mathcal{S}$ has Whitney (a)-property).

Let $\psi_f \mathbb{Z}_{\mathbf{X}}^{\bullet}$ be the sheaf of neighbouring cycles and let $i_x : \{x\} \to \mathbf{X}$ denote the inclusion. We prove the following result:

1.1 Theorem. *Let $(\mathbf{X}, x)$ and $f : (\mathbf{X}, x) \to (\mathbb{C}, 0)$ be as above and let $k = \dim \mathrm{Sing}_{\mathcal{S}} f$. For a general system of coordinates $(x_1, \ldots, x_m)$ on $\mathbb{C}^m$ and for $N_i \gg 0$ and small enough $|\varepsilon_i| > 0$, $\forall i \in \{1, \ldots, k\}$, the germ $g : (\mathbf{X}, x) \to (\mathbb{C}, 0)$, $g := f + \varepsilon_1 x_1^{N_1} + \cdots + \varepsilon_k x_k^{N_k}$ has an isolated singularity (i.e. $\mathrm{Sing}_{\mathcal{S}} g = \{x\}$) and there is a continuous stratified embedding of Milnor fibres $p : F_f \hookrightarrow F_g$ which induces a morphism at the level of neighbouring cycles at x:*

$$p^* : i_x^* \psi_g \mathbb{Z}_{\mathbf{X}}^{\bullet} \to i_x^* \psi_f \mathbb{Z}_{\mathbf{X}}^{\bullet}.$$

We shall embed the monodromy fibration of f into the monodromy fibration of g by using a special geometric monodromy and show that this stratified embedding induces a morphism at the level of neighbouring cycles (see 2.8). Our construction is inspired by the techniques introduced by Lê D.T. [Lê-4], [Lê-2], which we have already used and developed in a series of papers [Ti-1,2,3]. We discuss further on several consequences of the main result for the homotopy type, zeta-function of the monodromy, spectrum.

2 The main construction

Let $l : (\mathbf{X}, x) \to (\mathbb{C}, 0)$ be the restriction of a linear function on $\mathbb{C}^m$ and denote by φ the map $(l, f) : (\mathbf{X}, x) \to (\mathbb{C}^2, 0)$. We denote by $\mathrm{Sing}_{\mathcal{S}} \varphi$ the closed set $\bigcup_{j \in \Lambda} \mathrm{Sing}(\varphi_{|\mathcal{S}_j})$ and by $\Gamma_{\mathcal{S}}(l, f)$ the *polar locus* of f with respect to l, relative to $\mathcal{S}$, i.e. the closure of the set $\mathrm{Sing}_{\mathcal{S}} \varphi \setminus \mathrm{Sing}_{\mathcal{S}} f$.

By a Bertini type theorem (see e.g. [Lê-3]), there is a Zariski-open subset $\Omega_f \subset \check{\mathbb{P}}^{m-1}$ such that, for any $l \in \Omega_f$ we have the following:

(a) $\Gamma_{\mathcal{S}}(l, f)$ is a curve or it is void (therefore called *polar curve*) and the restriction of φ to $\Gamma_{\mathcal{S}}(l, f)$ is one to one.

(b) $\dim l^{-1}(0) \cap \mathrm{Sing}_{\mathcal{S}} f = \max\{0, \dim \mathrm{Sing}_{\mathcal{S}} f - 1\}$,

(c) $l^{-1}(0)$ is transversal to all Thom strata in a fixed (a_f)-stratification of $f^{-1}(0)$ which refines $\mathcal{S}$.

2.1 Lemma. *Let $f_N := f + \varepsilon l^N$, for $l \in \Omega_f$, $N \in \mathbb{N}$, $N \geq 2$ and $\varepsilon \in \mathbb{C}$. Then, for all except a finite number of values of ε, we have $\mathrm{Sing}_{\mathcal{S}} f_N = \mathrm{Sing}_{\mathcal{S}} f \cap l^{-1}(0)$. In particular $\varphi(\mathrm{Sing}_{\mathcal{S}} f_N) \subset \{0\}$ and $\dim \mathrm{Sing}_{\mathcal{S}} f_N = \max\{0, \dim \mathrm{Sing}_{\mathcal{S}} f - 1\}$.*

Proof. Let $p \in \mathcal{S}_i$. By a local change of coordinates at p, one may assume that l is the first coordinate x_1 on $\mathcal{S}_i$. If $p \notin \Gamma_{\mathcal{S}}(l, f) \cup \operatorname{Sing}_{\mathcal{S}} f$ then the germ at p of $f_{N|\mathcal{S}_i}$ is clearly nonsingular, hence we only have to look at points $p \in \Gamma(l, f) \cup \operatorname{Sing} f$, arbitrarily close to x. If $p \in \operatorname{Sing} f_{N|\mathcal{S}_i} \cap \operatorname{Sing}_{\mathcal{S}} f$ then $p \in \operatorname{Sing}_{\mathcal{S}} f \cap l^{-1}(0)$. On the other hand, if $p \in \Gamma_{\mathcal{S}}(l, f) \setminus \operatorname{Sing}_{\mathcal{S}} f$ then $(\partial f_N / \partial x_1)(p) = (\partial f / \partial x_1 + \varepsilon N x_1^{N-1})(p)$ can be equal to 0 for at most one value of ε and cannot be equal to zero if N is greater than all the Puiseux ratios ρ_i defined below. (If not so, then $x_1(p) = 0$, which would contradict the condition (c) above.) This value ε depends on the point p, hence it is constant on each component of $\Gamma_{\mathcal{S}}(l, f)$. Therefore $\operatorname{Sing}_{\mathcal{S}} f_N \cap \Gamma_{\mathcal{S}}(l, f) \subset \{x\}$, for all ε except a finite number of values. It follows that $\operatorname{Sing}_{\mathcal{S}} f_N = \operatorname{Sing}_{\mathcal{S}} f \cap l^{-1}(0)$. The equality of dimensions is immediate, by using condition (b) above. $\qquad\square$

We shall omit from now on the subscript $\mathcal{S}$ for the objects depending on $\mathcal{S}$, whenever this dependence is clear from the context.

For the map $\varphi_N := (l, f_N) : \mathbf{X} \to \mathbb{C}^2$, for $l \in \Omega_f$, one notices that $\Gamma(l, f_N) = \Gamma(l, f) \cup \operatorname{Sing} f$, hence $\dim \Gamma(l, f_N) \leq 1$ if and only if $\dim \operatorname{Sing} f \leq 1$.

Let $\Delta(l, f) := \varphi(\Gamma(l, f))$ denote the *Cerf diagram* of f, with respect to l, relative to $\mathcal{S}$. It is a germ of a curve in $\mathbb{C}^2$. Then $\Delta(l, f_N)$ is a curve too. Let $\Delta(l, f) = \cup_{i \in K} \Delta_i$ be the decomposition into irreducible components. The *Puiseux ratio* of Δ_i is defined as the quotient $\rho_i = \frac{m_i}{n_i}$, where $m_i := \operatorname{mult}_0 \Delta_i$ and $n_i := \operatorname{mult}_0(\Delta_i, \{\lambda = 0\})$. Notice that $\rho_i < 1$, since l is general. If (u, λ) denote the coordinates of $\mathbb{C} \times \mathbb{C}$, then a parametrization of Δ_i looks like $u = \sum_{j \geq m_i} c_{i,j} t^j$, $\lambda = t^{n_i}$, where $c_{i,m_i} \neq 0$.

In order to relate the Milnor fibres of f and f_N at x, we start from an idea due to Lê D.T, used in his proof of Iomdin's theorem [Io], [Lê-1], see also [Si], [Ti-3]. We develop in the next a general construction.

2.2 Let $l \in \Omega_f$. Following [Lê-4], [Lê-3, Theorem 2.4], [Lê-5, (2.7)], we recall that there is a fundamental system of privileged open polydiscs in $\mathbb{C}^m$, centred at 0, of the form $(D_\alpha \times P_\alpha)_{\alpha \in A}$ and a corresponding fundamental system $(D_\alpha \times D'_\alpha)_{\alpha \in A}$ of 2-discs at 0 in $\mathbb{C}^2$, such that φ induces, for any $\alpha \in A$, a map

$$\varphi_\alpha : \mathbf{X} \cap (D_\alpha \times P_\alpha) \to D_\alpha \times D'_\alpha$$

which is a topological fibration over $D_\alpha \times D'_\alpha \setminus (\Delta(l, f) \cup \{\lambda = 0\})$. Moreover, f induces a topological fibration

$$f_\alpha : \mathbf{X} \cap (D_\alpha \times P_\alpha) \cap f^{-1}(D'_\alpha \setminus \{0\}) \to D'_\alpha \setminus \{0\},$$

respectively

$$f'_\alpha : \mathbf{X} \cap (\{0\} \times P_\alpha) \cap f^{-1}(D'_\alpha \setminus \{0\}) \to D'_\alpha \setminus \{0\},$$

which is fibre homeomorphic to the Milnor fibration of f, respectively to the Milnor fibration of $f_{|\{l=0\}}$. The disc D'_α has been chosen small enough such that $\Delta(l, f) \cap \partial \overline{D_\alpha} \times D'_\alpha = \emptyset$.

2.3 The construction of the polydisc $D_\alpha \times P_\alpha$, as done in the proof of [Lê-4, Lemme 1.3.4], shows that this polydisc is privileged also with respect to f_N, in the sense that, for $|\eta| \neq 0$ small enough, $f^{-1}(\eta)$ is transversal to the natural stratification of $\mathbf{X} \cap (D_\alpha \times P_\alpha)$. Indeed, by the property (c) of Ω_f, the Thom stratification of $f^{-1}(0)$ induces a Thom stratification of $f^{-1}(0) \cap l^{-1}(0)$. Since $\mathrm{Sing}\, f_N = \mathrm{Sing}\, f \cap l^{-1}(0)$ and since f_N is of the very special form $f + \varepsilon l^N$, it follows that the stratum $\{f_N = 0\} \setminus \mathrm{Sing}\, f_N$ together with the strata included in $\mathrm{Sing}\, f \cap l^{-1}(0)$ of the Thom stratification of $f^{-1}(0) \cap l^{-1}(0)$ are a Thom stratification of $\{f_N = 0\}$.

The first consequence of this fact is that, for small enough $|\eta| \neq 0$, the set

$$F_N := f_N^{-1}(\eta) \cap \mathbf{X} \cap (D_\alpha \times P_\alpha)$$

is homeomorphic to the Milnor fibre of f_N at x. Let $F := f^{-1}(\eta) \cap \mathbf{X} \cap (D_\alpha \times P_\alpha)$ denote the Milnor fibre of f at x.

2.4 Let $s > 0$ be small enough such that $\{\lambda = \eta\}$ is transversal to $\Delta(l, f)$ within $D \times D'$, for all η with $0 < |\eta| \leq s$. Let $0 < s_0 < s$. Since $N \gg 0$ (in particular $N > 1/\rho_i$, where ρ_i are the Puisseux ratios of $\Delta(l, f)$, there exists $r_0 > 0$ such that the curve $\{\lambda + \varepsilon u^N = \eta\}$ is transversal to $\Delta(l, f)$ within $D \times D'$, for all η with $0 < |\eta| \leq s_0$ and all ε with $|\varepsilon| \leq r_0$.

Let us fix some η with $0 < |\eta| < s_0$. The map $j : D^* \times D' \to \mathbb{C}$, $(u, \lambda) \mapsto \frac{\eta - \lambda}{u^N}$ is submersive on $D^* \times (D' \setminus \overline{D_0}) \setminus \Delta$ and on Δ, where D_0 is a small enough open disc at 0. We may lift the vector field $\partial/\partial t$ on $[0, r_0]$ to a continuous vector field on $H := D^* \times (D' \setminus \overline{D_0}) \cap j^{-1}([0, r_0])$, tangent to $\Delta(l, f) \cap H$, then further lift if by φ to a controlled, "rugueux" vector field on $\varphi^{-1}(H)$, tangent to the strata of $\mathcal{S}$ and to $\Gamma(l, f)$. Notice that φ is a Thom map with respect to the stratification $\mathcal{S} \cup \Gamma(l, f)$, by property (a). By Thom's Second Isotopy Lemma (see [Ma], [GLPW] or [Ve]), we might integrate the latter vector field to construct a homeomorphism from $F = \varphi^{-1}(\{\lambda = \eta\})$ to a subset of $F_N = \varphi^{-1}(\{\lambda = \eta - \varepsilon u^N\})$, $|\varepsilon| < r_0$. This already gives an embedding $F \hookrightarrow F_N$, but to prove our theorem we need to show that it behaves well with respect to monodromy, see 2.8.

2.5 Let $S_\rho := \{z \in \mathbb{C} \mid |z| = \rho\}$. We have to construct a fibrewise stratified homeomorphism of the two fibrations over $S_{|\eta|}$. Let

$$D \times (D' \setminus \overline{D_0}) \times S_{|\eta|} \xrightarrow{\gamma} \mathbb{C} \times S_{|\eta|} \xrightarrow{\pi} \mathbb{C},$$

$\gamma(u, \lambda, \theta) = (j(u, \lambda), \theta)$, $\pi(z, \theta) = z$. The stratification on $D \times (D' \setminus \overline{D_0}) \times S_{|\eta|}$ is given by the products of the strata $D \times (D' \setminus \overline{D_0}) \setminus \Delta$, $\Delta \setminus \{0\}$, $\{0\}$ by the circle $S_{|\eta|}$.

We lift the vector field $\partial/\partial t$ by $\pi \circ \gamma$ on $D \times (D' \setminus \overline{D_0}) \times S_{|\eta|} \cap (\pi \circ \gamma)^{-1}([0, r_0])$ to a continuous vector field $\mathbf{v}$ tangent to the strata. By integrating $\mathbf{v}$, one gets a fibrewise diffeomorphism of fibrations over $S_{|\eta|}$.

We finally apply Thom's Second Isotopy Lemma for the maps $\hat{\varphi}$ and γ, where $\hat{\varphi} = (\varphi, \mathrm{id}) : \mathbf{X} \cap (D_\alpha \times P_\alpha) \times S_{|\eta|} \to D_\alpha \times D'_\alpha \times S_{|\eta|}$. (Since φ is a Thom map, $\hat{\varphi}$ will be a Thom map too.) We get a corresponding stratified, fibrewise homeomorphism of fibrations over the circle $S_{|\eta|}$. One may summarize the results by the following commutative diagram:

$$
\begin{array}{ccc}
\mathbf{X} \cap (D_\alpha \times P_\alpha) \cap f^{-1}(S_{|\eta|}) & \xrightarrow{\ \text{homeo}\ } & \mathbf{W} \subset \mathbf{X} \cap (D_\alpha \times P_\alpha) \cap f_N^{-1}(S_{|\eta|}) \\
\downarrow & & \downarrow \qquad\qquad \downarrow \\
\{(u,\lambda) \in D \times D',\ \theta \in S_{|\eta|} \mid \lambda = \theta\} & \xrightarrow{\ \text{diffeo}\ } & \mathbf{W} \subset \quad C_\varepsilon \\
& \searrow & \downarrow \qquad \swarrow \\
& & S_{|\eta|}
\end{array}
$$

where $C_\varepsilon := \{(u,\lambda) \in D \times D',\ \theta \in S_{|\eta|} \mid \lambda = \theta - \varepsilon_0 u^N\}$.

2.6 Let $\mathbf{w}$ be a smooth vector field on $D_\alpha \times S_{|\eta|}$, tangent to $\Delta(l,f) \cap (D_\alpha \times S_{|\eta|})$ and lifting the unit vector field of $S_{|\eta|}$ by the projection $D_\alpha \times S_{|\eta|} \to S_{|\eta|}$.

By our diffeomorphism, one can map it to a vector field $\mathbf{w}'$ on W. This is again a lift of the unit vector field on $S_{|\eta|}$. One may clearly extend $\mathbf{w}'$ to a vector field $\mathbf{u}$ on the whole "cylinder" C_ε such that it is still a lift of the unit vector field on $S_{|\eta|}$ and tangent to $\{\lambda = 0\} \cap C_\varepsilon$. Now $\mathbf{u}$ can be lifted by φ then integrated to give a characteristic homeomorphism h_{f_N} of F_N. We call it a *carrousel monodromy*.

Finally, by our stratified homeomorphism $\mathbf{W} \to \mathbf{X} \cap (D_\alpha \times P_\alpha) \cap f^{-1}(S_{|\eta|})$, we induce a geometric monodromy h_f of the fibre F. Altogether we have proven the following:

2.7 Proposition. *There is an embedding $i : F \hookrightarrow F_N$ and a geometric carrousel monodromy h_{f_N} of F_N such that its restriction to $i(F)$ is a geometric monodromy h_f of $i(F)$.* $\qquad\square$

2.8 Proof of Theorem 1.1

Proving that an embedding $p : F_f \hookrightarrow F_g$ induces a morphism at the level of neighbouring cycles amounts to proving that the induced morphism in cohomology $p^* : h^*(F_g, \mathbb{Z}) \to H^*(F_f, \mathbb{Z})$ is equivariant with respect to the algebraic monodromy, i.e. $h_f^* \circ p^* = p^* \circ h_g^*$.

Let then $f_N = f + \varepsilon l^N$ be as above. Consider the map $\varphi_N = (l, f_N)$. Then $\Delta(l, f_N) = \Delta(l,f) \cup \Delta_N$, where Δ_N is the germ of the curve $\{\lambda + \varepsilon u^N = 0\}$. By choosing $N \gg 0$ and in particular $1/N < \rho_i$, $\forall i \in K$, we insure that there are discs $\mathbf{D} \subset \overline{\mathbf{D}} \subset \mathbf{D}_N \subset \overline{\mathbf{D}_N} \subset D_\alpha$ such that:

(i) $\Delta(l,f) \cap D_\alpha \times D'_\alpha = \Delta(l,f) \cap \mathbf{D} \times D'_\alpha$,

(ii) $\Delta_N \cap D_\alpha \times D'_\alpha = \Delta_N \cap (\mathbf{D}_N \setminus \mathbf{D}) \times D'_\alpha$,

(iii) $W \subset \mathbf{D} \times D'_\alpha$.

This is a consequence of the existence of the *polar filtration* defined by Lê D.T., see [Lê-2] or [Ti-1]. As shown above, one can define a carrousel on $\mathbf{D}_N \times \{\eta\}$ which restricts to a carrousel on $\mathbf{D} \times \{\eta\}$ and an isomorphism of $W \cap \mathbf{D} \times \{\eta\}$. By lifting it, one gets the statement of Proposition 2.7.

It is now clear that we may proceed by induction. So let $f_i := f_{i-1} + \varepsilon_i l_i^{N_i}$, where $f_0 = f$, $l_i \in \Omega_{f_{i-1}}$, $N_i \gg 0$ and $|\varepsilon_i|$ sufficiently small (as described above), $\forall i \in \{1, \ldots, k\}$. Denoting by F_i the Milnor fibre of f_i at x, we obtain the following chain of commutative diagrams (by applying Proposition 2.7 at each step):

$$
\begin{array}{ccccccc}
F_0 & \hookrightarrow & F_1 & \hookrightarrow & \cdots & \hookrightarrow & F_k \\
\downarrow{\scriptstyle h_{f_0}} & & \downarrow{\scriptstyle h_{f_1}} & & & & \downarrow{\scriptstyle h_{f_k}} \\
F_0 & \hookrightarrow & F_1 & \hookrightarrow & \cdots & \hookrightarrow & F_k.
\end{array}
$$

By taking cohomology over $\mathbb{Z}$ in this diagram, our theorem is proved, for $f := f_0$ and $g := f_k$. $\qquad\Box$

3 Homotopy type of the Milnor fibre

3.1 Keeping the previous notations, we consider from now on the map $\varphi_N = (l, f_N) : \mathbf{X} \cap D \times P \cap \varphi_N^{-1}(D \times \hat{D}) \to D \times \hat{D}$. The Milnor fibre of f_N is stratified homeomorphic to $\varphi_N^{-1}(\mathbf{D}_N \times \{\eta\})$. On the other hand, we have shown that there is a subset $W \subset \mathbf{D} \subset \mathbf{D}_N$ such that W is diffeomorphic to a disc and $\varphi_N^{-1}(W \times \{\eta\})$ is stratified homeomorphic to the Milnor fibre F of f at x. Hence F is homotopy equivalent to $\varphi_N^{-1}(\mathbf{D} \times \{\eta\})$.

Let us define the carrousel $\mathbf{h}$ of $\mathbf{D}_N \times \{\eta\}$ in more detail. We have a lot of liberty in defining the vector field $\mathbf{u}$ from 2.6 and therefore we may assume that the circles $S := \partial\overline{\mathbf{D}} \times \{\eta\}$ and $S_N := \partial\overline{\mathbf{D}_N} \times \{\eta\}$ are pointwise fixed by the carrousel. Furthermore, by our assumption, there is a thin annulus A_N centered at 0 such that $A_N \subset \mathbf{D}_N \setminus \mathbf{D}$, $A_N \supset \Delta_N \cap \mathbf{D}_N$ and that the carrousel $\mathbf{h}$ rotates each point of $A_N \times \{\eta\}$ by $2\pi/N$. The latter property follows from the construction of the carrousel [*loc.cit.*].

We consider small enough discs $\delta_i \subset A_N \times \{\eta\}$ of the same radius, one at each point a_i of the set $\Delta_N \cap A_N \times \{\eta\} := \{a_1, \ldots, a_N\}$. Assuming that the points a_i are clockwise ordered, we have $\mathbf{h}(\delta_i) = \delta_{i+1}$. Now divide the disc $\mathbf{D}_N \times \{\eta\}$ into equal radial sectors R_i, $i \in \{1, \ldots, N\}$, such that $\delta_i \subset R_i$ and fix a point $b \in \partial\overline{\mathbf{D}} \times \{\eta\} \cap R_1$. Take a non selfintersecting path $\gamma_1 : [0, 1] \to R_1$ from b to some point c_1 of $\partial\overline{\delta_1}$ and fix a trivialization τ_1 from $\varphi_N^{-1}(c_1)$ to $\varphi_N^{-1}(b)$. Then $\gamma_i := \mathbf{h}^{i-1}(\gamma_1)$ is a path from b to $c_i := \mathbf{h}^{i-1}(c_1) \subset \partial\overline{\delta_i}$ and $\tau_i := h_{f_N}^{i-1}(\tau_1)$ is a trivialization from $\varphi_N^{-1}(c_i)$ to $\varphi_N^{-1}(b)$.

Let us fix a continuous stratified *specialization map*, as defined by M. Goresky, see [GM-1, §6]:

$$\mathrm{sp}_1 : \varphi_N^{-1}(\delta_1) \to \varphi_N^{-1}(a_1)$$

and denote $\mathrm{sp}_i := \mathbf{h}^{i-1}(\mathrm{sp}_1)$, for $i \in \{2, \dots, N\}$.

3.2 Theorem. *The Milnor fibre F_N is homotopy equivalent to the Milnor fibre $F = \varphi_N^{-1}(\overline{\mathbf{D}} \times \{\eta\})$, to which one attaches over $\varphi_N^{-1}(b)$, by the trivializations τ_i, the mapping cylinders $\mathrm{Cyl}(\mathrm{sp}_{i|\varphi_N^{-1}(c_i)})$, for $i \in \{1, \dots, N\}$.*

Proof. F_N is homotopy equivalent to $\varphi_N^{-1}(\overline{\mathbf{D}} \cup \cup_{i=1}^{N}(\mathrm{Im}\,\gamma_i \cup \delta_i))$. Since $\varphi_N^{-1}(\delta_i)$ contracts to $\varphi_N^{-1}(a_i)$ by the specialization map, we have the homotopy equivalence $\varphi_N^{-1}(\delta_i \cup \mathrm{Im}\,\gamma_i) \overset{\mathrm{ht}}{\simeq} \mathrm{Cyl}(\mathrm{sp}_{i|\varphi_N^{-1}(c_i)})$. $\qquad\square$

Notice that $\varphi_N^{-1}(a_i)$ is homotopy equivalent to the complex link $\mathrm{CL}(\{f = 0\}, x)$ of the germ $\{f = 0\}$ at x. In case $(\mathbf{X}, x)$ is smooth or a complete intersection (more generally, if the *rectified homotopical depth* $\mathrm{rhd}(\mathbf{X}, x)$ is maximal, see [HL] for the definition) then $\mathrm{CL}(\{f = 0\}, x)$ is homotopy equivalent to a bouquet $\vee_{\mu_0} S^{n-2}$ of spheres of dimension $n - 2 = \dim\{f = 0\} - 1$.

3.3 Corollary. (a) *If $l \in \Omega_f$ then:*

$$\chi(F_N) - \chi(F) = N[\chi(\mathrm{CL}(\{f = 0\}, x)) - \chi(F_{f_{|l=0}})].$$

(b) *Let $\mathrm{rhd}(\mathbf{X}, x) \geq \dim(\mathbf{X}, x) = n$. If $l \in \Omega_f$ then:*

$$\chi(F_N) = (1 - N)\chi(F) + N\mathrm{CL}(\mathbf{X}, x) + (-1)^{n-1}N\lambda^0,$$

where $\lambda^0 = \mathrm{mult}_x(\Gamma(l, f), \{f = 0\}) - \mathrm{mult}_x(\Gamma(l, f), \{l = 0\})$ is an invariant of f, not depending on $l \in \Omega_f$.

Proof. Since (a) is clear from Theorem 3.2, let us prove (b). Considering the map $\varphi = (l, f)$, we have that $\varphi^{-1}(\{\alpha\} \times D')$ is homeomorphic to $\mathrm{CL}(\mathbf{X}, x)$. If one takes small enough discs $\beta_j \subset \{\alpha\} \times D'$ centered at the points $p_j \in \Delta(l, f) \cap \{\alpha\} \times D'$ and β_0 at $\{0\} \times D'$ then $\varphi^{-1}(\beta_0) \overset{\mathrm{ht}}{\simeq} \mathrm{CL}(\{f = 0\}, x)$. For some $s_j \in \partial\overline{\beta}_j$, the pair $(\varphi^{-1}(\beta_j), \varphi^{-1}(s_j))$ is $(n - 2)$-connected, for any j (since p_j is the image of a stratified Morse point and $\mathrm{rhd}(\mathbf{X}, x) \geq \dim \mathbf{X}$), more precisely, $\varphi^{-1}(\beta_j)$ is obtained from $\varphi^{-1}(s_j)$ by attaching one $(n - 1)$-cell. It follows that $\mathrm{CL}(\mathbf{X}, x)$ is homotopy equivalent to the space obtained by attaching $\gamma := \mathrm{mult}_x(\Gamma(l, f), \{l = 0\})$ cells of dimension $n - 1$ to $\mathrm{CL}(\{f = 0\}, x)$. Therefore $\chi(\mathrm{CL}(\mathbf{X}, x)) = \chi(\mathrm{CL}(\{f = 0\})) - (-1)^{n-2}\gamma$.

On the other hand, F is homeomorphic to $\varphi^{-1}(D \times \{\eta\})$ and this is obtained from

$$\varphi^{-1}(\{0\} \times \{\eta\}) \overset{\mathrm{homeo}}{\simeq} F_{f_{|l=0}},$$

by attaching $b_x := \mathrm{mult}_x(\Gamma(l,f),\{f=0\})$ cells of dimension $n-1$ (by a similar argument as above, using that $\mathrm{rhd}(\mathbf{X},x) \geq n$). We get in particular $\chi(F) = \chi(F_{f_{|l=0}}) + (-1)^{n-1}b_x$. Then the claimed relation in (b) follows from (a) by the two proven relations among Euler numbers. $\qquad\square$

Let us now assume that $\mathrm{Sing}_{\mathcal{S}} f$ is one-dimensional. Then let $\mathrm{Sing}_{\mathcal{S}} f = \cup_{i \in I}\Sigma_i$ be the decomposition into irreducible components and let $d_i := \mathrm{mult}_x(\Sigma_i)$. In this case $\mathrm{CL}(\{f=0\},x)$ has a number of isolated singularities equal to $\sum_{i \in I} d_i$. Let us denote by F_i the local Milnor fibre at such a transversal isolated singularity, at some point of $\Sigma_i \setminus \{x\}$. This fibre does not depend on the point, see e.g. [St]. Then the attaching of $\mathrm{Cyl}(\mathrm{sp}_{j|\varphi_N^{-1}(c_j)})$ to $\varphi_N^{-1}(c_j)$ amounts to attaching one cone over each of the d_i transversal Milnor fibres $F_i \subset \varphi_N^{-1}(c_j)$. We have the following rather surprising bouquet result.

3.4 Theorem. *[Ti-3] If* $\dim \mathrm{Sing}_{\mathcal{S}} f = 1$*, then there is a homotopy equivalence:*

$$F_N \overset{\mathrm{ht}}{\simeq} (F \cup E) \bigvee_{\substack{i \in I \ \#M_i}} \bigvee S(F_i), \qquad (1)$$

where $S(F_i)$ denotes the topological suspension over F_i, $\#M_i = Nd_i - 1$, $E := \cup_{i \in I}\mathrm{Cone}(F_i)$ and $F \cup E$ is the attaching to F of one cone over $F_i \subset F$ for each $i \in I$.

Proof. From Theorem 1.1 and the above observations it already follows that

$$F_N \overset{\mathrm{ht}}{\simeq} (F \cup_{i \in I} \cup_{d_i - \mathrm{times}} \mathrm{Cone}(F_i)) \bigvee_{\substack{i \in I \ \#N_i}} \bigvee S(F_i), \qquad (2)$$

where $\#N_i = (N-1)d_i$. (This formula appeared in [Ti-2, Proposition 4.3] but contained an obvious error which should be corrected as we state above.) Notice that $\#M_i = \#N_i + d_i - 1$. To get the improvement from (2) to (1), we need a further introspection within the attaching process, for which we send the reader to the source [Ti-3]. $\qquad\square$

Taking Euler numbers, we easily get:

3.5 Corollary. *If* $\dim \mathrm{Sing}_{\mathcal{S}} f = 1$*, then*

$$\chi(F_N) = \chi(F) - N \sum_{i \in I} d_i \chi(F_i) + N\mathrm{mult}_x(\mathrm{Sing}_{\mathcal{S}} f).$$

3.6 Note. Let us restrict to the particular case $\mathbf{X} = \mathbb{C}^n$. Then Corollary 3.5 becomes the well known Iomdin-Lê formula [Io], [Lê-1]: $\chi(F_N) = \chi(F) + (-1)^{n-1}N\sum_{i \in I} d_i\mu_i$, where μ_i is the Milnor number of F_i. Our Corollaries 3.5 and 3.3 (b) may be considered as generalizations of this formula to the non smooth case. Other generalizations of the Iomdin-Lê formula, from different points of view but for smooth $\mathbf{X}$, have been obtained e.g. in [Mas-1], [MS], [Mas-2, pag. 49].

Our Theorem 3.4 is far more than a cell attaching result. For instance take the smooth case $\mathbf{X} = \mathbb{C}^n$. Then the suspension $S(F_i)$ is just a bouquet of spheres of dimension $n - 1$, since $F_i \overset{\mathrm{ht}}{\simeq} \vee_{\mu_i} S^{n-2}$. The previously known result (see [Va]) is that F_N is obtained from F by attaching $N \sum_{i \in I} d_i \mu_i$ cells of dimension $n - 1$. To be able to single out a number of spheres from an amorphous attaching one needs some more refined information on how the attaching is done. This is what we actually get by using the geometry of the monodromy.

4 Zeta-function of the monodromy

Let $\mathcal{F} \hookrightarrow E \to S^1$ be a locally trivial fibration with monodromy τ. The zeta-function of the monodromy τ is defined as:

$$\zeta(\mathcal{F}, \tau)(t) := \prod_{i \geq 0} \det[\mathbf{I} - t\tau^*; H^i(\mathcal{F}, \mathbb{Q})]^{(-1)^{i+1}}.$$

With this notation, the zeta-function of f at x is $\zeta_f(t) := \zeta(F, h_f)(t)$.

We have constructed in 2.5 an embedding of the monodromy fibration of f into the monodromy fibration of f_N. This enables one to investigate the relation among the corresponding zeta-functions, as follows. We remind from 3.1 that the carrousel $\mathbf{h}$ has the property that $\mathbf{h}^N(c_i) = c_i$, $\mathbf{h}^N(a_i) = a_i$, $\forall i \in \{1, \dots, N\}$. This implies that $h_i := h_{f_N}^N$ acts on the pair $(\varphi_N^{-1}(a_i), \varphi_N^{-1}(c_i))$, where h_{f_N} is the carrousel monodromy defined at 2.6.

Let now K denote the toric knot $\Delta_N \cap \mathbf{D}_N \times S_\eta$, where S_η is the circle of radius $|\eta|$. The counterclockwise orientation of S_η induces a direct orientation on K. We have $\{a_1, \dots, a_N\} \subset K$. Let N(K) be a closed tubular neighbourhood of K in $\mathbf{D}_N \times S_\eta$ such that $\mathrm{N(K)} \cap \mathbf{D}_N \times \{\eta\} = \cup_{j=1}^{N} \overline{\delta}_j$ and let $M, L \subset \partial\mathrm{N(K)}$ be the standard *meridian*, resp. the *longitude* of the knot K. We identify M with the circle $\partial\overline{\delta}_j$. The orientation on L is induced by the one on K and M has the counterclockwise orientation of $\partial\overline{\delta}_j$.

4.1 Definition. We call *meridian monodromy* the monodromy h_M of the fibration $\varphi_N^{-1}(M) \to M$, resp. *longitude monodromy* the monodromy h_L of the fibration $\varphi_N^{-1}(L) \to L$. We also denote by h_K the monodromy of the locally trivial fibration $\varphi_N^{-1}(K) \to K$.

Then we have the following Thom-Sebastiani type formula for the zeta-function.

4.2 Theorem.

$$\zeta_{f_N}(t) \cdot \zeta_f^{-1}(t) = \zeta^{-1}(\varphi_N^{-1}(c_1), h_L \circ h_M^N)(t^N) \cdot \zeta(\varphi_N^{-1}(a_1), h_K)(t^N).$$

Proof. The carrousel monodromy h_{f_N} restricts to monodromy actions on $\varphi_N^{-1}(\mathbf{D}_N \times \{\eta\})$, $\varphi_N^{-1}(\mathbf{D} \times \{\eta\})$, $\varphi_N^{-1}(A_N \times \{\eta\})$. Now $\varphi_N^{-1}(\mathbf{D}_N \setminus \mathbf{D} \times \{\eta\})$ is homotopy equivalent to $\varphi_N^{-1}(A_N \times \{\eta\})$ and by a Mayer-Vietoris argument as in [Ti-1, §3], we get:

$$\zeta_{f_N}(t) = \zeta_f(t) \cdot \zeta(\varphi_N^{-1}(A_N \times \{\eta\}), h_{f_N})(t).$$

Since the carrousel $\mathbf{h}$ rotates each point of $A_N \times \{\eta\}$ by $2\pi/N$, it follows that

$$[\zeta(\varphi_N^{-1}(A_N \times \{\eta\}), h_{f_N})(t)]^N = \zeta(\varphi_N^{-1}(A_N \times \{\eta\}), h_{f_N}^N)(t^N),$$

by the general theory of the zeta-function of the carrousel monodromy [Ti-1, §3, 3.3(6)].

On the other hand, by applying [Ti-1, Theorem 3.4], we get $\zeta(\varphi_N^{-1}(A_N \times \{\eta\}), h_{f_N}^N)(t) = [\zeta_1(t)]^N$, where $\zeta_1(t)$ is the zeta-function associated to the action of h_1^* on $H^\bullet(\varphi_N^{-1}(a_1), \varphi_N^{-1}(c_1))$. Altogether these give $\zeta(\varphi_N^{-1}(A_N \times \{\eta\}), h_{f_N})(t) = \zeta_1(t^N)$. By using the exact sequence

$$H^\bullet(\varphi_N^{-1}(a_1), \varphi_N^{-1}(c_1)) \to H^\bullet(\varphi_N^{-1}(a_1)) \to H^\bullet(\varphi_N^{-1}(c_1)) \overset{+1}{\to},$$

we get $\zeta_1(t) = \zeta(\varphi_N^{-1}(a_1), h_{f_N}^N)(t) \cdot \zeta^{-1}(\varphi_N^{-1}(c_1), h_{f_N}^N)(t)$. Now $(h_{f_N}^N)_{|\varphi_N^{-1}(a_1)}$ is by definition h_K and we also have

$$(h_{f_N}^N)_{|\varphi_N^{-1}(c_1)} = h_L \circ h_M^N, \tag{3}$$

since the trajectory of c_1 within $\partial N(K)$ under the action of the carrousel is a $(1, N)$-cable on K. $\square$

4.3 Remark. The formula for the above theorem can be written also as:

$$\zeta_{f_N}(t) \cdot \zeta_f^{-1}(t) = \zeta_1(t^N),$$

where $\zeta_1(t) = \zeta((\varphi_N^{-1}(a_1), \varphi_N^{-1}(c_1)), h_1^*)(t)$ is the zeta-function of the operator h_1^* on the relative cohomology $H^\bullet((\varphi_N^{-1}(a_1), \varphi_N^{-1}(c_1); \mathbb{Q})$.

4.4 Example. Let $f : (\mathbb{C}^n, 0) \to (\mathbb{C}, 0)$ be homogeneous of degree d and let $l \in \Omega_f$. Since $\mathrm{Sing}\, f$ is a union of linear subspaces, the only Puiseux ratio is $1/d$. Moreover, we have that $h_L \circ h_M^d$ is the identity. One can see this by using the action $s \cdot x = \exp(2\pi i s)x$ on $\varphi^{-1}(\partial N(K))$, for $s \in [0, 1]$; projecting this action on $\partial N(K)$ we get LM^d. On the other hand we have $1 \cdot x = x$, so the equality follows.

By a similar geometric action we see that $h_K = \mathrm{id}$. Therefore we get, for $N > d$:

$$\zeta_{f_N}(t) = \zeta_f(t)(1 - t^N)^{-\chi(\varphi_N^{-1}(a_1))}\zeta^{-1}(\varphi_N^{-1}(c_1), h_M^{N-d})(t^N).$$

4.5 Note. Specializing to the case of one-dimensional singular locus, we show how to derive from Theorem 4.2 Siersma's result [Si] on the zeta-function of the Iomdin series. More precisely, we shall prove an extension of Siersma's formula for the case of non-smooth underlying space. We may also refer the reader to [Ne, §8] for the zeta-function of a Iomdin type series, in case $\mathbf{X}$ is smooth and $\dim \operatorname{Sing} f = 1$.

As in §3, let $\operatorname{Sing}_{\mathcal{S}} f = \cup_{i \in I} \Sigma_i$. Then the toric knot $K_i := \varphi_N^{-1}(K) \cap \Sigma_i$ is a d_i-cyclic covering of K. The fibre $\varphi_N^{-1}(a_1)$ has $\sum_{i \in I} d_i$ singular points at $\cup_{i \in I} \varphi_N^{-1}(a_1) \cap \Sigma_i$. Let $B_{i,k}$ be a Milnor ball at some point $b_{i,k} \in \varphi_N^{-1}(a_1) \cap \Sigma_i$, $k \in \{1, \dots, d_i\}$, with Milnor fibre F_i (not depending on k). Notice that the homotopy type of F_i need not be a bouquet of spheres since $\mathbf{X}$ may not have maximal rectified homotopical depth. For small enough δ_1, we have the direct sum decomposition:

$$H^\bullet(\varphi_N^{-1}(a_1), \varphi_N^{-1}(c_1)) = \bigoplus_{i \in I} \bigoplus_{k=1}^{d_i} H^\bullet(B_{i,k} \cap \varphi_N^{-1}(\delta_1), F_i),$$

where $B_{i,k} \cap \varphi_N^{-1}(\delta_1)$ is contractible. Now the monodromy h_1^* acts in fact on $\bigoplus_{k=1}^{d_i} H^\bullet(B_{i,k} \cap \varphi_N^{-1}(\delta_1), F_i)$, for each fixed $i \in I$, since $K_{i_1} \cap K_{i_2} = \emptyset$, if $i_1 \neq i_2$. Moreover, each of the following three monodromies $(h_1^*)^{d_i}$, $(h_L^*)^{d_i}$, h_M^* acts on $H^\bullet(B_{i,k} \cap \varphi_N^{-1}(\delta_1), F_i) \simeq \tilde{H}^{\bullet-1}(F_i, \mathbb{Z})$. The action of $(h_L^*)^{d_i}$, resp. h_M^*, on $H^\bullet(F_i, \mathbb{Z})$ is what one usually calls (after Steenbrink [St]) *vertical monodromy*, also denoted by A_i, resp. *horizontal monodromy*, also denoted by T_i.

From (3) we get:

$$(h_1^*)^{d_i} = (h_L^*)^{d_i} \circ (h_M^*)^{N d_i} = A_i T_i^{N d_i}.$$

Then, by Theorem 4.2 and by elementary properties of the zeta-function, we get the following:

4.6 Corollary. *If* $\dim \operatorname{Sing}_{\mathcal{S}} f = 1$ *then*

$$\zeta_{f_N}(t) = \zeta_f(t) \cdot \prod_{i \in I} \tilde{\zeta}(F_i, A_i T_i^{N d_i})(t^{N d_i}),$$

where $\tilde{\zeta}$ denotes the zeta-function on the reduced cohomology.

4.7 Note on the spectrum. In [Sa-1], M. Saito defined the spectrum of $(\mathcal{M}, f, x)$, where $\mathcal{M}$ is a *polarizable MHM (i.e. mixed Hodge module)* (see [Sa-2], [Sa-3]). This extends the definition of the spectrum of a hypersurface singularity introduced by Steenbrink [St].

We assume for the remainder that $(\mathbf{X}, x)$ is a complete intersection, so then $\mathbb{Q}_\mathbf{X}^\bullet$ is a perverse sheaf. One denotes by $\mathbb{Q}_\mathbf{X}^H$ the MHM with $\mathbb{Q}_\mathbf{X}^\bullet$ as underlying sheaf. The functors φ_f and $\mathcal{H}^j i_x^*$ are considered in the category of MHM

114 $\qquad\qquad\qquad\qquad\qquad\qquad\qquad\qquad\qquad\qquad\qquad$ M. Tibăr

on **X**. Following Saito's notation in [Sa-1], the spectrum is an alternating sum

$$\mathrm{Sp}(\mathbb{Q}_\mathbf{X}^H, f, x) := \sum_{j \in \mathbb{Z}} (-1)^j \mathrm{Sp}(\mathcal{H}^j i_x^* \varphi_f \mathbb{Q}_\mathbf{X}^H, T_s)$$

of spectra of the mixed Hodge structures $\mathcal{H}^j i_x^* \varphi_f \mathbb{Q}_\mathbf{X}^H$ on the cohomology of the Milnor fibre $\tilde{H}^j(F_f, \mathbb{Q})$. The semi-simple part T_s of the monodromy h_f^* preserves the Hodge filtration F. The spectrum of a MHS H with an automorphism T is defined as $\mathrm{Sp}(H, T) := \sum_{\alpha \in \mathbb{Q}} n_\alpha(\alpha) \in \mathbb{Z}[\mathbb{Q}]$, where $n_\alpha = \dim \mathrm{Gr}_F^p H_{\mathbb{C}, \lambda}$, for $\lambda = \exp(2\pi i \alpha)$ and $p = [\alpha]$. The space $H_\mathbb{C}$ is the underlying complex vector space of H and $H_{\mathbb{C}, \lambda}$ denotes the eigenspace of T for the eigenvalue λ.

In case $\dim(\mathrm{Sing}_\mathcal{S} f) = 1$, M. Saito [Sa-1] proved a formula (previously conjectured by Steenbrink [St]) which relates $\mathrm{Sp}(\mathbb{Q}_\mathbf{X}^H, f + \varepsilon l^N, x)$ to $\mathrm{Sp}(\mathbb{Q}_\mathbf{X}^H, f, x)$. This formula reduces, by forgetting the Hodge structures, to the zeta-function formula from Corollary 4.6 (see Siersma's remark [Si, p. 184]). Saito's method of proof does not extend to the cases $\dim(\mathrm{Sing}_\mathcal{S} f) > 1$, but if such a formula still exists, then it must be compatible to the zeta-function formula which we have proven in Theorem 4.2. This latter formula, together with Remark 4.3, may suggest how a spectrum formula should look like; however, a spectrum formula cannot be derived directly from Theorem 1.1 since our theorem does not hold in the category of mixed Hodge modules.

References

[GLPW] C.G. Gibson; K. Wirthmüller; A.A. du Plessis; E.J.N. Looijenga: *Topological Stability of Smooth Mappings*, Lect. Notes in Math. **552**, Springer Verlag (1976).

[GM-1] M. Goresky; R. MacPherson: *Morse theory and intersection homology*, Analyse et Topologie sur les Espaces Singuliers, Astérisque **101** (1983), 135–192.

[HL] H. Hamm; Lê D.T.: *Rectified homotopical depth and Grothendieck conjectures*, in: Grothendieck Festschrift, vol. 2, Birkhäuser, Boston 1991, 311–351.

[Io] I.N. Iomdin: *Variétés complexes avec singularités de dimension un*, Sibirskii Mat. Zhurnal **15** (1974), 1061–1082.

[Lê-1] Lê Dũng Tráng: *Ensembles analytiques avec lieu singulier de dimension 1 (d'après Iomdine)*, Séminaire sur les singularités, Publ. Math. de l'Université Paris VII (1980), 87–95.

[Lê-2] Lê D.T.: *The geometry of the monodromy theorem*, C.P. Ramanujam – a tribute, Tata Institute, Springer-Verlag (1978).

[Lê-3] Lê D.T.: *Some remarks on the relative monodromy*, Real and Complex Singularities Oslo 1976, Sijhoff en Nordhoff, Alphen a.d. Rijn (1977), pp. 397–403.

[Lê-4]   Lê D.T.: *La monodromie n'a pas des points fixes*, J. Fac. Sc. Univ. Tokyo, sec.1A, **22** (1973), 409–427.

[Lê-5]   Lê D.T.: *Le théorème de la monodromie singulier*, C.R. Acad. Sc. , t. **288** (1979), 985–988.

[Lê-6]   Lê D.T.: *Complex Analytic Functions with Isolated Singularities* , J. Algebraic Geometry, **1** (1992), 83–100.

[Mas-1]   D.B. Massey: *The Lê Varieties, I*, Invent. Math. **99** (1990), 357–376.

[Mas-2]   D.B. Massey: *Lê Cycles and Hypersurface Singularities*, Lect. Notes in Math. **1615**, Springer Verlag1996.

[MS]   D.B. Massey; D. Siersma: *Deformations of polar methods*, Ann. Inst. Fourier **42** (1992), 737–778.

[Ma]   J. Mather: *Notes on topological stability*, Harvard University, (1970).

[Ne]   A. Némethi: *The zeta function of singularities*, J. Algebraic Geometry **2** (1993), 1–23.

[Sa-1]   M. Saito: *On Steenbrink's conjecture*, Math. Annalen **289** (1991), 703–716.

[Sa-2]   M. Saito: *Modules de Hodge polarisables*, Publ. RIMS Kyoto Univ. **24** (1988), 849–995.

[Sa-3]   M. Saito: *Mixed Hodge modules*, Publ. RIMS Kyoto Univ. **26** (1990), 221–333.

[Si]   D. Siersma: *The monodromy of a series of hypersurface singularities*, Comment. Math. Helvetici, **65** (1990), 181–197.

[St]   J. H. M. Steenbrink: *The spectrum of hypersurface singularities*, Astérisque **179-180** (1989), 163–184.

[Ti-1]   M. Tibăr: *Carrousel monodromy and Lefschetz number of singularities*, l'Enseignement Math. **37** (1993), 233–247.

[Ti-2]   M. Tibăr: *Bouquet decomposition of the Milnor fibre*, Topology **35**, 1 (1996), 227–241.

[Ti-3]   M. Tibăr: *A supplement to Iomdin-Lê theorem for singularities with one-dimensional singular locus*, in Singularities and Differential Equations, Banach Center Publications vol. **33** (1996), 411–419.

[Va]   J.-P. Vannier: *Familles à un paramètre de fonctions analytiques à lieu singulier de dimension un*, C.R. Acad. Sci. Paris, Ser. I, **303**, 8 (1986), 367–370.

[Ve]   J-L. Verdier: *Stratifications de Whitney et théorème de Bertini-Sard*, Invent. Math. **36** (1976), 295–312.

Chapter 2

Deformation Theory

Progress in Mathematics, Vol. 162, © 1998 Birkhäuser Verlag Basel/Switzerland

Discriminants and Vector Fields

Andrew A. du Plessis
Mathematisk Institut
Aarhus Universiteit
Ny Munkegade
800 Århus C
DENMARK

Charles T.C. Wall
Dept. of Math. Sci.
University of Liverpool
Liverpool L69 3BX
U. K.

It is just 20 years since the theory of the discriminant of a map began to be investigated as a topic in its own right. The year is clearly defined by important papers [2, 36, 29] from different authors. Numerous papers have appeared since then in which this topic plays an important role.

The theme of this paper is that the theory of the discriminant is now so well developed as to be a research tool in its own right, with applications to problems in which the role of the discriminant may not be apparent. Indeed, the discriminant gives detailed insight into – and usually determines the entire structure of – the map F; while on the other hand effective calculations of it may be made. Important varieties may be viewed as discriminants, ranging from the original example of the space of polynomials in one variable with a repeated root to the dual of an algebraic hypersurface. There are indeed a number of important papers from about 1850 on which may be seen in retrospect as using the concept of discriminant: we will not attempt to survey these here.

In this article we present a review of vector fields related to the discriminant from a geometrical viewpoint. Many authors have contributed to this theory, including Arnold, Bruce, Gaffney, Goryunov, Looijenga, K. Saito, Teissier, Zakalyukin, ... We will give below what appear to be the key references. However, as in the development of mathematics in other areas, many results have been discovered independently, or the original discovery was incomplete and the account superseded, so that precise attributions are difficult.

In Chapter 10 of our book [9] we used the discriminant matrix as a crucial ingredient of a computer algorithm, and presented the theory of the discriminant with this end in view. Although the presentation here is more geometrical, we also seek to emphasise the effective computability of the objects constructed.

1 Introduction

Suppose that F is a differentiable map $\mathbb{R}^n \to \mathbb{R}^p$ or, preferably, $\mathbb{C}^n \to \mathbb{C}^p$, with $n \geq p$: then the *critical set* $\Sigma F := \{\mathbf{x} \in \mathbb{C}^n \mid \text{rank } dF_x < p\}$, the set of points at which F is not submersive. The *discriminant* is its image $\Delta F := F(\Sigma F)$.

For arbitrary maps it is desirable to be more precise and define local sheaf structures on these sets, but under very mild conditions on F (which will be satisfied below) it suffices to regard them as reduced subspaces of $\mathbb{C}^n$ and $\mathbb{C}^p$ respectively. The terminology 'discriminant' is well established, but other terms are occasionally used (particularly in the real case) such as 'apparent contour' and 'profile'.

Suppose given tangent vector fields ξ on $\mathbb{C}^n$, η on $\mathbb{C}^p$ such that for each $\mathbf{x} \in \mathbb{C}^n$, $dF(\xi_x) = \eta_{F(x)}$. Then we say following Arnold [2] that the vector field η is *liftable*, ξ is *lowerable*, and ξ is a lift of η. The study of such vector fields is closely related to the theory of the discriminant, and they will provide a key to the geometry.

We begin with a very simple example, which will illustrate some of the main results without requiring hard computations. Consider the map

$$F(x, u) = (y, v) = (x^3 + xu, u).$$

The critical set is given by

$$0 = \begin{vmatrix} 3x^2 + u & 0 \\ x & 1 \end{vmatrix} = 3x^2 + u,$$

so we can take x as parameter, with $u = -3x^2$. On the discriminant we have $y = x^3 + xu = -2x^3$ and hence a parametrisation $(-2x^3, -3x^2)$ and an equation $27y^2 + 4v^3 = 0$. The map may be illustrated by the well known swallowtail diagram. This example, and its generalisation

$$F(x, u_1, \ldots, u_{n-1}) = (x^{n+1} + \sum_{1}^{n-1} u_i x^{n-i}, u_1, \ldots, u_{n-1}),$$

which may be treated similarly, were analysed in Arnold's paper [2]. The results were extended to the other simple singularities by Arnold [3]; see also Lyashko [20]. We proceed directly to a further generalisation. This seems to have been first stated in Zakalyukin [36] and Saito [28]; the details were completed in [4] and [37]; the paper [30] was also important.

Let $f_0 : \mathbb{C}^r \to \mathbb{C}$ have an isolated singular point at the origin. As is well known, this is equivalent to the condition that (f_0 is singular there and) the ideal $Jf_0 := \langle \partial f_0/\partial x_1, \ldots, \partial f_0/\partial x_r \rangle$ has finite codimension in the ring $\mathcal{O}_x$ of convergent power series in the x variables. Choose a basis $\phi_0 = 1, \phi_1, \ldots, \phi_{\mu-1}$ of the quotient vector space $\mathcal{O}_x/Jf_0$: it is traditional, and usually convenient, to choose the ϕ_i to be monomials. Now define the unfolding $F(\mathbf{x}, \mathbf{u}) = (y, \mathbf{v}) = (f(\mathbf{x}, \mathbf{u}), \mathbf{u})$, where $f(\mathbf{x}, \mathbf{u}) = f_0(\mathbf{x}) + \sum_{i=1}^{\mu-1} u_i \phi_i(\mathbf{x})$.

This is the usual construction of a versal unfolding of f_0, and is miniversal if f is weighted homogeneous: otherwise, it is not minimal, a point we will return to later. We will not discuss the theory of unfoldings here (it is well catered for in other survey articles – see e.g. [33]) but recall that the resulting map F has a stable germ at the origin, and hence – by openness of versality – has stable (multi-)germs on some neighbourhood of the origin, so on such a neighbourhood, F is *locally stable* in the sense of [9].

Now write $\mathcal{O}_{x,u}$ for the ring of convergent power series in the x and u variables, and $J_x f := \langle \partial f/\partial x_1, \dots, \partial f/\partial x_r \rangle$ for the indicated ideal in it. It follows from Nakayama's lemma that the ϕ_i form a free basis of the quotient $\mathcal{O}_u$-module $\mathcal{O}_{x,u}/J_x f$. Hence there exist uniquely determined elements $a_{i,j} \in \mathcal{O}_u$ such that

$$f\phi_i \equiv \sum_{j=0}^{\mu-1} a_{i,j}(\mathbf{u})\phi_j \mod J_x f \quad (0 \le i < \mu). \tag{1}$$

We then call $A = (a_{i,j}(\mathbf{v}) - y\,\delta_{i,j})$ the *discriminant matrix*. This is a key definition; for example, we will see that the equation of ΔF is given by $\det A = 0$.

We illustrate this with the above example. We take $f_0 = x^3$, $\phi_0 = 1$ and $\phi_1 = x$; this yields the map F of the example. The ideal $J_x f$ is generated by $3x^2 + u$ and we have

$$[(x^3 + xu)1 = 0.1 + (\tfrac{2}{3}u)x + \tfrac{1}{3}x(3x^2 + u),$$
$$[(x^3 + xu)x = (-\tfrac{2}{9}u^2).1 + 0.x + (\tfrac{1}{3}x^2 + \tfrac{2}{9}u)(3x^2 + u),$$

so the discriminant matrix is $\begin{pmatrix} -y & \tfrac{2}{3}v \\ -\tfrac{2}{9}v^2 & -y \end{pmatrix}$, and its determinant is $y^2 + \tfrac{4}{27}v^3$ — which does indeed give the correct equation.

However, the main importance of the discriminant matrix is in its relation to vector fields. Replacing the congruence (1) by an equation gives

$$f(\mathbf{x}, \mathbf{u})\phi_i \equiv \sum_{j=0}^{\mu-1} a_{i,j}(\mathbf{u})\phi_j + \sum_{k=1}^{r} b_{i,k}(\mathbf{x}, \mathbf{u})\frac{\partial f}{\partial x_k}. \tag{2}$$

This leads us to define the vector field ξ_i on $\mathbb{C}^n$ by

$$\xi_i = \sum_{k=1}^{r} b_{i,k}(\mathbf{x}, \mathbf{u})\frac{\partial}{\partial x_k} + \sum_{j=1}^{\mu-1}(a_{i,j}(\mathbf{u}) - f(\mathbf{x}, \mathbf{u})\delta_{i,j})\frac{\partial}{\partial u_j}.$$

Applying ξ_i to the components of F yields

$$\xi_i(f) = \sum_k b_{i,k}\frac{\partial f}{\partial x_k} + \sum_{j>0}(a_{i,j} - f\delta_{i,j})\phi_j = -(a_{i,0} - f\delta_{i,0}),$$

using (2), whereas

$$\xi_i(u_l) = \sum_j (a_{i,j} - f\delta_{i,j})\delta_{j,l} = a_{i,l} - f\delta_{i,l}.$$

Thus the vector field ξ_i lifts the vector field η_i defined on $\mathbb{C}^\mu$ by

$$\eta_i = \sum_{j=0}^{\mu-1} (a_{i,j}(\mathbf{v}) - y\delta_{i,j})\frac{\partial}{\partial v_j},$$

where $\partial/\partial v_0$ denotes $-\partial/\partial y$. These vector fields give real insight into the geometry: the following results hold, and will be discussed more fully below in a more general context.

- The vector fields η_i are tangent to ΔF.

- They form a free basis of the $\mathcal{O}_{u,y}$-module of vector fields tangent to ΔF.

- A vector field is tangent to ΔF if and only if it is liftable under F.

In the case where f_0 is weighted homogeneous, we can give a rather explicit procedure for finding the discriminant matrix. If the coordinates x_i have weights w_i and f_0 has weight 1, the *Euler vector field* $\epsilon_x = \sum w_i x_i \partial/\partial x_i$ satisfies $\epsilon_x f_0 = f_0$. We make f homogeneous by defining wt $u_i =$ wt $f_0 -$ wt ϕ_i (note that although in cases of interest the x coordinates have strictly positive weight, this is no longer the case for the u_j), and set $\epsilon_u = \sum$ wt $u_j \, u_j \partial/\partial u_j$ and $\epsilon_{x,u} = \epsilon_x + \epsilon_u$ (we will use corresponding notation for target coordinates also). Then

$$f = \epsilon_{x,u}f = \epsilon_x f + \sum \text{wt } u_j \, u_j \, \phi_j,$$

thus giving the first row of the discriminant matrix.

Choose a basis of unfolding monomials, such that each monomial on the list is obtained from one of lower degree by multiplying by a coordinate: say $\phi_r = x_i\phi_s$. Suppose the sth row already constructed, with an appropriate identity (2) (the induction starts with the above construction of the first row). Multiply through by x_i. Not all the products $x_i\phi_j$ will be unfolding monomials, but there exist relations

$$x_i\phi_j = \sum a_{i,j}^k \phi_k + \sum b_{i,j}^l \frac{\partial f_0}{\partial x_l}$$

(usually only 3 or 4 terms are non-zero) with each $a_{i,j}^k \in \mathbb{C}$ and $b_{i,j}^l \in \mathcal{O}_x$. Now replacing the term $x_i\phi_j$ by $\sum a_{i,j}^k \phi_k + \sum b_{i,j}^l \frac{\partial f}{\partial x_l}$ introduces an error of lower weight than the term in question. The procedure thus converges after a few iterations.

It is quite feasible to implement this procedure on a computer, but the size of the calculation increases rapidly with μ, and even more so if f_0 depends on auxiliary parameters. A number of devices to speed up the computation are described in [9, Chapter 10.4]. The procedure is illustrated in [9, Proposition 10.5.35].

2　General theory of the discriminant

In the above, we started with a single function f_0 of r variables and obtained F by versally unfolding. The theory of the discriminant, however, applies much more generally.

We first introduce a more systematic notation. For a map (or map-germ) F, its source will usually be denoted N and its target P, with respective dimensions n and p: we suppose throughout that $n \geq p$. In fact we will mainly deal with map-germs, but one need not think of results as valid only in a very small neighbourhood: indeed in the algebraic case they will apply on a Zariski-open, hence dense subset. The ring of germs of functions in the source will be denoted $\mathcal{O}_N$ and in the target $\mathcal{O}_P$, with respective maximal ideals $\mathfrak{m}_N, \mathfrak{m}_P$ corresponding to the source and target of the germ. The germs of vector fields in N and P define modules θ_N and θ_P; whereas the vector fields along F form the $\mathcal{O}_N$-module $\theta(F) = F^*\theta_P$.

With respect to a choice of local coordinates $\mathbf{x} = \{x_i \mid 1 \leq i \leq n\}$ in N and $\mathbf{y} = \{y_j \mid 1 \leq j \leq p\}$ in P, the $\partial/\partial x_i$ form a free $\mathcal{O}_N$-basis of θ_N and the $\partial/\partial y_j$ a free $\mathcal{O}_N$-basis of $\theta(F)$ and a free $\mathcal{O}_P$-basis of θ_P. It is very often convenient to identify θ_N, $\theta(F)$ and θ_P with $\mathcal{O}_N^n$, $\mathcal{O}_N^p$ and $\mathcal{O}_P^p$ respectively via such a choice.

We recall that the *critical set* ΣF is the set of points $\mathbf{x}$ in the source with $dF_{\mathbf{x}}$ not surjective, and that the *discriminant* is $\Delta F = F(\Sigma F)$. For a point $\mathbf{y}$ in the target, we set $\Sigma(F, \mathbf{y}) := \Sigma F \cap F^{-1}(\mathbf{y})$.

The above is fairly standard, as are Mather's notations

$$tF : \theta_N \longrightarrow \theta(F), \quad \omega F : \theta_P \longrightarrow \theta(F)$$

for the maps induced by F. We also write

$$\begin{aligned}
M(F) &:= \theta(F)/tF(\theta_N), \\
N(F) &:= \theta(F)/tF(\theta_N) + F^*\mathfrak{m}_P.(\theta_F).
\end{aligned}$$

Other notation is standard from a different viewpoint, e.g. sheaf theory or commutative algebra. Here $N(F)$ is usually denoted by T_F^1 or some variant thereof; instead of vector fields one considers the dual modules of differential forms, and the map $F^*\Omega_P^1 \to \Omega_N^1$ dual to tF, with cokernel Ω_F^1, so that $M(F) = \mathrm{Ext}_{\mathcal{O}_x}(\Omega_F^1, \mathcal{O}_x)$.

Although we will be mainly interested in germs which are either stable or close to being so, a number of important results are known in much greater generality. The hypothesis that will be in force throughout is that F has finite singularity type. For the case of germs, this is the condition that $N(F)$ has finite dimension, or equivalently, that the fibre $F^{-1}(\mathbf{y})$ has an isolated singularity. Globally, the condition implies that the restriction of F to its critical set ΣF is proper and finite (precise conditions are discussed in [9, Section 2.4]. It is proved in [15] that (under weaker assumptions) it follows that ΣF is Cohen-Macaulay.

The map (or germ) F is said to be *Thom transversal* (or TT for short) if its 1-jet $j^1 F$ is transverse to the stratification of 1-jet space by the rank of the first derivative (i.e. by the Thom manifolds Σ^k). The following are proved by [18], Chapter 4 for such maps (for stable maps they appear already in [29]): ΣF is reduced and normal; ΔF is irreducible; and F gives a normalisation $\Sigma F \to \Delta F$: this also follows from the (earlier) above result of [15].

Using the uniqueness of normalisation, it was shown in [35] that a stable map-germ is determined by its discriminant up to right equivalence. A far reaching generalisation of this was announced in part in [13] and (although preprints existed back in 1982) finally appeared as [8]: we confine ourselves here to the complex case. Then F is called a *critical normalisation* (CN for short) if ΣF is normal, and $F|\Sigma F$ is finite and generically injective. The normality is equivalent to F being TT off a codimension 2 subset of ΣF; generic injectivity is equivalent to a generic multijet transversality condition, so these hypotheses are very weak. The main theorem (in the complex case) states that two CN maps with the same discriminant are right equivalent. In these references the restriction $n \geq p$ is not made. Indeed, the cases $n \leq p$ are already to be found in Gaffney [11]: for the case $n = p$ (where the above definition of CN requires strengthening), this was in part inspired by [34].

A general (very tentative) principle states that (under reasonable hypotheses) two maps with homeomorphic discriminants are topologically equivalent. In some cases results of this nature may be deduced from those in our book [9]; further support is given by [12].

A second general principle which holds in great generality is that a vector field on P is liftable if and only if it is tangent to the discriminant ΔF.

If ξ lifts η, then integrating ξ and η gives local flows in N and P compatible with F, hence preserving the local geometry. It thus follows that η is tangent to ΔF.

The converse statement is due in the equidimensional case to Arnold [2]: lifts can be constructed locally at points where F is submersive or has a simple fold, and are then unique: since the complement of such points has codimension ≥ 2, the result follows by Hartogs' theorem. Results at various levels of generality have been obtained by a number of authors. It is shown by Looijenga [18], 6.13 that the conclusion holds if F is TT. The ultimate result is due to Bruce et al. [5], who give a necessary and sufficient condition for all vector fields tangent to ΔF to lift.

Now consider $M(F) := \theta(F)/tF(\theta_N)$. Although this is a module over $\mathcal{O}_N$, we next consider it (*via* F^*) as a module over $\mathcal{O}_P$: geometrically, this step corresponds to the mapping F from the critical set $\Sigma(F)$ (the support in N of the sheaf corresponding to $M(F)$) to the discriminant $\Delta(F)$ (its support in P).

A key result, due to Teissier [29] in part, and Greuel [16] in general, states that $M(F)$ has projective dimension 1 as $\mathcal{O}_P$-module, and we have a free presentation

$$0 \longrightarrow \mathcal{O}_P^\tau \xrightarrow{\ \alpha F\ } \mathcal{O}_P^\tau \xrightarrow{\ \overline{\phi F}\ } M(F) \longrightarrow 0; \tag{3}$$

here $\tau = \dim_{\mathbb{C}} M(F)/\mathfrak{m}_P M(F)$ is the Tjurina number of F, and $\overline{\phi F}$ is given by $\overline{\phi F}(a_1, \ldots, a_\tau) = \sum_{i=1}^\tau a_i \phi_i$, where $\phi_1, \ldots, \phi_\tau \in \theta(F)$ project to a $\mathbb{C}$-basis for $M(F)/\mathfrak{m}_P M(F)$, so, by Nakayama's lemma, to an $\mathcal{O}_P$-basis for $M(F)$.

The argument is a beautiful application of commutative algebra: the essential point is that $M(F)$ is a Cohen-Macaulay module over the relevant rings. We will not repeat it here. The argument is described briefly [9, Section 10.5.1] in the terminology used here; other accounts can be found in [18], [8].

Now consider the map $\sigma F = tF - \omega F : \theta_N \oplus \theta_P \to \theta(F)$. The definition above of liftability may be reformulated: ξ lifts η if and only if $tF(\xi) = \omega F(\eta)$, so if and only if $(\xi, \eta) \in \mathrm{Ker}\,\sigma F$. We shall suppose for now that σF is surjective. We recall that Mather [21] termed such map-germs F 'infinitesimally stable'; the condition is equivalent to a more geometrical notion of stability, as discussed in [9, p.28], so we shall just call them stable.

When σF is surjective, ωF defines a surjective map $\overline{\omega F}$ from $\theta_P \equiv \mathcal{O}_P^p$ to $M(F)$. The arguments just discussed above show that we have a free presentation

$$0 \longrightarrow \mathcal{O}_\Gamma^p \xrightarrow{\ \alpha F\ } \mathcal{O}_P^p \xrightarrow{\ \overline{\omega F}\ } M(F) \longrightarrow 0;$$

for $\overline{\omega F}$ is $\overline{\phi F}$ for $\phi_i = (0 \ldots, i, \ldots, 0)$, the i'th standard basis vector. (The ϕ_i's project to a spanning set, not necessarily a basis, for $M(F)/\mathfrak{m}_P M(F)$, but it is easy to see that this does not affect the conclusion.)

As in [9], we embed the above sequence in a diagram of short exact sequences

$$
\begin{array}{ccccccccc}
& & 0 & & 0 & & 0 & & \\
& & \downarrow & & \downarrow & & \downarrow & & \\
0 & \longrightarrow & D_0 F & \longrightarrow & \theta_N & \longrightarrow & \theta_N/D_0 F & \longrightarrow & 0 \\
& & \downarrow & & \downarrow & & \downarrow & & \\
0 & \longrightarrow & \mathrm{Ker}\,\sigma F & \longrightarrow & \theta_N \oplus \theta_P & \xrightarrow{\ \sigma F\ } & \theta(F) & \longrightarrow & 0 \\
& & \downarrow & & \downarrow & & \downarrow & & \\
0 & \longrightarrow & \mathrm{Ker}\,\overline{\omega F} & \longrightarrow & \theta_P & \xrightarrow{\ \overline{\omega F}\ } & M(F) & \longrightarrow & 0 \\
& & \downarrow & & \downarrow & & \downarrow & & \\
& & 0 & & 0 & & 0 & &
\end{array}
$$

where D_0F denotes the kernel of $tF : \theta_N \to \theta(F)$. It follows that the $\mathcal{O}_P$-module of liftable vector fields is identified with $\mathrm{Ker}\ \overline{\omega F}$, and hence with the free module $\mathcal{O}_P^p$.

We have seen that a vector field is liftable if and only if it is tangent to the discriminant. A hypersurface D such that the module Der log D of vector fields tangent to D is a free module is called by Saito [28] a *free divisor*. In this case, if a free basis is expressed as $\sum a_{i,j}\partial/\partial y_j$, then $det\,(a_{i,j}) = 0$ is an equation for D. It now follows that this applies to the discriminant of any stable map.

3   Construction of discriminant matrices and vector fields

We next present a generalization of the construction of Section 1.

Let $f : \mathbb{C}^n \to \mathbb{C}^p$ (with $n > p$) be a map-germ of finite singularity type, i.e. $N(f)$ has finite dimension $\tau = \tau(f)$, the *Tjurina number* of f. Choose a set of elements $\{\phi_j \mid 1 \leq j \leq \tau\}$ of $\theta(f)$ projecting to a basis of $N(f)$. We may suppose that df vanishes at the origin, and then choose $\phi_j = \partial/\partial y_j$ for $j \leq p$. We use the remaining ϕ_j to construct an unfolding $F : N \to P$ of f; as before by writing $F(\mathbf{x}, \mathbf{u}) = (\mathbf{y}, \mathbf{v}) = (F_1(\mathbf{x}, \mathbf{u}), \mathbf{u})$, where $F_1(\mathbf{x}, \mathbf{u}) = f(\mathbf{x}) + \sum_{i=1}^{\tau} u_i\phi_i(\mathbf{x})$. This is the standard construction for a stable map-germ. We have $N(F) \cong N(f)$; the $\partial/\partial y_j$ for $1 \leq j \leq \tau$ project to a basis of it, and dim $P = \tau$ is the least possible target dimension for a stable germ in the given extended contact class. Such a germ F we may call *ministable*. The formulation of the theory of discriminants is simplest for such maps.

However here we must be yet more precise. The discriminant Δf is a hypersurface (codimension 1) in the target, which we locally identify with $\mathbb{C}^p$. Choose a linear projection $\pi : \mathbb{C}^p \to \mathbb{C}^{p-1}$ whose restriction to Δf is a finite map-germ. We choose local coordinates such that this projection is defined by the first coordinate y_1, and write $\mathbf{y}'$ collectively for the remaining y coordinates. Then $M(f)$ is a free module over $\mathcal{O}_{y'}$: this follows by the same arguments as before (we have a Cohen-Macaulay module of depth zero).

Choose a free basis of this module, given by the classes $\overline{\psi_j}$ in $M(f)$ of $\psi_j \in \theta(f)$: we may suppose that the ϕ_i occur among the ψ_j, and in particular that for $j \leq p$, $\psi_j = \phi_j$ is the unit vector in the jth coordinate direction. Using the ψ_j instead of the ϕ_i we construct an unfolding $\tilde{F}$ of f and indeed of F: since F is stable, this is a trivial unfolding of F. To avoid overburdening notation, we will still write u, v for the unfolding variables in the present situation. Since the ψ_j give a free base of $M(\tilde{F})$ as $\mathcal{O}_{y',v}$-module, we can imitate the procedure of Section 1 by considering multiplication by the first component g of $\tilde{F}$ with respect to this basis. This yields congruences

$$y_1\psi_i \equiv \sum_{j \geq 1} a_{i,j}(\mathbf{y}', \mathbf{u})\psi_j \quad \mathrm{mod\ Im}\ t\tilde{F} \tag{4}$$

and equations

$$g(\mathbf{x}, \mathbf{u})\psi_i(\mathbf{x}) = \sum_{j \geq 1} a_{i,j}(\tilde{F}_1'(\mathbf{x}, \mathbf{u}), \mathbf{u})\psi_j(\mathbf{x}) + \sum_{k=1}^{n} b_{i,k}(\mathbf{x}, \mathbf{u})\frac{\partial \tilde{F}_1}{\partial x_k}. \tag{5}$$

leading to the vector fields given by

$$\xi_i = \sum_{k=1}^{n} b_{i,k}(\mathbf{x}, \mathbf{u})\frac{\partial}{\partial x_k} + \sum_{j \geq p}(a_{i,j}(\tilde{F}_1(\mathbf{x}, \mathbf{u}), \mathbf{u}) - g(\mathbf{x}, \mathbf{u})\delta_{i,j})\frac{\partial}{\partial u_j},$$

$$\eta_i = \sum_{j \geq 1}(a_{i,j}(\mathbf{y}', \mathbf{v}) - y_1\delta_{i,j})\frac{\partial}{\partial v_j},$$

(where $\partial/\partial v_i$ denotes $-\partial/\partial y_i$ for each $i \leq p$) such that ξ_i lifts η_i. As before, it can be seen that the η_i form a free basis of the module of liftable vector fields on the target.

We can, however, re-express this in terms of the original unfolding F. This construction is essentially due to Goryunov [14]. The fact that we have a trivial unfolding of a stable germ now reveals itself in the fact that there are liftable vector fields that do not vanish at the origin. From a computational point of view, we observe that multiplication by y_1 on $M(F)$ becomes nilpotent when reduced modulo $\mathfrak{m}_{y',v}$. We may thus bring the reduction modulo $\mathfrak{m}_{y',v}$ of the matrix of this multiplication to Jordan normal form.

Equivalently (see [9, p.452]) choose the above basis in the form $\{\overline{\psi_{i,r}} \mid 1 \leq r \leq e_i\}$ such that $y_1\overline{\psi_{i,r}} = \overline{\psi_{i,r+1}}$ for $r < e_i$ and the $\psi_{i,1} = \phi_i$ form a basis of $N(F)$. We may then set $y_1^{e_i}\phi_i \equiv \sum_{j,s} b_{i,j,s}y_1^{s-1}\phi_j$, where the $b_{i,j,s}$ are functions of $(\mathbf{y}', \mathbf{v})$. Then the matrix with entries $D_{i,j} = y_1^{e_i}\delta_{i,j} - \sum_{r=1}^{e_j} b_{i,j,r}y_1^{r-1}$ is a discriminant matrix for F.

This construction is the basis of many of the calculations in [9, Chapter 11].

4   Vector fields on the target of stable maps

If a map has a stable germ at some points, it also has stable germs at nearby points, but to build effectively on this it is essential to consider points in the target.

We recall that for any germ F of finite singularity type, the critical set ΣF has dimension $p - 1$, and the restriction to it of F defines a finite map. Thus for any $\mathbf{y} \in P$, the set $\Sigma(F, \mathbf{y}) = \Sigma F \cap F^{-1}(\mathbf{y})$ is finite.

A basic result of Mather states that for stable germs or multi-germs, $\mathcal{K}$- and $\mathcal{A}$-equivalence are the same. In the complex-analytic context, this can be stated as follows. The class of $F^{-1}(\mathbf{y})$ (in a neighbourhood of $\Sigma(F, \mathbf{y})$) up to complex-analytic equivalence determines the germ of F at $\Sigma(F, \mathbf{y})$ up to local coordinate change. In the case when $\Sigma(F, \mathbf{y})$ consists of a single point, this tells us that we can find local coordinates to put F in a convenient normal form, such as that given in a special case in the introductory section and generalised above. In the case when there are several points, we can combine the corresponding normal forms, as follows.

Suppose that $\Sigma(F, \mathbf{y}) = \{\mathbf{x}_1, \dots, \mathbf{x}_r\}$, and that the several singularities at the $\mathbf{x}_i$ correspond to normal forms $F_i(\mathbf{u}_i) = \mathbf{v}_i$, with F_i ministable. Then we can choose local coordinates in N and $(\mathbf{v}_0, \mathbf{v}_1, \dots, \mathbf{v}_r)$ in P such that near $\mathbf{x}_i$, F is given by the normal form

$$(\mathbf{v}_0, \mathbf{v}_1, \dots, \mathbf{u}_i, \dots, \mathbf{v}_r) \mapsto (\mathbf{v}_0, \mathbf{v}_1, \dots, F_i(\mathbf{u}_i), \dots, \mathbf{v}_r).$$

In such a coordinate system, a point belongs to the discriminant of F if and only if, for some i, its coordinate $\mathbf{v}_i$ belongs to the discriminant of F_i. Thus ΔF consists of several components meeting transversely. Moreover, a vector field of the form $(\eta_0, \cdots, \eta_r)$ is liftable if and only if η_i is liftable under F_i for $1 \leq i \leq r$.

We define two points $\mathbf{y}$ and $\mathbf{y}'$ in P to be *equivalent* if the germs of F at the sets $\Sigma(F, \mathbf{y})$ and $\Sigma(F, \mathbf{y}')$ are $\mathcal{K}$- (and hence $\mathcal{A}$-) equivalent). The equivalence classes form a partition of P which is, in fact, a foliated stratification. The parts of this partition are called *strict presentations of K-classes* in [9]; here we shall refer to them as *leaves*. The discriminant ΔF is the complement of the leaf where $\Sigma(F, \mathbf{y})$ is empty.

Proposition 4.1 *(i) The leaf containing $\mathbf{x}$ is smooth there, with tangent space given by the values at $\mathbf{x}$ of the liftable vector fields.*
(ii) The codimension of this leaf equals the sum of the Tjurina numbers of the singularities of F at the points of $\Sigma(F, \mathbf{x})$.

Proof. Part (i) follows from the remarks earlier in this section: we may choose local coordinates in which the map is in the above normal form, and in this case if the vector field $(\eta_1, \cdots, \eta_r)$ is liftable, then for $i > 0$ as η_i is liftable under F_i it vanishes at the origin. On the other hand, the leaf can be locally identified with the submanifold where all coordinates save $\mathbf{v}_0$ are zero.

(ii) We see from the discussion preceding the lemma that the codimension of the leaf is equal to the sum of the codimensions of the leaves corresponding to the germs F_i, so we may suppose F has only one singular point in the fibre.

We recall that the module of liftable vector fields can be identified with the kernel of $\overline{\omega F} : \theta_P \to M(F)$. By definition, τ is the dimension of $N(F) = M(F)/\mathfrak{m}_y.M(F)$. Hence the induced map $\mathbb{C}^p \cong \theta_P/\mathfrak{m}_y.\theta_P \to \mathfrak{m}_y.M(F)$ has kernel of dimension $p - \tau$. But this coincides with the evaluation at the point (the support of $\mathfrak{m}_p$) of the kernel, the liftable vector fields. Hence by (i) the dimension of the leaf is $p - \tau$, and its codimension is τ. $\qquad\square$

It follows from (i) that this partition coincides with what Saito [28] terms the 'logarithmic stratification' and Bruce and Roberts [6] term the 'holonomic stratification'.

5 Vector fields in the source

We now turn to vector fields in the source. We have seen that (under the hypothesis $n \geq p$) lowerings of vector fields are unique, and that the indeterminacy of liftings is not, and is given by

$$D_0 F = \operatorname{Ker} tF = \{\xi \; : \; tF(\xi) = 0\}.$$

We may also interpret $D_0 F$ as the module of vector fields tangent to all fibres of F. The structure of this module may be described as follows.

For each sequence $I \; : \; 1 \leq i_0 \leq \ldots \leq i_p \leq n$ we define the *Hamiltonian* vector field δ_I by $\delta_I = \sum_{j=1}^{n} D(I, j)\partial/\partial x_j$, where $D(I, j) = 0$ if $j \notin I$ and

$$D(I, i_k) = (-1)^{k-1} \left| \left(\frac{\partial f_r}{\partial x_{i_s}} : 1 \leq r \leq p, \; 0 \leq s \leq p, \; s \neq k \right) \right|.$$

Alternatively, we have

$$\delta_I(g) := \frac{\partial(g, f_1, \ldots, f_p)}{\partial(x_{i_0}, \ldots, x_{i_p})}.$$

Proposition 5.1 *The vector fields δ_I generate $D_0 F$ as $\mathcal{O}_N$-module.*

Proof. This follows from exactness of the Eagon-Northcott complex (see e.g. [25]), as observed in [9, p.422]. Indeed, provided only that $\dim \Sigma F = p - 1$, this sequence ([9, p.459]) shows that $\operatorname{Ker} tF$ is generated by the image of $\wedge^{p+1}G \otimes \tilde{R}$. Writing out the explicit form of this map, we see that the natural generators of $\wedge^{p+1}G$ map to the Hamiltonian vector fields. $\qquad\square$

Indeed, the exactness of this complex yields a projective resolution of $D_0 F$ over $\mathcal{O}_N$.

If we write $D_L F$ for the space of lowerable vector fields, then lowering gives an isomorphism of the quotient $D_L F / D_0 F$ onto the space of liftable vector fields, which we recall is in turn a free module over $\mathcal{O}_P$. It is, however, more convenient to study $\mathcal{O}_N$-modules, and we introduce

$$\begin{aligned} DF &= \{\xi \; : \; tF(\xi) \in F^*\mathfrak{m}_P.\theta(F)\}, \\ D_1 F &= D_0 F + F^*\mathfrak{m}_P.\theta_N; \end{aligned}$$

we have $D_0 F \subset D_1 F \subset DF$. While $D_0 F$ consists of the vector fields tangent to all fibres of F; DF consists of those tangent to the fibre over the base point. The relation of these to $D_L F$ will be discussed in the next section.

We now need some further notation. The *Jacobian ideal $J(F)$* is the ideal in $\mathcal{O}_N$ generated by the maximal minors of the Jacobian matrix: thus its support is ΣF. We have the local rings

$$Q(F) := \mathcal{O}_N / F^* \mathfrak{m}_P . \mathcal{O}_N, \quad \overline{Q(F)} := \mathcal{O}_N / J(F) + F^* \mathfrak{m}_P . \mathcal{O}_N.$$

Recall that the dimension of $N(F)$ is the Tjurina number τ: we denote by τ' the dimension of $\overline{Q(F)}$. There is also the Milnor number μ, defined as the Betti number of the Milnor fibre: for complete intersections there is no simple formula for this, though it can be expressed [15] as an alternating sum using partial maps of F. If F is weighted homogeneous, then [16] we have $\mu = \tau = \tau'$; the result $\mu = \tau$ is very useful and we use it below without further reference.

It was shown in [16] that $\tau' \leq \mu$, and that for curve singularities, $\tau \leq \mu$. That $\tau \leq \mu$ in general was shown in [19]. Also, $\tau = \tau'$ for trivial reasons if F unfolds a function germ, and (less trivially) if the fibre dimension $n - p = 1$; but not in general, as is shown by the example, pointed out to us by Vosegaard, of the cusp singularity $T_{2,2,2,3}$. Other expressions involving these numbers are given in [9, Proposition 10.5.3].

We have:

Proposition 5.2 *Let $\phi_1 \ldots \phi_\tau$ project to a basis for $N(F)$. Then there is a natural isomorphism*

$$\lambda \; : \; \frac{\operatorname{Ker} \overline{\phi F}}{\mathfrak{m}_P \operatorname{Ker} \overline{\phi F}} \longrightarrow \frac{DF}{D_1 F}.$$

In particular,

$$\dim_{\mathbb{C}} \frac{DF}{D_1 F} = \tau.$$

Proof. We define λ as follows: if $(\alpha_1 \ldots \alpha_\tau) \in \operatorname{Ker} \overline{\phi F} \subset \mathcal{O}_P^p$, then there exists a vector field ξ such that $tF(\xi) = \sum_{i=1}^{\tau} F^* \alpha_i \phi_i$. Since $\phi_1, \ldots, \phi_\tau$ project to a basis for $N(F)$, $\alpha_1, \ldots, \alpha_\tau$ must be contained in $\mathfrak{m}_P$, so $\xi \in DF$; we set $\lambda([(\alpha_1, \ldots, \alpha_\tau)]) = [\xi]$. It is easy to see that λ is a well-defined vector-space homomorphism.

We show first that λ is surjective. Let $\xi \in DF$. Then there exist $\beta_1, \ldots, \beta_p \in \theta(F)$ with $tF(\xi) = \sum_{i=1}^{p} F_i \beta_i$. Since $\phi_1, \ldots, \phi_\tau$ project to an $\mathcal{O}_P$-basis for $M(F)$, we can find, for $i = 1, \ldots, p$, $z_{ij} \in \mathcal{O}_P$, $j = 1, \ldots, \tau$, and $\psi_i \in \theta_N$ such that $\beta_i = \sum_{j=1}^{\tau} F^* z_{ij} \phi_j + tF(\psi_i)$. Thus, writing $\alpha_j = \sum_{i=1}^{p} y_i z_{ij} \in \mathfrak{m}_P$ for $j = 1, \ldots, \tau$, we see that $tF(\xi - \sum_{i=1}^{p} F_i \psi_i) = \sum_{j=1}^{\tau} F^* \alpha_j \phi_j$. Hence $(\alpha_1, \ldots, \alpha_\tau) \in \operatorname{Ker} \overline{\phi F}$, and $\lambda([(\alpha_1, \ldots, \alpha_\tau)]) = [\xi - \sum_{i=1}^{p} F_i \psi_i] = [\xi]$. Thus λ is indeed surjective.

Now observe that it follows from the exact sequence (3) of Section 2 that

$$\dim_{\mathbb{C}} \frac{\operatorname{Ker} \overline{\phi F}}{\mathfrak{m}_P \operatorname{Ker} \overline{\phi F}} = \tau;$$

thus to complete the proof, we need only show that

$$\dim_{\mathbb{C}} \frac{DF}{D_1 F} \geq \tau.$$

Choose $g \in \mathcal{O}_N$ such that $\Phi = (F, g) : N \to P \times \mathbb{C}$ has finite singularity type (this is a transversality condition off the base point, so holds in general). Applying vector fields to g induces a map of $DF/D_1 F$ onto

$$(DF(g) + F^* \mathfrak{m}_P.\mathcal{O}_N)/(D_1 F(g) + F^* \mathfrak{m}_P.\mathcal{O}_N).$$

It will thus suffice to show that

$$\dim \frac{\mathcal{O}_N}{D_1 F(g) + F^* \mathfrak{m}_P.\mathcal{O}_N} - \dim \frac{\mathcal{O}_N}{DF(g) + F^* \mathfrak{m}_P.\mathcal{O}_N} \geq \tau.$$

Now if ξ is a Hamiltonian vector field, $\xi(g)$ is a maximal minor in the Jacobian matrix of Φ. Since the Hamiltonian vector fields, together with $F^* \mathfrak{m}_P.\theta_N$, span $D_1 F$ we have $D_1 F(g) \subset J(\Phi) + F^* \mathfrak{m}_P.\mathcal{O}_N$. For the other term we have the following.

Lemma 5.3 *There is an exact sequence*

$$0 \longrightarrow \frac{\mathcal{O}_N}{DF(g) + F^* \mathfrak{m}_P.\mathcal{O}_N} \longrightarrow \frac{\theta(\Phi)}{t\Phi(\theta_N) + F^* \mathfrak{m}_P.\theta(\Phi)}$$

$$\longrightarrow \frac{\theta(F)}{tF(\theta_N) + F^* \mathfrak{m}_P.\theta(F)} \longrightarrow 0.$$

We are indebted to Mond and Montaldi for this result (private communication; see also [23]).

Proof (of Lemma). The penultimate map is induced by the projection of $\theta(\Phi) \cong \mathcal{O}_N^{p+1}$ on to $\theta(F) \cong \mathcal{O}_N^p$; it is easily seen to be well defined, and is clearly surjective. Since $t\Phi(\theta_N)$ projects onto $tF(\theta_N)$, any element of the kernel is represented by a vector parallel to $\mathbf{0} \times \mathbb{C}$ in $N \times \mathbb{C}$: say $(0, h)$ with $h \in \mathcal{O}_N$. Suppose this vector belongs to $t\Phi(\theta_N) + F^* \mathfrak{m}_P.\theta(\Phi)$; let the vector in θ_N be ξ. Equating the first p coordinates, we see that $tF(\xi) \in F^* \mathfrak{m}_P.\theta(F)$, so that $\xi \in DF$; equating the last now gives $\xi g - h \in F^* \mathfrak{m}_P.\mathcal{O}_N$, which proves the lemma. $\square$

It follows from the lemma that

$$\dim \frac{\mathcal{O}_N}{DF(g) + F^* \mathfrak{m}_P.\mathcal{O}_N} = \dim \frac{\theta(\Phi)}{t\Phi(\theta_N) + F^* \mathfrak{m}_P.\theta(\Phi)}$$

$$- \dim \frac{\theta(F)}{tF(\theta_N) + F^* \mathfrak{m}_P.\theta(F)}.$$

The last term here is just the dimension τ of $N(F)$. The theorem thus follows from the equality, proved by Greuel in [16],

$$\dim \frac{\mathcal{O}_N}{J(\Phi) + F^*(\mathfrak{m}_P).\mathcal{O}_N} = \dim \frac{\theta(\Phi)}{t\Phi(\theta_N) + F^*\mathfrak{m}_P.\theta(\Phi)}$$

which is also proved (in this notation) in [9, Proposition 10.5.3]. $\qquad\qquad\square$

If F is weighted homogeneous, there are induced actions of $\mathbb{C}^*$ on all the above spaces and modules, with which the maps are compatible. Further, the tangent spaces to the $\mathbb{C}^*$ actions on source and target define vector fields ϵ_N and ϵ_P such that the former lifts the latter.

Theorem 5.4 *If F is weighted homogeneous, the Euler vector field ϵ_x generates DF/D_1F as an $\mathcal{O}_N$-module, and its annihilator is $J(F) + F^*\mathfrak{m}_P.\mathcal{O}_N$.*

This theorem was stated in [32]; proofs appeared in [17] (excluding the case of curves) and [1]. As we find these difficult to follow, we offer a further proof here.

Proof. By [9, Lemma 10.5.6], there exists a weighted homogeneous function g such that $\Phi = (F, g)$ has finite singularity type. We define a map

$$\frac{\mathcal{O}_N.\epsilon_N + D_1F}{D_1F} \longrightarrow \frac{J(\Phi) + \Phi^*\mathfrak{m}_{P\times\mathbb{C}}.\mathcal{O}_N}{J(\Phi) + F^*\mathfrak{m}_P.\mathcal{O}_N}$$

as induced by $\xi \mapsto \xi g$. To check that this is well defined note first that D_0F is generated by the Hamiltonian vector fields δ_I, and that $\delta_I g$ is the determinant of a $(p+1)^2$ minor of the Jacobian matrix of Φ, hence belongs to $J(\Phi)$. Next, if $\xi \in F^*\mathfrak{m}_P.\theta_N$, we have $\xi g \in F^*\mathfrak{m}_P.\mathcal{O}_N$. Finally, taking $\xi = \epsilon_N$ gives $\epsilon_N g = \mathrm{wt}(g)g$, which belongs to the numerator, and generates the quotient, so the map is well defined and surjective.

Now $\mathcal{O}_N/\{J(\Phi) + \Phi^*\mathfrak{m}_{P\times\mathbb{C}}.\mathcal{O}_N\}$ has dimension $\tau(\Phi)$, and we know by [9, Proposition 10.5.3] that $\mathcal{O}_N/\{J(\Phi) + F^*\mathfrak{m}_P.\mathcal{O}_N\}$ has dimension $\mu(F) + \mu(\Phi)$. Recalling that $\tau(\Phi) = \mu(\Phi)$ and $\tau(F) = \mu(F)$, we see that the target of the above map has dimension $\tau(F)$. Thus the submodule of DF/D_1F generated by ϵ_N has dimension at least $\tau(F)$, which is the dimension of the whole space. Thus indeed ϵ_N is a generator.

Since $J(F) + F^*\mathfrak{m}_P.\mathcal{O}_N$ also has codimension $\tau(F)$ in $\mathcal{O}_N$, the second assertion will follow once we verify that this annihilates the class of ϵ_N in DF/D_1F. Now this is clear for $F^*\mathfrak{m}_P.\mathcal{O}_N$. Consider then a minor

$$\frac{\partial(f_1, \ldots, f_p)}{\partial(x_{i_1}, \ldots, x_{i_p})} = M_I,$$

say. Form a vector field Ξ by formally evaluating the $(p+1)\times(p+1)$ determinant whose first row consists of the vector fields $\epsilon_N, \partial/\partial x_{i_1}, \ldots, \partial/\partial x_{i_p}$, and whose

$(k+1)$st row $(1 \leq k \leq p)$ consists of the result of applying those vector fields to f_k. Then applying Ξ to $f_k(1 \leq k \leq p)$ yields a determinant whose first and $(k+1)$st rows are equal. Thus $\Xi.f_k = 0$ for $1 \leq k \leq p$, and $\Xi \in D_0F$.

Note now that the first column of the determinant defining Ξ has entries ϵ_N and $\epsilon_N.f_k = (\operatorname{wt} f_k)f_k$, $1 \leq k \leq p$; so that, expanding the determinant by its first column, we obtain $\Xi - M_I\epsilon_N \in F^*\mathfrak{m}_P.\theta_N$. Thus $M_I\epsilon_N \in D_1F$, and so M_I does indeed annihilate the class of ϵ_N in DF/D_1F, as required. $\qquad\square$

We saw in Proposition 4.1 that the Tjurina number of the singularities on a fibre of F is determined by vector fields in the target. Given a sufficiently explicit vector field on the fibre itself, we can effectively determine just what these singularities are.

Suppose in fact that F unfolds a function f_0, and that in some other fibre this is deformed to a function f_v (so that the fibre of F is given by $f_v(\mathbf{x}) = 0$). Let ξ be a vector field satisfying the condition defining DF but with respect to the fibre over v, so that by restriction we have $\xi f_v = \phi(\mathbf{x})f_v(\mathbf{x})$.

A first result is that at a critical point of f_v, all components of ξ vanish: indeed we have the

Lemma 5.5 *Write $f_v : \mathbb{C}^n \to \mathbb{C}^p$ for the map giving the fibre of F over v, so that $F(\mathbf{x}, v) = (f_v(\mathbf{x}), v)$. Let $\xi \in Df_v$. Then ξ vanishes on $\Sigma(f_v, 0)$.*

In the applications we have in mind, ξ will be the restriction of a lowerable vector field for F.

Proof. Let $\{h_t\}$ be the germ of flow for ξ near $\Sigma(f_v, 0)$. Since $\xi \in Df_v$, then $\xi(f_{v,i})$ vanishes on $f_v^{-1}(0)$ for each component $f_{v,i}$ $(1 \leq i \leq p)$ of f_v; and hence $f_v^{-1}(0)$ is invariant by h_t.

Since h_t is a germ of $\mathbb{C}$-analytic diffeomorphism, it must then map $\operatorname{Sing}(f_v^{-1}(0))$ to itself.

Now since f_v has finite singularity type, $\operatorname{Sing}(f_v^{-1}(0)) = \Sigma(f_v, 0)$ is finite. It follows that h_t must preserve this set pointwise. Thus the vector field ξ generating the flow vanishes at these points. $\qquad\square$

This locates the singular points. We can also, at least when F is an unfolding of a function, often use the vector fields to determine the singularity type.

Proposition 5.6 *Suppose that F is an unfolding of a function and that $\xi \in Df_v$; so that there exists $\phi \in \mathcal{O}_N$ with $\xi.f_v = \phi f_v$. Let $\mathbf{x} \in \Sigma(f_v, 0)$ and suppose $\phi(\mathbf{x}) \neq 0$: write $\eta = \xi/\phi$ near $\mathbf{x}$. Then the germ of f_v at $\mathbf{x}$ is quasi-homogeneous of weight 1 with respect to weights λ_i given by the eigenvalues of the Jacobian matrix $(\partial\eta_i/\partial x_j)$ (evaluated at $\mathbf{x}$); with the proviso that any pair of eigenvalues (μ_i, μ_j) with $\mu_i + \mu_j = 1$ is replaced by $\lambda_i = \lambda_j = \frac{1}{2}$.*

Proof. This proof refers throughout to germs at $\mathbf{x}$, so we suppress this restriction from our notation. Since $\phi(\mathbf{x}) \neq 0$, $\eta = \xi/\phi$ is analytic and $\eta.f_v = f_v$, so that f_v belongs to $J(f_v)$. It follows by the celebrated criterion of Saito [27] that f_v is indeed quasi-homogeneous, i.e. weighted homogeneous of weight 1 with respect to weights $w_1, \cdots, w_n$ in suitable coordinates y_i.

Let E be the Euler vector field in these coordinates and for these weights, so that $E.f_v = f_v$. Then $E - \eta$ belongs to $D_0(f_v)$, which by Proposition 5.1 is generated by the $\delta_{i,j} = \frac{\partial f_v}{\partial y_i}\frac{\partial}{\partial y_j} - \frac{\partial f_v}{\partial y_j}\frac{\partial}{\partial y_i}$. If f_v has zero 2-jet at $\mathbf{x}$, then each of these vector fields, and hence any element of $D_0(f_v)$, has zero 1-jet. Thus the 1-jet of η equals that of E, and its Jacobian matrix, given by the coefficients in this 1-jet, equals that of E, namely the diagonal matrix with entries w_i, so the result follows in this case.

If f_v has non-zero 2-jet at $\mathbf{x}$, then by the splitting theorem we may write it as $y_1^2 + \cdots + y_k^2 + g(y_{k+1}, \ldots, y_n)$ in appropriate coordinates – which, indeed, we may take as coordinates with respect to which f_v is weighted homogeneous, so that $w_1 = \cdots = w_k = \frac{1}{2}$. Then the 1-jet of $\delta_{i,j}$ is $2(y_i\frac{\partial}{\partial y_j} - y_j\frac{\partial}{\partial y_i})$ if $1 \le i < j \le k$, $2y_i\frac{\partial}{\partial y_j}$ if $1 \le i \le k < j$, and 0 if $k \le i < j$. Arguing as above, we see that the 1-jet of η has matrix of the form

$$\begin{pmatrix} \frac{1}{2}I + A & C \\ 0 & D \end{pmatrix},$$

where A is skew-symmetric, C is arbitrary, and D the diagonal matrix with entries $w_{k+1}, \cdots, w_n$. Since the eigenvalues of a skew-symmetric matrix are either 0 or occur in pairs $\pm\nu$, the eigenvalues of the Jacobian matrix of η are, besides $w_{k+1}, \cdots, w_n$, $\frac{1}{2}$ and pairs $\frac{1}{2} + \nu, \frac{1}{2} - \nu$. The result follows. $\qquad\square$

6 The relation between vector fields in source and target

The relation in question is given by the following key result. For the weighted homogeneous case, it is given by Goryunov [14]. The general result seems to be new in this form, but in Proposition 9.3 (p 170) of [18], Looijenga proves that Ker $\overline{\omega F}/\mathfrak{m}_P.$Ker $\overline{\omega F}$ is isomorphic to a certain local cohomology module, while Vosegaard has proved (unpublished) that this module is isomorphic to DF/D_1F.

Theorem 6.1 *If F is ministable, there is a natural isomorphism*

$$\lambda \ : \ \mathrm{Ker}\ \overline{\omega F}/\mathfrak{m}_P.\mathrm{Ker}\ \overline{\omega F} \longrightarrow DF/D_1F$$

induced by lifting vector fields.

Proof. Since F is ministable, the map $\mathcal{O}_P^p/\mathfrak{m}_P^p \to N(F)$ induced by ωF is a vector-space isomorphism, so that Ker $\overline{\omega F} \subset \mathfrak{m}_P^p$. Thus if $\eta \in$ Ker $\overline{\omega F}$, so that there exists a source vector field ξ such that $tF(\xi) = \omega F(\eta)$, then $\xi \in DF$. We thus define λ by $\lambda([\eta]) = [\xi]$.

To see that λ is well-defined, and an isomorphism, we note that, since F is miniversal, the standard basis elements $\phi_i = (0, \dots, 1, \dots, 0)$, $i = 1, \dots, p$, in $\mathcal{O}_P^p \equiv \theta(F)$ project to a basis for $N(F)$, that $\overline{\omega F} \equiv \overline{\phi F}$ for this basis, and that with this identification the λ defined above is the natural isomorphism of 5.2. $\square$

In Theorem 6.1 we have supposed F ministable. If we just assume F stable, the result becomes incorrect as stated. To see this, it is sufficient to observe that if we take a trivial unfolding $F \times 1_{\mathbb{C}}^k : N \times \mathbb{C}^k \to P \times \mathbb{C}^k$, the module $DF/D_1 F$ is unaltered (up to isomorphism) whereas Ker $\overline{\omega F}/\mathrm{m}_P.$Ker $\overline{\omega F}$ increases in dimension by k.

Since any stable map-germ is isomorphic to a trivial unfolding of a ministable map-germ, it follows that to repair the result we need only replace Ker $\overline{\omega F}$ by the module of liftable vector fields vanishing at the base-point. With this modification, we may apply the result to any point in the target of any stable map.

7 The instability locus and the discriminant matrix

Let f be any map-germ possessing a stable unfolding F. Then inclusion induces an isomorphism $N(f) \cong N(F)$, and indeed we may regard the unfolding as obtained by choosing suitable 'unfolding monomials' ϕ_j lifting a basis of $N(f)$ to $\theta(f)$. Write $\mathcal{O}_p, \mathcal{O}_P$ for the rings of germs of functions on the respective targets of f and F. Write x for the source variables of f, y for the target variables; for F we require further variables, and write x, u and y, v. Write p, a for the respective numbers of y and v variables.

The *instability locus* of f is the set of points y in the target such that the germ of f at $\Sigma(f, y)$ is *not* a stable germ. (Damon [7] calls this the *versality discriminant*). We are particularly interested in describing this set. It is central to the considerations of [9, Chapter 10], and is also the key to some interesting applications of the theory of discriminants.

We have discussed above the exact sequence

$$0 \longrightarrow \mathcal{O}_{y,v}^{p+a} \longrightarrow \mathcal{O}_{y,v}^{p+a} \xrightarrow{\overline{\omega F}} M(F) \longrightarrow 0,$$

where the middle term is identified with θ_P and the previous one with the module of liftable vector fields.

We know from [9, Proposition 10.4.2] that M deforms flatly, so that the exact sequence yields, on tensoring over $\mathcal{O}_{y,v}$ with $\mathcal{O}_y$, the sequence

$$0 \longrightarrow \mathcal{O}_y^{p+a} \longrightarrow \mathcal{O}_y^{p+a} \xrightarrow{\overline{\phi f}} M(f) \longrightarrow 0.$$

Thus the matrix of the first map in this sequence, which we may call the discriminant matrix for f, is obtained from the discriminant matrix for F by applying the homomorphism $\mathcal{O}_{y,v} \to \mathcal{O}_y$ to the entries. We suppose the first order derivatives of f to vanish at the origin, so that we may take the $\partial/\partial y_j$ as the first p of the ϕ_j. The corresponding summand $\mathcal{O}_y^p$ of $\mathcal{O}_y^{p+a}$ may then be identified with θ_P and the restriction of $\overline{\phi f}$ to it with $\overline{\omega f}$.

By Mather's criterion [22] for stable map-germs, the instability locus of f is the support of the module $\theta(f)/tf(\theta_N) + \omega f(\theta_P)$, i.e. of the cokernel of $\overline{\omega f}$. It follows from the above exact sequence that we can identify this with the cokernel of the induced map

$$\mathcal{O}_y^{p+a} \xrightarrow{\alpha f} \mathcal{O}_y^{p+a} \longrightarrow \mathcal{O}_y^a,$$

whose matrix consists of the last a columns of the discriminant matrix A. This is called the *instability matrix* and its image the *instability module*. The stage is thus set for application of the theory of the discriminant matrix .

The next result we require concerns the symmetry of the discriminant matrix. This was first observed by Arnold [2] for the discriminants of singularities of type A_n, and generalised by Mond and Pellikaan [24] as below. The restriction to weighted homogeneous germs has been removed by Palamodov [26]. A similar symmetry result has recently been obtained by Vosegaard [31] for the case when the fibres of F have dimension 2.

Theorem 7.1 *Let $f : (\mathbb{C}^s, 0) \to (\mathbb{C}, 0)$ be a weighted homogeneous germ of finite singularity type. Then any homogeneous unfolding of f admits a symmetric homogeneous discriminant matrix.*

We have seen above that the instability locus of f coincides with the locus where the rank of the submatrix consisting of a certain a columns of the discriminant matrix drops below a. In the case when the matrix is symmetric, we can replace this by a submatrix consisting of a of the rows. This is a clear advantage since, as we have seen, the rows give liftable vector fields, so have geometric content. To make use of the result, we need to identify the rows in question. This can be done in general by using the perfect pairing that underlies the proof of the theorem. This generality is not required for the next result, which gives the most important applications.

Let $f_0(\mathbf{x})$ be a weighted homogeneous function-germ, and choose homogeneous unfolding monomials ϕ_i, arranged in some order of increasing weight. We consider the partial unfoldings

$$F^k(\mathbf{x}, u_2, \ldots, u_{\mu-k}) = (f_u^k(\mathbf{x}), u_2, \ldots, u_{\mu-k}),$$

where

$$f_u^k(\mathbf{x}) = f_0(\mathbf{x}) + \sum_2^{\mu-k} u_i \phi_i(\mathbf{x}).$$

Thus $F = F^0$ is the versal, hence (C^∞-)stable unfolding, and F^k is obtained by omitting the last k unfolding monomials: we assume that the remainder have strictly lower weight.

Proposition 7.2 *([9, Theorem 10.5.32]) If F^k is the unfolding obtained by omitting the k unfolding monomials of highest weight, the instability module for F^k is the submodule of $\mathcal{O}^k$ spanned by the k columns of lowest weight of any discriminant matrix for F.*

For the k rows of lowest weight of the discriminant matrix must correspond by symmetry to the k columns of lowest weight.

We next observe the

Lemma 7.3 *The linear relation $\sum c_j a_{j,k} = 0$ between the columns of the discriminant matrix holds at a point $\mathbf{v}$ of the target of F if and only if the function $\psi = \sum c_j \phi_j$ satisfies the condition $\psi f_v \in J(f_v)$, where f_v is the restriction of f to the fibre over $\mathbf{v}$.*

Proof. We recall the equations (2) from Section 1:

$$f(\mathbf{x}, \mathbf{u})\phi_i \equiv \sum_{j=0}^{\mu-1} a_{i,j}(\mathbf{u})\phi_j + \sum_{k=1}^{r} b_{i,k}(\mathbf{x}, \mathbf{u})\frac{\partial f}{\partial x_k} \quad \text{for each} \quad i. \tag{6}$$

We now have the relations

$$\sum c_i(a_{i,j}(v) - y\delta_{i,j}) = 0 \quad \text{for each} \quad j. \tag{7}$$

Add c_i times the ith equation (6), and subtract the sum of ϕ_j times the jth relation (7). The terms in $a_{i,j}$ then all cancel, and, writing $f_v(x) = f(x, v) - y$ and (as above) $\psi = \sum c_j \phi_j$, we find $\psi(x)f_v(x) = \sum_{i,k} c_i b_{i,k} \partial f_v/\partial x_k$.

This establishes the forward implication: for the converse, if any relation of the form $\psi(x)f_v(x) = \sum \alpha_k \partial f_v/\partial x_k$ holds, then reversing the argument shows that at the point in question, $\sum c_i(a_{i,j}(v) - y\delta_{i,j})\phi_j$ lies in $J_x f$, whence the conclusion. $\qquad \Box$

Corollary 7.4 *For f as above, a point (y, v) lies in the instability locus if and only if there exists a linear combination $\psi(x)$ of the k unfolding monomials of least weight such that $\psi f_v \in J(f_v)$.*

We may again think more geometrically. The condition may be regarded as stating that the liftable vector field $\sum c_i \eta_i$ has a zero at the point in question. It follows that its lift $\xi = \sum c_i \xi_i$ is tangent to the fibre over this point. This vector field may be used as in Lemma 5.5 to locate and (under favourable circumstances) as in Proposition 5.6 to characterise the singularities of the fibre.

References

[1] Aleksandrov, A.G.: *Cohomology of a quasi-homogeneous complete intersection.* Math. USSR Izv. **26iii**, 437–477, (1986).

[2] Arnold, V.I.: *Wave front evolution and equivariant Morse lemma.* Comm. Pure Appl. Math. **29**, 557–582, (1976).

[3] Arnold, V.I.: *Indices of singular points of 1-forms on manifolds with boundary, convolution of invariants of reflection groups and singular projections of smooth surfaces.* Russian Math. Surveys **34ii**, 1–42, (1979).
Reprinted in *Singularity Theory: selected papers*, London Math. Soc. Lecture Notes no. **53**, Cambridge University Press, (1981).

[4] Bruce, J.W.: *Functions on discriminants.* Jour. London Math. Soc. (2) **30**, 551–567, (1984).

[5] Bruce, J.W.; A.A. du Plessis; L.C. Wilson: *Discriminants and liftable vector fields.* Jour. Alg. Geom. **3**, 725–753, (1994).

[6] Bruce, J.W. ; R.M. Roberts: *Critical points of functions on analytic varieties.* Topology **27**, 57–90, (1988).

[7] Damon, J.N.: *On the Pham example and the universal topological stratification of singularities.* pp. 161–168 in *Singularities* (ed. S. Lojaciewicz), Banach Center Publications **20**, PWN, Warsaw, (1988).

[8] du Plessis, A.A.; T. Gaffney; L.C. Wilson: *Map-germs determined by their discriminants.* to appear.

[9] du Plessis, A.A.; C.T.C. Wall: *The geometry of topological stability*, London Math. Soc. monographs, new series **9** Oxford University Press. (1995).

[10] du Plessis, A.A.; C.T.C. Wall: *Instabilities in the unfoldings of singularities in the E, Z and Q series.* Preprint, Universitet Aarhus.

[11] Gaffney, T.: *Properties of finitely determined germs*, Ph.D. thesis, Brandeis University, (1975).

[12] Gaffney, T.: *Polar multiplicities and the equisingularity of map-germs.* Topology **32**, 185–223, (1993).

[13] Gaffney, T.;L.C. Wilson: *Equivalence theorems in global singularity theory.* pp. 439–447 in *Proc. Symp. in Pure Math. 40 part 1: Singularities* (ed. P. Orlik), Amer. Math. Soc., (1983).

[14] Goryunov, V.V.: *Vector fields and functions on discriminants of complete intersections and bifurcation diagrams of projections.* Jour. Soviet Math. **52**, 3231–3245, (1990).

[15] Greuel, G.-M.: *Der Gauss-Manin Zusammenhang isolierter Singularitäten von vollständigen Durchschnitten.* Math. Ann. **214**, 235–266, (1975).

[16] Greuel, G.-M.: *Dualität in der lokalen Kohomologie isolierter Singularitäten.* Math. Ann. **250**, 157–173, (1980).

[17] Kersken, M.: *Reguläre Differentialformen.* Manuscripta Math. **46**, 1–26, (1984).

[18] Looijenga, E.J.N.: *Isolated singular points on complete intersections*, London Math. Soc. Lecture Notes no. **77**, Cambridge University Press, (1984).

[19] Looijenga, E.J.N.; J.H.M. Steenbrink: *Milnor numbers and Tjurina numbers of complete intersections.* Math. Ann. **271**,121–124, (1985).

[20] Lyashko, O.: *Geometry of bifurcation diagrams.* Jour. Soviet Math. **27**, 2735–2759, (1984).

[21] Mather, J.N.: *Stability of C^∞-mappings II: infinitesimal stability implies stability.* Ann. of Math. **89**, 254–291, (1969).

[22] Mather, J.N.: *Stability of C^∞-mappings IV: classification of stable germs by $\mathbb{R}$-algebras.* Publ. Math. IHES **37**, 223–248, (1970).

[23] Mond, D.M.Q.; J. Montaldi: *Deformations of maps on complete intersections, Damon's $\mathcal{K}_V$-equivalence and bifurcations.* pp 263–284 in *Singularities: Lille 1991* (ed. J.-P. Brasselet), London Math. Soc. Lecture Notes no. **201**, Cambridge University Press, (1994).

[24] Mond, D.M.Q.; R. Pellikaan: *Fitting ideals and multiple points of analytic mappings.* Springer lecture notes in Math. **1414**, 107–161, (1989).

[25] Northcott, D.G.: *Finite free resolutions*, Cambridge University Press, (1976).

[26] Palamodov, V.P.: *Tangent fields on deformations of complex spaces.* Math. USSR Sbornik **71**, 163–182, (1992).

[27] Saito, K.: *Quasihomogene isolierte Singularitäten von Hyperflächen.* Invent. Math. **14**, 123–142, (1971).

[28] Saito, K.: *Theory of logarithmic differential forms and logarithmic vector fields.* Jour. Fac. Sci. Tokyo, sec. 1A **27** 265–291, (1980).

[29] Teissier, B.: *The hunting of invariants in the geometry of discriminants.* pp. 565–677 in *Real and complex singularities, Oslo, 1976*, (ed. P. Holm), Sijthoff and Noordhoff, (1977).

[30] Terao, H.: *Discriminant of a holomorphic map and logarithmic vector fields*. Jour. Fac. Sci. Univ. Tokyo, sec. IA (Math.) **30**, 379–391, (1983).

[31] Vosegaard, H.: *Perfect pairings and discriminants of ICIS*. Preprint, Universitet Aarhus.

[32] Wahl, J.M.: *Derivations, automorphisms and deformations of quasi-homogeneous singularities*. pp 613–624 in *Proc. Symp. in Pure Math. 40 part 2: Singularities* (ed. P. Orlik), Amer. Math. Soc., (1983).

[33] Wall, C.T.C.: *Finite determinacy of smooth map-germs*. Bull. London Math. Soc. **13**, 481–539, (1981).

[34] Wilson, L.C.: *Equivalence of stable mappings between 2-dimensional manifolds*. Jour. Diff. Geom. **11**, 1–14, (1976).

[35] Wirthmüller, K.: *Singularities determined by their discriminant*. Math. Ann. **252**, 237–245, (1980).

[36] Zakalyukin, V.M.: *Reconstructions of wave fronts depending on one parameter*. Funct. Anal. Appl. **10**, 139–140, (1976).

[37] Zakalyukin, V.M.: *Perestroikas of fronts and caustics depending on parameters, versality of mappings*. Jour. Soviet Math. **27**, 2713–2735, (1984).

Progress in Mathematics, Vol. 162, © 1998 Birkhäuser Verlag Basel/Switzerland

Suspensions of Fat Points and Their Intersection Forms

Wolfgang Ebeling
Institut für Mathematik
Universität Hannover
Postfach 6009
30060 Hannover, GERMANY

Sabir M. Gusein-Zade *
Department of Geography
Moscow State University
Moscow 119899
RUSSIA

Dedicated to Egbert Brieskorn

Introduction

In this paper a fat point in $\mathbb{C}^2$ is a 0-dimensional isolated complete intersection singularity (*icis*) in $(\mathbb{C}^2, 0)$. Such a fat point X is defined by a system of equations $\{f = g = 0\}$ (f and g are function germs $(\mathbb{C}^2, 0) \to (\mathbb{C}, 0)$). We assume this system of equations to be generic. In particular it means that the function germs f and g define isolated curve singularities in $(\mathbb{C}^2, 0)$ with possibly minimal Milnor numbers. The p-fold suspension of X is the 1-dimensional *icis* $X^{(p)}$ in $(\mathbb{C}^3, 0)$ defined by the equations $\{f(x, y) + z^p - g(x, y) = 0\}$. The analytic type of the *icis* $X^{(p)}$ depends on the choice of the equations f and g. However (for a generic choice of the equations) $X^{(p)}$ is well-defined up to deformations with constant μ and μ_1 (for an n-dimensional *icis* $(F^{-1}(0), 0) \subset (\mathbb{C}^{n+k}, 0)$ defined by a map-germ $F = (f_1, \ldots, f_k) : (\mathbb{C}^{n+k}, 0) \to (\mathbb{C}^k, 0)$, μ_1 is the Milnor number of the $(n + 1)$-dimensional *icis* $((F')^{-1}(0), 0) \subset (\mathbb{C}^{n+k}, 0)$ defined by the map-germ $F' = (f_1, \ldots, f_{k-1}) : (\mathbb{C}^{n+k}, 0) \to (\mathbb{C}^{k-1}, 0)$ for a generic choice of the equations $f_1, \ldots, f_{k-1}, f_k$ of the *icis* $(F^{-1}(0), 0)$). Thus the Milnor fibre, the vanishing lattice, the monodromy group, (the set of) Coxeter-Dynkin diagrams,... of the suspension $X^{(p)}$ are well-defined. The relation between the monodromy operator of an *icis* and the monodromy operator of its p-fold suspension has been described in [ES]. The 2-fold suspension of the *icis* X is called its stabilization. Coxeter-Dynkin diagrams of stabilizations of fat points in $\mathbb{C}^2$, obtained by a version of the method of real morsifications, have been described in [EG2]. Here we give a similar description for p-fold suspensions.

---

*Supported by INTAS–94–4373 and Deutsche Forschungsgemeinschaft (436 RUS 17/171 /95).

At the same time we offer an equivariant analogue for the Picard-Lefschetz formula for the ($\mathbb{Z}_p$-equivariant) A_{p-1} singularity. We compute Coxeter-Dynkin diagrams for the 1-dimensional unimodular *icis* which are suspensions ([W]).

The authors are grateful to the referee of the paper for numerous useful suggestions.

1 p-fold suspensions of *icis*

Let $(X,0) \subset (\mathbb{C}^{n+k},0)$ be an n-dimensional *icis* defined by equations $\{f_1 = f_2 = \ldots = f_k = 0\}$ ($f_i : (\mathbb{C}^{n+k},0) \to (\mathbb{C},0)$, $F = (f_1, \ldots, f_k) : (\mathbb{C}^{n+k},0) \to (\mathbb{C}^k,0)$). The Milnor fibre of the *icis* $(X,0)$ is the manifold $V_\varepsilon = F^{-1}(\varepsilon) \cap B_\delta$ for $\delta > 0$ small enough and a generic (i.e. not belonging to the discriminant set $\mathcal{D}(F)$ of the map F) $\varepsilon \in \mathbb{C}^k$ with $0 < \|\varepsilon\| \ll \delta$ (B_δ is the ball of radius δ with the centre at the origin in $\mathbb{C}^k$). The Milnor fibre V_ε has the homotopy type of the bouquet of n-dimensional spheres (see [AGV]). The number μ of these spheres is called the Milnor number of the *icis* $(X,0)$. The (reduced modulo a point if $n = 0$) homology group $H_n(V_\varepsilon; \mathbb{Z}) \cong \mathbb{Z}^\mu$ of the Milnor fibre V_ε is called the vanishing homology group (or the Milnor lattice) of the *icis* $(X,0)$.

Systems of generators of the vanishing homology group of a special sort (so called *distinguished* ones) can be constructed in the following way (see [AGV], [E]). For a generic choice of a system of coordinates in $(\mathbb{C}^k,0)$ the germ $F' = (f_1, \ldots, f_{k-1})$ defines an $(n+1)$-dimensional *icis* $(X',0) = (F')^{-1}(0)$, the Milnor number μ_1 of which does not depend on the choice of a generic (!) system of equations (i.e. of coordinates in $(\mathbb{C}^k,0)$). Let $V'_{\varepsilon'} = (F')^{-1}(\varepsilon') \cap B_\delta$ ($\varepsilon' = (\varepsilon_1, \ldots, \varepsilon_{k-1})$, $0 < \|\varepsilon'\| \ll \delta$) be the Milnor fibre of the *icis* $(X',0)$. The restriction of a generic deformation $\tilde{f}_k$ of the kth function f_k to $V'_{\varepsilon'}$ is a Morse function, i.e., it has only non-degenerate critical points with different critical values u_i ($i = 1,2,\ldots,\nu$). Let u_0 be a non-critical value with $\|u_0\| > \|u_i\|$ ($i = 1,\ldots,\nu$). The level set $\{x \in V'_{\varepsilon'} : \tilde{f}_k(x) = u_0\}$ is diffeomorphic to the Milnor fibre of the *icis* $(X,0)$. Let s_i, $i = 1,\ldots,\nu$, be non-self-intersecting paths connecting the critical values u_i with the non-critical value u_0 in such a way that they lie inside the disc $D_{\|u_0\|} = \{u \in \mathbb{C} : \|u\| \leq \|u_0\|\}$ and every two of them intersect only at the point u_0. Let us suppose that the paths s_i (and thus the critical values u_i) are numbered clockwise according to the order in which they arrive at u_0 beginning from the boundary of the disc $D_{\|u_0\|}$. Each path s_i defines up to orientation a vanishing cycle δ_i in the homology group of the non-singular level set (isomorphic to the vanishing homology group of the *icis* $(X,0)$). The system $\{\delta_i\}$ is a system of generators of this group. A system of generators obtained in such a way is called *distinguished*. The number ν of the critical values (and of the cycles δ_i) is equal to the sum $\mu + \mu_1$ of the Milnor number $\mu = \mu(F)$ and of the Milnor number μ_1 of the *icis* defined by the germ $F' = (f_1, \ldots, f_{k-1})$. On the vanishing homology group there is the intersection form $(\cdot, \cdot)$. The matrix (δ_i, δ_j) of the intersection numbers of the (numbered, i.e. ordered) cycles δ_i is represented as usual by a graph — the Coxeter-Dynkin diagram of the *icis* $(X,0)$. In this graph, the multiplicity of the edge between the vertices corresponding to the cycles δ_i and δ_j is equal to $(-1)^{\frac{n(n+1)}{2}+1}(\delta_i,\delta_j)$ for $i < j$. Thus for $n = 1$ this multiplicity coincides with the intersection number (δ_i,δ_j) with $i < j$.

DEFINITION   *The p-fold suspension of the n-dimensional icis*

$$(X,0) = (F^{-1}(0),0)\,, \quad F = (f_1,\dots,f_k) : (\mathbb{C}^{n+k},0) \to (\mathbb{C}^k,0)$$

is the $(n+1)$-dimensional icis $(X^{(p)},0) \subset (\mathbb{C}^{n+k+1},0) = (\mathbb{C}^{n+k} \times \mathbb{C}^1, 0 \times 0)$ defined by the map-germ $(f_1 + a_1 z^p,\dots,f_k + a_k z^p)$ for generic coefficients a_i or (equivalently) by the map-germ $(f_1 + z^p, f_2,\dots,f_k)$ for a generic choice of equations $f_1,\dots,f_k$ (that is for a generic choice of the coordinate system in $(\mathbb{C}^k,0)$; z is a new variable).

The Milnor number μ and the number $\nu = \mu + \mu_1$ of vanishing cycles in a distinguished set of generators of the p-fold suspension $(X^{(p)},0)$ of an *icis* $(X,0)$ can be expressed in terms of invariants of the initial *icis* $(X,0)$. Namely:

$$\begin{aligned}
\mu(f_1 + z^p, f_2,\dots,f_k) &= (p-1)\mu(f_1,f_2,\dots,f_k) + p\cdot\mu(f_2,\dots,f_k), \\
\nu(f_1 + z^p, f_2,\dots,f_k) &= (p-1)\nu(f_1,f_2,\dots,f_k) + p\cdot\nu(f_2,\dots,f_k).
\end{aligned}$$

See also [D]. A refinement of a particular case of the first formula (for $k = 2$) can be found in [P].

2   Convenient equations and the corresponding real picture

This section is a short survey of results from [EG2], which is the reference for proofs. Most of the constructions which will be described in the sequel are of local nature. In order not to overload the paper by technical details we sometimes permit ourselves to be not very precise in descriptions of considered neighbourhoods or/and values of parameters.

In principle an arbitrary Coxeter-Dynkin diagram of a fat point does not determine a Coxeter-Dynkin diagram of its suspension (see [EG2] for an example of two fat points with the same sets of Coxeter-Dynkin diagrams but different Coxeter-Dynkin diagrams of their stabilizations). A Coxeter-Dynkin diagram of the suspension can be described in terms of a real picture corresponding to the fat point as we will point out in the sequel.

Let $(X,0) = \{f = g = 0\} \subset (\mathbb{C}^2,0)$ ($f = f(x,y)$, $g = g(x,y)$) be a fat point defined by a convenient system of equations in the sense of [EG2]. It means that:

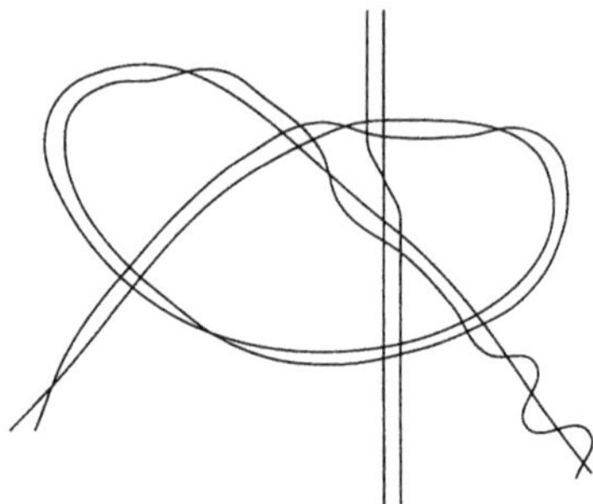

Figure 1: An example of a real picture

1) $\{f = g = 0\}$ is a generic system of equations of $(X, 0)$, i.e. the equation $\{f = 0\}$ defines an isolated curve singularity with possibly minimal Milnor number (equal to μ_1);

2) both f and g are real;

3) the curve $\{f = 0\}$ is real, i.e. it does not have pairs of complex conjugate components;

4) g can be assumed to be a small deformation of f, i.e. there is a real h such that $g = f + \lambda h$ for real λ small enough, $\{f = h = 0\}$ is a real system of equations of $(X, 0)$ (it implies that the curve $\{g = 0\}$ is also real and its Milnor number $\mu(\{g = 0\})$ is equal to the Milnor number $\mu(\{f = 0\}) = \mu_1$ of the curve $\{f = 0\}$).

Each fat point in $\mathbb{C}^2$ is up to a deformation with constant μ and μ_1 equivalent to a fat point defined by a convenient system of equations ([EG2]). Let $\tilde{f}$ and $\tilde{g}$ be convenient (real) deformations of the functions f and g in the sense of [EG2] again. It means that:

1) all critical points of the functions $\tilde{f}$ and $\tilde{g}$ (in an appropriate neighbourhood of the origin) are non-degenerate and real;

2) the values of $\tilde{f}$ and $\tilde{g}$ at all their saddle points are equal to zero;

3) all intersection points of the complex curves $\{\tilde{f} = 0\}$ and $\{\tilde{g} = 0\}$ are real and simple (double);

4) the branches of the real curve $\{\tilde{g} = 0\}$ go close and "almost parallel" to the branches of the real curve $\{\tilde{f} = 0\}$, intersecting the last one at smooth points (see Fig. 1).

It is possible to show that in this case all critical points of the restriction of the function $\tilde{g}$ to the zero-level set of the function $\tilde{f}$ in $\mathbb{C}^2$ are real and non-degenerate and that they are organized in the following way. The intersection points with the (real) curve $\{\tilde{g} = 0\}$ divide the real curve $\{\tilde{f} = 0\}$ into several smooth connected pieces (some of which can intersect at the self-intersection points of the curve $\{\tilde{f} = 0\}$). Each self-intersection point of the curve $\{\tilde{f} =$

$0\}$ corresponds to two intersection points with the curve $\{\tilde{g} = 0\}$. We shall say that the self-intersection point is *close* to them. The number ν of (non-degenerate) critical points of the function $\tilde{g}$ on a smoothing of the curve $\{\tilde{f} = 0\}$ is equal to the number of such bounded pieces (segments) plus twice the number of self-intersection points of the curve $\{\tilde{f} = 0\}$. There is one and only one non-degenerate critical point on each segment (there must be at least one critical point on such a segment since $\tilde{g}$ is equal to zero at its end points; the computation of the total number of the critical points shows that there is only one critical point and it is non-degenerate). Each self-intersection point of the curve $\{\tilde{f} = 0\}$ has to be regarded as a critical point of the function $\tilde{g}$ on this curve of multiplicity two, because it corresponds to a pair of non-degenerate critical points of $\tilde{g}$ on a smoothing of it. On one o f the two natural real smoothings of the self-intersection point these two critical points are real (a maximum and a minimum). On the other one they are imaginary.

The intersection forms (or Coxeter-Dynkin diagrams) of the fat point and of its stabilization can be described in terms of the geometry of real curves $\{\tilde{f} = 0\}$ and $\{\tilde{g} = 0\}$ ([EG2]).

Let $(X^{(p)}, 0) \subset (\mathbb{C}^3, 0)$ be the p-fold suspension of the fat point $(X, 0)$ which is defined by $(X^{(p)}, 0) = \{f + z^p = g = 0\}$ (z is a new variable). The number $\nu(X^{(p)})$ of non-degenerate critical points of a generic deformation of the function-germ g on the Milnor fibre of the singularity $f + z^p$ and hence the number of critical points of the function $\tilde{g}(x, y)$ on the (singular) level manifold $\{\tilde{f}(x, y) + z^p = 0\}$ counted with appropriate multiplicities is equal to $(p-1)\nu(X) + p \cdot \mu(g)$, where $\nu(X)$ is the number of critical points of the function $\tilde{g}(x, y)$ on the level curve $\{\tilde{f}(x, y) = 0\}$ (counted with multiplicities again) and $\mu(g)$ is the Milnor number of the singularity g (in our case $\mu(g) = \mu(f)$).

3 A distinguished set of vanishing cycles for the *icis* $\{x + z^p = 0, \ x \pm y^2 = 0\}$

At a (non-degenerate) critical point of the function $\tilde{g}(x, y)$ on the smooth part of the curve $\{\tilde{f} = 0\}$ the equations $\tilde{f} = 0$ and $\tilde{g} = 0$ can be reduced (by a real change of coordinates and multiplication of one of the equations by a positive constant) to the form $x = 0$, $x \pm y^2 + c = 0$. This also takes place at one of the two real critical points of the function $\tilde{g}$ which originate from a double point of the curve $\{\tilde{f} = 0\}$ after its (appropriate) smoothing. In order to define vanishing cycles corresponding to the point $(x_0, y_0, 0) \in \{\tilde{f} + z^p = 0, \ \tilde{g} = 0\} \subset \mathbb{R}^3$ in a distinguished set of generators of the vanishing homology group of the *icis* $\{f + z^p = 0, \ g = 0\}$, we have to describe a distinguished set of generators for the singularity $\{x + z^p = 0, x \pm y^2 = 0\}$. In fact this *icis* is an isolated hypersurface (and even plane curve) singularity of type A_{p-1}: the change of coordinates $\tilde{x} = x + z^p$, $\tilde{y} = y$, $\tilde{z} = z$ (with a linear change of the equations) reduces it to $\{\tilde{x} = 0, \ -- \tilde{z}^p \pm \tilde{y}^2 = 0\}$.

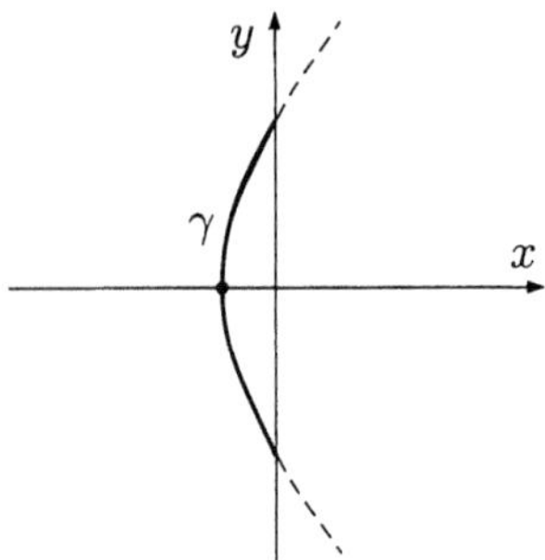

Figure 2: The arc γ

Its (standard) Coxeter-Dynkin diagram is well-known: it is the standard diagram of type A_{p-1} (see [AGV]). A description of a distinguished basis of its vanishing homology group can be obtained from the description of the distinguished basis of the A_{p-1} singularity of one variable in [AGV, pp. 62–66]. We will fix it in the following way. For $\varkappa > 0$ (small enough) the intersection of the projection of the curve $\{x + z^p = 0,\ x \pm y^2 = \pm\varkappa\}$ to the (complex) (x, y)-plane with the real plane contains the arc γ shown in Fig. 2 (for the "$-$" sign; for the "$+$" sign the picture is symmetric to this one). Let us orient the arc γ in the direction of decrease of y (i.e. counterclockwise as a part of the boundary of the region $x \pm y^2 \geq \pm\varkappa$). This means that to the right of γ there are smaller values of the (second) function $x \pm y^2$ and to the left larger ones. The preimage of the arc γ in the curve $\{x + z^p = 0,\ x \pm y^2 = \pm\varkappa\}$ consists of p arcs γ_q with $\operatorname{Arg} z = q \cdot \frac{2\pi}{p}$ for the "$-$" sign and with $\operatorname{Arg} z = \left(q - \frac{1}{2}\right)\frac{2\pi}{p}$ for the "$+$" sign ($q = 0, 1, \ldots, (p-1)$). Let us orient each of them according to the described orientation of its projection γ. Let δ be the *cycle* $(\gamma_0 - \gamma_1)$. From the description in [AGV] it is not difficult to understand (or to show) that the system of vanishing cycles $\delta, \sigma\delta, \ldots, \sigma^{p-2}\delta$ is a distinguished basis of the vanishing homology group of the *icis* $\{x + z^p = 0,\ x \pm y^2 = 0\}$ ($\sigma = \exp\left(2\pi i/p\right)$ — the generator of the cyclic group $\mathbb{Z}_p \subset S^1 \subset \mathbb{C}$ which acts on the Milnor fibre of the *icis* by the formula $\sigma(x, y, z) = (x, y, \exp\left(2\pi i/p\right) z)$). We will consider this basis as the standard one. There is the relation $(1 + \sigma + \ldots + \sigma^{p-1})\delta = 0$. One has $(\sigma^i\delta, \sigma^j\delta) = 0$ for $|j - i| > 1$, $(\sigma^i\delta, \sigma^{i+1}\delta) = 1$ for $0 \leq i \leq p - 3$.

4 Enumeration of vanishing cycles and the definition of their orientations

Each level set of the function $\tilde{f} + z^p$ (and in particular the (singular) zero level set) has the natural action of the cyclic group $\mathbb{Z}_p \subset S^1 \subset \mathbb{C}$: $\sigma^q(x, y, z) =$

$(x, y, \exp\left(\frac{2\pi q i}{p}\right) z)$. The function $\tilde{g}$ does not depend on z and thus it is invariant with respect to this action. Hence the set of critical points of the function $\tilde{g}$ on the (singular complex) surface $\{\tilde{f} + z^p = 0\}$ and the set of vanishing cycles are invariant with respect to this action as well. The vanishing homology group of the *icis* $X^{(p)}$ is a $\mathbb{Z}[\mathbb{Z}_p]$-module ($\mathbb{Z}[\mathbb{Z}_p]$ is the (integral) group ring of the group $\mathbb{Z}_p$). The intersection form on it is $\mathbb{Z}_p$-invariant as well.

The critical points of the restriction of the function $\tilde{g} = \tilde{g}(x, y)$ to the zero level set of the function $\tilde{f}(x, y) + z^p$ can be found from the system of equations

$$\begin{cases} \dfrac{\partial \tilde{g}}{\partial x} = \zeta \dfrac{\partial \tilde{f}}{\partial x}, \\[2mm] \dfrac{\partial \tilde{g}}{\partial y} = \zeta \dfrac{\partial \tilde{f}}{\partial y}, \\[2mm] 0 = p\zeta z^{p-1}, \\[2mm] f + z^p = 0, \end{cases}$$

where ζ is the Lagrange multiplier. We have either $z = 0$ (with multiplicity $(p-1)$) or $\lambda = 0$. If $z = 0$ we have just the equations for the critical points of the function $\tilde{g}$ on the (singular) curve $\{\tilde{f} = 0\} \subset \mathbb{C}^2$. There are $\nu(X)$ of them (including pairs of critical points corresponding to double points of the curve $\{\tilde{f} = 0\}$). All these points are real (and each of them has to be counted with multiplicity $(p-1)$ since it defines an A_{p-1}-singularity). If $\lambda = 0$ we have the system of equations for critical points of $\tilde{g}$ in $\mathbb{C}^2$. There are $\mu(g)$ solutions of this system. All the critical points of the function $\tilde{g}$ in $\mathbb{C}^2$ are real as well. They are in one-to-one correspondence with the double points of the curve $\{\tilde{g} = 0\} \subset \mathbb{R}^2$ and with the bounded components of its complement. To each critical point of the function $\tilde{g}$ in $\mathbb{C}^2$ there correspond p non-degenerate critical points of $\tilde{g}$ on the level surface $\{\tilde{f} + z^p = 0\}$ with $z = \sqrt[p]{-\tilde{f}(x, y)}$. All the critical values of the function $\tilde{g}$ on the surface $\{\tilde{f} + z^p = 0\}$ lie on the real line in the complex plane. To define a distinguished set of vanishing cycles, we will use a system of paths which connect critical values with a non-critical one from the upper half-plane $\{\operatorname{Im} u > 0\}$ and which lie completely in the upper half-plane with the only exception of their end points which coincide with the critical values. Such paths define (up to orientations) vanishing cycles corresponding to the critical points of the function $\tilde{g}$ in $\mathbb{C}^2$. However they do not define vanishing cycles corresponding to the critical points of the function $\tilde{g}$ on the curve $\{\tilde{f} = 0\}$ (including double points of this curve). To do so we can describe an appropriate system of paths for a small perturbation of the function $\tilde{g}$ (as a function of three variables), which splits the critical points of it on the surface $\{\tilde{f}(x, y) + z^p = 0\}$ with $z = 0$ into $(p-1)$ non-degenerate ones (and we have to make also a smoothing of the surface $\{\tilde{f}(x, y) + z^p = 0\}$ in order to do so for double points of this surface). We will not do it explicitly, supposing that we use a system of paths which corresponds to the "natural" system of vanishing cycles, which has been described in the previous section and about which we know that it is a distinguished one for the corresponding singularity (of type A_{p-1}). We will define one vanishing cycle for each critical point of the function $\tilde{g}$ on $\mathbb{R}^2$ or on the smooth part of the (singular) curve $\{\tilde{f} = 0\}$ and two vanishing cycles for each singular point of this curve in such a way that all other vanishing cycles from a distinguished set of generators of the vanishing homology group of the suspension $X^{(p)} = \{f + z^p = 0, \; g = 0\}$ are obtained from the defined ones by actions of (some) elements of the cyclic group $\mathbb{Z}_p$.

We have the following list of the critical points of the function $\tilde{g}$ on the surface $\{\tilde{f} + z^p = 0\}$ and of the corresponding vanishing cycles (compare with [EG2]).

1) The critical points corresponding to maxima or minima of the function $\tilde{g}$ in $\mathbb{R}^2$ (i.e. to bounded components of the complement to the curve $\{\tilde{g} = 0\} \subset \mathbb{R}^2$ or to the curve $\{\tilde{f} = 0\} \subset \mathbb{R}^2$). Let p_i^{++} (respectively p_n^{--}) be the critical points corresponding to maxima (respectively to minima) of the function $\tilde{g}$ in $\mathbb{R}^2$ with $\operatorname{Arg} z = -\frac{\pi}{p}$ (respectively with $\operatorname{Arg} z = 0$). Other critical points corresponding to maxima and minima are obtained from these ones by the action of the group $\mathbb{Z}_p$. E.g. the points corresponding to maxima with $\operatorname{Arg} z = \left(q - \frac{1}{2}\right)\frac{2\pi}{p}$ are $\sigma^q p_i^{++}$ ($\sigma = \exp\left(2\pi i/p\right)$ is the generator of the group $\mathbb{Z}_p$). The vanishing cycle corresponding to the critical point p_i^{++} or p_n^{--} will be denoted by δ_i^{++} or δ_n^{--} respectively. We define the orientation of the vanishing cycle δ_i^{++} (respectively of δ_n^{--}) on the lower (respectively higher) level set of the function $\tilde{g}$ (close to the critical one). On this level set it is represented by the preimage of an oval in the (x, y)-plane — a component of the real curve $\tilde{g}(x, y) = \tilde{g}(p_i^{++}) - \varkappa$ (respectively $\tilde{g}(x, y) = \tilde{g}(p_n^{--}) + \varkappa$), $\varkappa > 0$, and we take the counterclockwise orientation of it.

REMARKS   *1. Probably it seems to be more natural to take the point with $\operatorname{Arg} z = +\frac{\pi}{p}$ for p_i^{++}. However: a) we would like to have notations which correspond to those for stabilizations in [EG2]; b) it is reasonable to choose the sheets for the points p_i^{++} and p_n^{--} in such a way that they are transformed into each other while going in the set of values of $f + z^p$ from the positive half-line to the negative one in the upper half-plane, and a "natural" choice for p_i^{++} forces (in the sense of the described coordination) to have a "non-natural" one for p_n^{--}.*
2. Here and in the notations below the signs in the upper-right index of a point (and of the corresponding vanishing cycle) indicate the signs of the value of the function $\tilde{f}$ (the first sign) and of the value of $\tilde{g}$ (the second one).

2) The critical points corresponding to the saddle points of the function $\tilde{g}$ in $\mathbb{R}^2$, i.e. to the self-intersection points of the real curve $\{\tilde{g} = 0\}$ (they are in one-to-one correspondence with the self-intersection points of the curve $\{\tilde{f} = 0\}$). We will denote by p_k^{+0} or by p_l^{-0} (depending on the sign of the value

of $\tilde{f}$: see Remark 2 above) the one with $\operatorname{Arg} z = -\frac{\pi}{p}$ or with $\operatorname{Arg} z = 0$. The corresponding vanishing cycles will be denoted by δ_k^{+0} and δ_l^{-0}. The orientation of the cycle δ_k^{+0} or of the cycle δ_l^{-0} will be defined in the level set $\{\tilde{g} = \varkappa\}$ in $\mathbb{C}^2$ with $\varkappa > 0$ small enough. Let $(\tilde{x}, \tilde{y})$ be a positively oriented local system of real coordinates in $\mathbb{R}^2$ centred at the corresponding critical point and such that $\tilde{g}(\tilde{x}, \tilde{y}) = \tilde{x}\tilde{y}$. In the level set $\{\tilde{g}(\tilde{x}, \tilde{y}) = \varkappa\}$ $(\varkappa > 0)$ we define the orientation of the vanishing cycle by its parameterization of the form $\tilde{x} = \sqrt{\varkappa}e^{it}$, $\tilde{y} = \sqrt{\varkappa}e^{-it}$ (the value of z has to be determined from the equation $\tilde{f}(\tilde{x}, \tilde{y}) + z^p = 0$; $\operatorname{Arg} z = -\frac{\pi}{p}$ or 0; compare with [GZ]).

All other critical points correspond to the critical points of the function $\tilde{g}$ on the curve $\{\tilde{f} = 0\} \subset \mathbb{R}^2$ (and they lie on this curve).

3) The critical points corresponding to maxima and minima of $\tilde{g}$ on the (smooth part of the real) curve $\{\tilde{f} = 0\}$ will be denoted by p_j^{0+} and p_m^{0-} respectively ($\tilde{g}$ is positive in maxima and negative in minima). At each of these points the pair of functions $\tilde{f}(x, y)$ and $\tilde{g}(x, y)$ can be reduced to x and $x \pm (-y^2 + c)$ ("+" for a maximum and "−" for a minimum). The corresponding vanishing cycle δ_j^{0+} or δ_m^{0-} has been described in the previous section. Other basic vanishing cycles corresponding to this point are $\sigma^q \delta_j^{0+}$ or $\sigma^q \delta_m^{0-}$ with $q = 1, \ldots, p - 2$.

4) The double points of the curve $\{\tilde{f} = 0\} \subset \mathbb{R}^2$ are in one-to-one correspondence with the double points of the curve $\{\tilde{g} = 0\} \subset \mathbb{R}^2$. We will denote the double point of the curve $\{\tilde{f} = 0\}$ corresponding to the double point p_k^{+0} (respectively to the double point p_l^{-0}) of the curve $\{\tilde{g} = 0\}$ by ${}^?p_k^{0+}$ respectively by ${}^?p_l^{0-}$; the value of the function $\tilde{g}$ at a double point of the curve $\{\tilde{f} = 0\}$ has the same sign as the value of $\tilde{f}$ at the corresponding double point of the curve $\{\tilde{g} = 0\}$.

To each such point of the curve $\{\tilde{f} = 0\}$ there correspond $2(p - 1)$ basic vanishing cycles. They can be described in the following way. There are two natural real smoothings of the curve $\{\tilde{f} = 0\}$ at this point. On one of them the function $\tilde{g}$ has two real non-degenerate critical points (on the other one the critical points are non-degenerate but imaginary).

To define the vanishing cycles corresponding to the point ${}^?p_k^{0+}$ (respectively to ${}^?p_l^{0-}$) we take the first smoothing (i.e. with real critical points of the restriction of the function $\tilde{g}$ to this smoothing) and we take the critical point with greater (respectively smaller) value of the function $\tilde{g}$ (i.e. the one which is further from the corresponding self-intersection point of the curve $\{\tilde{g} = 0\}$). If $\{\tilde{\tilde{f}} = 0\}$ is the equation of the smoothing ($\tilde{\tilde{f}}$ is a deformation of $\tilde{f}$), then at the chosen point the pair $(\tilde{\tilde{f}} + z^p, \tilde{g})$ can be reduced to $(x + z^p, x \pm (y^2 - c))$ ("+" for ${}^?p_k^{0+}$ and "−" for ${}^?p_l^{0-}$; $c > 0$). The corresponding vanishing cycle (which has been described in the previous section) will be denoted by ${}'\delta_k^{0+}$ (respectively by ${}'\delta_l^{0-}$). We define ${}''\delta_k^{0+}$ (respectively ${}''\delta_l^{0-}$) to be equal to ${}'\delta_k^{0+}$ (respectively to ${}'\delta_l^{0-}$). The set ${}'\delta_k^{0+}, \sigma {}'\delta_k^{0+}, \ldots, \sigma^{(p-2)} {}'\delta_k^{0+}, {}''\delta_k^{0+}, \sigma {}''\delta_k^{0+}, \ldots, \sigma^{(p-2)} {}''\delta_k^{0+}$ (respectively the set ${}'\delta_l^{0-}, \sigma {}'\delta_l^{0-}, \ldots, \sigma^{(p-2)} {}'\delta_l^{0-}, {}''\delta_l^{0-}, \sigma {}''\delta_l^{0-}, \ldots, \sigma^{(p-2)} {}''\delta_l^{0-})$

is a distinguished set of generators of the Milnor lattice of the corresponding singularity.

REMARK *This definition of the cycles $^?\delta_k^{0+}$ and $^?\delta_l^{0-}$ (? =$'$ or $''$) seems to be different from that used in [EG2] for the corresponding cycles. However it is possible to check that (for $p = 2$) they just coincide with each other.*

5 The equivariant intersection form. The equivariant Picard-Lefschetz formula for the A_{p-1} singularity

Let $X^{(p)} = \{f + z^p = 0,\ g = 0\} \subset (\mathbb{C}^3, 0)$ be the p-fold suspension of the fat point $X = \{f = 0,\ g = 0\} \subset (\mathbb{C}^2, 0)$. It is convenient to describe the intersection form on the Milnor lattice of $X^{(p)}$ in terms of the equivariant intersection form on it (with values in the group ring $\mathbb{Z}[\mathbb{Z}_p]$ of the cyclic group $\mathbb{Z}_p$). The cyclic group $\mathbb{Z}_p \subset S^1 \subset \mathbb{C}$ acts on the Milnor fibre of the *icis* $X^{(p)}$ by the formula $\sigma^q(x, y, z) = (x, y, \exp(2\pi qi/p)\, z)$, where $\sigma = \exp(2\pi i/p) \in \mathbb{C}$ is the generator of the group $\mathbb{Z}_p$. Thus the group $\mathbb{Z}_p$ acts on the Milnor lattice H of the *icis* $X^{(p)}$ and this action makes H a $\mathbb{Z}[\mathbb{Z}_p]$-module. The intersection form on H is $\mathbb{Z}_p$-invariant. Let us define the equivariant intersection number $\langle \alpha, \beta \rangle$ of the cycles α and β of H as

$$\sum_{q=0}^{p-1} (\alpha, \sigma^q \beta) \sigma^q \in \mathbb{Z}[\mathbb{Z}_p]$$

$((\cdot, \cdot)$ is the usual ($\mathbb{Z}$-valued) intersection form on H).

The equivariant intersection form $\langle \cdot, \cdot \rangle$ is $\mathbb{Z}[\mathbb{Z}_p]$-sesquilinear. This means the following. For $c = \sum c_q \sigma^q \in \mathbb{Z}[\mathbb{Z}_p]$ let $\bar{c} = \sum c_q \sigma^{-q}$. The operation $c \mapsto \bar{c}$ is an involution of the group ring $\mathbb{Z}[\mathbb{Z}_p]$. Then one has $\langle c\alpha, \beta \rangle = c\langle \alpha, \beta \rangle$, $\langle \alpha, c\beta \rangle = \bar{c}\langle \alpha, \beta \rangle$, $\langle \beta, \alpha \rangle = -\overline{\langle \alpha, \beta \rangle}$. A description in terms of the equivariant intersection form permits, in particular, to save space for the table of intersection numbers. E.g. the equality of the form $\langle \delta_1, \delta_2 \rangle = 1 - \sigma$ means that $(\delta_1, \delta_2) = 1$, $(\delta_1, \sigma\delta_2) = -1$, $(\delta_1, \sigma^q\delta_2) = 0$ for $q = 2, \ldots, p-1$ (and also that $(\sigma^{-1}\delta_1, \delta_2) = -1$, $(\sigma^q\delta_1, \delta_2) = 0$ for $q = -2, \ldots, -(p-1); \ldots)$.

We have mentioned that if (x_0, y_0) is a non-degenerate critical point of the function $\tilde{g}(x, y)$ on (the smooth part of) the curve $\{\tilde{f} = 0\}$, then near the point $(x_0, y_0, 0) \in \{\tilde{f} + z^p = 0\}$ the restriction of the function $\tilde{g}(x, y)$ to the curve $\{\tilde{f} + z^p = 0\} \subset \mathbb{C}^3$ has an equivariant singularity of type A_{p-1}. Going around the corresponding critical value in $\mathbb{C}$ in the positive direction (i.e. counterclockwise) one gets the corresponding transformation of the Milnor fibre (the monodromy transformation), which can be assumed to be equivariant, and thus a $\mathbb{Z}[\mathbb{Z}_p]$-linear automorphism of the Milnor lattice (of the *icis* $\{f + z^p = 0,\ g = 0\}$). Let us describe this automorphism in terms of the equivariant intersection form on the Milnor lattice. It is essentially used in the course of the calculations (for the proof of Theorem 2).

As we know (see Section 3) the vanishing homology group of the singularity under consideration as a $\mathbb{Z}[\mathbb{Z}_p]$-module is generated by one vanishing cycle δ (and is isomorphic to $\mathbb{Z}[\mathbb{Z}_p]/(1+\sigma+\ldots+\sigma^{p-1}))$. The set $\delta, \sigma\delta, \ldots, \sigma^{p-2}\delta$ is a distinguished basis of this vanishing homology group. It is not difficult to see that $\langle \delta, \delta \rangle = \sigma - \sigma^{-1}$. A path s from the critical value to the non-critical one (corresponding to the Milnor fibre) determines a vanishing cycle (which will be denoted by δ also) in the Milnor lattice H of the *icis* $X^{(p)}$. Let h be the monodromy transformation of the Milnor lattice H induced by the (simple) loop corresponding to the path s. Let us consider the lattice H as embedded into $H_\mathbb{Q} = H \otimes_\mathbb{Z} \mathbb{Q}$ ($\mathbb{Q}$ is the field of rational numbers). $H_\mathbb{Q}$ is a $\mathbb{Q}[\mathbb{Z}_p]$-module, where $\mathbb{Q}[\mathbb{Z}_p]$ is the rational group ring of the group $\mathbb{Z}_p$.

THEOREM 1 *For $a \in H$ one has*

$$h(a) = a - \langle a, \lambda \rangle \delta, \tag{1}$$

$$h^{-1}(a) = a + \langle a, (\delta - \lambda) \rangle \delta,$$

where $\lambda = \frac{1}{p}((p-1) + (p-2)\sigma + \ldots + \sigma^{p-2})\delta \in (\mathbb{Q}[\mathbb{Z}_p])\delta$.

Of course these formulae can also be applied in cases different from those for suspensions.

Proof. Let h_i be the usual Picard-Lefschetz transformation of the lattice H corresponding to the cycle $\delta_i = \sigma^i \delta$, $i = 0, 1, \ldots, (p-2)$: $h_i(a) = a - (a, \delta_i)\delta_i$. Since $\delta_0, \delta_1, \ldots, \delta_{p-2}$ is a distinguished basis of the corresponding lattice, one has $h = h_0 \circ h_1 \circ \ldots \circ h_{p-2}$. Therefore

$$
\begin{aligned}
h(a) &= a - \sum_{q=1}^{p-2} \left(\sum_{r=q}^{p-2} (a, \delta_r) \right) \delta_q \\
&= a - \sum_{q=1}^{p-2} \left(\sum_{r=q}^{p-2} (a, \sigma^r \delta) \right) \sigma^q \delta + \left(a, \frac{1}{p} \sum_{i=1}^{p-1} i\sigma^{i-1}\delta \right) \left(\sum_{j=0}^{p-1} \sigma^j \delta \right).
\end{aligned}
$$

The last term could be added because of the relation $(1+\sigma+\ldots+\sigma^{p-1})\delta = 0$. By repeated application of this relation, one can easily compute that the coefficient of $\sigma^q\delta$, $q = 0, \ldots, p-1$, in the above formula for $h(a)$ is equal to

$$-\left(a, \frac{1}{p}((p-1) + (p-2)\sigma + \ldots + \sigma^{p-2})\sigma^q\delta \right).$$

The first of the two formulas in Theorem 1 now follows from the definition of the equivariant intersection form. The proof of the second one is similar. $\quad\square$

REMARKS   *1. It is reasonable to consider h as an equivariant Picard-Lefschetz transformation and thus to call the equation (1) the equivariant Picard-Lefschetz formula (for the singularity A_{p-1}). The equivariant analogue of the Picard-Lefschetz formula for a free $\mathbb{Z}_p$-orbit consisting of non-degenerate critical points (in our case — for a set of critical points corresponding to a critical point of the function $\tilde{g}$ in $\mathbb{C}^2$) looks just like the usual (non-equivariant) one: $h(a) = a - \langle a, \delta \rangle \delta$.*

2. It is possible to write the formulae of Theorem 1 in the form

$$h(a) = a - c\langle a, \delta \rangle \delta,$$

$$h^{-1}(a) = a + (1 - c)\langle a, \delta \rangle \delta,$$

where

$$c := \frac{1}{p} \sum_{i=1}^{p-1} (p - i)\sigma^{1-i} \in \mathbb{Q}[\mathbb{Z}_p].$$

Note that $c\sigma - c = -\frac{1}{p}(1 + \cdots + \sigma^{p-1}) + \sigma$, and therefore

$$(1 - c)(1 - \sigma) \equiv 1 \; mod \; \mathbb{Q}(1 + \cdots + \sigma^{p-1}).$$

However we prefer the other form believing that it is more generalizable.

3. For calculations it is useful to know the (equivariant) intersection number $\langle \delta, \lambda \rangle$. It is equal to $1 + \sigma - \frac{2}{p}(1 + \ldots + \sigma^{p-1})$. If $\langle a, \delta \rangle = (1 - \sigma)P(\sigma)$ $(P(\sigma) \in \mathbb{Q}[\mathbb{Z}_p])$, then $\langle a, \lambda \rangle \delta = -\sigma P(\sigma)\delta$, $\langle a, \delta - \lambda \rangle \delta = P(\sigma)\delta$.

4. For $a = \delta$, one has $h(\delta) = -\sigma \delta$.

6   The intersection form of the suspension

In order to formulate the main statement let us use the following conventions. Each vanishing cycle is indexed by two indices, q and another one. We group the vanishing cycles together by using two pairs of brackets which can be curly or square brackets according to the following rules:

1) Interior brackets (curly or square) mean the set of cycles, indicated inside them, with different q. Inside curly brackets the index q runs from 0 to $p - 1$ and inside square ones from 0 to $p - 2$.

2) Exterior brackets mean the set of all (groups of) cycles with all possible values of the other index (i.e. the index different from q). Exterior brackets can only be curly ones.

3) Inside any pair of curly brackets the order of cycles (for interior brackets) or of groups of cycles (for exterior ones) is not essential.

4) In a pair of square brackets the order of cycles corresponds to the order $0, 1, \ldots, (p - 2)$ of the index q.

REMARK *There is a little redundancy in this notation: if and only if the first of the two upper right indices of the cycles is zero, we have interior square brackets.*

The self-intersection points and the components of the complement to the curve $\{\tilde{f} = 0\} \subset \mathbb{R}^2$ are in one-to-one correspondence with those for the curve $\{\tilde{g} = 0\} \subset \mathbb{R}^2$. This fact permits us to use a phrase of the sort "the self-intersection point of the curve $\{\tilde{f} = 0\}$ corresponding to the cycle δ_k^{+0}" (the genuine critical point corresponding to a cycle always can be understood from the notations).

THEOREM 2 *The cycles*

$$\{\{\sigma^q \delta_i^{++}\}\}, \ \{[\sigma^q \delta_j^{0+}]\}, \ \{[\sigma^{q\,\prime} \delta_k^{0+}], [\sigma^{q\,\prime\prime} \delta_k^{0+}]\}, \ \{\{\sigma^q \delta_k^{+0}\}, \{\sigma^q \delta_l^{-0}\}\},$$

$$\{[\sigma^{q\,\prime} \delta_l^{0-}], \ [\sigma^{q\,\prime\prime} \delta_l^{0-}]\}, \ \{[\sigma^q \delta_m^{0-}]\}, \ \{\{\sigma^q \delta_n^{--}\}\}$$

with the indicated order form a distinguished set of generators of the vanishing homology group of the suspension $X^{(p)} = \{f + z^p = 0, \ g = 0\}$ of the fat point $X = \{f = g = 0\}$. The equivariant intersection numbers between these cycles are equal to zero except in the following cases.

I. Between the cycles corresponding to the critical points of the function $\tilde{g}$ in $\mathbb{R}^2$ (i.e. δ_i^{++}, δ_k^{+0}, δ_l^{-0}, and δ_n^{--}).

1. *$\langle \delta_i^{++}, \delta_k^{+0} \rangle = 1$ (respectively $\langle \delta_i^{++}, \delta_k^{+0} \rangle = \sigma$) if the double point of the curve $\{\tilde{g} = 0\}$ corresponding to the cycle δ_k^{+0} is on the boundary of the component of the complement to the curve $\{\tilde{g} = 0\}$ corresponding to the cycle δ_i^{++} and inside (respectively outside) the corresponding component of the complement to the curve $\{\tilde{f} = 0\}$.*

2. *$\langle \delta_i^{++}, \delta_l^{-0} \rangle = 1$ if the double point of the curve $\{\tilde{g} = 0\}$ corresponding to the cycle δ_l^{-0} is on the boundary of the component of the complement to the curve $\{\tilde{g} = 0\}$ corresponding to the cycle δ_i^{++}.*

3. *$\langle \delta_k^{+0}, \delta_n^{--} \rangle = 1$ if the double point of the curve $\{\tilde{g} = 0\}$ corresponding to the cycle δ_k^{+0} is on the boundary of the component of the complement to the curve $\{\tilde{g} = 0\}$ corresponding to the cycle δ_n^{--}.*

4. *$\langle \delta_l^{-0}, \delta_n^{--} \rangle = 1$ (respectively $\langle \delta_l^{-0}, \delta_n^{--} \rangle = \sigma$) if the double point of the curve $\{\tilde{g} = 0\}$ corresponding to the cycle δ_l^{-0} is on the boundary of the component of the complement to the curve $\{\tilde{g} = 0\}$ corresponding to the cycle δ_n^{--} and inside (respectively outside) the corresponding component of the complement to the curve $\{\tilde{f} = 0\}$.*

5. *$\langle \delta_i^{++}, \delta_n^{--} \rangle = -1$ if the components of the complement to the curve $\{\tilde{g} = 0\}$ corresponding to the cycles δ_i^{++} and δ_n^{--} have a common side.*

II. Between the cycles corresponding to the critical points of the function $\tilde{g}$ on the curve $\{\tilde{f} = 0\} \subset \mathbb{R}^2$ (i.e. δ_j^{0+}, ${}^?\delta_k^{0+}$, ${}^?\delta_l^{0-}$ and δ_m^{0-}; $? ='$ or $''$).

0. For each of these cycles the self-intersection number is equal to $\sigma - \sigma^{-1}$.

1. $\langle \delta_j^{0+}, \delta_m^{0-} \rangle = 1 - \sigma$ (respectively $\langle \delta_j^{0+}, \delta_m^{0-} \rangle = -1 + \sigma$) if the segment corresponding to the cycle δ_m^{0-} is a continuation of the segment corresponding to the cycle δ_j^{0+} and if there is no self-intersection point (respectively if there is a self-intersection point) of the curve $\{\tilde{f} = 0\}$ near their common end.

2. $\langle \delta_j^{0+}, {}^?\delta_k^{0+} \rangle = 1 - \sigma$ (respectively $\langle \delta_j^{0+}, {}^?\delta_l^{0-} \rangle = 1 - \sigma$) if the self-intersection point of the curve $\{\tilde{f} = 0\}$ corresponding to the cycle(s) ${}^?\delta_k^{0+}$ (respectively to ${}^?\delta_l^{0-}$) is close to one end point of the segment corresponding to the cycle δ_j^{0+}.

3. $\langle {}^?\delta_k^{0+}, \delta_m^{0-} \rangle = 1 - \sigma$ (respectively $\langle {}^?\delta_l^{0-}, \delta_m^{0-} \rangle = 1 - \sigma$) if the self-intersection point of the curve $\{\tilde{f} = 0\}$ corresponding to the cycle(s) ${}^?\delta_k^{0+}$ (respectively to ${}^?\delta_l^{0-}$) is close to one end point of the segment corresponding to the cycle δ_m^{0-}.

III. Between the cycles corresponding to the critical points of $\tilde{g}$ in $\mathbb{R}^2$ and the cycles corresponding to the critical points of $\tilde{g}$ on the curve $\{\tilde{f} = 0\} \subset \mathbb{R}^2$.

1. $\langle \delta_i^{++}, \delta_j^{0+} \rangle = - - 1 + \sigma^{-1}$ (respectively $\langle \delta_i^{++}, \delta_m^{0-} \rangle = 1 - \sigma^{-1}$) if the critical point corresponding to the cycle δ_j^{0+} (respectively δ_m^{0-}) is on the boundary of the component of the complement to the curve $\{\tilde{f} = 0\} \subset \mathbb{R}^2$ corresponding to the cycle δ_i^{++}.

2. $\langle \delta_i^{++}, {}^?\delta_k^{0+} \rangle = 1 - \sigma^{-1}$ if the self-intersection point of the curve $\{\tilde{f} = 0\} \subset \mathbb{R}^2$ corresponding to the cycle ${}^?\delta_k^{0+}$ is inside the component of the complement to the curve $\{\tilde{g} = 0\} \subset \mathbb{R}^2$ corresponding to the cycle δ_i^{++} (the last cycle corresponds both to a component of the complement to the curve $\{\tilde{f} = 0\}$ and to one of the complement to the curve $\{\tilde{g} = 0\}$).

3. $\langle \delta_i^{++}, {}^?\delta_l^{0-} \rangle = 0.$

4. $\langle \delta_j^{0+}, \delta_k^{+0} \rangle = - - \sigma^2 + \sigma$ (respectively $\langle \delta_k^{+0}, \delta_m^{0-} \rangle = -1 + \sigma^{-1}$) if an end point of the segment corresponding to the cycle δ_j^{0+} (respectively δ_m^{0-}) is close to the self-intersection point of the curve $\{\tilde{g} = 0\} \subset \mathbb{R}^2$ corresponding to the cycle δ_k^{+0}.

5. $\langle \delta_j^{0+}, \delta_l^{-0} \rangle = 1 - \sigma$ (respectively $\langle \delta_l^{-0}, \delta_m^{0-} \rangle = 1 - \sigma$) if an end point of the segment corresponding to the cycle δ_j^{0+} (respectively δ_m^{0-}) is close to the self-intersection point of the curve $\{\tilde{g} = 0\} \subset \mathbb{R}^2$ corresponding to the cycle δ_l^{-0}.

6. $\langle {}^?\delta_k^{0+}, \delta_k^{+0}\rangle = -\sigma + \sigma^2$, $\langle \delta_l^{-0}, {}^?\delta_l^{0-}\rangle = -1 + \sigma$.

7. $\langle \delta_j^{0+}, \delta_n^{--}\rangle = -1 + \sigma$ *(respectively* $\langle \delta_m^{0-}, \delta_n^{--}\rangle = 1 - \sigma$*) if the critical point corresponding to the cycle* δ_j^{0+} *(respectively* δ_m^{0-}*) is on the boundary of the component of the complement to the curve* $\{\tilde{f} = 0\} \subset \mathbb{R}^2$ *corresponding to the cycle* δ_n^{--}.

8. $\langle {}^?\delta_k^{0+}, \delta_n^{--}\rangle = 0$.

9. $\langle {}^?\delta_l^{0-}, \delta_n^{--}\rangle = -1 + \sigma$ *if the self-intersection point of the curve* $\{\tilde{f} = 0\} \subset \mathbb{R}^2$ *corresponding to the cycle* ${}^?\delta_l^{0-}$ *is inside the component of the complement to the curve* $\{\tilde{g} = 0\} \subset \mathbb{R}^2$ *corresponding to the cycle* δ_n^{--}.

REMARK *In some cases (for particular curves* $\{\tilde{f} = 0\}$ *and* $\{\tilde{g} = 0\}$*) there can be a situation when, say, an intersection point of the curve* $\{\tilde{f} = 0\}$ *corresponding to the cycle* ${}^?\delta_k^{0+}$ *is close to both end points of the segment corresponding to the cycle* δ_j^{0+}*. In this case the connection between them has to be counted twice and we have* $\langle \delta_j^{0+}, {}^?\delta_k^{0+}\rangle = 2(1 - \sigma)$*. We do not indicate such ("evident") modifications of the formulae.*

7 Sketch of the proof of Theorem 2

As in [EG2] the greater part of the calculations of the (equivariant) intersection numbers of the cycles from the described distinguished set consists in bringing them to the zero level set of the function $\tilde{g}$ (or close to it if the double points of it are of interest) along the real axis in the complex line of values of $\tilde{g}$ and analyzing their intersections there. Some additional considerations are needed if the cycle meets a critical point in the course of this deformation: it is necessary to go around the corresponding critical value *in the upper half-plane*. In the zero-level set all these cycles are represented by parts of the preimage of the real curve $\{\tilde{g} = 0\}$ (under the projection $(x, y, z) \mapsto (x, y)$ to the (x, y)-plane). The cycles can either intersect at a point or coincide along a certain common part. To calculate the intersection number one has to keep track of the behavior of the cycles at the intersection point or at the end (divergence) points of the common part. The intersection points or/and the divergence points of the cycles can be the preimages either of intersection points of the (real) curves $\{\tilde{f} = 0\}$ and $\{\tilde{g} = 0\}$ or of self-intersection points of the curve $\{\tilde{g} = 0\}$.

Near (the preimage of) an intersection point of the curves $\{\tilde{f} = 0\}$ and $\{\tilde{g} = 0\}$ the complex curve $\{\tilde{f} + z^p = 0, \tilde{g} = 0\}$ is non-singular and z is a (regular) local coordinate on it. The preimage of the real curve $\{\tilde{g} = 0\}$ in the surface $\{\tilde{f} + z^p = 0\}$ (i.e. the set $\{\tilde{f} + z^p = 0, \tilde{g} = 0, x \text{ and } y \text{ are real}\}$) coincides with the union of p lines (see Fig. 3; $p = 5$) and thus consists of $2p$ rays $\{\text{Arg } z = q\pi/p\}$, $q = 0, 1, \ldots, 2p - 1$. Near such a point each cycle from the distinguished set goes along two of these $2p$ rays. Therefore the behavior

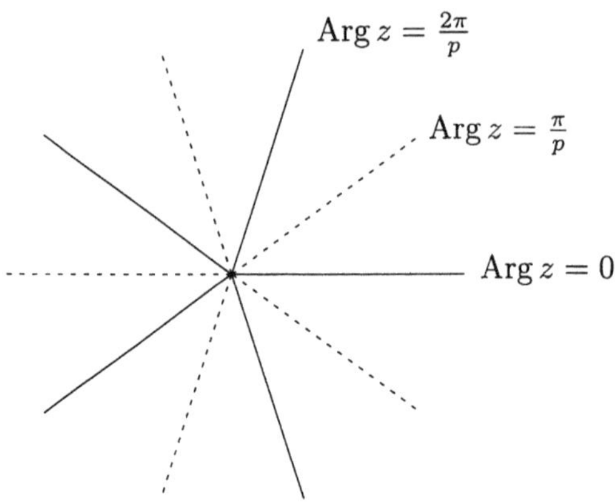

Figure 3: Planar structure at an intersection point of $\{\tilde{f} = 0\}$ and $\{\tilde{g} = 0\}$

of cycles near this point can be described (and understood) by considering the natural planar structure at this point shown in Fig. 3.

The situation near a preimage of a self-intersection point of the curve $\{\tilde{g} = 0\}$ can be described (and understood) in terms of (two) planar structures corresponding to such a point (see the description in [EG2]). In fact each vanishing cycle can be considered as consisting of some arcs — preimages of arcs of the real curve $\{\tilde{g} = 0\}$ and some halves of vanishing cycles corresponding to the self-intersection points of this curve.

If in the course of the deformation to the zero-level set along the real axis a cycle δ_i^{++} or δ_j^{0+} (respectively a cycle δ_n^{--} or δ_m^{0-}) meets the critical point corresponding to cycles $^?\delta_k^{0+}$ (respectively to $^?\delta_l^{0-}$) it is more convenient to use the following trick. At first we make the smoothing of the curve $\{\tilde{f} = 0\}$ which makes the critical points of the function $\tilde{g}$ (and its critical values) corresponding to the cycles $^?\delta_k^{0+}$ (or $^?\delta_l^{0-}$) imaginary (non-real). After that in the course of the deformation of the cycle δ_i^{++}, δ_j^{0+}, δ_n^{--}, or δ_m^{0-} along the real axis it does not meet the (corresponding) critical points and hence the cycle in the zero-level set obtained in this way can be easily described. Thus its (equivariant) intersection numbers with other cycles can be calculated by considerations of the described planar structures at the intersection or divergence points. However the cycle obtained in this way is not δ_i^{++} or δ_j^{0+} (respectively δ_n^{--} or δ_m^{0-}) itself but the image of the respective cycle under the inverse Picard-Lefschetz transformation corresponding to the cycle $'\delta_k^{0+}$ (respectively under the Picard-Lefschetz transformation corresponding to the cycle $'\delta_l^{0-}$). Using the equivariant Picard-Lefschetz formula (Section 5) one can determine the desired equivariant intersection number (the intersection numbers of the cycles in the pair with the cycle $'\delta_k^{0+}$ or $'\delta_l^{0-}$ should be already determined).

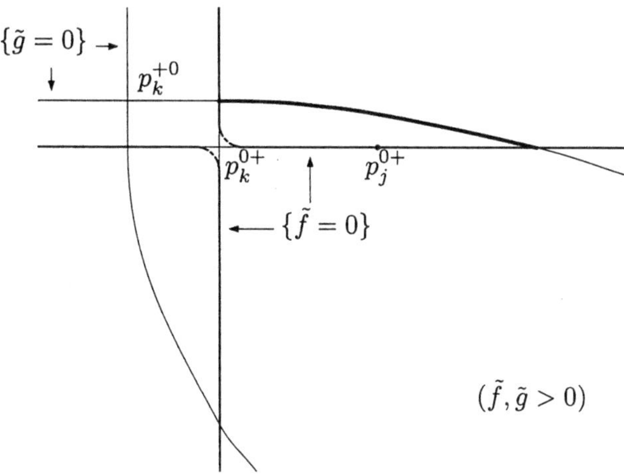

Figure 4: The picture for calculating $\langle\delta_j^{0+},\delta_k^{+0}\rangle$

An example of a calculation by considering planar structures at intersection and/or divergence points can be obtained from one in [EG2] (for $\langle\delta_j^{0+},\delta_n^{--}\rangle$) without essential changes.

As an example which uses the method of "artificial" removing of critical points (and of critical values) from the real domain, we calculate the intersection number $\langle\delta_j^{0+},\delta_k^{+0}\rangle$ if an end point of the segment corresponding to the cycle δ_j^{0+} is close to the self-intersection point p_k^{+0} of the curve $\{\tilde{g}=0\}\subset\mathbb{R}^2$ corresponding to the cycle δ_k^{+0} (Fig. 4).

The intersection numbers $\langle{}'\delta_k^{0+},\delta_k^{+0}\rangle=-\sigma+\sigma^2$ and $\langle\delta_j^{0+},{}'\delta_k^{0+}\rangle=1-\sigma$ have to be calculated in advance. Both of them can be calculated directly (by considering planar structures at intersection and divergence points). In the first case the method of removing of critical points from the real domain can make the calculations shorter. However in this case only the "degenerate" version of the equivariant Picard-Lefschetz formula (see Remark 4 at the end of Section 5) has to be used. The situation after the smoothing of the self-intersection point ${}^?p_k^{0+}$ of the curve $\{\tilde{f}=0\}$, for which the corresponding critical points of the function $\tilde{g}$ on this curve are non-real, can be understood from the same figure. The cycle, vanishing at the point p_j^{0+}, being deformed to the zero-level set of the function $\tilde{g}$ along the real axis, consists of preimages of the arc marked in the figure.

Hence (near the zero-level set) it has no (geometrical) intersections with the cycle δ_k^{+0} and the corresponding equivariant intersection number is equal to zero. The cycle thus obtained is not δ_j^{0+} but the image of δ_j^{0+} under the inverse (equivariant) Picard-Lefschetz transformation h^{-1} corresponding to the cycle

$'\delta_k^{0+}$. So, $\langle h^{-1}\delta_j^{0+}, \delta_k^{+0}\rangle = 0$. By Remark 2 following Theorem 1 we have

$$
\begin{aligned}
h^{-1}\delta_j^{0+} &= \delta_j^{0+} + (1-c)\langle\delta_j^{0+}, {}'\delta_k^{0+}\rangle\,{}'\delta_k^{0+} \\
&= \delta_j^{0+} + (1-c)(1-\sigma)\,{}'\delta_k^{0+} \\
&= \delta_j^{0+} + {}'\delta_k^{0+}.
\end{aligned}
$$

Thus $\langle\delta_j^{0+}, \delta_k^{+0}\rangle = -\langle{}'\delta_k^{0+}, \delta_k^{+0}\rangle = \sigma - \sigma^2.$ $\square$

8   Relations between vanishing cycles of the distinguished set

In the vanishing homology group of the *icis* $\{f + z^p = 0, g = 0\}$ there are $\mu(f + z^p) = (p-1)\mu(f)$ linearly independent (over $\mathbb{Z}$) relations between th e elements of a distinguished set of generators. On the other hand the set of relations is a $\mathbb{Z}[\mathbb{Z}_p]$-submodule in $(\mathbb{Z}[\mathbb{Z}_p])^\nu$. Its rank over the group ring $\mathbb{Z}[\mathbb{Z}_p]$ is equal to $\mu(f)$ and thus over $\mathbb{Z}[\mathbb{Z}_p]$ there are $\mu(f)$ linearly independent relations.

THEOREM 3   *The linearly independent (over $\mathbb{Z}[\mathbb{Z}_p]$) relations between the cycles of the distinguished set in the Milnor lattice of the p-fold suspension $X^{(p)} = \{f + z^p = 0,\ g = 0\}$ of the icis (fat point) $\{f = 0,\ g = 0\}$ correspond to the self-intersection points of the real curve $\{\tilde f = 0\} \subset \mathbb{R}^2$ and to the bounded components of its complement. A relation of the first type has the form ${}'\delta_k^{0+} = {}''\delta_k^{0+}$ or ${}'\delta_l^{0-} = {}''\delta_l^{0-}$. A relation of the second type has the form*

$$
(1-\sigma)\left(\delta_i^{++} + \sum \delta_k^{+0}\right) -
$$
$$
- \left(\sum \delta_j^{0+} + \sum \sigma^{-s}\,{}'\delta_k^{0+} + \sum {}'\delta_l^{0-} + \sum \delta_m^{0-}\right) = 0
$$

for a component of the complement corresponding to the cycle δ_i^{++} (i.e. with $\tilde f > 0$) and the form

$$
(1-\sigma)\left(\delta_n^{--} + \sum \delta_l^{-0}\right) +
$$
$$
+ \left(\sum \delta_j^{0+} + \sum {}'\delta_k^{0+} + \sum \sigma^s\,{}'\delta_l^{0-} + \sum \delta_m^{0-}\right) = 0
$$

for a component corresponding to the cycle δ_n^{--} (i.e. with $\tilde f < 0$). In each formula the first sum is over k (respectively l) for which the self-intersection points of the curve $\{\tilde g = 0\}$ corresponding to the cycle δ_k^{+0} (respectively to δ_l^{-0}) lie inside the component under consideration. The other sums are over those j, k, l, and m, for which the points of the (real) curve $\{\tilde f = 0\}$ corresponding to the cycles δ_j^{0+}, ${}'\delta_k^{0+}$, ${}'\delta_l^{0-}$, and δ_m^{0-} lie on the boundary of the component. Here

$s = 1$ *if the self-intersection point of the curve* $\{\tilde{g} = 0\} \subset \mathbb{R}^2$ *corresponding to the cycle* δ_k^{+0} *(respectively to the cycle* δ_l^{-0}*) is inside the component (and thus participates in the first sum) and* $s = 0$ *otherwise.*

Proof. The relations of the first type are evident. The form of the relations corresponding to the bounded components of the complement to the curve $\{\tilde{f} = 0\}$ can be guessed with the help of the following statement (which we formulate more precisely than in [EG2]).

Let $h : M^2 \to \mathbb{C}$ be a holomorphic function on a non-singular complex surface M^2, $S \subset M^2$ — a compact smooth real oriented subsurface in M such that $h|_S$ is real, it has only non-degenerate critical points, and the critical values of $h|_S$ at all saddle points are equal to zero. Let u_0 be a small non-critical value of h in the upper half-plane: the absolute value of u_0 should be smaller than that of any non-zero critical value of h. Let us take a system of paths connecting the critical values of $h|_S$ with u_0 which go in the upper half-plane along the real axis in $\mathbb{C}$ and very close to it (with smaller imaginary part than all non-real critical values of h). Let $\delta_1, \dots, \delta_s$ be the system of vanishing cycles in the level manifold (curve) $h^{-1}(u_0)$ defined by this system of paths. We orient them in the following way. We define the orientation of the vanishing cycle corresponding to a maximum (minimum) of $h|_S$ with the (real) critical value u in the level set $\{h = u - \varkappa\}$ (respectively in the level set $\{h = u + \varkappa\}$) with $\varkappa > 0$ small enough. In this level set the vanishing cycle is represented by a (real) oval in the surface S and we orient it counterclockwise (according to the orientation of S). At a saddle point of $h|_S$ we choose a positively oriented system $(\tilde{x}, \tilde{y})$ of real local coordinates on S such that $h(\tilde{x}, \tilde{y}) = \tilde{x}\tilde{y}$ and define the orientation of the corresponding vanishing cycle in the level set $\{h = \varkappa\}$) ($\varkappa > 0$ small enough) by its parameterization $\tilde{x} = \sqrt{\varkappa}e^{it}$, $\tilde{y} = \sqrt{\varkappa}e^{-it}$.

LEMMA 1 *The sum* $\delta_1 + \dots + \delta_s$ *of the described vanishing cycles* δ_i *is equal to zero.*

The situation we have is far from the one in the Lemma. However an appropriate smoothing of the level set $\{\tilde{f} + z^p = 0\}$ leads to a situation close to the described one. The difference is that the values of the restriction of function $\tilde{g}$ to (the smoothing of) the level set $\{\tilde{f} + z^p = 0\}$ at the saddle points are not equal to zero. However they are close to zero and hence it is possible to assume that one has a relation of the described form. When the relations have been written down it is not difficult to verify them. $\qquad\square$

9 Examples

In order not to overload figures with Dynkin diagrams we will use the following conventions (which can be applied only to suspensions and moreover only to diagrams obtained by the described method).

1. Vertices of a diagram correspond to vanishing cycles from the distinguished set which were chosen as generators of the Milnor lattice H as a $\mathbb{Z}[\mathbb{Z}_p]$-module.

2. If a vanishing cycle (from the distinguished set) generates a free $\mathbb{Z}[\mathbb{Z}_p]$-submodule in H, i.e. if it is one of δ_i^{++}, δ_k^{+0}, δ_l^{-0}, or δ_n^{--}, then the corresponding vertex is indicated by $\circ$. If a vanishing cycle is one of δ_j^{0+} or δ_m^{0-}, then the corresponding vertex is indicated by $\bullet$. We show only one vertex for the pair of (equal) cycles $({}'\delta_k^{0+}, {}''\delta_k^{0+})$ or $({}'\delta_l^{0-}, {}''\delta_l^{0-})$. Such a vertex is indicated by $\star$. Cycles corresponding to $\bullet$ or $\star$ vertices generate submodules in H isomorphic to $\mathbb{Z}[\mathbb{Z}_p]/(1+\sigma+\ldots+\sigma^{p-1})$. Therefore they have similar properties and are treated in the same way (as $\bullet$ vertices) in the next convention.

3. The (equivariant) intersection number of cycles corresponding to two $\circ$ vertices (respectively corresponding to two $\bullet$ vertices or to one $\bullet$ and one $\star$ vertex), taken in their order in the distinguished set (!), is equal to $\pm\sigma^w$ (respectively to $\pm\sigma^w(1-\sigma)$). It is shown by a solid (for the "+" sign) or by a dotted (for the "$-$" sign) edge with the weight w associated to it. If the weight w is equal to zero, it (the weight, but not the edge) is not indicated in the diagram.

1. The most simple suspension is the suspension over the intersection of two quadrics (i.e. of two pairs of lines) in $\mathbb{C}^2$: $\{x^2 - y^2 + z^p = xy = 0\}$. For $p = 3$ this is the simple (in the sense of M. Giusti [G]) *icis* Z_9. Its Coxeter-Dynkin diagram (obtained by a different method) has been described in [EG1]. The real picture corresponding to this singularity (one of several similar ones) is shown in Fig. 5. Its (equivariant) Coxeter-Dynkin diagram obtained from Theorem 2 is shown in Fig. 6. This diagram gives rise to the usual (non-equivariant) Coxeter-Dynkin diagram shown in Fig. 7.

2. In Wall's list of unimodular curve *icis* in $\mathbb{C}^3$ ([W]) there are only *icis* , the first (generic) equations of which have singularities not more complicated than A_2. Thus there can be suspensions (with $p > 2$, i.e. not stabilizations) only over fat points of the form $\{xy = 0, (x+y^r)(y+x^s) = 0\}$ (moreover there can be only 3-suspensions over such *icis*). A real picture corresponding to this fat point is shown in Fig. 8 ($r = 3$, $s = 4$). The corresponding Coxeter-Dynkin diagram is shown in Fig. 9. Instead of numbering vertices according to their order in the distinguished basis here and in the example below we add arrows to the edges in the direction of the increase of the numbering. In fact any numbering coherent with the arrows determines a distinguished set of generators.

3. The examples 1 and 2 are somewhat degenerate: there are only few possible types of connections (points of Theorem 2) used for them (e.g. the function $\tilde{g}$ has no maxima or minima in the plane $\mathbb{R}^2$). As a "non-degenerate" example we

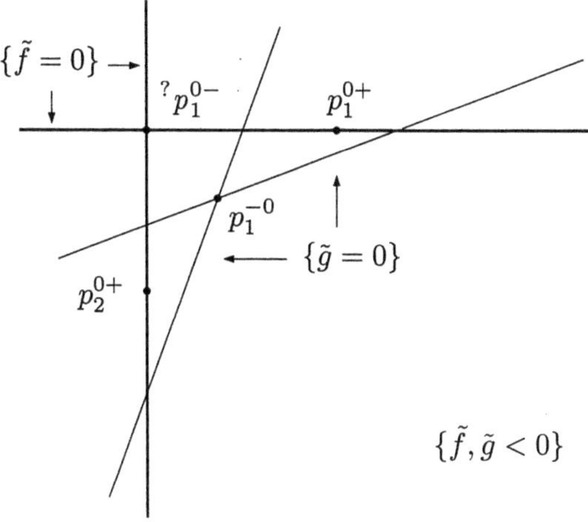

Figure 5: Real picture for Z_9

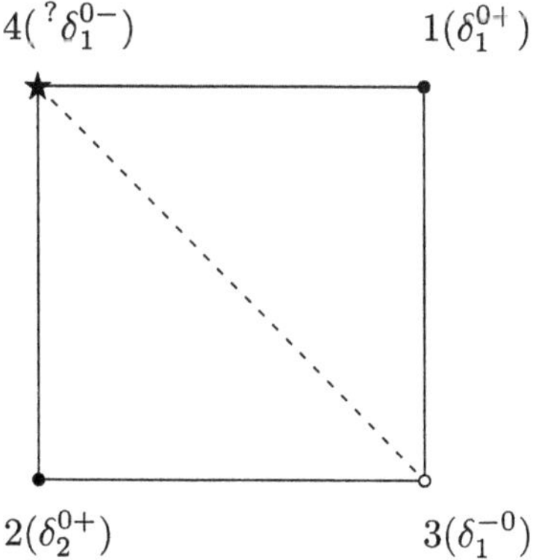

Figure 6: Equivariant Coxeter-Dynkin diagram of Z_9

                    W. Ebeling and S.M. Gusein-Zade

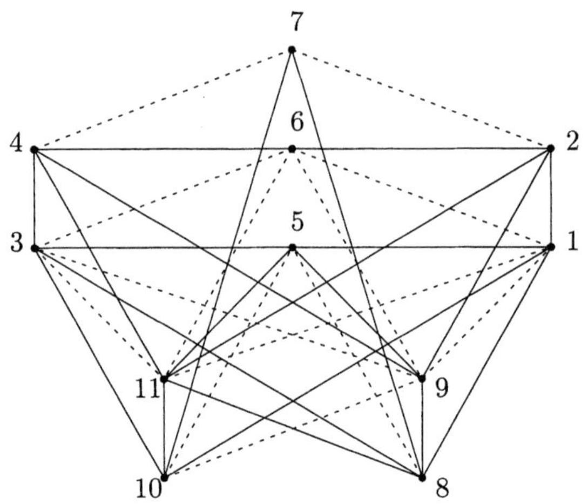

Figure 7: Coxeter-Dynkin diagram of Z_9

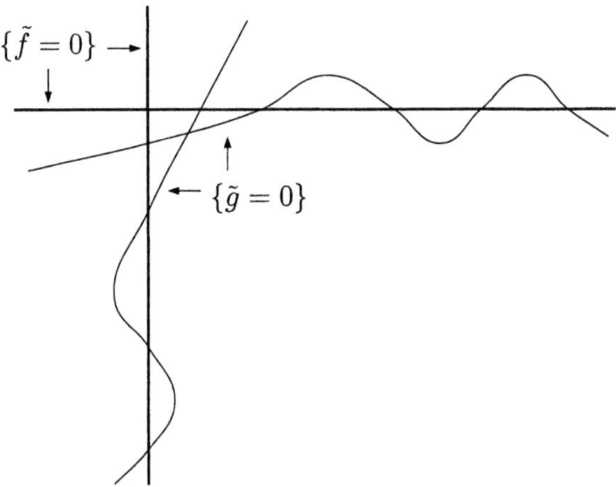

Figure 8: Real picture for $\{xy = (x + y^r)(y + x^s) = 0\}$; $r = 3$, $s = 4$

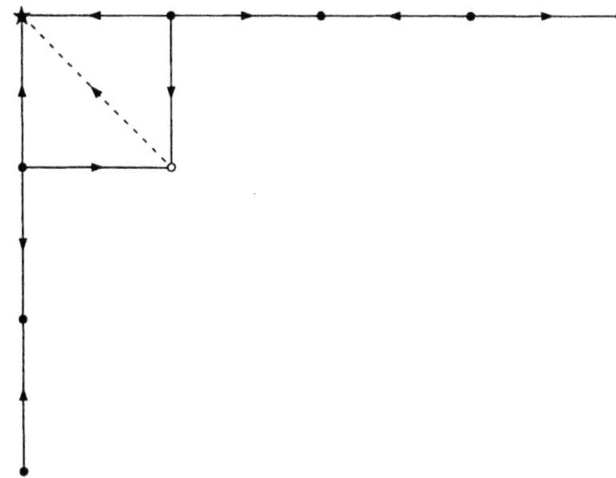

Figure 9: Equivariant Coxeter-Dynkin diagram for $\{xy = (x+y^r)(y+x^s) = 0\}$

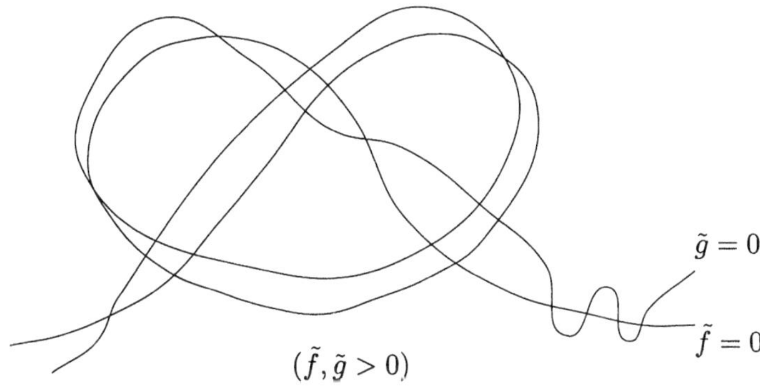

Figure 10: Example 3

show the equivariant Coxeter-Dynkin diagram for the suspension over the fat point considered in [EG2]: $x^3 - y^4 = 0$, $y^5 = 0$, using the same real picture — Fig. 10.

REMARK *This picture is not the most convenient one if one wants to have a less complicated diagram. For this it is reasonable to pull all (possible) intersection points of the curves $\{\tilde{f} = 0\}$ and $\{\tilde{g} = 0\}$ to one "tail". It can be done either by admissible modifications ([EG2]) or by choosing appropriate analytic deformations of the equations.*

Since it is (almost) impossible to show the Coxeter-Dynkin diagram in one figure and for better visuality we show it in three figures (Fig. 11–13) corresponding to the three parts of Theorem 2.

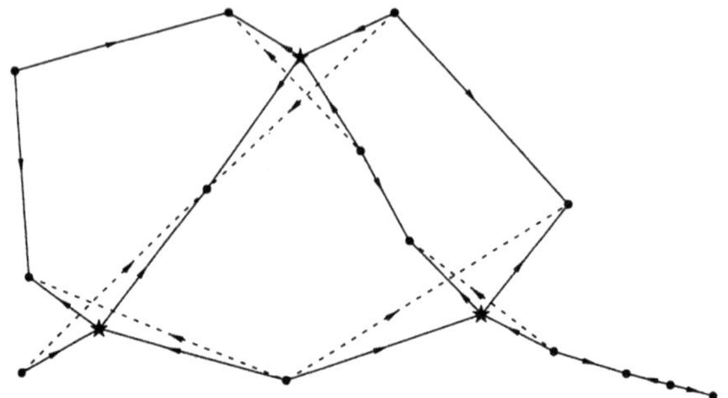

Figure 11: Equivariant Coxeter-Dynkin diagram: part I

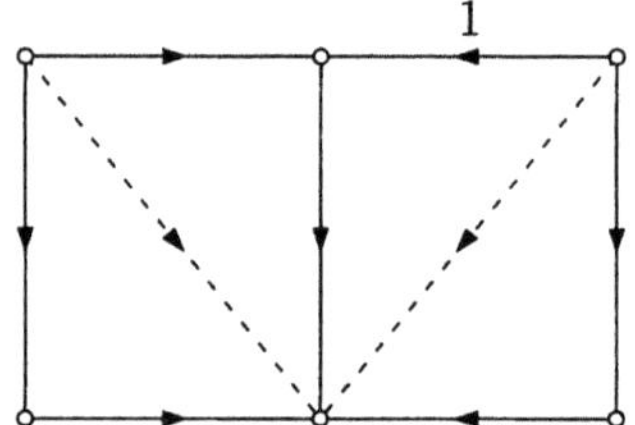

Figure 12: Equivariant Coxeter-Dynkin diagram: part II

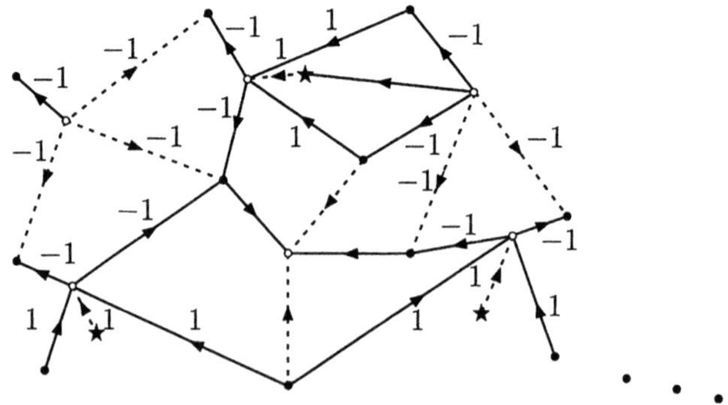

Figure 13: Equivariant Coxeter-Dynkin diagram: part III

References

[AGV] Arnold, V. I., Gusein-Zade, S. M., Varchenko, A. N.: Singularities of Differentiable Maps, Volume II, Birkhäuser, Boston, Basel, Berlin 1988.

[D] Damon, J.: Topological triviality of versal unfoldings of complete intersections. Ann. Inst. Fourier, 34, 225–251 (1984).

[E] Ebeling, W.: The Monodromy Groups of Isolated Singularities of Complete Intersections. Lecture Notes in Mathematics, Vol. 1293, Springer-Verlag, Berlin etc., 1987.

[EG1] Ebeling, W., Gusein-Zade, S. M.: Coxeter-Dynkin diagrams of the complete intersection singularities Z_9 and Z_{10}. Math. Z. 218, 549–562 (1995).

[EG2] Ebeling, W., Gusein-Zade, S. M.: Coxeter-Dynkin diagrams of fat points in $\mathbb{C}^2$ and of their stabilizations. Math. Annalen, to be published (1996).

[ES] Ebeling, W., Steenbrink, J.H.M.: Spectral pairs for isolated complete intersection singularities. Preprint University of Hannover, 272 (1996).

[G] Giusti, M.: Classification des singularités isolées d'intersections complètes simples. C.R. Acad. Sc. Paris, Sér. A, t. 284, 167–170 (1977).

[GZ] Gusein-Zade, S. M.: Intersection matrices for certain singularities of functions of two variables. Funktsional Anal. i Prilozhen. 8:1, 11–15 (1974) (Engl. translation in Functional Anal. Appl. 8, 10–13 (1974)).

[P] Pickl, A.: Die Homologie der Einhängung eines vollständigen Durchschnitts mit isolierter Singularität, Mathematica Gottingensis, Heft 15, SFB Geometrie und Analysis (1985).

[W] Wall, C. T. C.: Classification of unimodal isolated singularities of complete intersections, Proc. Symp. in Pure Math. Vol. 40, Part 2, 625–640 (1983).

Progress in Mathematics, Vol. 162, © 1998 Birkhäuser Verlag Basel/Switzerland

Brieskorn Lattices and Torelli Type Theorems for Cubics in $\mathbb{P}^3$ and for Brieskorn-Pham Singularities with Coprime Exponents

Claus Hertling
Universität Bonn
Mathematisches Institut
Beringstraße 3
D – 53115 Bonn,
GERMANY

Abstract

The first part is a survey about polarized mixed Hodge structures and Brieskorn lattices for hypersurface singularities. It describes wellknown properties of these objects and contains new results about classification spaces and moduli spaces for these objects. In the second part the Brieskorn lattices of cubics in $\mathbb{C}^4$ and of Brieskorn-Pham singularities with coprime exponents are studied. A nice application is a global Torelli theorem for cubics in $\mathbb{P}^3$ by some pure Hodge structure.

1 Introduction

1970, in his paper on the complex monodromy of isolated hypersurface singularities, Brieskorn laid the foundations for the study of the local Gauß-Manin connection for hypersurface singularities [Br]. Since then, many people have extended and applied this theory of the Gauß-Manin connection. We refer to [AGV] for a comprehensive introduction, many applications and a long list of references.

Studying the Gauß-Manin connection means essentially studying the integrals of holomorphic forms over vanishing cycles. This classical viewpoint is adopted in [AGV]. Other references, [Ph1][SK2][SM1], which also give some survey, are written in a more technical language and stress different aspects.

Many of the applications refer to deformations of singularities. Central for these applications is the relation between the Gauß-Manin connection and the mixed Hodge structure on the middle cohomology, which was found by Varchenko [Va1]. Starting with the Gauß-Manin connection of a holomorphic function germ $f : (\mathbb{C}^{n+1}, 0) \to (\mathbb{C}, 0)$ with an isolated singularity, he constructed a mixed Hodge structure on the middle cohomology $H^n(X_\infty, \mathbb{C})$ of a Milnor fibre of the singularity. His construction was modified [Ph2][SchSt] to obtain Steenbrink's mixed Hodge structure [St]. M. Saito greatly extended this to his theory of mixed Hodge modules.

The mixed Hodge structure contains much information about the singularity. Most famous are the spectral pairs [St][AGV], which are equivalent to the Hodge numbers, and the spectral numbers, which are a bit weaker. Again, we refer to [AGV] for their properties and applications.

Here we want to take into account all the information of the mixed Hodge structure and study Torelli type questions. Within a μ-constant family of singularities, the spectral pairs are constant [Va2], and the mixed Hodge structure varies holomorphically. It is a quite fine analytic invariant and is sensitive to many of the moduli of a μ-constant family, but, as it turns out, in general not to all. So, in general, it is not fine enough to give Torelli theorems for μ-homotopy classes of singularities. For example, in the case of semiquasihomogeneous singularities, the mixed Hodge structure depends only on the quasihomogeneous part.

One finds some kind of explanation for this and a refinement, if one looks at the construction of the mixed Hodge structure from the Gauß-Manin connection [Va1]: The mixed Hodge structure is determined only by the principle parts of the asymptotic expansions of the integrals of holomorphic n-forms over n-cycles in the Milnor fibres. If we take into account also the higher parts of the expansion of the integrals, we obtain richer analytic information, which, in fact, is sensitive to all moduli in μ-constant families [SM2].

Using this we will define an invariant BL of the right equivalence class of a singularity and study Torelli type questions, classifying spaces, and period mappings for the invariant.

To give a precise definition of this invariant, we need the Brieskorn lattice $H_0'' = \Omega^{n+1}_{\mathbb{C}^{n+1}, 0} / df \wedge d\Omega^{n-1}_{\mathbb{C}^{n+1}, 0}$ [Br](cf. [SM1]). It is central for the algebraic description of the Gauß-Manin connection and in some sense combines all the information given by the integrals of holomorphic n-forms over n-cycles in the Milnor fibres. It is a free $\mathbb{C}\{t\}$-module of rank μ, the meromorphic extension $\mathbb{C}\{t\}[t^{-1}] \cdot H_0'' = \mathcal{G}_0$ is a regular singular $\mathbb{C}\{t\}[\partial_t]$-module, the Gauß-Manin connection [Br][Ph1]. The Brieskorn lattice determines Steenbrink's mixed Hodge structure on $H^n(X_\infty, \mathbb{C})$ [Va1][Ph2][SchSt], it varies holomorphically within a μ-constant family of singularities, and it satisfies an infinitesimal Torelli theorem if the μ-constant stratum is smooth [SM2].

It yields the invariant BL of the right equivalence class of a hypersurface singularity as follows. $G_\mathbb{Z}$ is the group of automorphisms of the Milnor lattice $H_n(X_\infty, \mathbb{Z})$ which respect monodromy, intersection form, and Seifert form. $G_\mathbb{Z}$ acts on $\mathcal{G}_0$. This group action induces an equivalence relation for subspaces of $\mathcal{G}_0$. The invariant BL is the equivalence class of the Brieskorn lattice.

In [He1][He2] this invariant BL was studied together with two weaker invariants; one of them is the Picard-Fuchs singularity [AGV]. We conjectured that BL satisfies global Torelli theorems for all isolated hypersurface singularities. We proved this for the unimodal and bimodal singularities: BL determines the right equivalence class for nearly all unimodal and bimodal hypersurface singularities (cf. §4).

In §§5 and 6 of this paper such global Torelli theorems will be obtained for two other classes of singularities, in §§2, 3, and 4 we will put together the general framework for such investigations: known and new results for polarized mixed Hodge structures, Brieskorn lattices, classifying spaces, and period mappings.

Now let us give a more detailed description of the contents of this paper. To obtain a classifying space D_{BL} for Brieskorn lattices, we first need a classifying space D_{PMHS} for mixed Hodge structures. Here it is important that Steenbrink's mixed Hodge structure is polarized. §2 is devoted to this polarization and the space D_{PMHS}. The definition of polarized mixed Hodge structure, which is used here, is motivated by [Schm]. The polarization of the given mixed Hodge structure is a consequence of results of Schmid [Schm], Steenbrink [St], and Scherk [Sche].

§3 is a quite dense account of the properties of the Brieskorn lattice. Using the polarization, a pairing will be defined, which is essentially K. Saito's higher residue pairing. We will define a classifying space D_{BL} for Brieskorn lattices and review results from [He3] on the spaces D_{PMHS} and D_{BL}. There exists a double fibration

$$D_{\mathrm{BL}} \to D_{\mathrm{PMHS}} \to D_{\mathrm{prim}},$$

where D_{prim} is a product of classifying spaces for polarized pure Hodge structures. The fibres of the holomorphic fibre bundles $D_{\mathrm{BL}} \to D_{\mathrm{PMHS}}$ and $D_{\mathrm{PMHS}} \to D_{\mathrm{prim}}$ are isomorphic to $\mathbb{C}^{N_{\mathrm{BL}}}$ and $\mathbb{C}^{N_{\mathrm{PMHS}}}$ for some $N_{\mathrm{BL}}, N_{\mathrm{PMHS}} \in \mathbb{N}$. The group $G_{\mathbb{Z}}$ acts on D_{PMHS} and D_{BL}. The actions are properly discontinuous [He3]. The quotient $D_{\mathrm{BL}}/G_{\mathbb{Z}}$ is the moduli space for the invariant BL. Analogously, the quotient $D_{\mathrm{PMHS}}/G_{\mathbb{Z}}$ parametrizes the isomorphism classes $PMHS$ of the polarized mixed Hodge structures on $H^n(X_\infty, \mathbb{C})$.

In §4 the invariant BL and known results on period mappings and Torelli theorems are given.

In §6 the invariant BL will be studied for all families of semiquasihomogeneous singularities with weights $(\frac{1}{a_0}, \ldots, \frac{1}{a_n})$, where $a_0, \ldots, a_n$ are pairwise coprime. These are the μ-constant strata of the Brieskorn-Pham singularities with coprime exponents. They can have arbitrary high modality. The global Torelli theorem for the invariant BL can be derived with some general arguments from the statement that there the group $G_{\mathbb{Z}}$ is $G_{\mathbb{Z}} = \{\pm M_h^k \mid k \in \mathbb{Z}\}$. Here M_h is the monodromy on the Milnor lattice. The proof of this statement uses a result of Thom and Sebastiani [SeTh] and properties of unit roots and cyclotomic polynomials.

In §5 the semiquasihomogeneous singularities with weights $(\frac{1}{3}, \frac{1}{3}, \frac{1}{3}, \frac{1}{3})$ will be studied. The proof of the global Torelli theorem for BL uses [St], [Schm], the global Torelli theorem for cubics in $\mathbb{P}^4$ of Tjurin, Clemens and Griffiths [Tju][ClGr], and the cancellation property of Hauser and Müller [HaMu]. As an application, we obtain a global Torelli theorem for smooth cubics in $\mathbb{P}^3$ in the following way.

There exists a four-dimensional coarse moduli space for cubics in $\mathbb{P}^3$. But the Hodge structures on the cohomology groups are trivial. They do not give a global or generic or infinitesimal Torelli theorem. Thereby the cubics are an exception within the smooth projective hypersurfaces. But the cone of a smooth cubic is a quasihomogeneous singularity with weights $(\frac{1}{3}, \frac{1}{3}, \frac{1}{3}, \frac{1}{3})$. The cohomology $H^3(X_\infty, \mathbb{C})$ of the Milnor fibre of this singularity carries a mixed Hodge structure, and the sum $H^3(X_\infty, \mathbb{C})_{\neq 1}$ of the eigenspaces of the monodromy with eigenvalues $\neq 1$ carries a polarized pure Hodge structure, which is invariant with respect to the monodromy. The moduli space of such polarized Hodge structures is four-dimensional. A consequence of the global Torelli theorem 5.1 for BL is corollary 5.2: The period mapping from the coarse moduli space for cubics in $\mathbb{P}^3$ to this moduli space for polarized Hodge structures is injective.

2 Hypersurface singularities and polarized mixed Hodge structures

In §2 we will fix notations for the classical topological data of a hypersurface singularity. Furthermore, we will describe a canonical, but less well known nondegenerate bilinear form S on the rational cohomology $H^n(X_\infty, \mathbb{Q})$ of the Milnor fibre. S gives a polarization for Steenbrink's mixed Hodge structure [St]. We will define a classifying space D_{PMHS} for such polarized mixed Hodge structures and indicate some results on D_{PMHS} from [He3]. S will also be used in §3.

Let $f : (\mathbb{C}^{n+1}, 0) \to (\mathbb{C}, 0)$ be a holomorphic function with an isolated singularity at 0 and Milnor number μ. If we choose a sufficiently small open ball B around 0 in $\mathbb{C}^{n+1}$ and a sufficiently small punctured disc T' around 0 in $\mathbb{C}$, then

$$f : X' = f^{-1}(T') \cap B \to T'$$

is a C^∞ fibre bundle, the Milnor fibration. The fibres $X_t = f^{-1}(t) \cap B, t \in T'$, have the homotopy type of a bouquet of μ n-spheres [Mi]. If $u : T_\infty \to T'$ is the universal covering and $X_\infty = X' \times_{T'} T_\infty$, then for any $\tau \in T_\infty$ the natural inclusion $X_{u(\tau)} \hookrightarrow X_\infty$ is a homotopy equivalence. $H_n(X_\infty, \mathbb{Z}) \cong \mathbb{Z}^\mu$ is the Milnor lattice.

We have the monodromy M_h, the intersection form q and the Seifert form l on $H_n(X_\infty, \mathbb{Z})$ [AGV](Ch 2). The variation $\mathrm{Var} : H^n(X_\infty, \mathbb{Z}) \to H_n(X_\infty, \mathbb{Z})$

is an isomorphism. Seifert form l and Variation Var determine one another by

$$l(a,b) = \langle \mathrm{Var}^{-1}(a), b \rangle \qquad \text{for } a, b \in H_n(X_\infty, \mathbb{Z}).$$

The Seifert form l determines monodromy M_h and intersection form q by

$$l(M_h a, b) = (-1)^{n+1} l(b, a) \quad \text{and} \quad q(a, b) = -l(a, b) + (-1)^{n+1} l(b, a).$$

Thus any automorphism of the Milnor lattice $H_n(X_\infty, \mathbb{Z})$, which respects the Seifert form, also respects monodromy and intersection form. The group of all automorphisms of the Milnor lattice, which respect the Seifert form, will be denoted by $G_\mathbb{Z}$.

We have the monodromy $M = M_u \cdot M_s = M_s \cdot M_u$ with unipotent part M_u and semisimple part M_s on the cohomology $H^n(X_\infty, \mathbb{Q})$. The nilpotent part $N = \log M_u$ satisfies $N^{n+1} = 0$. We set

$$
\begin{aligned}
H^n(X_\infty, \mathbb{C})_\lambda &= \ker(M_s - \lambda \cdot id) \subset H^n(X_\infty, \mathbb{C}), \\
H^n(X_\infty, \mathbb{C})_{\neq 1} &= \bigoplus_{\lambda \neq 1} H^n(X_\infty, \mathbb{C})_\lambda, \\
H^n(X_\infty, \mathbb{Z})_{\neq 1} &= H^n(X_\infty, \mathbb{Z}) \cap H^n(X_\infty, \mathbb{C})_{\neq 1},
\end{aligned}
$$

and similarly $H^n(X_\infty, \mathbb{Z})_1$, $H^n(X_\infty, \mathbb{Q})_{\neq 1}$, $H^n(X_\infty, \mathbb{Q})_1$.

The intersection form q is $(-1)^n$-symmetric, i.e. symmetric for even n, skewsymmetric for odd n. The intersection form is nondegenerate on $H_n(X_\infty, \mathbb{Q})_{\neq 1}$ and induces an isomorphism $H^n(X_\infty, \mathbb{Q})_{\neq 1} \cong H_n(X_\infty, \mathbb{Q})_{\neq 1}$ and a nondegenerate bilinear form q^* on $H^n(X_\infty, \mathbb{Q})_{\neq 1}$.

There is a canonical extension S of $(-1)^{n(n-1)/2} \cdot q^*$ to $H^n(X_\infty, \mathbb{Q})$. The best way to introduce and understand S, is to follow a construction of Brieskorn [Br] and Scherk [Sche]. We can embed the Milnor fibration into a fibration $Y' \to T'$ with nonsingular projective fibres Y_t, $t \in T'$, with $Y_\infty = Y' \times_{T'} T_\infty$ analogously to X_∞, and with monodromy M_Y and nondegenerate pairing q_Y^* on the primitive cohomology $P^n(Y_\infty, \mathbb{Q})$. Now, by a result of Scherk [Sche], we can choose the projective fibration $Y' \to T'$ in such a way that the embedding $i : X_\infty \to Y_\infty$ induces a surjective mapping $i^* : P^n(Y_\infty, \mathbb{Q}) \to H^n(X_\infty, \mathbb{Q})$ with kernel $\ker i^* = \ker(M_Y - id)$.

Let $M_{Y,s}$, $M_{Y,u}$, $N_Y = \log M_{Y,u}$ be the semisimple, unipotent and nilpotent part of M_Y and $P^n(Y_\infty, \mathbb{Q})_1 = \ker(M_{Y,s} - id)$. There is a well defined and uniquely determined bilinear form S on $H^n(X_\infty, \mathbb{Q})$ with the following properties. S is invariant with respect to the monodromy, the restriction of S to $H^n(X_\infty, \mathbb{Q})_{\neq 1}$ is equal to $(-1)^{n(n-1)/2} \cdot q^*$, the restriction of S to $H^n(X_\infty, \mathbb{Q})_1$ is given by

$$S(a, b) = (-1)^{n(n-1)/2} q_Y^*(\tilde{a}, N_Y \tilde{b})$$

for $a, b \in H^n(X_\infty, \mathbb{Q})_1$, $\tilde{a}, \tilde{b} \in P^n(Y_\infty, \mathbb{Q})_1$, $i^*(\tilde{a}) = a$, $i^*(\tilde{b}) = b$. It is easy to prove Lemma 2.1. The properties in Lemma 2.1 show that S is determined by the variation and the monodromy and is independent of the choice of the projective fibration $Y' \to T'$.

LEMMA 2.1 *The bilinear form S on $H^n(X_\infty, \mathbb{Q})$ is invariant with respect to the monodromy, the restriction of S to $H^n(X_\infty, \mathbb{Q})_{\neq 1}$ is equal to $(-1)^{n(n-1)/2} \cdot q^*$, the restriction of S to $H^n(X_\infty, \mathbb{Q})_1$ is given by*

$$S(a, b) = (-1)^{n(n-1)/2} \langle (a, \text{Var} \circ (\textstyle\sum_{l \geq 1} \frac{1}{l}(-1)^{l-1}(M - id)^{l-1})b \rangle$$

for $a, b \in H^n(X_\infty, \mathbb{Q})_1$. The form S is nondegenerate, S is $(-1)^n$-symmetric on $H^n(X_\infty, \mathbb{Q})_{\neq 1}$ and $(-1)^{n+1}$-symmetric on $H^n(X_\infty, \mathbb{Q})_1$. S induces an isomorphism $H^n(X_\infty, \mathbb{Q}) \cong H_n(X_\infty, \mathbb{Q})$ and a nondegenerate bilinear form S_* on $H_n(X_\infty, \mathbb{Q})$. If $a, b \in H_n(X_\infty, \mathbb{Q})_1$ then $S_*(N\,a, b) = (-1)^{n(n-1)/2} \cdot q(a, b)$.

The bilinear form S gives a polarization for Steenbrink's mixed Hodge structure on $H^n(X_\infty, \mathbb{C})$ in the sense of §6 in [Schm]. This is made precise in Proposition 2.3. But first the properties of the weight filtration and the bilinear form S are given in Lemma 2.2, which is a direct consequence of Lemma (6.4) in [Schm].

LEMMA 2.2 *The weight filtration $W_\bullet$ on $H^n(X_\infty, \mathbb{Q})$ is invariant with respect to M_s. If $(H_\mathbb{Q}, m)$ is one of the two pairs $(H^n(X_\infty, \mathbb{Q})_{\neq 1}, n)$ and $(H^n(X_\infty, \mathbb{Q})_1, n + 1)$, then the restriction of $W_\bullet$ to $H_\mathbb{Q}$ has the following properties.*

a) $W_\bullet$ is the unique increasing filtration such that $N(W_l) \subset W_{l-2}$ and such that $N^l : Gr^W_{m+l} \to Gr^W_{m-l}$ is an isomorphism.

b) $S(W_l, W_{l'}) = 0$ if $l + l' < 2m$.

c) A nondegenerate $(-1)^{m+l}$-symmetric bilinear form S_l is well defined on Gr^W_{m+l} for $l \geq 0$ by the requirement: $S_l(a, b) = S(\tilde{a}, N^l \tilde{b})$ if $\tilde{a}, \tilde{b} \in W_{m+l}$ represent $a, b \in Gr^W_{m+l}$.

d) The primitive subspace $P_{m+l}(H_\mathbb{Q})$ of Gr^W_{m+l} is defined by

$$P_{m+l} = \ker(N^{l+1} : Gr^W_{m+l} \to Gr^W_{m-l-2})$$

if $l \geq 0$ and $P_{m+l} = 0$ if $l < 0$. Then

$$Gr^W_{m+l} = \bigoplus_{i \geq 0} N^i P_{m+l+2i},$$

and this decomposition is orthogonal with respect to S_l if $l \geq 0$.

The primitive subspaces of Gr^W_k restricted to $H^n(X_\infty, \mathbb{C})_{\neq 1}$ and $H^n(X_\infty, \mathbb{C})_1$ will be denoted by $P_{\neq 1, k}$ and $P_{1, k}$. We have the decomposition $P_{\neq 1, k} \oplus P_{1, k} = \bigoplus_\lambda P_{\lambda, k}$ into the eigenspaces $P_{\lambda, k}$ of the semisimple part M_s of the monodromy.

A description of Steenbrink's Hodge filtration $F^\bullet$ on $H^n(X_\infty, \mathbb{C})$ is given in §3. Proposition 2.3 follows from [St] and [Schm]: If $Y' \to T'$ is a projective fibration as above, then Schmid's limit Hodge filtration on $P^n(Y_\infty, \mathbb{C})$ together with N_Y and $(-1)^{n(n-1)/2} \cdot q_Y^*$ gives a polarized mixed Hodge structure on

$P^n(Y_\infty)$ in the sense of theorem (6.16) in [Schm]. Following Steenbrink [St], the mapping $P^n(Y_\infty) \to H^n(X_\infty)$ is a morphism of mixed Hodge structures. Proposition 2.3 follows with the definition of S. More details are given in [He3].

PROPOSITION 2.3

a) *$F^\bullet$ and $W_\bullet$ give a mixed Hodge structure on $H^n(X_\infty, \mathbb{C})$. $F^\bullet$ and $W_\bullet$ are invariant with respect to M_s. N is a $(-1,-1)$-morphism, thus $N(F^p) \subset F^{p-1}$ and $F^p N^j P_{\lambda,k} = N^j F^{p+j} P_{\lambda,k}$ and*

$$F^p Gr_k^W H^n(X_\infty, \mathbb{C})_\lambda = \bigoplus_{j \geq 0} F^p N^j P_{\lambda,k+2j}.$$

b) $S(F^p H^n(X_\infty, \mathbb{C})_{\neq 1}, F^{n+1-p} H^n(X_\infty, \mathbb{C})_{\neq 1}) = 0,$
$S(F^p H^n(X_\infty, \mathbb{C})_1, F^{n+2-p} H^n(X_\infty, \mathbb{C})_1) = 0.$

c) *The pure Hodge structures $F^\bullet P_{\neq 1, n+l}$ on $P_{\neq 1, n+l}$ and $F^\bullet P_{1, n+1+l}$ on $P_{1, n+1+l}$ are polarized by S_l.*

In [He3] such polarized mixed Hodge structures and their classifying spaces are studied. To get a classifying space D_{PMHS}, we have to fix the lattice $H^n(X_\infty, \mathbb{Z})$, the endomorphisms N and M_s on $H^n(X_\infty, \mathbb{Q})$, and some numbers $f_{\lambda,k}^p$. Then the classifying space D_{PMHS} consists of all filtrations $F^\bullet$ on $H^n(X_\infty, \mathbb{C})$ with the Hodge numbers $\dim F^p P_{\lambda,k} = f_{\lambda,k}^p$ and the properties in proposition 2.3. If we consider only the pure Hodge structures on the quotients $Gr_\bullet^W$, we get a classifying space D_{prim} in the following way. D_{prim} is the product of classifying spaces for the pure polarized Hodge structures on the primitive subspaces $P_{\lambda,k} + P_{\bar{\lambda},k}$ (we need only the real structure on $P_{\lambda,k} + P_{\bar{\lambda},k}$, not a lattice, for the definition of these classifying spaces). D_{prim} is a complex manifold and a homogeneous space with respect to a real Lie group [Schm]. We have a canonical projection $\pi_{\text{PMHS}} : D_{\text{PMHS}} \to D_{\text{prim}}$.

The group $G_{\mathbb{Z}}$ of automorphisms of the Milnor lattice which respect the Seifert form is canonically isomorphic to the group of automorphisms of $H^n(X_\infty, \mathbb{Z})$ which respect S, N and M_s. The group $G_{\mathbb{Z}}$ acts on D_{PMHS} and on D_{prim}. The quotient $D_{\text{PMHS}}/G_{\mathbb{Z}}$ is the moduli space for the isomorphism classes $PMHS$ of polarized mixed Hodge structures on $H^n(X_\infty)$. In [He3] the following is proved.

PROPOSITION 2.4

a) *D_{PMHS} is a complex manifold and a homogeneous space with respect to a real Lie group. π_{PMHS} is a locally trivial fibre bundle with fibres isomorphic to $\mathbb{C}^{N_{PMHS}}$, $N_{PMHS} \in \mathbb{N}$.*

b) *The group $G_{\mathbb{Z}}$ acts properly discontinuously on D_{PMHS}. The moduli space $D_{PMHS}/G_{\mathbb{Z}}$ for polarized mixed Hodge structures on $H^n(X_\infty)$ is a normal complex space and has only quotient singularities.*

REMARK 2.5 Varchenko [Va2] proved that the spectral numbers and spectral pairs are constant within a μ-constant family of singularities. Thus also the Hodge numbers $f^p_{\lambda,k}$ are constant. We have a period mapping from the parameter space of the μ-constant family to the moduli space $D_{\mathrm{PMHS}}/G_{\mathbb{Z}}$. This period mapping is locally liftable to D_{PMHS}.

3 Brieskorn lattice

As in §2 let $f : (\mathbb{C}^{n+1}, 0) \to (\mathbb{C}, 0)$ be a hypersurface singularity and $f : X' \to T'$ a Milnor fibration. The Brieskorn lattice of f is $H_0'' = H_0''(f) = \Omega^{n+1}_{X,0}/df \wedge d\Omega^{n-1}_{X,0}$ [Br]. Brieskorn lattice and Gauß-Manin connection determine Steenbrink's Hodge filtration on $H^n(X_\infty, \mathbb{C})$ [Va1][Ph2][SchSt]. The Brieskorn lattice induces an invariant of the right equivalence class of f, which is finer than the polarized mixed Hodge structure on $H^n(X_\infty)$. In §3 we will state the essential properties of Gauß-Manin connection and Brieskorn lattice.

The cohomology bundle $H^n = \bigcup_{t \in T'} H^n(X_t, \mathbb{C})$ is a flat complex vector bundle. $H^n(X_\infty, \mathbb{C})$ can be identified with the space of the global flat many-valued sections in H^n. If $A \in H^n(X_\infty, \mathbb{C})_\lambda$ and $\alpha \in \mathbb{Q}$ such that $e^{-2\pi i\alpha} = \lambda$, then

$$s(A, \alpha)(t) = t^\alpha \exp\left(\log t \, \tfrac{-N}{2\pi i}\right) A(t)$$

is a unique holomorphic section in H^n. Let $\mathcal{H}^n$ be the sheaf of germs of holomorphic sections in H^n and $i : T' \to T$ the inclusion. The germs $s(A, \alpha)_0 \in (i_*\mathcal{H}^n)_0$ in 0 of the sections $s(A, \alpha)(t)$ span a $\mathbb{C}\{t\}[t^{-1}]$-vectorspace $\mathcal{G}_0$ of dimension μ . This space $\mathcal{G}_0$ is invariant with respect to the differential operator $\partial_t : (i_*\mathcal{H}^n)_0 \to (i_*\mathcal{H}^n)_0$, which is induced by the covariant derivative. $\mathcal{G}_0$ is a regular singular $\mathbb{C}\{t\}[t^{-1}]$-module; it is called the Gauß-Manin connection [Ph1]. The mapping

$$\psi_\alpha : H^n(X_\infty, \mathbb{C})_\lambda \to \mathcal{G}_0, \quad \psi_\alpha(A) = s(A, \alpha)_0$$

is injective, the image $\psi_\alpha(H^n(X_\infty, \mathbb{C})_\lambda)$ is

$$C_\alpha = \ker(t\partial_t - \alpha)^{n+1} \subset \mathcal{G}_0.$$

These subspaces C_α are the key to understand the structure of $\mathcal{G}_0$. The mapping ψ_α satisfies

$$(t\partial_t - \alpha) \circ \psi_\alpha = \psi_\alpha \circ \left(\tfrac{-N}{2\pi i}\right) \quad \text{and} \quad t \circ \psi_\alpha = \psi_{\alpha+1}.$$

$t : C_\alpha \to C_{\alpha+1}$ is bijective, and $\partial_t : C_\alpha \to C_{\alpha-1}$ is bijective if $\alpha \neq 0$. The eigenspaces C_α induce the decreasing $V^\bullet$-filtration on $\mathcal{G}_0$,

$$
\begin{aligned}
V^\alpha &= \sum_{\beta \geq \alpha} \mathbb{C}\{t\}C_\beta = \bigoplus_{\alpha \leq \beta < \alpha+1} \mathbb{C}\{t\}C_\beta, \\
V^{>\alpha} &= \sum_{\beta > \alpha} \mathbb{C}\{t\}C_\beta = \bigoplus_{\alpha < \beta \leq \alpha+1} \mathbb{C}\{t\}C_\beta.
\end{aligned}
$$

The ring

$$R = \mathbb{C}\{\{\partial_t^{-1}\}\} = \left\{ \sum_{i\geq 0} a_i \partial_t^{-i} \mid \sum_{i\geq 0} a_i t^i/i! \in \mathbb{C}\{t\} \right\}$$

is the ring of microdifferential operators with constant coefficients [Ph1]. It is easy to see that the subspace $\mathbb{C}\{t\} \cdot C_\alpha$ is a free R-module of rank $\dim_{\mathbb{C}} C_\alpha$ if $\alpha \notin \mathbb{Z}_{<0}$. The subspace $V^{>-1}$ is a free R-module of rank μ,

$$V^{>-1} = \bigoplus_{-1<\alpha\leq 0} R \cdot C_\alpha = \bigoplus_{-1<\alpha\leq 0} \mathbb{C}\{t\} \cdot C_\alpha.$$

The mappings ψ_α, $-1 < \alpha \leq 0$, are put together to give an isomorphism

$$\psi : \bigoplus_{-1<\alpha\leq 0} \psi_\alpha : H^n(X_\infty, \mathbb{C}) \to \bigoplus_{-1<\alpha\leq 0} C_\alpha$$

of vectorspaces. We use the structure of $V^{>-1}$ as R-module, the isomorphism ψ, and the bilinear form S on $H^n(X_\infty, \mathbb{C})$ to define a pairing on $V^{>-1}$.

DEFINITION 3.1 The pairing $P_S : V^{>-1} \times V^{>-1} \to R \cdot \partial_t^{-1}$ is defined by the following properties. Let $\alpha, \beta \in (-1, 0]$, $a \in C_\alpha$, $b \in C_\beta$, $g_1(\partial_t^{-1}), g_2(\partial_t^{-1}) \in R$, and

$$\begin{aligned}
P_S(a,b) &= 0, \quad \text{if } \alpha + \beta \notin \mathbb{Z}, \\
P_S(a,b) &= \tfrac{1}{(2\pi i)^n} S(\psi^{-1}(a), \psi^{-1}(b)) \cdot \partial_t^{-1}, \quad \text{if } \alpha + \beta = -1, \\
P_S(a,b) &= \tfrac{1}{(2\pi i)^{n+1}} S(\psi^{-1}(a), \psi^{-1}(b)) \cdot \partial_t^{-2}, \quad \text{if } \alpha = \beta = 0, \\
P_S(g_1(\partial_t^{-1})a, g_2(\partial_t^{-1})b) &= g_1(\partial_t^{-1})g_2(-\partial_t^{-1})P_S(a,b).
\end{aligned}$$

$P_S^{(-l)}$ is the part of P_S in $\mathbb{C} \cdot \partial_t^{-l}$, i.e. $P_S(a,b) = \sum_{l\geq 1} P_S^{(-l)}(a,b)$ and $P_S^{(-l)}(a,b) \in \mathbb{C} \cdot \partial_t^{-l}$.

LEMMA 3.2

i) $P_S : C_\alpha \times C_\beta \to 0$ *if $\alpha + \beta \notin \mathbb{Z}$, $\alpha, \beta > -1$,*
    $P_S : C_\alpha \times C_\beta \to \mathbb{C} \cdot \partial_t^{-\alpha-\beta-2}$ *is a perfect pairing if $\alpha + \beta \in \mathbb{Z}$, $\alpha, \beta > -1$.*

ii) $P_S^{(-l)}$ *is $(-1)^{n+1+l}$-symmetric.*

iii) $[t, P_S(a,b)] = P_S(ta, b) - P_S(a, tb)$,
    i.e. $(l-1)P_S^{(-l+1)}(a,b) \cdot \partial_t^{-1} = P_S^{(-l)}(ta, b) - P_S^{(-l)}(a, tb)$.

Proof. i) and ii) follow from lemma 2.1; iii) follows by an easy calculation from $(t\partial_t - \alpha) \circ \psi_\alpha = \psi_\alpha \circ (\frac{-N}{2\pi i})$. $\qquad\square$

REMARK 3.3 P_S is the restriction of K. Saito's higher residue pairing [SK1], [SK2] to $V^{>-1}$. This follows from proposition 3.4 and [SM1] §2. This residue pairing is defined on the Gauß-Manin system $(\int_f^{n+1} \mathcal{O}_X)_0$ [SM1] §2. The space $V^{>-1}$ is canonically embedded in the Gauß-Manin system. But here we use the Gauß-Manin connection $\mathcal{G}_0$, definition 3.1, and proposition 3.4 instead of the Gauß-Manin system.

If $\omega \in \Omega_X^{n+1}$ is a holomorphic $(n+1)$-form, then the Gelfand-Leray form $\frac{\omega}{df}|_{X_t}$ gives a holomorphic section $s[\omega](t)$ in the cohomology bundle H^n,

$$s[\omega](t) = \left[\frac{\omega}{df}\Big|_{X_t} \right] \in H^n(X_t, \mathbb{C}), \quad t \in T'.$$

For any $\omega \in \Omega_{X,0}^{n+1}$ the germ $s[\omega]_0 \in (i_* \mathcal{H}^n)_0$ of the section $s[\omega](t)$ is in $\mathcal{G}_0$ [Br] and even in $V^{>-1}$ [Ma]. The kernel of the mapping $\Omega_{X,0}^{n+1} \to V^{>-1}$, $\omega \mapsto s[\omega]_0$ is $df \wedge d\Omega_{X,0}^{n-1}$ [Ma]. The Brieskorn lattice $H_0'' = \Omega_{X,0}^{n+1}/df \wedge d\Omega_{X,0}^{n-1}$ will be identified with its image in $V^{>-1}$.

PROPOSITION 3.4

    i) $\mathbb{C}\{t\}[t^{-1}] \cdot H_0'' = \mathcal{G}_0, \qquad H_0'' \subset V^{>-1}.$

    ii) $tH_0'' \subset H_0''$, H_0'' *is a free* $\mathbb{C}\{t\}$*-module of rank* μ.

    iii) $\partial_t^{-1} H_0'' \subset H_0''$, H_0'' *is a free R-module of rank* μ.

Proof. i) [Br] and [Ma]. ii) follows from i) and $t \cdot s[\omega]_0 = s[f\omega]_0$. iii) follows from i) and $\partial_t^{-1} s[d\eta]_0 = s[df \wedge \eta]_0$ for $\eta \in \Omega_{X,0}^n$ [Br].   □

The Grothendieck residue on the Jacobi algebra induces a nondegenerate pairing Res_f on $\Omega_{X,0}^{n+1}/df \wedge \Omega_{X,0}^n = H_0''/\partial_t^{-1} H_0''$ [SK1][SK2][Va4].

PROPOSITION 3.5

    i) $P_S(H_0'', H_0'') \subset R \cdot \partial_t^{-n-1}$, *i.e.* $P_S^{(-l)}(H_0'', H_0'') = 0$ *if* $1 \leq l \leq n$.

    ii) $P_S^{(-n-1)}(s[\omega_1]_0, s[\omega_2]_0) = \mathrm{Res}_f(\omega_1, \omega_2) \cdot \partial_t^{-n-1}$ *if* $\omega_1, \omega_2 \in \Omega_{X,0}^{n+1}$.

Proof. Statements of this type can be found in [SM1] 2.7. But they are not specific about the constants in the definition 3.1 of P_S. Explicit calculations which take into account all the constants can be found in [Va4] §3.3. Varchenko uses a projective fibration $Y' \to T'$ like the one which we used to define S in §2. He gives results on the sections in the bundle $\bigcup_{t \in T'} P^n(Y_t)$ and on the pairing q_Y^* in $P^n(Y_t)$. One has to translate these results into statements on H_0'' and P_S and calculate all the constants in [Va4] §3.3. That gives i) and ii).   □

COROLLARY 3.6   *i) H_0'' is isotropic of maximal size with respect to the antisymmetric bilinear form $P_S^{(-n)}$, i.e. $P_S^{(-n)}(h, H_0'') = 0 \iff h \in H_0''$.*

ii) $H_0'' \supset V^{n-1}$, $\dim H_0''/V^{n-1} = \frac{1}{2}\dim V^{>-1}/V^{n-1}$.

Proof. i) This follows easily from $P_S^{(-n)}(H_0'', H_0'') = 0$ (3.4 i)) and from the fact that $P_S^{(-n-1)} = \mathrm{Res}_f$ is well defined and nondegenerate on $H_0''/\partial_t^{-1}H_0''$.

ii) $P_S^{(-n)}(V^{>-1}, V^{n-1}) = 0$ (3.2 i)), $H_0'' \subset V^{>-1}$ (3.3 i)), and i) imply $H_0'' \supset V^{n-1}$ and

$$H_0'' = (H_0'' \cap \bigoplus_{-1<\alpha<n-1} C_\alpha) \oplus V^{n-1}.$$

Now, $P_S^{(-n)}$ is nondegenerate on $\bigoplus_{-1<\alpha<n-1} C_\alpha$ (cf. 3.2 i)), and $H_0'' \cap \bigoplus_{-1<\alpha<n-1} C_\alpha$ is a maximal isotropic subspace. $\qquad\square$

Varchenko [Va1] used the Gauß-Manin connection and the Brieskorn lattice to construct a mixed Hodge structure on $H^n(X_\infty, \mathbb{C})$. His construction was modified later [Ph2][SchSt] (cf. [SM1]) to obtain Steenbrink's [St] mixed Hodge structure. The modified version can be given as follows.

If a subspace $K \subset V^{>-1}$ satisfies $\mathbb{C}\{t\}[t^{-1}] \cdot K = \mathcal{G}_0$ and $\partial_t^{-1}K \subset K$, then K induces a decreasing filtration $F_K^\bullet$ on $H^n(X_\infty, \mathbb{C})$, which is invariant with respect to M_S, by

$$F_K^p H^n(X_\infty, \mathbb{C})_\lambda = \psi_\alpha^{-1}(V^\alpha \cap \partial_t^{n-p}K/V^{>\alpha}), \qquad \alpha \in (-1, 0],\ e^{-2\pi i\alpha} = \lambda.$$

PROPOSITION 3.7 $F_{H_0''}^\bullet$ *is Steenbrink's Hodge filtration $F^\bullet$.*

REMARK 3.8 $N(F^p) \subset F^{p-1}$ follows from $tH_0'' \subset H_0''$; $S(F^p, F^{n+1-p}) = 0$ on $H^n(X_\infty)_{\neq 1}$ and $S(F^p, F^{n+2-p}) = 0$ on $H^n(X_\infty)_1$ follow from $P_S(H_0'', H_0'') \subset R \cdot \partial_t^{-n-1}$.

The Hodge numbers of $(F^\bullet, W_\bullet)$ and the spectral pairs, which are defined in [St] [AGV], determine one another. A little coarser are the spectral numbers: μ rational numbers α with multiplicities $d(\alpha)$,

$$
\begin{aligned}
d(\alpha) &= \dim Gr_F^{[n-\alpha]} H^n(X_\infty)_\lambda \qquad \text{if } \lambda = e^{-2\pi i\alpha} \\
&= \dim V^\alpha \cap H_0''/V^{>\alpha} - \dim V^\alpha \cap \partial_t^{-1}H_0''/V^{>\alpha}.
\end{aligned}
$$

They satisfy the symmetry $d(\alpha) = d(n - 1 - \alpha)$. On the one hand, this follows from the mixed Hodge structure. On the other hand, it can also be derived from proposition 3.4, in the spirit and as an extension of corollary 3.5.

EXAMPLE 3.9 In the case of semiquasihomogeneous singularities of degree 1, the weights $(w_0, \ldots, w_n)$ determine the spectral numbers by [St]

$$\sum_\alpha d(\alpha) t^{\alpha+1} = \prod_{i=0}^{n} (t^{w_i} - t)(1 - t^{w_i})^{-1}.$$

In §5 the semiquasihomogeneous singularities with weights $(\frac{1}{3}, \frac{1}{3}, \frac{1}{3}, \frac{1}{3})$ will be discussed. For these singularities we get with $\zeta = e^{2\pi i/3}$, $k \in \mathbb{Z}$,

$$\begin{aligned}
\sum_\alpha d(\alpha) t^\alpha &= t^{1/3} + 4t^{2/3} + 6t + 4t^{4/3} + t^{5/3}, \\
\dim C_{k+\frac{1}{3}} &= \dim C_{k+\frac{2}{3}} = \dim H^3(X_\infty, \mathbb{C})_\zeta = \dim H^3(X_\infty, \mathbb{C})_{\bar\zeta} = 5, \\
\dim C_k &= \dim H^3(X_\infty, \mathbb{C})_1 = 6, \\
\dim C_\alpha &= 0 \qquad \text{if } \alpha \notin \tfrac{1}{3}\mathbb{Z}.
\end{aligned}$$

From the spectral numbers and from proposition 3.5 we get the following statements on the polarized mixed Hodge structure. They are illustrated by the next picture.

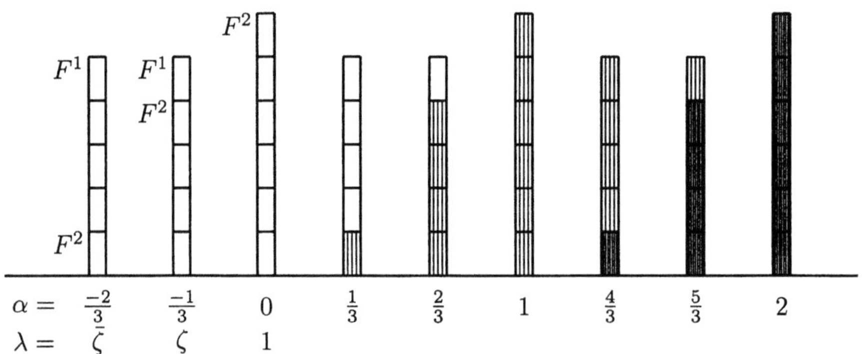

$$W_3 = F^3 = 0 \text{ and } W_4 = F^2 = H^3(X_\infty, \mathbb{C})_1 \text{ on } H^3(X_\infty, \mathbb{C})_1,$$

the polarized Hodge structure on $H^3(X_\infty, \mathbb{C})_1$ is trivial.

$$\begin{aligned}
W_2 = F^3 = 0 \text{ and } W_3 &= F^1 = H^3(X_\infty, \mathbb{C})_{\neq 1} \text{ on } H^3(X_\infty, \mathbb{C})_{\neq 1}, \\
\dim F^2 H^3(X_\infty, \mathbb{C})_{\bar\zeta} &= 1, \quad \dim F^2 H^3(X_\infty, \mathbb{C})_\zeta = 4, \\
S(F^2 H^3(X_\infty, \mathbb{C})_{\bar\zeta}, F^2 &H^3(X_\infty, \mathbb{C})_\zeta) = 0,
\end{aligned}$$

and the polarized Hodge structure on $H^3(X_\infty, \mathbb{C})_{\neq 1}$ is determined by $F^2 H^3(X_\infty, \mathbb{C})_{\bar\zeta}$.

In §2 the classifying space D_{PMHS} of all polarized mixed Hodge structures on $H^n(X_\infty)$ with fixed Hodge numbers was defined. Propositions 3.3, 3.4 and 3.6

show how to define a classifying space D_{BL} of Brieskorn lattices.

$$D_{\mathrm{BL}} = \big\{ \text{ subspaces } K \subset V^{>-1} \mid \mathbb{C}\{t\}[t^{-1}] \cdot K = \mathcal{G}_0, \ \partial_t^{-1} K \subset K,$$
$$tK \subset K, \ P_S^{(-l)}(K,K) = 0 \text{ if } 1 \le l \le n, \ F_K^{\bullet} \in D_{\mathrm{PMHS}} \big\},$$

$\pi_{\mathrm{BL}} : D_{\mathrm{BL}} \to D_{\mathrm{PMHS}}$ is the canonical projection. In [He3] this classifying space is studied carefully. A main result is

PROPOSITION 3.10 $\pi_{BL} : D_{BL} \to D_{PMHS}$ *is a locally trivial bundle with fibres isomorphic to* $\mathbb{C}^{N_{BL}}$,

$$N_{BL} = \sum_{\alpha+\beta<n-2,\alpha<\beta} d(\alpha)d(\beta) + \sum_{2\alpha<n-2} \tfrac{1}{2}d(\alpha)(d(\alpha)+1) < \tfrac{1}{4}\mu^2.$$

There is a canonical $\mathbb{C}^*$*-action with negative weights on the fibres of* π_{BL} *and therefore a canonical zero section of the bundle* π_{BL}.

Via ψ any $g \in G_{\mathbb{Z}}$ induces an automorphism of the Gauß-Manin connection $\mathcal{G}_0$. The group $G_{\mathbb{Z}}$ acts on D_{BL}.

COROLLARY 3.11 [He3] *The group* $G_{\mathbb{Z}}$ *respects the fibres and the* $\mathbb{C}^*$*-action on the fibres of the bundle* π_{BL}. *The group* $G_{\mathbb{Z}}$ *acts properly discontinuously on* D_{BL}. *The quotient* $D_{BL}/G_{\mathbb{Z}}$ *is a normal complex space and has only quotient singularities.*

EXAMPLE 3.12 *In the case of the semiquasihomogeneous singularities with weights* $(\frac{1}{3}, \frac{1}{3}, \frac{1}{3}, \frac{1}{3})$, *the classifying space* D_{PMHS} *is canonically isomorphic to the open subset*

$$\big\{ \langle v \rangle \subset H^3(X_\infty, \mathbb{C})_{\bar{\zeta}} \mid iS(v, \bar{v}) > 0 \big\} \subset \mathbb{P}H^3(X_\infty, \mathbb{C})_{\bar{\zeta}} \cong \mathbb{P}^4$$

of the four-dimensional projective space $\mathbb{P}H^3(X_\infty, \mathbb{C})_{\bar{\zeta}}$. *The classifying space* D_{BL} *is a vector bundle with base* D_{PMHS} *and one-dimensional fibre,*

$$D_{BL} = \big\{ \langle c_{1/3} + s^- \cdot c_{2/3} \rangle \oplus \tilde{C}_{2/3} \oplus V^1 \mid s^- \in \mathbb{C}, \text{ there exists}$$
$$v \in H^3(X_\infty, \mathbb{C})_{\bar{\zeta}} \text{ such that } iS(v, \bar{v}) > 0,$$
$$c_{1/3} = \partial_t^{-1}\psi_{-2/3}v \in C_{1/3}, \ c_{2/3} = \partial_t^{-1}\psi_{-1/3}\bar{v} \in C_{2/3},$$
$$\tilde{C}_{2/3} = \{c \in C_{2/3} \mid P_S(c_{1/3}, c) = 0\} \big\}.$$

4 The invariant BL

The elements of $D_{\mathrm{BL}}/G_{\mathbb{Z}}$ are equivalence classes of subspaces of $\mathcal{G}_0$ with respect to the operation of $G_{\mathbb{Z}}$ on $\mathcal{G}_0$. The equivalence class in $D_{\mathrm{BL}}/G_{\mathbb{Z}}$ of a Brieskorn lattice $H_0'' = H_0''(f)$ is an invariant of the right equivalence class of f. We call this invariant $BL(f)$. The space $D_{\mathrm{BL}}/G_{\mathbb{Z}}$ is a moduli space for such invariants. There is another description of the elements of $D_{\mathrm{BL}}/G_{\mathbb{Z}}$: We can consider $G_{\mathbb{Z}}$ as the automorphism group of the tuple

$$(H^n(X_\infty,\mathbb{Z}),\ M_s,\ N,\ S,\ \psi,\ \mathcal{G}_0)$$

and $D_{\mathrm{BL}}/G_{\mathbb{Z}}$ as the set of isomorphism classes of tuples

$$(H^n(X_\infty,\mathbb{Z}),\ M_s,\ N,\ S,\ \psi,\ \mathcal{G}_0,\ K\subset\mathcal{G}_0).$$

The invariant BL was defined and studied first in [He1] [He2], under the name LBL and together with two weaker invariants. One of them is the Picard-Fuchs singularity [AGV]. The invariant BL contains very fine analytic information and is a good candidate for Torelli theorems for hypersurface singularities. In [He2] the following conjecture is formulated.

CONJECTURE. *The invariant $BL(f)$ of a hypersurface singularity f determines the right equivalence class of f.*

The main result in [He2], a global Torelli theorem for the unimodal and bimodal singularities, confirms the conjecture.

THEOREM 4.1 *The invariant BL determines the right equivalence class for all unimodal and bimodal singularities, possibly with the exception of the subseries* $Z_{1,14k},\ S_{1,10k}, S_{1,10k}^{\sharp}\ (k\geq 1)$.

In the case of those three subseries, the methods which were used were not strong enough to determine the invariant BL. In §5 and §6 such global Torelli theorems will be proved for the semiquasihomogeneous singularities with weights $(\frac{1}{3},\frac{1}{3},\frac{1}{3},\frac{1}{3})$ and for all Brieskorn Pham singularities (and their μ-constant deformations) with pairwise coprime exponents.

Apart from these global Torelli theorems, the conjecture is confirmed by an infinitesimal Torelli theorem for all hypersurface singularities [SM2](3.1 and 3.2). Let f_0 be a hypersurface singularity and S a sufficiently small open subset of the μ-constant stratum in some semiuniversal unfolding of f_0. Within this μ-constant stratum the topological data like Milnor lattice and Seifert form and also the Hodge numbers [Va2] are constant. We obtain a period mapping

$$\Phi : S \to D_{\mathrm{BL}},\quad s \mapsto H_0''(f_s)$$

for sufficiently small S. The period mapping Φ is holomorphic.

PROPOSITION 4.2 [SM2]

a) *If S is smooth then $\Phi : S \to D_{BL}$ is an immersion.*

b) *Even if S is not smooth any fibre of Φ is finite.*

In the case of semiquasihomogeneous singularities, we know more about S and the period mapping Φ. Let f_0 be a quasihomogeneous singularity with weights $(w_0, \ldots, w_n)$ and degree 1. Then $D_{\mathrm{PMHS}} = D_{\mathrm{prim}}$ and $N_{\mathrm{PMHS}} = 0$, because the monodromy is finite. The μ-constant stratum S in a suitable semiuniversal unfolding is a product $S = S^0 \times S^-$ with $S^- = \mathbb{C}^{\dim S^-}$. Here

$$f_s = f_{(s^0,0)} + \sum_i s_i^- d_i \,, \qquad s = (s^0, s^-) = (s^0, (s_i^-)_i) \in S^0 \times S^-,$$

$f_{(s^0,0)}$ is quasihomogeneous, the d_i are the monomials of degree > 1 in a monomial basis of the Jacobi algebra,

$$\begin{aligned}
\dim S^0 &= d(\alpha_1 + 1) \\
\dim S^- &= \sum_{\alpha > \alpha_1 + 1} d(\alpha), \\
\alpha_1 &= \sum w_i - 1 = \text{smallest spectral number.}
\end{aligned}$$

This follows from [Va3]. Setting $\deg s_i^- = 1 - \deg d_i < 0$, we obtain a $\mathbb{C}^*$-action with negative weights on the fibres S^- of the trivial bundle $S^0 \times S^-$. From proposition 3.8 we have a $\mathbb{C}^*$-action with negative weights on the fibres of the bundle π_{BL}. From [He2](2.4) proposition 4.3 follows.

PROPOSITION 4.3 *In the case of semiquasihomogeneous singularities, the period mapping* $\Phi : S \to D_{BL}$ *is a fibre preserving* $\mathbb{C}^*$-*equivariant embedding of bundles.*

The following table gives the dimensions $\dim S$ (=modality), $\dim D_{\mathrm{prim}}$, N_{PMHS}, and N_{BL} for the unimodal and bimodal singularities and for the singularities in §5 and §6. In the case of semiquasihomogeneous singularities, we write $\dim S = \dim S^0 + \dim S^-$.

Singularities	$\dim S$	$\dim D_{\mathrm{prim}}$	N_{PMHS}	N_{BL}
$\tilde{E}_6,\ \tilde{E}_7,\ \tilde{E}_8$	$1 = 1 + 0$	1	0	0
$T_{pqr},\ \frac{1}{p} + \frac{1}{q} + \frac{1}{r} > 1$	1	0	1	0
14 exceptional unimodal	$1 = 0 + 1$	0	0	1
$E_{3,0}, Z_{1,0}, Q_{2,0}, W_{1,0}, S_{1,0}, U_{1,0}$	$2 = 1 + 1$	1	0	1
14 exceptional bimodal	$2 = 0 + 2$	0	0	2
8 bimodal series	2	0 or 1	0	≥ 2
semiqh. with weights $(\frac{1}{3},\frac{1}{3},\frac{1}{3},\frac{1}{3})$	$5 = 4 + 1$	4	0	1
semiqh. with weights $(\frac{1}{a_0},..,\frac{1}{a_n})$ and pairwise coprime a_i	$\dim S = 0 + \dim S^-$	0	0	$\geq \dim S^-$

The table shows that any level of the double fibering $D_{\mathrm{BL}} \to D_{\mathrm{PMHS}} \to D_{\mathrm{prim}}$ can contain geometric information. For 6 of the 8 listed classes $\dim S = \dim D_{\mathrm{BL}}$. That is not typical. In general, one can expect that $\dim D_{\mathrm{BL}}$ is much bigger than the dimension of the μ-constant stratum, and that $\dim D_{\mathrm{prim}}$, N_{PMHS} and N_{BL} are not 0.

This chapter ends with some remarks on the unimodal and bimodal singularities. Details can be found in [He1][He2]. The families which are considered there contain representatives for all right equivalence classes of the unimodal and bimodal singularities. The base spaces S are not small. The period mapping $S \to D_{\mathrm{BL}}$ is manyvalued in most cases. It is univalued only for the exceptional unimodal and bimodal singularities. It induces a mapping $S/\text{right equivalence} \to D_{\mathrm{BL}}/G_{\mathbb{Z}}$. This mapping is always injective (possibly with the exception of the subseries $Z_{1,14k}$, $S_{1,10k}, S^{\sharp}_{1,10k}$ $(k \geq 1)$) It is even bijective for all unimodal and the exceptional bimodal singularities. To get these results one has to determine the right equivalence classes, the manyvalued period mapping $S \to D_{\mathrm{BL}}$, and the action of $G_{\mathbb{Z}}$ on D_{BL}. Often the last part is most difficult. But also calculating the period mapping is not easy. In theorem 4.1 the mentioned subseries are excepted because there the period mapping $S \to D_{\mathrm{BL}}$ could not be determined precisely enough.

5 Semiquasihomogeneous singularities with weights $\left(\frac{1}{3}, \frac{1}{3}, \frac{1}{3}, \frac{1}{3}\right)$

This chapter is devoted to the discussion and proof of the following two statements.

THEOREM 5.1 *In the case of the semiquasihomogeneous singularities with weights $\left(\frac{1}{3}, \frac{1}{3}, \frac{1}{3}, \frac{1}{3}\right)$, the invariant BL determines the right equivalence class.*

COROLLARY 5.2 *(A global Torelli theorem for cubics in $\mathbb{P}^3$)*
Up to isomorphism a smooth cubic in $\mathbb{P}^3$ is determined by the polarized pure Hodge structure on the subspace $H^3(X_\infty, \mathbb{C})_{\neq 1}$ of the cohomology of the Milnor fibre of the respective homogeneous singularity in $\mathbb{C}^4$.

The Hodge structures on the cohomology groups are trivial for cubics in $\mathbb{P}^3$ – in contrast to nearly all other projective hypersurfaces. The Hodge structures do not give a global or generic or infinitesimal Torelli theorem. The primitive second cohomology of a cubic in $\mathbb{P}^3$ is canonically isomorphic to the subspace $H^3(X_\infty, \mathbb{C})_1$ of the cohomology of the Milnor fibre of the homogeneous singularity in $\mathbb{C}^4$. The corollary shows that the other subspace $H^3(X_\infty, \mathbb{C})_{\neq 1}$ of the cohomology of the singularity satisfies all wishes.

A smooth cubic in $\mathbb{P}^3$ determines an isolated quasihomogeneous singularity with weights $\left(\frac{1}{3}, \frac{1}{3}, \frac{1}{3}, \frac{1}{3}\right)$ and vice versa. Two quasihomogeneous singularities with these weights are right equivalent if and only if the respective cubics are analytically isomorphic. The set of all homogeneous polynomials of degree 3 in 4 variables is a 20 dimensional vectorspace. The subset V_{isol} of the polynomials which give a smooth cubic in $\mathbb{P}^3$ is the complement of the zero set of a resultant. The group $GL(4, \mathbb{C})$ is reductive and acts on V_{isol}. For any $f \in V_{\text{isol}}$ the dimension of the orbit $GL(4, \mathbb{C}) \cdot f$ in V_{isol} is

$$\dim GL(4, \mathbb{C}) \cdot f = \dim \sum_{i,j} \mathbb{C} \cdot x_i \frac{\partial f}{\partial x_j} = 16 = \dim GL(4, \mathbb{C}).$$

Thus the quotient $V_{\text{isol}}/GL(4, \mathbb{C}) = \mathcal{M}_{\text{cubics}}$ is a four-dimensional affine variety and a coarse moduli space for cubics in $\mathbb{P}^3$.

For any $f \in V_{\text{isol}}$ we can choose a semiuniversal unfolding and a μ-constant stratum $S = S^0 \times S^-$ with small S^0 and $S^- = \mathbb{C}^{\dim S^-}$ as in §4. Then $\dim S^0 = 4 = \dim \mathcal{M}_{\text{cubics}}$ and $\dim S^- = 1$. That follows from the spectral numbers. In example 3.7 and 3.10 we discussed the spectral numbers, the Hodge numbers and the spaces D_{PMHS} and D_{BL}. We have $\dim H^3(X_\infty, \mathbb{C})_1 = 6$ and $\dim H^3(X_\infty, \mathbb{C})_\zeta = \dim H^3(X_\infty, \mathbb{C})_{\bar{\zeta}} = 5$ with $\zeta = e^{2\pi i/3}$. The polarized mixed Hodge structure is determined by the one-dimensional space $F^2 H^3(X_\infty, \mathbb{C})_{\bar{\zeta}}$. The classifying space D_{PMHS} is canonically isomorphic to an open subset of the four-dimensional projective space $\mathbb{P} H^3(X_\infty, \mathbb{C})_{\bar{\zeta}}$. The space D_{BL} is a trivial vector bundle with one-dimensional fibre and base D_{PMHS}. Corollary 5.2 is equivalent to the following proposition.

PROPOSITION 5.3 *The period mapping $\mathcal{M}_{cubics} \to D_{PMHS}/G_{\mathbb{Z}}$ is injective.*

Proof. The proof is an application of the results of Schmid [Schm] and Steenbrink [St] on mixed Hodge structures, of the global Torelli theorem for cubics in $\mathbb{P}^4$ of Tjurin, Clemens and Griffiths [Tju][ClGr], and of the cancellation property for complex space germs of Hauser and Müller [HaMu].

We fix a polynomial $f - f(x_0, x_1, x_2, x_3) \in V_{\text{isol}}$. Similar to §2, we embed the Milnor fibration in a projective fibration. The zero set

$$Y_t = \{(x_0 : \ldots : x_4) \mid f(x) - tx_4^3 = 0\} \subset \mathbb{P}^4$$

is smooth if $t \neq 0$. The only singular point of Y_0 is $(0 : 0 : 0 : 0 : 1)$. We define Y, Y', $u : T_\infty \to T'$ and $Y_\infty = Y' \times_{T'} T_\infty$ as in §2. The fibres Y_t, $t \neq 0$, are all analytically isomorphic. The monodromy M_Y on $H^3(Y_\infty) = \mathbb{P}^3(Y_\infty)$ is finite. Therefore the variation of Hodge structures is trivial. The isomorphism $H^3(Y_{u(\tau)}, \mathbb{C}) \cong H^3(Y_\infty, \mathbb{C})$ induces an isomorphism of the Hodge filtration $F^\bullet_{u(\tau)}$ on $H^3(Y_{u(\tau)}, \mathbb{C})$ and the limit Hodge filtration $F^\bullet_\infty$ on $H^3(Y_\infty, \mathbb{C})$. Following [St], the sequence

$$0 \to H^3(Y_0) \to H^3(Y_\infty) \to H^3(X_\infty) \to H^4(Y_0) \to H^4(Y_\infty) \to 0$$

is an exact sequence of mixed Hodge structures and equivariant with respect to the monodromy. Here $H^\bullet(Y_0)$, $H^\bullet(Y_\infty)$, and $H^\bullet(X_\infty)$ are equipped with Deligne's, Schmid's, and Steenbrink's mixed Hodge structure respectively. Let

Z be the cubic in $\mathbb{P}^3 = \{x_4 = 0\} \subset \mathbb{P}^4$, which is given by f, so $Z = Y_t \cap \mathbb{P}^3$. Then

$$H^\bullet(Y_0) \cong H^\bullet(Y_t, X_t) \cong H^{\bullet-2}(Z),$$
$$H^3(Y_0) = 0, \quad H^4(Y_0) \cong H^2(Z).$$

From the exact sequence above we obtain isomorphisms

$$\mathbb{P}^3(Y_\infty) \cong H^3(Y_\infty) \cong H^3(X_\infty)_{\neq 1}$$

of polarized pure Hodge structures and an isomorphism

$$H^3(X_\infty)_1 \cong P^2(Z),$$

which is a $(-1,-1)$-morphism of trivial Hodge structures. The polarized Hodge structure on $H^3(X_\infty)_{\neq 1}$ is isomorphic to the polarized Hodge structure on $H^3(Y_t) = P^3(Y_t)$. The global Torelli theorem of Tjurin, Clemens and Griffiths [Tju][ClGr] shows that this polarized Hodge structure determines the cubic $Y_t \subset \mathbb{P}^4$ up to isomorphism. Therefore it determines the polynomial $f + x_4^3$ up to homogeneous coordinate changes. It remains to show:

Let $f_1, f_2 \in V_{isol}$ be given. If there is a homogeneous coordinate change φ such that $(f_1 + x_4^3) \circ \varphi = f_2 + x_4^3$, then there is a homogeneous coordinate change ψ such that $f_1 \circ \psi = f_2$.

This can be shown by elementary calculations. But the following argument is more elegant. Professor Hauser kindly communicated it to me.

Suppose there are $f_1, f_2 \in V_{\text{isol}}$ and a homogeneous coordinate change φ such that $(f_1 + x_4^3) \circ \varphi = f_2 + x_4^3$. Then the complex space germs of the Jacobi ideals $\left(\frac{\partial(f_1+x_4^3)}{\partial x_i}\right)$ and $\left(\frac{\partial(f_1+x_4^3)}{\partial x_i}\right)$ are isomorphic. The cancellation property of Hauser and Müller [HaMu] shows that the space germs of the Jacobi ideals $\left(\frac{\partial f_1}{\partial x_i}\right)$ and $\left(\frac{\partial f_2}{\partial x_i}\right)$ are isomorphic. Thus the Jacobi algebras of f_1 and f_2 are isomorphic. Because of a result of Mather and Yau, [MaYa] f_1 and f_2 are contact equivalent. As f_1 and f_2 are homogeneous, there is a homogeneous coordinate change ψ such that $f_1 \circ \psi = f_2$. $\qquad\square$

The rest of §5 is devoted to the single semiquasihomogeneous μ-constant parameter and to the proof of theorem 5.1. Apart from proposition 5.3, we need the following property of the group $G_\mathbb{Z}$.

LEMMA 5.4 *If $\gamma \in G_\mathbb{Z}$ acts on $H^3(X_\infty, \mathbb{C})_{\bar\zeta}$ like $\lambda \cdot id$, then $\lambda \in \{\pm 1, \pm\zeta, \pm\bar\zeta\}$.*

Proof. The reason for this property is the structure of the lattice $H^3(X_\infty, \mathbb{Z})$ as a monodromy module. This can be computed from a Coxeter Dynkin diagram. But it is easier to use the property that $\mathbb{Z}[\zeta]$ is a principal domain. The $\mathbb{Z}[\zeta]$-module

$$H^3(X_\infty, \mathbb{Z}[\zeta])_{\bar\zeta} = H^3(X_\infty, \mathbb{C})_{\bar\zeta} \cap H^3(X_\infty, \mathbb{Z}[\zeta])$$

has no torsion as a $\mathbb{Z}$-module and thus no torsion as a $\mathbb{Z}[\zeta]$-module. It is a free $\mathbb{Z}[\zeta]$-module of rank 5. Any $\gamma \in G_{\mathbb{Z}}$ acts as a $\mathbb{Z}[\zeta]$-module automorphism on $H^3(X_\infty, \mathbb{Z}[\zeta])_{\bar{\zeta}}$. If $\gamma \in G_{\mathbb{Z}}$ acts like $\lambda \cdot id$, then $\lambda \in \mathbb{Z}[\zeta]$ and $|\lambda| = 1$, so $\lambda \in \{\pm 1, \pm \zeta, \pm \bar{\zeta}\}$. $\qquad\qquad\square$

It is well known that the Hessian $H(f) = (\frac{\partial^2 f}{\partial x_i \partial x_j})$ of $f \in V_{\mathrm{isol}}$ is not an element of the Jacobi ideal $(\frac{\partial f}{\partial x_0}, .., \frac{\partial f}{\partial x_3})$. The Jacobi algebra of f is graded by the weights $(\frac{1}{3}, \frac{1}{3}, \frac{1}{3}, \frac{1}{3})$, and $H(f)$ is a generator of the homogeneous subspace with degree > 1. Therefore any semiquasihomogeneous singularity with quasihomogeneous part f is right equivalent to a singularity $f + s \cdot H(f)$, $s \in \mathbb{C}$. Let

$$\tilde{V}_{\mathrm{isol}} = \{f + s \cdot H(f) \mid f \in V_{\mathrm{isol}}, \ s \in \mathbb{C}\}.$$

The projection $\pi_{hess} : \tilde{V}_{\mathrm{isol}} \to V_{\mathrm{isol}}$ is a trivial vector bundle. The group $GL(4, \mathbb{C})$ acts on $\tilde{V}_{\mathrm{isol}}$ as a vector bundle, because $H(f \circ \varphi) = H(f) \circ \varphi \cdot (\det \varphi)^2$ if $f \in V_{\mathrm{isol}}$, $\varphi \in GL(4, \mathbb{C})$.

LEMMA 5.5 *The orbits of $GL(4, \mathbb{C})$ in $\tilde{V}_{isol}$ are the right equivalence classes in $\tilde{V}_{isol}$. The quotient $\tilde{V}_{isol}/GL(4, \mathbb{C})$ is an affine variety and a coarse moduli space for the semiquasihomogeneous singularities with weights $(\frac{1}{3}, \frac{1}{3}, \frac{1}{3}, \frac{1}{3})$.*

Proof. Any biholomorphic coordinate change $\varphi : (\mathbb{C}^4, 0) \to (\mathbb{C}^4, 0)$ has a unique linear part $\varphi_0 \in GL(4, \mathbb{C})$. If $\tilde{f} = f + sH(f) \in \tilde{V}_{\mathrm{isol}}$ then

$$\tilde{f} \circ \varphi \circ \varphi_0^{-1} - \tilde{f} \in \left(\frac{\partial f}{\partial x_0}, \ldots, \frac{\partial f}{\partial x_3} \right).$$

If $\tilde{f} \circ \varphi \in \tilde{V}_{\mathrm{isol}}$ then $\tilde{f} \circ \varphi = \tilde{f} \circ \varphi_0$ follows. This proves the first part. The rest of the lemma is a consequence of the following statements. $GL(4, \mathbb{C})$ is reductive. $\tilde{V}_{\mathrm{isol}}$ is affine. The orbits of $GL(4, \mathbb{C})$ in $\tilde{V}_{\mathrm{isol}}$ have the same dimension as the orbits of $GL(4, \mathbb{C})$ in V_{isol}; all the orbits have dimension 16. $\qquad\square$

We fix $f \in V_{\mathrm{isol}}$. A small open neighborhood of f in V_{isol} with special properties will be constructed. We choose a four-dimensional disc T around f, which is transversal to the orbit $GL(4, \mathbb{C})f$, and an open neighborhood U of id in $GL(4, \mathbb{C})$. Both are supposed to be sufficiently small. Then the mappings

$$U \times T \ \to \ U \cdot T \subset V_{\mathrm{isol}} \quad \text{and}$$
$$U \times \pi_{hess}^{-1}(T) \ \to \ U \cdot \pi_{hess}^{-1}(T) = \pi_{hess}^{-1}(U \cdot T) \subset \tilde{V}_{\mathrm{isol}}$$

are biholomorphic; $U \cdot T$ is an open neighborhood of f in V_{isol}; and $\pi_{hess} : U \cdot \pi_{hess}^{-1}(T) \to U \cdot T$ is a vector bundle.

There exist period mappings $\tilde{\Phi} : U \cdot \pi_{hess}^{-1}(T) \to D_{\mathrm{BL}}$ and $\Phi : U \cdot T \to D_{\mathrm{PMHS}}$ with $\Phi \circ \pi_{hess} = \pi_{\mathrm{BL}} \circ \tilde{\Phi}$. These period mappings $\tilde{\Phi}$ and Φ are U-invariant. Because of proposition 4.3 the pair $(\tilde{\Phi}, \Phi)$ is a morphism of vector bundles. The restriction of $(\tilde{\Phi}, \Phi)$ to $(\pi_{hess}^{-1}(T), T)$ is an isomorphism from the vector bundle $\pi_{hess}^{-1}(T) \to T$ to an open subbundle of the vector bundle $D_{\mathrm{BL}} \to D_{\mathrm{PMHS}}$.

LEMMA 5.6 *Suppose* $f \in V_{isol}$ *is fixed and* $(\tilde{\Phi}, \Phi)$ *is as above. For any* $\gamma \in G_{\mathbb{Z}}^{\Phi(f)} = \{g \in G_{\mathbb{Z}} \mid g \ \Phi(f) = \Phi(f)\}$ *there exists a linear coordinate change* $\varphi \in GL(4, \mathbb{C})^f = \{\psi \in GL(4, \mathbb{C}) \mid f \circ \psi = f\}$ *such that*

$$\gamma \ \tilde{\Phi}(\tilde{f}) = \tilde{\Phi}(\tilde{f} \circ \varphi) \ \text{ for all } \ \tilde{f} \in \pi_{hess}^{-1}(f) = \{f + s \cdot H(f) \mid s \in \mathbb{C}\}.$$

Theorem 5.1 follows at once from proposition 5.3 and lemma 5.6: Suppose that g_1 and g_2 are two singularities such that $BL(g_1) = BL(g_2)$ and $PMHS(g_1) = PMHS(g_2) = PMHS(f)$. Because of proposition 5.3 there exist singularities f_1 and f_2 such that

$$f_i \text{ and } g_i \text{ are right equivalent, } \ f_1, f_2 \in \{f + s \cdot H(f) \mid s \in \mathbb{C}\},$$
$$\text{there exists } \gamma \in G_{\mathbb{Z}}^{\Phi(f)} \text{ such that } \gamma \ \tilde{\Phi}(f_1) = \tilde{\Phi}(f_2).$$

Because of lemma 5.6 the singularities f_1 and f_2 and thus also g_1 and g_2 are right equivalent. $\qquad\Box$

Proof of Lemma 5.6. Any $\varphi \in GL(4, \mathbb{C})^f$ induces an automorphism of the Milnor lattice of f and an element $\gamma_\varphi \in G_{\mathbb{Z}}$. This γ_φ satisfies $\gamma_\varphi \in G_{\mathbb{Z}}^{\Phi(f)}$ and

$$\gamma_\varphi \ \tilde{\Phi}(\tilde{h}) = \tilde{\Phi}(\tilde{h} \circ \varphi)$$

for any $\tilde{h} \in \pi_{hess}^{-1}(U')$, where U' is a small neighborhood of f in V_{isol}.

Let $\gamma \in G_{\mathbb{Z}}^{\Phi(f)}$. The group $GL(4, \mathbb{C})^f$ is finite. Because of proposition 5.3 there exists a $\varphi \in GL(4, \mathbb{C})^f$ such that

$$\gamma \ \Phi(h) = \Phi(h \circ \varphi)$$

for any $h \in V_{\mathrm{isol}}$ in a small neighborhood of f. Hence the actions of γ and of γ_φ on D_{PMHS} coincide on an open subset of D_{PMHS}, thus they coincide on D_{PMHS}. Because of lemma 5.4 the composition $\gamma_\varphi \circ \gamma^{-1}$ acts like $\lambda \cdot id$ with $\lambda \in \{\pm 1, \pm\zeta, \pm\bar{\zeta}\}$ on $H^3(X_\infty, \mathbb{C})_{\bar{\zeta}}$ and like $\bar{\lambda} \cdot id$ on $H^3(X_\infty, \mathbb{C})_\zeta$. Therefore the composition $\gamma_\varphi \circ \gamma^{-1}$ acts as multiplication by 1, ζ or $\bar{\zeta}$ on the fibres of the bundle $D_{\mathrm{BL}} \to D_{\mathrm{PMHS}}$. If $\psi = \zeta \cdot id \in GL(4, \mathbb{C})$ then

$$(h + s \cdot H(h)) \circ \psi = h + (\zeta \cdot s) \cdot H(h), \quad h \in V_{\mathrm{isol}},$$

and γ_ψ acts as multiplication by ζ on the fibres of the bundle $D_{\mathrm{BL}} \to D_{\mathrm{PMHS}}$. After choosing a suitable power ψ^k of ψ, the composition $\gamma_{\psi^k} \circ \gamma_\varphi \circ \gamma^{-1} = \gamma_{\varphi \circ \psi^k} \circ \gamma^{-1}$ acts trivial on D_{BL}. Now lemma 5.6 follows. $\qquad\Box$

6 Brieskorn-Pham singularities with pairwise coprime exponents

In this section the following result will be proved.

THEOREM 6.1 *Let $a_i \in \mathbb{N}$, $i = 0, \ldots, n$ be pairwise coprime numbers. In the case of the semiquasihomogeneous singularities with weights $(\frac{1}{a_0}, \ldots, \frac{1}{a_n})$, the invariant BL determines the right equivalence class.*

Any such singularity is right equivalent to a μ-constant deformation of the Brieskorn-Pham singularity $f_0 = \sum_{i=0}^{n} x_i^{a_i}$. The dimension of the μ-constant stratum of f_0 increases together with n and $a_0, \ldots, a_n$ to arbitrary large numbers.

The Brieskorn-Pham singularities with coprime exponents can be dealt with so well because of two reasons: They are quasihomogeneous, and all eigenspaces of the monodromy are one-dimensional. The second holds, for the monodromy of f_0 is the tensorproduct of the monodromies of the singularities $x_i^{a_i}$ [SeTh], and the a_i are pairwise coprime. Thus

$$D_{\mathrm{prim}} = D_{\mathrm{PMHS}} = \{pt\} \text{ and } D_{\mathrm{BL}} \cong \mathbb{C}^{N_{\mathrm{BL}}}.$$

By choosing μ monomials which represent a basis of the Jacobi algebra, we get a semiuniversal unfolding. The monomials with degree > 1 are denoted by d_l, $l \in L$. The μ-constant stratum $S \cong \mathbb{C}^{|L|}$ parametrizes the semiquasihomogeneous singularities

$$f_s = f_0 + \sum s_l \cdot m_l.$$

The coordinates s_l are equipped with the negative weights $\deg s_l = 1 - \deg d_l$. The μ-constant stratum $S = S^0 \times S^-$ satisfies

$$S^0 = \{pt\}, \ S = S^- \cong \mathbb{C}^{\dim S^-}.$$

Because of proposition 4.3 the period mapping $\Phi : S \to D_{\mathrm{BL}}$, $s \mapsto H_0''(f_s)$, is a $\mathbb{C}^*$-equivariant embedding. In general, $\dim S$ is much smaller than $\dim D_{\mathrm{BL}}$.

Let $a = a_0 \cdots a_n$. The $\mathbb{C}^*$-actions on $\mathbb{C}^{n+1}$ and on S are given by

$$c * x = c * (x_0, \ldots, x_n) = (c^{a/a_0} \cdot x_0, \ldots, c^{a/a_n} \cdot x_n),$$
$$c * s = c * (s_l)_l = (c^{a \cdot \deg s_l} \cdot s_l)_l,$$

and satisfy $f_{c*s}(c*x) = c^a \cdot f_s(x)$. If $c^a = 1$ then f_s and f_{c*s} are right equivalent. Theorem 6.1 follows from Proposition 6.2.

PROPOSITION 6.2 *The following statements are equivalent.*

i) f_{s_1} and f_{s_2} are right equivalent.

ii) There exists a number $c \in \mathbb{C}^*$ such that $c^a = 1$ and $s_2 = c * s_1$.

iii) $BL(f_{s_1}) = BL(f_{s_2})$.

Proof. It remains to show iii) $\Rightarrow$ ii). The $\mathbb{C}^*$-action on D_{BL} is given in the following way. If $c \in \mathbb{C}^*$, $K \in D_{\mathrm{BL}}$, $\sigma \in K$, $\sigma = \sum \sigma_\alpha$ such that $\sigma_\alpha \in C_\alpha$, then

$$
\begin{aligned}
c * \sigma &= \sum c^{a(-\alpha)} \sigma_\alpha, \\
c * K &= \{c * \sigma \mid \sigma \in K\}.
\end{aligned}
$$

The monodromy M_h as element of $G_{\mathbb{Z}}$ acts on D_{BL} like the unit root $\bar{\zeta} = \exp(2\pi i \frac{-1}{a})$ as element of $\mathbb{C}^*$, and $-id$ as element of $G_{\mathbb{Z}}$ acts trivial on D_{BL}. Thus the $\mathbb{C}^*$-equivariant embedding $S \to D_{\mathrm{BL}}$ induces an injective mapping $S/\langle \zeta \rangle \to D_{\mathrm{BL}}/\langle \pm M_h \rangle$.

Now for the proof of iii) $\Rightarrow$ ii) it is sufficient to show $G_{\mathbb{Z}} = \{\pm M_h^k \mid k \in \mathbb{Z}\}$, which will be done in proposition 6.3. $\qquad\square$

PROPOSITION 6.3 *In the case of a Brieskorn-Pham singularity* $\sum x_i^{a_i}$ *with pairwise coprime exponents, the group* $G_{\mathbb{Z}}$ *of automorphisms of the Milnor lattice which respect the Seifert form is* $G_{\mathbb{Z}} = \{\pm M_h^k \mid k \in \mathbb{Z}\}$.

Proof. The Milnor lattice of the A_{a_i-1} singularity x^{a_i} will be denoted by $\tilde{H}_0(A_{a_i-1})$, the monodromy by M_i. Then $M_i^{a_i} = id$, and there exists a cycle δ_i such that

$$
\bigoplus_{j=1}^{a_i-1} \mathbb{Z} \cdot M_i^{j-1} \cdot \delta_i = \tilde{H}_0(A_{a_i-1}).
$$

Following Thom and Sebastiani [SeTh], the Milnor lattice and the monodromy of the Brieskorn-Pham singularity are

$$
\begin{aligned}
H_n(X_\infty, \mathbb{Z}) &\cong \bigotimes_{i=0}^{n} \tilde{H}_0(A_{a_i-1}) \quad \text{and} \\
M_h &= \bigotimes_{i=0}^{n} M_i.
\end{aligned}
$$

Obviously $M_h^{a/a_i} = M_i^{a/a_i}$ and $\gcd(a_i, \frac{a}{a_i}) = 1$. Thus

$$
\bigoplus_{k=0}^{\mu-1} \mathbb{Z} \cdot M_h^k \cdot \delta_0 \otimes \cdots \otimes \delta_n = H_n(X_\infty \mathbb{Z}).
$$

Hence any automorphism C of the Milnor lattice which commutes with the monodromy M_h is determined by the image $C \cdot \delta_0 \otimes \cdots \otimes \delta_n$ of $\delta_0 \otimes \cdots \otimes \delta_n$. If

$$
C \cdot \delta_0 \otimes \cdots \otimes \delta_n = \sum_{k=0}^{\mu-1} c_k \cdot M_h^k \cdot \delta_0 \otimes \cdots \otimes \delta_n
$$

and $c(x) = \sum c_k x^k \in \mathbb{Z}[x]$ then $C = c(M_h)$.

All eigenspaces of the monodromy are one-dimensional. The set Eiw of the μ eigenvalues is

$$Eiw = \left\{ \exp\left(2\pi i \tfrac{c}{a}\right) \;\middle|\; 0 < c < a, \; c \not\equiv 0 \;(mod\; a_i) \right\}.$$

The intersection form is nondegenerate. The set of automorphisms of the Milnor lattice which respect monodromy and intersection form is

$$\{C = c(M_h) \mid c(x) \in \mathbb{Z}[x], \; |c(\lambda)| = 1 \; \forall \; \lambda \in Eiw\}.$$

For the proof of proposition 6.3 it remains to show that this set is $\{\pm M_h^k \mid k \in \mathbb{Z}\}$. This statement can be seen as a generalisation of the number theoretic fact [Wa](Ch 1,2): *For any unit root λ the numbers of absolute value 1 in $\mathbb{Z}[\lambda]$ are*

$$\{g(\lambda) \in \mathbb{Z}[\lambda] \mid |g(\lambda)| = 1\} = \{\pm \lambda^k \mid k \in \mathbb{Z}\}.$$

The proof of the statement uses this fact, the number theoretic lemma 6.4, and some property of Eiw, which allows to proceed inductively. We need some notations:

- the order $\operatorname{ord}\lambda$ of a unit root λ is $\operatorname{ord}\lambda = \min(k \in \mathbb{N} \mid \lambda^k = 1)$,

- the mth cyclotomic polynomial ϕ_m is $\phi_m = \prod_{ord\lambda=m}(x - \lambda)$,

- the set of orders of Eiw is $Ord := \{\operatorname{ord}\lambda \mid \lambda \in Eiw\}$,

- $e(q) := \exp(2\pi i q)$ for any $q \in \mathbb{C}$, $\zeta := e(\tfrac{1}{a})$.

We can choose three sequences $(m_i)_{i=1,\ldots,|Ord|}$, $(j(i))_{i=2,\ldots,|Ord|}$, $(p_i)_{i=2,\ldots,|Ord|}$ of natural numbers with the properties

- $Ord = \{m_i \mid i = 1,\ldots,|Ord|\}$, $\quad m_1 = a$,

- $j(i) < i$, $\quad p_i$ is a prime number,

- $m_{j(i)} = m_i \cdot p_i$, $\quad p_2 \le p_3 \le \cdots \le p_{|Ord|}$.

The sets $Ord_i := \{m_j \mid j = 1,\ldots,i\}$ form a chain $Ord_1 \subset Ord_2 \subset \cdots \subset Ord_{|Ord|} = Ord$ such that

$$Ord_i = Ord_{i-1} \cup \left\{ \frac{m_{j(i)}}{p_i} \right\} \quad \text{if } i \ge 2.$$

Suppose a is even. The number i_0 such that $2 = p_2 = \cdots = p_{i_0} < p_{i_0+1} \le \cdots \le p_{|Ord|}$ is $i_0 \ge 2$. Then $j(i) = i - 1$ and $m_i = \frac{a}{2^{i-1}}$ if $i \le i_0$. The proof for odd a is a simplification of the following reasoning.

Let $c(x) \in \mathbb{Z}[x]$ such that $|c(\lambda)| = 1 \; \forall \; \lambda \in Eiw$, and let $C = c(M_h)$. There exists a $k \in \mathbb{Z}$ such that $\pm \zeta^k \cdot c(\zeta) = 1$.

Lemma 6.4 b) implies $\pm e(\frac{2}{a})^k \cdot c(e(\frac{2}{a})) = \pm 1$.

$$c_{new}(x) := \begin{cases} \pm x^k c(x) & \text{if} \quad \pm e(\frac{2}{a})^k \cdot c(e(\frac{2}{a})) = 1 \\ (-x^{a/2})(\pm x^k)c(x) & \text{if} \quad \pm e(\frac{2}{a})^k \cdot c(e(\frac{2}{a})) = -1 \end{cases}$$

c_{new} satisfies

$$c_{new}(\zeta) = c_{new}(e(\frac{1}{m_1})) = 1, \quad c_{new}(e(\frac{2}{a})) = c_{new}(e(\frac{1}{m_2})) = 1.$$

Lemma 6.4 c) implies $c_{new}(e(\frac{1}{m_i})) = 1$ if $3 \leq i \leq i_0$.

If $i > i_0$ then $c_{new}(e(\frac{1}{m_i})) = 1$ follows inductively from lemma 6.4 a) and from $c_{new}(e(\frac{1}{m_{j(i)}})) = 1$. Hence $c_{new}(\lambda) = 1 \; \forall \; \lambda \in Eiw$ and $c_{new}(M_h) = id$ and $C \in \{\pm M_h^k \mid k \in \mathbb{Z}\}$. $\qquad\qquad\qquad\qquad\qquad\qquad\qquad\qquad\qquad\square$

REMARK 6.4 Two facts are crucial for the proof of proposition 6.3:
1. The Milnor lattice $H_n(X_\infty, \mathbb{Z})$ is a cyclic $\mathbb{Z}[M_h]$-module.
2. We obtain the whole set Ord, if we start with $a = ord\ \zeta$ and repeatedly divide an already given order by a (power of a) prime number to get another order.

LEMMA 6.5 *Let p be a prime number, $k, m \in \mathbb{N}$, $c(x) \in \mathbb{Z}[x]$ such that $c(e(\frac{1}{p^k m})) = 1$ and $|c(e(\frac{1}{m}))| = 1$.*

   a) If $p \geq 3$ then $c(e(\frac{1}{m})) = 1$.

   b) If $p = 2$ then $c(e(\frac{1}{m})) = \pm 1$.

   c) If $p = 2$ and $c(e(\frac{1}{p^l m})) = 1$ for some $l \neq k$ then $c(e(\frac{1}{m})) = 1$.

Proof. The proof is based on elementary properties of the rings $\mathbb{Z}[\lambda]$ for unit roots λ. In the following, p, q are always prime numbers, k, l, m are natural numbers ≥ 1 and λ is a unit root. Together with the notations from 6.3 like $ord\ \lambda$ and cyclotomic polynomial ϕ_m, we need the norm

$$Norm_m : \mathbb{Z}[e(\tfrac{1}{m})] \to \mathbb{Z}, \quad g(e(\tfrac{1}{m})) \mapsto \prod\nolimits_{ord\ \lambda = m} g(\lambda)$$

and the Euler function $\varphi : \mathbb{N} \to \mathbb{N}$, $\varphi(m) = \deg \phi_m$. In the following, "unit" always means an invertible element in some suitable ring $\mathbb{Z}[\lambda]$.

    If p does not divide m, then obviously

$$\phi_{p^{k+1}m}(x) = \phi_{p^k m}(x^p) \quad \text{and} \quad \phi_{pm}(x) = \frac{\phi_m(x^p)}{\phi_m(x)}.$$

$\phi_{p^k}(1) = p$ and $1 + x + \cdots + x^{m-1} = \prod_{l|m, l \neq 1} \phi_l(x)$ imply inductively $\phi_m(1) = 1$ for any m which is not a power of a prime number. Hence $1 - \lambda$ is a unit in $\mathbb{Z}[\lambda]$ if $ord\ \lambda$ is not a power of a prime number.

Statement 1. $\phi_{p^k m}(e(\frac{1}{m})) = p \cdot \text{unit}.$

Proof. If p_i are different prime numbers and $k_i \geq 1$, then

$$\phi_{p_1^{k_1} \cdots p_l^{k_l}}(x) = \phi_{p_1 \cdots p_l}(x^{p_1^{k_1-1} \cdots p_l^{k_l-1}}).$$

Hence statement 1 can be reduced to the statement

$$\phi_{p_1 \cdots p_l}\left(e\left(\frac{1}{p_2 \cdots p_l}\right)\right) = p_1 \cdot \text{unit}.$$

If p, q, m are given such that $p \neq q$ and p and q do not divide m, then

$$\phi_{pm}\left(e\left(\frac{1}{qm}\right)\right) = e\left(\frac{\varphi(pm)}{qm}\right) \cdot \prod_{\text{ord}(\lambda)=pm} \left(1 - \lambda \cdot e\left(\frac{-1}{qm}\right)\right)$$

is a unit, because the order $\text{ord}\,(\lambda \cdot e(\frac{-1}{qm}))$ is not a power of a prime number. Using $\phi_{pm}(x^q) = \phi_{pqm}(x)\phi_{pm}(x)$, we get

$$\phi_{pqm}\left(e\left(\frac{1}{qm}\right)\right) = \phi_{pm}\left(e\left(\frac{1}{m}\right)\right) \cdot \text{unit}.$$

Thus the statement can be reduced to the trivial case $\phi_p(1) = p$. $\qquad\square$

Statement 2. Let $m = \text{ord}\,\lambda$. The set $\{Norm_m(1 - \lambda^k) \mid k \in \mathbb{Z}\}$ is the union of the set $\{0\}$, the set

$$\left\{ p^{\frac{\varphi(m)}{p^{l-1}(p-1)}} \;\middle|\; l \geq 1,\ p \text{ prime number such that } p^l | m \right\},$$

and, if and only if m is not a power of a prime number, the set $\{1\}$.

Proof. If $\text{ord}\,(\lambda^k)$ is not a power of a prime number, then $Norm_m(1 - \lambda^k) = 1$ because $\phi_{\text{ord}\,(\lambda^k)}(1) = 1$. If $\text{ord}\,(\lambda^k) = p^l$ then

$$Norm_m(1 - \lambda^k) = (\phi_{p^l}(1))^{\varphi(m)/\varphi(p^l)} = p^{\varphi(m)/p^{l-1}(p-1)}. \qquad\square$$

Now we will restrict to the proof of a). The proofs of b) and c) are very similar. Let $p,\ k,\ c(x)$ be given such that

$$p \geq 3, \quad c\left(e\left(\frac{1}{p^k m}\right)\right) = 1, \quad \left|c\left(e\left(\frac{1}{m}\right)\right)\right| = 1.$$

There exists a polynomial $r(x) \in \mathbb{Z}[x]$ such that $1 - c(x) = \phi_{p^k m}(x) \cdot r(x)$. Then

$$Norm_m\left(1 - c\left(e\left(\frac{1}{m}\right)\right)\right) = Norm_m\left(\phi_{p^k m}\left(e\left(\frac{1}{m}\right)\right)\right) \cdot Norm_m\left(r\left(e\left(\frac{1}{m}\right)\right)\right).$$

Statement 1 implies

$$Norm_m\left(\phi_{p^k m}\left(e\left(\frac{1}{m}\right)\right)\right) = (\pm 1) \cdot p^{\varphi(m)}.$$

From [Wa](Ch 1,2) and $|c(e(\frac{1}{m}))| = 1$ we obtain

$$c(e(\tfrac{1}{m})) \in \left\{ \pm e(\tfrac{l}{m}) \mid l \in \mathbb{Z} \right\}.$$

Case 1: m is odd. Then $\mathbb{Z}[e(\frac{1}{m})] = \mathbb{Z}[e(\frac{1}{2m})]$, $Norm_m = Norm_{2m}$, $\{\pm e(\frac{l}{m}) \mid l \in \mathbb{Z}\} = \{e(\frac{l}{2m}) \mid l \in \mathbb{Z}\}$. Because of statement 2 and $p^{l-1}(p-1) > 1$, the only number in $\{Norm_m(1 - e(\frac{l}{2m})) \mid l \in \mathbb{Z}\}$, which is divisible by $p^{\varphi(2m)} = p^{\varphi(m)}$, is 0. Thus $Norm_m(1 - c(e(\frac{1}{m}))) = 0$ and $c(e(\frac{1}{m})) = 1$.

Case 2: m is even. Then $\{\pm e(\frac{l}{m}) \mid l \in \mathbb{Z}\} = \{e(\frac{l}{m}) \mid l \in \mathbb{Z}\}$. With m instead of $2m$ we can argue as in case 1. This proves a). $\qquad\square$

References

[AGV] Arnold, V.I.; Gusein-Zade, S.M.; Varchenko, A.N.: *Singularities of Differentiable Maps.* Vol. II, Boston Basel Berlin: Birkhäuser (1988).

[Br] Brieskorn, E.: *Die Monodromie der isolierten Singularitäten von Hyperflächen.* Manuscripta Math. **2**, 103–161, (1970).

[ClGr] Clemens, H.; Griffiths, P.: *The intermediate Jacobian of the cubic threefold.* Ann. Math. **95**, 281–356, (1972).

[De] Deligne, P.: *Equations différentielles à points singuliers réguliers.* Lecture Notes Math. vol. **163**, Berlin Heidelberg New York: Springer (1970).

[HaMu] Hauser, H; Müller, G.: *The cancellation property for direct products of analytic spaces.* Math. Ann. **286**, 209–223, (1990).

[He1] Hertling, C.: *Analytische Invarianten bei den unimodularen und bimodularen Hyperflächensingularitäten.* Dissertation. Bonner Math. Schriften **250**, Bonn (1992).

[He2] Hertling, C.: *Ein Torellisatz für die unimodalen und bimodularen Hyperflächensingularitäten.* Math. Ann. **302**, 359–394, (1995).

[He3] Hertling, C.: *Classifying spaces and moduli spaces for polarized mixed Hodge structures and for Brieskorn lattices.* Preprint.

[Ma] Malgrange, B.: *Intégrales asymptotiques et monodromie.* Ann. Sci. École Norm. Sup. **7**, 405–430 (1974).

[MaYa] Mather, J.; Yau, St.S.-T.: *Classification of isolated hypersurface singularities by their moduli algebras.* Invent. Math. **69**, 243–252, (1982).

[Mi]   Milnor, J.: *Singular points of complex hypersurfaces.* Ann. Math. Stud. vol. **61**, Princeton University Press (1968).

[Ph1] Pham, F.: *Singularités des systèmes différentielles de Gauss-Manin.* Prog. Math. 2, Boston Basel Stuttgart: Birkhäuser (1979).

[Ph2] Pham, F.: *Structure de Hodge mixte associée à point critique isolé.* Astérisque **101–102**,268–285, (1983).

[SK1] Saito, K.: *The higher residue pairings $K_F^{(k)}$ for a family of hypersurface singular points.* In: Singularities, Proc. of symp. in pure math. **40.2**, 441–463, (1983).

[SK2] Saito, K.: *Period mapping associated to a primitive form.* Publ. Res. Inst. Math. Sci. Kyoto Univ. **19**, 1231–1264, (1983).

[SM1] Saito, M.: *On the structure of Brieskorn lattices.* Ann. Inst. Fourier Grenoble **39**, 27–72, (1989).

[SM2] Saito, M.: *Period mapping via Brieskorn modules.* Bull. Soc. math. France **119**, 141–171, (1991).

[SchSt] Scherk, J.; Steenbrink, J.H.M.: *On the mixed Hodge structure on the cohomology of the Milnor fibre.* Math. Ann. **271**, 641–665, (1985).

[Sche] Scherk, J.: *On the monodromy theorem for isolated hypersurface singularities.* Invent. Math. **58**, 289–301, (1980).

[Schm] Schmid, W.: *Variation of Hodge structure: The singularities of the period mapping.* Invent. Math. **22**, 211–319, (1973).

[SeTh] Sebastiani, M.; Thom, R.: *Un resultat sur la monodromie.* Invent. Math. **13**, 90–96, (1971).

[St]   Steenbrink, J.H.M.: *Mixed Hodge structure on the vanishing cohomology.* In: Real and complex singularities, Oslo (1976), Holm, P.(ed.).Alphen aan den Rijn: Sijthoff and Noordhoff, pp. 525–562, (1977).

[Tju] Tjurin, A.N.: *The geometry of the Fano surface of a nonsingular cubic $F \subset \mathbb{P}^4$ and Torelli's theorem for Fano surfaces and cubics.* Math. USSR Izvestiya **5**, 517–546, (1971).

[Va1] Varchenko, A.N.: *The asymptotics of holomorphic forms determine a mixed Hodge structure.* Sov. Math. Dokl. **22**, 772–775,(1980).

[Va2] Varchenko, A.N.: *The complex singular index does not change along the stratum $\mu = $constant.* Functional Anal. Appl. **16**, 1–9, (1982).

[Va3]   Varchenko, A.N.: *A lower bound for the codimension of the stratum* μ =*constant in terms of the mixed Hodge structure.* Moscow Univ. Math. Bull. **37**, 30–33, (1982).

[Va4]   Varchenko, A.N.: *On the local residue and the intersection form on the vanishing cohomology.* Math. USSR Izvestiya **26**, 31–52,(1986).

[Wa]   Washington, L.C.: *Introduction to cyclotomic fields.* New York: Springer (1982).

Progress in Mathematics, Vol. 162, © 1998 Birkhäuser Verlag Basel/Switzerland

Equiclassical Deformation of Plane Algebraic Curves

Eugenii Shustin*
School of Mathematical Sciences
Tel Aviv University
Ramat Aviv
Tel Aviv 69978
ISRAEL

Abstract

We study the geometry of equiclassical families of plane algebraic curves, that means the sets of irreducible curves of a given degree d with given geometric genus g and class c (degree of the dual curve). It is known that the generic member of an equigeneric family (set of curves of given degree with given genus) is a nodal curve, or, in other words, any curve can be deformed into a nodal curve of the same degree and genus. We give sufficient conditions for the existence of a deformation of a plane irreducible curve into a curve of the same degree, genus and class, having only nodes and cusps. For instance, if

$$c \geq 2g - d + 2,$$

then the generic member of an equiclassical family is a curve with nodes and ordinary cusps, that strengthens the Diaz-Harris sufficient condition $c \geq 2g - 1$. In particular, any curve of degree ≤ 10 can be equiclassically deformed into a curve with nodes and cusps.

1 Introduction

Let $\mathbf{P}^{d(d+3)/2}$ be the space of plane algebraic curves of degree d over an algebraically closed field of characteristic zero, $\Delta \subset \mathbf{P}^{d(d+3)/2}$ be the discriminant, and $\Delta' \subset \Delta$ be the set of singular reduced irreducible curves. We shall concentrate on the *stratification problem* by introducing a certain equivalence relation of singular curves and describing the geometry of equivalence classes and their

---

*The author was partially supported by the grant no. 6836-1-9 of the Ministry of Science and Arts, Israel, and by the Minerva Center for Geometry at the Tel Aviv University.

adjacency. The *equisingular* and *equigeneric* stratifications have been studied after Severi [17]. The equisingular stratification (locally, μ-constant stratification) is complicated: its strata may be smooth or not, irreducible or not, and so on (see, for instance, [8], [13], [15], [17], [18], [23]). On the other hand, the set $V_{d,g} \subset \Delta'$ of curves of degree d and genus g is an irreducible variety of dimension $3d - 1 + g$ [10], [17], whose generic point corresponds to a nodal curve [17] (see also [1], [16]), and which is smooth at any point corresponding to a nodal curve [17]. For other properties see [4], [6], [20].

In this article we consider an intermediate stratification – *equiclassical*: a subdivision of Δ' into the sets $V_{d,g,c}$ of curves of degree d, genus g and class c (degree of the dual curve). It was studied in [3], [5]. In particular, an equiclassical deformation of a curve $F \in V_{d,g,c}$ means an equiclassical deformation of any singular point of F, i.e. in the stratum $\{\delta = \text{const}, \ \kappa = \text{const}\}$, where δ is the δ-invariant, and κ is the intersection number of a curve with its generic polar at the point considered, by the classical formulae [5] (Definition 3.12), [11], [22]:

$$g = \frac{(d-1)(d-2)}{2} - \sum_{z \in F} \delta(z), \quad c = d(d-1) - \sum_{z \in F} \kappa(z). \tag{1}$$

Diaz and Harris [5] proved that

- any plane algebraic curve isolated singular point can be deformed in the stratum $\{\delta = \text{const}, \ \kappa = \text{const}\}$ into a collection of ordinary nodes and cusps (further simply, *nodes* and *cusps*); more precisely, into $3\delta - \kappa$ nodes and $\kappa - 2\delta$ cusps, since [22]

$$\delta(\text{node}) = \delta(\text{cusp}) = 1, \quad \kappa(\text{node}) = 2, \quad \kappa(\text{cusp}) = 3;$$

- if $c \geq 2g - 1$ then any curve in $V_{d,g,c}$ can be equiclassically deformed into a curve with only nodes and cusps, i.e. a generic member of any component of $V_{d,g,c}$ is a curve with nodes and cusps.

Here we strengthen the second result:

THEOREM 1.1 *If*

$$c \geq 2g - d + 2, \tag{2}$$

then a generic member of any irreducible component of $V_{d,g,c}$ is a curve with only nodes and cusps.

Plücker's formulae

$$g = \frac{(d-1)(d-2)}{2} - n - k, \quad c = d(d-1) - 2n - 3k$$

give us the numbers n, k of nodes and cusps of these generic members

$$n = \frac{d^2 - 7d + 6}{2} + c - 3g, \quad k = 2d - 2 - c + 2g.$$

Under the condition (2), $k \leq 3d - 4$; hence (see [8], [23] (Chapter VIII)) the dimension of any component of $V_{d,g,c}$ is

$$\frac{d(d+3)}{2} - n - 2k = c - g + d + 1,$$

and $V_{d,g,c}$ is smooth at any point corresponding to a curve with nodes and cusps. But $V_{d,g,c}$ may be reducible, for instance, the classical Zariski example of a reducible set of sextic curves with six cusps [23] (Chapter VIII) shows that $V_{6,4,12}$ satisfies (2) and is reducible.

COROLLARY 1.2 *If $2g \leq d+1$ then $V_{d,g,c}$ is an irreducible variety of dimension $c - g + d + 1$, whose generic member is a curve with nodes and cusps.*

Indeed, for $d \geq 3$, the inequality $2g \leq d + 1$ implies (2). Hence an open dense subset of $V_{d,g,c}$ consists of curves with nodes and cusps. But then $V_{d,g,c}$ is irreducible by [12], where the inequality $2g \leq d+1$ is proven to be sufficient for the irreducibility of the corresponding variety of curves with nodes and cusps.

COROLLARY 1.3 *Any irreducible plane curve of degree ≤ 10 can be equiclassically deformed into a curve with nodes and cusps.*

We give one more sufficient criterion for the existence of such equiclassical deformations, applicable to individual curves. For a singular point z of a curve $F \in V_{d,g,c}$, put

$$\widetilde{\kappa}(z) = \kappa(z) - \delta(z),$$

and denote by k_m, $m \geq 1$, the number of singular points $z \in F$ with $\widetilde{\kappa}(z) = m$.

THEOREM 1.4 *Let*

$$\sum_{i \geq m} \sqrt{2ik_i} \leq \frac{d}{m+1} + 2, \quad m \geq 1. \tag{3}$$

Then a generic member of any irreducible component of $V_{d,g,c}$, containing F, is a curve with nodes and cusps.

It can easily be seen that the condition (3) is not covered by (2).

I am very grateful to C. Lossen for his remarks that allowed me to correct mistakes.

2 Preliminaries

Here we introduce notations and recall some notions and facts used below in
the proof.

If S is a zero-dimensional subscheme of $\mathbf{P}^2$ (or a curve T) then by $\mathcal{J}_S \subset \mathcal{O}_{\mathbf{P}^2}$
(resp. $\mathcal{J}_{S/T} \subset \mathcal{O}_T$) we denote the ideal sheaf of S on $\mathbf{P}^2$ (resp. on T). Let F be
a reduced irreducible plane curve of degree d. Denote by Con_F the conductor
sheaf on F – the annulator of the sheaf $n_* \mathcal{O}_{\widetilde{F}} / \mathcal{O}_F$, where $n : \widetilde{F} \to F$ is the
normalization. It can be expressed as

$$\mathrm{Con}_F = \mathcal{O}_F(-\Delta(F)),$$

where $\Delta(F)$ is the so-called "adjoint", or "double point divisor" (see [7], section
9.3, and also [2], Appendix A). It is known that

$$\deg \Delta = 2 \sum_{z \in \mathrm{Sing}(F)} \delta(z),$$

where

$$\delta(z) = \dim(\mathcal{O}_{F,z}/\mathrm{Con}_{F,z})$$

is the δ-invariant (see, for instance [5] (Definition 3.12)). More precisely ([2]
(Appendix A)),

$$\Delta(F) = \sum_P ((\Pi \cdot P) - m(P)) \cdot P, \quad m(P) = \mathrm{mult}(P) - 1,$$

where P runs through all local branches of F centered at singular points, Π is
a generic polar curve of F, $(\Pi \cdot P)$ is the intersection number, and $\mathrm{mult}(P)$ is
the multiplicity of the branch P (intersection number with a generic straight
line through the center of P). Let D be an effective Cartier divisor on F.
The sheaf $\mathcal{O}_F(-D - \Delta(F))$ is an ideal sheaf of some zero-dimensional scheme
$X(D) \subset F$. By [5] (sec. 4.3), the tangent cone to $V_{d,g,c}$ at $F \in V_{d,g,c}$ is contained
in $|\mathcal{J}_{X(D_{ec})}(d)|$, where

$$D_{ec} = \sum_P m(P) \cdot P, \tag{4}$$

with the sum taken over all local branches P of F at its singular points. Below
we use the formulae

$$\deg D_{ec} = \sum_{w \in \mathrm{Sing}(F)} (\kappa(w) - 2\delta(w)) , \tag{5}$$

$$\deg X(D_{ec}) = \sum_{w \in \mathrm{Sing}(F)} (\kappa(w) - \delta(w)) . \tag{6}$$

LEMMA 2.1 *In the above notation, if* $\deg D \leq 3d - 1$ *then*

$$h^0(\mathbf{P}^2, \mathcal{J}_{X(D)}(d)) = g + 3d - \deg D, \tag{7}$$

where g is the geometric genus of F.

Proof of Lemma 2.1. From the exact sequence

$$0 \to \mathcal{O}_{\mathbf{P}^2} \overset{\times F}{\to} \mathcal{J}_{X(D)}(d) \to \mathcal{O}_F(-D - \Delta(F))(d) \to 0$$

one obtains

$$h^0(\mathbf{P}^2, \mathcal{J}_{X(D)}(d)) = h^0(F, \mathcal{O}_F(-D - \Delta(F))(d)) + 1 \ .$$

To compute the latter term we apply the Riemann-Roch theorem to the curve F in the form [21] (Chapter 8):

$$\begin{aligned}
h^0(\mathcal{O}_F(-D - \Delta(F))(d)) &= (d^2 - \deg\Delta(F) - \deg D) - g + 1 \\
&\quad + h^1(\mathcal{O}_F(-D - \Delta(F))(d)).
\end{aligned} \tag{8}$$

Under the condition $\deg D \leq 3d - 1$,

$$d^2 - \deg D - \deg\Delta(F) = 2g + 3d - 2 - \deg D > 2g - 2 \ ;$$

hence

$$h^1(\mathcal{O}_F(-D - \Delta(F))(d)) = 0,$$

which implies (7), due to (8). $\qquad\square$

3 Proof of Theorem 1.1

Step 1. By [5] (sec. 5.2), the dimension of any component of $V_{d,g,c}$ is at least $c - g + d + 1$. Hence we have only to show that under condition (2), the dimension of any equisingular stratum in $V_{d,g,c}$, consisting of curves with at least one singularity different from node or cusp, is less than $c - g + d + 1$.

Let $F \in V_{d,g,c}$, and $V(F)$ be a germ at F of the equisingular stratum (i.e. consisting of curves with the same collection of singularities).

Step 2. Assume that F has a singular point z of multiplicity ≥ 3. Consider the set $V(F, z) \subset V(F)$ of curves having the same singularity at z as F. Clearly,

$$\dim V(F) \leq \dim V(F, z) + 2 \ .$$

By [9] (Theorem 2 and its proof) the tangent cone to $V(F, z)$ at F is contained in the linear system $|\mathcal{J}_{X(D')}(d)|$, where

$$D' = \sum_P m(P) \cdot P + \sum_Q \mathrm{mult}(Q) \cdot Q = D_{ec} + \sum_Q \mathrm{mult}(Q) \cdot Q,$$

with P running through all local branches of F centered at all singular points, Q running through all local branches of F centered at z. Since the multiplicity of z is ≥ 3, we can choose a divisor $D_{ec} \leq D \leq D'$ such that

$$D = D_{ec} + \sum_Q m'(Q) \cdot Q, \quad \sum_Q m'(Q) = 3 \ .$$

In particular, by (5),

$$\deg D = \sum_{w \in \mathrm{Sing}(F)} (\kappa(w) - 2\delta(w)) + 3 \ .$$

Since, by (1), (2),

$$\sum_{w \in \mathrm{Sing}(F)}(\kappa(w) - 2\delta(w)) = d(d-1) - c - (d-1)(d-2) + 2g$$
$$= 2d - 2 - c + 2g \le 3d - 4, \tag{9}$$

we obtain $\deg D \le 3d - 1$; hence by Lemma 2.1,

$$\dim V(F, z) \le \dim |\mathcal{J}_{X(D)}(d)| = h^0(\mathbf{P}^2, \mathcal{J}_{X(D)}(d)) - 1$$
$$\le g + 3d - 1 - \deg D \le g + 3d - 1 - (2d + 1 - c + 2g) = c - g + d - 2$$

and

$$\dim V(F) \le \dim V(F, z) + 2 \le c - g + d < \dim V_{d,g,c} \ .$$

Step 3. Assume that F has a double point z different from node and cusp, that is, of type $A_n, n \ge 3$. If $n = 2k + 1$, $k \ge 1$, then F has two non-singular branches P_1, P_2, centered at z and intersecting with multiplicity $k+1$. If $n = 2k$, $k \ge 2$, then F has one branch P of multiplicity 2, centered at z. We recall that $\delta(A_{2k+1}) = k + 1$, $\delta(A_{2k}) = k$. The equisingular ideal $I^{es} \subset \mathcal{O}_{\mathbf{P}^2,z}$ (unlike [5] we consider an ideal in the local ring of the plane, not of the curve) coincides with the Jacobian ideal and can be represented as follows:

$$I = \{f \in \mathcal{O}_{\mathbf{P}^2,z} \ : \ (f \cdot \Pi) \ge n\} \ ,$$

where Π is a generic (clearly, non-singular at z) polar curve of F.

LEMMA 3.1 *If $n = 2k + 1$, then the ideal*

$$I' = \{f \in \mathcal{O}_{\mathbf{P}^2,z} \ : \ (f \cdot \Pi) \ge k + 2\}$$

contains the ideal $I'' \subset \mathcal{O}_{\mathbf{P}^2,z}$ defined by

$$I'' = \{f \in \mathcal{O}_{\mathbf{P}^2,z} \ : \ (f \cdot P_i) \ge k + 2, \quad i = 1, 2\}.$$

If $n = 2k$, then the ideal

$$I' = \{f \in \mathcal{O}_{\mathbf{P}^2,z} \ : \ (f \cdot \Pi) \ge k + 2\}$$

contains the ideal $I'' \subset \mathcal{O}_{\mathbf{P}^2,z}$ defined by

$$I'' = \{f \in \mathcal{O}_{\mathbf{P}^2,z} \ : \ (f \cdot P) \ge 2k + 3\}.$$

In both the cases

$$\dim I'/I'' = 1 \ . \tag{10}$$

Proof. Since all the relations are invariant with respect to a local smooth coordinate change, one can assume that

$$F = \{y^2 - x^{n+1} = 0\}, \quad \Pi = \{y = 0\},$$

and

$$P_{1,2} = \{y = \pm x^{k+1}\}, \quad \text{if} \quad n = 2k + 1,$$
$$P = \{y^2 - x^{n+1} = 0\}, \quad \text{if} \quad n = 2k.$$

Then one easily sees that

$$f \in I'' \iff f(x,y) = \sum_{i+(k+1)j>k+1} a_{ij} x^i y^j, \quad n = 2k+1, \quad (11)$$
$$f \in I'' \iff f(x,y) = \sum_{2i+(2k+1)j>2k+2} a_{ij} x^i y^j, \quad n = 2k, \quad (12)$$

which implies $I' \supset I''$. Since $\dim \mathcal{O}_{\mathbf{P}^2,z}/I' = k+2$ and, by (11), (12), $\dim \mathcal{O}_{\mathbf{P}^2,z}/I'' = k+3$, one obtains (10). $\qquad \Box$

Let us introduce the following zero-dimensional schemes $X, X', X'' \subset \mathbf{P}^2$, concentrated at the singular points of F. At any point $w \in \mathrm{Sing}(F)$, $w \neq z$, put

$$(\mathcal{J}_X)_w = (\mathcal{J}_{X'})_w = (\mathcal{J}_{X''})_w = (\mathcal{J}_{X(D_{ec})})_w \ ,$$

where D_{ec} is defined by (4). At the point z, put

$$(\mathcal{J}_X)_z = I^{es}, \quad (\mathcal{J}_{X'})_z = I', \quad (\mathcal{J}_{X''})_w = I'' \ .$$

By [5], [9] (Theorem 1), the tangent cone to $V(F)$ at F is contained in $|\mathcal{J}_X(d)|$; hence

$$\dim V(F) \leq h^0(\mathbf{P}^2, \mathcal{J}_X(d)) - 1 \ .$$

Since $I' \supset I^{es}$ as $n > 3$, we have $X \supset X'$ and

$$\dim V(F) \leq h^0(\mathbf{P}^2, \mathcal{J}_{X'}(d)) - 1 \ .$$

On the other hand, by Lemma 3.1, $X' \subset X''$ and $\deg X' = \deg X'' - 1$; hence

$$\dim V(F) \leq h^0(\mathbf{P}^2, \mathcal{J}_{X''}(d)) \ .$$

To compute the latter term, we note that $X'' = X(D)$, where $D = D_{ec} + P_1 + P_2$ if $n = 2k+1$, and $D = D_{ec} + 2P$ if $n = 2k$. By (2), (5), as in the second step, one obtains

$$\deg D = \deg D_{ec} + 2 = \sum_{w \in \mathrm{Sing}(F)} (\kappa(w) - 2\delta(w)) + 2$$
$$= 2d - 2 - c + 2g + 2 = 2d - c + 2g \leq 3d - 2 \ .$$

Hence, by Lemma 2.1

$$\dim V(F) \leq h^0(\mathbf{P}^2, \mathcal{J}_{X''}(d)) \leq g + 3d - (2d - c + 2g)$$
$$= d + c - g < \dim V_{d,g,c} \ ,$$

and we are done.

4   Proof of Corollary 1.3

LEMMA 4.1 *For any curve $F \in V_{d,g,c}$ the inequality*

$$c \geq 2g - \frac{d^2 - 8d + 6}{3} - \varepsilon, \quad \varepsilon = \begin{cases} 1, & \text{if } d = 4, 6 \\ 0, & \text{if } d \neq 4, 6 \end{cases} \tag{13}$$

holds true.

One can easily verify that (13) implies (2) for $d \leq 10$, and derive Corollary 1.3 from Theorem 1.1.

Proof of Lemma 4.1. Denote by F^* the dual curve of F. For a singular point z of a curve, put

$$m(z) = \kappa(z) - 2\delta(z) = \sum_P m(P),$$

where P runs through all local branches centered at z (cf. [5], Definition 3.12). Since (see, for instance, [14], Remark 10.10)

$$\delta(z) \geq \frac{\mathrm{mult}(z) \cdot (\mathrm{mult}(z) - 1)}{2} \geq m(z),$$

one derives

$$m(z) \leq \frac{\kappa(z)}{3} . \tag{14}$$

Hence

$$\sum_{z \in \mathrm{Sing}(F)} m(z) \leq \frac{1}{3} \sum_{z \in \mathrm{Sing}(F)} \kappa(z) = \frac{1}{3}(d(d-1) - c). \tag{15}$$

On the other hand, from Plücker's formula

$$d = c(c-1) - \sum_{w \in \mathrm{Sing}(F^*)} \kappa(w)$$

and from

$$\sum_{z \in \mathrm{Sing}(F)} m(z) - \sum_{w \in \mathrm{Sing}(F^*)} m(w) = 3(d - c)$$

(see [21], §26, formula (10)), it follows that

$$\begin{aligned}
\sum_{z \in \mathrm{Sing}(F)} m(z) &= 3(d - c) - \sum_{w \in \mathrm{Sing}(F^*)} m(w) \\
&= 3c(c-2) - \sum_{w \in \mathrm{Sing}(F^*)} (3\kappa(w) - m(w)) \\
&\leq 3c(c-2) - \frac{8}{3} \sum_{w \in \mathrm{Sing}(F^*)} \kappa(w) \\
&= 3c(c-2) - \frac{8}{3}(c(c-1) - d) = \frac{1}{3}(c^2 - 10c + 8d) .
\end{aligned}$$

Combining this with (15), one obtains

$$\sum_{z \in \mathrm{Sing}(F)} m(z) \le \frac{1}{3} \max_c \min\{d^2 - d - c,\ c^2 - 10c + 8d\} = \frac{1}{3}d(d-2) + \varepsilon,$$

which by (9),

$$\sum_{z \in \mathrm{Sing}(F)} m(z) = \sum_{z \in \mathrm{Sing}(F)} (\kappa(z) - 2\delta(z)) = 2d - 2 - c + 2g$$

implies (13). $\qquad\qquad\square$

5 Proof of Theorem 1.4

Without loss of generality assume that $d > 10$.

We shall show that if $F \in V_{d,g,c}$ satisfies (3) and has singularities different from nodes and cusps, then

$$\dim V(F) < c - g + d + 1 \ .$$

Hence, as in Section 3, the germ of the equisingular stratum $V(F)$ does not contain the germ of any component of $V_{d,g,c}$. Thus, in any component of $V_{d,g,c}$ the curve F can be deformed into a curve F' with another collection of singularities. Since the left-hand side in (3) does not increase in deformations of curves, F' satisfies (3) as well, and we proceed in this manner, until we obtain a curve with only nodes and cusps.

As in the proof of Theorem 1.1, assuming that F has a singular point of multiplicity ≥ 3 or of type $A_k, k \ge 3$, we introduce the divisor D defined in the same manner, and reduce the problem to the relation

$$h^1(\mathbf{P}^2, \mathcal{J}_{X(D)}(d)) = 0 \ . \tag{16}$$

The proof of (16) is a word-by-word copy of the argument in the proof of the similar h^1-vanishing relations (15), (16) in [19], Section 4.3, and we omit it here.

References

[1] Arbarello, E. and Cornalba, M.: A few remarks about the variety of irreducible plane curves of given degree and genus. *Ann. sci. Ecole norm. sup.* **16** (1983), no. 3, 467–488.

[2] Arbarello, E., Cornalba, M., Griffiths, Ph., and Harris, J.: *Geometry of algebraic curves*, vol. 1, Springer, New York, 1985.

[3] Diaz, S.: Irreducibility of the equiclassical locus. *J. Diff. Geom.* **29** (1989), 489–498.

[4] Diaz, S. and Harris, J.: Geometry of the Severi variety. *Trans. Amer. Math. Soc.* **309** (1988), no. 1, 1–34.

[5] Diaz, S. and Harris, J.: Ideals associated to deformations of singular plane curves. *Trans. Amer. Math. Soc.* **309** (1988), no. 2, 433–468.

[6] Diaz, S. and Harris, J.: *Geometry of Severi varieties, II: Independence of divisor classes and examples.* Preprint, 1992.

[7] Fulton, W.: *Intersection Theory (Ergebnisse der Mathematik und ihrer Grenzgebiete)*, Springer, Berlin etc., 1984.

[8] Greuel, G.-M. and Karras, U.: Families of varieties with prescribed singularities. *Compos. math.* **69** (1989), no. 1, 83–110.

[9] Gudkov, D. A. and Shustin E. I.: On the intersection of the close algebraic curves. In: *Topology (Proc. Leningrad Internat. Topological Conf., Leningrad, Aug. 1982)/ Lect. Notes in Math., vol. 1060*, Springer, Berlin etc., 1984, pp. 278–289.

[10] Harris, J.: On the Severi problem. *Invent. Math.* **84** (1985), 445–461.

[11] Hironaka, H.: Arithmetic genera and effective genera of algebraic curves. *Mem. Coll. Sci. Univ. Kyoto. Sect. A30* (1956), 177–195.

[12] Kang, P.-L.: A note on the variety of plane curves with nodes and cusps. *Proc. Amer. Math. Soc.* **106** (1989), no. 2, 309–312.

[13] Luengo, I.: The μ-constant stratum is not smooth. *Invent. Math.* **90** (1987), 139–152.

[14] Milnor, J.: *Singular points of complex hypersurfaces (Annals of Math. Studies 61)*, Princeton Univ. Press, Princeton, 1968.

[15] Nobile, A.: On families of singular plane projective curves. *Ann. mat. pura ed appl.* **138** (1984), 341–378.

[16] Nobile, A.: On specialization of curves, I. *Trans. Amer. Math. Soc.* **282** (1984), no. 2, 739–748.

[17] Severi, F.: *Vorlesungen über Algebraische Geometrie* (Anhang F). Teubner, Leipzig, 1921.

[18] Shustin, E.: Geometry of equisingular families of plane algebraic curves. *J. Algebraic Geometry* **5** (1996), no. 2, 209–234.

[19] Shustin, E.: Smoothness of equisingular families of plane algebraic curves. *International Math. Research Notices* **2** (1997), 67–82.

[20] Treger, R.: Local properties of families of plane curves. *J. Diff. Geom.* **39** (1994), 51–55.

[21] Van der Waerden, B.L.: *Einführung in die algebraische Geometrie*, 2nd edition, Springer, Berlin etc., 1973.

[22] Walker, R.: *Algebraic curves.* Dover, New York, 1950.

[23] Zariski, O.: *Algebraic surfaces*, 2nd ed. Springer, Heidelberg, 1971.

Progress in Mathematics, Vol. 162, © 1998 Birkhäuser Verlag Basel/Switzerland

Monodromy of Complete Intersections and Surface Potentials

Victor A. Vassiliev*
Steklov Math. Institute
Vavilova Str. 42
Moscow
RUSSIA

To Egbert Brieskorn with admiration

Abstract

Following Newton, Ivory and Arnold, we study the Newtonian potentials of algebraic hypersurfaces in $\mathbf{R}^n$. The ramification of (analytic continuations of) these potential depends on a monodromy group, which can be considered as a proper subgroup of the local monodromy group of a complete intersection (acting on a *twisted* vanishing homology group if n is odd). Studying this monodromy group we prove, in particular, that the attraction force of a hyperbolic layer of degree d in $\mathbf{R}^n$ coincides with appropriate *algebraic* vector-functions everywhere outside the attracting surface if $n = 2$ or $d - 2$, and is non-algebraic in all domains other than the hyperbolicity domain if the surface is generic and $(d \geq 3)\&(n \geq 3)\&(n + d \geq 8)$.

Recently W. Ebeling removed the last restriction $d + n \geq 8$, see his Appendix to this article.

1 Introduction

Two famous theorems of Newton assert that

a) a homogeneous spherical layer in Euclidean space does not attract bodies inside the sphere, and

b) exterior bodies are attracted by it to the center of the sphere as by the point-wise particle whose mass is equal to the mass of the entire sphere.

---

*Research supported by the Russian Fund of Basic Investigations (project 95-01-00846a) and INTAS grant (Project # 4373)

Ivory [I] extended both these theorems to the attraction of ellipsoids, and Arnold [A 82] extended the first of them to the attraction of arbitrary hyperbolic hypersurfaces: such a surface does not attract the particles inside the hyperbolicity domain; see also [G 84].

In any component of the complement of the attracting surface this attraction force coincides with a real analytic vector-function; we investigate the ramification of this function, in particular (following one another famous theory of Newton, see [A 87], [AV]) the question if it is algebraic or not.

We describe the monodromy group responsible for the ramification and identify it as a subgroup of the local monodromy group of a complex complete intersection of codimension 2 in $\mathbf{C}^n$. Unlike the usual local monodromy action, this monodromy representation is reducible: e.g. the Newton–Ivory–Arnold theorem depends on the fact that the homology class of the set of real points of a hyperbolic surface defines an invariant element of this action (although this element is not equal to zero: indeed, otherwise even the potential function of the force would be zero, and not only its gradient field, which is wrong already in the Newton's case).

Although we consider mainly the orbit of a very special cycle, formed by all real points of a hyperbolic polynomial, all our calculations can be applied to more general situations, e.g. when the integration cycle is an arbitrary linear combination of real components (maybe non-compact) of an algebraic hypersurface in $\mathbf{R}^n$.

In the case of odd n, this group acts in a vanishing homology group with twisted coefficients (so that the corresponding kernel form $r^{2-n}ds$ of the potential function can be integrated correctly along its elements). In § 2.3 we extend the standard facts concerning vanishing homology of complete intersections to this group, cf. [Ph 65], [G 88].

There is a (non-formal) partition of all classes of isolated singularities of complete intersections into series with varying dimension n of the ambient space $\mathbf{C}^n$ (but with the constant codimension p of the complete intersection), see [E], [AGLV]; e.g., all singularities given by p generic quadrics in the spaces $\mathbf{C}^n$ with different n and fixed p form such a series. To any such series there corresponds a series of reflection groups, also depending on the parameter n; for such n that $n-p$ is even, these groups coincide with the (standard) local monodromy groups of corresponding singularities. The homology groups described in § 2.3 fill in the gap: for $n-p$ odd, the reflection group of the natural series coincides with the monodromy action on such a twisted homology group of the corresponding singularity. (In the marginal case $p = 1$, all the reflection groups of the series coincide, see [GZ], [G 88].)

This is a reason why the qualitative behavior of attraction forces in the spaces of any dimension is essentially the same, unlike the usual situation (see e.g. [P], [ABG], [A 87], [AV], [V 94]) when the functions given by similar integral representations behave in very different way in the spaces of dimensions of different parity.

For $n = 2$ and arbitrary d, our monodromy group is finite, thus the analytic continuation of the attraction force is finitely-valued, in particular (by the Riemann's existence principle) algebraic, see § 5.1 below. A realistic estimate of the number of values of this continuation is given by Theorem 4. In particular, we get a new series of examples when the attraction force coincides with a *single-valued* (rational) vector-function outside the hyperbolicity domain, see the Corollary to Theorem 4.

For $d = 2$ and arbitrary $n > 2$, the monodromy group is infinite, and the orbit of any integration cycle lies on an ellipsoidal cylinder in the vanishing homology space. Fortunately, the integral of the attracting charge takes zero value on the directing plane of this cylinder, thus the number of its values along the elements of any orbit again is finite, see § 5.2.

In all the other cases (when $d \geq 3$ and $n \geq 3$) it seems likely that the monodromy group defined by the *generic* algebraic surface of degree d in $\mathbf{R}^n$ is large enough to ensure that the Newton's integral (and any other non-zero linear form on the space of vanishing cycles) takes an infinite number of values on the orbit of any non-invariant vector (and the unique invariant vector is presented by the integration cycle corresponding to the hyperbolicity domain of an hyperbolic charge). I can prove this conjecture only if the additional restriction $d + n \geq 8$ is satisfied[1].

Everywhere below all the homology groups $H_*(\cdot)$ are reduced modulo a point.

2 Vanishing homology and local monodromy of complete intersections

Here we recall the basic facts about the local Picard–Lefschetz theory of isolated singularities of complete intersections (see e.g. [H], [E], [AGLV]) and extend them to the case of twisted vanishing homology groups.

2.1 Classical theory

Let $f : (\mathbf{C}^n, 0) \to (\mathbf{C}^p, 0)$ be a holomorphic map, $f = (f_1, \ldots, f_p)$, and suppose that the variety $f^{-1}(0)$ is an *isolated complete intersection singularity* (ICIS) at 0 (i.e. it is a smooth $(n - p)$-dimensional variety in a punctured neighborhood of 0). Suppose that the coordinates in $\mathbf{C}^p$ are chosen generically, then the map $\tilde{f} \equiv (f_1, \ldots, f_{p-1}) : \mathbf{C}^n \to \mathbf{C}^{p-1}$ also defines an ICIS at 0. Let B be a sufficiently small *closed* disc centered at the origin in $\mathbf{C}^n$, and $c = (c_1, \ldots, c_p)$ a generic point very close to the origin in $\mathbf{C}^p$. The corresponding manifolds $X_f \equiv f^{-1}(c) \cap B$ and $\tilde{X}_f \equiv \tilde{f}^{-1}(c_1, \ldots, c_{p-1}) \cap B$ are called the *Milnor fibres*

---

[1]For a complete proof, removing this restriction, see the Appendix to this article, written by W. Ebeling

of f and $\tilde{f}$. Their homology groups are connected by the exact sequence

$$\cdots \to H_{n-p+1}(\tilde{X}_f) \to H_{n-p+1}(\tilde{X}_f, X_f) \xrightarrow{\partial} H_{n-p}(X_f) \to \cdots . \qquad (1)$$

Proposition 1 (see [M], [H]). *The sequence (1) is trivial outside the fragment presented here. All groups in (1) are free Abelian. Moreover, the spaces X_f and $\tilde{X}_f$ are homotopy equivalent to the wedges of spheres of dimensions $n - p$ and $n - p + 1$ respectively.*

The rank of $H_{n-p}(X_f)$ is called the *Milnor number* of the complete intersection f and is denoted by $\mu(f)$. The Milnor numbers of all *quasihomogeneous* complete intersections are calculated in [GH] (in [MO] for $p = 1$); we need the following special case of this calculation.

Proposition 2. *1. The Milnor number of a homogeneous function $f : \mathbf{C}^n \to \mathbf{C}^1$ of degree d with isolated singularity at 0 is equal to $(d - 1)^n$.*
2. The Milnor number of a complete intersection $f = (f_1, f_2)$ with isolated singularity at 0, where the functions f_1 and f_2 are homogeneous of degrees a and b respectively, is equal to $((a - 1)^n b - (b - 1)^n a)/(a - b)$ if $a \neq b$, and to $(a - 1)^n(an - a + 1)$ if $a = b$.

The rank $\mu(f) + \mu(\tilde{f})$ of the middle group $H_{n-p+1}(\tilde{X}_f, X_f)$ of (1) is equal to the number of (Morse) critical points of the restriction of f_p on $\tilde{X}_f$. The generators of this group are represented by the *Lefschetz thimbles* defined by the (non-intersecting) paths in $\mathbf{C}^1$ connecting the non-critical value c_p of this restriction with all critical values, namely, any of these thimbles is an embedded disc swept out by the one-parametric family of *vanishing spheres* lying in the varieties $f^{-1}(c_1, \ldots, c_{p-1}, \tau)$, where τ runs over the corresponding path in $\mathbf{C}^1$: when τ tends to the endpoint (i.e. to a critical value of this restriction) the cycles of this family contract to the corresponding critical point. These vanishing spheres in the variety X_f (which corresponds to the common starting point c_p of these paths) generate the group $H_{n-p}(X_f)$, while the elements of $H_{n-p+1}(\tilde{X}_f)$ define relations among them.

2.2 Picard–Lefschetz formula for standard homology

Let $s \subset \mathbf{C}^1$ be the set of all these critical values, then the group $\pi_1(\mathbf{C}^1 \setminus s)$ acts naturally on all groups of (1). This action commutes with all arrows in (1) and is trivial on the left-hand group $H_{n-p+1}(\tilde{X}_f)$. The action on the middle and right-hand groups $H_{n-p+1}(\tilde{X}_f, X_f)$, $H_{n-p}(X_f)$ is determined by the Picard–Lefschetz formula: a class $\delta \in H_{n-p+1}(\tilde{X}_f, X_f)$, being transported along a *simple loop* (see [Ph 67], [V 94]) ω_i, corresponding to the path connecting c_p with the i-th critical value, becomes $\delta + (-1)^{(n-p+1)(n-p+2)/2}\langle \partial\delta, \partial\delta_i\rangle\delta_i$, where δ_i is the class of the thimble defined by this path, ∂ is the boundary operator in (1), and $\langle \cdot, \cdot \rangle$ is the intersection form in $H_{n-p}(X_f)$. In particular, a similar formula describes the monodromy action of the same loop on $H_{n-p}(X_f)$: it sends an element Δ of this group to

$$\Delta + (-1)^{(n-p+1)(n-p+2)/2} \langle \Delta, \Delta_i \rangle \Delta_i, \tag{2}$$

where $\Delta_i \equiv \partial \delta_i$ is the sphere vanishing along this path.

Proposition 3 (see e.g. [AGV]). *The intersection form $\langle \cdot, \cdot \rangle$ is symmetric if $n-p$ is even and skew-symmetric if $n - p$ is odd. The self-intersection index of any vanishing sphere is equal to 2 if $n-p \equiv 0 (mod\ 4)$ and to -2 if $n-p \equiv 2 (mod\ 4)$.*

In particular, if $n - p$ is even, then any transportation along a simple loop ω_i acts on the group $H_{n-p}(X_f)$ (respectively, $H_{n-p+1}(\tilde{X}_f, X_f)$) as the reflection in the hyperplane orthogonal to the vector Δ_i (respectively, δ_i) with respect to the intersection form in the homology of X_f (respectively, the form induced by the boundary operator from this intersection form). The latter action is a central extension of the former one.

More generally, let F be a k-parametric deformation of f, i.e. a map $F : \mathbf{C}^n \times \mathbf{C}^k \to \mathbf{C}^p$ such that $F(\cdot, 0) \equiv f$. For any $\lambda \in \mathbf{C}^k$ lying in a sufficiently small neighborhood D^k of the origin, denote by f_λ the map $F(\cdot, \lambda)$ and by $\tilde{f}_\lambda$ the map $\mathbf{C}^n \to \mathbf{C}^{p-1}$ given by first $p-1$ coordinate functions of f_λ. Set $X_{f,\lambda} = f_\lambda^{-1} \cap B$ and $\tilde{X}_{f,\lambda} = \tilde{f}_\lambda^{-1} \cap B$. If F is "not very degenerate" then for almost all values of λ these varieties are smooth (with boundaries) and have the same topological type; e.g. the varieties $X_f, \tilde{X}_f$ participating in (1) appear in the p-parametric deformation consisting of maps $f_\lambda \equiv (f_1 - \lambda_1, \ldots, f_p - \lambda_p)$ and correspond to the particular value $(\lambda_1, \ldots, \lambda_p) = (c_1, \ldots, c_p)$.

Definition 1. The *discriminant variety* $\Sigma(F)$ of F is the set of such $\lambda \in D^k$ that the topological type of the pair of varieties $(\tilde{X}_{f,\lambda}, X_{f,\lambda})$ does not coincide with that for all neighboring λ, i.e., either the origin in $\mathbf{C}^{p-1}$ is a critical value of $\tilde{f}_\lambda$ or the origin in $\mathbf{C}^1$ is a critical value of $f_p|_{\tilde{X}_{f,\lambda}}$. An exact sequence similar to (1) appears for any $\lambda \in D^k \setminus \Sigma(F)$, as well as the monodromy action of the group $\pi_1(D^k \setminus \Sigma(F))$ on this sequence.

Now suppose that the deformation F keeps $\tilde{f}$ undeformed, i.e., $\tilde{f}_\lambda \equiv \tilde{f}$ for any λ; in particular the action of this group on the left-hand group in (1) is trivial. A standard speculation with the Zariski's theorem (see e.g. [AGV], [V]) allows us to reduce this action to the above-considered action of the group $\pi_1(\mathbf{C}^1 \setminus s)$, and thus to the Picard–Lefschetz operators.

There is a natural map, *Leray tube operation*

$$t : H_{n-p}(X_f) \to H_{n-p+1}(\tilde{X}_f \setminus X_f) \tag{3}$$

described e.g. in [Ph 67], [AGLV], [V 94]: for any cycle γ in X_λ the cycle $t(\gamma)$ is swept out by the small circles in $\tilde{X}_f \setminus X_f$ which are the boundaries of the fibres of the natural fibration of the tubular neighborhood of X_f.

2.3   Twisted vanishing homology of complete intersections

Let L_{-1} (respectively, $\pm\mathbf{Z}$) be the local system on $\tilde{X}_f \setminus X_f$ with the fibre $\mathbf{C}^1$ (respectively, $\mathbf{Z}^1$) such that any loop having an odd linking number with X_f acts on this fibre as multiplication by -1. In particular, $L_{-1} \equiv \pm\mathbf{Z} \otimes \mathbf{C}$.

Consider the obvious homomorphism

$$j : H_{n-p+1}(\tilde{X}_f \setminus X_f, L_{-1}) \to H^{lf}_{n-p+1}(\tilde{X}_f \setminus X_f, L_{-1}), \tag{4}$$

where $H^{lf}_*(\cdot)$ denotes the homology of locally finite chains.

The Lefschetz thimbles define elements also in the right-hand group of (4) and in the similar group $H^{lf}_{n-p+1}(\tilde{X} \setminus X_f, \pm\mathbf{Z})$: indeed, they are embedded discs in $\tilde{X}_f \setminus X_f$ with boundary in X_f, and thus their interior parts can be lifted to an arbitrary leaf of the local system L_{-1} or $\pm\mathbf{Z}$. For any such thimble $\delta_i \in H^{lf}_{n-p+1}(\tilde{X} \setminus X_f, \pm\mathbf{Z})$ there is an element $\kappa_i \in H_{n-p+1}(\tilde{X} \setminus X_f, \pm\mathbf{Z})$, the *vanishing cycle* defined by the same path in $\mathbf{C}^1$, such that $j(\kappa_i) = 2\delta_i$, see [Ph 65] and Fig. 1, where such a cycle in one-dimensional $\tilde{X}$ is shown.

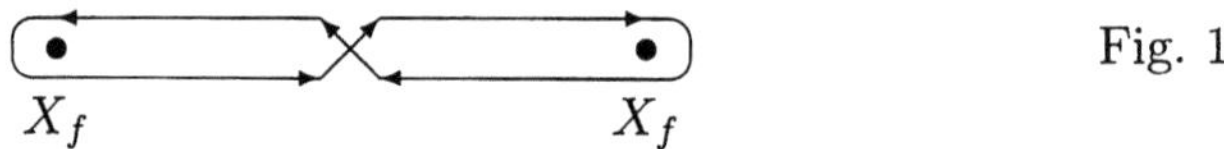

Fig. 1

Theorem 1. *a) The homomorphism (4) is an isomorphism, as well as the similar homomorphism of homology groups reduced mod $\partial\tilde{X}_f$,*

$$H_{n-p+1}(\tilde{X}_f \setminus X_f, \partial\tilde{X}_f \setminus \partial X_f; L_{-1}) \to H^{lf}_{n-p+1}(\tilde{X}_f \setminus X_f, \partial\tilde{X}_f \setminus \partial X_f; L_{-1});$$

b) the dimensions of both groups (4) are equal to $\nu(f) \equiv \mu(f) + \mu(\tilde{f})$, and similar homology groups in all other dimensions are trivial;
c) the right-hand group in (4) is freely generated by the Lefschetz thimbles specified by an arbitrary distinguished (see e.g. [AGV], [AGLV]) system of paths connecting the noncritical value c_p of $f_p|_{\tilde{X}}$ with all critical values.

Corollary. *The left-hand group in (4) is generated by the* vanishing cycles *defined by the same paths.*

Proof of the theorem. The fact that the map (4) (and also its relative version) is isomorphic is a general algebraic fact, which is true for all local systems L_α with monodromy indices $\alpha \neq 1$: this follows from the comparison of the *Leray spectral sequences* (see e.g. [GrH], § III.5) calculating the indicated homology groups and applied to the identical embedding $\tilde{X}_f \setminus X_f \to \tilde{X}_f$.

The assertion of statement b) concerning the right-hand group in (4) follows from the similar assertion concerning the non-twisted vanishing homology group $H_{n-p+1}(\tilde{X}_f, X_f; \mathbf{Z}) \equiv H^{lf}_{n-p+1}(\tilde{X}_f \setminus X_f, \mathbf{Z})$ (see Proposition 1), the fact that $\pm\mathbf{Z} \otimes \mathbf{Z}_2 = \mathbf{Z} \otimes \mathbf{Z}_2 = \mathbf{Z}_2$ (the constant local system with fibre $\mathbf{Z}_2$) and from the formula of universal coefficients. The same reasons prove that the $\mathbf{Z}_2$-torsion of the group $H^{lf}_*(\tilde{X}_f \setminus X_f, \pm\mathbf{Z})$ is trivial in all dimensions.

Statement c) follows now from the fact that the images of thimbles are linearly independent already in the group $H^{lf}_{n-p+1}(\tilde{X}_f \setminus X_f, \pm\mathbf{Z}) \otimes \mathbf{Z}_2$. $\qquad\square$

Let $\mathfrak{I}$ be the subgroup in $H_{n-p+1}(\tilde{X}_f \setminus X_f, \pm\mathbf{Z})$ generated by vanishing cycles κ_i defined by all possible paths (probably it coincides with entire $H_{n-p+1}(\tilde{X}_f \setminus X_f, \pm\mathbf{Z}))$.

Lemma 1. *For any elements $\alpha, \beta \in \mathfrak{I}$, their intersection index is even.*

Indeed, this index is equal to the (well-defined) intersection index of α and $j(\beta)$, and $j(\beta) \in 2H^{lf}_{n-p+1}(\tilde{X}_f \setminus X_f, \pm\mathbf{Z})$.

Define the bilinear form $\langle \cdot, \cdot \rangle$ on $\mathfrak{I}$ equal to half this intersection index.

Proposition 4. *The form $\langle \cdot, \cdot \rangle$ is symmetric if $n-p$ is odd and is skew-symmetric if $n - p$ is even. For any basis vanishing cycle κ_i, $\langle \kappa_i, \kappa_i \rangle$ is equal to 2 if $n - p \equiv 3 (mod\ 4)$ and to -2 if $n - p \equiv 1 (mod\ 4)$.*

In the terms of this form, the monodromy action on the group $\mathfrak{I}$ is defined by the same Picard–Lefschetz formula as before: the monodromy along the simple loop ω_i takes a cycle κ to

$$\kappa + (-1)^{(n-p+1)(n-p+2)/2} \langle \kappa, \kappa_i \rangle \kappa_i. \tag{5}$$

3 Surface potentials and Newton–Ivory–Arnold theorem

3.1 Potential function of a surface

Denote by dV the volume differential form in $\mathbf{R}^n$, i.e. the form $dx_1 \wedge \ldots \wedge dx_n$ in the Euclidean positively oriented coordinates $x_1, \ldots, x_n$.

Denote by r the Euclidean norm in $\mathbf{R}^n$, $r = (x_1^2 + \cdots + x_n^2)^{1/2}$, and by C_n the area of the unit sphere in $\mathbf{R}^n$.

Definition 2. The *elementary Newton–Coulomb potential function*, or, which is the same, the *standard fundamental solution of the Laplace operator* in $\mathbf{R}^n$, is the function equal to $\frac{1}{2\pi}\ln r$ if $n = 2$, and to $-r^{2-n}/((n - 2)C_n)$ if $n \geq 3$. Denote this function by G.

This function can be interpreted as the potential of the force of attraction by a particle of unit mass placed at the origin, i.e., the attraction force of this particle is equal to $-\operatorname{grad} G$.

The attraction force with which a body K with density distribution P attracts a particle of unit mass placed at the point $x \in \mathbf{R}^n$ is equal to minus the gradient of the corresponding *potential function*, whose value at the point

x is equal to the integral over K of the differential form $G(x - z)P(z)dV(z)$ (if such an integral exists).

Let F be a smooth function in Euclidean space $\mathbf{R}^n$, and M_F the hypersurface $\{F = 0\}$. Suppose that grad $F \neq 0$ at the points of M_F, so that M_F is smooth.

Definition 3. The *standard charge* ω_F on the surface M_F is the differential form dV/dF, i.e. the $(n-1)$-form such that for any tangent frame $(l_2, \ldots, l_n)$ of M_F and a transversal vector l_1 the product of the values $\omega_F(l_2, \ldots, l_n)$ and (dF, l_1) is equal to the value $dV(l_1, \ldots, l_n)$. The *natural orientation* of the surface M_F is the orientation defined by this differential form.

In particular, the value at a point $x \notin M_F$ of the limit of potential functions of homogeneous (with density $1/\epsilon$) distributions of charges between the surfaces $F = 0$ and $F = \epsilon$ is equal to the integral of the standard charge form

$$G(x - z)\omega_F(z) \tag{6}$$

along the naturally oriented surface M_F.

In a similar way, any function P on the surface M_F defines the charge $P \cdot \omega_F$, which is called the *standard charge with density P*; the potential at the point x of this charge is equal to the integral of the form

$$G(x - z)P(z)\omega_F(z) \tag{7}$$

along the naturally oriented surface M_F. The attraction force of this charge is equal to minus the gradient of this potential function.

In these terms, the theorems of Newton and Ivory look as follows.

Theorem. *The potential of the standard charge of the sphere (respectively, an ellipsoid) in $\mathbf{R}^n$ given by the canonical equation (i.e. by a polynomial F of degree 2) is equal to a constant inside the sphere (the ellipsoid), while outside it coincides (up to multiplicative constant) with the potential function defined by any smaller ellipsoid confocal to ours.*

Arnold extended the "interior" part of this theorem to all *hyperbolic* layers.

Definition 4. An algebraic hypersurface M of degree d in $\mathbf{R}P^n$ is *strictly hyperbolic* with respect to a point $x \in \mathbf{R}P^n \setminus M$ if any real line through x intersects M at exactly d different real points. A polynomial $F : \mathbf{R}^n \to \mathbf{R}$ is strictly hyperbolic with respect to the point $x \in \mathbf{R}^n$ if the projective closure $\bar{M}_F$ of the corresponding surface M_F is.

Proposition 5 (see e.g. [ABG]). *If a hypersurface $M \subset \mathbf{R}P^n$ is strictly hyperbolic with respect to a point x, then it is also strictly hyperbolic with respect to any point in the same component of the complement of M. Any strictly hyperbolic hypersurface is smooth.*

Definition 5. The *hyperbolicity domain* of a surface M is the union of points x such that M is hyperbolic with respect to x.

Proposition 6 (see [N]). *The set of all hypersurfaces M of given degree d in $\mathbf{R}P^n$, which are strictly hyperbolic with respect to a given point x, is contractible (or, equivalently, the set of all polynomials of degree d defining them consists of two contractible components).*

In particular, all the strictly hyperbolic surfaces M of a given degree d in $\mathbf{R}P^n$ are situated topologically in the same way: if d is even, then M is ambient (and even rigid) isotopic to the union of $[d/2]$ concentric spheres lying in an affine chart in $\mathbf{R}P^n$; if d is odd, then M is isotopic to the union of $[d/2]$ concentric spheres plus the improper projective hyperplane. The hyperbolicity domain consists of the interior points of the "most interior" spheroid. This spheroid is always convex in $\mathbf{R}P^n$, in particular, the hyperbolicity domain in $\mathbf{R}^n$ may consist of at most two connected components.

The hyperbolic surface M_F separates the space $\mathbf{R}^n$ into *zones*: the k-th zone consists of all points $x \in \mathbf{R}^n \setminus M_F$ such that the minimal number of intersection points of M_F with segments connecting x and points of the hyperbolicity domains is equal to k. In particular, the maximal index k of a zone is equal to $[d/2] + 1$ if d is odd and the hyperbolicity domain in $\mathbf{R}^n$ consists of one component, and is equal to $[d/2]$ otherwise.

Given a strictly hyperbolic polynomial F, let us fix some path-component of its hyperbolicity domain in $\mathbf{R}^n$, and number the components of M_F starting from the boundary of this component (which becomes number 1), its neighboring component gets number 2, etc.

Definition 6. The *Arnold cycle* of F is the manifold $\bar{M}_F$, oriented in such a way that in the restriction to its finite part M_F all odd components are taken with the natural orientation (see Definition 3), while all even components are taken with the reversed orientations.

The *hyperbolic potential* (respectively, *hyperbolic potential with density P*) of the surface M_F at a point $x \in \mathbf{R}^n \setminus M_F$ is the integral of the form (6) (respectively, (7)) along the Arnold cycle. As usual, the *attraction forces* defined by these potentials are equal to minus the gradients of the potential functions.

Lemma 2. *This definition of the Arnold cycle is correct, i.e. the orientations of different non-compact components of M_F thus defined are the restrictions of the same orientation of the corresponding components of $\bar{M}_F$.*

The proof is immediate.

Theorem (see [A 82]). *The hyperbolic potential of the surface M_F (and moreover any hyperbolic potential with density P, where P is a polynomial of degree $\leq d - 2$) is constant inside the hyperbolicity domain.*

(In other words, the points of the hyperbolicity domain are not attracted by the standard charge on M_F taken with sign 1 or -1 depending on the parity of the number of the component on which this charge is distributed.)

The proof follows Newton's original proof: for any infinitesimally narrow cone centred at the point x, whose direction is not asymptotic for the surface M_F, the forces of attraction to the pieces of M_F cut by the cone annihilate one another. Indeed, let us restrict the polynomial F to the line L in $\mathbf{R}^n$ through x contained in this cone; then this attraction force is equal to the solid angle of our cone multiplied by the sum of the numbers $P(A_i)/F'(A_i)$ over all zeros A_i of the polynomial $F|_L$. The last sum is zero because it is the sum of the residues of a rational function over all its complex poles.

The restriction deg $P \le d - 2$ from the Arnold's theorem ensures that the integration form (7) is "regular at infinity", i.e. extends to a holomorphic form on the projective hypersurface $\bar{M}_F$. Givental [G 84] remarked that a similar statement is true for polynomial potentials of arbitrary degree if the integration cycle M_F is compact in $\mathbf{R}^n$: in this case the potential function in the hyperbolicity domain coincides with a polynomial of degree $\le$ deg $P - d + 2$.

In other domains the potential also coincides with real analytic functions; in the next sections we study the global behavior of these functions, in particular their algebraicity. The ramification of these functions is defined by the action of certain monodromy group on a certain homology group; in the next § 4 we define these objects, and in § 5 we calculate this monodromy group.

4 Monodromy group responsible for the ramification of potentials

4.1 Homology groups

For any point $x \in \mathbf{C}^n$, $x = (x_1, \dots, x_n)$, denote by $S(x)$ the cone in $\mathbf{C}^n$ given by the equation

$$(z_1 - x_1)^2 + \cdots + (z_n - x_n)^2 = 0. \tag{8}$$

Denote by $@ \equiv @(x)$ a local system over $\mathbf{C}^n \setminus S(x)$ with fibre $\mathbf{Z}$ such that the corresponding representation $\pi_1(\mathbf{C}^n \setminus S(x)) \to \mathrm{Aut}(\mathbf{Z})$ maps the loops whose linking numbers with $S(x)$ are odd to the multiplication by -1.

We specify this local system in such a way that integrals of the form $r(\cdot - x)dz_1 \wedge \dots \wedge dz_n$ along the $(n-1)$-dimensional cycles with coefficients in it are well defined. Namely, we consider the two-fold covering over $\mathbf{C}^n \setminus S(x)$, on which this form is single-valued, and the direct image in $\mathbf{C}^n \setminus S(x)$ of this bundle under the obvious projection of this covering. The trivial $\mathbf{Z}$-bundle over $\mathbf{C}^n \setminus S(x)$ is naturally included in this direct image as a subbundle; the desired local system is the quotient bundle of these two local systems. Obviously, integrals of the form $r(\cdot - x)dz_1 \wedge \ldots \wedge dz_n$ (and of its products by all single-valued functions) along the piecewise smooth n-chains with coefficients in this local system are well-defined, and if these chains are cycles, these integrals depend only on their homology classes.

Let $F : \mathbf{C}^n \to \mathbf{C}$ be a polynomial, $W_F \subset \mathbf{C}^n$ the set of its zeros, and $\bar{W}_F$ the projective closure of W_F. For any $x \in \mathbf{C}^n$ we denote by $\mathcal{H}(x)$ the group

$$H_{n-1}(W_F \setminus S(x), \mathbf{Z}) \tag{9}$$

in the case of even n, and the group

$$H_{n-1}(W_F \setminus S(x), \ @(x)) \tag{10}$$

if n is odd. Similarly, denote by $\mathcal{PH}(x)$ the group

$$H_{n-1}(\bar{W}_F \setminus \bar{S}(x), \mathbf{Z}) \tag{11}$$

in the case of even n, and the group

$$H_{n-1}(\bar{W}_F \setminus \bar{S}(x), \ @(x)) \tag{12}$$

in the case of odd n.

Definition 7. If the polynomial F is real (i.e., $F(\mathbf{R}^n) \subset \mathbf{R}$) and strictly hyperbolic, then the Arnold cycle defines correctly an element of the group $\mathcal{PH}(x)$ (and even of the group $\mathcal{H}(x)$ if M_F is compact); these elements are called the *Arnold homology classes* and are denoted by $PA(x)$ and $A(x)$ respectively.

In the case of odd n, integrals of the form (7) along $(n-1)$-chains in $W_F \setminus S(x)$ with coefficients in $@(x)$ are well defined, and the values of these integrals along the cycles depend only on their homology classes in the group (10). Moreover, if $\deg P \leq d - 2$, and hence the form (7) is regular at infinity, then it can be integrated along the chains in $\bar{W}_F \setminus \bar{S}(x)$, and the integrals along the cycles depend only on their classes in the group (12).

In the case of even $n > 2$ the form (7) is single-valued, and no problems with the definition of similar integrals along the elements of the group (9) (or even (11) if $\deg P \leq d-2$) arise, and in the exceptional case $n = 2$, when (7) is logarithmic, we remember that we are interested not in the potential, but in its first partial derivatives with respect to the parameter x (i.e. in the components of the attraction force vector). Therefore we integrate not the form (7) but its partial derivatives

$$\frac{x_i - z_i}{(x_1 - z_1)^2 + (x_2 - z_2)^2} P(z)\omega_F, \ i = 1, 2\,;$$

these forms are already single-valued and there is no problem in integrating them along the elements of the group (9) (or (11) if $\deg P \leq d - 2$).

4.2 Homological bundles

For almost all $x \in \mathbf{C}^n$ the groups $\mathcal{H}(x)$ (respectively, $\mathcal{PH}(x)$) are naturally isomorphic to one another. The set of exceptional x (for which the pair $(\bar{W}_F \setminus \bar{S}(x), W_F \setminus S(x))$ is not homeomorphic to these for all neighboring x') belongs to a proper algebraic subvariety in $\mathbf{C}^n$ consisting of three components:

a) W_F itself,

b) the set of such x that $S(x)$ and W_F are tangent outside x in $\mathbf{C}^n$, and

c) the set of such x that the projective closure of $S(x)$ in $\mathbf{C}P^n$ is "more nontransversal" to the closure of W_F at their infinitely distant points.

For a generic F the last component is empty, and the second is irreducible provided additionally that $n \geq 3$. Denote this algebraic set of all exceptional $x \in \mathbf{C}^n$ by $\Sigma(F)$.

Consider two fibre bundles over $\mathbf{C}^n \setminus \Sigma(F)$ whose fibres over a point x are the spaces $W_F \setminus S(x)$, $\bar{W}_F \setminus \bar{S}(x)$, and associate with them the homological bundles whose fibres over the same point are the groups $\mathcal{H}(x)$ and $\mathcal{PH}(x)$. As usual, the Gauss–Manin connection in these bundles defines the monodromy representations

$$\pi_1(\mathbf{C}^n \setminus \Sigma(F)) \;\rightarrow\; \text{Aut } \mathcal{H}(x), \tag{13}$$

$$\pi_1(\mathbf{C}^n \setminus \Sigma(F)) \;\rightarrow\; \text{Aut } \mathcal{PH}(x). \tag{14}$$

These representations obviously commute with the natural map $\mathcal{H}(x) \rightarrow \mathcal{PH}(x)$.

Let u be the potential function of the polynomial charge $P \cdot \omega_F$, i.e. the function defined for any x by the integral of the form (7) along the Arnold cycle. The ramification of (the analytic continuation of) the function u depends on the monodromy action (13) (respectively, (14)) on the Arnold element in $\mathcal{H}(x)$ (respectively, in $\mathcal{PH}(x)$).

Namely, for any multiindex $\nu \in \mathbf{Z}^n_+$ ($\nu \neq 0$ if $n = 2$) consider the linear forms

$$N^{(\nu)} : \mathcal{H}(x) \rightarrow \mathbf{C}, \quad PN^{(\nu)} : \mathcal{PH}(x) \rightarrow \mathbf{C}, \tag{15}$$

whose values on the cycle γ are equal to the integral along γ of the ν-th partial derivative of the form (7) with respect to the parameter x.

Proposition 7. *For any ν ($\neq 0$ if $n = 2$) and $x \in \mathbf{R}^n \setminus \Sigma(F)$, the ν-th partial derivative of the potential function of the standard charge of the compact hyperbolic surface M_F with density P is finite-valued at x if and only if the linear form $N^{(\nu)}$ takes finitely many values on the orbit of the cycle $A(x)$ under the action of the monodromy group (13). If P is a polynomial of degree $\leq d-2-|\nu|$, then the same is true for non-compact hyperbolic surfaces if we replace $A(x)$ by $PA(x)$, $N^{(\nu)}$ by $PN^{(\nu)}$, and the action (13) by (14).*

This is a tautology.

4.3 The invariant cycle

In this subsection we show that for any F and $x \in \mathbf{C}^n \setminus \Sigma(F)$ the representation (14) has an invariant vector; if F is a real hyperbolic polynomial and x lies in its hyperbolicity domain, then this cycle coincides with the Arnold homology class.

Denote by $PS \subset \mathbf{C}P^{n-1}$ the common "infinite" part of all cones $\bar{S}(x) \subset \mathbf{C}P^n$ and by @ the local system over $\mathbf{C}P^{n-1} \setminus PS$ such that any system @(x) is induced from it by the obvious projection with center x.

Proposition 8. *The groups*

$$H_{n-1}(\mathbf{C}P^{n-1} \setminus PS) \tag{16}$$

(if n is even) and

$$H_{n-1}(\mathbf{C}P^{n-1} \setminus PS, @) \tag{17}$$

(if n is odd) are one-dimensional. The generators of all these groups are presented by the class of the submanifold $\mathbf{R}P^{n-1} \subset \mathbf{C}P^{n-1} \setminus PS$.

The proof is elementary.

The obvious map $\Pi : \bar{W}_F \setminus \bar{S}(x) \to \mathbf{C}P^{n-1} \setminus PS$ (projection from the center x) is a d-fold ramified covering of complex (and thus oriented) manifolds. The variety $\Pi^{-1}(\mathbf{R}P^{n-1})$ admits thus an orientation (@(x)-orientation if n is odd) induced from the chosen orientation of $\mathbf{R}P^{n-1}$; denote by $\Omega(x)$ the class of this variety in the group $\mathcal{P}\mathcal{H}(x)$.

Proposition 9. *1. The classes $\Omega(x)$ for different x constitute a section of the homology bundle over $\mathbf{C}^n \setminus \Sigma(F)$ with fibres $\mathcal{P}\mathcal{H}(x)$, which is invariant under the Gauss–Manin connection, in particular these classes are invariant under the representation (14).*
2. If F is a real hyperbolic polynomial and x lies in its hyperbolicity domain, then $\Omega(x)$ coincides with the Arnold homology class $PA(x)$.

This follows immediately from the construction.

4.4 Reduced Arnold class

For an arbitrary element γ of the group $\mathcal{P}\mathcal{H}(x)$, the corresponding potential function $u_\gamma(x)$ can be defined as the integral of the form (7) along the cycle γ (if this integral exists), in particular the usual potential $u(x)$ coincides with $u_{PA(x)}(x)$.

In this subsection we for any point $x \in \mathbf{R}^n \setminus \Sigma(F)$ replace the corresponding Arnold class $PA(x)$ by another class $P\tilde{A}(x)$, whose potential function $u_{P\tilde{A}(x)}$ ramifies in exactly the same way, but which is more convenient because (as we shall see later)

a) it is represented by a cycle lying in the "finite" part $W_F \backslash S(x)$ of $\bar{W}_F \backslash \bar{S}(x)$ and thus defining an element $\tilde{A}(x)$ of the (much better studied) group $\mathcal{H}(x)$, and

b) if n is even, then this element $\tilde{A}(x)$ can be obtained by the "Leray tube operation" (3) from a certain homology class $\alpha(x) \in H_{n-2}(W_F \cap S(x))$, so that the action (13) on it is reduced to the similar action on this more standard group.

Indeed, it follows from Proposition 9, that if the class $\gamma' \in \mathcal{PH}(x)$ is obtained by the Gauss–Manin connection over some path in $\mathbf{C}^n \setminus \Sigma(F)$ from the Arnold cycle $PA(\mathbf{x})$, where $\mathbf{x}$ is a point in the hyperbolicity domain of a compact hyperbolic surface, then the potential function $u_{\gamma'}(x)$ is a single-valued holomorphic function in $\mathbf{C}^n \setminus \Sigma(F)$. Therefore the ramification of our integrals defined by the class γ coincides with that defined by the class $\gamma - \gamma'$ (if both integrals are well-defined).

For any point $x \in \mathbf{R}^n \setminus M_F$ we choose canonically some class γ' obtained in this way. Namely, we choose an arbitrary point $\mathbf{x} \in \mathbf{R}^n$ in the hyperbolicity domain (if this domain has two components in $\mathbf{R}^n$, then in the component closest to x, i.e. such that the segment connecting x and $\mathbf{x}$ has $\leq [d/2]$ intersections with M_f). Then connect x with $\mathbf{x}$ by a complex line and take the path in this line that goes from $\mathbf{x}$ to x along the real segment and misses any point of W_F along a small arc in the *lower* complex half-line with respect to this direction (i.e. the half-line into which the vector $i \cdot (\mathbf{x} - x)$ is directed). See Fig. 2.

Fig. 2

For any $x \in \mathbf{R}^n \setminus M_F$, denote by $PA_{\mathrm{hyp}}(x)$ the class in $\mathcal{PH}(x)$ obtained from $PA(\mathbf{x})$ by the Gauss–Manin connection over this path. We are interested in the monodromy of the class $PA(x) - PA_{\mathrm{hyp}}(x)$, which will be called the *reduced Arnold class* and denoted by $P\tilde{A}(x)$.

4.5 Groups $\mathcal{H}(x)$ and the vanishing homology of complete intersections

We shall consider especially carefully the case when the attracting surface W_F satisfies certain genericity conditions, namely, the following ones.

We say that two holomorphic hypersurfaces in $\mathbf{C}^n$ are *simple tangent* at their common point, if in some local holomorphic coordinates with origin at this point one of them is given by the equality $z_n = 0$, and the second by $z_n = z_1^2 + \cdots + z_{n-1}^2$.

Definition 8. The polynomial F (and the corresponding hypersurface W_F) is *S-generic* if the projective closure $\bar{W}_F$ of W_F is smooth and transversal to the improper hyperplane $\mathbf{C}P^n \setminus \mathbf{C}^n$, its "infinite part" $\bar{W}_F \setminus \mathbf{C}^n$ is transversal in the improper hyperplane $\mathbf{C}P^{n-1}$ to the standard quadric $\{z_1^2 + \cdots + z_n^2 = 0\}$, i.e. to the boundary of any cone $S(x)$, and additionally the set of points at which W_F is *simple* tangent to appropriate cones $S(x)$ is dense in the set of all points of tangency of W_F and these cones at their nonsingular points.

The transversality conditions from this definition can be reformulated as follows: let $\bar{F}$ be the principal (of degree d) homogeneous part of F, and $r^2 \equiv z_1^2 + \cdots + z_n^2$, then the function $\bar{F}$ has an isolated singularity at 0, and also the pair of functions $(\bar{F}, r^2)$ defines a (homogeneous) complete intersection with an isolated singularity at 0.

Theorem 2. *Suppose that the algebraic surface $W_F = \{F = 0\}$ in $\mathbf{C}^n$ is S-generic, $\deg F = d$. Then for a generic x the ranks of both groups (9), (10) (in particular, of the group $\mathcal{H}(x)$) are equal to $(d-1)^n + (2(d-1)^n - d)/(d-2)$ if $d > 2$, and to $2n$ if $d = 2$.*

Indeed, the pair of functions $(F, r^2(\cdot - x))$ defining the manifolds $W_F, S(x)$ is a perturbation of the complete intersection $(\bar{F}, r^2)$, changing only terms of lower degree of these polynomials. Thus the pair $(W_F, W_F \cap S(x))$ for smooth W_F and nondiscriminant x is homeomorphic to the pair $(\tilde{X}_f, X_f)$ from (1), and the local system $@(x)$ is isomorphic to the system $\pm\mathbf{Z}$ on $\tilde{X}_f \setminus X_f$, see § 2.3. For the group (10) the assertion of the theorem follows now from Theorem 1 and Propositions 1 and 2.

Denote by ∂W_F the "infinite part" $\bar{W}_F \setminus \mathbf{C}^n$ of $\bar{W}_F$. Then the group (9) is Poincaré–Lefschetz dual to the group $H_{n-1}(\bar{W}_F, \partial W_F \cup (\bar{W}_F \cap \bar{S}(x)))$. Consider the homological exact sequence of the triple $(\bar{W}_F, \partial W_F \cup (\bar{W}_F \cap \bar{S}(x)), \partial W_F)$. By Proposition 1 and Poincaré duality in the manifolds W_F, $W_F \cap S(x)$, the only nontrivial fragment in this sequence is

$$0 \to H_{n-1}(\bar{W}_F, \partial W_F) \to H_{n-1}(\bar{W}_F, \partial W_F \cup (\bar{W}_F \cap \bar{S}(x))) \to$$
$$\to H_{n-2}(\bar{W}_F \cap \bar{S}(x), \partial W_F \cap \bar{S}(x)) \to 0, \qquad (18)$$

and the assertion of our theorem about the group (9) follows from Proposition 2.

Remark. It is easy to see that the map

$$H_{n-2}(W_F \cap S(x)) \to H_{n-1}(W_F \setminus S(x)), \qquad (19)$$

conjugate with respect to Poincaré dualities to the third arrow in (18), coincides with the Leray tube operation (3), in particular in this case this operation is monomorphic.

So we have identified the pair $(W_F, W_F \cap S(x))$ with a standard object of the theory of singularities of complete intersections. The pair of functions $(F, r^2(\cdot - x))$ defining this complete intersection participates in three important families, which depend on $n, 1$ and $n+1$ parameters respectively. Since all of them keep the first function F unmoved, we describe only the corresponding families of second components. The first family consists of all functions $r^2(\cdot - \tilde{x})$, $\tilde{x} \in \mathbf{C}^n$; the second of all functions $r^2(\cdot - x) - \tau$, $\tau \in \mathbf{C}$, and the third of all functions

$$\rho_\lambda \equiv z_1^2 + \cdots + z_n^2 + \lambda_1 z_1 + \cdots + \lambda_n z_n + \lambda_0. \tag{20}$$

Denote the parameter space of the third deformation by T; the parameter spaces $\mathbf{C}^n$ and $\mathbf{C}^1$ of the first and second families are obviously included in it.

Define the set Σ_T as the set of all such points $\lambda \in T$ that the variety $W_F \cap \{\rho_\lambda = 0\}$ is not smooth; the intersection of Σ_T with the parameter space of the first (respectively, the second) subfamily coincides with $\Sigma(F)$ (respectively, the set s of critical values of the restriction of $r^2(\cdot - x)$ on W_F, see § 2.2).

By the Zariski theorem, the obvious homomorphism $\pi_1(\mathbf{C}^1 \setminus s) \to \pi_1(T \setminus \Sigma_T)$ is monomorphic, in particular the monodromy group generated by the action of the latter group in $\mathcal{H}(x)$ coincides with the standard monodromy group of the complete intersection $(\bar{F}, r^2)$ considered in § 2.2, 2.3.

Definition 9. The monodromy group defined by the Gauss–Manin representation $\pi_1(\mathbf{C}^1 \setminus s) \to Aut(\mathcal{H}(x))$ (or, equivalently, $\pi_1(T \setminus \Sigma_T) \to Aut(\mathcal{H}(x))$) is called the *big* monodromy group, while the similar monodromy group defined by the natural action (13) is the *small* one.

Below we shall see that the small monodromy group actually is a proper subgroup of the big one. To describe it we need several more reductions and notions.

The subgroup $\mathcal{J}(x) \subset \mathcal{H}(x)$ for any n is defined as that generated by all vanishing cycles in $W_F \setminus S(x)$ defined by all paths in $\mathbf{C}^1 \setminus s$ connecting 0 with all the points of s, see § 2: for even n it coincides with the image of the Leray tube map (19), for odd n it is just the group $\mathfrak{F}$ described in the end of § 2.3.

On this subgroup there is a symmetric bilinear form $\langle \cdot, \cdot \rangle$: in the case of odd n it was defined before Proposition 4 (as half the intersection index), and in the case of even n it is induced by the tube *mono*morphism (19) from the intersection index on the group $H_{n-2}(W_F \cap S(x))$. By Propositions 3 and 4, for any vanishing cycle $\alpha \in \mathcal{H}(x)$ $\langle \alpha, \alpha \rangle$ is equal to 2 if $[\frac{n+1}{2}]$ is odd and to -2 if $[\frac{n+1}{2}]$ is even.

Lemma 3. *For any n, the action of the big monodromy group on $\mathcal{H}(x)$ preserves the subgroup $\mathcal{J}(x)$ and the bilinear form $\langle \cdot, \cdot \rangle$ on it.*

This follows immediately from the Picard–Lefschetz formulae (2), (5).

Now suppose that the polynomial F is real and hyperbolic.

Theorem 3. *For any point x from the k th zone of $\mathbf{R}^n \setminus \Sigma(F)$, $k \le [d/2]$, the reduced Arnold class $P\tilde{A}(x) = PA(x) - PA_{\mathrm{hyp}}(x)$ can be represented by a cycle with support in $W_F \setminus S(x)$ which is homological in $\mathcal{H}(x)$ to the sum of k pairwise orthogonal vanishing cycles. In particular, its homology class $\tilde{A}(x)$ belongs to* *the subgroup $\mathcal{J}(x)$, and its self-intersection index $\langle \tilde{A}(x), \tilde{A}(x) \rangle$ is equal to $2k$ if $[\frac{n+1}{2}]$ is odd, and to $-2k$ if $[\frac{n+1}{2}]$ is even.*

Indeed, these vanishing cycles are constructed as follows. If the point $y \in \mathbf{R}^n \setminus M_F$ is sufficiently close to a component of M_F, then in a small disc $B \subset \mathbf{C}^n$ centered at y the pair $(W_F, S(y))$ is diffeomorphic to the pair consisting of the plane $\{x_1 = 1\}$ and the cone $S(0)$; it is easy to see that both groups $H_{n-1}(B \cap W_F \setminus S(y))$ and $H_{n-1}(B \cap W_F \setminus S(y), @(y))$ are isomorphic to $\mathbf{Z}$ and generated by vanishing cycles defined by the one-parametric family of maps $(F, r^2(\cdot - y) - \tau)$, $\tau \in \mathbf{C}^n$ (in the first case this cycle is equal to the tube around the vanishing cycle in $W_F \cap S(y)$).

Lemma 4 (see [V 94], Lemma 2 in § III.3.4). *If we go from the hyperbolicity domain along a line in $\mathbf{R}^n$ and traverse a component of M_F, then the Arnold class corresponding to the point after the traversing is equal to the sum of this vanishing cycle and of the similar Arnold cycle for the point before it transported by the Gauss–Manin connection over the arc of the path from Fig. 2 connecting them.*

In particular, the difference $PA(x) - PA_{hyp}(x)$ for x from the k-th zone is homologous to the sum of k vanishing cycles; by construction all these cycles lie in the finite domain $W_F \setminus S(x)$. The homology class of this sum in $\mathcal{H}(x)$ is exactly the promised reduced Arnold class $\tilde{A}(x)$, see § 4.4. It remains only to prove that these cycles are pairwise orthogonal. To do it, consider a model hyperbolic surface: the union of $[d/2]$ concentric close spheres of radii $1, 1 + \varepsilon, \ldots, 1 + ([d/2] - 1)\varepsilon$ (which do not intersect one another even in the complex domain) and, if d is odd, one plane distant from these spheres.

Although this surface is not S-generic, the above-described construction of the cycle $\tilde{A}(x)$ can be accomplished for any point x in the k-th zone where $k \leq [d/2]$ and, if d is odd and $k = [d/2]$, then x lies much closer to the exterior ovaloid than to the additional plane. Then any of our k vanishing cycles lies on the complexification of its own sphere, in particular they do not intersect one another, and our assertion is proved for the (very degenerate) model hyperbolic surface. We can change this surface arbitrarily weakly so that its closure $\bar{W}_F$ becomes S-generic and transversal to $S(x)$, but the topological shape of the pair $(W_F, S(x))$ does not change in a large ball in $\mathbf{C}^n$ containing all our k vanishing cycles. Therefore they have zero intersection indices also for a certain generic hyperbolic polynomial. Finally, the set of nongeneric real hyperbolic polynomials, all whose "nongenericity" lies in the complex domain, has codimension at least 2 in the space of all strictly hyperbolic polynomials, and, by Proposition 6, the space of pairs of the form {a strictly hyperbolic polynomial F of degree d in $\mathbf{R}^n$; a point x of its k-th zone with $k \leq [d/2]$} is open and path-connected; this gives our assertion also for arbitrary generic F.

5 Description of the small monodromy group and finiteness theorems in the cases $n = 2$ and $d = 2$

5.1 The two-dimensional case

Let $n = 2$. Denote by $\eta(F)$ the number of factors $x_1^2 + x_2^2$ in the decomposition of the principal part $\bar{F}$ of the polynomial F into the simplest real factors. (Of course, if $\eta(F) > 0$ then F is not S-generic.)

Theorem 4. *The attraction force of the standard charge, distributed on a hyperbolic curve $\{F = 0\}$ of degree d in $\mathbf{R}^2$ coincides in the k-th zone with the sum of two algebraic vector-functions, any of which is $\leq \binom{d - \eta(F)}{k}$-valued. The same is true for the standard charge with polynomial density P of degree $\leq d - 2$.*

If the hyperbolic curve $\{F = 0\}$ is compact and the density function P is holomorphic, then the corresponding attraction force coincides in the k-th zone with the sum of two analytic finite-valued (and even algebraic if P is a polynomial) vector-functions, any of which also is $\leq \binom{d - \eta(F)}{k}$-valued.

Corollary. *If d is even and $\bar{F} \equiv (x_1^2 + x_2^2)^{d/2}$, then the attraction force coincides with a rational vector-function in the "most nonhyperbolic" $(d/2)$-th zone.*

Example. If $d = 2$, then $\eta(F) \neq 0$ only in the Newtonian case (when M_F is a circle). In this case the attraction force is single-valued, in all the other irreducible cases it is 4-valued in the 1-st zone.

Proof of Theorem 4. If $n = 2$, then the surface $S(x)$ consists of two complex lines through x, collinear to the lines $\{x_1 = \pm i \cdot x_2\}$. The reduced Arnold class $\tilde{A}(x)$ corresponding to a point x from the k-th zone is represented by $2k$ small circles in $W_F \setminus S(x)$ around the intersection points of these two lines with W_F: k circles around the points of any line. It follows from the construction of Arnold cycles that all these circles close to one line are oriented in accordance with the complex structure of the normal bundle of this line, while close to all points of the other they are oriented clockwise. The total number of such intersection points in the finite domain for any line is equal to $d - \eta(F)$. Moving the point x in $\mathbf{C}^2 \setminus \Sigma(F)$ we can only permute these $d - \eta(F)$ circles (and, if W_F is smooth, all permutations can be realized). Therefore the orbit of the monodromy group consists of $\binom{d - \eta(F)}{k}^2$ elements; this implies Theorem 4. $\square$

Remark. Already in this case we see that the small monodromy group actually is smaller than the big one. Indeed, the standard ("big") monodromy group of the complete intersection $(\bar{F}, r^2)$ in $\mathbf{R}^2$ is just the permutation group of all $2d$ points of the Milnor fibre. In particular, the orbit of the reduced Arnold class from the k-th zone under this action consists of $\binom{2d}{k, k, 2d - 2k}$ points, which is much more than $\binom{d}{k}^2$ provided by Theorem 4 in the case $\eta(F) = 0$.

Remark about Ivory's second theorem. Given a hyperbolic surface, do there exist other surfaces defining the same attraction force in some exterior zone? If yes, these surfaces define the same ramification locus of the analytic continuations of these forces. In the case of irreducible plane curves this locus consists of $d(d-1)$ lines tangent to W_F and parallel to the line $x_1 = i \cdot x_2$ plus $d(d-1)$ lines parallel to the line $x_1 = -i \cdot x_2$. If $d = 2$, the set of curves for which these ramification loci coincide consists of all conics inscribed in a given rectangle whose sides are parallel to these two directions. It is easy to see that this set is one-parametric and coincides with the family of confocal conics. For larger d, such copotential families do not exist or at least are exceptional, because the number $2d(d-1)$ of conditions that the curves of such a family should satisfy becomes much greater than the dimension of the space of curves.

5.2 Reduction of the kernel of the form $\langle \cdot, \cdot \rangle$ and the case of conical sections

Denote by $\mathrm{Ker}\mathcal{J}(x)$ the kernel of the bilinear form $\langle \cdot, \cdot \rangle$ on the group $\mathcal{J}(x)$, i.e. the set of all $\gamma \in \mathcal{J}(x)$ such that $\langle \gamma, \alpha \rangle = 0$ for any α. By the Picard–Lefschetz formula, this subspace is invariant under the monodromy action, and hence this action on the quotient lattice $\tilde{\mathcal{J}}(x) \equiv \mathcal{J}(x)/\mathrm{Ker}\mathcal{J}(x)$ is well defined.

Theorem 5. *If F is S-generic, $x \in \mathbf{C}^n \setminus \Sigma(F)$, and P a polynomial of degree p, then any form $N^{(\nu)}$ (see (15)) with $|\nu| \geq p+2-d$ takes zero value on* $\mathrm{Ker}\mathcal{J}(x)$.

Proof. Let n be even, so that $\mathcal{J}(x) = t(H_{n-2}(W_F \cap S(x_0)))$, see (19). By Poincaré duality in $W_F \cap S(x_0)$, the condition $\gamma \in \mathrm{Ker}\ \mathcal{J}(x_0)$ implies that the cycle $t^{-1}(\gamma) \in H_{n-2}(W_F \cap S(x_0))$ is homologous in the projective closure $\bar{W}_F \cap \bar{S}(x_0) \subset \mathbf{C}P^n$ of $W_F \cap S(x_0)$ to a cycle which lies in the improper subspace $\bar{W}_F \cap \bar{S}(x_0) \cap (\mathbf{C}P^n \setminus \mathbf{C}^n)$. The tube around this homology provides the homology of γ to some cycle belonging to $\partial W_F \setminus \bar{S}(x_0) \equiv (\bar{W}_F \setminus \bar{S}(x_0)) \cap (\mathbf{C}P^n \setminus \mathbf{C}^n)$. The last space is an $(n-2)$-dimensional Stein manifold, thus γ is homologous to zero in $\bar{W}_F \setminus \bar{S}(x_0)$. On the other hand, the forms $D_x^{(\nu)}|_{x=x_0} G(x-y)P(y)\omega_F(y)$ with $|\nu| \geq 2+p-d$ can be extended to holomorphic forms on $\bar{W}_F \setminus \bar{S}(x_0)$, thus their integrals along γ are equal to zero.

In the case of odd n, the condition $\gamma \in Ker\ \mathcal{J}(x_0)$ also implies that γ is homologous in $\bar{W}_F \setminus \bar{S}(x_0)$ (as a cycle with coefficients in $@(x_0) \otimes \mathbf{C}$) to a cycle in the improper subspace: indeed, by Poincaré duality this condition implies that γ defines a trivial element of the group $H_{n-1}^{lf}(\bar{W}_F \setminus S(x_0), \partial W_F \setminus S(x_0); @(x_0))$, and hence, by the relative part of Theorem 1a), also of the group $H_{n-1}(W_F \setminus S(x_0), \partial W_F \setminus S(x_0); @(x_0) \otimes \mathbf{C}))$. The rest of the proof is the same as for even n.

Corollary. *In the conditions of Theorem 5, the linear form $N^{(\nu)}$ induces a form on the quotient lattice $\tilde{\mathcal{J}}(x)$, and the number of different values of this form on any orbit of the monodromy action on $\mathcal{J}(x)$ coincides with similar number for the induced form and induced monodromy action on $\tilde{\mathcal{J}}(x)$.*

Theorem 6. *For any $n \geq 3$ the potential of the standard charge (6) distributed on a strictly hyperbolic surface $\{F = 0\}$ of degree 2 in $\mathbf{R}^n$ coincides in the 1-st zone with an algebraic function.*

Proposition 10. *If n is even, $n > 2$, and F is a generic quadric in $\mathbf{C}^n$, then the pair consisting of the corresponding lattice $\mathcal{J}(x)$ and the bilinear form $\langle \cdot, \cdot \rangle$ on it coincides with that defined by the extended root system $\tilde{D}_{n+1}$. For odd n this pair is a direct sum of the lattice $\tilde{D}_{n+1}$ and the $(n-1)$-dimensional lattice with zero form on it.*

This fact in the case of even n and non-twisted homology is proved in [E], and the calculation for odd n is essentially the same.

Proof of Theorem 6. If F is a *generic* quadric, then by the Proposition 10 the lowered form $\langle \cdot, \cdot \rangle$ on the quotient lattice $\tilde{\mathcal{J}}(x)$ is isomorphic to the canonical form on the lattice D_{n+1}, in particular is elliptic. Hence the orbit of any class in this lattice (in particular of the coset of the reduced Arnold class) under the reduced monodromy action is finite, and any linear form takes finitely many values on it.

Finally, the *non-generic* quadric F can be approximated by a one-parameter family F_τ, $\tau \in (0, \epsilon]$, of generic quadrics. The analytic continuation of the potential function $u = u(F)$ is equal to the limit of similar continuations of potentials $u(F_\tau)$. Hence the number of leaves of $u(F)$ is majorized by the (common) number of leaves of any of the $u(F_\tau)$. $\qquad\square$

This proof estimates the number of leaves of potential functions of quadrics by the numbers of elements of length $\sqrt{-2}$ in the lattice D_{n+1}. As we shall see in the next subsection, this majorization is not sharp: a more precise upper bound is the number of integer points in the intersection of the sphere of radius $\sqrt{-2}$ with a certain affine sublattice of corank 1 that does not pass through the origin.

5.3   Principal theorem on the small monodromy group

The obvious map $\Pi : \bar{W}_F \setminus \bar{S}(x) \to \mathbf{C}P^{n-1} \setminus PS$ (see § 4.3) induces a homomorphism Π_* of the group $\mathcal{J}(x)$ to the group (16) (if n is even) or (17) (if n is odd). Denote by $\mathcal{M}(x)$ the kernel of this homomorphism.

Theorem 7. *Suppose that the polynomial $F : \mathbf{C}^n \to \mathbf{C}$ is S-generic. Then for any $x \in \mathbf{C}^n \setminus \Sigma(F)$*

a) *the map Π_* is epimorphic, in particular $\mathcal{M}(x)$ is a sublattice of corank 1 in $\mathcal{J}(x)$;*

b) *$\mathcal{M}(x)$ is spanned by all vectors but one of some basis of vanishing cycles in $\mathcal{J}(x)$;*

c) *the small monodromy group in $\mathcal{J}(x)$ is generated by reflections (with respect to the form $\langle \cdot, \cdot \rangle$) in all the basis vanishing cycles generating $\mathcal{M}(x)$;*

> *d) the set of these basis vanishing cycles in $\mathcal{M}(x)$ is transitive under the action of this small monodromy group;*
>
> *e) the subgroup $\mathrm{Ker}\,\mathcal{J}(x) \subset \mathcal{J}(x)$ belongs to $\mathcal{M}(x)$. If F is a real hyperbolic polynomial of degree d, x a point from the k-th zone, $1 \leq k \leq [d/2]$, and $\tilde{A}(x)$ the corresponding reduced Arnold cycle, then additionally*
>
> *f) $\tilde{A}(x)$ belongs to k times the generator of the quotient group $\mathcal{J}(x)/\mathcal{M}(x) \sim \mathbf{Z}$, in particular it does not belong to $\mathcal{M}(x)$;*
>
> *g) the linear form $\langle \tilde{A}(x), \cdot \rangle$ on $\mathcal{M}(x)$ is not trivial.*

For the proof of this theorem and of Theorem 8 see § 6.

Corollary. *The orbit of any element of $\mathcal{J}(x)$ under the small monodromy group lies in some affine hyperplane parallel to $\mathcal{M}(x)$.*

Indeed, this follows from Theorem 7c) and Picard–Lefschetz formula.

Definition 10. A polynomial $P : \mathbf{C}^n \to \mathbf{C}$ is *very degenerate* with respect to W_F if it is equal to 0 at all points $y \in W_F$ at which appropriate surfaces of the form $S(x)$ are tangent to W_F at their smooth points.

Theorem 8. *Suppose that W_F is S-generic, and the polynomial P is not very degenerate with respect to W_F. Then there exist multiindices $\nu \in \mathbf{Z}_+^n$ with arbitrarily large $|\nu|$ such that for a generic $x \in \mathbf{C}^n \setminus \Sigma(F)$, the restriction on $\mathcal{M}(x)$ of the linear form $N^{(\nu)}$ (see (15)) is not trivial.*

5.4 Main conjectures

Conjecture 1. *If the hyperbolic polynomial F of degree $d \geq 3$ in $\mathbf{C}^n$, $n \geq 3$, is S-generic, then the potential function of the standard charge (7) with not very degenerate P does not coincide with algebraic functions in the components of $\mathbf{R}^n \setminus \Sigma(F)$ other than the hyperbolicity domain; moreover, the same is true for some arbitrarily high partial derivatives of this potential function.*

Theorem 7 reduces this conjecture to the following Conjecture 2 (proved recently by W. Ebeling, see the Appendix).

Definition 11. A triple $(A; \langle \cdot, \cdot \rangle; g)$ consisting of an integer lattice A, an *even* integer-valued symmetric bilinear form on it and a group $g \subset \mathrm{Aut}(A)$ generated by the reflections in hyperplanes orthogonal to several elements a_i of length $\sqrt{-2}$ in A, is called *completely infinite* if for any element $a \in A$ such that not all numbers $\langle a, a_i \rangle$ are equal to 0, any nonzero linear form $A \otimes \mathbf{C} \to \mathbf{C}$ takes infinitely many values on the orbit of a under the action of the group g.

Conjecture 2. *For any S-generic polynomial F of degree $d \geq 3$ in $\mathbf{C}^n$, $n \geq 3$, the triple consisting of the group $\mathcal{M}(x)$, the bilinear form equal (up to sign if $[\frac{n+1}{2}]$ is odd) to the form $\langle \cdot, \cdot \rangle$ defined before Lemma 3, and the "small" monodromy group on $\mathcal{M}(x)$, is completely infinite.*

In [V 94] this conjecture was proved if additionally $n + d \geq 8$.

Proposition 11. *Conjecture 2 implies Conjecture 1.*

Proof. Let x be a nondiscriminant point in the k-th zone, $1 \leq k \leq [d/2]$, for which the assertion of Theorem 8 with a certain ν is satisfied. By Theorem 7b), g) there is a vanishing cycle $\Gamma \in \mathcal{M}(x)$ such that $\langle \tilde{A}, \Gamma \rangle \neq 0$. By the Picard–Lefschetz formula, the monodromy along the corresponding simple loop takes $\tilde{A}$ to $\tilde{A} + \lambda\Gamma$, $\lambda \neq 0$. By Conjecture 2, for generic x the form $N^{(\nu)}$ takes infinitely many values on the orbit of the added term $\lambda\Gamma$ under the action of the small monodromy group. On the other hand, this infinite number is estimated from above by the number $q(q-1)$, where q is the number of values of the form N^ν on the orbit of $\tilde{A}(x)$, in particular this number q is also infinite.

Finally, for the points x from the $([d/2]+1)$-th zone (if it exists) the assertion of the Conjecture 1 follows from the fact that the potential function defined by the charge (7) obviously extends to an analytic function on $\mathbf{R}P^n \setminus M_F$, hence its algebraicity in the $([d/2]+1)$-th zone is equivalent to that in the zone separated from it by a piece of the improper subspace in $\mathbf{R}P^n$; the number of the latter zone is surely less than $[d/2]+1$. $\square$

6   Proof of Theorems 7, 8

All the main characters of statements a)–e) of Theorem 7 corresponding to all S-generic F of the same degree in $\mathbf{C}^n$ and all $x \notin \Sigma(F)$ are isomorphic to one another, therefore we can assume that F is a real hyperbolic polynomial and x a real point.

The proof of statement e) follows immediately from that of Theorem 5.

Any induction step from the proof of Theorem 3 obviously increases the image of $\tilde{A}(x)$ under the map Π_* by a generator of the target homology group; all such k steps are locally topologically equivalent, and hence add a fixed generator of this target group with the same sign. This proves statement f) of Theorem 7, and statement a) is a direct corollary of it.

For any $k = 1, 2, \ldots, [d/2]$, and any point x in the k-th zone, consider the difference of the projective Arnold class $PA(x)$ and the element in $\mathcal{PH}(x)$ obtained as in the definition of the reduced Arnold cycles (i.e. by transportation along an arc in the lower complex half-line) from a similar class $PA(x')$, x' in the $(k-1)$-st zone. By Lemma 4, if x and x' are sufficiently close to one another and to the k-th component of M_F separating them, then this class can be realized by a cycle lying in a small disc B containing both these points x, x'. Denote by $a(x)$ the class of this cycle in the group $\mathcal{H}(x)$; by continuity this class $a(x)$ is well defined also for arbitrary x from the same zone (not necessarily close to M_F). By Lemma 4, for all x not in the hyperbolicity domain the corresponding maps Π_* send the elements $a(x)$ into the same element of the group (16) or (17).

Theorem 9. *If $n > 2$, then*

> a) *all classes $a(x)$, corresponding to all points $x \in \mathbf{R}^n \setminus \Sigma(F)$, x not in the hyperbolicity domain or in the $([d/2]+1)$-th zone, can be obtained from one another by the Gauss–Manin connection in the homology bundle $\{\mathcal{H}(x) \to x\}$ over some path in $\mathbf{C}^n \setminus \Sigma(F)$. These classes $a(x)$ do not belong to $\mathcal{M}(x)$, and any of them, being added to the set of $\dim \mathcal{J}(x) - 1$ basis elements of $\mathcal{M}(x)$, mentioned in statements b), c) of Theorem 7, completes this set to a basis in $\mathcal{J}(x)$;*

> b) *for arbitrary x in the k-th zone, $1 \leq k \leq [d/2]$, the linear form $\langle a(x), \cdot \rangle$ on the group $\mathcal{M}(x)$, defined by our bilinear form, is nontrivial.*

6.1 Comparison of big and small monodromy groups

Now we compare the fundamental groups of $\mathbf{C}^n \setminus \Sigma(F)$ and of the complement of the discriminant variety Σ_T of the deformation (20) of the complete intersection $(\bar{F}, r^2)$. Since F is S-generic, the set $\Sigma(F)$ consists of only two components, W_F and the set of $x \notin W_F$ such that $S(x)$ is tangent to W_F; if $n > 2$, then the latter component is irreducible.

Let us choose the distinguished point $\mathbf{x}$ of the space $T \setminus \Sigma_T$ in the hyperbolicity domain of the subspace $\mathbf{R}^n \setminus \Sigma(F)$. The group $\pi_1(T \setminus \Sigma_T)$ acts in the usual way on the group $\mathcal{H}(\mathbf{x})$ and generates the "big" monodromy group, see § 4.5.

Let Λ be a generic 2-plane in T, and $L = \Lambda \cap \mathbf{C}^n$; $\bar{U}$ a small neighbourhood of L in the projective compactification of T, and $U = \bar{U} \cap T$ the affine part of $\bar{U}$. Let L' be a generic line in Λ through $\mathbf{x}$ sufficiently close to L, so that $L' \subset U$ and L' intersects Σ_T transversally.

Lemma 5. *The obvious maps $\pi_1(L \setminus \Sigma(F)) \to \pi_1(\mathbf{C}^n \setminus \Sigma(F))$ and $\pi_1(L' \setminus \Sigma_T) \to \pi_1(U \setminus \Sigma_T) \to \pi_1(T \setminus \Sigma_T)$ are epimorphic.*

The proof follows directly from the generalized Lefschetz theorem (see [GM]).

Thus the small and big monodromy groups are generated by simple loops lying in $L \setminus \Sigma_T$ and $L' \setminus \Sigma_T$, respectively. Let us compare these collections of loops.

Lemma 6. *The group $\mathcal{J}(\mathbf{x})$ is generated by the cycles vanishing along the paths of an arbitrary distinguished system in L' connecting the distinguished point $\mathbf{x}$ with all points of $L' \cap \Sigma_T$.*

Indeed, the group $\pi_1(L' \setminus \Sigma_T)$ acts on the group $\mathcal{J}(\mathbf{x})$; this monodromy action is described by the Picard–Lefschetz formulae, see § 2. Lemma 6 follows from these formulae, from Lemma 5, and from the fact that the group $\mathcal{J}(\mathbf{x})$ coincides with the linear hull of the orbit of any vanishing cycle under the action of the big monodromy group, see [Gab], [E].

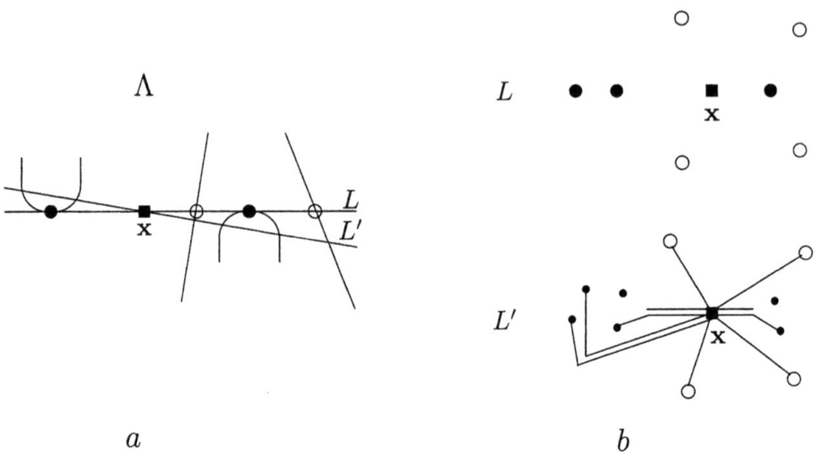

Fig. 3. Lines L and L' and discriminant points in them

The set $L \cap \Sigma(F)$ consists of several points of two kinds: the points of transversal intersection of L and W_F and points $x \notin W_F$ such that $S(x)$ is tangent to W_F.

Lemma 7. a) *Close to a generic point y of the submanifold $W_F \subset \mathbf{C}^n \subset T$ (i.e. to a point at which the generating lines of the cone $S(y)$ are transversal to W_F) the variety Σ_T is smooth and has simple tangency with $\mathbf{C}^n$ along W_F. In particular, the intersection of Σ_T with any 2-plane Λ transversal to W_F coincides close to the points of $\Lambda \cap W_F$ with a smooth curve having simple tangency with the line $\Lambda \cap \mathbf{C}^n \equiv L$;*

b) *if F is S-generic, then close to a generic point of the variety $(\Sigma(F) \setminus W_F) \subset \mathbf{C}^n \subset T$ the variety Σ_T is smooth and intersects $\mathbf{C}^n$ transversally along $(\Sigma(F) \setminus W_F)$.*

The proof is immediate.

Thus the cardinality of $L' \cap \Sigma_T$ is equal to the cardinality of $L \cap \Sigma(F)$ plus $\deg F$: to any point of $L \cap (\Sigma(F) \setminus W_F)$ there corresponds one close point of $L' \cap \Sigma_T$, while to any point of $L \cap W_F$ there correspond two such points; see Fig. 3a.

Since the point $\mathbf{x}$ lies in the hyperbolicity domain, all points of $L \cap W_F$ are real. For any such point y belonging to the k-th component of M_F, let $y_+ \in \mathbf{R}^n \setminus \Sigma(F)$ be a close point in the k-th zone. For such a point y_+, the class $a(y_+)$ was defined before Theorem 9.

Let us agree to choose the distinguished system of paths in L' in such a way that the paths connecting $\mathbf{x}$ with any two points of $L' \cap \Sigma_T$ arising from the same point y of $L \cap W_F$ go together up to a small common neighbourhood of these two points and are close to the real segment in L connecting $\mathbf{x}$ and

y, while the paths in L' connecting $\mathbf{x}$ with any other points of $L' \cap \Sigma_T$ do not touch this small neighbourhood; see Fig. 3b.

Definition 12. A point of $L' \cap \Sigma_T$ is of the first kind (respectively, of the second kind) if it arises from a close point of W_F (respectively, of $\Sigma(F) \setminus W_F$) in L after the move $L \to L'$. A cycle in $\mathcal{J}(\mathbf{x})$ vanishing over a path of our distinguished system in L' that connects $\mathbf{x}$ with a point $y \in \Sigma_T$ is called a cycle of the first kind (respectively, of the second kind) if this point y is of the first (respectively, the second) kind.

In Fig. 3b the points of $L \cap W_F$ and the points of the first kind in L' are shown by small black circles, while the points of $L \cap (\Sigma(F) \setminus W_F)$ and the points of the second kind in L' are shown by white circles.

Lemma 8. *a) Two cycles of the first kind in $\mathcal{H}(\mathbf{x})$, vanishing over two distinguished paths connecting $\mathbf{x}$ with two points of $L' \cap \Sigma_T$ arising from the same close point y of $L \cap W_F$, coincide (maybe up to sign);*

b) this cycle coincides (maybe up to sign) with the cycle $a(y_+)$ transported from the point y_+ to $\mathbf{x}$ along the path described in the definition of the reduced Arnold class. In particular, the map Π_* sends the homology class of any such cycle into a generator of the corresponding group (16) or (17);

c) the monodromy action in the group $\mathcal{H}(\mathbf{x})$, defined by any simple loop in $L \setminus \Sigma(F)$ going around some point of $L \cap W_F$, is trivial;

d) any cycle in $\mathcal{H}(\mathbf{x})$ vanishing over a path in $L \setminus \Sigma(F)$ connecting $\mathbf{x}$ with a point of $\Sigma(F) \setminus W_F$ belongs to the subspace $\mathcal{M}(\mathbf{x})$. In particular, the same is true for any cycle of the second kind defined by a path of our distinguished system in $L' \setminus \Sigma_T$ connecting $\mathbf{x}$ with a point (of the second kind) of Σ_T.

Proof. Consider the space of complex lines through $\mathbf{x}$ transversal to Σ_T in the plane Λ. Obviously this space is a projective line with several points removed, one of which is the point $\{L\}$. Consider a small loop in this space, which starts and finishes at the point $\{L'\}$ and goes once around the point $\{L\}$. This loop takes one of the two distinguished paths from statement a) of the lemma into the other, thus this statement follows.

Statement c) is a direct consequence of a). Indeed, the loop considered there is homotopic in $\Lambda \setminus \Sigma_T$ to a loop in $L' \setminus \Sigma_T$ which turns around two discriminant points defining the same vanishing cycle, thus its monodromy action is equal to the square of the reflection in the hyperplane orthogonal to this vanishing cycle.

Statement b) follows from Lemma 4 and the local shape of the pair $(W_F, \mathcal{S}(\lambda))$ where $\mathcal{S}(\lambda)$ is the variety of zeros of the polynomial (20) defined by the discriminant point λ of the first kind. The way in which the pairs of distinguished paths connecting x with different pairs of points of the first kind miss one another is not important, because by the proof of Theorem 3 all the cycles of the first kind that vanish over the paths going from $\mathbf{x}$ to the points arising from different points of $L \cap W_F$ on the same side of $\mathbf{x}$ in $\operatorname{Re} L$ are pairwise orthogonal.

Statement d) of the lemma follows immediately from the constructions. $\qquad\square$

Thus, the vanishing cycles of the first (respectively, second) kind are exactly those that are sent by the map Π_* into a generator of the group (16) or (17) (respectively, into a zero class).

Lemma 9. *Any vanishing cycle of the first kind in $\mathcal{J}(\mathbf{x})$ can be transformed into any other by a sequence of reflections in the hyperplanes orthogonal to cycles of the second kind and to this cycle itself.*

By the Picard–Lefschetz formula, this lemma follows from the next one.

Lemma 9′. *There exists a distinguished system of paths in $L' \setminus \Sigma_T$ connecting $\mathbf{x}$ with all points of $L' \cap \Sigma_T$, such that all vanishing cycles of the first kind defined by this system are equal to each other.*

Proof. (This proof simulates that of the well-known fact that the fundamental group of the complement of a smooth irreducible algebraic hypersurface in $\mathbf{C}^n$, $n \geq 2$ is isomorphic to $\mathbf{Z}$.)

Let y_1 be any point of $L \cap W_F$. Let us fix an arbitrary path γ_1 in $L \setminus \Sigma_T$ connecting $\mathbf{x}$ with y_1. Denote by A^n the space of complex lines in $\mathbf{C}^n$, and by $\operatorname{Reg}(\Sigma(F))$ the subset of A^n consisting of lines transversal to $\Sigma(F)$. Consider a path $\chi_1 : [0,1] \to A^n$ such that $\chi_1(0) = L$, $\chi_1([0,1)) \subset \operatorname{Reg}(\Sigma(F))$, the last point $\chi_1(1)$ is a line transversal to $\Sigma(F)$ everywhere except for one point of simple tangency with W_F, and one of the two points of $\chi_1(\tau) \cap W_F$, $\tau = 1 - \varepsilon$ that coalesce at this tangency point is obtained from the point y_1 of the similar set corresponding to the value $\tau = 0$ during the deformation of the set $\chi_1(\tau) \cap W_F$, $\tau \in [0, 1 - \varepsilon]$.

Consider the continuous deformation $\gamma_1[\tau]$, $\tau \in [0, 1]$, of the path γ_1 such that $\gamma_1[0] = \gamma_1$, $\gamma_1[\tau] \subset \chi_1(\tau)$, and for any τ the path $\gamma_1[\tau]$ connects in $\chi_1(\tau) \setminus \Sigma(F)$ a point of $\chi_1(\tau) \cap W_F$ with some distinguished point $\mathbf{x}(\tau) \in \gamma_1[\tau] \setminus \Sigma(F)$, $\mathbf{x}(0) = \mathbf{x}$. At almost the final instant $\tau = 1 - \varepsilon$, the endpoint $\gamma_1[1 - \varepsilon](1)$ of the path $\gamma_1[1 - \varepsilon]$ lies very close to some other point of $\chi_1(1 - \varepsilon) \cap W_F$ (with which it coalesces at the instant $\tau = 1$). Connect this new point with $\mathbf{x}(1 - \varepsilon)$ by a path $\gamma_2[1 - \varepsilon]$ in $\chi_1(1 - \varepsilon) \setminus \Sigma(F)$ that goes very close to $\gamma_1[1 - \varepsilon]$ but does not intersect it except for the initial point. Then construct a continuous family of paths $\gamma_2[\tau] \subset \chi_1(\tau)$, $\tau \in [0, 1 - \varepsilon]$, such that for any τ the corresponding path $\gamma_2[\tau]$ connects a point of $\chi_1[\tau] \cap W_F$ with $\mathbf{x}(\tau)$ and does not intersect other points of $\chi_1(\tau) \cap \Sigma(F)$ or of the path $\gamma_1[\tau]$. At the instant $\tau = 0$ we get a path $\gamma_2 \equiv \gamma_2[0] \subset L$ connecting $\mathbf{x}$ with some point y_2 of W_F.

Then consider a new path $\chi_2 : [0, 1] \to A^n$, $\chi_2([0, 1)) \subset \operatorname{Reg}(\Sigma(F))$, connecting L with some new simple tangent line to W_F and having no extra nontransversalities with $\Sigma(F)$, in such a way that at the last instant $\tau = 1$ one of the two points of $\chi_2(\tau) \cap W_F$ that coalesce at the tangency point is obtained by deformation along our path χ_2 from one of the points y_1 or y_2, and the other two points of these two pairs do not coincide. Arguing as before, we construct a third path in $L \setminus \Sigma(F)$, connecting $\mathbf{x}$ with some third point of $L \cap W_F$, and so on.

After the $(d-1)$-th step we get a system of d nonintersecting paths in $L \setminus \Sigma(F)$, connecting $\mathbf{x}$ with all points of $L \cap W_F$. Complete this family to any distinguished collection of paths connecting $\mathbf{x}$ with all points of $L \cap \Sigma(F)$. For the close perturbation $L' \subset T$ of L, take a close distinguished system of paths in L', connecting the point $\mathbf{x}$ with all points of $L' \cap \Sigma_T$ in such a way that to any path in L connecting $\mathbf{x}$ with W_F there correspond two paths connecting $\mathbf{x}$ with two close points of the first kind. This system of paths is the desired one. For instance, the cycles vanishing along the (perturbed) paths γ_1 and γ_2 define the same vanishing homology class in $\mathcal{J}(\mathbf{x})$: indeed, a similar assertion for the cycles in the group $\mathcal{H}(\mathbf{x}(1-\varepsilon)) \equiv H_{n-1}(W_F \setminus S(\mathbf{x}(1-\varepsilon)), \mathbf{Z})$ or $H_{n-1}(W_F \setminus S(\mathbf{x}(1-\varepsilon)), @(\mathbf{x}(1-\varepsilon)))$ is proved just as the statement a) of Lemma 8, and for other values of $\tau \in [0, 1-\varepsilon]$ it follows by continuity. Lemmas $9'$ and 9 are thus proved. $\square$

Now we are ready to prove statement b) of Theorem 7. Indeed, by Lemma 6 the group $\mathcal{J}(\mathbf{x})$ is generated by the vanishing cycles of the first and second kind. By Lemma 9 and the Picard–Lefschetz formula, all vanishing cycles of the first kind lie in the linear span of an arbitrary one of them (for which we can take the class obtained by the Gauss–Manin connection from $a(x)$, x from the $\mathbf{k}$-th zone, $1 \le k \le [d/2]$, see statement b) of Lemma 8) and the vanishing cycles of the second kind (which lie in $\mathcal{M}(\mathbf{x})$, see statement d) of Lemma 8).

Statement c) of Theorem 7 follows immediately from statement c) of Lemma 8, and statement d) follows from the fact that the variety $\Sigma(F) \setminus W_F$ is irreducible.

Proof of Theorem 9a). We can assume that the points y_1 and y_2, whose classes $a(y_1)$ and $a(y_2)$ we want to transfer to each other, lie very close to the "interior" (i.e. closest to the hyperbolicity domain) components of M_F bounding corresponding zones. For such y_i the class $a(y_i)$ is realized by a cycle generating the group $H_{n-1}(W_F \cap B \setminus S(y_i))$ or $H_{n-1}(W_F \cap B \setminus S(y_i), @(y_i))$, where B is a small neighbourhood of y_i; see Lemma 4. Thus, for the desired path connecting y_1 and y_2 we can take the path that goes very close to the set of generic points of W_F (i.e. of such points y close to which all the generating lines of the cones $S(y)$ are transversal to W_F and hence the pairs $(W_F, S(y))$ have locally the same topological structure). $\square$

Statement b) of Theorem 9 follows from Theorem 3 and the connectedness of Dynkin diagrams of isolated singularities of complete intersections.

Proof of the statement g) of Theorem 7. First of all, this statement is true in the case when M_F is an ellipsoid with different eigenvalues. Indeed, by Theorem 3 in this case $\tilde{A}(x)$ is a vanishing cycle, and the assertion follows from the connectedness of the Dynkin diagram and the fact that the group $\mathcal{M}(x)$ is nontrivial for such F, see e.g. [E].

For arbitrary d, consider the model (not S-generic) hyperbolic surface M_F consisting of $[d/2]$ ellipsoids $\alpha_1 x_1^2 + \cdots + \alpha_n x_n^2 = j$, $\quad j = 1, 1+\varepsilon, \dots, 1 + ([d/2] - 1)\varepsilon$, where all α_i are positive and distinct, plus, if d is odd, a distant hyperplane. The class $\tilde{A}(x)$ for x from the k-th zone, $1 \leq k \leq [d/2]$, is then equal to the sum of k vanishing cycles, each of which lies in the complexification of its own ellipsoid; see the proof of Theorem 3. By the previous special case of a single ellipsoid, in each of these k complexified ellipsoids $\mathcal{E}_i$ there is a compact cycle Γ defining an element of the group $H_{n-1}(\mathcal{E}_i \setminus S(x))$ if n is even, or in $H_{n-1}(\mathcal{E}_i \setminus S(x), @(x))$ if n is odd, such that $\langle \tilde{A}(x), \Gamma \rangle \neq 0$ and the map Π_* sends the homology class of Γ into the zero homology class.

Consider a perturbation of our model hyperbolic polynomial F which replaces it by a S-generic one and is so weak that it does not change the topology of the variety $W_F \cup S(x)$ inside a sufficiently large disc, in which the cycles Γ and $\tilde{A}(x)$ lie. The cycle $\tilde{\Gamma}$ close to Γ in the moved manifold W_F satisfies all the above conditions, and statement g) of Theorem 7 is proved for *some* S-generic hyperbolic polynomial. For an arbitrary such polynomial this statement follows from the fact that all the generic surgeries separating different path-components of the space of all strictly hyperbolic S-generic surfaces of given degree in $\mathbf{R}^n$ (these surgeries correspond to the smooth hyperbolic surfaces in $\mathbf{R}P^n$ simple tangent to the non-proper plane) preserve the homology classes $\tilde{A}(x)$ (provided that the corresponding point x and the distinguished point in the hyperbolicity domain do not change in this surgery). (In formal terms, this preservation means that these homology classes corresponding to the polynomials before and after the surgery are transposed into one another by the natural connection over any *short* connecting them path in the space of all *complex* S-generic polynomials.) $\hfill\square$

6.2   Proof of Theorem 8

Let c be a point of simple tangency of a cone $S(x_0)$ and W_F such that $P(c) \neq 0$. Let Υ be an affine complex line through x_0 in $\mathbf{C}^n$, transversal to the common tangent hyperplane of $S(x_0)$ and W_F at c; let ξ be an affine coordinate on it with the origin at x_0. Consider the one-parametric family of surfaces $S(x(\xi))$, $x(\xi) \in \Upsilon$. The elements $S(x(\xi))$ of this family with ξ from a small *punctured* neighborhood of the origin are transversal to W_F in a small disc B centred at c, and the vanishing element $\gamma(\xi)$ of the group $H_{n-1}(B \cap W_F \setminus S(x(\xi)))$ (if n is even) or $H_{n-1}(B \cap W_F \setminus S(x(\xi)), @(x(\xi)))$ (if n is odd) is well defined (up to sign) by this family. By the Picard–Lefschetz formulae of § 2, in both cases the rotation of ξ around 0 sends $\gamma(\xi)$ to $-\gamma(\xi)$.

Define the function $\Xi(\xi)$, $\xi \in \mathbf{C}$, as the integral of the form (7) with $x = x(\xi) \in \Upsilon$ along the cycle $\gamma(\xi)$. It is sufficient to prove that there are arbitrarily high derivatives of this function not equal identically to 0. This follows from the next lemma.

Lemma 11. *The function $\Xi(\xi)$ is represented by a power series of the variable $\sqrt{\xi}$, whose leading (of smallest degree) term with non-zero coefficient has degree 1. (Of course, all even powers of this series vanish.)*

Proof. Using the Leray residue theorem, we can replace the integral (7) along the cycle $\gamma(\xi)$ by the integral of the form $G(x(\xi) - z)P(z)/F(z)dz_1 \wedge \ldots \wedge dz_n$ along the Leray tube $t\gamma(\xi) \in H_n(B \setminus (W_F \cup S(x(\xi))))$ or $\in H_n(B \setminus (W_F \cup S(x(\xi))), @(x(\xi)))$. Close to c the holomorphic function $\mathbf{C}^n \to \mathbf{C}$ is defined, which assigns to any point the coordinate ξ of the origin $x(\xi)$ of the cone $S(x(\xi))$ containing it. Choose this function for the last local coordinate w_n at c; by the Morse lemma we can choose the remaining coordinates $w_1, \ldots, w_{n-1}$ in such a way that W_F is locally given by $w_n = w_1^2 + \cdots + w_{n-1}^2$. In these coordinates our differential form becomes

$$(w_n - \xi)^{-(n-2)/2}(w_n - w_1^2 - \cdots - w_{n-1}^2)^{-1}I(w_1, \ldots, w_n)dw_1 \wedge \ldots \wedge dw_n, \qquad (21)$$

where the function I does not vanish at c. Let $I = I_0 + I_1 + \cdots$ be the expansion of I into the sum of quasihomogeneous polynomials of degrees $0, 1, \ldots$ respectively with respect to the weights $\deg w_1 = \cdots = \deg w_{n-1} = 1$, $\deg w_n = 2$. Using the corresponding group of quasihomogeneous dilations $(w_1, \ldots, w_{n-1}, w_n) \to (\tau w_1, \ldots, \tau w_{n-1}, \tau^2 w_n)$ we see, that the integral along $t\gamma(\xi)$ of the form similar to (21), in which I_m is substituted instead of I, is a homogeneous function in ξ of degree $(m+1)/2$. It is easy to calculate that this function corresponding to the constant polynomial $I_0 \neq 0$ is not the identical zero function; this proves our lemma. $\qquad \square$

References

[A 82]   V.I. Arnold: *On the Newtonian potential of hyperbolic layers.* Trudy Tbilisskogo Universiteta, Ser. Math./Mekh./Astron., **V.** 232–233, # 13–14, 23–29,(1982). English transl. in: Selecta Math. Soviet., **4:2**, 103–106, (1985).

[A 83]   V.I. Arnold: *Magnetic field analogues of the theorems of Newton and Ivory.* Uspekhi Mat. Nauk **38:5**, 145–146,(1983).

[A 87]   V.I. Arnold: *Kepler's second law and the topology of Abelian integrals.* Kvant, No. **12**, 17–21 (Russian); *A topological proof of transcendence of Abelian integrals in Newton's Principia*, Quant, No. **12**, 1–15, (1987); *The 300-th anniversary of mathematical natural philosophy and celestial mechanics*, Priroda, No. **8**, 5–15, (1987) (in Russian).

[ABG]   M.F. Atiyah; R. Bott; L. Gårding: *Lacunas for hyperbolic differential operators with constant coefficients.* Acta Math., **124**, 109–189, (1970) and **131**, 145–206, (1973).

[AV]   V.I. Arnold; V.A. Vassiliev: *Newton's Principia read 300 years later.* Notices Amer. Math. Soc., **36:9**, 1148–1154, (1989).

[AVG]   V.I. Arnold; A.N. Varchenko; S.M. Gusein-Zade: *Singularities of differentiable maps.* **V 1, 2**; "Nauka", Moscow, (1982, 1984); Engl. transl.: Birkhäuser, Basel, (1985, 1988).

[AVGL]   V.I. Arnold; V.A. Vassiliev; V.V. Goryunov; O. V. Lyashko: *Singularities, 1 and 2.* Dynamical systems, **6; 39**, VINITI, Moscow, (1988; 1989); English transl.: Encyclopaedia Math. Sci., **V. 6; 39**, Springer-Verlag, Berlin, New York, (1993).

[E]   W. Ebeling: *The monodromy groups of isolated singularities of complete intersections.* Lect. Notes Math., **vol.1293**, Springer, Berlin a.o., (1987).

[G 84]   A.B. Givental: *Polynomiality of the electrostatic potentials.* Uspekhi Mat. Nauk, **39:5**, 253–254 (in Russian), (1984).

[G 88]   A.B. Givental: *Twisted Picard–Lefschetz formulae.* Funkts. Anal. i Prilozh., 22:1, 12–22, (1988); Engl. translation in Functional Anal. Appl., 22:1, 10–18 (1988).

[Gab]   A.M. Gabrielov, *Bifurcations, Dynkin diagrams and modality of isolated singularities,* Funkts. Anal. i Prilozh., **8:2**, 7–12, (1974); Engl. transl. in Funct. Anal. Appl., **8**, 94–98, (1974).

[GH]   G.-M. Greuel; H.A. Hamm: *Invarianten quasihomogener vollständiger Durchschnitte.* Invent. Math., **49:1**, 67–86, (1978).

[GrH]   Ph. Griffiths; J. Harris: *Principles of algebraic geometry.* John Wiley & Sons, New York a.o., (1978).

[GM]   M. Goresky; R. MacPherson: *Stratified Morse theory.* Springer-Verlag, Berlin and New York, (1986).

[GR]   H. Grauert; R. Remmert: *Komplexe Räume.* Math. Ann., **136:2**, 245–318, (1958).

[GZ]   S.M. Gusein–Zade: *Monodromy groups of isolated singularities of hypersurfaces.* Uspekhi Mat. Nauk **32, no.2**, 23–65, (1977); Engl. transl. in Russian Math. Surveys **32, No.2**, 23–69, (1977).

[H]   H. Hamm: *Locale topologische Eigenschaften komplexer Räume.* Math. Ann., **191**, 235–252, (1971).

[I]   J. Ivory: *On the attraction of homogeneous ellipsoids.* Philos. Trans., **99**, 345–372, (1809).

[M]   J. Milnor: *Singular points of complex hypersurfaces.* Princeton Univ. Press, Princeton, NJ, and Univ. of Tokyo Press, Tokyo, (1968).

[MO]   J. Milnor; P. Orlik: *Isolated singularities, defined by weighted isolated polynomials.* Topology, **9:2**, 385–393, (1970).

[N]   W. Nuij: *A note on hyperbolic polynomials.* Math. Scand. **23**, 69–72, (1968).

[New]   I. Newton: *Philosophiae Naturalis Principia Mathematica.* London, (1687).

[P]   I.G. Petrovsky: *On the diffusion of waves and the lacunas for hyperbolic equations.* Matem. Sbornik, **17(59)**, 289–370, (1945).

[Ph 65] F. Pham: *Formules de Picard–Lefschetz génèralisées et ramification des intégrales.* Bull. Soc. Math. France, **93**, 333–367, (1965).

[Ph 67] F. Pham: *Introduction à l'étude topologique des singularités de Landau.* Gauthier-Villars, Paris, (1967).

[VSh]   A.D. Vainshtein; B.Z. Shapiro: *Multidimensional analogues of the Newton and Ivory theorems.* Funkts. Anal. i Prilozh., **19:1**, 20–24, (1985); Engl. translation in Functional Anal. Appl., **19:1**, 17–20, (1985).

[V 86]   V.A. Vassiliev: *Sharpness and the local Petrovskii condition for hyperbolic equations with constant coefficients.* Izv. Akad. Nauk SSSR Ser. Mat. **50**, 242–283, (1986); Engl. transl. in Mat. USSR Izv. **28**, 233–273, (1987).

[V 94]   V.A. Vassiliev: *Ramified Integrals, Singularities and Lacunas.* Kluwer, (1994).

[Z]   O. Zariski: *On the Poincaré group of a projective hypersurface.* Ann. Math. **38**, 131–141, (1937).

Appendix to the paper of V.A. Vassiliev

Wolfgang Ebeling

In this appendix we prove Conjecture 2 of the paper "Monodromy of complete intersections and surface potentials" (cf. [V]) of V. A. Vassiliev. We use without further reference the definitions and notations of this paper.

Theorem [V, Conjecture 2] *For any S-generic polynomial F of degree $d \geq 3$ in $\mathbf{C}^n$, $n \geq 3$, the triple consisting of the group $\mathcal{M}(x)$, the bilinear form equal (up to sign if $\left[\frac{n+1}{2}\right]$ is odd) to the form $\langle \cdot, \cdot \rangle$ defined before Lemma 3 of [V], and the "small" monodromy group on $\mathcal{M}(x)$, is completely infinite.*

Proof. By [V, Theorem 7], $\mathcal{M}(x)$ is a sublattice of corank 1 in the lattice $\mathcal{J}(x)$. Let ρ be the rank of $\mathcal{J}(x)$ and ρ_+, ρ_- the inertia indices of the symmetric bilinear form on $\mathcal{J}(x)$.

In the case n even, $\mathcal{J}(x)$ is isomorphic to the Milnor lattice $H_{n-2}(X_f)$ of the ICIS defined by $f = (F, r^2(\cdot - x))$. Under the assumptions $n \geq 3$, $d \geq 3$, this singularity is not hyperbolic, so $\min\{\rho_+, \rho_-\} \geq 2$.

In the case when n is odd, $\mathcal{J}(x)$ contains the group $H_{n-2}(X_f)$ with a symmetric bilinear form obtained in the following way: Write the skew-symmetric intersection matrix with respect to a basis of vanishing cycles in the form $S = A - A^t$, where A is an upper triangular matrix with -1 on the diagonal. Then consider the symmetric bilinear form defined by the symmetric matrix $A + A^t$. Denote by μ_+ the positive inertia index of this form. For $n = 3$, $d = 3$, the ICIS $f = (F, r^2(\cdot - x))$ is $\mathcal{K}$-equivalent to the singularity $(xy, (x+y^2)(y+x^2)+z^3)$. This is a 3-fold suspension of a fat point. A Coxeter-Dynkin diagram with respect to a distinguished set of vanishing cycles for this singularity is calculated in [EGZ, Sect. 9], a $\mathbf{Z}_3$-equivariant Coxeter-Dynkin diagram is the graph of [EGZ, Fig. 9] with $r = s = 2$. Then one can compute that $\mu_+ = 2$. It follows that $\min\{\rho_+, \rho_-\} \geq 2$ for all $n \geq 3$, $d \geq 3$.

We multiply the symmetric bilinear form by -1. So we may assume that $\langle \delta, \delta \rangle = 2$ for any vanishing cycle $\delta \in \mathcal{J}(x)$ and $2 \leq \rho_- \leq \rho_+$.

Now consider the lattice $\mathcal{M}(x)$. Denote by ρ' the rank and by ρ'_+, ρ'_- the corresponding inertia indices of $\mathcal{M}(x)$. Since $\mathcal{M}(x)$ has corank 1 in $\mathcal{J}(x)$, it follows that $\rho'_+, \rho'_- \geq 1$. We denote by $\mathrm{Ker}\,\mathcal{M}(x)$ the kernel of the bilinear form $\langle \cdot, \cdot \rangle$ on $\mathcal{M}(x)$. Let $L = \mathcal{M}(x)/\mathrm{Ker}\,\mathcal{M}(x)$, Δ be the image of the set of vanishing cycles in L, and Γ be the subgroup of $O(L)$ generated by the reflections corresponding to the elements of Δ. By [V, Theorem 7], $\langle \delta, \delta \rangle = 2$ for all $\delta \in \Delta$, Δ generates L, and Δ is a single orbit under the action of Γ. Let $V = L \otimes \mathbf{C}$. Then the pair (V, Δ) satisfies the conditions (A), (B), and (C) of [EO, Sect. 1]. Since $\rho'_- \geq 1$, it follows from that paper that Δ is an infinite set (cf. also [D, Lemme $(4.4.2^s)$]). Then also the set $\{\langle \delta, \delta' \rangle | \delta, \delta' \in \Delta\}$ is infinite (cf. [FM, 6.2.2, Claim 2.7]). For suppose that this set is finite. Since Δ generates V, there is a basis $\delta_1, \ldots, \delta_{\rho'}$ of V with $\delta_i \in \Delta$. Let δ_i^* be the dual element to δ_i. Since every $\delta \in \Delta$ can be written as

$$\delta = \sum_i \langle \delta, \delta_i \rangle \delta_i^*,$$

there are only finitely many possibilities for δ. So Δ would be finite, a contradiction.

In order to prove the theorem, it suffices to show that the triple $(L; \langle \cdot, \cdot \rangle; \Gamma)$ is completely infinite. Let $v \in V$ be an arbitrary vector such that not all numbers $\langle v, \delta \rangle$ for $\delta \in \Delta$ are equal to 0. We have to show that any nonzero linear form $V \to \mathbf{C}$ takes infinitely many values on the orbit $\Gamma \cdot v$ of v under the action of the group Γ. Assume the contrary. Let $l : V \to \mathbf{C}$ be a nonzero linear form which takes only finitely many values on $\Gamma \cdot v$. Let $\delta \in \Delta$ be such that $\langle v, \delta \rangle = a \neq 0$. Then $s_\delta(v) = v - a\delta$ and hence

$$l(\delta) = \frac{1}{a}(l(v) - l(s_\delta(v))).$$

Since l takes only finitely many values on the orbit $\Gamma \cdot v$, it follows that l also takes only finitely many values on the orbit of δ which is Δ.

Let $\delta' \in \Delta$ be an element with $l(\delta') \neq 0$. Such an element exists, since otherwise l would vanish on Δ which would force l to be zero since Δ generates V. Since the set $\{\langle \delta_1, \delta_2 \rangle | \delta_1, \delta_2 \in \Delta\}$ is infinite, there exist $\delta_1, \delta_2 \in \Delta$ with $\langle \delta_1, \delta_2 \rangle \notin \{-1, 0, 1\}$. Since Γ acts transitively on Δ, $\delta' = \gamma \delta_1$ for some $\gamma \in \Gamma$. Put $\delta'' = \gamma \delta_2$. Then $\langle \delta', \delta'' \rangle = \langle \delta_1, \delta_2 \rangle \notin \{-1, 0, 1\}$. Let W be the linear plane in V spanned by δ' and δ''. Then W is not contained in the hyperplane $l^{-1}(0)$. The affine hyperplane $l^{-1}(c)$ corresponding to a value c of l on Δ cuts W in a line. Let q be the quadratic form corresponding to the symmetric bilinear form $\langle \cdot, \cdot \rangle$, i.e., q is defined by $q(y) = \langle y, y \rangle$ for all $y \in V$. Then $\Delta \cap W$ is contained in the intersection of the non-degenerate complex plane quadric $q^{-1}(2) \cap W$ with a finite set of lines. This intersection is finite. On the other hand, $s_{\delta'} s_{\delta''}$ is an element of infinite order in $O(W)$ (cf. [D, Lemme $(4.4.3^s)$]). Hence $\Delta \cap W$ must have infinitely many elements, a contradiction. This proves the theorem. $\quad\square$

References

[D]   P. Deligne: *La conjecture de Weil.* II, Publ. Math. IHES **52**, 137–252, (1980).

[EGZ] W. Ebeling; S.M. Gusein-Zade: *Suspensions of fat points and their intersection forms.* These proceedings.

[EO]   W. Ebeling; Ch. Okonek: *Donaldson invariants, monodromy groups, and singularities.* Intern. J. Math. **1**, 233–250, (1990).

[FM]   R. Friedman; J.W. Morgan: *Smooth Four-Manifolds and Complex Surfaces.* Springer, Berlin etc., (1994).

[V]   V.A. Vassiliev: *Monodromy of complete intersections and surface potentials.* These proceedings.

Chapter 3

Resolution

Progress in Mathematics, Vol. 162, © 1998 Birkhäuser Verlag Basel/Switzerland

P-Resolutions of Cyclic Quotients from the Toric Viewpoint

Klaus Altmann
Institut für reine Mathematik
Humboldt-Universität zu Berlin
Ziegelstraße 13A
D-10099 Berlin
GERMANY

1 Introduction

(1.1) The break through in deformation theory of (two-dimensional) quotient singularities Y was Kollár/Shepherd-Barron's discovery of the one-to-one correspondence between so-called P-resolutions, on the one hand, and components of the versal base space, on the other (cf. [KS], Theorem (3.9)). It generalizes the fact that all deformations admitting a simultaneous (RDP-) resolution form one single component, the Artin component.

According to definition (3.8) in [KS], P-resolutions are partial resolutions $\pi : \tilde{Y} \to Y$ such that

- the canonical divisor $K_{\tilde{Y}|Y}$ is ample relative to π (a minimality condition) and

- $\tilde{Y}$ contains only mild singularities of a certain type (so-called T-singularities).

Despite their definition as those quotient singularities admitting a $\mathbb{Q}$-Gorenstein one-parameter smoothing ([KS], (3.7)), there are at least three further descriptions of the class of T-singularities: An explicit list of their defining group actions on $\mathbb{C}^2$ ([KS], (3.10)), an inductive procedure to construct their resolution graphs ([KS], (3.11)), and a characterization using toric language ([Al], (7.3)).

The latter one begins with the observation that affine, two-dimensional toric varieties (given by some rational, polyhedral cone $\sigma \subseteq \mathbb{R}^2$) provide exactly the two-dimensional cyclic quotient singularities. Then, T-singularities come from cones over rational intervals of integer length placed in height one (i.e. contained in the affine line $(\bullet, 1) \subseteq \mathbb{R}^2$).

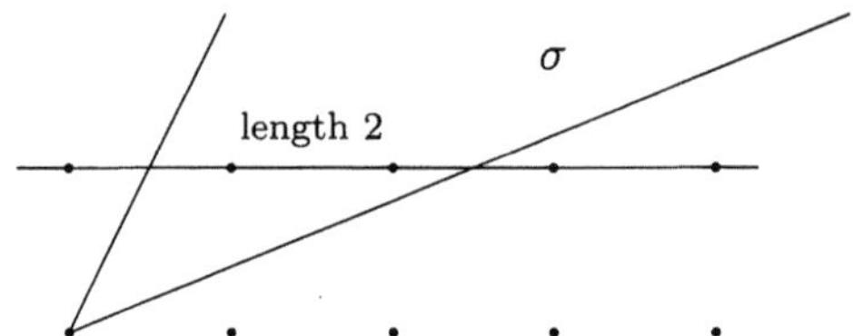

If the affine interval is of length $\mu + 1$, then the corresponding T-singularity will have Milnor number μ (on the $\mathbb{Q}$-Gorenstein one-parameter smoothing).

(1.2) In [Ch] and [St] Christophersen and Stevens gave a combinatorial description of all P-resolutions for two-dimensional, cyclic quotient singularities. Using an inductive construction method (going through different cyclic quotients with step-by-step increasing multiplicity), they have shown that there is a one-to-one correspondence between P-resolutions, on the one hand, and certain integer tuples $(k_2, \ldots, k_{e-1})$ yielding zero if expanded as a (negative) continued fraction (cf. (4.2)), on the other hand.

The aim of the present paper is to provide an elementary, direct method for constructing the P-resolutions of a cyclic quotient singularity (i.e. a two-dimensional toric variety) Y_σ. Given a chain $(k_2, \ldots, k_{e-1})$ representing zero, we will give a straight description of the corresponding polyhedral subdivision of σ. (In particular, the bijection between those 0-chains and P-resolutions will be proved again by a different method.)

2   Cyclic quotient singularities

In the following we remind the reader of basic notions concerning continued fractions and cyclic quotients as well as fix notation. References are [Od] (§1.6) or the first sections in [Ch] and [St], respectively.

(2.1) Definition: To integers $c_1, \ldots, c_r \in \mathbb{Z}$ we will assign the continued fraction $[c_1, \ldots, c_r] \in \mathbb{Q}$ if the following inductive procedure is well-defined (i.e. if no division by 0 occurs):

- $[c_r] := c_r$,
- $[c_i, \ldots, c_r] := c_i - 1/[c_{i+1}, \ldots, c_r]$.

If $c_i \geq 2$ for $i = 1, \ldots, r$, then $[c_1, \ldots, c_r]$ is always defined and yields a rational number greater than 1. Moreover, all these numbers may be represented by those continued fractions in a unique way.

(2.2) Let $n \geq 2$ be an integer and $q \in (\mathbb{Z}/n\mathbb{Z})^*$ be represented by an integer between 0 and n. Each q provides a group action of $\mathbb{Z}/n\mathbb{Z}$ on $\mathbb{C}^2$ via the matrix

$\begin{pmatrix} \xi & 0 \\ 0 & \xi^q \end{pmatrix}$ (with ξ a primitive n-th root of unity). The quotient is denoted by $Y(n,q)$.

In toric language, $Y(n,q)$ equals the variety Y_σ assigned to the polyhedral cone $\sigma := \langle (1,0); (-q,n) \rangle$ contained in $\mathbb{R}^2$. Y_σ is defined as Spec $\mathbb{C}[\sigma^\vee \cap \mathbb{Z}^2]$ with

$$\sigma^\vee := \{ r \in (\mathbb{R}^2)^* \mid r \geq 0 \text{ on } \sigma \} = \langle [0,1]; [n,q] \rangle \subseteq (\mathbb{R}^2)^* \cong \mathbb{R}^2 .$$

Notation: Just to distinguish between $\mathbb{R}^2$ and its dual $(\mathbb{R}^2)^* \cong \mathbb{R}^2$, we will denote these vector spaces by $N_{\mathbb{R}}$ and $M_{\mathbb{R}}$, respectively. (Hence, $\sigma \subseteq N_{\mathbb{R}}$ and $\sigma^\vee \subseteq M_{\mathbb{R}}$.) Elements of $N_{\mathbb{R}} \cong \mathbb{R}^2$ are written in parentheses; elements of $M_{\mathbb{R}} \cong \mathbb{R}^2$ are written in brackets. The natural pairing between $N_{\mathbb{R}}$ and $M_{\mathbb{R}}$ is denoted by $\langle \, , \, \rangle$, which should not be confused with the symbol indicating the generators of a cone. Finally, all these remarks apply for the lattices $N \cong \mathbb{Z}^2$ and $M \cong \mathbb{Z}^2$, too.

(2.3) Let n, q be as before. We may write $n/(n-q)$ and n/q (both are greater than 1) as continued fractions

$$n/(n-q) = [a_2, \ldots, a_{e-1}] \text{ and } n/q = [b_1, \ldots, b_r] \quad (a_i, b_j \geq 2).$$

The a_i's and the b_j's are mutually related by Riemenschneider's point diagram (cf. [Ri]).

Denote by $w^1, w^2, \ldots, w^e$ the lattice points on the compact edges of the convex hull of $(\sigma^\vee \cap M) \setminus \{0\}$. If ordered the right way, we obtain $w^1 = [0,1]$ and $w^e = [n,q]$ for the first and the last point, respectively.

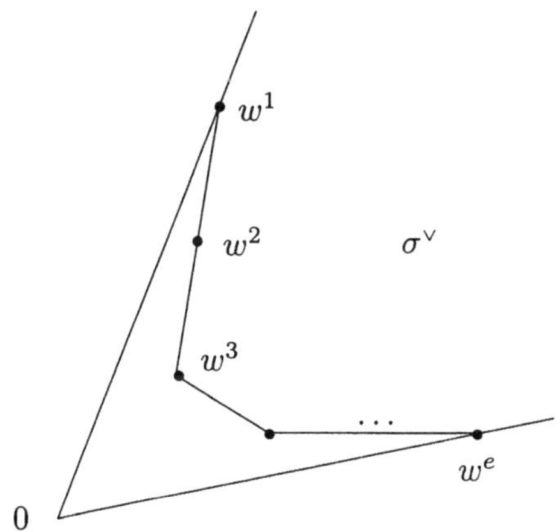

Then, $E := \{ w^1, \ldots, w^e \}$ is the minimal generating set (the so-called Hilbert basis) of the semigroup $\sigma^\vee \cap M$. These point are related to our first continued fraction by

$$w^{i-1} + w^{i+1} = a_i \, w^i \quad (i = 2, \ldots, e-1).$$

Remark: The surjection $\mathbb{N}^E \longrightarrow \sigma^\vee \cap M$ provides a minimal embedding of Y_σ. In particular, e equals its embedding dimension.

In a similar manner we can define $v^0, \ldots, v^{r+1} \in \sigma \cap N$ in the original cone; now we have $v^0 = (1,0)$, $v^{r+1} = (-q, n)$, and the relation to the continued fractions is $v^{j-1} + v^{j+1} = b_j\, v^j$ (for $j = 1, \ldots, r$).

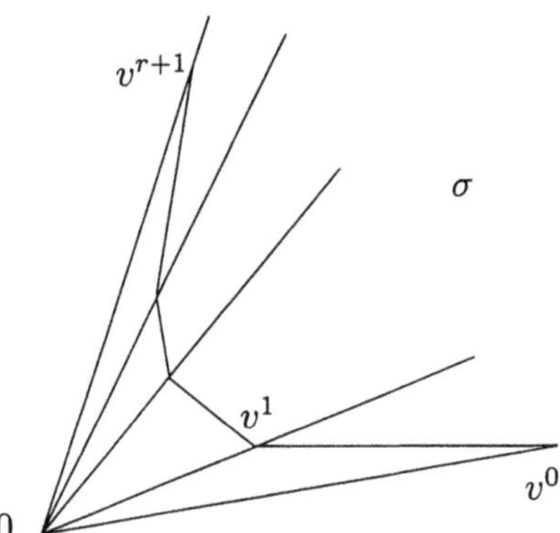

Drawing rays through the origin and each point v^j, respectively, provides a polyhedral subdivision Σ of σ. The corresponding toric variety Y_Σ is a resolution of our singularity Y_σ. The numbers $-b_j$ equal the self intersection numbers of the exceptional divisors; since $b_j \geq 2$, the resolution is *minimal*.

3   The maximal resolution

(3.1) Definition: ([KS], (3.12)) For a resolution $\pi : \tilde{Y} \to Y$ we may write $K_{\tilde{Y}|Y} := K_{\tilde{Y}} - \pi^* K_Y = \sum_j (\alpha_j - 1) E_j$, where the E_j denote the exceptional divisors, and $\alpha_j \in \mathbb{Q}$. Then, π will be called *maximal* if it is maximal with respect to the property $0 < \alpha_j < 1$.

The maximal resolution is uniquely determined and dominates all the P-resolutions. Hence, for our purpose, it is more important than the minimal one. It can be constructed from the minimal resolution by successively blowing up points $E_i \cap E_j$ with $\alpha_i + \alpha_j \geq 0$ (cf. Lemma (3.13) and Lemma (3.14) in [KS]).

(3.2) Proposition: *The maximal resolution of Y_σ is toric. It can be obtained by drawing rays through* 0 *and all interior lattice points (i.e. $\in N$) of the triangle* $\Delta := \mathrm{conv}\,(0, v^0, v^{r+1})$, *respectively.*

Proof. In order to keep track of the rational numbers α_j, we will show how they can be "seen" in an arbitrary toric resolution of Y_σ. Let $\Sigma < \sigma$ be a subdivision generated by one-dimensional rays through the points $u^0, \ldots, u^{s+1} \in \sigma \cap N$. In particular, $u^0 = v^0 = (1,0)$ and $u^{s+1} = v^{r+1} = (-q, n)$; moreover, for the minimal resolution we have $s = r$ and $u^j = v^j$ $(j = 0, \ldots, r+1)$. Denote by $c_1, \ldots, c_s$ the integers given by the relations

$$u^{j-1} + u^{j+1} = c_j \, u^j \qquad (j = 1, \ldots, s).$$

In particular, $c_j = b_j$ for the minimal resolution. As usual, the numbers $-c_j$ equal the self intersection numbers of the exceptional divisors E_j in Y_Σ. Indeed, $D := \sum_i u^i E_i$ is a principal divisor (if you do not like coefficients u^i from N, evaluate them using arbitrary elements of M); hence,

$$
\begin{aligned}
0 = E_j \cdot D &= E_j \cdot (u^{j-1} E_{j-1} + u^j E_j + u^{j+1} E_{j+1}) \\
&= u^{j-1} + (E_j)^2 \, u^j + u^{j+1} \\
&= (c_j + (E_j)^2) \cdot u^j \qquad (j = 1, \ldots, s).
\end{aligned}
$$

On the other hand, we can use the projection formula to obtain

$$
\begin{aligned}
-2 = 2\,g(E_j) - 2 &= K_{\tilde{Y}|Y} \cdot E_j + (E_j)^2 \\
&= \sum_i (\alpha_i - 1)\,(E_i \cdot E_j) + (E_j)^2 \\
&= (\alpha_{j-1} - 1) + (\alpha_j - 1)(E_j)^2 + (\alpha_{j+1} - 1) + (E_j)^2 \,,
\end{aligned}
$$

hence

$$\alpha_{j-1} + \alpha_{j+1} = c_j \, \alpha_j \qquad (j = 1, \ldots, s;\ \alpha_0, \alpha_{s+1} := 1).$$

Looking at the definition of the c_j's via relations among the lattice points u^j, there has to be some $R \in M_{\mathbb{R}}$ such that

$$\alpha_j = \langle u^j, R \rangle \qquad (j = 0, \ldots, s+1).$$

The conditions $\langle u^0, R \rangle = \alpha_0 = 1$ and $\langle u^{s+1}, R \rangle = \alpha_{s+1} = 1$ determine R uniquely. Now, we can see that α_j measures exactly the quotient between the length of the line segment $\overline{0\,u^j}$ and the length of the Δ-part of the line through 0 and u^j. In particular, $\alpha_j < 1$ if and only if u^j lies below the line connecting u^0 and u^{s+1}.

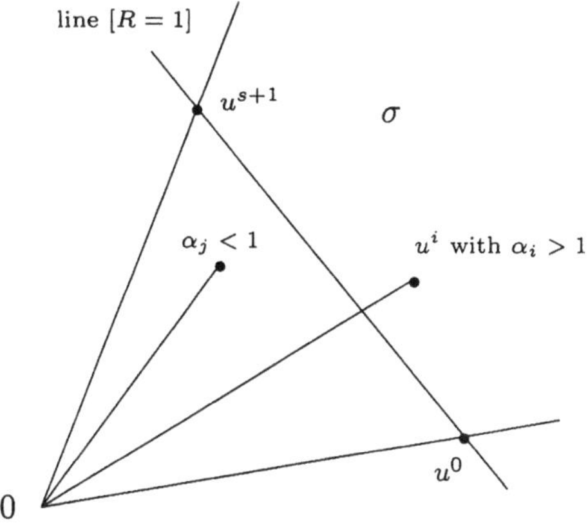

This explains how to construct the maximal resolution:
Starting with the minimal one, continue subdividing each small cone $\langle u^j, u^{j+1} \rangle$ into $\langle u^j, u^j + u^{j+1} \rangle \cup \langle u^j + u^{j+1}, u^{j+1} \rangle$ as long as it contains interior lattice points below the line $[R = 1]$, i.e. belonging to int Δ. $\square$

Corollary: *Every P-resolution is toric.*

Proof. P-resolutions are obtained by blowing down curves in the maximal resolution. $\square$

(3.3) Example: We take the example $Y(19, 7)$ from [KS], (3.15). Since $\sigma = \langle (1, 0), (-7, 19) \rangle$, the interior of Δ is given by the three inequalities

$$y > 0, \quad 19x + 7y > 0, \quad \text{and} \quad 19x + 8y < 19 \quad \text{(corresponding to } R = [1, 8/19]).$$

The only primitive lattice points (i.e. generating rays) contained in int Δ are

$$u^1 = (0, 1), \quad u^2 = (-1, 4), \quad u^3 = (-2, 7),$$
$$u^4 = (-1, 3), \quad u^5 = (-5, 14), \quad u^6 = (-4, 11).$$

They provide the maximal resolution. The corresponding α's can be obtained by taking the scalar product with $R = [1, 8/19]$, which yields 8/19, 13/19, 18/19, 5/19, 17/19, and 12/19.

The minimal resolution uses only the rays through $u^1 = (0, 1)$, $u^4 = (-1, 3)$, and $u^6 = (-4, 11)$, respectively.

4   P-resolutions

(4.1) In this section we will speak about *partial* toric resolutions $\pi : Y_\Sigma \to Y_\sigma$. Nevertheless, we use the same notation as we did for the maximal resolution: The fan Σ subdividing σ is generated by rays through $u^0, \ldots, u^s \in \sigma \cap N$; each ray u^j corresponds to an exceptional divisor $E_j \subseteq Y_\Sigma$. However, since $u^{j-1} + u^{j+1}$ need not to be a multiple of u^j, the numbers c_j no longer make sense.

Lemma: ([Re], (4.3)) *For $K := K_{Y_\Sigma}$ or $K := K_{Y_\Sigma | Y_\sigma}$ the intersection number $(E_j \cdot K)$ is positive, zero, or negative if the line segments $\overline{u^{j-1}u^j}$ and $\overline{u^j u^{j+1}}$ form a strict concave, flat, or strict convex "roof" over the two cones, respectively.*

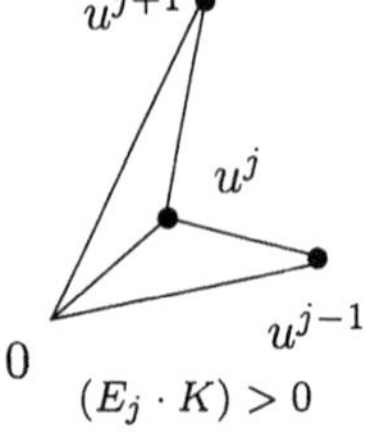

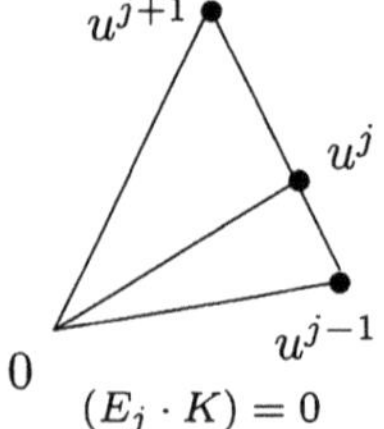

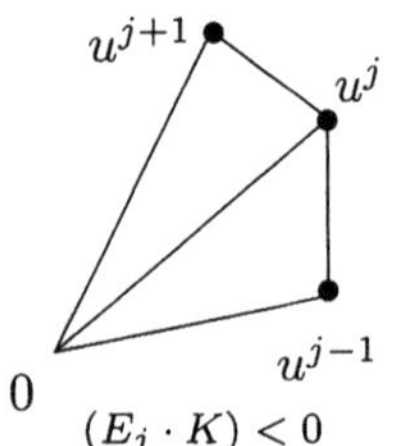

Proof. Using $K := K_{Y_\Sigma} = -\sum_{i=0}^{s+1} E_i$ (cf. [Od], (2.1)) we have

$$(E_j \cdot K) = -(E_j \cdot E_{j-1}) - (E_j)^2 - (E_j \cdot E_{j+1}) \,.$$

On the other hand, as in the proof of Proposition (3.2), we know that

$$0 = (E_j \cdot E_{j-1})\, u^{j-1} + (E_j)^2\, u^j + (E_j \cdot E_{j+1})\, u^{j+1} \,.$$

Combining both formulas yields the final result

$$(E_j \cdot K)\, u^j = (E_j \cdot E_{j-1})\, (u^{j-1} - u^j) + (E_j \cdot E_{j+1})\, (u^{j+1} - u^j) \,. \qquad \square$$

Remark: The previous lemma together with Proposition (3.2) illustrate again the fact that all P-resolutions are dominated by the maximal resolution.

(4.2) In [Ch] Christophersen has defined the set

$$K_{e-2} := \{(k_2,\ldots,k_{e-1}) \in \mathbb{N}^{e-2} \mid [k_2,\ldots,k_{e-1}] \text{ is well defined and yields } 0\,\}$$

of chains representing zero. To every such chain, non-negative integers $q_1,\ldots,q_e$ are assigned. They are characterized by the following equivalent properties:

- $q_1 = 0$, $q_2 = 1$, and $q_{i-1} + q_{i+1} = k_i\, q_i$ $(i = 2,\ldots,e-1)$;

- $q_{e-1} = 1$, $q_e = 0$, and $q_{i-1} + q_{i+1} = k_i\, q_i$ $(i = 2,\ldots,e-1)$;

- $q_e = 0$ and $[k_i,\ldots,k_{e-1}] = q_{i-1}/q_i$
  with $\gcd(q_{i-1},q_i) = 1$ $(i - 2,\ldots,c \quad 1)$.

(The two latter properties do not even use the fact that the continued fraction $[k_2,\ldots,k_{e-1}]$ yields zero.)

Remark: The elements of K_{e-2} are in one-to-one correspondence with triangulations of a (regular) $(e-1)$-gon with vertices $P_2,\ldots,P_{e-1},P_*$. The numbers k_i tell how many triangles are attached to P_i, and the q_i have an easy meaning in this language, too.

Finally, for a given Y_σ with embedding dimension e, Christophersen defines

$$K(Y_\sigma) := \{(k_2,\ldots,k_{e-1}) \in K_{e-2} \mid k_i \le a_i\} \,.$$

Theorem: *Each P-resolution of Y_σ (i.e. the corresponding subdivision Σ of σ) is given by some $\underline{k} \in K(Y_\sigma)$ in the following way:*

(1) Σ is built from the rays that are orthogonal to $w^i/q_i - w^{i-1}/q_{i-1} \in M_\mathbb{R}$ (for $i = 3,\ldots,e-1$). In some sense, if the occurring divisions by zero are

interpreted properly, Σ may be seen as dual to the Newton boundary generated by $w^i/q_i \in \sigma^\vee$ $(i = 1, \ldots, e)$.

(2) The affine lines $[\langle \bullet, w^i \rangle = q_i]$ form the "roofs" of the Σ-cones. In particular, the (possibly degenerate) cones $\tau^i \in \Sigma$ correspond to the elements $w^1, \ldots, w^e \in E$. The "roof" over the cone τ^i has length $\ell_i := (a_i - k_i)\, q_i$, where the length is defined via the metric induced by the lattice structure $M \subseteq M_\mathbb{R}$ on rational lines. In particular, τ^i is degenerate if and only if $k_i = a_i$. The Milnor number of the T-singularity Y_{τ^i} equals $(a_i - k_i - 1)$.

(4.3) *Proof.* According to the notation introduced in (4.1), the fan Σ consists of (non-degenerate) cones $\tau^j := \langle u^{j-1}, u^j \rangle$ with $j = 1, \ldots, s+1$. (Except for $u^0 = (1, 0)$ and $u^s = (-q, n)$, their generators u^j are primitive lattice points (i.e. $\in N$) contained in $\mathrm{int}\,\Delta \subseteq \sigma$.)

Step 1: For each τ^j there are $w \in E, d \in \mathbb{N}$ such that $\langle u^{j-1}, w \rangle = \langle u^j, w \rangle = d$.

First, it is very clear that there is a primitive lattice point $w \in M$ and a non-negative number $d \in \mathbb{R}_{\geq 0}$ admitting the desired properties. Moreover, since $u^j \in N$, d has to be an integer, and Reid's Lemma (4.1) tells us that $w \in \sigma^\vee$. It remains to show that w belongs to the Hilbert basis $E \subseteq \sigma^\vee \cap M$ as well. Denote by ℓ the length of the line segment $\overline{u^{j-1}u^j}$ on the "roof" line $[\langle \bullet, w \rangle = d]$. Since τ^j represents a T-singularity, we know from (7.3) of [Al] (cf. (1.1) of the present paper) that $d | \ell$. In particular, $\overline{u^{j-1}u^j}$ contains the d-th multiple $d \cdot u$ of some lattice point $u \in \tau^j \cap M$ (w.l.o.g. not belonging to the boundary of σ). Hence, $\langle u, w \rangle = 1$ and $u \in \mathrm{int}\,\sigma \cap M$, and this implies $w \in E$.

Step 2: As each of the cones $\tau^1, \ldots, \tau^{s+1} \in \Sigma$ is assigned to some element $w \in E$, a renumbering will be used to make the notation more obvious: Let $\tau^i = \langle u^{i-1}, u^i \rangle$ be the cone assigned to $w^i \in E$, and denote by d_i, ℓ_i the height and the length of its "roof" $\overline{u^{i-1}u^i}$, respectively. Some of these cones might be degenerated, i.e. $\ell_i = 0$. This it at least true for the extremal τ^1 and τ^e coinciding with the two rays spanning σ. Here we even have $d_1 = d_e = 0$; in particular $u^0 = u^1 = (1, 0)$ and $u^{e-1} = u^e = (-q, n)$.

Since $d_i | \ell_i$, we may introduce integers $k_i \leq a_i$ yielding $\ell_i = (a_i - k_i)\, d_i$. For $i = 2, \ldots, e-1$ they are even uniquely determined.

Step 3: Using the following three ingredients

(i) $\langle u^{i-1}, w^i \rangle = \langle u^i, w^i \rangle = d_i$ $(i = 1, \ldots, e)$,

(ii) $w^{i-1} + w^{i+1} = a_i\, w^i$ $(i = 2, \ldots, e-1;$ cf. (2.3)), and

(iii) $\langle u^i - u^{i-1}, w^{i-1} \rangle = \ell_i = (a_i - k_i)\, d_i$

     (since $\{w^{i-1}, w^i\}$ forms a $\mathbb{Z}$-basis of M), we obtain

$$
\begin{aligned}
d_{i-1} + d_{i+1} &= (a_i\,d_i + d_{i-1}) - (a_i\,d_i - d_{i+1}) \\
&= (a_i\,d_i + \langle u^{i-1}, w^{i-1}\rangle) - \langle u^i,\, a_i\,w^i - w^{i+1}\rangle \\
&= a_i\,d_i + \langle u^{i-1}, w^{i-1}\rangle - \langle u^i, w^{i-1}\rangle \\
&= a_i\,d_i + \langle u^{i-1} - u^i,\, w^{i-1}\rangle \\
&= a_i\,d_i - (a_i - k_i)\,d_i \;=\; k_i\,d_i \quad (\text{for } i = 2,\dots,e-1).
\end{aligned}
$$

In particular, $k_i \geq 0$ (and in fact ≥ 1 for $e > 3$). Moreover, because $\{w^{i-1}, w^i\}$ forms a basis of M and $u^{i-1} \in N$ is primitive, we have $\gcd(d_{i-1}, d_i) = 1$. It follows that $d_i = q_i$, since both sequences of integers satisfy the second of the three properties mentioned in the beginning of (4.2). Finally, the third of these properties yields $[k_2, \dots, k_{e-1}] = q_1/q_2 = d_1/d_2 = 0$, i.e. $\underline{k} \in K_{e-2}$.

The reverse direction, i.e. the fact that each $K(Y_\sigma)$-element indeed yields a P-resolution, follows from the above calculations in a similar manner. $\qquad\square$

Remark: Subdividing each τ^i further into $(a_i - k_i)$ equal cones (with each "roof" length q_i) yields the so-called M-resolution (cf. [BC]) assigned to a P-resolution. It is defined to contain only T_0-singularities, i.e. T-singularities with Milnor number 0; in exchange, $K_{\tilde{Y}|Y}$ does not need to be relatively ample anymore. This property is replaced by "relatively nef".

Examples: (1) The continued fraction $[1, 2, 2, \dots, 2, 1] = 0$ yields $q_1 = q_e = 0$ and $q_i = 1$ otherwise. In particular, the "roof" lines equal $[\langle \bullet, w^i\rangle = 1]$ (for $i = 2, \dots, e-1$), describing the RDP-resolution of Y_σ. The assigned M-resolution equals the minimal resolution mentioned at the end of (2.3).

(2) Let us return to Example (3.3): The embedding dimension e of Y_σ is 6, the vector $(a_2, \dots, a_{e-1})$ equals $(2, 3, 2, 3)$, and, except the trivial RDP element mentioned in (1), $K(Y_\sigma)$ contains only $(1, 3, 1, 2)$ and $(2, 2, 1, 3)$.

In both cases we already know that $q_1 = q_6 = 0$ and $q_2 = q_5 = 1$. The remaining values are given by the equation $q_3/q_4 = [k_4, k_5]$, i.e. we obtain $q_3 = 1$, $q_4 = 2$ or $q_3 = 2$, $q_4 = 3$, respectively.

Hence, in case of $(1, 3, 1, 2)$ the fan Σ is given by the additional rays through $(0, 1)$ and $(-4, 11)$. For $\underline{k} = (2, 2, 1, 3)$ we only need the one through $(-1, 4)$.

References

[Al] Altmann, K.: *Minkowski sums and homogeneous deformations of toric varieties.* Tôhoku Math. J. **47**, 151–184, (1995).

[BC] Behnke, K., Christophersen, J.A.: *M-resolutions and deformations of quotient singularities.* Amer. J. Math. **116**, 881–903, (1994).

[Ch] Christophersen, J.A.: *On the Components and Discriminant of the Versal Base Space of Cyclic Quotient Singularities.* In: Singularity Theory and its Applications, Warwick 1989, Part I: Geometric Aspects of Singularities, pp. 81–92, Springer-Verlag Berlin Heidelberg, (LNM 1462), (1991).

[KS] Kollár, J., Shepherd-Barron, N.I.: *Threefolds and deformations of surface singularities.* Invent. math. **91**, 299–338, (1988).

[Od] Oda, T.: *Convex bodies and algebraic geometry.* Ergebnisse der Mathematik und ihrer Grenzgebiete (3/15), Springer-Verlag, (1988).

[Re] Reid, M.:*Decomposition of Toric Morphisms.* In: Arithmetic and Geometry, papers dedicated to I.R. Shavarevich, ed. M. Artin and J. Tate, pp. 395–418, Birkhäuser (1983).

[Ri] Riemenschneider, O.: *Deformationen von Quotientensingularitäten (nach zyklischen Gruppen).* Math. Ann. **209**, 211–248, (1974).

[St] Stevens, J.: *On the versal deformation of cyclic quotient singularities.* In: Singularity Theory and its Applications, Warwick 1989, Part I: Geometric Aspects of Singularities, pp. 302–319, Springer-Verlag Berlin Heidelberg, (LNM 1462), (1991).

Progress in Mathematics, Vol. 162, © 1998 Birkhäuser Verlag Basel/Switzerland

On Characteristic Cones, Clusters and Chains of Infinitely Near Points

Antonio Campillo *
Dep. de Algebra, Geometria y Topologia
Facultad de Ciencias
Prado de la Magdalena sn.
47005 Valladolid
SPAIN

Gérard González-Sprinberg
Institut Fourier
Université de Grenoble I
BP 74, 38402 St. Martin d'Hères
FRANCE

1 Introduction

Let $\pi : Z \to X$ be a projective birational morphism of smooth surfaces. Assume π is an isomorphism outside $\pi^{-1}(Q)$ for some closed point $Q \in X$. For each factorization $Z \to Y \to X$, where Y is a normal surface, one has that Y is the blowing up of a complete (i.e. integrally closed) ideal I on X with $I\mathcal{O}_Z$ invertible and support at Q. Since π is the composition of the successive blowing ups of a finite set $\mathcal{C}$ of infinitely near points (we will call $\mathcal{C}$ a constellation) to Q, the theorem of Zariski on unique factorization of complete ideals allow to describe all these sandwiched surfaces Y as well as the contractions $Z \to Y$. Such a contraction becomes the minimal resolution of the singularities of Y (a class of rational singularities called sandwiched).

Zariski asked the question of extending to higher dimensions the theory of complete ideals. Unique factorization is not true in general. Lipman in [7] has showed that one has unique factorization with integer (maybe negative) exponents if the morphism π is obtained by blowing up a constellation of i.n.p. to a smooth point in dimension $d \geq 3$. In this case complete ideals are said to be finitely supported and they are studied in [1] by means of clusters, i.e.

---

*Partially supported by DGICYT n. PB94-1111-C02-01

integer weighted constellations. For toric constellations, i.e. when X is a toric variety and the points of $\mathcal{C}$ are closed orbits, all the sandwiched varieties can be described in terms of $\mathcal{C}$. If, furthermore, the toric constellation is a chain, i.e. a sequence of i.n.p. such that each one is in the exceptional divisor created by its predecesor, then one has unique factorization with non negative exponents and Zariski's result and conclusions follow in the same way.

Cutkosky [2] approaches the general case (of projective birational morphisms $\pi : Z \to X$ where Z, X are normal) by means of the characteristic cone $\tilde{P}(Z/X)$, i.e. the cone in $N^1(Z/X) \otimes_{\mathbf{Z}} \mathbf{R}$ spanned by the classes of the π-generated line bundles. Since sandwiched varieties correspond one to one to topological cells of $\tilde{P}(Z/X)$ (a result which follows from Kleiman [3]), one can study factorizations properties of complete ideals from the cone structure of $\tilde{P}(Z/X)$ as, according to [6], complete ideals are nothing but sets of global sections of π-generated line bundles. In Zariski's situation the cone $\tilde{P}(Z/X)$ is regular polyhedral (a consequence of unique factorization). Cutkosky gives two examples showing that the cone $\tilde{P}(Z/X)$ could be non polyhedral and even non closed. Both examples are cases of non chain constellations with respective cardinalities 10 and 17. In the first case one has infinitely many sandwiched varieties.

Two natural questions arise. First, to give conditions in order that the number of sandwiched varieties is finite. Second, in the case of constellations which are chains, to investigate if Zariski result holds.

This paper deals with the above two questions. In Section 2 we prove that the number of sandwiched varieties is finite if the cone $NE(Z/X)$ of relative nef curves is (finite) polyhedral. For the case of constellations, we prove, in Section 3, that $NE(Z/X)$ is polyhedral if $NE(E_i)$ is so for every component E_i of the exceptional divisor of π. If $d = 3$, we derive that $NE(E_i)$ is polyhedral if the set of points blown up to create E_i from the projective plane is either toric or it has cardinality at most 8. Finally in Section 4, we show examples of chains proving that the characteristic cone can be polyhedral non simplicial, or regular with semigroup of complete ideals smaller than that of lattice points of $\tilde{P}(Z/X)$, or even non polyhedral and non closed (the cardinalities of involved constellations is at most 10). Thus, Zariski's results of dimension 2 fail in higher dimension even for chains.

2 Characteristic cones, complete ideals and sandwiched varieties

Consider a projective birational morphism $T \to S$ where T and S are normal algebraic varieties over an algebraically closed field k. Let $Q \in S$ be a closed point and set $R = \mathcal{O}_{S,Q}$, $X = \operatorname{Spec} R$, $Z = T \times_S X$ and $\pi : Z \to X$ the induced projective birational morphism. Denote by $N_1(Z/X)$ (resp. $N^1(Z/X)$) the abelian group of 1-dimensional cycles on Z whose support contracts to Q

(resp. Cartier divisors on Z) modulo numerical equivalence. Here a one dimensional cycle C (resp. Cartier divisor D) is numerically equivalent to 0 iff one has $C \cdot D = 0$ for all Cartier divisors D (resp. all complete curves contracted to Q) on Z.

Set $A_1(Z/X) = N_1(Z/X) \otimes_{\mathbf{Z}} \mathbf{R}$, $A^1(Z/X) = N^1(Z/X) \otimes_{\mathbf{Z}} \mathbf{R}$. Since $\pi^{-1}(Q)$ is a projective scheme over k and the vector space $A^1(Z/X)$ maps injectively into $A^1(\pi^{-1}(Q))$, the dimension $\rho(Z/X)$ of $A^1(Z/X)$ is finite and, therefore, the intersection pairing makes $A_1(Z/X)$ and $A^1(Z/X)$ dual vector spaces.

Let $NE(Z/X)$ be the convex cone in $A_1(Z/X)$ generated by the classes of the effective curves in Z which contract to Q. In the dual space $A^1(Z/X)$ we will consider two cones $P(Z/X)$ and $\tilde{P}(Z/X)$. $P(Z/X)$ is the dual cone of $-NE(Z/X)$, i.e. the cone consisting of the vectors l such that $c \cdot l \leq 0$ for every class c of a contracted effective curve in Z. In other words the cone $P(Z/X)$ is minus the semiample relative cone for the morphism π at Q (see [3]). $\tilde{P}(Z/X)$ is the convex cone generated by the classes of the Cartier semiample divisors D such that $\mathcal{O}_Z(-D)$ is generated by their global sections. The cone $\tilde{P}(Z/X)$ is called the characteristic cone for π at Q, the terminology being due to Hironaka.

Since a divisor whose ideal sheaf is generated by its global sections is numerically effective, one has $\tilde{P}(Z/X) \subset P(Z/X)$. On the other hand, according to [3], one has that the topological interior $P^\circ(Z/X)$ of $P(Z/X)$ is minus the relative ample cone (i.e., the convex cone generated by the classes of divisors such that $\mathcal{O}_Z(-D)$ is ample or, equivalently such that $C \cdot D < 0$ for every effective contracted to Q curve C on Z). Since some multiple of an ample divisor is generated by its global sections, it follows that $P^\circ(Z/X) \subset \tilde{P}(Z/X)$ and hence $P^\circ(Z/X) = \tilde{P}^\circ(Z/X)$.

The characteristic cone can also be seen as the convex cone generated by the Cartier divisors D such that $\mathcal{O}_Z(-D) = I\mathcal{O}_Z$ for some ideal $I \subset R$ (see [6]). Among the ideals I with the above property there is a largest one, namely the integral closure $\bar{I}$ of I. Integrally closed ideals are also called complete ideals since Zariski showed that they are the local analogous to complete linear systems. Thus, $\tilde{P}(Z/X)$ can be understood as the convex cone generated by the divisor classes corresponding (by the correspondence $I \mapsto D$ given by $I\mathcal{O}_Z = \mathcal{O}_Z(-D)$) to the complete ideals of R such that $I\mathcal{O}_Z$ is invertible. Notice that the set of such complete ideals is a semigroup for the $*$-operation given by $I * J = \overline{IJ}$ and that the above correspondence takes the operation $*$ to the summation of divisors. Denote this semigroup, or its additive image in $A^1(Z/X)$, by $S(Z/X)$. If $\tilde{S}(Z/X)$ is the additive semigroup of lattice points of $\tilde{P}(Z/X)$, i.e. $\tilde{S}(Z/X) = \tilde{P}(Z/X) \cap N^1(Z/X)$, one obviously has $S(Z/X) \subset \tilde{S}(Z/X)$ and both semigroups generate the cone $\tilde{P}(Z/X)$.

For a sandwiched variety we will mean a normal scheme for which π factorizes as a product of birational projective maps $Z \to Y \to X$. From Kleiman [3], it follows that there is a one to one correspondence between sandwiched varieties and topological cells of the characteristic cone (see [2, theorem 13]).

Moreover, for two sandwiched varieties Y and Y', the cell associated to Y' is included in the one associated to Y if and only if there is a birational morphism $Y' \to Y$. The correspondence works as follows: the relative interior of the cell associated to Y contains exactly the classes of those divisors corresponding to complete ideals I such that Y is the normalized blowing up of I.

For a convex cone K generating a vector space $\mathbf{R}^m$, the cells are defined in the following way. The only m-dimensional cell is K and, by descending induction, the other cells are those of the maximal convex cones contained in $K \setminus K^\circ$, where the upper index $^\circ$ means the relative interior. For the cones $P(Z/X)$ and $\tilde{P}(Z/X)$, which both generate $A^1(Z/X)$, the only $\rho(Z/X)$-dimensional relative interior of a cell is $P^\circ(Z/X)$. The cells are uniquely determined by their relative interior.

Theorem 2.1 *The inclusion of relative interiors of cells gives an injective map from the set of cells of $\tilde{P}(Z/X)$ into that of $P(Z/X)$.*

Proof. Let $\tilde{U}$ be a cell of $\tilde{P}(Z/X)$ with associated sandwiched variety Y. Then the morphism $Z \to Y$ contracts exactly those curves C such that $C \cdot D = 0$ where D is any divisor with class in $\tilde{U}^\circ$. Thus, all such divisors D have classes contained in the same cell U of $P(Z/X)$ and, therefore since those divisor classes generate $\tilde{U}^\circ$, one has $\tilde{U}^\circ \subset U^\circ$ and the map is well defined. Assume that for a second cell $\tilde{U}'$ of $\tilde{P}(Z/X)$ one has $\tilde{U}'^\circ \subset U^\circ$ and take complete ideals I, J with respective divisor classes in $\tilde{U}^\circ, \tilde{U}'^\circ$. Then the divisor class of $I * J$ is in $\tilde{U}''^\circ \subset \tilde{U}^\circ$ for a third cell $\tilde{U}''$.

For the associated sandwiched varieties Y, Y', Y'' one has birational morphisms $Y'' \to Y$ and $Y'' \to Y'$. Moreover, the curves of Z in $\pi^{-1}(Q)$ contracted in the three varieties Y, Y', Y'' are exactly the same. On the other hand, since $Y'' \neq Y$ and both Y'', Y are normal, there exists a complete curve C'' in Y'' which is contracted in Y. Take a curve C in the inverse image of C'' in Z dominating C'' (it always exists because we are dealing with algebraic varieties). Now, C is contracted in Y but it is not in Y'' which is a contradiction. This completes the proof. $\qquad\qquad\square$

Corollary 2.2 *If the cone $NE(Z/X)$ is polyhedral then the set of sandwiched varieties relative to π is finite.*

Proof. Since $-NE(Z/X)$ is polyhedral, its dual cone $P(Z/X)$ is also polyhedral, so it has finitely many cells and, hence, there are finitely many sandwiched varieties. $\qquad\qquad\square$

3   Cones and constellations of infinitely near points

From now on, we will consider the case in which X is smooth and π is the composition of a sequence of blowing ups at closed points. The semigroup $S(Z/X)$ is studied in [7] and [1]. We will use here the description in [1].

Assume $X = \operatorname{Spec} R$ is smooth and $\dim X = d \geq 2$. For a constellation of infinitely near points (i.n.p. in short) to Q we mean a set $\mathcal{C} = \{Q_0, Q_1, \ldots, Q_n\}$ where $Q_0 = Q$ and each Q_i is a closed point in the blown up variety of the variety containing Q_{i-1} with center at Q_{i-1} which maps to Q in X. Let $\pi : Z \to X$ be the composition of the successive blowing ups of the points of $\mathcal{C}$. Denote by B_i the exceptional divisor of the blowing up with center at Q_i, by E_i (resp. E_i^*) the strict (resp. total) transform of B_i in Z.

Both, the classes of $\{E_0, E_1, \ldots, E_n\}$ and those of $\{E_0^*, E_1^*, \ldots, E_n^*\}$ are basis of the lattice $N^1(Z/X)$. The basis change is given by

$$E_i = E_i^* - \sum_{j \to i} E_j^*$$

where $j \to i$ means that Q_j is proximate to Q_i, i.e. that Q_j belongs to the strict transform of B_i in the variety containing Q_j.

For each i, one has E_i dominates B_i and the restriction $\pi : E_i \to B_i$ is a map obtained by composition of the successive blowing ups at the points of the set $\mathcal{C}_i$ of proximate points to Q_i ($\mathcal{C}_i$ can be considered as union of $(d-1)$-dimensional constellations). Since $Pic(Z/X) \to Pic(E_i)$ is surjective, one has an injective linear map $A_1(E_i) \to A_1(Z/X)$, where $A_1(E_i) = N_1(E_i) \otimes_{\mathbf{Z}} \mathbf{R}$ and $N_1(E_i)$ is the group of 1-cycles modulo numerical equivalence on E_i. The cone $NE(E_i)$ generated by the classes in $N_1(E_i)$ of effective curves on E_i is mapped, by the above linear map, into the cone $NE(Z/X)$. It is clear that $NE(Z/X)$ is nothing but the convex sum of the images of the cones $NE(E_i)$ in $A_1(Z/X)$.

Proposition 3.1 *If $NE(E_i)$ is a polyhedral cone for every i, then the number of sandwiched varieties relative to π is finite.*

Proof. The convex sum $NE(Z/X)$ is polyhedral, so the result follows from Corollary 2.2. $\qquad\square$

For each index i, $NE(E_i)$ is a polyhedral cone in each of the two following cases. First, if the set $\mathcal{C}_i$ is toric, i.e. if it consists of i.n.p. which are 0-dimensional orbits of some structure of toric variety on the projective space $B_i \cong \mathbf{P}^{d-1}$, then $NE(E_i)$ is the cone generated by 1-dimensional orbits, hence it is rational polyhedral (see [5], [1]). Second, one can apply Kawamata's theorem [4] which guarantees that $NE(E_i)$ is rational polyhedral if it is contained in the half space $c \cdot K_{E_i} < 0$ where K_{E_i} is the class of the canonical divisor of E_i, i.e. if the anticanonical bundle of E_i is ample. Thus one obtains the following result.

Corollary 3.2 *If for each i one has either $\mathcal{C}_i$ is toric or the anticanonic bundle of E_i is ample, then the number of sandwiched varieties relative to π is finite.*

In particular, if the whole constellation $\mathcal{C}$ is toric (i.e. if Q_0 is a closed orbit of a toric structure on the affine d-dimensional space and all the i.n.p. Q_i are also closed orbits for the derived toric structures on the blow up spaces) the cones $NE(Z/X)$ and $P(Z/X)$ are rational polyhedral. Moreover, as shown in [1], in this case one has $P(Z/X) = \tilde{P}(Z/X)$, $S(Z/X) = P(Z/X) \cap N^1(Z/X)$, and the extremal rays in $NE(Z/X)$ can be described explicitly in terms of the combinatorics of the constellation. Thus, one can characterize [1, 2.20] those toric constellations for which the cone $NE(Z/X)$ is simplicial. One sees that in this case $NE(Z/X)$, and so also $P(Z/X) = \tilde{P}(Z/X)$, is a regular cone and $S(Z/X)$ is a free semigroup. This characterization includes the case of chains, i.e. constellations such that $Q_{i+1} \in B_i$ for each $i \geq 0$. Later on, we will show with some examples that these results are not true in general for non toric chains.

One can use Kawamata's theorem with some weaker assumptions than in Corollary 3.2. For fixed i and $j \to i$, denote by $E_{ij} = E_i \cap E_j$, $E_{ij}^* = E_i \cap E_j^*$ (here $\cap$ means the cycle given by the proper intersection). The canonical divisor of E_i is given by $-dH_i^* + (d-2) \sum_{j \to i} E_{ij}^*$, where H_i^* is the total transform in E_i of a general hyperplane in B_i. Assume that the linear system of effective divisors F' on B_i such that $\pi_i^* F' \geq (d-2) \sum_{j \to i} E_{ij}^*$ has a base point set $\mathcal{S}_i$ of dimension at most one. Then, if $C' \subset B_i$ is an irreducible curve not contained in $supp(\mathcal{S}_i)$, it follows from Bézout's theorem (applied to C' and some convenient member of the above linear system) that the class c in $N_1(E_i)$ of the strict transform of C' in E_i satisfies $c \cdot K_{E_i} \leq 0$. This means that $NE(E_i)$ is generated by the curves in the region $c \cdot K_{E_i} \leq 0$, the classes of the curves in $supp(\mathcal{S}_i)$ and the exceptional curves in the region $c \cdot K_{E_i} > 0$ (exceptional means contracted by π_i). Kawamata's theorem gives information on the intersection of $N_1(E_i)$ with the region $c \cdot K_{E_i} < 0$ (the set of extremal rays in this region is discrete). Thus, since the classes of the exceptional curves will appear also in others $NE(E_j)$, it is possible to know the contribution of $NE(E_i)$ to $NE(Z/X)$ if one controls the curves with class in the hyperplane $c \cdot K_{E_i} = 0$.

We will precise the above situation for $d = 3$. Assume that the linear system $\mathcal{F}_i$ of curves F' in $B_i \cong \mathbf{P}^2$ with $\pi_i^* F' \geq \sum_{j \to i} E_{ij}^*$ is non-empty (notice that this is always true if card $(\mathcal{C}_i) \leq 9$), i.e. one has $\dim(\mathcal{S}_i) \leq 1$. Thus $NE(E_i)$ is generated by its intersection with $c \cdot K_{E_i} \leq 0$ and finitely many more classes of curves, namely those in $supp(\mathcal{S}_i)$ and the exceptional ones. Furthermore, if the linear system F_i contains a pencil (e.g. if card $(\mathcal{C}_i) \leq 8$), then $NE(E_i)$ is generated by its intersection with $c \cdot K_{E_i} < 0$ and finitely many more classes, namely those of the exceptional curves and those of the irreducible components of the members of the pencil. Notice that this set of classes is finite as all the general curves of the pencil have the same class in $N_1(E_i)$ and, hence, this also happens for the classes of their irreducible components.

Now, the extremal rays of $N_1(E_i)$ in the region $c \cdot K_{E_i} < 0$ are those corresponding to irreducible curves $C \subset E_i$ which can be contracted on a smooth surface, i.e. those irreducible curves such that $p_a(C) = 0$ and $C \cdot C = -1$, or equivalently $C \cdot C = C \cdot K_{E_i} = -1$. Consider on the lattice $N_1(E_i)$ the basis given by the classes of the cycles $-\tilde{H}_i^*$ and $-E_{ij}^*$ for $j \to i$. Thus, if the class of C has coordinates $(-n, \{e_j\}_{j \to i})$ in the above basis, then the conditions $C \cdot C = C \cdot K_{E_i} = -1$ are written in the following way

$$\sum_{j \to i} e_j^2 = n^2 + 1, \quad \sum_{j \to i} e_j = 3n - 1.$$

Notice that, if the irreducible curve C is not exceptional then n is the degree of its image C' in B_i and e_j is the multiplicity at Q_j of the strict transform of C'. If C is exceptional, then $n = 0$ and C, being irreducible should be one of the curves E_{ij} with j maximal (i.e. such that there is no index l with $l \to i$ and $l \to j$).

Lemma 3.3 *With notations as above, keep the assumption $d = 3$. For each i denote by $\mathcal{R}_i$ the set of rays in $NE(E_i)$ which are either extremal for $NE(E_i)$ in the region $c \cdot K_{E_i} < 0$ or generated by classes of irreducible curves in the hyperplane $c \cdot K_{E_i} = 0$. Then one has:*

(i) If $\operatorname{card}(\mathcal{C}_i) \le 8$ the set $\mathcal{R}_i$ is finite.

(ii) If $\operatorname{card}(\mathcal{C}_i) = 9$ the set $\mathcal{R}_i$ has at most one limit point, namely the ray generated by the class of $C_0 = 3H_i^* - \sum_{j \to i} E_{ij}^*$.

Proof. Any ray in $\mathcal{R}_i$ is generated by a vector of coordinates $(-n, \{e_j\}_{j \to i})$ where either $\sum_{j \to i} e_j = 3n - 1$ and $\sum_{j \to i} e_j^2 = n^2 + 1$ (extremal rays in $c \cdot K_{E_i} < 0$) or $\sum_{j \to i} e_j = 3n$ and $\sum_{j \to i} e_j^2 = n^2 + 2$ (classes of curves with $C \cdot K_{E_i} = 0$ and $p_a(C) \ge 0$). Since for any value of n there are only finitely many possible values of $\{e_j\}_{j \to i}$ fitting in one of two above arithmetical situations, any limit ray of $\mathcal{R}_i$ should be a limit of rays generated by vectors as above with $n \to \infty$. Such a limit is generated by a vector of type $(-1, \{x_j\}_{j \to i})$ with $x_j \ge 0$, $\sum_{j \to i} x_j = 3$ and $\sum_{j \to i} x_j^2 = 1$. Now, if $h = \operatorname{card}(\mathcal{C}_i)$, the h-variable function $\sum_{j \to i} x_j^2$ has an absolute minimum at $x_j = 3/h$ for every $j \to i$, the minimum value being $9/h$. Thus, if $h \le 8$ the equality $\sum_{j \to i} x_j^2 = 1$ is impossible and therefore the set $\mathcal{R}_i$ is discrete and hence finite. If $h = 9$, the equality $\sum_{j \to i} x_j^2 = 1$ implies that $x_j = 1/3$ for each $j \to i$, so the ray generated by C_0 is the only possible limit point of $\mathcal{R}_i$. $\qquad\square$

Theorem 3.4 *Let Q be a smooth closed point of a 3-dimensional variety and $\pi : Z \to X$ a morphism obtained by blowing up a constellation C of i.n.p. to Q. Assume that for every $Q_i \in C$ either Q_i is toric or $\operatorname{card}(\mathcal{C}_i) \le 8$. Then The cone $NE(Z/X)$ is polyhedral and the number of sandwiched varieties relative to π is finite.*

Proof. If $\mathcal{C}_i$ is toric, the cone $NE(E_i)$ is polyhedral. If $\operatorname{card}(\mathcal{C}_i) \le 8$, then by Lemma 3.3 $NE(E_i)$ is generated by the finite set $\mathcal{R}_i$ and finitely many other curves (the linear system F_i contains a pencil in this case), so the cone $NE(E_i)$ is also polyhedral. Thus $NE(Z/X)$ is also a polyhedral cone and, hence, by Corollary 2.2 the set of sandwiched varieties is finite. $\qquad\square$

Remark. If card $(\mathcal{C}_i) = 9$, the cone $NE(E_i)$ could be non polyhedral as shown, for instance, in example 1 [2, p. 37] when $\mathcal{C} = \{Q_0, Q_1, \ldots, Q_9\}$ and $Q_1, \ldots, Q_9$ are the intersection points of two general cubics in B_0. The method to discuss the examples in next section shows us how in practice, even for nine points, in many cases, one can decide if the cone $NE(E_i)$ is polyhedral or not.

4 Clusters and chains of infinitely near points

Let $\pi : Z \to X = \mathrm{Spec}\,(R)$ the morphism obtained by blowing up a constellation $\mathcal{C}$ of i.n.p. to the smooth closed point $Q \in \mathrm{Spec}\,(R)$. The classes of the divisors E_i^* are a basis for the lattice $N^1(Z/X) = Pic(Z/X)$. Thus to give a relative divisor $D = \sum m_i E_i^*$ is equivalent to give an integer weight on the points of $\mathcal{C}$ by assigning to Q_i the weight m_i. Such a weighted constellation is called a cluster. A cluster is called idealistic if the divisor D comes from a complete ideal I such that $I\mathcal{O}_Z$ is invertible, i.e. if it belongs to the semigroup $S(Z/X)$. Thus, $S(Z/X)$ can be considered as the additive semigroup of the idealistic clusters and $\tilde{P}(Z/X)$ as the cone generated by those clusters. The cone $P(Z/X)$ is given by the so called proximity inequalities (see [1]), i.e., for each irreducible exceptional curve C and i the only index such that $C \subset E_i$ and its image C' in B_i is not a point, the inequality

$$\deg(C')m_i \geq \sum_{j \to i} e_j(C')m_j \,,$$

where $\deg(C')$ is the degree of C' in the projective space B_i and $e_j(C')$ the multiplicity at Q_j of the strict transform of C. Corollary 3.2 and Theorem 3.4 give conditions under which the cone $P(Z/X)$ is given by finitely many proximity inequalities.

A classic result by Zariski, which has given rise to the theory of complete ideals, asserts that if $d = 2$ the cone $\tilde{P}(Z/X)$ is polyhedral regular and that one has $\tilde{P}(Z/X) = P(Z/X)$ and $S(Z/X) = P(Z/X) \cap N^1(Z/X)$. This follows from the obvious fact that $NE(Z/X)$ is the regular cone generated by the classes of the curves E_i and the fact that any cluster satisfying the proximity inequalities is idealistic. In [1] it is shown that the same is true if $d \geq 3$ and the constellation is toric and it is a chain. By a chain we mean that $\mathcal{C} = \{Q_0, Q_1, \ldots, Q_n\}$ and $Q_{i+1} \in B_i$ for every $i \geq 0$. The example quoted in Remark 3 shows that $\tilde{P}(Z/X)$ could be non polyhedral for a suitable constellation and therefore $S(Z/X)$ is not a finitely generated semigroup. Even $\tilde{P}(Z/X)$ could be non closed, and hence $\tilde{P}(Z/X) \neq P(Z/X)$ as shown in example 3 in [2, p. 37], where $\mathcal{C} = \{Q_0, Q_1, \ldots, Q_{16}\}$ the sixteen last points being in general position in B_0.

The result of Zariski in dimension two implies that, in this case, the semigroup $S(Z/X)$ is free, i.e. that one has unique factorization with non negative exponents in terms of the irreducible elements. Zariski proposed to extend to higher dimensions this kind of results. The discussion in terms of the structure of the various cones can provide several types of generalizations of the Zariski's above factorization property. Thus, to be $\tilde{P}(Z/X)$ a simplicial cone means that one has semiunique factorization, i.e. unique factorization with rational exponents in terms of the primitive extremal vectors of $S(Z/X)$. To be $\tilde{P}(Z/X)$ polyhedral but not simplicial means non unique semifactorization and $\tilde{P}(Z/X) \neq P(Z/X)$ or $P(Z/X)$ non polyhedral which means non unique semifactorization in terms of infinitely many primitive extremals. Lipman in [7] showed that for any constellation, the semigroup $S(Z/X)$ contains a concrete lattice basis of $N^1(Z/X)$, so that in terms of the basis one has unique factorization with integral exponents.

A natural question is to ask if Zariski's result is true for constellations which are chains. Next examples show that this question has a negative answer. For the all three examples we assume $d = 3$.

Example 4.1 Consider the chain $\mathcal{C} = \{Q_0, Q_1, \ldots, Q_5\}$ where $Q_1, \ldots, Q_5$ are five consecutive points on a smooth conic G in B_0, i.e. $Q_{i+1} \in B_i$ for $i \geq 0$ and Q_1 is on G and Q_i on the strict transform of G for $i \geq 2$. In particular $i \to 0$ for $i \geq 1$, the embedding $N_1(E_0) \to N_1(Z/X)$ is an isomorphism and it takes the cone $NE(E_0)$ to $NE(Z/X)$. Take the basis $\{-H_0^*, -E_{01}^*, \ldots, -E_{05}^*\}$ on $N_1(E_0)$ and represent the vectors in $A_1(E_0)$ by their 6-uple of coordinates.

The linear system $\mathcal{F}_0$ contains the pencil generated by the cubics $G + L$, $G + L'$, where L, L' are generic lines in B_0. Thus after the comments in Section 3, the cone $NE(E_0)$ is generated by the class $g = (-2, 1, 1, 1, 1, 1)$, the exceptional classes $f_1 = (0, -1, 1, 0, 0, 0)$, $f_2 = (0, 0, -1, 1, 0, 0)$, $f_3 = (0, 0, 0, -1, 1, 0)$, $f_4 = (0, 0, 0, 0, -1, 1)$, $f_5 = (0, 0, 0, 0, 0, -1)$ and the vectors $(-n, e_1, \ldots, e_5)$ with $\sum_{i=1}^5 e_i^2 = n^2 + 1$, $\sum_{i=1}^5 e_i = 3n - 1$ and $n > 0$ (all these vectors are classes of effective curves in E_0, may be non irreducible ones, as the number of imposed conditions by the multiplicities e_i is $(1/2) \sum e_i(e_i + 1)$ which is one unit less than the dimension of the space of n forms). Since $\operatorname{card}(\mathcal{C}_0) = 5$, the only possibilities for these vectors are $l = (-1, 1, 1, 0, 0, 0)$ and $g = (-2, 1, 1, 1, 1, 1)$.

Thus $NE(E_0)$ is generated by the seven vectors $l, f_1, \ldots, f_5, g$ and it is not a simplicial cone. The dual cone $P(Z/X)$ of $-NE(Z/X)$ is given by the following proximity inequalities $m_0 \geq m_1 + m_2$, $2m_0 \geq m_1 + \cdots + m_5$, $m_1 \geq m_2 \geq m_3 \geq m_4 \geq m_5 \geq 0$. By looking to solutions with equality at least in 5 of the above inequalities one obtains the following 9 extremal vectors: $(1, 0, 0, 0, 0, 0)$, $(1, 1, 0, 0, 0, 0)$, $(2, 1, 1, 0, 0, 0)$, $(2, 1, 1, 1, 0, 0)$, $(2, 1, 1, 1, 1, 0)$, $(3, 2, 1, 1, 1, 1)$, $(4, 2, 2, 2, 1, 1)$, $(5, 2, 2, 2, 2, 2)$, $(6, 3, 3, 2, 2, 2)$, the six first ones being the Lipman basis. Since in this case one has $\tilde{P}(Z/X) = P(Z/X)$ one has non unique semifactorization: The cone $\tilde{P}(Z/X)$ is not simplicial, hence it is not regular.

Example 4.2 Consider the chain $\mathcal{C} = \{Q_0, Q_1, \ldots, Q_9\}$ where $Q_1, \ldots, Q_9$ are consecutive points on an inflection point Q_1 of a rational cubic C_0 in B_0. As above one has $NE(E_0) = NE(Z/X)$ and the vectors $A_1(E_0)$ can be represented by a 10-uple of coordinates.

The linear system $\mathcal{F}_0$ contains the pencil generated by $\mathcal{C}_0$ and $3L'$ where L' is the tangent line to $\mathcal{C}_0$ at Q_1. From Section 3, $NE(E_0)$ is generated by the class $l = (-1, 1, 1, 1, 0, 0, 0, 0, 0, 0)$ and the exceptional classes $f_1, \ldots, f_9$ as above (i.e., f_i has -1 as i-th entry, 1 as $(i+1)$-th entry for $i \leq 8$ and 0 as entry otherwise). In fact, notice that the class of the transform of $\mathcal{C}_0$ and those of the effective curves with $C \cdot C = C \cdot K_{E_0} = -1$ are in the cone generated by $l, f_1, \ldots, f_9$ (since $C \cdot (C - K_{E_0}) = 0$ and $L \cdot (C - K_{E_0}) < 0$, where L is the strict transform of L', it follows from Bézout theorem that L should be a component of C, so C is not irreducible).

One has $\tilde{P}(Z/X) = P(Z/X)$ so $\tilde{P}(Z/X)$ is a regular cone. If Q_1 is the origin of the curve $y = x^3$, then there is no cubic having intersection multiplicity 8 with $\mathcal{C}_0$ at Q_1, so the cluster with weight $m = (3, 1, 1, 1, 1, 1, 1, 1, 1, 0)$ satisfies the proximity relations but it is not idealistic (otherwise the tangent cone of the hypersurface given by a general element of the ideal would achieve the intersection multiplicity 8). Thus one has $S(Z/X) \neq \tilde{S}(Z/X)$. One has unique semifactorization and the Lipman basis contains the vector $(4, 1, 1, 1, 1, 1, 1, 1, 1, 0)$, so in Lipman factorization there are clusters (for instance the cluster with weight $2m$) with negative exponents. The semigroup $S(Z/X)$ has more than 10 irreducible elements, so if one wants non negative integral coefficients one has non unique factorization.

Finally, if we consider only 8 points Q_i instead of 9, one gets an alternative example with identical characteristics.

Example 4.3 Consider the chain $\mathcal{C} = \{Q_0, Q_1, \ldots, Q_9\}$ where $Q_1, \ldots, Q_9$ are consecutive points on a non inflection smooth point Q_1 of a rational cubic $\mathcal{C}_0$ in B_0. Take, for instance, Q_1 the origin of $y = x^2 + x^3$. Since the only irreducible curve A in B_0 having intersection multiplicity with $\mathcal{C}_0$ greater or equal than $3 \deg(A)$ is the same curve $\mathcal{C}_0$, it follows that $NE(E_0)$ is included in $c \cdot K_{E_0} \leq 0$ and its intersection with $c \cdot K_{E_0} = 0$ is the cone generated by the classes $c_0 = (-3, 1, 1, 1, 1, 1, 1, 1, 1, 1)$ and $f_1, \ldots, f_8$ (as in Example 4.2). Thus $NE(E_0)$ is generated besides $c_0, f_1, \ldots, f_8$ by f_9 and those $c = (-n, e_1, \ldots, e_9)$ such that $n > 0$, $\sum_{i=1}^{9} e_i^2 = n^2 + 1$, $\sum_{i=1}^{9} e_i = 3n - 1$ and $e_1 \geq e_2 \geq \cdots \geq e_9 \geq 0$. One can see that each such a c is the class of an irreducible curve, so those c are extremal vectors for $NE(E_0)$. Finally, notice that there are infinitely many values of c (take, for instance the sequence $(-(3t^2 + 3), t^2 + t, t^2 + 2, t^2 + 1, t^2 + 1, t^2 + 1, t^2 + 1, t^2 + 1, t^2 + 1, t^2 - t))$, so $NE(E_0)$ is not a polyhedral cone. Thus the dual cone $P(Z/X)$ is also not a polyhedral cone. Moreover, since $NE(E_0)$ is included in $c \cdot K_{E_0} \leq 0$, the cluster with weight $m = (3, 1, 1, 1, 1, 1, 1, 1, 1, 1)$ satisfies the proximity inequalities but sm is not an idealistic cluster for every $s \geq 1$ (otherwise the tangent cone to a general element of the ideal will be the curve of degree $3s$ with intersection multiplicity $9s$ with $\mathcal{C}_0$ at Q_1 and not containing $\mathcal{C}_0$ in its support). It follows that one has $P(Z/X) \neq \tilde{P}(Z/X)$ and $S(Z/X)$ is a non finitely generated semigroup.

Finally, we remark that this is an example of non closed characteristic cone obtained by blowing up only ten points (in a chain).

References

[1] A. Campillo, G. González-Sprinberg, M. Lejeune-Jalabert: *Clusters of infinitely near points.* Math. Annalen 306(1), 169–194 (1996).

[2] S.D. Cutkosky: *Complete ideals in Algebra and Geometry.* Contemporary Math. Vol 159, 27–39 (1994).

[3] S. Kleiman: *Toward a numerical theory of ampleness.* Ann. Math. 84, 293–344 (1966).

[4] Y. Kawamata. *The cone of curves of algebraic varieties.* Ann. Math. 119, 603–623 (1984).

[5] G. Kempf, F. Knudsen, D. Mumford and B. Saint Donat. *Toroidal embeddings.* Lect. Notes in Math. 339. Springer-Verlag (1973).

[6] J. Lipman: *Rational singularities with applications to algebraic surfaces and unique factorization.* Publ. Math. I.H.E.S. 36, 195–279 (1969).

[7] J. Lipman: *On complete ideals in regular local rings.* Algebraic Geometry and Commutative Algebra in Honour to Nagata, 203–231 (1987).

Progress in Mathematics, Vol. 162, © 1998 Birkhäuser Verlag Basel/Switzerland

On Kleinian Singularities and Quivers

Heiko Cassens
Reuters A.G.
Graf-Adolf-Str. 35–37
40210 Düsseldorf
GERMANY

Peter Slodowy
Mathematisches Seminar
Universität Hamburg
Bundesstraße 55
20146 Hamburg
GERMANY

Introduction

Starting from McKay's observation on the description of (an essential part of) the representation theory of binary polyhedral groups Γ in terms of extended Coxeter-Dynkin-Witt diagrams $\underline{\tilde{\Delta}}(\Gamma)$ and working in the differential geometric framework of Hyper-Kähler-quotients P.B. Kronheimer was able to give a new construction of the semiuniversal deformations of the Kleinian singularities $X = \mathbb{C}^2/\Gamma$ as well as of their simultaneous resolutions ([24], [25], [26]). As far as the deformations were concerned, he already gave a purely algebraic geometric formulation of his results in terms of representations of certain quivers naturally attached to the diagrams $\underline{\tilde{\Delta}}(\Gamma)$. By making use of the invariant-theoretic notion of "linear modification" (cf. Section 6, below) and applying it to Kronheimer's quiver construction we show here how to obtain a purely algebraic geometric simultaneous resolution as well (Section 7). On the way, we shall take the opportunity to remind the reader of various facts about

Kleinian singularities	(Section 1),
McKay's observation	(Section 2),
Symplectic geometry	(Section 3),
Kronheimer's work	(Section 4), and
Quivers	(Section 5).

This article covers the main results of the doctoral dissertation [9] written at Hamburg university under the guidance of the second named author and supported by a DFG-grant (Ri 303/3-2). More details and worked out examples may be found there.

1   Reminder on Kleinian singularities (cf. e.g. [37])

Let $\Gamma \subset SU(2) \subset SL_2(\mathbb{C})$ denote a finite subgroup. Up to conjugacy there are five classes of such groups:

- $\mathcal{C}_n$,   cyclic of order $n, n \geq 2$,
- $\mathcal{D}_n$,   binary dihedral of order $4n, n \geq 2$,
- $\mathcal{T}$,   binary tetrahedral of order 24,
- $\mathcal{O}$,   binary octahedral of order 48,
- $\mathcal{J}$,   binary icosahedral of order 120

The *Kleinian singularity* attached to Γ ist the quotient singularity $S = \mathbb{C}^2/\Gamma$, which may be viewed (by the invariant theory of Γ, F. Klein, [22]) as a hypersurface in $\mathbb{C}^3$

$$S = \{(x, y, z) \in \mathbb{C}^3 \mid R(x, y, z) = 0\}$$

with an isolated singularity at 0. Here, R is the relation between three fundamental generators of the invariant ring $\mathbb{C}[u, v]^\Gamma$ of $\mathbb{C}^2$:

Γ	$\mathcal{C}_n$	$\mathcal{D}_n$	$\mathcal{T}$	$\mathcal{O}$	$\mathcal{J}$
R	$X^n + YZ$	$X(Y^2 - X^n) + Z^2$	$X^4 + Y^3 + Z^2$	$X^3 + XY^3 + Z^2$	$X^5 + Y^3 + Z^2$

The minimal resolution $\pi : \tilde{S} \to S$ of S had essentially been studied by Du Val [11]. He obtained an exceptional fibre $\pi^{-1}(0)$ of the form

$$\pi^{-1}(0) = C_1 \cup \cdots \cup C_r \quad,$$

where the C_i are smooth rational curves with self-intersection $C_i \cdot C_i = -2$ and pairwise transversal intersection according dually to a Coxeter-Dynkin-Witt diagram $\underline{\Delta} = \underline{\Delta}(\Gamma)$ of type A, D, E:

Γ	$\mathcal{C}_n$	$\mathcal{D}_n$	$\mathcal{T}$	$\mathcal{O}$	$\mathcal{J}$
$\underline{\Delta}$	A_{n-1}	D_{n+2}	E_6	E_7	E_8

Another way to describe this information is as follows. The homology group $H_2(\tilde{S}, \mathbb{Z})$ is freely generated by the classes of the irreducible exceptional components

$$H_2(\tilde{S}, \mathbb{Z}) = \bigoplus_{i=1}^{r} \mathbb{Z}[C_i] \quad .$$

This group is equipped with an intersection product $\langle \ , \ \rangle$ and the lattice $(H_2(\tilde{S}, \mathbb{Z}), -\langle \ , \ \rangle)$ identifies with the root lattice Q of type $\underline{\Delta}(\Gamma)$ equipped with the (normalized) Killing form, the classes $[C_i]$ corresponding to a system of simple roots (cf. [4] for the basic definitions).

Deformations and resolutions of deformations of Kleinian singularities were intensively studied in the period 1966 - 1970 by Brieskorn, Grothendieck, and Tjurina, cf. [5], [6], [7], [39]. The semiuniversal deformation of S may be easily constructed

$$\begin{array}{ccc} S & \hookrightarrow & \mathcal{X} \\ \downarrow & & \downarrow \chi \\ 0 & \in & \mathcal{U} \end{array}.$$

Here, $\mathcal{U}$ is a smooth r-dimensional space ($r =$ number of exceptional components of $\pi : \tilde{S} \to S =$ rank of the corresponding root system of type $\underline{\Delta}\,(\Gamma)$) which may be identified, quite naturally, with the quotient $\mathbf{h}/W$ of a Cartan subalgebra $\mathbf{h}$ in a simple complex Lie algebra $\mathbf{g}$ of type $\underline{\Delta}\,(\Gamma)$ by the action of the finite Weyl group W. Under this identification the discriminant locus of χ, $\{u \in \mathcal{U} \mid \mathcal{X}_u = \chi^{-1}\,(u)$ is singular $\}$, corresponds to the ramification locus of the ramified covering $\mathbf{h} \to \mathbf{h}/W$ in a "type-preserving" way:

$$\left(\begin{array}{c} \text{types of singularities} \\ \text{in the fibre } \mathcal{X}_u \end{array} \right) = \left(\begin{array}{c} \text{types of irreducible Weyl} \\ \text{group factors in the isotropy} \\ \text{group } W_{\tilde{u}} \end{array} \right)$$

$$(\tilde{u} \text{ a preimage of } u \in \mathcal{U} = \mathbf{h}/W \text{ in } \tilde{U} = \mathbf{h}).$$

A simultaneous resolution of χ can be obtained after pull-back by $\mathbf{h} \to \mathbf{h}/W$:

$$\begin{array}{ccccc} \mathcal{X} & \leftarrow & \mathcal{X} \times_{\mathcal{U}} \tilde{\mathcal{U}} & \leftarrow & \mathcal{Y} \\ \chi \downarrow & & \downarrow & & \downarrow \ominus \\ \mathcal{U} & \leftarrow & \tilde{\mathcal{U}} & = & \tilde{\mathcal{U}} \\ \parallel & & \parallel & & \parallel \\ \mathbf{h}/W & & \mathbf{h} & & \mathbf{h} \end{array}$$

Here, every fibre $\ominus^{-1}\,(\tilde{u})$ of $\ominus$ is a minimal resolution of the corresponding fibre $\chi^{-1}\,(u)$ of χ, $\tilde{\mathcal{U}} \ni \tilde{u} \mapsto u \in \mathcal{U}$.

In Brieskorn's theory ([7]) relating Kleinian singularities to simple Lie algebras, the above diagram is induced by the following one

$$\begin{array}{ccccc} \mathbf{g} & \leftarrow & \mathbf{g} \times_{\mathbf{h}/W} \mathbf{h} & \leftarrow & G \times^B \mathbf{b} \\ \downarrow & & \downarrow & & \downarrow \\ \mathbf{h}/W & \leftarrow & \mathbf{h} & = & \mathbf{h} \end{array}$$

where G is the adjoint group $\mathrm{Aut}^0\,(\mathbf{g})$ of $\mathbf{g}$ and B a Borel subgroup with Lie algebra $\mathbf{b}$ containing $\mathbf{h}$.

2   McKay's observation

Whereas the construction of Kleinian singularities started in a uniform way from the finite subgroups $\check{\Gamma} \subset SL_2(\mathbb{C})$, this group didn't play any role in [7]. A key step in the re-introduction of Γ is McKay's observation of 1979 (cf. e.g. [14], [31], [37]).

Let $R_0, R_1, \ldots, R_r$ denote the irreducible complex representations of Γ, R_0 the trivial one and N the "natural" one obtained from the inclusion $\Gamma \subset SL_2(\mathbb{C}) \subset GL_2(\mathbb{C})$. (In the following, we tacitly identify these representations with their equivalence classes, i.e. their characters.) Then there exists a bijection

$$\check{\Gamma} = \{R_0, \ldots, R_r\} \quad \longleftrightarrow \quad \text{(vertices of the extended diagram } \underline{\tilde{\Delta}}(\Gamma))$$

$$R_i \quad \longleftrightarrow \quad \overset{i}{\bullet}$$

$$\text{such that} \quad \mathrm{Hom}_\Gamma(R_i, R_j \otimes N) = \begin{cases} 0 \\ \mathbb{C} \\ \mathbb{C}^2 \end{cases} \iff \quad \begin{matrix} \overset{i}{\bullet} \ \ \overset{j}{\bullet} \\[4pt] \overset{i}{\bullet}\!\!-\!\!\overset{j}{\bullet} \\[4pt] \overset{i}{\bullet}\!\!=\!\!\overset{j}{\bullet} \end{matrix}$$

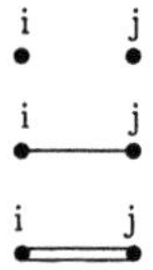

i.e. R_i occurs with multiplicity m in $R_j \otimes N$ exactly when the nodes i and j are connected by m edges.

Examples.

$$\Gamma = \mathcal{C}_2 \quad , \quad \underline{\tilde{\Delta}}(\Gamma) = \tilde{A}_1$$

$$\Gamma = \mathcal{C}_3 \quad , \quad \underline{\tilde{\Delta}}(\Gamma) = \tilde{A}_2$$

$$\Gamma = \mathcal{J} \quad , \quad \underline{\tilde{\Delta}}(\Gamma) = \tilde{E}_8$$

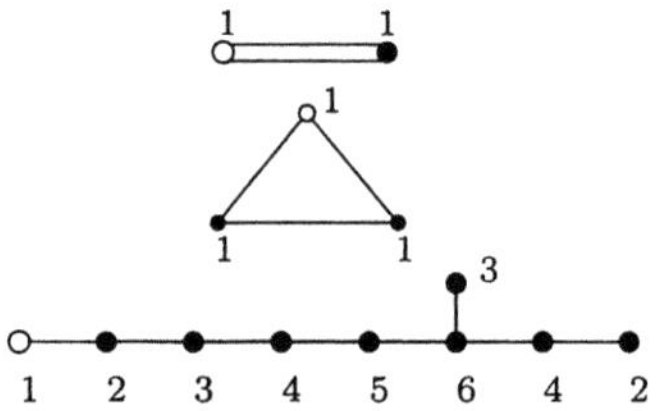

The attached numbers are the dimensions (degrees) $d_i = \dim_\mathbb{C} R_i$. The trivial representation can always be chosen to correspond to the extra node of $\underline{\tilde{\Delta}}(\Gamma)$. The case $\Gamma = \mathcal{C}_2$ is the only one with multiplicities $\neq 0, 1$.

We may rephrase McKay's observation in many ways:

Let $m_{ij} := \dim_\mathbb{C} \mathrm{Hom}_\Gamma(R_i, R_j \otimes N)$ and $c_{ij} := 2\,\delta_{ij} - m_{ij}$ $(i, j = 0, \ldots, r)$. Then $C = ((c_{ij}))$ is an extended Cartan matrix of type $\underline{\tilde{\Delta}}(\Gamma)$. Or, if

$$R(\Gamma) = \bigoplus_{i=0}^{r} \mathbb{Z}\, R_i$$

denotes the integral representation ring of Γ with standard scalar product $\langle R_i, R_j \rangle = \delta_{ij}$, then $(R(\Gamma), (\ ,\ ))$ with

$$(R, R') := 2\,\langle R, R' \rangle - \langle R, R' \otimes N \rangle = \langle R, R' \otimes (2\,R_0 - N) \rangle$$

may be identified with the root lattice of an affine root system of type $\tilde{\Delta}(\Gamma)$, $\{R_0, \dots, R_r\}$ corresponding to a system of simple roots and $R = \bigoplus_{i=0}^{r} d_i R_i$ (the regular representation) corresponding to a minimal isotropic vector (imaginary root), cf. e.g. [19] Chap. 6.

A direct relation of McKay's observation to the resolution of Kleinian singularities was found by Gonzales-Sprinberg and Verdier ([16]). Their result was developed later in many directions (cf. [23], [3], [12], [13]).

3 Symplectic geometry and momentum maps

Kronheimer's work makes serious use of ideas and results from symplectic geometry. In line with our later applications we shall deal with these matters directly in the framework of complex analytic or algebraic geometry. (We might in fact choose to work over an arbitrary field with some mild restrictions on its characteristic. In the literature one usually finds treated the real differentiable case, cf. e.g. [1], [2], [17], $\dots$.)

Let M be a complex analytic (resp. algebraic) manifold. A *symplectic form* ω on M is a holomorphic (resp. algebraic) 2-form on M, i.e. $\omega \in \Omega^2(M) = \Gamma(M, \wedge^2 T^* M)$, such that

- ω is closed, $d\omega = 0$,

- ω is nondegenerate on $T_p M$ for all $p \in M$.

A pair (M, ω) with ω as above is called a *symplectic manifold*. Note that $\dim_{\mathbb{C}} M$ is even because of the nondegeneracy condition. The form ω induces an isomorphism

$$TM \xrightarrow{\ \sharp\ } T^* M$$

sending a vector field $X \in \mathcal{X}(M) = \Gamma(M, TM)$ to the 1-from $X^\sharp = i_X \omega = \omega(X, ?) \in \Omega^1(M) = \Gamma(M, T^* M)$. The inverse of $\sharp$ will be denoted $\flat$.

Let $f \in \mathcal{O}(M) = \Gamma(M, \mathcal{O})$ be a global function on M and $df \in \Omega^1(M)$ its differential. Then we obtain a vector field $X_f := (df)^\flat \in \mathcal{X}(M)$. The vector field X_f is *symplectic* (or a canonical infinitesimal transformation), i.e. it satisfies

$$L_{X_f} \omega = 0$$

where L_X denotes the Lie derivative with respect to the vector field X. (This follows from the formula $L_X = d \circ i_X + i_X \circ d$ since $d\omega = 0$ and $i_{X_f}\omega = df$.) Any vector field of the form X_f for some $f \in \mathcal{O}(M)$ is called *Hamiltonian*. Let $\mathrm{Can}(X)$ denote the Lie subalgebra of $\mathcal{X}(M)$ of symplectic vector fields and $\mathcal{H}(M)$ the subspace of Hamiltonian vector fields. $\mathcal{H}(M)$ is a Lie subalgebra, too. In fact, $[X_f, X_g] = X_{\{f,g\}}$ where $\{f, g\} = \omega((df)^\flat, (dg)^\flat)$ denotes the *Poisson bracket* of f and g in $\mathcal{O}(M)$, and we have an obvious exact sequence of Lie algebras

$$0 \to \mathbb{C} \to (\mathcal{O}(M), \{\ \}) \to \mathcal{H}(M) \to 0.$$

Let G be an analytic (resp. algebraic) group acting morphically and symplectically on M (i.e. $G \times M \to M, (g, m) \mapsto gm$, is analytic (resp. algebraic) and $g^*\omega = \omega$ for all $g \in G$). Then the Lie algebra $\mathbf{g}$ of G "acts infinitesimally" on M by means of symplectic vector fields, i.e. we obtain a homomorphism of Lie algebras $\kappa : \mathbf{g} \to \mathrm{Can}(M)$. The action of G on M is called *Hamiltonian* if the image of $\mathbf{g}$ lies in $\mathcal{H}(M)$ and if the homomorphism $\kappa : \mathbf{g} \to \mathcal{H}(M)$ can be lifted to a G-equivariant homomorphism $\tilde{\kappa} : \mathbf{g} \to \mathcal{O}(M)$ of Lie algebras

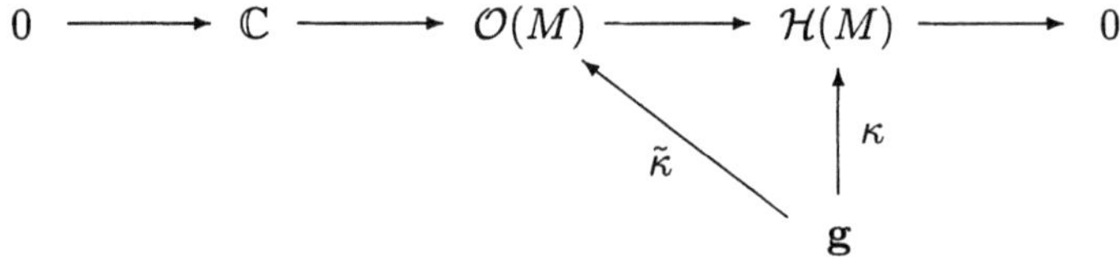

(note that G acts naturally on all terms involved). There exist general conditions on M and G which guarantee that any symplectic G-action is Hamiltonian (i.e. $H^1(M, \mathbb{C}) = 0$ for the appropriate de Rham cohomology and $H^2(\mathbf{g}, \mathbb{C}) = 0$, e.g. $\mathbf{g}$ semisimple), but we will be interested here only in very special cases.

Any Hamiltonian G-action gives rise to a *momentum map* $\mu : M \to \mathbf{g}^*$ of M into the dual of $\mathbf{g}$. It corresponds to the map $\tilde{\kappa} : \mathbf{g} \to \mathcal{O}(M)$ under the identification $\mathrm{Hom}_{\mathrm{vectorspace}}(\mathbf{g}, \mathcal{O}(M)) = \mathbf{g}^* \otimes \mathcal{O}(M) = \mathrm{Mor}(M, \mathbf{g}^*)$ and, up to a shift by a constant from the co-centre $(\mathbf{g}^*)^G$, it is characterized by the following two properties:

- μ is G-equivariant,

- for any $A \in \mathbf{g}, \kappa(A) = X_{A \circ \mu}$ (or, equivalently, $d(A \circ \mu) = i_{\kappa(A)}\omega$).

Lemma 1. *Assume there is a G-invariant 1-from $\alpha \in \Omega^1(M)$ such that $\omega = d\alpha$. Then the action of G on M is Hamiltonian and a momentum map*

$$\mu : M \to \mathbf{g}^*$$

is given by $\mu(p)(A) = -(i_{\kappa(A)}\alpha)(p)$ for all $p \in M, A \in \mathbf{g}$.

Proof. Note that, by the G-invariance of α,

$$d(A \circ \mu) = -d \circ i_{\kappa(A)}\alpha = -L_{\kappa(A)}\alpha + i_{\kappa(\alpha)}d\alpha = i_{\kappa(A)}\omega$$

for all $A \in \mathbf{g}$. The G-equivariance of μ follows from that of α (and κ). $\qquad\square$

Example. Let $M = T^*X$ denote the cotangent bundle of a manifold X, Then M carries a natural symplectic form $\omega = d\alpha$ where α denotes the canonical 1-form

$$\alpha_p(v) = p(\pi_* v).$$

Here $\pi : M \to X$ denotes the bundle projection, $v \in T_p M, \pi_* v \in T_{\pi(p)} X, p \in T^*_{\pi(p)} X \subset M$. The form α (and thus ω) is preserved under all "point transformations" of M, i.e. under all transformations of T^*X induced by isomorphisms of X.

In the sequel we shall encounter only the following situation.

Lemma 2. *Let (M, ω) denote a complex symplectic vector space and let $G \subset Sp(M, \omega)$ be an algebraic subgroup of the symplectic group of (M, ω) with Lie algebra $\mathbf{g}$. Then a momentum map $\mu : M \to \mathbf{g}^*$ is given by*

$$\mu(x)(A) \;=\; (A \circ \mu)(x) \;=\; \tfrac{1}{2}\,\omega(Ax, x) \quad (x \in M, A \in \mathbf{g}).$$

Proof. Since $G \subset Sp(M, \omega)$ we have for all $x \in M, v \in T_x M, A \in \mathbf{g}$,

$$
\begin{aligned}
d_x(A \circ \mu)(v) &= \tfrac{1}{2}(\omega(Av, x) + \omega(Ax, v)) \\
&= \tfrac{1}{2}(-\omega(v, Ax) + \omega(Ax, v)) \\
&= (i_A \cdot \omega)(v) \quad .
\end{aligned}
$$

The G-equivariance is also immediate. $\qquad\square$

We are mainly interested in the following two geometric properties of momentum maps.

Lemma 3. *Let $\mu : M \to \mathbf{g}^*$ be a momentum map for a Hamiltonian G-action on (M, ω). Then a point $p \in M$ is regular for μ, i.e. $d_p\mu : T_p M \to \mathbf{g}^*$ is surjective, if and only if the stabilizer $G_p = \{g \in G | gp = p\}$ of p in G is finite.*

Proof. By the basic property of μ we have

$$A(d_p\mu(v)) = d_p(A \circ \mu)(v) = \omega(\kappa(A)_p, v)$$

for all $v \in T_p M, A \in \mathbf{g}$. We now have the obvious equivalences:

$$
\begin{aligned}
d_p\mu \text{ is surjective} \quad &\Longleftrightarrow \quad \text{there is no } A \in \mathbf{g} \text{ such that} \\
& \qquad\quad A(d_p\mu(T_p M)) = \omega(\kappa(A)_p, T_p M) = 0 \\
&\Longleftrightarrow \quad \text{(by the nondegeneracy of } \omega) \\
& \qquad\quad \text{for all } A \in \mathbf{g}, \; \kappa(A)_p \neq 0 \\
&\Longleftrightarrow \quad G_p \text{ is finite.} \qquad\square
\end{aligned}
$$

Again, let $\mu : M \to \mathbf{g}^*$ be a momentum map and let $c \in (\mathbf{g}^*)^G$. Then G acts on the fiber $\mu^{-1}(c)$ because of the G-equivariance of μ. We can thus form the quotient $\mu^{-1}(c)//G$ in the category of algebraic varieties.

Proposition. *Assume G acts freely on $\mu^{-1}(c)$. Then $\mu^{-1}(c)$ and $\mu^{-1}(c)//G$ are manifolds. Moreover, $\mu^{-1}(c)//G$ carries a unique symplectic form $\overline{\omega}$ such that $\tilde{\omega} = q^*\overline{\omega}$, where $q : \mu^{-1}(c) \to \mu^{-1}(c)//G$ is the quotient map and $\tilde{\omega}$ the restriction of ω to $\mu^{-1}(c)$.*

Proof. The first statement follows from Lemma 3 and the freeness of the G-action on $\mu^{-1}(c)$. More precisely, as a consequence of Luna's slice theorem (cf. [27]), the quotient map $q : \mu^{-1}(c) \to \mu^{-1}(c)//G$ is a principal G-fiber bundle, now. To prove the second statement note first that for any $p \in \mu^{-1}(c), T_p(\mu^{-1}(c)) = \operatorname{Ker} d_p\mu = \{v \in T_pM | \omega(\kappa(A)_p, v) = 0 \text{ for all } A \in \mathbf{g}\}$. Since ω is nondegenerate on T_pM and since $T_p(G.p) = \{\kappa(A)_p | A \in \mathbf{g}\} \subset T_p(\mu^{-1}(c))$ this implies that $T_p(G.p)$ is the radical for the form $\tilde{\omega} = \omega_p|_{T_p(\mu^{-1}(c))}$ and that the induced form on the normal space $N_p = T_p(\mu^{-1}(c))/T_p(G.p)$ to $G.p$ in $\mu^{-1}(c)$ is nondegenerate. For any $\overline{p} \in \mu^{-1}(c)//G$ and tangent vectors $\overline{v}, \overline{w} \in T_{\overline{p}}(\mu^{-1}(c)//G)$ define

$$\overline{\omega}_{\overline{p}}(\overline{v}, \overline{w}) := \omega_p(v, w)$$

for some $p \in q^{-1}(\overline{p})$ and some lifts $v, w \in T_p(\mu^{-1}(c)), D_pq(v) = \overline{v}, D_pq(w) = \overline{w}$. The above developments show that this definition does not depend on the choices of the lifts v, w and that it defines a nondegenerate symplectic form on $T_{\overline{p}}(\mu^{-1}(c)//G)$. The G-invariance of ω yields its independence of the choice of $p \in q^{-1}(\overline{p})$. We finally have to show that $\overline{\omega}$ defines a regular and closed form on $\mu^{-1}(c)//G$. These are both local properties. For simplicity let us first argue in the complex-analytic category. By Luna's slice theorem (loc. cit.) we can find, for any $\overline{p} \in \mu^{-1}(c)//G$ and $p \in q^{-1}(\overline{p})$, a sufficiently small transversal slice S_p to $G \cdot p$ in $\mu^{-1}(c)$ (i.e. $T_pS_p \oplus T_p(G.p) = T_p(\mu^{-1}(c))$) such that q induces an analytic isomorphism $\overline{q} : S_p \to V_{\overline{p}}$ onto a neighborhood $V_{\overline{p}}$ of $\overline{p}$ in $\mu^{-1}(c)//G$. Moreover, $\overline{q}^*\overline{\omega}$ is just the restriction of ω to S_p which proves the (complex analytic) regularity and closedness of $\overline{\omega}$. In the algebraic category, similar spaces $S_p, V_{\overline{p}}$ are available, however, $\overline{q} : S_p \to V_{\overline{p}}$ is only étale (and without loss of generality, surjective) (cf. loc. cit.). The regularity of $\overline{\omega}$ follows from that of $\omega|_{S_p}$ by étale descent, i.e. by using the exactness of

$$\mathcal{O}(V_p) \xrightarrow[\overline{q}^*]{} \mathcal{O}(S_p) \underset{pr_i^*}{\rightrightarrows} \mathcal{O}(S_p \times_{V_p} S_p)$$

(note that tangent bundles, and thus all of their associated bundles, are locally trivial in the Zariski topology). The closedness of $\overline{\omega}$ follows from that of $\omega|_{S_p}$ by the injectivity and naturality ($\overline{q}^* \circ d = d \circ \overline{q}^*$) of $\overline{q}^*$. $\qquad\square$

4 Kronheimer's work

A new approach to the deformation and resolution theory of Kleinian singularities was given in the Oxford thesis [25] of P.B. Kronheimer, cf. also [24], [26]. His construction starts directly from the finite group Γ and uses quotient constructions in a "Hyper-Kähler-setting". (In the following, but only in this section, we shall have to make use of the analogues of the developments of Section 3 in the real-differentiable category.)

Let N denote the natural and R the regular representation of Γ, as before. N carries a natural (i.e. Γ-invariant) quaternionic structure, and by choosing a Γ-invariant hermitian scalar product on R we obtain a (Γ-invariant) real structure on $\operatorname{End}(R) = \operatorname{Hom}_{\mathbb{C}}(R, R)$ (i.e. the set $\operatorname{End}(R)_{\mathbb{R}}$ of real points of this structure consists of the hermitian (or antihermitian) matrices with respect to this scalar product), thus a (Γ-invariant) quaternionic structure on

$$M = \operatorname{End}(R) \otimes_{\mathbb{C}} N = \operatorname{End}(R)_{\mathbb{R}} \otimes_{\mathbb{R}} \mathbb{H} \quad .$$

Let $U(R)$ denote the unitary subgroup of $\operatorname{End}(R)$ acting on $\operatorname{End}(R)$ by conjugation, and let

$$\begin{aligned} U(\Gamma) &= \{g \in U(R) \mid g\,\gamma = \gamma\,g \quad \forall \gamma \in \Gamma\} \\ SU(\Gamma) &= U(\Gamma) \cap SL(R) \quad . \end{aligned}$$

Because of $R \cong \bigoplus_{i=0}^{r} R_i \otimes \mathbb{C}^{d_i}$ we have

$$U(\Gamma) \cong \prod_{i=0}^{r} U(d_i)$$

$$SU(\Gamma) \cong \left\{(A_0, \dots, A_r) \in U(\Gamma) \mid \prod_{i-0}^{r} \det(A_i) = 1\right\}.$$

The compact group $U(\Gamma)$ acts on M and on

$$M(\Gamma) = M^{\Gamma} = (\operatorname{End}(R) \otimes N)^{\Gamma}$$

preserving the quaternionic structures. In this situation we obtain three $U(\Gamma)$-invariant real symplectic structures on $M(\Gamma)$. Namely, we can choose a $U(\Gamma)$-invariant scalar product $\langle \ , \ \rangle$ on $M(\Gamma)$ such that the quaternionic operators I, J, K are anti-self-adjoint with respect to $\langle \ , \ \rangle$. The real symplectic forms are then given by

$$\begin{aligned} \omega_I(v, w) &= \langle v, Iw \rangle \\ \omega_J(v, w) &= \langle v, Jw \rangle \\ \omega_K(v, w) &= \langle v, Kw \rangle. \end{aligned}$$

Accordingly we obtain a $U(\Gamma)$-equivariant Hyper-Kähler-momentum map (now in the real-differentiable category)

$$\mu_{\mathbb{H}} : M(\Gamma) \longrightarrow \underline{u}(\Gamma)^* \otimes_{\mathbb{R}} \mathbb{H}_0$$

where $\mathbb{H}_0 = \mathbb{R}\, I \oplus \mathbb{R}\, J \oplus \mathbb{R}\, K \cong \mathbf{su}\,(2)$ denotes the pure quaternions, and where $\mathbf{u}\,(\Gamma)^*$ is the dual of the Lie algebra $\mathbf{u}\,(\Gamma)$ of $U\,(\Gamma)$ acted on by $U\,(\Gamma)$ via the co-adjoint action. Since the diagonal scalars

$$\mathbb{T} = \{(\lambda, \dots, \lambda) \in \prod_{i=0}^{r} U\,(d_i)\,|\,\lambda \in U\,(1)\}$$

act trivially on $M\,(\Gamma)$ and since $\mathbf{u}\,(\Gamma)^*$ may be identified with $\mathbf{u}\,(\Gamma)$ by means of the trace form $(A, B) \mapsto tr\,(A\,B^*)$, the target of $\mu_{\mathbb{H}}$ may be identified with $\mathbf{su}\,(\Gamma) \otimes_{\mathbb{R}} \mathbb{H}_0$, where $\mathbf{su}\,(\Gamma)$ is the Lie algebra of $SU\,(\Gamma)$. Let $\mathbf{c} = \{(\mu_0, \dots, \mu_r) \in \bigoplus_{i=0}^{r} \mathbf{u}\,(d_i)\,|\,\mu_i \in \mathbf{u}\,(1),\ \sum_{i=0}^{r} d_i\mu_i = 0\}$ denote the r-dimensional centre of $\mathbf{su}\,(\Gamma)$, i.e.

$$\mathbf{c} = (\mathbf{su}\,(\Gamma))^{U\,(\Gamma)} \quad .$$

Then $U\,(\Gamma)$ acts on any fibre of $\mu_{\mathbb{H}}^{-1}\,(\vec{\zeta})$, $\vec{\zeta} = \zeta_1 I + \zeta_2 J + \zeta_3 K \in \mathbf{c} \otimes \mathbb{H}_0$, and the real differential-geometric quotient $\mu_{\mathbb{H}}^{-1}\,(\vec{\zeta})/U\,(\Gamma)$ is a *Hyper-Kähler-quotient* (at least at its smooth points it carries the structure of a *Hyper-Kähler-manifold*).

Remark. The whole approach of Kronheimer may be motivated to some extent from the point of view of moduli spaces for certain types of "singular connections" on the real orbifold $\mathbb{C}^2/\Gamma$. For an elaboration of that, cf. [35] Chapter IV.

Kronheimer shows that for all $\vec{\zeta} \in \mathbf{c} \otimes \mathbb{H}_0$, the quotient $\mu_{\mathbb{H}}^{-1}\,(\vec{\zeta})/U\,(\Gamma)$ is a complex-analytic surface with at most isolated (Kleinian) singularities, in particular

$$\mu_{\mathbb{H}}^{-1}\,(\vec{0})/U\,(\Gamma) \cong \mathbb{C}^2/\Gamma \quad .$$

Moreover, there is a natural identification of $\mathbf{c} \otimes \mathbb{C} = \mathbf{c} \otimes J \oplus \mathbf{c} \otimes K$ with a Cartan subalgebra $\mathbf{h}$ of type $\underline{\Delta}\,(\Gamma)$ such that the complex r-parameter family

$$\mu_{\mathbb{H}}^{-1}\,(\mathbf{c} \otimes \mathbb{C})/U\,(\Gamma) \longrightarrow \mathbf{c} \otimes \mathbb{C}$$

realizes a pull-back (via $\mathbf{h} \to \mathbf{h}/W$) of the semiuniversal deformation of $\mathbb{C}^2/\Gamma$, and shifting this family "into the I-direction" provides a simultaneous resolution

$$\mu_{\mathbb{H}}^{-1}\,(\mathbf{c} \otimes \mathbb{C})/U\,(\Gamma) \quad \longleftarrow \quad \mu_{\mathbb{H}}^{-1}\,(\zeta_1 I + \mathbf{c} \otimes \mathbb{C})/U\,(\Gamma)$$
$$\downarrow \qquad\qquad\qquad\qquad\qquad\qquad \downarrow$$
$$\mathbf{c} \otimes \mathbb{C} \qquad \overset{\sim}{\longleftarrow} \qquad \zeta_1 I + \mathbf{c} \otimes \mathbb{C}$$

(for any fixed generic, i.e. regular $\zeta_1 \in \mathbf{c}$).

Remark. Due to the differential-geometric context, Kronheimer's original results [26] are formulated in a slightly weaker form when concerned with the algebro-geometric structure of singular points. However, he already devised a more algebraic approach, to which we turn in Section 5, and which yields these results in full strength. The crucial step is to view the composition

$$M\left(\Gamma\right) \xrightarrow{\mu_{\mathbb{H}}} \mathbf{u}^*\left(\Gamma\right) \otimes \mathbb{H}_0 \xrightarrow{pr} \mathbf{u}^*\left(\Gamma\right) \otimes \mathbb{C}$$

(where pr is induced by the projection $I, J, K \mapsto 0, 1, i$ of $\mathbb{H}_0$ onto $\mathbb{C}$) as a complex momentum map for the action of the reductive group

$$GL\left(\Gamma\right) = \{g \in GL\left(R\right) \,|\, g\,\gamma = \gamma\,g \quad \forall_{\gamma \in \Gamma}\}$$

on the complex symplectic vector space $M\left(\Gamma\right)$. (Note that $U\left(\Gamma\right) \otimes_{\mathbb{R}} \mathbb{C} = GL\left(\Gamma\right)$, that N is a symplectic Γ-module and that $GL\left(\Gamma\right)$ acts complex orthogonally on $\mathrm{End}\left(R\right)$.)

5 Quivers

A *quiver* (cf. [15]) is the same as an oriented graph consisting of two (finite) sets I and A, where I is the set of vertices and A the set of arrows (oriented edges) between vertices, e.g.

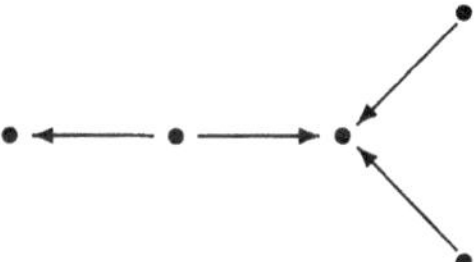

To any quiver (I, A) is associated the abelian category $\mathrm{Rep}(I, A)$ of its *representations* over a field k, say $\mathbb{C}$ in our context.

A (complex) *representation* of a quiver associates to any vertex $i \in I$ a (complex) vector space V_i and to any arrow $i \xrightarrow{a} j$ a linear map $f_a : V_i \to V_j$. The vector $\vec{d} = (d_i, i \in I)$, $d_i = \dim_{\mathbb{C}} V_i$, is called the *dimension* of the representation.

A *morphism* $\Psi : (V_i, f_a) \to (W_i, g_a)$ of two representations of (I, A) consists of a family $\Psi = (\varphi_i, i \in I)$ of linear maps $\varphi_i : V_i \to W_i$ such that the diagram

$$
\begin{array}{ccc}
V_i & \xrightarrow{\varphi_i} & W_i \\
\downarrow f_a & & \downarrow g_a \\
V_j & \xrightarrow{\varphi_j} & W_j
\end{array}
$$

commutes for any arrow $a \in A$. There are obvious notions of isomorphisms, sub- and quotient respresentations, direct sum, decomposition and indecomposability, making $\mathrm{Rep}(I, A)$ into an abelian category. Identifying any complex vector space of dimension $\vec{d}$ with $\mathbb{C}^d$, we may view

$$M(I, A, \vec{d}) = \bigoplus_{a \in A} \mathrm{Hom}(\mathbb{C}^{d_i(a)}, \mathbb{C}^{d_j(a)})$$

as the set of all representations of (I, A) of given dimension $\vec{d} = (d_i, i \in I)$. The group $GL(I, A, \vec{d}) = \prod_{i \in I} GL_{d_i}(\mathbb{C})$ acts on $M(I, A\vec{d})$ by simultaneous conjugation (base change in all V_i), and two representations in $M(I, A, \vec{d})$ are isomorphic exactly when they lie in the same $GL(I, A, \vec{d})$-orbit.

By means of McKay's observation Kronheimer interprets $M(\Gamma)$ in terms of quivers:

$$
\begin{aligned}
M(\Gamma) \;&=\; (\mathrm{End}\,(R) \otimes N)^\Gamma = (R^* \otimes R \otimes N)^\Gamma \\[2mm]
&=\; \mathrm{Hom}_\Gamma\,(R, R \otimes N) = \mathrm{Hom}_\Gamma\,\Big(\bigoplus_{i=0}^{r} R_i \otimes \mathbb{C}^{d_i}, \bigoplus_{j=0}^{r} R_j \otimes N \otimes \mathbb{C}^{d_j} \Big) \\[2mm]
&=\; \bigoplus_{i,j=0}^{r} \mathrm{Hom}_\Gamma\,(R_i, R_j \otimes N) \otimes \mathrm{Hom}\,(\mathbb{C}^{d_i}, \mathbb{C}^{d_j}) \\[2mm]
&=\; \bigoplus_{i \;\; j} \mathrm{Hom}\,(\mathbb{C}^{d_i}, \mathbb{C}^{d_j}) \\[2mm]
&=\; M(I, A, \vec{d})
\end{aligned}
$$

for the quiver (I, A) consisting of the vertices $I = \{0, 1, \dots, r\}$ of the extended diagram $\tilde{\underline{\Delta}}\,(\Gamma)$, two arrows (in both directions) for any edge of $\tilde{\underline{\Delta}}\,(\Gamma)$, and the dimension $\vec{d} = (d_0, \dots, d_r)$, $d_i = \dim_\mathbb{C} R_i$.

Example. $\Gamma = \mathcal{J}$, $\tilde{\underline{\Delta}}\,(\Gamma) = \tilde{E}_8$

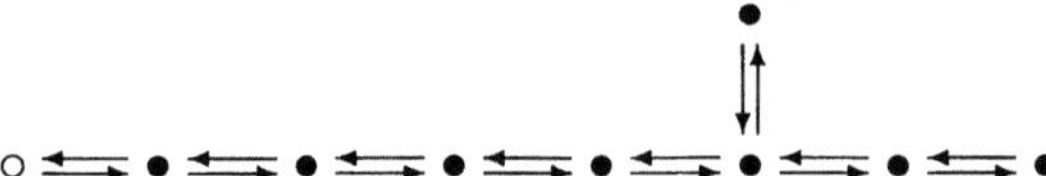

Henceforth, we shall call (I, A) as above the *McKay quiver of type $\underline{\Delta}\,(\Gamma)$.*

Let $G(\Gamma) = (\prod_{i=0}^{r} GL_{d_i}(\mathbb{C}))/\mathbb{C}^*$ denote the group effectively operating on $M(\Gamma)$ (i.e. $\mathbb{C}^*$ is the diagonal scalar subgroup). A $G(\Gamma)$-invariant symplectic form and corresponding momentum map on $M(\Gamma)$ can now easily be written out in terms of this interpretation (cf. [29] §12).

Fix a function $\varepsilon : A \to \mathbb{C}^*$ such that $\varepsilon(a) + \varepsilon(\bar{a}) = 0$ for all arrows

$$
\overset{i \;\;\; a \;\;\; j}{\bullet \!\!\longrightarrow\!\! \bullet}
$$

and opposite arrows

$$
\overset{i \;\;\; \bar{a} \;\;\; j}{\bullet \!\!\longleftarrow\!\! \bullet}
$$

(e.g. $\varepsilon(a) = +1 = -\varepsilon(\bar{a})$ for arrows a belonging to a fixed orientation of the edges of $\tilde{\underline{\Delta}}\,(\Gamma)$). For any pair $\varphi = (\varphi_a, a \in A), \psi = (\psi_a, a \in A)$ we put

$$
\langle \varphi, \psi \rangle := \sum_{a \in A} \varepsilon(a)\, tr\,(\varphi_a \, \psi_{\bar{a}})
$$

Then $\langle\ ,\ \rangle$ is a non-degenerate $G(\Gamma)$-invariant symplectic form on $M(\Gamma)$ with momentum map

$$\mu_{\mathbb{C}} : M(\Gamma) \longrightarrow (\operatorname{Lie} G(\Gamma))^* \subset \bigoplus_{i=0}^{r} M_{d_i}(\mathbb{C})$$

given by

$$\mu_{\mathbb{C}}(\varphi) = \Big(\ \ldots\ , \ \overbrace{\sum \varepsilon(a)\,\varphi_a\,\varphi_{\bar{a}}}^{i\text{-th entry}} , \ \ldots\ \Big)$$

Example.

$$\varepsilon = \pm 1 \ (\text{according to orientation})$$

The i-th entry of $\mu_{\mathbb{C}}(\varphi)$, $\quad \varphi = (\varphi_{\alpha\beta} : V_\alpha \to V_\beta)$, is now

$$\varphi_{ji}\,\varphi_{ij} - \varphi_{ki}\,\varphi_{ik} - \varphi_{li}\,\varphi_{il} \quad .$$

In quiver-theoretic terms, $\mu_{\mathbb{C}}^{-1}(0)$ may be viewed as the set of all representations of $(I, A, \vec{d})$ with certain relations.

The dual Z of the centre of $\operatorname{Lie} G(\Gamma)$ may, again, be identified with

$$\mathbf{c} \otimes \mathbb{C} = \Big\{ (\mu_0, \mu_1, \ldots, \mu_r) \in \prod M_{d_i}(\mathbb{C}) \,\big|\, \mu_i \in \mathbb{C}, \ \sum_{i=0}^{r} d_i \mu_i = 0 \Big\},$$

and for any $z \in Z$ the fibre $\mu_{\mathbb{C}}^{-1}(z)$ is acted on by $G(\Gamma)$. It is a consequence of the results of Kempf-Ness (cf. [32] Appendix) that

$$\mu_{\mathbb{C}}^{-1}(z) \,//\, G(\Gamma) \cong \mu_{\mathbb{H}}^{-1}(0, \zeta_2, \zeta_3)/U(\Gamma)$$

for all $z = \zeta_2 + \zeta_3 i \in \mathbf{c} \otimes \mathbb{C} = Z$ (as topological spaces, or as complex manifolds when restricted to the regular points). Thus we obtain a purely invariant theoretic construction of the pull-back of the semiuniversal deformation

$$
\begin{array}{ccc}
\mathcal{X} \times_{\mathbf{h}/W} \mathbf{h} & \xleftarrow{\ \sim\ } & \mu_{\mathbb{C}}^{-1}(Z)//G(\Gamma) \\
\downarrow & & \downarrow \\
\mathbf{h} & \xleftarrow{\ \sim\ } & Z
\end{array}
$$

Remark. To establish the above isomorphism in the category of analytic spaces or algebraic varieties it is necessary to leave the differential-geometric frame work. A crucial result in that respect is the flatness of the momentum map $\mu_{\mathbb{C}} : M\,(\Gamma) \to (\text{Lie}\,G\,(\Gamma))^*$ which follows from a determination of certain fibre dimensions by Lusztig (cf. [28], [29], [9]).

Example. Let $\Gamma = \mathcal{C}_{n+1}$. Then $\underline{\tilde{\Delta}}\,(\Gamma)$ is of type $\tilde{A}_n$, and the McKay quiver (I, A) has the following form:

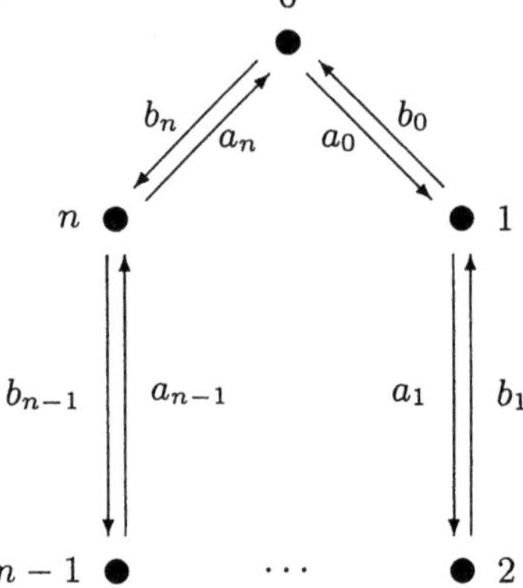

We have

$$
\begin{aligned}
M\,(\Gamma) &= \mathbb{C}^{n+1} \oplus \mathbb{C}^{n+1} \\
&= \{(a_0, \dots, a_n, b_0, \dots, b_n)\,|\,a_i, b_i \in \mathbb{C}\}
\end{aligned}
$$

since $\vec{d} = (1, 1, \dots, 1)$, and

$$
G\,(\Gamma) = (\mathbb{C}^*)^{n+1}/\mathbb{C}^* \cong (\mathbb{C}^*)^n,
$$

in particular, $Z = \text{Lie}\,(G\,(\Gamma))^*$ since $G\,(\Gamma)$ is commutative, and $\mu_{\mathbb{C}}^{-1}\,(Z) = M\,(\Gamma)$ itself. The action of an element $t = (t_0, \dots, t_n) \in (\mathbb{C}^*)^{n+1}$ on $M\,(\Gamma)$ is given by

$$
t \cdot (a_0, \dots, a_n, b_0, \dots, b_n) = (t_1\,t_0^{-1}\,a_0, \dots, t_0\,t_n^{-1}\,a_n, t_0\,t_1^{-1}\,b_0, \dots, t_n\,t_0^{-1}\,b_n)
$$

with (obvious) fundamental invariants

$$
\begin{aligned}
z_i &= a_i\,b_i & (i = 0, \dots, n) \\
x &= a_0\,a_1\,\dots\,a_n \\
y &= b_0\,b_1\,\dots\,b_n
\end{aligned}
$$

and relation

$$
z_0\,z_1\,\dots\,z_n = xy .
$$

Thus $M\,(\Gamma)//G\,(\Gamma)$ is given as the hypersurface

$$
\{(z_0, \dots, z_n, x, y) \in \mathbb{C}^{n+3}\,|\,z_0\,z_1\,\dots\,z_n = xy\},
$$

and the map to $Z = \{(\mu_0, \ldots, \mu_n) \in \mathbb{C}^{n+1} \mid \sum_{i=0}^n \mu_i = 0\}$ is given by

$$(z_0, \ldots, z_n, x, y) \mapsto (z_0 - z_1, z_1 - z_2, \ldots, z_n - z_0)$$

(here we have fixed the clockwise orientation on (I, A) and $\varepsilon = \pm 1$). Introducing new coordinates on $M(\Gamma)//G(\Gamma)$ by putting

$$
\begin{aligned}
z &:= \frac{1}{n+1} \left(\sum_{i=0}^n z_i \right) \\
\lambda_i &= z - z_i \qquad (i = 0, \ldots, n)
\end{aligned}
$$

and identifying Z with the standard Cartan subalgebra

$$\mathbf{h} = \left\{ (\lambda_0, \ldots, \lambda_n) \mid \sum_{i=0}^n \lambda_i = 0 \right\}$$

in $\mathbf{sl}_{n+1}$ by means of the isomorphism

$$
\begin{aligned}
\mathbf{h} &\to Z \\
(\lambda_0, \ldots, \lambda_n) &\mapsto (\lambda_1 - \lambda_0, \lambda_2 - \lambda_1, \ldots, \lambda_0 - \lambda_n)
\end{aligned}
$$

we obtain the standard form

$$
\begin{aligned}
M(\Gamma)//G(\Gamma) &= \{(x, y, z, \lambda_0, \ldots, \lambda_n) \in \mathbb{C}^3 \times \mathbf{h} \mid \prod_{i=0}^n (z - \lambda_i) = xy\} \\
\downarrow \qquad\qquad & \qquad\qquad \downarrow pr_2 \\
\mathbf{h} \qquad &= \qquad\qquad \mathbf{h}
\end{aligned}
$$

for the pull-back

$$
\begin{aligned}
\mathcal{X} \times_{\mathbf{h}/W} \mathbf{h} \\
\downarrow \\
\mathbf{h}
\end{aligned}
$$

of the semiuniversal deformation of the Kleinian singularity $S = \{z^{n+1} = xy\}$ of type A_n (cf. e.g. [20], [39]).

6 Linear modifications

As we have seen in the last section there is a purely algebraic, invariant-theoretic construction of (a pull-back of) the semiuniversal deformation of $\mathbb{C}^2/\Gamma$ in terms of the associated quiver. One of the main objectives of the doctoral thesis [9] is a similar construction for the simultaneous resolution. For that we make use of a quite basic operation in invariant theory which we call "linear modification" and which has recently been studied independently and in different contexts by other authors (e.g. [8], [10], [21], [34], [38]).

Let V denote an affine algebraic variety over $\mathbb{C}$ acted upon (morphically) by a reductive algebraic group G, and let $V//G = \mathrm{Specmax}\,(\mathbb{C}\,[V]^G)$ denote the categorical quotient of V by G. As a set we may identify $V//G$ with the set of closed orbits of G in V ([32], [33]).

Let $\chi : G \to \mathbb{C}^*$ denote a multiplicative character of G (an element of the additively written group $X^*\,(G) = \mathrm{Hom}\,(G, \mathbb{C}^*)$). Then we consider the graded ring

$$\mathbb{C}\,[V]^{G,\chi} := \bigoplus_{m=0}^{\infty} \mathbb{C}\,[V]_{m\,\chi}$$

where $\mathbb{C}\,[V]_{m\,\chi}$ denotes the subspace of G-semi-invariants transforming according to the character $m\,\chi$ and we put

$$V//^{\chi} G := \mathrm{Projm}\,(\mathbb{C}\,[V]^{G,\,\chi}),$$

the homogeneous (maximal) spectrum of $\mathbb{C}\,[V]^{G\chi}$ (cf. [18] II 7).

There is a natural projective morphism

$$V//^{\chi} G \to V//G$$

which can be shown to be birational provided all components of V admit G-stable points. (In fact, it induces an isomorphism above the G-stable orbits, $V\,(G-\mathrm{stable})//G.$) This map is called the χ-*linear modification of* $V//G$.

The space $V//^{\chi} G$ itself can be viewed, in terms of Mumford's geometric invariant theory [32], as a quotient of V by G with respect to a G-linearization of V given by the trivial line bundle $V \times \mathbb{C}$ with nontrivial action

$$g\,(v, \lambda) = (g\,v, \chi\,(g)\,\lambda)\quad ,\quad v \in V, \lambda \in \mathbb{C}\quad ,\quad g \in G.$$

In this situation, a point $v \in V$ is called

- χ-*semistable* $\iff$ there exists an $m > 0$ and an $f \in \mathbb{C}\,[V]_{m\,\chi}$ such that $f\,(v) \neq 0$.

- χ-*stable* $\iff$ the isotropy group G_v is finite and there exists an $f \in \mathbb{C}\,[V]_{m\,\chi}$, $m > 0, f\,(v) \neq 0$, such that the orbit $G.v$ is closed in the affine open subset $V_f = \{w \in V \mid f\,(w) \neq 0\}$.

As a set, $V//^{\chi} G$ may be identified with the set of closed orbits of G in the set $V\,(\chi\text{-semistable})$ of χ-semistable points of V.

Let us have a look at some examples:

1) Let $V = \mathbb{C}^2$ with typical element (x, y), $G = \mathbb{C}^*$ with typical element λ, and action $\lambda\,(x, y) = (\lambda\,x, \lambda\,y)$. We have $V//G = \mathrm{Specmax}\,(\mathbb{C}\,[V]^G) = \mathrm{Specmax}\,\mathbb{C} = $ a point. Let $\chi : G \to \mathbb{C}^*$ be given by $\lambda \mapsto \lambda^{-1}$. Then

$V//^\chi G = \mathrm{Projm}\,(\mathbb{C}\,[V]^{G,\chi}) = \mathrm{Projm}\,(\mathbb{C}\,[x,y]) = \mathbb{P}^1\,(\mathbb{C})$, and $V\,(\chi\text{-semistable})$ $= V\backslash\{0\}$.

If instead of χ we would have used $\chi^{-1} = id : \lambda \mapsto \lambda$, we would have obtained $V//^{\chi^{-1}} G = \mathrm{Projm}\,(\mathbb{C}) = \emptyset$.

2) Let $V = \mathbb{C}^3$ with typical element (x,y,z), $G = \mathbb{C}^*$ with typical element λ, and action $\lambda\,(x,y,z) = (\lambda\,x, \lambda\,y, \lambda^{-1}\,z)$. Now $\mathbb{C}\,[V]^G = \mathbb{C}\,[u,v]$ with $u = xz, v = yz$, and $V//G = \mathbb{C}^2$. Let χ be as in 1). Then $\mathbb{C}\,[V]^{G,\chi} = \mathbb{C}\,[u,v][x,y]/(yu - xv)$ and $V//^\chi G$ is the blow-up of 0 in $\mathbb{C}^2$. We have $V\,(\chi\text{-semistable}) = \mathbb{C}^3\backslash z\text{-axis}$ and $V//^\chi G = \mathbb{C}^2\backslash\{0\} \cup \mathbb{P}^1\,(\mathbb{C})$. Starting from χ^{-1} would give no change:

$$V//^{\chi^{-1}} G = \mathrm{Projm}\,(\mathbb{C}[u,v][z]) = \mathrm{Specmax}\,\mathbb{C}[u,v] = V//^\chi G.$$

3) The next example encompassing the previous two as "degenerate" limit cases has been amply discussed in the context of "flips" (cf. e.g. [34], [38], or [9]). Let $p,q > 0$, and $V = \mathbb{C}^{p+1} \times \mathbb{C}^{q+1} \ni (x_0,\dots,x_p,y_0,\dots,y_q), G = \mathbb{C}^* \ni \lambda$,

$$\lambda((x_i),(y_j)) = ((\lambda x_i),(\lambda^{-1} y_j)).$$

Now $\mathbb{C}[V]^G = \mathbb{C}[u_{ij}; i = 0,\dots,p; j = 0,\dots,q]/I$, where $u_{ij} = x_i x_j$ and where I is the ideal generated by all 2×2-minors of the matrix $U = ((u_{ij}))$, i.e. $V//G$ is the affine cone over the Segre-embedding of $\mathbb{P}^p(\mathbb{C}) \times \mathbb{P}^q(\mathbb{C})$ into $\mathbb{P}^{pq+p+q}(\mathbb{C}) = \mathbb{P}(\mathbb{C}^{p+1} \otimes \mathbb{C}^{q+1})$. Let χ be as in 1) and 2). Then

$$\mathbb{C}[V]^{G,\chi} = (\mathbb{C}[u_{ij}]/I)[x_0,\dots,x_p]/J,$$

where J is the ideal generated by all

$$x_k u_{ij} - x_i u_{kj}; \quad i,k = 0,\dots,p; \quad j = 0,\dots,q.$$

Viewing the blow-up of $0 \in \mathbb{C}^{p+1}$ as the bundle $\mathcal{O}_{\mathbb{P}^p}(-1)$ we may identify $V//^\chi G$ as the $(q+1)$-fold sum

$$V//^\chi G = \bigoplus_{s=0}^{q} \mathcal{O}_{\mathbb{P}^p}(-1)_{(s)}$$

of $\mathcal{O}_{\mathbb{P}^p}(-1)$ over $\mathbb{P}^p$. The set of χ-semistable points consists of

$$V\backslash(\{0\} \times \mathbb{C}^{q+1}).$$

In direct geometric terms the map $V//^\chi G \to V//G$ may either be described as the "blow-down" of the zero-section $\mathbb{P}^p \subset \bigoplus_{s=0}^{q} \mathcal{O}_{\mathbb{P}^p}(-1)_{(s)}$ to $0 \in V//G \subset \mathbb{C}^{p+1} \otimes \mathbb{C}^{q+1}$ or the "blow-up by $(q+1)$-planes" of $0 \in V//G$, i.e. $V//^\chi G$ may be viewed as the closure in $(V//G) \times Gr$ of the incidence variety

$$\{(p,W) \in ((V//G)\backslash\{0\}) \times Gr | p \in W\}$$

(here Gr denotes the Grassmannian $Gr_{q+1}(\mathbb{C}^{p+1} \otimes \mathbb{C}^{q+1})$ of all $(q+1)$-planes in $\mathbb{C}^{(p+1)(q+1)} = \mathbb{C}^{(p+1)} \otimes \mathbb{C}^{(q+1)}$ through 0), and $V//^{\chi}G \to V//G$ is realized by the first projection.

A similar picture is obtained by using χ^{-1}, the roles of the x_i and y_j being interchanged now. If one denotes by

$$\widetilde{V//G}$$

the ordinary blow-up of $0 \in V//G \subset \mathbb{C}^{p+1} \otimes \mathbb{C}^{q+1}$ with exceptional fibre $\mathbb{P}^p \times \mathbb{P}^q$, one obtains the following cartesian "flip diagram"

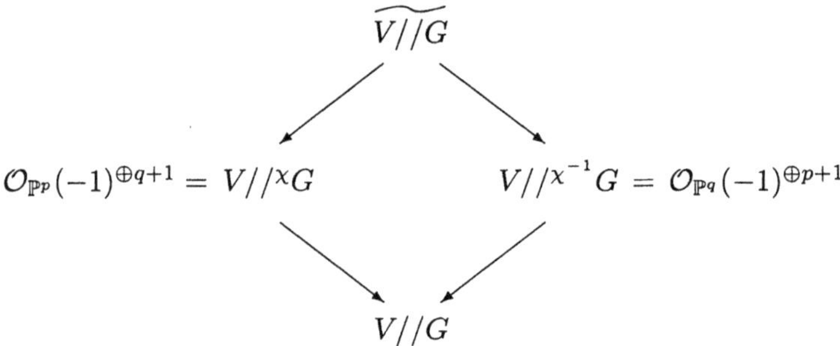

containing

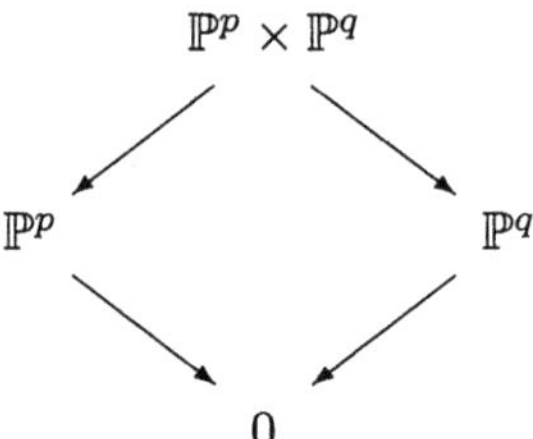

Note that in the examples above non-closed (or unstable) G-orbits in V have become "χ-stabilized" and thus contribute to the quotient space $V//^{\chi}G$.

7   Simultaneous resolution

We can now apply a linear modification to Kronheimer's construction (write $\mu_{\mathbb{C}}^{-1}(Z) = M(\Gamma)_Z$)

$$
\begin{array}{ccc}
\mathcal{X} \times_{\mathbf{h}/W} \mathbf{h} & \xleftarrow{\ \sim\ } & M(\Gamma)_Z //G(\Gamma) \\
\downarrow & & \downarrow \\
\mathbf{h} & \xleftarrow{\ \sim\ } & Z
\end{array}
$$

i.e. for any $\chi \in X^* \left(G \left(\Gamma \right) \right)$ we obtain a relative projective morphism

$$
\begin{array}{ccc}
M \left(\Gamma \right)_Z // G \left(\Gamma \right) & \longleftarrow & M \left(\Gamma \right)_Z //^{\chi} G \left(\Gamma \right) \\
\downarrow & & \downarrow \\
\mathbf{c} \otimes \mathbb{C} & = & Z
\end{array}
$$

Theorem: *For generic χ this diagram is a simultaneous minimal resolution, in particular, $\mu_{\mathbb{C}}^{-1} \left(0 \right) //^{\chi} G \left(\Gamma \right)$ is a minimal resolution of $\mathbb{C}^2 / \Gamma$.*

Before indicating details of the proof of this result we have to define the notion of genericity for characters $\chi \in X^* \left(G \left(\Gamma \right) \right)$.

Let us first consider $GL \left(\Gamma \right) = \prod_{i=0}^r GL_{d_i} \left(\mathbb{C} \right)$ whose group of characters is free abelian of rank $r + 1$

$$
X^* \left(GL \left(\Gamma \right) \right) = \tilde{X}^* := \bigoplus_{i=0}^r \mathbb{Z} \tilde{\omega}_i \quad .
$$

Here, $\tilde{\omega}_i$ is given by the "i-th determinant"

$$
\tilde{\omega}_i \left(A_0, \ldots A_r \right) = \det A_i \quad .
$$

The dual lattice $\tilde{X}_* = \mathrm{Hom}_{\mathbb{Z}} \left(\tilde{X}^*, \mathbb{Z} \right)$ is the group of one-parameter-groups of $\mathbb{C}^*$ into the factor commutator group

$$
GL \left(\Gamma \right) / \left(GL \left(\Gamma \right), GL \left(\Gamma \right) \right) = \prod_{i=0}^r GL_{d_i} \left(\mathbb{C} \right) / \left(GL_{d_i} \left(\mathbb{C} \right), GL_{d_i} \left(\mathbb{C} \right) \right) \quad ,
$$

$$
\tilde{X}_* = \bigoplus_{i=0}^r \mathbb{Z} \tilde{\alpha}_i \quad \text{with} \quad \tilde{\omega}_i \left(\tilde{\alpha}_i \right) = \delta_{ij} \quad .
$$

We shall view $\tilde{X}_*$ as the root lattice of an affine root system (of type $\tilde{\underline{\Delta}} \left(\Gamma \right)$) with simple roots $\tilde{\alpha}_0, \ldots, \tilde{\alpha}_r$ and $\tilde{X}^*$ as the corresponding weight lattice with fundamental dominant weights $\tilde{\omega}_0, \ldots, \tilde{\omega}_r$.

The character group $X^* \left(G \left(\Gamma \right) \right)$ is embedded into $\tilde{X}^*$ as the set of characters $\omega \in \tilde{X}^*$ vanishing on the scalar subgroup

$$
\sigma \; : \; \mathbb{C}^* \to \prod_{i=0}^r GL_{d_i} \left(\mathbb{C} \right)
$$

$$
\lambda \; \mapsto \left(\lambda, \ldots, \lambda \right) \quad .
$$

Since σ corresponds to $\delta = \sum_{i=0}^r d_i \, \tilde{\alpha}_i$ in $\tilde{X}_*$, the minimal imaginary root, we may view $X_* = \tilde{X}_* / \mathbb{Z} \, \delta$ as a root lattice Q of type $\underline{\Delta} \left(\Gamma \right)$ (i.e. A_r, D_r, E_r) and $X^* = X^* \left(G \left(\Gamma \right) \right) = \{ \omega \in \tilde{X}^* \, | \, \omega \left(\delta \right) = 0 \}$ as the corresponding weight lattice P. We can now define:

A character $\chi \in X^* (G (\Gamma))$ is *generic* exactly when χ lies inside a Weyl chamber, i.e. when $\chi (\alpha) \neq 0$ for all roots α in X_*. (In the frame-work of affine root systems we may rephrase this condition as: $\chi (\tilde{\alpha}) \neq 0$ for all real roots $\tilde{\alpha}$ in $\tilde{X}_*$, when χ is considered as an element of $\tilde{X}^*$.)

In the proof of the above theorem we shall profit from a "direct" interpretation, due to A.D. King (cf. [21]), of χ-stability of representations in terms of subrepresentations. This will be similar to the interpretation of stability for vector bundles on curves in terms of subbundles.

Note that the character group $\tilde{X}^* = \bigoplus_{i=0}^r \mathbb{Z} \tilde{\omega}_i$ has a "universal" meaning for the given McKay quiver (I, A) independent of a fixed dimension vector (i.e. $\tilde{X}^* = X^* (GL (I, A, \tilde{\alpha}) = X^* (\prod_{i=0}^r GL_{a_i} (\mathbb{C}))$ for any dimension vector $\tilde{\alpha} = (a_0, \dots, a_r)$ with all $a_i \neq 0$). More naturally, the dual group $\tilde{X}_* = \bigoplus_{i=0}^r \mathbb{Z} \tilde{\alpha}_i$ may be viewed as the group generated by all dimension vectors $\tilde{\alpha} = \sum_{i=0}^r a_i \tilde{\alpha}_i$ of representations of (I, A). Thus any $\chi = \sum_{i=0}^r x_i \tilde{\omega}_i$ in $\tilde{X}^*$ defines a "numerical" character (denoted by $\chi^\sharp$) on the Grothendieck group $K (I, A)$ of representations of (I, A) :

$$\chi^\sharp : K (I, A) \xrightarrow{\dim} \tilde{X}_* \xrightarrow{\chi} \mathbb{Z},$$

$$\chi^\sharp (\mathcal{A}) = \chi (\dim \mathcal{A}) = \sum_{i=0}^r x_i a_i, \quad \text{if } \dim \mathcal{A} = \sum_{i=0}^r a_i \tilde{\alpha}_i.$$

We now have the following

Characterization Theorem (King, [21]). *Let $\mathcal{A} \in M (I, A, \alpha)$ be a representation of (I, A) of dimension $\tilde{\alpha}$ and let χ be a character of $G(I, A, \alpha) = GL(I, A, \alpha)/\mathbb{C}^*$, i.e. $\chi \in \tilde{X}^*$ such that $\chi^\sharp (\mathcal{A}) = \chi (\tilde{\alpha}) = 0$. Then $\mathcal{A}$ is χ-stable (resp. χ-semistable) with respect to the action of $G (I, A, \tilde{\alpha})$ on $M (I, A, \tilde{\alpha})$ exactly when $\chi^\sharp (\mathcal{A}') > 0$ (resp. $\chi^\sharp (\mathcal{A}') \geq 0$) for all proper subrepresentations $\mathcal{A}'$ of $\mathcal{A}$.*

Remark. Let $\mathcal{A}$ be a vector bundle of rank r and degree d on a curve C. Then one can reformulate the (semi-) stability of $\mathcal{A}$ in similar terms when considering the numerical character $\mathcal{A}' \mapsto d \cdot \text{rank} (\mathcal{A}') - r \cdot \deg (\mathcal{A}')$ on subbundles $\mathcal{A}'$ of $\mathcal{A}$ (cf. [32], [33]).

The following Root Lemma plays already a crucial role in Kronheimer's work (cf. [25] Section 4.4 for its disguise in terms of Hyper-Kähler-quotients).

Let $M (I, A, \tilde{\alpha})$ denote the representations of a McKay quiver of type $\underline{\tilde{\Delta}} (\Gamma)$ with arbitrary dimension vector $\tilde{\alpha} = \sum_{i=0}^r a_i \tilde{\alpha}_i \in \tilde{X}_*$. As in Section 5 we can define a complex $G (I, A, \tilde{\alpha})$-equivariant momentum map with respect to a similar symplectic form (i.e. using the same ε)

$$\mu_{\mathbb{C}} : M (I, A, \tilde{\alpha}) \rightarrow (\text{Lie } G (I, A, \tilde{\alpha}))^*.$$

Root Lemma. *Let $\mathcal{A} \in M (I, A, \tilde{\alpha})$ be such that*

$$\mu_{\mathbb{C}} (\mathcal{A}) = z \in ((\text{Lie } G (I, A, \tilde{\alpha}))^*)^{G (I, A, \tilde{\alpha})}$$

and such that the isotropy group of $\mathcal{A}$ in $G(I, A, \tilde{\alpha})$ is finite . Then $\tilde{\alpha}$ is a real or imaginary affine root in $\tilde{X}_*$.

Proof. Lemma 3 of Section 3 shows that, because of the isotropy condition, $\mu_{\mathbb{C}}$ is submersive at $\mathcal{A}$, thus $\mu_{\mathbb{C}}^{-1}(z)$ is a complex manifold of dimension

$$\dim M(I, A, \tilde{\alpha}) - \dim G(I, A, \tilde{\alpha}) = \sum_{i \neq j} m_{ij} a_i a_j - \left(\sum_i a_i^2 - 1 \right)$$

at $\mathcal{A}$. Since z is co-central, the whole $G(I, A, \tilde{\alpha})$-orbit of $\mathcal{A}$ lies in $\mu_{\mathbb{C}}^{-1}(z)$. Thus $\dim G(I, A, \tilde{\alpha}) = \dim G(I, A, \tilde{\alpha}).\mathcal{A} \leq \dim \mu_{\mathbb{C}}^{-1}(z)$ (at $\mathcal{A}$). And thus

$$2 \left(\sum_{i=0}^r a_i^2 - 1 \right) \leq \sum_{i \neq j} m_{ij} a_i a_j \qquad \text{or} \qquad (\tilde{\alpha}, \tilde{\alpha}) = \sum_{i,j=0}^r c_{ij} a_i a_j \leq 2$$

where $c_{ij} = 2\,\delta_{ij} - m_{ij}$ are the Cartan coefficients (of type $\underline{\tilde{\Delta}}(\Gamma)$) and $(\ ,\ )$ is the (affine) Killing form on $\tilde{X}_*$ (cf. Section 2). However, any element $\tilde{\alpha} \in \tilde{X}_*$ with $(\tilde{\alpha}, \tilde{\alpha}) \leq 2$ is either a real root $((\tilde{\alpha}, \tilde{\alpha}) = 2)$ or an imaginary root $((\tilde{\alpha}, \tilde{\alpha}) = 0)$ (cf. e.g. [19] Prop. 5.10). $\qquad \square$

Remarks. 1) Assume $\tilde{\alpha} = \sum_{i=0}^r a_i \tilde{\alpha}_i$ with $0 \leq a_i \leq d_i$ for all i, $a_j < d_j$ for at least one j, and $(\tilde{\alpha}, \tilde{\alpha}) \leq 2$. Since $\sum_{i=0}^r d_i \tilde{\alpha}_i$ is the minimal (positive) imaginary root we now obtain either $\tilde{\alpha} = 0$ or $(\tilde{\alpha}, \tilde{\alpha}) = 2$.

2) It is a standard fact in quiver theory that isotropy groups of representations are connected (they are of the form $U/\mathbb{C}^*$ where U is the Zariski open subset of units of their endomorphism algebra). Thus the isotropy group is finite exactly when it is trivial.

The following lemma shows how the genericity condition enters.

Stability Lemma. *Let $\chi \in X^*(G(\Gamma))$ be generic. Then*

$$M(\Gamma)_Z(\chi\text{-semistable}) = M(\Gamma)_Z(\chi\text{-stable}).$$

Proof. Let $\mathcal{A} \in M(\Gamma)_Z \subset M(I, A, \delta)$ be χ-semistable but not χ-stable. Then there is a proper subrepresentation $\mathcal{A}'$ of $\mathcal{A}, \mathcal{A}' \in M(I, A, \tilde{\alpha})$ such that $\chi^\sharp(\mathcal{A}') = \chi(\tilde{\alpha}) = 0$. If we assume $\mathcal{A}'$ to be minimal with this property, then $\mathcal{A}'$ is χ-stable. Thus the isotropy group of $\mathcal{A}'$ in $G(I, A, \tilde{\alpha})$ is finite. Moreover, computing with the explicit form of the momentum map $\mu_{\mathbb{C}}$ for $M(V, A, \alpha)$ shows that $\mu_{\mathbb{C}}(a')$ is co-central, too. According to the Root Lemma and Remark 1 the dimension $\tilde{\alpha}$ of $\mathcal{A}'$ is a real root and thus we get a contradiction to the genericity assumption: $\chi(\tilde{\alpha}) \neq 0$ for all real roots $\tilde{\alpha}$ in $\tilde{X}_*$. $\qquad \square$

We have now assembled the essential facts to show that the diagram

$$M_Z(\Gamma)//G(\Gamma) \quad \longleftarrow \quad M_Z(\Gamma)//{}^\chi G(\Gamma)$$
$$\downarrow \qquad\qquad\qquad\qquad\qquad \downarrow$$
$$Z \qquad \stackrel{=}{\longleftarrow} \qquad Z$$

is a simultaneous minimal resolution for generic χ.

1) It is shown by Kronheimer ([25], [26]) that any fibre $\mu_{\mathbb{C}}^{-1}(z), z \in Z$, contains (a dense open set of) $G(\Gamma)$-stable points and that the singular points of $\mu_{\mathbb{C}}^{-1}(z)//G(\Gamma)$ correspond to non-stable (closed) orbits. According to the general properties of linear modifications alluded to in Section 6 it is sufficient to show that $M_Z//{}^\chi G(\Gamma)$ and the map $M_Z//{}^\chi G(\Gamma) \to Z$ are smooth.

2) $M_Z(\Gamma)//{}^\chi G(\Gamma)$ is the orbit space of $M_Z(\Gamma)(\chi$-semistable) $= M_Z(\Gamma)(\chi$-stable) (recall the Stability Lemma) by the action of $G(\Gamma)$. Since $G(\Gamma)$ acts with finite isotropy groups it acts freely, in fact (cf. Remark 2). Thus, $M_Z(\Gamma)//{}^\chi G(\Gamma)$ is smooth as soon as $M_Z(\Gamma)(\chi$-stable) is smooth. However, this follows from the fact (cf. Lemma 3, Section 3) that $\mu_{\mathbb{C}}$ is smooth at all points of $M(\Gamma)(\chi$-stable) and that Z is smooth in $(\mathrm{Lie}\, G(\Gamma))^*$.

3) Since $M_Z(\Gamma)(\chi$-stable) $\xrightarrow{\mu_{\mathbb{C}}} Z$ is smooth, the smoothness of

$$M_Z(\Gamma)//{}^\chi G(\Gamma) \to Z$$

follows by commutativity

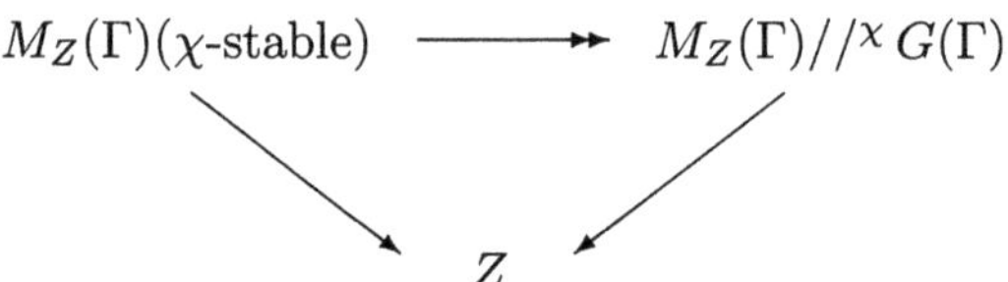

4) Finally, all the fibres $\mu_{\mathbb{C}}^{-1}(z)//{}^\chi G(\Gamma)$ are minimal resolutions of the singular fibres $\mu_{\mathbb{C}}^{-1}(z)//G(\Gamma)$. This follows from an interpretation of the $\mu_{\mathbb{C}}^{-1}(z)//{}^\chi G(\Gamma)$ as symplectic quotients of the symplectic manifold $M_Z(\Gamma)(\chi$-stable) by the freely acting group $G(\Gamma)$ (cf. Section 3, Proposition). As is well known, the adjunction formula then implies that all exceptional components must be rational curves of self-intersection -2, i.e. $\mu_{\mathbb{C}}^{-1}(z)//{}^\chi G(\Gamma)$ must be a minimal resolution.

Remarks. 1) The strategy for the above proof is different from the original one in [9] which is based on an analysis of exceptional components. In fact, candidates for exceptional components can be obtained as follows:

Let Ω denote an orientation of the quiver (V, A) (non-cyclic in the case $\tilde{A}_r$) and let

$$C_\Omega := \{(\varphi_a, a \in A) \in M(\Gamma) \mid \varphi_a = 0 \text{ for all } a \notin \Omega\}.$$

Then C_Ω is a vector space of dimension $|\Gamma| = \sum_{i=0}^{r} d_i^2$ contained in the "nilpotent" variety $\mathcal{N} = \mu_{\mathbb{C}}^{-1}(0)(G(\Gamma)\text{-unstable})$.

According to Lusztig, $\mathcal{N}$ is pure dimensional of dimension $|\Gamma|$ (cf. [28], [29]), and thus C_Ω is a component of $\mathcal{N}$. It follows from results of Happel (cf. [9] Kap. 5) that $C_\Omega//^\chi G(\Gamma)$ is either empty or a projective line. In the case of type $\tilde{A}_r$ all components of $\mathcal{N}$ are of this form, and it is possible to reconstruct the exceptional fibre of

$$\mu_{\mathbb{C}}^{-1}(0)//G(\Gamma) \longleftarrow \mu_{\mathbb{C}}^{-1}(0)//^\chi G(\Gamma)$$

by analysing the Ω with C_Ω (χ-stable) $\neq \emptyset$. However, in all other cases there are vastly more components of $\mathcal{N}$ (cf. e.g. [28]), and it is easily seen that additional components, different from the C_Ω, enter the resolution picture.

2) In case $\chi \in X^*(G(\Gamma))$ is not generic one obtains partial resolutions whose type depends on the positon of χ with respect to the walls (root hypersurfaces) in $X^*(G(\Gamma))$ (in this situation $\mu_{\mathbb{C}}^{-1}(0)//^\chi G(\Gamma)$ is a singular surface with Kleinian singularities whose types correspond to the types of the irreducible components of the sub-root-system $\Phi_\chi = \{\alpha \in X_* \,|\, \|\alpha\|^2 = 2, \langle \chi, \alpha \rangle = 0\}$; in case $\tilde{A}_r$ this can be checked explicitly, cf. [9]).

3) In case of the cyclic groups $\Gamma = \mathcal{C}_{n+1}, \tilde{\Delta}(\Gamma) = \tilde{A}_n$, the various linear modifications can be described quite explicitly. Among other things, one obtains Brieskorn's original formulae for the simultaneous resolution ([5], cf. also [20], [39]) if one chooses χ to be minimally generic (i.e. the half-sum of positive roots), cf. [9].

4) It should be clear that our proof above is of a purely algebraic-geometric nature and works at least over arbitrary algebraically closed fields of characteristic zero. Over the complex numbers, one is tempted to invoke a general comparison result between linear modifications and "shifted" symplectic quotients generalizing Kempf-Ness. On the set-theoretic level, such a result can be found in [21]. However, up to now this doesn't seem to have been extended to the topological level.

References

[1] ABRAHAM, R., MARSDEN, J.: *Foundations of Mechanics.* Benjamin/ Cummings, Menlo Park, 1978.

[2] ARNOLD, V.I.: *Mathematical Methods of Classical Mechanics.* Graduate Texts in Math. **60**, Springer Verlag, 1978.

[3] ARTIN, M., VERDIER, J.L.: *Reflexive modules over rational double points.* Math. Ann. **270**, 79–82 (1985).

[4] BOURBAKI, N.: *Groupes et algèbres de Lie, Chap. IV–VI.* Hermann, Paris, 1968.

[5] BRIESKORN, E.: *Über die Auflösung gewisser Singularitäten von holomorphen Abbildungen.* Math. Annalen **166** (1966), 76–102.

[6] BRIESKORN, E.: *Die Auflösung der rationalen Singularitäten holomorpher Abbildungen.* Math. Annalen **178** (1968), 255–270.

[7] BRIESKORN, E.: *Singular elements of semisimple algebraic groups.* Actes Congr. Int. Math. Nice 1970, t.2, 279–284.

[8] BRION, M., PROCESI, C.: *Action d'un tore dans une variété projective.* Progress in Math. **92**(1990), 509–539.

[9] CASSENS, H.: *Lineare Modifikationen algebraischer Quotienten, Darstellungen des McKay-Köchers und Kleinsche Singularitäten.* Dissertation, Fachbereich Mathematik, Universität Hamburg, June 1994.

[10] DOLGACHEV, I., HU, Y.: *Variation of geometric invariant theory quotients.* preprint, University of Michigan, Ann Arbor, 1993.

[11] DUVAL, P.: *On isolated singularities which do not affect the conditions of adjunction I, II, III.* Proc. Cambridge Phil. Soc. **30**, 453–459, 483–491 (1934).

[12] ESNAULT, H., KNÖRRER, H.: *Reflexive modules over rational double points.* Math. Ann. **272**, 545–548 (1985).

[13] ESNAULT, H.: *Reflexive modules on quotient singularities.* J. Reine Angew. Math. **362**, 63–71 (1985).

[14] FORD, D., McKAY, J.: *Representations and Coxeter graphs.* in "The Geometric Vein" (The Coxeter-Festschrift), Ed. Ch. Davis, B. Grünbaum, F.A. Sherk, Springer Verlag, 1981, pp. 549–554.

[15] GABRIEL, P.: *Unzerlegbare Darstellungen I.* Manuscripta Math. **6**(1972), 71–103.

[16] GONZALES-SPRINBERG, G., VERDIER, J.L.: *Construction géometrique de la correspondance de McKay.* Ann. Sci E.N.S. **16**, 409–449 (1983).

[17] GUILLEMIN, V., STERNBERG, S.: *Symplectic techniques in physics.* Cambridge University Press, 1984.

[18] HARTSHORNE, R.: *Algebraic Geometry.* Springer Graduate Text **52**, Springer Verlag, Berlin Heidelberg New York, 1977.

[19] KAC, V.G.: *Infinite dimensional Lie algebras.* 3rd ed., Cambridge University Press, 1990.

[20] KAS, A.: *On the resolution of certain holomorphic mappings.* in "Global Analysis, Papers in honour of K. Kodaira", Ed.P.C. Spencer, S. Iyanaga, Princeton Univ. Press, Princeton, 1969, 289–294.

[21] KING, A.D.: *Moduli of representations of finite dimensional algebras.* Quarterly Journal of Math. Oxford (2) **45** (1994), 515–530.

[22] KLEIN, F.: *Vorlesungen über das Ikosaeder und die Auflösung der Gleichungen vom fünften Grade.* Teubner, Leipzig, 1884, and Birkhäuser, Basel, 1993. English translation : Lectures on the icosahedron and the solution of equations of the fifth degree. Trübner and Co., London, 1888, 1913, Dover, New York, 1956.

[23] KNÖRRER, H. *Group representations and the resolution of rational double points.* Contemporary Mathematics (AMS) Vol. **45** (1985) 175–222.

[24] KRONHEIMER P.B.: *Instantons gravitationnels et singularités de Klein.* C. R. Acad. Sci. Paris, **303** Série I (1986), 53–55.

[25] KRONHEIMER P.B.: Oxford Ph. D. Thesis, 1987.

[26] KRONHEIMER P.B.: *The construction of ALE spaces as hyper-Kähler quotients.* Journ. of Diff. Geometry **29**(1989), 665–683.

[27] LUNA, D.: *Slices étales.* Bull. Soc. Math. France, Mémoire **33** (1973), 81–105.

[28] LUSZTIG, G.: *Canonical bases arising from quantized enveloping algebras II.* Progr. Theor. Phys. **102** (1990).

[29] LUSZTIG, G.: *Quivers, perverse sheaves, and quantized enveloping algebras.* J. AMS **4(2)**, 1991, 365–42.

[30] LUSZTIG, G.: *Affine Quivers and Canonical Bases.* Publ. Math IHES **76** (1992), 111–163.

[31] MCKAY, J.: *Graphs, singularities, and finite groups.* Proc. Symp. Pure Math. Vol. **37**, 183–186 (1980).

[32] MUMFORD, D., FOGARTY, J., KIRWAN, F.: *Geometric Invariant Theory.* 3rd ed., Springer Verlag, 1994.

[33] NEWSTEAD, P.: *Introduction to moduli problems and orbit spaces.* TIFR Lecture Notes, Springer Verlag, 1978.

[34] REID, M.: *What is a flip?* Preprint Warwick University, 1993.

[35] SARDO-INFIRRI, A.V.: *Resolutions of Orbifold Singularities and Representation Moduli of McKay Quivers*. Ph.D. dissertation, Oxford University, 1994, also: preprint No. **984**, RIMS, Kyoto-University, 1994.

[36] SLODOWY, P.: *Simple singularities and simple algebraic groups*. Springer Lecture Notes in Math. **815**, 1980.

[37] SLODOWY, P.: *Platonic solids, Kleinian singularities, and Lie groups.* in "Algebraic Geometry", Proc., Ann Arbor 1981, Ed. I. Dolgachev, Springer Lecture Notes in Math. **1008**, 102–138(1983).

[38] THADDEUS, M.: *Geometric invariant theory and flips.* preprint, Oxford University, 1994, *alg-geom* 9405004

[39] TJURINA, G.N.: *Resolutions of flat deformations of rational double points.* Functional Anal. Appl. **4**, (1), 68–73(1970).

Progress in Mathematics, Vol. 162, © 1998 Birkhäuser Verlag Basel/Switzerland

Seventeen Obstacles for Resolution of Singularities

Herwig Hauser
Mathematisches Institut
Universität Innsbruck
A-6020 Innsbruck
AUSTRIA

Introduction

This is the first of a series of papers related to resolution of singularities. We present here examples which explain why many arguments and proofs work in special situations, say small dimension or zero characteristic, but fail in general. This exhibits in particular the delicacy of resolution of singularity for arbitrary excellent schemes. The examples were originally assembled for the author's personal records. They might be of some interest to a larger audience, especially to readers for whom the flavour of resolution of singularities is concealed by technique.

We shall concentrate here on the classical approach developed by Zariski, Abhyankar, Hironaka and several other mathematicians towards a constructive proof of resolution of singularities by a sequence of well chosen monoidal transformations [A1, H1, BM1, Sp1, V1, Z1]. The basic idea is to construct sufficiently fine local invariants of singularities which determine the center of blowing up at each stage as the locus of points on the variety where the invariants take their maximal values and to measure the improvement the variety undergoes when passing to the blown up variety by comparing these and possibly further invariants before and after the blowup.

At present there is no completely satisfying answer to this objective. One reason is the lack of conceptuality of the proposed and studied invariants, already apparent in characteristic 0, the other, which is partly a consequence of the first, that the known invariants sometimes behave badly under blowup in positive characteristic.

We shall set up a catalogue of cautions one has to be aware of when using standard resolution invariants or searching new ones. In principle, there are two types of local arguments involved: a combinatorial one, which investigates resolution with respect to fixed coordinate systems and only treats the origins of the charts in the blown up varieties, and a reduction argument, which intends to reduce the general situation where neither coordinates nor the point of the exceptional divisor are specified to the combinatorial situation.

The combinatorial problem is also known as Hironaka's polyhedral game. In the context of valuations it has been solved by Zariski [Z2, p. 861]. The solution proposed by Spivakovsky [Sp2] relies on invariants already considered by Hironaka. However, these invariants cause difficulties for the reduction problem, especially in positive characteristic, which have not been overcome yet, see examples 16 and 17. We meet here a typical phenomenon of resolution of singularities: the various detail problems have to be solved *coherently* and *simultaneously* in the sense that each separate solution to one problem only serves if it has been suitably adjusted with respect to the solutions of the other problems.

Our exposition is confined to the local problem, i.e. the definition of local invariants of singularities and their behaviour under blowing up. This is the core of the problem, though global aspects are not to be neglected. We shall deprive our presentation from technical and sophisticated decoration. The main obstructions already appear in simple and elementary circumstances.

Recently, there has opened an entirely new approach to resolution of singularities through work of de Jong. His method combines induction based on semistable reduction with arguments from toric geometry and proves resolution of singularities up to a finite map in any characteristic [J]. See Berthelot's Bourbaki note [Be] for another account on this as well as various applications. There are subsequent papers of Abramovich, de Jong, respectively, Bogomolov and Pantev extending these techniques and proving on a few pages a weak version of Hironaka's theorem in characteristic zero [AJ, BP]. The difference to Hironaka's version is that centers are blown up which are not necessarily contained in the singular locus of the variety and hence the variety is also modified in regular points.

Let us formulate the strong version of resolution of singularities in the context of varieties over a field:

Embedded Resolution of Singularities. *Given a reduced variety X over a field $\mathbf{K}$ and embedded in a smooth variety W, construct a smooth variety $\tilde{W}$ and a proper birational morphism $\pi : \tilde{W} \to W$ such that the inverse image $\tilde{X}$ of X under π has only normal crossing singularities and such that the restriction of π over the smooth points of X is an isomorphism.*

Recall that a variety is said to have normal crossing singularities if it consists of a union of smooth components intersecting transversally, more precisely, such that at any intersection point there is a coordinate system for which the variety is defined locally by monomials. The most explicit way to construct $\tilde{W}$ is by a finite sequence of well chosen blowups. This requires to determine the centers of each blowup as a subvariety of the singular locus Sing X of X and to show that after a finite number of steps only normal crossing singularities are left.

It is convenient, even though a priori more difficult, to treat the embedded situation. This allows to work with regular systems of parameters which are very helpful to define local invariants. Then to prove embedded resolution of singularities of varieties it suffices to solve the following problem:

Inductive Resolution Problem. *Given a field* $\mathbf{K}$ *and a smooth variety* W *over* $\mathbf{K}$*, construct the following objects:*

(1) *Centers: A map*

$$Z : \{reduced\ subvarieties\ X\ of\ W\} \to \{subvarieties\ of\ \operatorname{Sing} X\}$$

such that blowing up W *in* Z_X *gives a smooth variety* $\tilde{W}$*.*

(2) *Resolution invariant: A well-ordered set* Γ *and a map*

$$i : \{reduced\ subvarieties\ X\ of\ W\ with\ a\ point\ p\ on\ X\} \to \Gamma$$

such that for $\tilde{X} = \pi^{-1}X$ *the following conditions are satisfied:*

(i) $i_p X = 0$ *if and only if* X *has normal crossing at* p*.*

(ii) $i_{p'}\tilde{X} < i_p X$ *for all* $p \in Z_X$ *and* $p' \in D = \pi^{-1}Z_X$*.*

(iii) i *is upper semicontinuous w.r.t.* p*.*

(iv) $\{p \in X,\ i_p X\ maximal\ on\ X\} \subset Z_X$*.*

As a variant, replace (i) by $i_p X = 0$ if and only if X is smooth at p and in (ii) the total transform $\tilde{X}$ by the strict transform X', cf. Section 1. Observe that in contrast to the classical notion of permissible centers as in [H1, III.5] we allow singular and even non-reduced centers Z_X, provided W' is smooth, see Section 1. and the discussion after example 3. Some authors require equality in (iv), i.e., that Z_X is the smallest stratum of the stratification of X induced by i [BM1].

The resolution invariant i is usually a vector of integer or rational invariants equipped with the lexicographic order. Typically it is shown that its first component does not increase under blowup. If it remains constant, the situation must be sufficiently specific to be able to show that the second component does not increase. Then the argument is repeated, until one arrives at a component which strictly decreases.

The definition of the centers is a global problem, whereas the invariants are of local nature. However, often local invariants are used to define the centers locally, and then patching arguments are needed to show that the local constructions expand to give global centers [V2]. Some authors restrict even further the nature of the invariants as to be defined along a valuation, cf. with the concept of uniformization [Z2, Sp1]. It then suffices to prove that the invariant has dropped only at the point of the exceptional divisor which is determined by the valuation.

The four conditions above allow to apply the following induction argument: The stratification of X by subvarieties along which i is constant is locally finite by (iii) and since X is Noetherian. By (ii) and (iv) each point of the worst stratum $\{p \in X,\ i_p X \text{ maximal on } X\}$ improves under the blowup. Hence the maximal value of i on X drops when passing to X'. As Γ is well ordered, it becomes 0 in a finite number of steps, and (i) applies.

Let us comment on the induction invariants usually considered. The most important one is the order of the defining ideal, i.e. the largest power of the maximal ideal whichcontains the ideal. For hypersurfaces it is just the order of the series expansion of f w.r.t. some coordinate system. By a theorem of Nagata-Zariski the order does not increase under localization, i.e. is upper semicontinuous [H1, p. 218]. Also it is known not to increase under permissible blowup in any dimension and characteristic (infinitesimal upper semicontinuity.) The concept of multiplicity has been extended by Hironaka to the non hypersurface case by introducing the finite sequence ν^* of integers given as the orders of the elements of a standard basis of the defining ideal, listed increasingly and ordered lexicographically, cf. [H1, III.1] and Section 2.(5) below. It does not increase under permissible blowing up [H1, III. Thm. 3], but may behave badly under localization, see example 11. Bennett suggested to replace ν^* by the Hilbert-Samuel function of the local ring and established its infinitesimal and ordinary semicontinuity for arbitrary excellent schemes [Bn, Si]. It is generally accepted today (though not the conviction of the author) that the Hilbert-Samuel function should be the first invariant to be considered, at least for the definition of the centers.

The next step is to investigate the situation where, say, the Hilbert-Samuel function remains constant under blowup. Further invariants are then necessary. There are several options appearing in the literature, depending on the context. Hironaka and Abhyankar propose, already in the hypersurface case, to take the minimal number of variables necessary to define the initial form of f, cf. Section 2.(2) and example 7. Other invariants can be constructed from the Newton polyhedron associated to the power series expansion of f. These are often generalizations of the notion of maximal slope for plane curves, sometimes known as weighted orders.

A very intricate problem consists in measuring the singularities of the equimultiple locus. For technical reasons we have to be vague here and must refer the reader to [H5, BM2, V1] for more details. Simplifying first the singularities of the equimultiple locus is crucial in order to be able to define sufficiently big centers for resolving the variety itself, cf. example 1. This requires to apply induction on invariants related to the equimultiple locus. Therefore the local invariants associated to X will also involve invariants of 'smaller' varieties. In characteristic 0, a precise meaning can be given to the comparative 'smaller' through the local embedding dimension and the theory of idealistic presentations, but this fails in positive characteristic and still remains unsettled, cf. example 8. We know of no appropriate method in characteristic $p > 0$ to handle this difficulty in arbitrary dimension. See [A2] for an explicit description of the problems in the curve case.

Most of the examples below illustrate situations where certain of these invariants do not to satisfy (ii) or (iii) in general circumstances. Other examples point at problems which occur in the definition of the centers and show why the construction of new invariants is very subtle and delicate. In [Ha1, Ha2, Ha3, Ha6] we develop techniques which allow the systematic construction of invariants and provide a uniform frame for many of the known invariants. This seems to be a prerequisite towards a conceptual approach to the inductive resolution problem. Up to now, many invariants and resolution arguments are established ad hoc by defining all kinds of distinguished coordinate systems and introducing some numerical data associated to them. You may find coordinates which are prepared, well-prepared, very well-prepared [H3], tangential, normal [H6], good, essential, nontangential [A4], essential [BM1], good, quasi-good [LO], maximal clean, Zariski clean [Mo1, Mo2], convenient [V2], just to list a few of the used labels. All these notions follow the same pattern. Namely, they can be characterized as coordinates which *maximize* some numerical datum attached to the singularity [Ha1].

We do not discuss in this article methods and topics as normalization, Nash modification, toroidal resolution, uniformization, rectilinearization, valuation theory, resolution of surfaces or quasi-ordinary singularities, see for instance [Br1, Br2, GT, KKMS, La, Lp2, Ok, OW, Su, Z2, Z5, Z6]. For further introductions and surveys we refer to [A2, A3, A4, Ar, Bd, BM3, CGO, Co3, Gi1, Gr1, H6, H7, Lp1, Od1, Od3, SG1, V3]. References for some historical papers are [Al, Ch, Ju, Lv1, Lv2, N, NB, W]. We apologize for any omission or incompleteness which might have occurred in citing results and techniques from the literature.

The preparation of this article was supported by a grants from the Austrian-Spanish cooperation program Acciones Integradas and the D. Swarovski Forschungsförderungsfonds. We thank O. Villamayor, V. Cossart and M. Spivakovsky for valuable suggestions and references.

1   Basics

We start by recalling some fundamental notions on blowups. References are e.g. [H1, III.2, H3, H6, Ha5, Hs, II.7, Gi1, HIO, Bb, Gr2, IV.7, K]. Let us place in some n-dimensional affine space $\mathbf{A}^n$ over a field $\mathbf{K}$ and fix a point p in $\mathbf{A}^n$. Since we are interested in local questions at p, we accordingly fix a regular local Noetherian ring $\mathbf{R}$ of dimension n and maximal ideal $\mathbf{M}$. Without to much loose of generality we suppose that $\mathbf{R}$ is complete. Moreover we shall assume $\mathbf{R}$ equicharacteristic, so that, by Cohen's Structure Theorem, $\mathbf{R}$ can be viewn as a formal power series ring in n variables over $\mathbf{K}$. With the exception of example 7, the reader may think of $\mathbf{K}$ as being algebraically closed.

Any non-zero ideal $\mathbf{P}$ of $\mathbf{R}$ gives rise to the Rees algebra or blowup algebra $\mathbf{S} = \oplus_{i\geq 0}\mathbf{P}^i$ of $\mathbf{R}$ with center $\mathbf{P}$. Here, $\mathbf{P}^0 = \mathbf{R}$, so that $\mathbf{R} \subset \mathbf{S}$. Localizing $\mathbf{S}$ at a maximal ideal $\mathbf{Q}'$ which contains $\mathbf{M}$ and passing to the completion gives a complete local ring $\mathbf{R}'$ and an inclusion $\pi^* : \mathbf{R} \to \mathbf{R}'$ which we shall call the blowup of $\mathbf{R}$ with center $\mathbf{P}$ considered locally at $\mathbf{Q}'$. Geometrically, this corresponds to blow up $\mathbf{A}^n$ in the subvariety Z defined by $\mathbf{P}$ and to look at $\tilde{\mathbf{A}}^n = \operatorname{Proj} \mathbf{S}$ locally at the point p' defined by $\mathbf{Q}'$ in the exceptional divisor $D = \pi^{-1}Z$, where $\pi : \tilde{\mathbf{A}}^n \to \mathbf{A}^n$ denotes the projection associated to $\mathbf{R} \subset \mathbf{S}$.

Any set of generators $a_1,\ldots,a_k$ of $\mathbf{P}$ induces a covering of $\tilde{\mathbf{A}}^n$ by affine charts $\operatorname{Proj} \mathbf{S} = \bigcup_{i=1}^{k} \operatorname{Spec} \mathbf{R}\left[\frac{\mathbf{P}}{a_i}\right] = \bigcup_{i=1}^{k} \operatorname{Spec} \mathbf{R}[\frac{a_1}{a_i},\ldots,\frac{a_k}{a_i}]$. Each chart is specified by selecting one generator a_j. Letting $\mathbf{Q}$ be the ideal of $\mathbf{R}$ generated by the remaining ones we have $\mathbf{R}\left[\frac{\mathbf{P}}{a_j}\right] = \mathbf{R}\left[\frac{\mathbf{Q}}{a_j}\right]$. In particular, $\mathbf{Q}$, or j, determine a point p' in D, namely the origin of this affine chart. More generally, any affine chart on $\tilde{\mathbf{A}}^n$ can be given by selecting an ideal $\mathbf{Q} \subset \mathbf{P}$ of $\mathbf{R}$ with $\mathbf{P}/\mathbf{Q}$ of height 1 in $\mathbf{R}/\mathbf{Q}$. Take any element $a \in \mathbf{P}$ whose class generates $\mathbf{P}/\mathbf{Q}$ and get the chart $\mathbf{R}\left[\frac{\mathbf{Q}}{a}\right]$. It turns out that changing $\mathbf{Q}$ by elements from $\mathbf{P}^2$ does not alter the chart, i.e. that the chart is determined by the image of $\mathbf{Q}$ in $\mathbf{P}/\mathbf{P}^2$. We shall always specify the point p' of the exceptional divisor or the chart we are looking at by giving the ideal $\mathbf{Q}$ of $\mathbf{R}$. Geometrically, this is reflected by the fact that any point of the exceptional divisor corresponds to a direction in the normal cone of Z in $\mathbf{A}^n$.

If $x_1,\ldots,x_n$ are local coordinates of $\mathbf{A}^n$ at p, i.e., a regular system of parameters of $\mathbf{R}$, then $\mathbf{R} \cong \mathbf{K}[[x]]$ and the j-th affine chart $\mathbf{K}[[x]][\frac{\mathbf{P}}{a_j}]$ gives rise to an explicit description when $\mathbf{P}$ is generated by part of the coordinates, say x_i with $i \in J$, where $J \subset \{1,\ldots,n\}$ and $j \in J$. Up to permutation, it suffices to consider $J = \{1,\ldots,j\}$ for some j, say $\mathbf{P} = (x_1,\ldots,x_j)$ and $\mathbf{Q} = (x_1,\ldots,x_{j-1})$. In this way $\mathbf{K}[[x]][\frac{\mathbf{P}}{a_j}]$ equals $\mathbf{K}[[x]][\frac{x_1}{x_j},\ldots,\frac{x_{j-1}}{x_j}]$. Localization and completion at the ideal $(\frac{x_1}{x_j},\ldots,\frac{x_{j-1}}{x_j},x_j,\ldots,x_n)$ gives $\mathbf{R}' = \mathbf{K}[[\frac{x_1}{x_j},\ldots,\frac{x_{j-1}}{x_j},x_j,\ldots,x_n]]$, which is again a regular complete Noetherian local ring of dimension n. Hence setting $y_i = \frac{x_i}{x_j}$ for $i < j$ and $y_i = x_i$ else defines a regular system of parameters for $\mathbf{R}'$. The map $\pi^* : \mathbf{R} \to \mathbf{R}'$ is then given by $x_i \to y_iy_j$ for $i < j$ and $x_i \to y_i$ for $i \geq j$.

In this situation we say that the local blowing up is *monomial of type* $j \in J$ w.r.t. the coordinates x. The coordinates y in $\mathbf{R}'$ are called the induced coordinates, often denoted again by x for convenience. The exceptional divisor D is defined by the principal ideal (y_j).

Observe that the map π^* is still very simple when $\mathbf{P}$ is generated by monomials in the variables, though $\mathbf{R}'$ may then fail to be regular, cf. example 3. Compare this with toric modifications and toric singularities [GT, KKMS, Ok].

Consider now a variety X in $\mathbf{A}^n$. Let $\mathbf{I} \subset \mathbf{R}$ be the ideal defining X locally at p. Its total transform $\tilde{X}$ is the inverse image of X under $\pi : \tilde{\mathbf{A}}^n \to \mathbf{A}^n$. In the j-th affine chart it is defined by the ideal $\tilde{\mathbf{I}} = (\pi^*\mathbf{I})\mathbf{R}'$ of $\mathbf{R}'$. We shall write $\tilde{\mathbf{I}} = \mathbf{IR}'$ for short. The strict transform X' of X is the closure in $\tilde{\mathbf{A}}^n$ of the inverse image of $X \setminus Z$ under π. Its ideal is $\mathbf{I}' = \bigcup_{m \geq 0} y_j^{-m}(\mathbf{I} \cap \mathbf{P}^m)\mathbf{R}'$. In particular, for hypersurfaces f, the strict transform equals $f' = y_j^{-o}f$ where o denotes the order of f w.r.t. $\mathbf{P}$, $o = \max\{m, f \in \mathbf{P}^m\}$. For arbitrary ideals, the strict transform can be computed in terms of generators of the ideal by means of standard bases [H1, III.2].

The center $\mathbf{P}$ is permissible for a hypersurface X if the order of f w.r.t. $\mathbf{P}$ and $\mathbf{M}$ is the same, i.e., if Z is contained in the equimultiple locus of X. In general, $\mathbf{P}$ is permissible if the Hilbert-Samuel function of X is constant along Z [Bn, Thm. 3]. In the literature, Z is assumed smooth, but we do not impose this restriction here.

Assume that we are in the monomial situation of type $j \in J$ w.r.t. coordinates x in $\mathbf{R}$, i.e. $\mathbf{P} = (x_1, \ldots, x_j)$ and $\mathbf{Q} = (x_1, \ldots, x_{j-1})$. Let y be the induced coordinates in $\mathbf{R}'$. Denote by $f = \sum c_\alpha x^\alpha$ the expansion of f w.r.t. x and let f' be its strict transform. Since π^* is given by substitution of the variables by monomials, f' has expansion $f' = \sum c_\alpha y^{\alpha^*}$ where the map $*$: supp $f \to$ supp f' is defined by the following formula:

$$\alpha_i^* = \alpha_i \ \text{ for } \ i \neq j, \qquad \alpha_j^* = \alpha_j + (\alpha_1 + \ldots + \alpha_{j-1} - o).$$

Here supp $f = \{\alpha \in \mathbb{N}^n, c_\alpha \neq 0\}$ denotes the support of f. In the examples below we shall always use this formula to compute the strict transform. Let N_f denote the Newton polyhedron of f, i.e., the convex hull in R_+^n of supp $f + \mathrm{R}_+^n$. As $*$ is an affine map of R^n the polyhedron $N_{f'}$ of the strict transform is computed by taking the convex hull of the images under $*$ of the vertices of N_f. The improvement of $N_{f'}$ relative to N_f can be read off in various ways. However, as the Newton polyhedron depends on the chosen coordinates such a measure need not be a local invariant of the singularity. It has to be made coordinate independent, at least to a certain extent. Among the various attempts to extract coordinate free information from the Newton polyhedron, e.g. [CP, H2, LT, Y], none suffices to construct an invariant i satisfying (i) to (iv) for arbitrary characteristic. We believe that this is one of the central difficulties in resolution problems. Note that the hypersurface $f = 0$ has become a normal crossing singularity at a point of the exceptional divisor if and only if the Newton polyhedron of the total transform has precisely one vertex α, and the strict transform of f has become smooth if and only if its Newton polyhedron has a vertex of total degree 1.

2 Examples

Several of the following examples can be found in the literature or belong to what some people call folklore. The remaining ones have been constructed to clarify the author's view on the subject. In the sequel, f' always denotes the strict transform of f. The reader is asked to check its asserted form in the examples by computing it in all charts according to the formula for $*$ given above. Occasionally we include *exercises* and *problems*. The first are mostly computational and shall consolidate the understanding. For the latter, a complete answer is not always known yet.

(1) Centers of Blowup. For varieties with isolated singularities there is no ambiguity how to choose the center. Only the singular points are permissible. For non-isolated singularities unfortunate choices of the center may yield the same or worse singularities.

EXAMPLE 1. Let $f = x^2 + yz$ be the rational double point with isolated singularity at the origin. Blowing up 0 resolves the singularity. For instance, taking the chart corresponding to $\mathbf{Q} = (x, y)$ gives $f' = x^2 + y$ smooth.

Let $f = x^2 + y^2 z$ be the Whitney umbrella. The singular locus is the z-axis. Since f has order two in each singular point the axis coincides with the equi-multiple locus of f ($=$ locus of points of maximal multiplicity.) Two centers are permissible, the origin and the z-axis (provided that the variety is considered locally at 0.) *Exercise*: Blowing up 0 reproduces the singularity in the chart $\mathbf{Q} = (x, y)$ whereas blowing up the z-axis resolves the singularity. This contradicts our intuition of stratifying the variety according to the complexity of the singularities and to take as center the worst stratum. On first view the origin of the Whitney umbrella seems to be worse than the singularities on the z-axis. The example indicates that this naive stratification is too subtle. We have to look for a coarser one. A good recipe is to blow up permissible centers of maximal possible dimension. Such choice is supported by the fact that each singular point has to belong to a center at least once in the resolution process, else it would remain singular until the end. This is an equilibrium problem. Roughly speaking, the centers should be as coarse, the invariants as fine as possible.

Modify the Whitney umbrella to $f = x^2 + y^m z$ with $m \geq 3$. Then for $\mathbf{P} = \mathbf{M}$ and $\mathbf{Q} = (x, y)$ we get $f' = x^2 + y^m z^{m-1}$ which is worse than f in whatever sense one may think of. Again we have to blow up the z-axis to improve the situation: in the chart $\mathbf{Q} = (y)$ the multiplicity drops, in the chart $\mathbf{Q} = (x)$ we get $f' = x^2 + y^{m-2} z$. *Problem*: Give a coordinate free description of the improvement of the singularity in the second chart.

EXAMPLE 2. The Theorem of Beppo Levi asserts that for surfaces whose equimultiple locus has only normal crossing singularities a finite sequence of blowups in smooth permissible centers of maximal dimension makes the multiplicity drop [Lv1, Lv2, Z1, p. 522]. The algorithm is not canonical since at certain stages several choices of centers may be possible. The same philosophy fails in higher dimension without further specification of the centers. Spivakovsky gave in [Sp4] an example for this in dimension 4, a variant of which is presented here: Let $f = x^3 + yz^2 w^4 + yz^4 w^2$. One checks by inspection that all coordinate axes except the x-axis are permissible. Blowing up the w-axis, i.e., $\mathbf{P} = (x, y, z)$, the strict transform of f at the point $\mathbf{Q} = (x, z)$ equals $f' = x^3 + z^2 w^4 + y^2 z^4 w^2$. Now the y- and z-axis are permissible. Blowing up $\mathbf{P} = (x, y, w)$ gives $f'' = f$ at the point corresponding to $\mathbf{Q} = (x, w)$. We have run into a cycle of length two.

Observe that each choice of center provides an improvement of the singularity w.r.t. certain variables and a deterioration w.r.t. others. It is then clear that the combination of two blowups may possibly cancel these changes, and that is what happens in the example. A first reaction would be to choose the second blowup according to the first one in order to add the improvements and the deteriorations and then to conclude that if the singularity has sufficiently improved w.r.t. to some variables, it has become smooth, regardless the effect w.r.t. the other variables. We know of no effective and intrinsic criterion for this, so let us formulate it as a *Problem*: measure asymmetric improvements and define the subsequent center depending on the prior blowups and this measure.

A second possible answer is as follows: In the first blowup the situation is symmetric w.r.t. z and w. Choosing arbitrarily one of the axes as center destroys this symmetry. The resulting algorithm is no longer canonical. Imagine that the local symmetry of the variety is not induced from a global symmetry. As we consider local invariants to define the center, we will never detect the global asymmetry and it is not reflected in our choice. To avoid this phenomenon we may blow up lower dimensional centers in presence of symmetries. In the example, instead of choosing one axis, we can blow up the origin. This choice is canonical, and keeping track of this preparatory blowing up the local symmetry will disappear along the points of the exceptional divisor, since at an intersection point p' of D with the strict transform of the z- or w-axis one equimultiple curve , namely the exceptional curve, is new and one, the strict transform of the equimultiple curve from below, is old, hence they can be distinguished. There may still exist a local or even global symmetry of the blown up variety which interchanges the two curves, but our records will tell us that it is not a symmetry of the resolution process. This allows to choose locally at p' the new center canonically.

This approach has been applied successfully in the past, cf. [BM1, V1]. It burdens enormously the amount of data to define the invariants and notation becomes quite cumbersome. The derived algorithm is usually far from being the most economic one.

Much more effective is to allow in such cases singular centers, with, say, at most normal crossing singularities. The first relevant situation occurs when the equimultiple locus of a surface in $\mathbf{A}^3$ consists of two transversal smooth curves as in the example. Classically, singular centers are not treated as permissible. One reason is that blowing up a smooth variety in a singular center generally creates singularities:

EXAMPLE 3. Blow up two lines in $\mathbf{A}^3$, say $\mathbf{P} = (x, yz)$. In the chart $\mathbf{Q} = (yz)$ we have $\mathbf{R}' = \mathbf{K}[[x, y, z, x/yz]] = \mathbf{K}[[y, z, x/yz]]$ smooth, whereas for $\mathbf{Q} = (x)$ the ring $\mathbf{R}' = \mathbf{K}[[x, y, z, yz/x]] \cong \mathbf{K}[[x, y, z, w]]/(xw - yz)$ is singular.

Pfeifle has determined non reduced structures for normal crossing singularities such that blowing up smooth varieties in the corresponding ideals produces smooth varieties [P]. Since the ideals can be chosen to be monomial, computations are very explicit and can be implemented. For the given example, one such non reduced structure on the two lines in $\mathbf{A}^3$ is the ideal $\mathbf{P}^\sharp = (x, yz)(x, y)(x, z) = (x^3, x^2y, x^2z, xyz, y^2z^2)$. It provides five smooth charts for the blown up variety. For instance, in the chart corresponding to $\mathbf{Q} = (x^3, x^2y, x^2z, y^2z^2)$ we get

$$\mathbf{R}' = \mathbf{K}[[x, y, z, \tfrac{x^3}{xyz}, \tfrac{x^2y}{xyz}, \tfrac{x^2z}{xyz}, \tfrac{y^2z^2}{xyz}]] = \mathbf{K}[[y, \tfrac{x}{y}, \tfrac{yz}{x}]]$$

which is regular. It is not clear whether there is a general pattern to find these non reduced structures on normal crossing singularities. The problem is related to the resolution of toric singularities (which appear when blowing up a reduced normal crossing ideal), cf. [KKMS]. *Exercise*: Find another non-reduced structure on two transversal lines in $\mathbf{A}^3$ so that the blowup $\tilde{\mathbf{A}}^3$ is smooth.

There is a fourth way how to treat multiple possible choices of centers in presence of symmetries. It consists in equipping the variety locally at each of its points with a flag $\mathbf{F}$ of smooth varieties in the ambient space. This corresponds to choosing a maximal chain of regular ideals in $\mathbf{R}$. Assume you blow up a smooth center transversal to all members of the flag. In [Ha2] it is shown that each point of the exceptional divisor then inherits in a natural way a flag $\mathbf{F}'$ induced from the flag below.

This new flag allows to untie symmetries when there appear various candidates as permissible centers by considering their position relative to the flag. Moreover flags allow to define the notion of subordinate coordinate systems. These are coordinates for which the ideals of the defining chain in $\mathbf{R}$ are generated by decreasing collections of them. The set of all subordinate coordinates is clearly an invariant of the pair $(\mathbf{R}, \mathbf{F})$. Using a generalized Gauss-Bruhat decomposition established in [Ha2] for the automorphism group $\mathrm{Aut}\,\mathbf{R}$ of $\mathbf{R}$ it is possible to show that a whole sector of the Newton polyhedron of a hypersurface f is fixed under automorphisms of $\mathbf{R}$ which preserve subordinate coordinates. Therefore, this portion of vertices is also an invariant of the flagged ambient space $\mathbf{R}$. Now it turns out that subordinate coordinates remain subordinate under blowing up, that is, w.r.t. the induced flag in $\mathbf{R}'$. It follows that the invariant part of the Newton polyhedron corresponding to this sector can be compared explicitly before and after the blowup. This is relevant for constructing components of a suitable resolution invariant i.

(2) Multiplicity. For $f \in \mathbf{R}$ let o respectively o' denote the multiplicity of f and of its strict transform f'. As mentioned above, $o' \leq o$ for any permissible blowing up. In general, equality may hold:

EXAMPLE 4. $f = x^2 + y^5$, $\mathbf{P} = (x, y)$, $\mathbf{Q} = (x)$, then $f' = x^2 + y^3$ has again multiplicity 2. In order to measure the improvement we are led to consider the exponent of the y-monomial. This is not intrinsic since there might be several monomials in the expansion of f to look at, even if we assume f given in Weierstrass form. Besides, the expansion depends on the coordinates. For plane curves as in the example, the classical second invariant after the multiplicity is the slope of the Newton polygon of f. Assume f is given in Weierstrass form $f = x^o + \sum_{i<o} c_i x^i$ with series c_i in y. Then the slope of f is defined as the slope of the first segment of the Newton polygon of f, i.e. $\mathrm{ord}(c_i)/(i - o)$ where i is the largest index for which c_i does not vanish identically. In the example the slope passes from $-2/5$ to $-2/3$. If, in general, it becomes < -1, the multiplicity has dropped. If the slope arrives precisely at -1 the multiplicity drops in the next blow up. This reasoning is not exact since the slope as defined above may vary under changes of coordinates. To make it coordinate independent, take the maximal slope which occurs for all coordinate systems. Then the same assertions hold and form the basis for inductive resolution of plane curve singularities, see [BK] for an extensive treatment, and [A2] for the case of positive characteristic. The notion of maximality is discussed in section 2.(6). There is another constructive way to resolve plane curve singularities via toric modifications defined through the fan associated to the Newton polygon [GT, Ok]. *Problem*: Prove inductively resolution of plane curve singularities by point blowups without using the multiplicity as induction invariant. Such proof becomes necessary when reducing the resolution of surfaces to the plane curve case, see [H3] and [Ha3] for a discussion and an answer. Blowups for which the multiplicity remains constant are characterized in [SG2].

EXAMPLE 5. Blow up the plane curve $f = x^2 - y^2 + y^5$ in the origin. The strict transform intersects the exceptional divisor D in $\tilde{\mathbf{A}}^2$ in two points corresponding to $\mathbf{Q}_1 = (x + y)$ and $\mathbf{Q}_2 = (x - y)$. Applying the coordinate change $(x, y) \rightarrow (x - y, y)$ in $\mathbf{R}$ transforms $\mathbf{Q}_1$ into $\mathbf{Q}_1 = (x)$ and f into $f = x^2 - 2xy + y^5$. The strict transform in this chart is $f' = x^2 - 2x + y^3$ which is smooth at 0 if the characteristic is different 2. For characteristic equal 2 we have $f = (x + y)^2 + y^5$ in the original coordinates and the multiplicity of the strict transform remains 2 at the point $\mathbf{Q}_1$. *Exercise*: Explain how this phenomenon fits into the preceding argument for the maximal slope (cf. [A2]).

EXAMPLE 6. Consider the isolated surface singularity $f = x^2 + y^3 + yz^3$ with the origin as the only permissible center. In the chart $\mathbf{Q} = (x, y)$ we get $f' = x^2 + y^3 z + yz^2$. One monomial has decreased, the other increased. *Exercise*: Prove that something has improved. *Problem*: Express the improvement without using coordinates. You may consult [H3] or [Ha3, Ha4] for a detailed discussion.

For fixed coordinates in $\mathbf{R}$, the multiplicity drops under monomial blowup of type $j \in J$ if and only if the variable x_j occurs in the initial form of f (= homogeneous polynomial of lowest degree of f) defining the tangent cone of the variety at the origin. Hence, for arbitrary blowup, it drops in all charts if and only if the tangent cone of f is not a cartesian product with a line transversal to the center. This observation led Hironaka to define the invariant τ as the minimal number of variables necessary to define the tangent cone of f. Abhyankar proceeds similarly, cf. e.g. [A4, p. 230]. Above we have seen an example (ex. 5) where τ equals 2 in characteristic $\neq 2$ and 1 in characteristic 2. If $\tau = n$ then the multiplicity drops at each point of the exceptional divisor, which suggests to apply induction on $n - \tau$. In characteristic zero, constant multiplicity implies that τ does not decrease [H1, III. Thm. 3]. In three variables, this holds in any characteristic, and allows Hironaka to prove resolution of excellent surfaces by studying further invariants deduced from the Newton polyhedron of f [H3, Co1]. In dimension four he gives the following example of decrease of τ, hence increase of $n - \tau$, an example which initiated a different definition of τ via additive forms in positive characteristic [H4, Gi2, Gi3, Od2, Od3].

EXAMPLE 7. [H4, Thm. 3, p. 331, Od2, p. 300] Let $\mathbf{K}$ be an imperfect field of characteristic 2 admitting elements a and b such that $\mathbf{K}^2(a, b)$ has degree 4 over $\mathbf{K}^2$. Let $f = x^2 + ay^2 + bz^2 + abw^2 + $ (higher order terms). Then $\tau = 4$. Blow up the maximal ideal $\mathbf{M}$. The ideal $\mathbf{Q}' = (x + yz, y^2 + b, z^2 + a)$ is maximal in the Rees algebra $\mathbf{S}$ and has residue field $\neq \mathbf{K}$. Then up to terms of degree ≥ 3 the strict transform at $\mathbf{Q}'$ equals $f' = x^2 + ay^2 + bz^2 + ab = (x+yz)^2 - (y^2+b)(z^2+a)$ and hence $\tau' = 3$. This type of phenomenon can of course be avoided by working with algebraically closed ground fields.

(3) Equimultiple locus. One important ingredient in resolution arguments is induction on the number of variables, i.e., on the local embedding dimension. As its definition involves derivatives, it is natural to expect complications in positive characteristic. This happens in particular in the theory of maximal contact. The basic observation in this theory is the simple fact that the equimultiple locus S of a hypersurface singularity in $\mathbf{A}^n$ has local embedding dimension less than n, provided the characteristic is zero. For given f, let $\mathbf{J}$ be the ideal of $\mathbf{R}$ generated by all partial derivatives of f up to degree less than the multiplicity of f. Its zero locus is S. By definition of the multiplicity of f, one of the derivatives appearing among the generators of $\mathbf{J}$ must have order 1 and therefore defines a smooth hypersurface. As S is contained in it, the assertion follows.

The theory of maximal contact and of idealistic presentations allows, by induction, to resolve first the singularities of S. Once S has only normal crossing singularities its components can be chosen as centers of further blowups in order to improve the variety itself. In general, the equimultiple locus of a hypersurface is not a hypersurface. This forces Hironaka to use resolution of non hypersurfaces of lower embedding dimension to resolve hypersurfaces [H5]. Bierstone and Milman have modified the argument as to get a proof for hypersurfaces in zero characteristic which remains inside this class [BM1].

The reasoning on the embedding dimension breaks completely down in positive characteristic.

EXAMPLE 8. Let $\mathbf{K}$ be a field of characteristic 2. Consider $f = x^2 + yz^3 + zw^3 + y^7w$ of order 2 at 0 and embedding dimension 4. The equimultiple locus S is given by the first order partial derivatives of f, say $\mathbf{J} = (0, z^3 + y^6w, yz^2 + w^3, zw^2 + y^7)$. It has embedding dimension 4 since the monomial curve C parametrized by t^{32}, t^7, t^{19}, t^{15} lies in S but cannot be embedded locally at 0 into $\mathbf{A}^3$.

Compare this with the induction on the number of variables based on the Weierstrass form of a series and the application of Tschirnhausen transformations [A1, A5, BM2, Co2, Mo1, Mo2, Su]. Again this fails in positive characteristic.

There are attempts by Giraud to extend the concept of maximal contact to positive characteristic [Gi2, Gi3], but his theory is not fully applicable. We are not aware of any explicit substitute for the local embedding dimension as induction invariant for arbitrary characteristic.

Instead of the local embedding dimension one could try with the dimension of the equimultiple locus. It turns out that it varies quite arbitrarily under blowing up and hence does not serve as a resolution invariant:

EXAMPLE 9. Take $f = x^2 + y^4 + z^4$ with isolated singularity at 0 in zero characteristic. Blow up the origin and consider the chart $\mathbf{Q} = (x, y)$. Then $f' = x^2 + y^4z^2 + z^2 = x^2 + (z\sqrt{y^4 + 1})$. This singularity is locally isomorphic to the cartesian product of the plane curve $x^2 + z^2$ with a line which forms its equimultiple locus.

For a detailed study of equimultiple curves and their behaviour under blowing up, see [Z1, Z6]. Zariski shows that for surfaces the equimultiple locus of the blown up variety can at most consist of the strict transform of the old equimultiple locus plus possibly some new but smooth equimultiple curve. This allows to reduce always by point blowups to a normal crossing equimultiple locus. This is false for threefolds:

EXAMPLE 10. The singular and equimultiple locus may become worse under blowing up. Consider the hypersurface $f = x^2 + y^3 + z^2w + w^4$ in characteristic zero. Its Jacobian ideal is $j(f) = (x, y^2, zw, z^2 + 4w^3)$. Hence the singularity at 0 is isolated of multiplicity 2. Blow up the origin and look at the chart corresponding to $\mathbf{Q} = (x, y, z)$. The strict transform of f is given by $f' = x^2 + y^3w + z^2w + w^2$. It has Jacobian ideal $j(f') = (x, y^2w, zw, y^3 + z^2 + 2w)$. The singular locus of X' is the singular curve $y^3 + z^2 = 0$ inside the plane $x = w = 0$. As the multiplicity is 2, it coincides with the two-fold locus of X'.

EXAMPLE 11. The number of strata appearing in the equimultiple stratification of X at 0 may increase. Consider the hypersurface $f = x^3 + y^4 + yz^2w + z^4w + w^6$ in zero characteristic with Jacobian ideal $j(f) = (x^2, 4y^3 + z^2w, (y + 2z^2)zw, yz^2 + z^4 + 6w^5)$. Any singular point close to 0 must satisfy $x = z = w = 0$ and then $y = 0$ follows immediately. So the singularity at 0 is isolated. Blow up the origin and look at the chart corresponding to $\mathbf{Q} = (x, y, z)$. The strict transform of f is given by $f' = x^3 + y^4w + yz^2w + z^4w^2 + w^3$ and has Jacobian ideal $j(f') = (x^2, (4y^3 + z^2)w, (y + 2z^2w)zw, y^4 + yz^2 + 2z^4w + 3w^2)$. A short computation shows that the singular locus of X' near 0 is contained in the plane $x = w = 0$. It then follows that it equals the curve $y(y^3 + z^2) = 0$. Hence the two-fold locus of X' is a reducible plane curve with one smooth and one singular component. The three-fold locus is again the origin.

(4) Hilbert-Samuel function. Hironaka's original generalization of the multiplicity to the non hypersurface case is an increasing sequence ν^* of integers defined as follows: given an ideal $\mathbf{I}$ in $\mathbf{R}$ consider a generator system $f_1, \ldots, f_m$ of $\mathbf{I}$ subject to the following condition: the initial forms of the f_i are a minimal generator system of the initial forms ideal of $\mathbf{I}$ (= ideal generated by all initial forms of elements of $\mathbf{I}$) and the f_i are numbered by increasing multiplicity. Then ν^* is the sequence $\nu_i = o(f_i)$ of multiplicities, completed to an infinite sequence by adding infinity. Two sequences are compared pairwise lexicographically. He called such generators a standard basis of $\mathbf{I}$, distinct to nowadays use of the term, where standard bases are defined w.r.t. monomial orders on $\mathbb{N}^n$ and referring to initial monomials instead of initial forms. Permissibility of the center (= normal flatness of the variety along it) can be expressed by the equimultiplicity of each element of the standard basis along the center [H1, II. Thm. 2]. This description allows the explicit construction of the strict transform of the ideal via the strict transforms of its standard basis. Hironaka showed that ν^* does not increase under permissible blowing up. Later on, ν^* was commonly replaced by the Hilbert-Samuel function of the variety, for which Bennett had proven semicontinuity under blowing up and localization in arbitrary characteristic [Bn, Si]. Hironaka himself observes that ν^* does not behave well under localization:

EXAMPLE 12. [H1, remark p. 220] Let $\mathbf{K}$ be an arbitrary field, and let $\mathbf{I} = (xy + (x + z^2)w, x(x + y) + (y + z)^2w, (x + z)^2(x + y) - (y + z)^2y)$ in $\mathbf{R} = \mathbf{K}[[x, y, z, w]]$. It has $\nu^* = (2, 2, 3, \infty, \ldots)$. Localizing in (x, y, z) gives $\mathbf{I}_{loc} = (xy + (x + z)^2w, x(x + y) + (y + z)^2w)$ with $\nu^*_{loc} = (2, 2, \infty, \ldots) > \nu^*$.

Any choice of monomial order allows to refine ν^* by taking instead of initial forms of $\mathbf{I}$ initial monomials w.r.t. to this order, e.g. the graded lexicographic order. The resulting ideal, the initial ideal w.r.t. the order, is a monomial ideal and hence a purely combinatorial object. In contrast to ν^* it depends on the choice of coordinates. It can be made coordinate free by considering the generic initial ideal, i.e., the initial ideal which occurs for a generic choice of coordinates. Genericity makes only sense for infinite ground fields. More conceptually, the set of monomial ideals can be totally ordered by comparing their vertices lexicographically w.r.t. the given monomial order, and then the generic initial ideal coincides with the minimal initial ideal, where the minimum is taken over all choices of coordinates [Ha1]. This follows from the upper semicontinuity of initial ideals in deformations [BM4].

If the order is compatible with total degre the initial ideal determines ν^* and the Hilbert-Samuel function, but not conversely, as the induced stratification is in general strictly finer than the Hilbert-Samuel stratification:

EXAMPLE 13. [Ga, p. 567] Take $\mathbf{I} = (x^2, xy, xz+y^2w, y^3)$. Its initial ideal w.r.t. the graded lexicographic order with $x < y < z < w$ is in $\mathbf{I} = (x^2, xy, xz, y^3)$ and is already the generic initial ideal. The initial ideal stratum is the origin, whereas the Hilbert-Samuel stratum coincides with the w-axis. To see this, it suffices to consider $\mathbf{I}_t = (x^2, xy, xz + y^2(w - t), y^3)$ with $t \in \mathbf{K}$ so that in $\mathbf{I}_t = (x^2, xy, y^2)$. The defect of the initial ideal stratification is the following: the origin is the only permissible center w.r.t. in $\mathbf{I}$, but blowing it up the strict transform $\mathbf{I}'$ at the point $\mathbf{Q} = (x, y, z)$ equals $\mathbf{I}$. Hence the stratification is to fine to describe suitable centers.

This shows that initial ideals might be used to define part of the resolution invariant i but they are not suited to define the centers. Actually, the generic or minimal initial ideal still does not contain sufficient information as to measure the improvement of a singularity under blowup: for hypersurfaces, and having a graded monomial order, it just gives the multiplicity. This is due to the fact that the Newton polyhedron degenerates completely in generic coordinates to a polyhedron with one compact face. The interesting information appears in very specific coordinates when the vertices of the Newton polyhedron are as remote from the origin as possible, see the discussion below.

(5) Automorphisms. Invariants as the slope, τ or the initial ideal are examples of a frequent construction of invariants: Choose coordinates x in $\mathbf{R}$, develop f or the generators of the defining ideal into its power series expansion and extract some numerical datum. It will depend on the coordinates. To make it independent take e.g. its minimal or maximal value over all choices of coordinates. This gives an invariant of the singularity. In order to study its behaviour under blowup one proceeds as follows [Ha1].

Let us restrict to hypersurfaces for simplicity. Let $q_x f$ be a numerical datum associated to the power series expansion of f in the coordinates x, for which the minimum $q_{\min} f = \min_x \{q_x f\}$ over all coordinate choices exists. If the blowup is monomial in the coordinates x and if y denote the induced coordinates in $\mathbf{R}'$ the change from $q_x f$ to $q_y f'$ is computed by substituting in the expansion of f the variables x by the corresponding monomials in y according to the formula for $*$. Assume that we can show in this way that $q_x f > q_y f'$ for monomial blowups. Now, coordinates for which the blowup is monomial need not realize the minimal value of q. Assume, however, that there exists a coordinate change $x \mapsto \tilde{x}$ in $\mathbf{R}$ which realizes the minimum and such that the blowup stays monomial w.r.t. $\tilde{x}$. We then obtain

$$q_{\min} f = q_{\tilde{x}} f > q_{\tilde{y}} f' \geq q_{\min} f'$$

where $\tilde{y}$ denote the coordinates in $\mathbf{R}'$ induced from $\tilde{x}$. This shows that the minimum $q_{\min} f$ has dropped. Observe that it was not necessary to realize the minimum in $\mathbf{R}'$. We only used that $q_x f > q_y f'$ under monomial blowup and that there is a coordinate change in $\mathbf{R}$ realizing the minimum and preserving monomiality. It turns out that if $q_x f$ is defined through a monomial order as a certain initial monomial of f the minimizing coordinate changes can be chosen from a product $S\mathbf{U}$ inside the automorphism group of $\mathbf{R}$, where S is the permutation group on the coordinates and where $\mathbf{U}$ is a generalized 'unipotent' subgroup [Ha2]. This relies on the Gauss-Bruhat decomposition of $\mathrm{Aut}\,\mathbf{R}$. The explicit description of these subgroups allows to determine the cases where it is possible to choose a minimizing coordinate change in $\mathbf{R}$ such that the monomiality of the blowup is preserved.

For maxima, the argument is upside down. Assume again that we have $q_x f > q_y f'$ for monomial blowups. Now suppose that there exists a maximizing coordinate change $y \to \tilde{y}$ in $\mathbf{R}'$ which is induced from a coordinate change in $\mathbf{R}$ and such that the monomiality of the blowup is preserved. Then

$$q_{\max} f \geq q_{\tilde{x}} f > q_{\tilde{y}} f' = q_{\max} f'$$

where $\tilde{x}$ denote the coordinates in $\mathbf{R}$ obtained from x by the coordinate change. The maximum $q_{\max} f$ has dropped. It was not necessary to realize it in $\mathbf{R}$.

Note that if $q_x f$ belongs to a well ordered set, its minimum always exist, whereas the maximum need not, even in case the set $\{q_x f\}$ is bounded from above. Nevertheless, if $q_x f$ is upper semicontinuous w.r.t. x the maximum, if it exists, is in general much more sensitive to improvements of the singularity because it corresponds to very special choices of coordinates. If $q_x \mathbf{I}$ is defined as the initial ideal of $\mathbf{I}$ w.r.t. a given monomial order the maximum exists as is shown by a double application of Artin's Approximation Theorem and using the standard basis criterion of Becker-Buchberger [Ha1].

To realize maxima in $\mathbf{R}'$ is more difficult then to realize minima in $\mathbf{R}$ when required to preserve the monomial situation. One of the reasons is that for automorphisms of $\mathbf{R}'$ to be induced from automorphisms of $\mathbf{R}$, it is necessary but not sufficient that they fix the exceptional divisor D of the blowup:

EXAMPLE 14. Let $\mathbf{R}'$ be the local ring obtained from $\mathbf{R} = \mathbf{K}[[x, y, z]]$ by blowing up the maximal ideal and looking at the chart $\mathbf{Q} = (x, y)$. Then automorphisms g of $\mathbf{R}'$ of form $g(x, y, z) = (x + y^2, y, z)$ fix the exceptional divisor $z = 0$ but are not induced from automorphisms in $\mathbf{R}$ preserving the monomial situation.

Maximal initial ideals seem to be very appropriate to be used as resolution invariants. They appear implicitly and in modified form for hypersurfaces in various papers of Abhyankar, Hironaka, Moh and others, and each time their existence is proven by hand. The technique of [Ha1] provides a simultaneous proof of these results. Yet there is another difficulty to apply initial ideals successfully. Under monomial blowup, the expansion of f transforms under the map $*$ into the expansion of f', but the initial monomial of f' need not be the transform of the initial monomial of f. Hence direct comparison is sometimes impossible.

EXAMPLE 15. Let $f = x^5 + x^4 y^2 + y^4 z^5$. Consider the monomial order on $\mathbb{N}^3$ given by $\alpha < \beta$ if $(\alpha_1 + \alpha_3, \alpha_2, \alpha_1) < (\beta_1 + \beta_3, \beta_2, \beta_1)$ lexicographically. Then $x^4 y^2$ is the initial monomial of f. Blowup the origin and look at the chart given by $\mathbf{Q} = (x, y)$. Then $f' = x^5 + x^4 y^2 z + y^4 z^4$ and its initial monomial is $y^4 z^4$.

Monomial orders are hence in general not compatible in all charts with monomial blowup. It might be possible to overcome this obstruction by considering a refinement of the notion of monomial orders, the so called monomial rotation orders introduced in [Ha6]. These are defined by rotating a hyperplane in $\mathbb{R}_+^n$ around a fixed vertex and taking as initial monomials those whose exponents are touched first by this hyperplane. One recovers all vertices which are adjacent to the selected one, cf. [Ha3] and the notion of critical tropism [LT].

(6) Relative Multiplicity. This is an invariant suggested by Abhyankar and Hironaka. It is used by Spivakovsky in his solution to Hironaka's polyhedral game and by Moh and Cossart for resolution of threefolds [Sp2, Co2, Mo1, Mo2]. It also appears in the constructive resolutions of Bierstone-Milman and Villamayor [BM1, V1]. We describe it in the simplest possible context. Consider a power series f in three variables of form $f(x, y, z) = x^o + g(y, z)$ with g of order $> o$. Factor from g the largest monomial in y and z, say $g(y, z) = y^i z^k h(y, z)$. The relative multiplicity of f w.r.t. the given coordinates is defined as the order of h w.r.t. y and z. It depends on x, y, z and will be denoted by $r_{xyz} f$. In order to get a coordinate independent invariant consider all coordinates x, y, z in which f has the form $f = x^o + g(y, z)$. Among these, take those x for which the Newton polygon of g is minimal set-theoretically. For fixed y and z, it can be checked that the minimal Newton polygon is unique. Hence it only depends on y and z. Now vary y and z. In the presence of components of the exceptional divisor, y and z are subject to define them by $y = 0$ and/or $z = 0$; else there is no condition. Next, choose y and z such that the monomial factored from g has highest possible degree $i + k$. In all such coordinates the order of h is the same and called the relative multiplicity $r(f)$ of f (in the literature, this number is usually divided by o and called the weighted order of f.)

   The definition is relatively involved and not as conceptual as one would wish. Moreover, in positive characteristic, r may increase under blowing up. Also it does not behave well under deformations:

EXAMPLE 16. [Mo2, ex. 3.2] Let $f = x^p + y^2 z^{3p-2} + yz^{3p+1} + z^{3p+2} = x^p + z^{3p-2}(y^2 + yz^3 + z^4)$ where p is the characteristic of $\mathbf{K}$. Here $r_{xyz}f = 2 \neq r(f)$. Apply the coordinate change $\varphi = (x - z^3, y - z, z)$ and get $f_1 = \varphi^* f = x^p + yz^{3p-2}(y - 2z + z^3)$. Then $r_{xyz}f_1 = r(f_1) = r(f) = 1$. Blow up the origin and consider the point corresponding to $\mathbf{Q} = (x, y)$. The exceptional divisor D is given by $z = 0$ and the strict transform of f equals $f' = x^p + y^2 z^{2p} + yz^{2p+2} + z^{2p+2} = x^p + z^{2p}(y^2 + yz^2 + z^2)$. We have $r_{xyz}f' = r(f') = 2 > r(f)$.

EXAMPLE 17. Let $f = x^p + y^m z^{2p}(z^p + y^{p+1} + z^{p+1})$ with $r(f) = p$. Consider the deformation $f_t = x^p + (y-t)^m z^{2p}(z^p + y^{p+1} + z^{p+1})$. Applying the automorphism $\varphi = (x - t^{m/p} z^3, y, z)$ gives $\varphi^* f_t = x^p + z^{2p}(yz^p u_t + y^{p+1} + z^{p+1})$ with some unity u_t of $\mathbf{R}$ and $r(f_t) = p + 1$ if $t \neq 0$. Hence the relative multiplicity is not upper semicontinuous under deformation.

References

[A1]   Abhyankar, S.: Local uniformization of algebraic surfaces over ground fields of characteristic $p \neq 0$. Ann. Math. **63** (1956), 491–526. [*One of the first contributions to the global resolution problem in positive characteristic.*]

[A2]   Abhyankar, S.: Desingularization of plane curves. In: Summer Institute on Algebraic Geometry. Arcata 1981, Proc. Symp. Pure Appl. Math. **40** Amer. Math. Soc. [*Discusses invariants as the maximal slope for plane curves and explains problems which are special for positive characteristic. Describes Tschirnhausen transformations.*]

[A3]   Abhyankar, S.: Current status of the resolution problem. Summer Institute on Algebraic Geometry 1964. Proc. Amer. Math. Soc. [*Surveys the situation before and around Hironaka's Annals paper.*]

[A4]   Abhyankar, S.: Algebraic Geometry for Scientists and Engineers. Math. Surveys and Monographs **35**. Amer. Math. Soc. 1990. [*Provides a description of the state of the art in resolution of singularities and related problems (p. 253).*]

[A5]   Abhyankar, S.: Analytic desingularization in characteristic zero. Preprint 1996. [*Local resolution of hypersurfaces, cf. with* [BM 2].]

[Al]   Albanese, G.: Transformazione birazionale di una superficie algebrica qualunque in un altra priva di punti multipli. Rend. Circ. Mat. Palermo **48** (1924).

[AJ]   Abramovich, D., de Jong, J.: Smoothness, semi-stability and toroidal geometry. Preprint 1996. [*Short proof a weak version of Hironaka's Theorem in characteristic zero, cf. with* [BP].]

[Ar]   Artin, M.: Lipman's proof of resolution of singularities. In: Arithmetic Geometry (eds. G. Cornell, J.H. Silverman). [*Surveys the paper* [Lp 2] *on resolution of two dimensional schemes.*]

[Bb]   Brandenberg, M.: Aufblasungen affiner Varietäten. Diplomarbeit Zürich 1992. [*Detailed introduction to blowup.*]

[Bd]   Brodmann, M.: Computerbilder von Aufblasungen. El. Math. **50** (1995), 149–163.

[Be]   Berthelot, P.: Altérations de variétés algébriques (d'après A.J. de Jong). Sém. Bourbaki, exp. 815 (1995/96). [*Reviews the proof and gives various applications of de Jong's result.*]

[BK]   Brieskorn, E., Knörrer, H.: Ebene algebraische Kurven. Birkhäuser 1981. [*Discusses in detail slopes of Newton polygon and maximal contact.*]

[BM1]   Bierstone, E., Milman, P.: Canonical desingularization in characteristic zero by blowing up the maximum strata of a local invariant. To appear in Invent. Math. [*Systematic presentation of resolution invariants and treatment of non hypersurface case.*]

[BM2]   Bierstone, E., Milman, P.: Uniformization of analytic spaces. J. Amer. Math. Soc. **2** (1989), 801–836. [*Establishes a local version of resolution in the analytic category, cf. with* [A 5].]

[BM3]   Bierstone, E., Milman, P.: A simple constructive proof of canonical resolution of singularities. In: Effective Methods in Algebraic Geometry (eds. T. Mora, C. Traverso). Progress in Math. **94**, Birkhäuser 1991, 11–30.

[BM4]   Bierstone, E., Milman, P.: Relations among analytic functions I. Ann. Inst. Fourier **37** (1987), 187–239. [*Includes study on initial ideals.*]

[Bn]   Bennett, B.-M.: On the characteristic function of a local ring. Ann. Math. **91** (1970), 25–87. [*Proves that the Hilbert-Samuel function does not increase under permissible blowing up in any characteristic and that it is upper-semicontinuous under localization.*]

[BP]   Bogomolov, F., Pantev, T.: Weak Hironaka Theorem. To appear in Math. Res. Letters. [*Short proof a weak version of Hironaka's Theorem in characteristic zero, cf. with* [AJ].]

[Br1]   Brieskorn, E.: Über die Auflösung gewisser Singularitäten von holomorphen Abbildungen. Math. Ann. **166** (1966), 76–102.

[Br2]   Brieskorn, E.: Die Auflösung der rationalen Singularitäten holomorpher Abbildungen. Math. Ann. **178** (1968), 255 – 270.

[Ch]   Chisini, O.: La risoluzione delle singolarità di una superficie mediante transformazioni birazionali dello spazio. Mem. Accad. Sci. Bologna VII. **8** (1921).

[CGO]   Cossart, V., Giraud, J., Orbanz, U.: Resolution of surface singularities. Lecture Notes in Math. vol. 1101, Springer 1984. [*Compares resolution methods for surfaces by Abhyankar, Hironaka, Jung and Zariski.*]

[Co1]   Cossart, V.: Desingularization of embedded excellent surfaces. Tôhoku Math. J. **33** (1981), 25–33. [*Fills a gap in* [H 3] *by extending H.'s argument to non rational points.*]

[Co2]   Cossart, V.: Polyèdre caractéristique d'une singularité. Thèse d'Etat, Orsay 1987. [*Normal form of functions in three variables under blowup in characteristic p.*]

[Co3]   Cossart, V.: Désingularisation des surfaces (d'après Zariski). Preprint Ecole Polytechnique 1995.

[CP]   Cano, F., Piedra, R.: Characteristic polygon of surface singularities. In: Géométrie algébrique et applications II (eds. J.-M. Aroca, T. Sanchez-Giralda, J.-L. Vicente), Proc. of Conference on Singularities, La Rábida 1984. Hermann 1987.

[Ga]   Galligo, A.: A propos du Théorème de Préparation de Weierstrass. Springer Lecture Notes in Math. 409 (1973), 543–579.

[Gi1]   Giraud, J.: Etude locale des singularités. Cours de 3e Cycle, Orsay 1971/72. [*Exposition of basic notions for resolution.*]

[Gi2]   Giraud, J.: Sur la théorie du contact maximal. Math. Z. **137** (1974), 285–310. [*Discusses definition of τ in positive characteristic.*]

[Gi3]   Giraud, J.: Contact maximal en caractéristique positive. Ann. Scient. E.N.S. **8** (1975), 201–234.

[Gr1]   Grothendieck, A.: Traveaux de Heisouke Hironaka sur la résolution des singularités. Actes Congrès International, Nice 1970. [*Short summary, admits that he has not got through the paper completely.*]

[Gr2]   Grothendieck, A.: EGA IV. Publ. Math. IHES **24** (1965).

[GT]   Goldin, R, Teissier, B.: Resolving singularities of plane analytic branches with one toric morphism. Preprint ENS Paris 1995. [*Uses toric modifications to resolve plane curve singularities.*]

[H1]   Hironaka, H.: Resolution of singularities of an algebraic variety over a field of characteristic zero. Ann. of Math. **79** (1964), 109–326. [*Comprehensive presentation of many concepts used in resolution arguments, such as blowup, strict and total transform, normal flatness, standard bases. To a large extent very clear and informative, the complicated part being the proper induction argument.*]

[H2]   Hironaka, H.: Characteristic polyhedra of singularities. J. of Math. Kyoto Univ. **7** (1967), 251–293. [*Develops coordinate free Newton polygon, no application to blowup.*]

[H3]   Hironaka, H.: Desingularization of excellent surfaces. Notes by B. Bennett at the Conference on Algebraic Geometry, Bowdoin 1967. Reprinted in: Cossart, V., Giraud, J., Orbanz, U.: Resolution of surface singularities. Lecture Notes in Math. 1101, Springer 1984. [*Explicit study of Newton polyhedron to make multiplicity drop in a finite number of blowups. The argument has to be complemented by* [Co 1] *for non rational points.*]

[H4]   Hironaka, H.: Additive groups associated with points of a projective space. Ann. Math. **92** (1970), 327–334. [*Discussion of* τ *in positive characteristic.*]

[H5]   Hironaka, H.: Idealistic exponents of singularity. In: Algebraic Geometry, the Johns Hopkins Centennial Lectures. Johns Hopkins University Press 1977. [*Development of division theorem to study standard bases and their role in resolution arguments. Treats non hypersurfaces through the notion of idealistic presentation.*]

[H6]   Hironaka, H.: Schemes etc. In: 5th Nordic Summer School in Mathematics (ed. F. Oort), Oslo 1970, 291–313. [*Uses differential operators to investigate* τ. *Contains various examples.*]

[H7]   Hironaka, H.: Desingularization of complex analytic varieties. Actes Congrès Intern. Math., Nice 1970, 627–631.

[Ha1]   Hauser, H.: Maximal and minimal initial ideals in resolution processes. Preprint Innsbruck 1996. [*Proves that the minimal and maximal initial ideal over all coordinate choices exists. Short proof of infinitesimal semicontinuity of Hilbert-Samuel function. Studies maximal initial ideals as resolution invariants.*]

[Ha2]   Hauser, H.: Gauss-Bruhat decomposition, flag varieties and blowing up. Preprint Innsbruck 1996. [*Establishes product decomposition for automorphisms of power series rings. Allows to determine maximizing and minimizing coordinate changes. Applies to construct flag invariants from the Newton polyhedron.*]

[Ha3]   Hauser, H.: Triangles, prisms and tetrahedra as resolution invariants. Preprint Innsbruck 1996. [*Resolves Hironaka's polyhedral game in three variables. Uses theory of convex polytopes to construct new resolution invariants.*]

[Ha4]   Hauser, H.: A critical reading of Hironaka's Bowdoin lecture on desingularization of excellent surface singularities. Preprint Innsbruck 1996. [*Provides conceptual foundation of* [H 3]. *Describes advantages and limitations of H.'s approach.*]

[Ha5]   Hauser, H.: The concept of blowing up in algebraic geometry. In preparation.

[Ha6]   Hauser, H.: Monomial rotation orders. In preparation.

[HIO]   Herrmann, M., Ikeda, S., Orbanz, U.: Equimultiplicity and blowing up. Springer 1988. [*Discusses in detail concepts as multiplicity, normal flatness, Hilbert-Samuel function, etc.*]

[Hs]   Hartshorne, R.: Algebraic Geometry. Springer 1977.

[J]   de Jong, A.: Smoothness, semi-stability and alterations. To appear in Publ. Math. IHES. [*'The' new approach to resolution of singularities through semi-stable reduction.*]

[Ju]   Jung, H.: Darstellung der Funktionen eines algebraischen Körpers zweier unabhängiger Veränderlicher x, y in der Umgebung einer Stelle $x = a$, $y = b$. J. Reine Angew. Math. **133** (1908), 289–314. [*Proves local resolution for complex surfaces. His method is reviewed in* [Lp1].]

[K]   Kunz, E.: Algebraische Geometrie IV. Vorlesung Regensburg. [*Lectures on Hironaka's Annals paper 1964.*]

[KKMS]   Kempf, G., Knudson, F., Mumford, D., Saint-Donat, B.: Toroidal Embeddings I. Lect. Notes Math. **339** Springer 1973. [*Theory of toric modification and resolution.*]

[La]   Laufer, H.: Normal two-dimensional singularities. Ann. Math. Studies 71. [*Explicit resolution of normal surface singularities. Includes examples.*]

[LO]   Lê, D.T., Oka, M.: On resolution complexity of plane curves. Kodai Math. J. **18** (1995), 1–36. [*Describes number of blowups necessary to resolve plane curve singularities.*]

[Lp1]   Lipman, J.: Introduction to resolution of singularities. Proceedings Symp. Pure Appl. Math. **29** Amer. Math. Soc. 1975, 187–230. [*Surveys various methods of resolution by Jung, Walker, Zariski and Abhyankar.*]

[Lp2]   Lipman, J.: Desingularization of 2-dimensional schemes. Ann. Math. **107** (1978), 151–207. [*Resolution based on homological methods, duality and differentials.*]

[LT]   Lejeune, M., Teissier, B.: Contribution à l'étude des singularités du point de vue du polygone de Newton. Thèse d'Etat 1973.

[Lv1]   Levi, B.: Sulla risoluzione delle singolarità puntuali delle superficie algebriche dello spazio ordinario per transformazioni quadratiche. Ann. Mat. Pura Appl. II **26** (1897).

[Lv2]   Levi, B.: Risoluzione delle singolarità puntuali delle superficie algebriche. Atti Acad. Sci. Torino **33** (1897), 66–86.

[Mo1]   Moh, T.-T.: Canonical uniformization of hypersurface singularities of characteristic zero. Comm. Alg. **20** (1992), 3207–3249.

[Mo2]   Moh, T.-T.: On a Newton polygon approach to the uniformization of singularities of characteristic p. In: Algebraic Geometry and Singularities (eds. A. Campillo, L. Narváez). Proc. Conf. on Singularities La Rábida. Birkhäuser 1996. [*Includes example that weighted order may increase in positive characteristic.*]

[N]   Noether, M.: Über einen Satz aus der Theorie der algebraischen Funktionen. Math. Ann. **6** (1873), 351–359.

[NB]   Noether, M., Brill, A.: Die Entwicklung der Theorie der algebraischen Funktionen in älterer und neuerer Zeit. Jahresber. Dt. Math. Verein. III (1892/93), 107–566. [*Resolution of complex curves. Compares the methods of Kronecker, Noether and Hamburger.*]

[Od1]   Oda, T.: Infinitely very near singular points. Adv. Studies Pure Math. **8** (1986), 363–404. [*Characteristic free survey on Hilbert-Samuel function and ν^* w.r.t. resolution problems.*]

[Od2]   Oda, T.: Hironaka's additive group scheme. In: Number Theory, Algebraic Geometry and Commutative Algebra, in honor of Y. Akizuki, Kinokuniya, Tokyo, 1973, 181–219.

[Od 3] Oda, T.: Hironaka group schemes and resolution of singularities. In: Proc. Conf. on Algebraic Geometry, Tokyo and Kyoto 1982. Lecture Notes in Math. 1016, Springer 1983, 295–312.

[Ok]   Oka, M.: Geometry of plane curves via toroidal resolution. In: Algebraic Geometry and Singularities (eds. A. Campillo, L. Narváez). Proc. Conf. on Singularities La Rábida. Birkhäuser 1996.

[OW]   Orlik, P., Wagreich, P.: Equivariant resolution of singularities with $\mathbf{C}^*$ action. In: Proceedings of the 2nd Conference on Compact Transformation Groups, Amherst 1971.

[P]   Pfeifle, J.: Blowing up non reduced normal crossing ideals. Diplomarbeit, Innsbruck 1996. [*Investigates non reduced monomial ideals which give smooth varieties when blown up in some ambient space.*]

[SG1]   Sanchez Giralda, T.: Teoria de singularidades de superficies algebroides sumergidas. Monografias y memorias de Matemática IX, Publ. del Instituto Jorge Juan de Matemáticas. Madrid 1976.

[SG2]   Sanchez Giralda, T.: Caractérisation des variétés permises d'une hypersurface algébroide. C.R. Acad. Sci. Paris **285** (1977), 1073–1075. [Determines the cases where the multiplicity does not drop under blowup.]

[Si]   Singh, B.: Effect of permissible blowing up on the local Hilbert function. Invent. Math. **26** (1974), 201–212. [*Refinement of Bennett's result on the infinitesimal semicontinuity of the Hilbert-Samuel function.*]

[Sp1]   Spivakovsky, M.: Resolution of Singularities. Preprint 1995. [*Preliminary version. Combines Zariski's valuation theory with Hironaka's approach.*]

[Sp2]   Spivakovsky, M.: A solution to Hironaka's Polyhedra Game. In: Arithmetic and Geometry. Papers dedicated to I.R. Shafarevich on the occasion of his sixtieth birthday, vol II (eds.: M. Artin, J. Tate). Birkhäuser 1983, 419–432.

[Sp3]   Spivakovsky, M.: A counterexample to Hironaka's 'hard' polyhedra game. Publ. RIMS, Kyoto University **18** (1983), 1009–1012.

[Sp4]   Spivakovsky, M.: A counterexample to the theorem of Beppo Levi in three dimensions. Invent. Math. **96** (1989), 181–183.

[Su]   Sussmann, H.: Real-analytic desingularization and subanalytic sets: an elementary approach. Trans. Amer. Math. Soc. **317** (1990), 417–461.

[V1]   Villamayor, O.: Constructiveness of Hironaka's resolution. Ann. Sci. Ec. Norm. Sup. Paris **22** (1989), 1–32. [*Develops a canonical resolution algorithm.*]

[V2]   Villamayor, O.: Patching local uniformizations. Ann. Sci. Ec. Norm. Sup. Paris **25** (1992), 629–677. [*Shows how to glue local constructions to get global ones.*]

[V3]   Villamayor, O.: An introduction to the algorithm of resolution. In: Algebraic Geometry and Singularities (eds. A. Campillo, L. Narváez). Proc. Conf. on Singularities La Rábida. Birkhäuser 1996.

[W]   Walker, R.J.: Reduction of singularities of an algebraic surface. Ann. Math. **36** (1935), 336–365. [*Resolution of surface singularities over* **C**.]

[Y]   Youssin, B.: Newton polyhedra without coordinates. Mem. AMS **433** (1990), 1–74, 75–99.

[Z1]   Zariski, O.: Reduction of singularities of algebraic three dimensional varieties. Ann. Math. **45** (1944), 472–542. [*Resolution of three-fold singularities over a field of characteristic* 0.]

[Z2]   Zariski, O.: Local uniformization theorem on algebraic varieties. Ann. Math. **41** (1940), 852–896.

[Z3]   Zariski, O.: The reduction of singularities of an algebraic surface. Ann. Math. **40** (1939), 639–689. [*Resolution of surface singularities over a field of characteristic* 0.]

[Z4]   Zariski, O.: A simplified proof for resolution of singularities of an algebraic surface. Ann. Math. **43** (1942), 583–593.

[Z5]   Zariski, O.: Exceptional singularities of an algebraic surface and their reduction. Atti Accad. Naz. Lincei Rend. Cl. Sci. Fis. Mat. Natur., serie VIII, **43** (1967), 135–196.

[Z6]   Zariski, O.: A new proof of the total embedded resolution theorem for algebraic surfaces. Amer. J. Math. **100** (1978), 411–442.

Chapter 4

Applications

Progress in Mathematics, Vol. 162, © 1998 Birkhäuser Verlag Basel/Switzerland

Sur la topologie des polynômes complexes

Enrique Artal-Bartolo *
Departamento de Matemáticas
Universidad de Zaragoza
50009 Zaragoza
ESPAGNE

Pierrette Cassou-Noguès and Alexandru Dimca
Laboratoire de mathématiques pures
Université Bordeaux I
350 Cours de la Libération
33405 Talence
FRANCE

1 Introduction

Soit $f \in \mathbb{C}[x_1, \ldots, x_n]$ un polynôme de degré d. On sait [P], [ST] qu'il existe un ensemble fini $\Lambda \subset \mathbb{C}$ tel que

$$f : \mathbb{C}^n \setminus f^{-1}(\Lambda) \longrightarrow \mathbb{C} \setminus \Lambda$$

est une fibration localement triviale. Dorénavant Λ désigne le plus petit ensemble qui possède cette propriété. Si $t \in \Lambda$, la fibre $F_t = f^{-1}(t)$ est appellée *fibre irrégulière* de f, sinon elle est dite *régulière* ou *générique*. Soit δ un nombre réel positif assez petit, $\delta \notin \Lambda$. On note

$$D_\delta = \{t \in \mathbb{C} \mid \ |t| < \delta\}, \ S_\delta = \partial \overline{D}_\delta \ \text{ et } \ T_\delta = f^{-1}(D_\delta).$$

Si la fibre F_0 est régulière, alors on a les deux faits suivants.

A. *Pour δ un nombre réel positif assez petit, la fibration autour de la fibre F_0, $f : f^{-1}(S_\delta) \longrightarrow S_\delta$, a un opérateur de monodromie $T_0 : H^*(F_\delta) \longrightarrow H^*(F_\delta)$ trivial (les coefficients pour les groupes de cohomologie sont toujours dans $\mathbb{C}$ sauf dans le Théorème 3 et sa preuve où ils sont dans $\mathbb{Z}$).*

---

*financé en partie par CAICYT PB94-0291

B. *L'inclusion $j : F_0 \hookrightarrow T_\delta$ est une équivalence homotopique.*

L'assertion **B** implique l'assertion suivante

B'. *Tous les nombres de Betti de F_0 et de T_δ coincident, c'est à dire*

$$b_j(F_0) = b_j(T_\delta) \quad \forall j \in \mathbb{N}.$$

Le but de cet article est d'étudier ce qui se passe pour les fibres irrégulières F_0 qui satisfont la condition suivante

     (CF) $(F_0)_{red}$ a seulement des singularités isolées,

tandis que le polynôme f doit satisfaire la condition

     (CP) $\tilde{b}_j(F_\delta) = 0$ pour tout $j \neq n - 1$.

Ici X_{red} est la variété réduite associée à la variété X et $\tilde{b}_j(M) = \dim \tilde{H}^j(M)$.

Exemple 1. Les deux classes principales d'exemples sont les suivantes:

(E1) $n = 2$ et f est un polynôme primitif (la fibre générique F_δ est irréductible), voir [A].

(E2) $n > 2$ et le polynôme f a seulement des singularités isolées dans $\mathbb{C}^n$ et à l'infini, voir [ST], [Pa] pour les définitions et les propriétés principales de ces polynômes.

Pour ces deux classes de polynômes, si la fibre F_s est à singularités isolées, soient $K_s = (F_s)_{sing}$, μ_s la somme des nombres de Milnor des singularités de F_s et λ_s la somme des différences des nombres de Milnor à l'infini de F_s et F_δ par rapport à une compactification convenable Z. Pour $n = 2$ on peut toujours prendre $Z = \mathbb{P}^2$. On sait alors qu'on a :

$$(\chi 1) \qquad \chi(F_s) - \chi(F_\delta) = (-1)^n(\mu_s + \lambda_s)$$

et $\chi(F_s) \neq \chi(F_\delta)$ si et seulement si $s \in \Lambda$.

     Il y a d'autres exemples de polynômes qui vérifient les conditions **(CP)** et **(CF)** (pour toute fibre F_0):

(E3) $n \geq 2$ et le polynôme f est modéré, voir [B], ou, plus généralement, M-modéré, voir [NZ].

Pour $n = 2$ et une fibre F_s quelconque, soit $K_s = ((F_s)_{red})_{sing}$.

Théorème 1. *Soit f un polynôme qui satisfait* **(CP)** *et F_0 une de ses fibres qui satisfait* **(CF)**. *Alors*

$$codim \, Ker \, (T_0 - Id) = (-1)^n(\chi(F_0) - \chi(F_\delta)) - \tilde{b}_{n-2}(F_0 \setminus K_0).$$

Soit $n(F_s)$ le nombre de composantes irréductibles de la fibre F_s. Remarquons que l'on a

$$n(F_s) - 1 = \tilde{b}_0(F_s \setminus K_s).$$

Il est intéressant de remarquer que, pour $n > 2$, l'invariant $b_{n-2}(F_0 \setminus K_0)$ dépend en général de la position de singularités sur la fibre F_0.

Exemple 2. Soient C_1, C_2 deux courbes de degré 6 dans $\mathbb{P}^2$, ayant chacune 6 cusps A_2 et telles que les cusps de C_1 sont situés sur une conique, tandis que les cusps de C_2 n'ont pas cette propriété, tout comme dans l'exemple classique de Zariski. Soient S_1, S_2 dans $\mathbb{P}^3$ les revêtements de degré 6 de $\mathbb{P}^2$ ramifiés le long de C_1 et, respectivement, de C_2. On sait alors que

$$b_3(S_1) = 2, \qquad b_3(S_2) = 0$$

voir [Z] ou[D2], p. 210. Soit $H \subset \mathbb{P}^3$ un plan tel que $S_i \cap H$ soit lisse pour $i = 1, 2$. Soit $\mathbb{C}^3 = \mathbb{P}^3 \setminus H$ et soient $f_i = 0$ l'équation de la surface $F_{0i} = S_i \setminus H$. Il est facile à voir que:

(i) f_i est un polynôme modéré [B], en particulier il a seulement des singularités isolées dans $\mathbb{C}^3$ et pas de singularités à l'infini.

(ii) F_{0i} est un bouquet de 2-spheres et $b_2(F_{0i}) = 65$.

(iii) $b_1(F_{01} \setminus (F_{01})_{sing}) = 2$ et $b_1(F_{02} \setminus (F_{02})_{sing}) = 0$. (utiliser la dualité de Lefschetz pour le couple $(S_i, (S_i \cap H) \cup (S_i)_{sing})$).

Des situations similaires ont été considerées par García-Nemethi [GN], mais chez eux ce sont les singularités dans l'hyperplan à l'infini qui jouent le rôle central.

Nous calculons à titre d'exemple l'opérateur de monodromie T_0 pour le polynôme suivant. Notons

$$s = xy + 1; \quad p = sx - 1; \quad g = p^3 + p^4 + p^2 x; \quad h = g/x^2;$$

$$f = h + 5/3\,ps - 1/3\,s$$

Ce polynôme a été introduit par Briançon dans un manuscrit non publié de 1985, et est étudié en détail dans [ACL]. Il a toutes ses fibres lisses et irréductibles et deux fibres irrégulières à l'infini, pour $t = 0$ et $t = -16/9$. Nous montrons que son opérateur de monodromie $T_0 \neq Id$ a toutes ses valeurs propres égales à 1.

Soit $\pi = \pi_1(\mathbb{C} \setminus \Lambda, \delta)$ et soit $H^{n-1}(F_\delta)^\pi$ la partie fixe par l'action de la monodromie de la fibration

$$f : \mathbb{C}^n \setminus f^{-1}(\Lambda) \longrightarrow \mathbb{C} \setminus \Lambda.$$

On a alors le théorème suivant des cycles invariants *affines*.

Théorème 2. *Si le polynôme f satisfait la condition* **(CP)** *et toutes ses fibres satisfont la condition* **(CF)** *alors*

$$dim\ H^{n-1}(F_\delta)^\pi = \sum_{s\in\Lambda} \tilde{b}_{n-2}(F_s \setminus K_s).$$

Corollaire 1. *La famille de sous-espace vectoriels $(Ker(T_s - Id))_{s\in\Lambda}$ est en position générale dans $H^{n-1}(F_\delta)$, i.e.*

$$\sum_s codim\ Ker\ (T_s - Id) = codim\ (\cap_s Ker\ (T_s - Id)).$$

Dans le cas $n = 2$, l'application $f : \mathbb{C}^2 \to \mathbb{C}$ peut être étendue en une application régulière $\overline{\phi} : X \to \mathbb{P}^1$. Alors $D_\infty = X\setminus\mathbb{C}^2$ est une courbe à croisements normaux dont toutes les composantes sont lisses et rationnelles. Les composantes D telles que $\overline{\phi}|D$ est surjective sont appelées composantes dicritiques. On note $\delta(f)$ le nombre de ces composantes. Ce nombre est indépendent de la résolution choisie et, d'après Kaliman [K], il vérifie l'inégalité suivante

$$(K) \qquad \delta(f) - 1 \geq \sum_s (n(F_s) - 1)$$

avec égalité si la fibre générique est rationnelle ou , comme nous le montrons, si $|\Lambda| = 1$. Nous allons retrouver ce résultat de Kaliman, en montrant sa relation avec le Théorème 2.

Corollaire 2. *Pour $n = 2$, l'action de la monodromie est triviale si et seulement si la fibre générique est rationnelle et le nombre de dicritiques de f est égal au nombre de places à l'infini de la courbe F_δ (chaque dicritique est alors de degré un).*

Les polynômes correspondant à ce cas ont été étudiés en detail par Miyanishi et Sugie [MS].

L'égalité suivante de caractéristiques d'Euler

$$\chi(T_\delta) = \chi(T_\delta \setminus F_0) + \chi(F_0)$$

voir Fulton [Fu] (où le premier terme est trivial, car $T_\delta \setminus F_0$ est fibré au-dessus de S_δ) combinée avec la condition **(CF)** sur F_0 implique que cette fibre satisfait la propriété **(B')** si et seulement si on a $\tilde{b}_{n-2}(F_0) = 0$. En effet, on a toujours $\tilde{b}_j(F_0) = 0$ pour tout $j \neq n - 2, n - 1$ (voir la preuve du Thm. 1).

Pour la condition **(B)** la situation est beaucoup plus subtile:

Théorème 3. *Dans le cas $n = 2$, soit F_0 une fibre réduite du polynôme primitif f. Alors les trois propriétés suivantes sont équivalentes:*

(a) $j : F_0 \to T_\delta$ *est une équivalence d'homotopie.*

(b) F_0 *est connexe et* $j_* \, H_1(F_0; \mathbb{Z}) \to H_1(T_\delta; \mathbb{Z})$ *est un isomorphisme.*

(c) F_0 *est connexe et la valeur* $t = 0$ *est régulière à l'infini pour le polynôme* f.

Pour la définition d'une fibre régulière à l'infini on peut voir la section 3 ci-dessous.

Il est évident que les implications $(c) \Rightarrow (a) \Rightarrow (b)$ restent vraies en toute dimension. Par contre, l'implication clé $(b) \Rightarrow (c)$ n'est plus vraie en dimension supérieure:

Exemple 3. Soit $f : \mathbb{C}^4 \longrightarrow \mathbb{C} : \quad f = x + x^4 y + y^2 z^3 + t^5$

Alors F_0 est lisse, irrégulière et difféomorphe à $\mathbb{C}^3$, voir [CD]. En outre, le polynôme f a seulement des singularités isolées dans $\mathbb{C}^4$ et à l'infini.

On peut utiliser le fait que f est quasihomogène pour montrer que T_δ est contractible. Cela donne (a), (b) mais $\Lambda = \{0\}$.

Dans la deuxième section nous démontrons les Théorèmes 1 et 2 par des calculs dans l'espace affine $\mathbb{C}^n$.

En utilisant une compactification de f pour $n = 2$, nous redémontrons l'inégalité de Kaliman en section 3 Nous donnons aussi des renseignements sur le calcul effectif de l'opérateur T_0 dans ce cas.

Finalement, en section 4 nous utilisons la résolution minimale de Lê-Weber et les propriétés des espaces lenticulaires pour démontrer le Théorème 3. Nous donnerons aussi des exemples où j_* n'est ni injective ni surjective et des exemples où elle est injective et l'image est d'indice fini plus grand que 1. Nous donnerons aussi des exemples qui montrent que le théorème n'est pas vrai en général pour les fibres non-réduites.

2 Démonstration des Théorèmes 1 et 2

Nous commençons par un résultat très simple.

Lemme 1.

(i) *Pour tout polynôme* $f : \mathbb{C}^n \to \mathbb{C}$ *on a* $b_j(T_\delta) = 0$ *pour tout* $j \geq n$.

(ii) *Si le polynôme* f *satisfait la condition (CP), alors on a* $\tilde{b}_j(T_\delta) = 0$ *pour tout* $j \neq n - 1$.

Preuve. Soit $S \subset \mathbb{C}$ l'image d'un plongement d'un disque fermé dans $\mathbb{C}$ tel que:

$$\text{(a)} \ \ \bar{D}_\delta \cap S = \{\delta\}, \quad \text{(b)} \ \ \Lambda \subset D_\delta \cup int(S).$$

Alors, soit $\tilde{S} = f^{-1}(S)$. La retraction $\mathbb{C} \to \bar{D}_\delta \cup S$ se relève et donne une retraction $\mathbb{C}^n \to \tilde{S} \cup \bar{T}_\delta$. C'est la suite de Mayer-Vietoris de cette réunion qui donne le résultat en passant par le fait que $b_j(F_\delta) = 0$ pour $j > \dim(F_\delta)$, voir [H]. $\qquad \square$

322   *E. Artal-Bartolo, P. Cassou-Noguès and A. Dimca*

Démonstration du Théorème 1. La suite de Gysin de l'hypersurface lisse $Z = (F_0)_{red} \setminus K_0$ dans $M = T_\delta \setminus K_0$ donne la suite exacte

$$(\alpha) \quad \ldots \to H^j(M) \to H^j(M \setminus Z) \to H^{j-1}(Z) \to H^{j+1}(M) \to \ldots$$

voir [D2], p. 46. On étudie tout d'abord le cas $n > 2$.

Alors, pour $0 < j \leq n - 3$ et pour $j = n$, les termes extrêmes sont égaux à 0. D'autre part, pour $j = n - 2$, le morphisme

$$H^{n-3}(Z) \to H^{n-1}(M)$$

est trivial, car on peut le relier au morphisme correspondant dans la suite exacte de Gysin de Z dans $\mathbb{C}^n \setminus K_0$.

La suite exacte de Wang de la fibration $f : M \setminus Z \to D_\delta^*$, voir [D2], p. 74, combinée avec les remarques ci-dessus donnent alors

$$(\beta) \quad \dim \operatorname{Ker}(T_0 - Id) = b_n(T_\delta \setminus F_0) = b_{n-1}(F_0 \setminus K_0),$$
$$\tilde{b}_j(F_0 \setminus K_0) = 0 \ \text{pour tout } j \leq n - 3.$$

D'autre part, la suite exacte du couple $(F_0, F_0 \setminus K_0)$ implique la suite exacte

$$(\gamma) \quad \begin{aligned} &\cdots \to \bigoplus_{k \in K_0} \tilde{H}^{n-2}(L_k) \to H^{n-1}(F_0) \to H^{n-1}(F_0 \setminus K_0) \to \\ &\to \bigoplus_{k \in K_0} H^{n-1}(L_k) \to \ldots \end{aligned}$$

où L_k sont les entrelacs des singularités de F_0.

Utilisant la $(n - 3)$-connectivité de ces entrelacs, voir [Mi], on obtient $\tilde{b}_j(F_0) = 0$ pour $j \leq n - 3$. En prennant les caractéristiques euleriennes dans la dernière suite exacte on obtient

$$\dim \operatorname{Ker}(T_0 - Id) = (-1)^{n-1}(\chi(F_0) - 1) + \tilde{b}_{n-2}(F_0 \setminus K_0).$$

Finalement, cela donne

$$\operatorname{codim} \operatorname{Ker}(T_0 - Id) = (-1)^n(\chi(F_0) - \chi(F_\delta)) - \tilde{b}_{n-2}(F_0 \setminus K_0)$$

vu la structure très simple de $H^*(F_\delta)$.

Pour $n = 2$, la suite exacte (α) et la relation (β) donnent

$$\dim H^1(F_0 \setminus K_0) = \dim \operatorname{Ker}(T_0 - Id) + |K_0|$$

car on a évidemment $\dim H^3(T_\delta \setminus K_0) = |K_0|$. Ceci implique le Thm. 1. dans ce cas.

D'autre part, encore pour $n = 2$ on a $\dim H^1(L_k) = n(F_0, k)$, le nombre de branches du germe de courbe $((F_0)_{red}, k)$. Soit $\Delta_k(t)$ le polynôme caractéristique de la monodromie locale associée au germe $((F_0)_{red}, k)$. Alors

$n(F_0, k) - 1$ est la multiplicité de 1 comme racine du polynôme $\Delta_k(t)$. La suite exacte (γ) donne alors l'inégalité suivante

$$\dim \mathrm{Ker}\,(T_0 - Id) \geq \sum_k (n(F_0, k) - 1).$$

Quand la fibre F_0 est réduite, cette inégalité, plus une localisation évidente de la monodromie T_0 autour des singularités (F_0, k) donne le résultat suivant.

Corollaire 3. *Pour $n = 2$, le produit des polynômes caractéristiques locaux $\Delta_k(t)$ divise le polynôme caractéristique $\Delta(t)$ de la monodromie globale T_0.*

Démonstration du Théorème 2. Soit $\Lambda = \{\lambda_1, \ldots, \lambda_m\}$. Pour tout $i \in \{1, \ldots, m\}$, soit l_i l'image d'un plongement $c_i : I \to \mathbb{C}$, $I = [0, 1]$, tel que

(i) $c_i(0) = \delta$, $c_i(1) = \lambda_i$

(ii) $l_i \cap l_j = \{\delta\}$ pour $i \neq j$.

Soit D_i un petit disque fermé dans $\mathbb{C}$, centré en λ_i et soit

$$L_k = f^{-1}\left(\bigcup_{1 \leq i \leq k} l_i \cup \bigcup_{1 \leq i \leq k} D_i \setminus \Lambda\right)$$

pour tout $k \in \{1, \ldots, m\}$. Avec un choix convenable des chemins l_i, il existe une retraction $\mathbb{C}^n \setminus f^{-1}(\Lambda) \to L_m$ donc

$$b_{n-1}(L_m) = b_{n-1}(\mathbb{C}^n \setminus f^{-1}(\Lambda)) = \sum_s b_{n-2}(F_s \setminus K_s)$$

(on peut utiliser à nouveau la suite de Gysin comme ci-dessus).

Pour finir la preuve dans le cas $n = 2$, il suffit de montrer le résultat suivant. Pour chaque $s \in \Lambda$ soit T_s l'operateur de monodromie sur $H^{n-1}(F_\delta)$ associé à un lacet élémentaire dans π qui tourne autour de s.

Lemme 2. (n=2) *Le morphisme $j_k^* : H^1(L_k) \to H^1(F_\delta)$ induit par l'inclusion pour $k = 1, \ldots, m$ satisfait les propriétés suivantes:*

(i) $\dim \mathrm{Ker}\, j_k^* = k$

(ii) $\mathrm{Im}\, j_k^* = H^1(F_\delta)^{<T_{\lambda_1}, \ldots, T_{\lambda_k}>}$ où $< T_{\lambda_1}, \ldots, T_{\lambda_k} >$ dénote le groupe engendré par les opérateurs de monodromie $T_{\lambda_1}, \ldots, T_{\lambda_k}$.

Preuve. Pour $k = 1$, on peut supposer $\lambda_1 = 0$ et alors le résultat se déduit de la suite exacte de Wang.

Le passage de k à $k + 1$ se fait en utilisant la suite de Mayer-Vietoris suivante (pour un recouvrement fermé):

$$0 \to H^1(L_{k+1}) \to H^1(L_k) \oplus H^1(f^{-1}(D_{k+1}^*)) \to H^1(F_\delta) \to \ldots$$

où $D_{k+1}^* = D_{k+1} \setminus \{\lambda_{k+1}\}$.

Soit i_{k+1} l'inclusion de F_δ dans $f^{-1}(l_{k+1} \cup D_{k+1}) \setminus F_{\lambda_{k+1}}$. S'il y a $x \in H^1(L_k)$ et $y \in H^1(f^{-1}(l_{k+1} \cup D_{k+1}) \setminus F_{\lambda_{k+1}}) = H^1(f^{-1}(D_{k+1}^*))$ tels que

$$z = j_k^*(x) = i_{k+1}^*(y),$$

alors on déduit que $z \in H^1(F_\delta)^{<T_{\lambda_1},\ldots,T_{\lambda_{k+1}}>}$.

De plus, pour tout élément z de ce type, la dimension de l'espace de couples (x,y) comme au-dessus est égale à $\dim \operatorname{Ker} j_k^* + \dim \operatorname{Ker} i_{k+1}^* = k+1$. Ceci achève la preuve du Lemme. $\qquad\square$

Dans le cas $n > 2$ on a un lemme similaire, sauf le fait que les morphismes j_k^* sont maintenant des monomorphismes.

Preuve du Corollaire 1 Il suffit de montrer qu'on a

$$(\chi 2) \qquad b_{n-1}(F_\delta) = (-1)^n \Big(\sum_{s \in \Lambda} (\chi(F_s) - \chi(F_\delta)) \Big).$$

Ceci provient de l'égalité

$$1 = \chi(\mathbb{C}^n) = \chi(\mathbb{C}^n \setminus f^{-1}(\Lambda)) + \sum_{s \in \Lambda} \chi(F_s)$$

car on a

$$\chi(\mathbb{C}^n \setminus f^{-1}(\Lambda)) = (1 - |\Lambda|)\chi(F_\delta). \qquad\square$$

3 L'inégalité de Kaliman et calcul de l'opérateur de monodromie T_0

Soit $f \in \mathbb{C}[x,y]$ un polynôme de degré d. Pour étudier l'application polynomiale $\mathbb{C}^2 \longrightarrow \mathbb{C}$ associée à f, il est utile d'introduire une compactification X de $\mathbb{C}^2$.

Soit $F(x,y,z) \in \mathbb{C}[x,y,z]$ l'homogénéisé de degré d de f. Si $f(x,y) = f_0 + f_1(x,y) + f_2(x,y) + \cdots + f_d(x,y)$ est la décomposition de f en formes homogènes, alors

$$F(x,y,z) = f_d(x,y) + f_{d-1}(x,y)z + \cdots + f_1(x,y)z^{d-1} + f_0 z^d.$$

Considérons le pinceau de courbes dans $\mathbb{P}^2$

$$C_t : \qquad F(x,y,z) - tz^d = 0$$

et l'application rationnelle associée

$$\phi : \mathbb{P}^2 ---> \mathbb{P}, \quad \phi(x,y,z) = (F(x,y,z), z^d)$$

Notons L_∞ la droite à l'infini $\{z = 0\}$. Soit

$$B = \{b_1, \ldots, b_{k(f)}\} = \{f_d(x,y) = z = 0\} \subset L_\infty$$

l'ensemble des points d'indétermination de ϕ. Il existe une suite finie minimale d'éclatements de points $\pi : X \longrightarrow \mathbb{P}^2$ telle que

(i) $\pi : X \setminus \pi^{-1}(B) \longrightarrow \mathbb{P}^2 \setminus B$ soit un isomorphisme,

(ii) La fonction rationnelle $\overline{\phi} = \phi \circ \pi$ est régulière sur X,

(iii) Si $\tilde{F}_t$ est la transformée stricte d'une courbe C_t, alors $\tilde{F}_t \cup \pi^{-1}(B)$ est un diviseur à croisements normaux, pour tout $t \in \mathbb{C}$.

Pour la construction de $\pi : X \longrightarrow \mathbb{P}^2$, on peut voir [F]. Notons

$$D_\infty = X \setminus \mathbb{C}^2 = \pi^{-1}(L_\infty).$$

Alors D_∞ est une courbe à croisements normaux, dont toutes les composantes irréductibles sont rationnelles et lisses. De plus, le graphe dual T_∞ de D_∞ est un arbre. L'ensemble des composantes de D_∞ est une réunion disjointe de trois sous ensembles:

(a) Les composantes irréductibles D telles que $\overline{\phi}|D$ est constante et la valeur atteinte est ∞.

(b) Les composantes irréductibles D telles que $\overline{\phi}|D$ est surjective. Ces composantes sont appellées *composantes dicritiques*. Le degré d'une telle composante est le degré de l'application $\overline{\phi}|D$. Soit $\delta(f)$ le nombre de ces dicritiques. On a évidemment $\delta(f) \geq k(f)$. En outre, on sait que pour un polynôme primitif f, le p.g.c.d. de $\{\deg(D)|D$ dicritique$\}$ est égal à 1, voir [A].

(c) Les composantes irréductibles D telles que $\overline{\phi}|D$ est constante et la valeur atteinte est finie.

On a ainsi une description de l'ensemble Λ. Soit $t \in \mathbb{C}$. Alors $t \in \Lambda$ si l'on est dans l'une des situations suivantes:

(i) Il existe un dicritique D tel que t est une valeur critique de $\overline{\phi}|D$.

(ii) Il existe une composante irréductible D de D_∞ de type (c) telle que t est la valeur atteinte par $\overline{\phi}|D$.

(iii) La fibre F_t est singulière.

Dans les conditions (i) et (ii), on dit que la fibre est *irrégulière à l'infini*. La démonstration de ce critère, se trouve dans [LW] et [F].

Soit $T_i = \pi^{-1}(b_i)$. Si Δ_i est un petit disque de $\mathbb{P}^2$ centré en b_i, alors $N_i = \pi^{-1}(\Delta_i)$ est un voisinage tubulaire de T_i et $\partial N_i \simeq \partial \Delta_i = \mathbb{S}^3$. Le graphe dual associé à T_i est un arbre.

Soient $D_{i,1}, \ldots, D_{i,k_i}$ les composantes dicritiques de T_i. Les composantes connexes de $T_i \setminus (\cup_j D_{i,j})$ sont encore des arbres.

On rappelle que δ est un nombre réel positif assez petit, $F_\delta = f^{-1}(\delta)$, est une fibre régulière de f, et $\tilde{F}_\delta$ est la transformée stricte de la courbe C_δ. Soit $A_{i,j} = \tilde{F}_\delta \cap D_{i,j}$, $k_{i,j} = Card(A_{i,j})$. On a donc

(i) F_δ a comme compactification lisse $\tilde{F}_\delta = \overline{\phi}^{-1}(\delta)$.

(ii) $k_{i,j} = \deg(\overline{\phi}_{|D_{i,j}})$.

(iii) $\sum_j k_{i,j}$ est le nombre de branches du germe (C_δ, b_i).

1) *Un diagramme fondamental.*

Soit $E = \overline{\phi}^{-1}(S_\delta)$ et notons $\mathcal{D} = \cup_{i,j} D_{i,j}$. Alors

$$\overline{\phi} : (E, E \cap \mathcal{D}) \longrightarrow S_\delta$$

est une fibration relative de fibre $(\tilde{F}_\delta, A)$ où $A = \cup_{i,j} A_{i,j}$. On a donc le diagramme suivant:

$$(\star) \quad \begin{array}{ccccccccc} 0 & \longrightarrow & H_1(\tilde{F}_\delta) & \longrightarrow & H_1(\tilde{F}_\delta, A) & \longrightarrow & \tilde{H}_0(A) & \longrightarrow & 0 \\ & & \tilde{T} \downarrow & & T_r \downarrow & & T_A \downarrow & & \\ 0 & \longrightarrow & H_1(\tilde{F}_\delta) & \longrightarrow & H_1(\tilde{F}_\delta, A) & \longrightarrow & \tilde{H}_0(A) & \longrightarrow & 0 \end{array}$$

où $\tilde{T}$, T_r et T_A sont les opérateurs de monodromie associés à la fibration $\overline{\phi}$. Via la dualité d'Alexander [S,p. 297]

$$H_1(\tilde{F}_\delta, A) \simeq H^1(F_\delta)$$

l'opérateur de monodromie relative

$$T_r : H_1(\tilde{F}_\delta, A) \longrightarrow H_1(\tilde{F}_\delta, A)$$

correspond à l'opérateur

$$T_0 : H^1(F_\delta) \longrightarrow H^1(F_\delta).$$

La dualité de Poincaré nous donne l'égalité

$$\dim H^1(\tilde{F}_\delta)^\pi = \dim H_1(\tilde{F}_\delta)^\pi.$$

En effet le cycle c est invariant si et seulement si le cocycle $T_c = \langle c, - \rangle$ est invariant, où $\langle -, - \rangle$ est la forme d'intersection des cycles sur $\tilde{F}_\delta$. Le théorème (projectif) des cycles invariants [De] donne un épimorphisme

$$H^1(X) \to H^1(\tilde{F}_\delta)^\pi.$$

D'autre part, on a $\mathbb{C}^2 \subset X$ et donc $\pi_1(X) = 0$, ce qui implique $H^1(X) = 0$. On peut utiliser le diagramme $(\star)$ et obtenir l'égalité

$$H_1(\tilde{F}_\delta, A)^\pi \cap H_1(\tilde{F}_\delta) = H_1(\tilde{F}_\delta)^\pi = 0.$$

Elle implique que

$$\partial : H_1(\tilde{F}_\delta, A)^\pi \to \tilde{H}_0(A)^\pi$$

est un monomorphisme, qui en général n'est pas un épimorphism, voir [Se], p. 119. D'autre part, chaque dicritique produit un élément évident dans $H_0(A)^\pi$, donc $\dim H_0(A)^\pi = \delta(f)$ et $\dim \tilde{H}_0(A)^\pi = \delta(f) - 1$.

On a donc montré l'inégalité

$$\dim H^1(F_\delta)^\pi \leq \delta(f) - 1$$

qui donne une démonstration de l'inégalité de Kaliman. Si $g = \mathrm{genre}\,(\tilde{F}_\delta) = 0$ alors il est clair que cette inégalité est une égalité car alors ∂ est un isomorphisme.

Remarque. La version cohomologique du diagramme $(\star)$ donne un monomorphisme

$$\tilde{H}^0(A)^\pi \to H^1(\tilde{F}_\delta, A)^\pi$$

donc on a

$$\dim H_1(F_\delta)^\pi = \dim H^1(\tilde{F}_\delta, A)^\pi \geq \delta(f) - 1.$$

Exemple 4. Soif f un polynôme tel que la forme $f_d(x, y)$ a toutes ses racines simples. Alors on a $\delta(f) = d$, car il suffit d'éclater une fois chaque point b_i. La fibre générique peut être identifiée dans ce cas à la fibre de Milnor de la singularité $(C, 0)$ définie par $f_d = 0$.

La partie fixe $H_1(F_\delta)^\pi$ correspond à la partie fixe par rapport à l'action de la monodromie locale qui, en vue des formules de Picard-Lefschetz, correspond au noyau $\mathrm{Rad}\,L$ du réseau de Milnor associé L.

En effet, pour toute singularité isolée d'hypersurface, soit T_{loc} l'opérateur de monodromie locale associé en homologie. Alors, la matrice associée à l'opérateur $T_{loc} - Id$ est essentiellement la matrice d'intersection du réseau L, voir [D2], p. 93. Donc, les cycles invariants par T_{loc} sont les mêmes que les cycles invariants par tout le groupe de monodromie. Dans notre situation, l'opérateur T_{loc} correspond à l'opérateur de monodromie à l'infini $T(f)_\infty$ du polynôme f, voir [NZ], [GN]. On a

$$\begin{aligned}
\dim H_1(F_\delta)^\pi &= \dim\,(\mathrm{Rad}\,L) = \dim L - b_1(\tilde{F}_\delta) \\
&= (d-1)^2 - (d-1)(d-2) = d - 1
\end{aligned}$$

voir [D2], p. 157. Donc, la dimension de l'espace des cycles invariants (en homologie) est indépendante des monômes de degré $< d$ dans f.

Par contre, pour $f = x^3 + y^3$ on a $\sum(n(F_s) - 1) = 2$ et pour $f = x^3 + y^3 + xy$ on a $\sum(n(F_s) - 1) = 0$, i.e. la dimension de l'espace des cocycles invariants (en cohomologie) dépend aussi des termes de degré $< d$.

De plus, la famille $f_t = x^3 + y^3 + txy$ est une famille de polynômes *M-tame* à nombre de Milnor global constant, pour laquelle la dimension de l'espace de cycles invariants en cohomologie n'est pas constante malgré le fait que la famille d'opérateurs $T(f_t)_\infty$ est constante d'après [HZ].

Preuve du Corollaire 2. Si l'action de la monodromie est triviale, on a

$$H_1(\tilde{F}_\delta) = H_1(\tilde{F}_\delta)^\pi = 0$$

donc $g = 0$. Mais alors l'égalité de Kaliman donne

$$\delta(f) - 1 = b_1(F_\delta).$$

On conclut en remarquant que $b_1(F_\delta) = |A| - 1$, donc $\delta(f) = |A|$. D'autre part $|A| - \delta(f) = \sum(k_{i,j} - 1)$.

2) Calcul des dimensions de $Ker(\tilde{T} - Id)$ et de $Ker(T_A - Id)$

Nous allons maintenant calculer les dimensions de Ker $(\tilde{T} - Id)$ et de Ker $(T_A - Id)$ pour obtenir encore une autre démonstration de l'inégalité de Kaliman (qui donne une égalité aussi pour $|\Lambda| = 1$) et pour déterminer la relation entre les blocs de Jordan pour la valeur propre 1 des opérateurs T_0, $\tilde{T}$ et T_A.

a) Calcul de $Ker(\tilde{T} - Id)$

Ce calcul utilise le théorème des cycles invariants [C] pour le morphisme projectif

$$\overline{\phi} : X_\delta \longrightarrow D_\delta$$

où $X_\delta = \overline{\phi}^{-1}(\bar{D}_\delta)$. Ce théorème nous donne une suite exacte

$$H^1(X_\delta) \xrightarrow{j^\star} H^1(\tilde{F}_\delta) \xrightarrow{Id-\tilde{T}} H^1(\tilde{F}_\delta).$$

On montre que l'on a le résultat suivant, qui est un cas particulier des résultats de de Jong et de Steenbrink, voir [JS].

Lemme 3. *Le morphisme induit par inclusion*

$$j^\star : H^1(X_\delta) \longrightarrow H^1(\tilde{F}_\delta)$$

est injectif.

Preuve. Grâce au théorème des cycles invariants, il suffit de prouver que $\dim H^1(X_\delta) = \dim \mathrm{Ker}\,(Id - \tilde{T})$. Soit $Y = \overline{\phi}^{-1}(0)$. Dans la suite exacte du couple $(X_\delta, X_\delta \setminus Y)$, on a la partie suivante

$$\ldots \longrightarrow H^1(X_\delta, X_\delta \setminus Y) \longrightarrow H^1(X_\delta) \longrightarrow H^1(X_\delta \setminus Y) \longrightarrow$$
$$\xrightarrow{\delta} H^2(X_\delta, X_\delta \setminus Y) \xrightarrow{S} H^2(X_\delta) \longrightarrow \ldots$$

On a les relations

$$H^1(X_\delta, X_\delta \setminus Y) \simeq H^1(X_\delta, \partial X_\delta) \simeq H^3(X_\delta) = 0,$$
$$H^2(X_\delta, X_\delta \setminus Y) \simeq H^2(X_\delta) \simeq H_2(X_\delta).$$

En outre, on peut considérer le morphisme S comme étant donné par la forme d'intersection des composantes de Y, voir [D2, p. 52]. La suite exacte de Wang de la fibration $X_\delta \setminus Y \longrightarrow D_\delta^*$, nous donne

$$\dim H^1(X_\delta \setminus Y) = \dim H^2(X_\delta \setminus Y) = \dim \operatorname{Ker}(Id - \tilde{T}) + 1.$$

Le théorème des cycles invariants implique

$$\dim H^1(X_\delta) \geq \dim \operatorname{Ker}(Id - \tilde{T}).$$

On en déduit donc que $\dim(\operatorname{Ker} S) \leq 1$.

Il suffit de démontrer que $\dim(\operatorname{Ker} S) = 1$, c'est à dire que $\dim(\operatorname{Ker} S) \neq 0$. Or $[\tilde{F}_\delta]$ est un élément de $\operatorname{Ker} S$ car $\tilde{F}_\delta \cap Y_i = \emptyset$, pour toute composante irréductible Y_i de Y. De plus $[\tilde{F}_\delta]$ est non nul dans $H_2(X_\delta)$, car il est non nul dans $H_2(X)$, comme tout cycle donné par une sous variété projective dans une variété projective. $\qquad\square$

Corollaire 4.
$$\dim \operatorname{Ker}(\tilde{T} - Id) = b_1(Y).$$

Preuve. Il suffit de remarquer que puisque $\overline{\phi}$ est propre, on a une rétraction de X_δ sur Y [L]. $\qquad\square$

b) Calcul de $b_1(Y)$

Puisque $\tilde{F}_\delta$ s'obtient à partir de F_δ, en ajoutant $\sum_{i,j} k_{i,j}$ points, on a

$$\chi(\tilde{F}_\delta) = \chi(F_\delta) + \sum_{i,j} k_{i,j}.$$

Notons l_i le nombre de branches du germe (C_0, b_i). Alors $\tilde{F}_0$ s'obtient à partir de F_0, en ajoutant $\sum_i l_i$ points. Donc

$$\chi(\tilde{F}_0) = \chi(F_0) + \sum_i l_i.$$

On a donc **Lemme 4.**

$$b_1(\tilde{F}_0) = n(F_0) + b_0(F_0) - 1 + b_1(F_\delta) - \sum_i l_i - -(\chi(F_0) - \chi(F_\delta)).$$

Pour chaque dicritique, $D_{i,j}$, soit $n_{i,j}$ le nombre d'arbres $T_1^{i,j}, \ldots, T_{n_{i,j}}^{i,j}$ rattachés à $D_{i,j}$, qui sont contenus dans la fibre $Y = \overline{\phi}^{-1}(0)$.

On a $|\tilde{F}_0 \cap D_{i,j}| = k_{i,j} - n_{i,j} - e_{i,j}$ où $e_{i,j}$ sont des entiers non négatifs qui sont egaux à 0 si et seulement si $\overline{\phi}|D_{i,j}$ a 0 comme valeur régulière. De plus, si $n_{i,j}^k = |\tilde{F}_0 \cap T_k^{i,j}|$, on a

$$l_i = \sum_j (\sum_k n_{i,j}^k + k_{i,j} - n_{i,j} - e_{i,j}).$$

Donc **Lemme 5.**

$$b_1(\tilde{F}_\delta) \;=\; 2 - n(F_0) - b_0(F_0) + b_1(\tilde{F}_0) + $$
$$+ \sum_{i,j,k}(n_{i,j}^k - 1) - \sum_{i,j} e_{i,j} + (\chi(F_0) - \chi(F_\delta)).$$

Lemme 6.

$$b_1(Y) = b_1(\tilde{F}_0) + \sum_{i,j,k}(n_{i,j}^k - 1) - b_0(F_0) + 1.$$

Preuve du Lemme 6. Puisque les arbres $T_k^{i,j}$ sont des bouquets de 2-spheres, on a une suite exacte de Mayer-Vietoris

$$0 \longrightarrow H_1(\amalg_{i,j,k} T_k^{i,j}) \oplus H_1(\tilde{F}_0) \longrightarrow H_1(Y) \longrightarrow H_0(F(\textstyle\sum_{i,j,k}(n_{i,j}^k))) \longrightarrow$$
$$\longrightarrow H_0(F(m)) \oplus H_0(\tilde{F}_0) \longrightarrow H_0(Y) \longrightarrow 0$$

où $F(l)$ est une ensemble fini à l éléments, et m est le nombre de triplets (i,j,k). On a aussi les égalités suivantes

$$b_0(F_0) = b_0(\tilde{F}_0), \quad b_0(Y) = 1.$$

En effet, pour la dernière, on peut utiliser le fait que le bord d'un voisinage tubulaire de Y est connexe, car il est difféomorphe à $f^{-1}(S_\delta)$.

Remarque. Si 0 n'est pas une valeur critique de $\overline{\phi}|\mathcal{D}$, le nombre de places à l'infini de la courbe F_0 en chaque point b_i est supérieur (ou égal) au nombre de places à l'infini en b_i de la courbe générique F_δ.

S'il n'existe pas d' arbre rattaché dans Y, le nombre de places à l'infini de la courbe F_0 en chaque point b_i est inférieur (ou égal) au nombre de places à l'infini en b_i de la courbe générique F_δ. En effet, la première condition équivaut à $e_{i,j} = 0$, qui implique $l_i \geq \sum_j k_{i,j}$.

c) Retour sur l'inégalité de Kaliman

Pour chaque $s \in \Lambda$ soient $Y_s = \overline{\phi}^{-1}(s)$ et T_s (resp. $\tilde{T}_s, T_{A,s}$) l'operateur de monodromie sur $H^1(F_\delta)$ (resp. sur $H_1(\tilde{F}_\delta), \tilde{H}_0(A))$ associé à un lacet élémentaire dans π qui tourne autour de s.

Dans [K] on montre, à partir d'un calcul simple de caractéristiques d'Euler similaire au calcul ci-dessus pour obtenir l'égalité $(\chi 2)$, l'égalité suivante

$$\sum_s k_s = 2g + \delta(f) - 1 - \sum_s (n(F_s) - 1)$$

où $k_s = 2g - b_1(Y_s)$ et g est le genre de la courbe $\tilde{F}_\delta$.

D'après le Corollaire 4, on a $k_s = \operatorname{codim}\operatorname{Ker}(\tilde{T}_s - Id)$. Pour obtenir l'inégalité (K) il suffit donc de montrer

$$(K') \qquad\qquad \sum_s k_s \geq 2g.$$

Tout d'abord on a une inégalité d'algèbre linéaire

$$\sum_s k_s \geq \operatorname{codim} H_1(\tilde{F}_\delta)^\pi \quad \text{car} \quad H_1(\tilde{F}_\delta)^\pi = \cap_s \operatorname{Ker}(\tilde{T}_s - Id).$$

On a une égalité ici si et seulement si la famille $(\operatorname{Ker}(\tilde{T}_s - Id))_{s \in \Lambda}$ est en position générale, ce qui est évidemment le cas si $g = 0$ ou si $|\Lambda| = 1$. D'autre part, nous avons vu ci-dessus que $H_1(\tilde{F}_\delta)^\pi = 0$, ce qui achève cette seconde preuve de l'inégalité de Kaliman. Pour le polynôme de Briançon, l'inégalité de Kaliman est stricte, donc la condition de position générale ci-dessus n'est pas toujours vraie (à comparer avec le Cor. 1).

3) *Calcul des opérateurs de monodromie*

Nous allons maintenant calculer des opérateurs de monodromie T_s pour les deux fibres irrégulières du polynôme de Briançon.

Rappelons d'abord quelques principes généraux. D'après le diagramme $(\star)$, on a

$$\Delta(t) = \Delta_A(t)\tilde{\Delta}(t)$$

où

$$
\begin{aligned}
\Delta(t) &= \det(tId - T_r) = \det(tId - T_0) \\
\Delta_A(t) &= \det(tId - T_A) \\
\tilde{\Delta}(t) &= \det(tId - \tilde{T})
\end{aligned}
$$

On peut calculer le polynôme $\tilde{\Delta}$ en utilisant la fonction Zêta de la monodromie opérant sur la courbe $\tilde{F}_\delta$. En effet on a

$$Z(t) = \frac{\tilde{\Delta}(t)}{(1-t)^2}.$$

La fonction Zêta de la monodromie est calculée en utilisant la formule [A'C]

$$Z(t) = \prod_{m \geq 1} (1 - t^m)^{-\chi(S_m)}$$

où S_m est la partie lisse de $Y = \overline{\phi}^{-1}(0)$ qui a la multiplicité m.

Pour calculer Δ_A, on utilise les remarques suivantes.

Remarquons tout d'abord que l'opérateur T_A est égal à l'identité si et seulement si $\overline{\phi}|\mathcal{D}$ n'a pas 0 comme valeur critique, donc si et seulement si $\sum_{i,j} e_{i,j} = 0$. En effet, soit $x_0 \in D_{i,j}$ un point singulier de $\overline{\phi}|D_{i,j}$ tel que $\overline{\phi}(x_0) = 0$. Soit $k = \deg_{x_0}(\overline{\phi}|D_{i,j})$. Alors, si δ tend vers 0, k points $y_1, \ldots, y_k$ de l'ensemble $A_{i,j}$ vont tendre vers x_0 et sur l'ensemble $\{y_1, \ldots, y_k\}$ la monodromie agit comme une permutation cyclique.

L'action de la monodromie sur $H_0(A)$ a une matrice qui contient $\sum_{i,j}(k_{i,j} - e_{i,j})$ blocs, chaque bloc correspondant à une permutation cyclique. Chaque bloc donne donc une contribution de 1 pour la dimension de $\mathrm{Ker}\,(T_A - Id)$ dans $H^0(A)$. Donc dans $\tilde{H}_0(A)$ on a

$$\dim \mathrm{Ker}\,(T_A - Id) = \sum_{i,j}(k_{i,j} - e_{i,j}) - 1.$$

En particulier, on a

Corollaire 5.

   (i) $T_A = Id$ si et seulement si $e_{i,j} = 0$ pour tout couple (i,j).

   (ii) $\dim \mathrm{Ker}\,(T_0 - Id) = \dim \mathrm{Ker}\,(\tilde{T}_0 - Id) + \dim \mathrm{Ker}\,(T_{A,0} - Id)$.

On peut maintenant préciser la structure de Jordan de T_0 correspondant à la valeur propre 1. Soient $m(T_0), m(\tilde{T}), m(T_A)$ les nombres des blocs de Jordan de $T_0, \tilde{T}, T_A$ pour la valeur propre 1.

Soient $\alpha_i, i = 1, ..., m(T_0)$ les dimensions des blocs de T_0 et soient β_j, γ_k les nombres correspondants pour $\tilde{T}$ et T_A (en effet $\gamma_k = 1$ pour tout k). La relation ci-dessus entre les polynômes caractéristiques donne

$$\sum_i \alpha_i = \sum_j \beta_j + \sum_k \gamma_k.$$

D'autre part, le Corollaire 5 implique

$$m(T_0) = m(\tilde{T}) + m(T_A).$$

A chaque bloc de Jordan $\tilde{B}_j$ de $\tilde{T}$ correspond un unique bloc de Jordan B_j de T_0 tel que

$$\beta_j = \dim B_j \geq \dim \tilde{B}_j = \alpha_j.$$

On déduit que cette inégalité doit être une égalité et que les blocs de Jordan de T_0 pour la valeur propre 1 sont précisement ceux de $\tilde{T}$ et de T_A.

Corollaire 6. *La partie $(T_0)_1$ de l'opérateur T_0 qui correspond à la valeur propre 1 est semisimple si et seulement si le graphe dual de Y est un arbre.*

Preuve. Les calculs ci-dessus montrent que la multiplicité de 1 comme racine du polynôme $\tilde{\Delta}(t)$ est précisement donnée par

$$mult_1(\tilde{\Delta}(t)) = b_1(Y) + (|Y_{sing}| + 1 - n(Y))$$

Puis, l'opérateur $(T_0)_1$ est semisimple si et seulement si l'opérateur $(\tilde{T}_0)_1$ est semi-simple, donc si et seulement si

$$|Y_{sing}| + 1 - n(Y) = 0$$

On voit facilement que l'on a cette dernière égalité si et seulement si le graphe dual de Y est un arbre. $\qquad\square$

Nous rappelons que dans la situation locale, pour une singularité de courbe on a toujours un opérateur semisimple $(T_{loc})_1$. L'exemple suivant montre que cela n'est pas le cas dans la situation globale.

Exemple 5. On rappelle que le polynôme de Briançon a toutes ses fibres lisses et irréductibles. Il a deux fibres irrégulières à l'infini, pour $t_1 = -16/9$ et $t_0 = 0$. On a $\lambda_0 = 1$ et $\lambda_1 = 3$, où λ_1 est l'analogue de λ_0 mais cette fois pour la fibre F_{t_1}. Pour la fibre générique on a $b_1(F_\delta) = \sum_i \lambda_i = 4$.

En calculant la fibre $\overline{\phi}^{-1}(t_1)$, on obtient

$$Z(t) = \frac{t^6 - 1}{(t^3 - 1)(t^2 - 1)(t - 1)} = \frac{t^2 - t + 1}{(t - 1)^2}$$

donc

$$\tilde{\Delta}(t) = t^2 - t + 1.$$

Dans ce cas on a $\mathrm{Card}\,(\dot{A}) = 3$. La monodromie T_A agit de façon non triviale car il y a un dicritique D_i tel que $\overline{\phi}|D_i$ a degré 2 et t_1 comme valeur critique. Donc

$$\Delta_A(t) = (t - 1)(t + 1).$$

Alors

$$\Delta(t) = \Delta_A(t)\tilde{\Delta}(t) = (t^3 + 1)(t - 1).$$

On peut donc énoncer:

Pour le polynôme de Briançon, la monodromie autour de la fibre F_{t_1} est semi-simple de polynôme caractéristique

$$\Delta(t) = (t^3 + 1)(t - 1).$$

En calculant la fibre $\overline{\phi}^{-1}(t_0)$ on a

$$Z(t) = 1 \quad \text{donc} \quad \tilde{\Delta}(t) = (t - 1)^2.$$

On est dans le cas où $T_A = Id$, donc

$$\Delta_A(t) = (t - 1)^2 \quad \text{et finalement} \quad \Delta(t) = \Delta_A(t)\tilde{\Delta}(t) = (t - 1)^4.$$

On voit donc que toutes les valeurs propres de la monodromie T_0 sont égales à 1. Puisque $b_1(F_0) = \lambda_1 = 3$, le Théorème 1 implique

$$\dim \mathrm{Ker}\,(T_0 - Id) = 3.$$

Pour le polynôme de Briançon, la matrice de la monodromie T_0 autour de la fibre F_0 est semblable à la matrice

$$A = \begin{pmatrix} 1 & 1 & 0 & 0 \\ 0 & 1 & 0 & 0 \\ 0 & 0 & 1 & 0 \\ 0 & 0 & 0 & 1 \end{pmatrix}.$$

4   Démonstration du Théorème 3

Dans cette section nous supposons que la fibre F_0 est réduite et connexe. Pour simplifier les notations, notons $F := F_0 = f^{-1}(0)$ et $V := \overline{T_\delta}$. Soit $j : F \hookrightarrow V$ l'inclusion.

Remarque. Dans cette section, nous travaillons avec la preimage d'un disque fermé pour pouvoir utiliser la suite exacte de Mayer-Viétoris pour un recouvrement fermé; les résultats seront valables si l'on remplace V par T_δ car l'interieur d'une variété à bord a le même type d'homotopie que la variété toute entière.

La partie (a)$\Rightarrow$(b) est évidente. En utilisant le cas propre, il est aussi facile de voir (c)$\Rightarrow$(a). La partie (b)$\Rightarrow$(c) est une conséquence du lemme suivant:

Lemme 7. *Supposons que F est irrégulière à l'infini et connexe. Alors, l'application* $j_* : H_1(F; \mathbb{Z}) \to H_1(V; \mathbb{Z})$ *n'est pas surjective.*

La preuve du lemme occupera la suite de cette section.

Rappel. Tous les groupes d'homologie dans cette section où le groupe de coefficients n'est pas explicite sont à coefficientes dans $\mathbb{Z}$.

Considérons la résolution $\overline{\phi} : X \to \mathbb{P}^1$ de la section 3. Soit $\pi : X \to \mathbb{P}^2$ la suite d'éclatements qui donne lieu à $\overline{\phi}$. Nous pouvons extraire une sous-suite minimale $\tilde{\pi} : \widetilde{X} \to \mathbb{P}^2$ telle que

  (i) $\tilde{\pi} : \widetilde{X} \setminus \tilde{\pi}^{-1}(B) \to \mathbb{P}^2 \setminus B$ soit un isomorphisme,

  (ii) La fonction rationnelle $\tilde{\phi} := \phi \circ \tilde{\pi}$ est régulière sur $\widetilde{X}$.

Cette résolution de f est celle de [LW]. Nous avons trois types de composantes irréductibles dans la courbe $\tilde{D}_\infty = \widetilde{X} \setminus \mathbb{C}^2$ qui sont définis de la même façon que pour D_∞. Dans ce cas, nous pouvons être plus explicites avec les composantes irréductibles de type (c). Nous allons décrire les composantes connexes de la réunion des composantes irréductibles de type (c) de $\tilde{D}_\infty$.

Soit C une telle composante connexe. Il s'agit d'une courbe à croisements normaux, qui est un bambou au sens de [LW]. C'est-à-dire, on peut ordonner les composantes irreductibles $D_1, \ldots, D_n$ de C de telle sorte que $D_i \cdot D_{i+1} = 1$ si $i = 1, \ldots, n-1$ et $D_i \cap D_j = \emptyset$ si $|i - j| \neq 1$(le graphe dual est linéaire). En plus, $D_i^2 \leq -2$, $i = 1, \ldots, n$.

    Il est aussi démontré dans [LW] qu'il existe une unique composante dicritique D_C telle que $D \cap C \neq \emptyset$; en plus, $D_C \cdot C = 1$ et le point d'intersection se trouve dans une composante au bout, par exemple, dans D_1. Finalement, étant donné un dicritique D il existe au plus un bambou C tel que $D = D_C$.

Nous utiliserons à nouveau le critère d'irrégularité à l'infini de la section 3, adapté à la résolution $\tilde{\pi}$. La fibre F est irrégulière à l'infini si l'une des deux situations suivantes a lieu:

  (i) Il existe un dicritique D tel que $t = 0$ est une valeur critique de $\tilde{\phi}_{|D}$.

  (ii) Il existe un bambou C tel que $\tilde{\phi}_{|C}$ est constante égale à $t = 0$.

Définition. La *partie dicritique* de F, notée Dic_0, est l'ensemble des points P dans les dicritiques tels que $\tilde{\phi}(P) = 0$.

Remarque. Un point $P \in \mathrm{Dic}_0$ peut être de deux types: soit P est dans l'adhérence de F dans $\tilde{X}$, soit P appartient à un bambou.

Nous allons étudier F près des points dans Dic_0.

Cas 1. *P est dans l'adhérence $\bar{F}$ de F dans $\tilde{X}$.*

Fixons d'abord la notation:

- D est le dicritique tel que $P \in D$.

- B_P est une *petite* boule de Milnor (fermée) de $D \cup \bar{F}$ en P.

- $F_P := F \cap B_P$ (F_P est contenu dans $\mathbb{C}^2$) et $\bar{F}_P := \bar{F} \cap B_P = F_P \cup \{P\}$.

- $V_P := V \cap B_P$.

- $D_P := D \cap B_P$.

- $\bar{F}_P := \bar{F}_P^1 \cup \cdots \cup \bar{F}_P^{n(P)}$ est la décomposition en composantes irréductibles (locales) de $\bar{F}_P$.

- Pour $j = 1, \ldots, n(P)$, soit $m_P^j := (\bar{F}_P^j \cdot D_P)_P$ et soit $m_P := \sum_{j=1}^{n(P)} m_P^j$.

- Pour $j = 1, \ldots, n(P)$, soit $K_P^j := \bar{F}_P^j \cap \partial B_P$; il s'agit d'un nœud dans $\partial B_P = S^3$. Soit $K_P := \bigcup_{j=1}^{n(P)} K_P^j = \bar{F}_P \cap \partial B_P$.

- $V_P^\partial := V_P \cap \partial B_P$; il s'agit d'une réunion de $n(P)$ tores pleins, voisinage tubulaire de l'entrelacs K_P dans $\partial B_P = S^3$.

- $S_P := \partial B_P \setminus D$; il s'agit du complément dans $\partial B_P = S^3$ du nœud trivial $D_P^\partial := D_P \cap \partial B_P$, c'est-à-dire, un tore plein ouvert.

- Soit M_P l'intersection avec ∂B_P d'une courbe lisse et transverse à D en P (de sorte que la réunion de M_P et D_P^∂ est un entrelacs de Hopf dans ∂B_P). Il s'agit de l'âme de S_P.

En utilisant un champs de vecteurs comme dans [Mi], nous avons le résultat suivant:

Lemme 8. *Les inclusions $V_P \hookrightarrow B_P \setminus D$ et $S_P \hookrightarrow B_P \setminus D$ sont des équivalences d'homotopie.*

Notons aussi $[M_P]$ l'image dans $H_1(V_P; \mathbb{Z})$ de $[M_P] \in H_1(B_P \setminus D; \mathbb{Z})$. Par conséquent $[M_P]$ engendre le groupes abeliens $H_1(S_P; \mathbb{Z})$, $H_1(B_P \setminus D; \mathbb{Z})$, $H_1(V_P; \mathbb{Z})$, qui sont tous isomorphes à $\mathbb{Z}$.

Nous rappelons que le coefficient d'enlacement de deux nœuds algébriques coïncide avec le nombre d'intersection des germes de courbes qui les définissent, voir [BK]. D'autre part, M_P définit un méridien positivement orienté du nœud D_P^∂.

La définition du coefficient d'enlacement implique que si L est un nœud contenu dans S_P, alors, sa classe d'homologie est égale à la classe de M_P multiplié par le coefficient d'enlacement de L et D_P^∂. Nous en déduisons:

Lemme 9. *Pour* $j = 1, \ldots, n(P)$, $[K_P^j] = m_P^j[M_P]$ *dans* $H_1(V_P; \mathbb{Z})$.

Cas 2. $P \in C$ *où* $C := D_1 \cup \cdots \cup D_{r(P)}$ *est un bambou.*

Fixons la notation comme dans le cas précedent:

- D est le dicritique tel que $P \in D$.

- B_P est un voisinage tubulaire (fermé) de C.

- $F_P := F \cap B_P$ (F_P est contenu dans $\mathbb{C}^2$) et $\bar{F}_P := \bar{F} \cap B_P$.

- $V_P := V \cap B_P$.

- $D_P := D \cap B_P$.

- $\bar{F}_P := \bar{F}_P^1 \cup \cdots \cup \bar{F}_P^{n(P)}$ est la décomposition en composantes irréductibles de $\bar{F}_P$.

- Pour $j = 1, \ldots, n(P)$, soit $m_P^j := (m_P^j(1), \ldots, m_P^j(r(P)))$ où $m_P^j(k) := \bar{F}_P^j \cdot D_k$, $k = 1, \ldots, r(P)$. Nous remarquons que dans le $r(P)$-tuple m_P^j, le nombre de coordonnées non nulles est égal à 1 ou 2.

- Pour $j = 1, \ldots, n(P)$, soit $K_P^j := \bar{F}_P^j \cup \partial B_P$; il s'agit d'un nœud dans ∂B_P qui est un espace lenticulaire non homéomorphe à S^3. Soit $K_P := \bigcup_{j=1}^{n(P)} K_P^j$.

- $V_P^\partial := V_P \cap \partial B_P$; il s'agit d'une réunion de $n(P)$ tores pleins, voisinage tubulaire de l'entrelacs K_P dans ∂B_P.

- $S_P := \partial B_P \setminus D$; il s'agit du complément dans $\partial B_P = S^3$ du nœud *trivial* $D_P^\partial := D_P \cap \partial B_P$, où *trivial* veut dire que S_P un tore plein ouvert.

- Pour chaque $k = 1, \ldots, r(P)$, soit C_k une curvette de D_k; notons $C_k^\partial := C_k \cap \partial B_P$. Soit $M_P := C_{r(P)}^\partial$, c'est-à-dire le bord de la dernière curvette. Il s'agit de l'âme de S_P.

On montre aisément:

Lemme 8'. *Les inclusions* $V_P \hookrightarrow B_P \setminus D$ *et* $S_P \hookrightarrow B_P \setminus D$ *sont des équivalences d'homotopie.*

Notons aussi $[M_P]$ l'image dans $H_1(V_P; \mathbb{Z})$ de $[M_P] \in H_1(B_P \setminus D; \mathbb{Z})$. Par conséquent $[M_P]$ engendre les groupes abeliens $H_1(S_P; \mathbb{Z})$, $H_1(B_P \setminus D; \mathbb{Z})$, $H_1(V_P; \mathbb{Z})$, qui sont tous isomorphes à $\mathbb{Z}$.

Pour chaque $k = 1, \ldots, r(P)$, $[C_k^\partial] \in H_1(S_P; \mathbb{Z})$; on pose $l_k \in \mathbb{Z}$ tel que $[C_k^\partial] = l_k[M_P]$. Nous allons calculer ces nombres-là en termes des fractions continues, en suivant les techniques de Mumford pour calculer le groupe fondamental d'une variété graphée dont le graphe dual est un arbre et toutes les composantes sont rationelles, voir [Mu]. Dans notre cas la variété graphée est un espace lenticulaire; la condition sur les self-intersections, implique qu'elle n'est pas homéomorphe à la sphère de dimension 3. Nous interprétons ces résultats en homologie. Il démontre que $H_1(S_P)$ est engendré par les classes d'homologie des bords des curvettes C_k^∂, $k = 1, \ldots, r(P)$ avec les relations suivantes:

$$
\begin{aligned}
[C_{r(P)-1}^\partial] &= -e_{r(P)}[C_{r(P)}^\partial], \\
-e_k[C_k^\partial] &= [C_{k+1}^\partial] + [C_{k-1}^\partial], \qquad k = 1, \ldots, r(P)-1,
\end{aligned}
\tag{M}
$$

où $-e_k$ est la self-intersection de D_k, $k = 1, \ldots, r(P)$.

Remarque. Bien que le résultat de Mumford soit valable pour les variétés graphées fermées, dans sa démonstration (voir par exemple la page 11 de [Mu]), il calcule l'homologie de la variété à bord associée à chaque sommet (un S^1-fibré de base une sphère trouée, dont le nombre de trous est égal à la valence du sommet dans le graphe). C'est de cette façon que nous obtenons les relations (M) qui sont évidemment valables dans S_P, dont nous savons par ailleurs qu'il s'agit d'un tore plein ouvert d'âme M_P.

Nous allons exprimer $[C_k^\partial]$, $k = 1, \ldots, r(P)$, en fonction du générateur $[M_P]$ de $H_1(S_P)$ en utilisant les fractions continues.

Définition. Soient $e_1, \ldots, e_r$ des entiers ≥ 2. On définit la fraction continue $[[e_1, \ldots, e_r]]$ comme suit:

(i) Si $r = 1$, $[[e_1]] := e_1$.

(ii) Si $r > 1$, alors, $[[e_1, \ldots, e_r]] := e_1 - \dfrac{1}{[[e_2, \ldots, e_r]]}$.

Remarque. Soient $e_1, \ldots, e_r$ des entiers ≥ 2. Alors il existe une suite d'entiers positifs $n_0, n_1, \ldots, n_r$, avec $n_r = 1$, tels que $\mathrm{pgcd}(n_j, n_{j-1}) = 1$, $j = 1, \ldots, r$, et

$$
[[e_j, \ldots, e_r]] = \frac{n_{j-1}}{n_j}, \quad j = 1, \ldots, r.
$$

Par exemple, si $r > 1$, $n_{r-1} = e_r$; si $r > 2$, $n_{r-2} = e_{r-1}e_{r-2}$. Ces entiers sont toujours plus grands que 1, sauf n_r.

Si nous utilisons les relations (M) pour calculer les l_j, nous trouvons une situation semblable à celle des n_j pour la fraction continue $[[e_j, \ldots, e_{r(P)}]]$. Plus précisément, notons $n_0, n_1, \ldots, n_{r(P)}$ la suite de la remarque précédente. Il est facile de montrer par recurrence descendante que $l_j = n_j$, $j = 1, \ldots, r(P)$. En utilisant l'unicité de cette décomposition en fractions continues, nous en déduisons:

Lemme 10. *Pour* $k = 1, \ldots, r(P)$ *on a* $l_j \geq 1$; *en plus,* $l_j = 1$ *si et seulement si* $j = r(P)$.

Il est facile de voir que $[K_P^j] = \sum_{k=1}^{r(P)} m_P^j(k)[C_k^\partial]$, $j = 1, \ldots, n(P)$. Cette égalité nous permet de démontrer:

Lemme 11. *Pour $j = 1, \ldots, n(P)$,*

$$[K_P^j] = \Big(\sum_{k=1}^{r(P)} m_P^j(k) l_k \Big) [M_P]$$

dans $H_1(V_P; \mathbb{Z})$.

La preuve du Lemme 7 passe par l'utilisation de deux suites exactes de Mayer-Viétoris associées aux décompositions fermées suivantes:

$$F_\infty := \coprod_{P \in \mathrm{Dic}_0} F_P, \quad F_a := \overline{F \setminus F_\infty},$$
$$V_\infty := \coprod_{P \in \mathrm{Dic}_0} V_P, \quad V_a := \overline{V \setminus V_\infty}.$$

Il est évident que:

$$F_\infty \cup F_a = F, \quad F_\infty \cap F_a = \coprod_{P \in \mathrm{Dic}_0} K_P,$$
$$V_\infty \cup V_a = V, \quad V_\infty \cap V_a = \coprod_{P \in \mathrm{Dic}_0} V_P^\partial.$$

Nous allons considérer le diagramme suivant, où les flèches horizontales sont celles des suites exactes de Mayer-Viétoris des décompositions ci-dessus et les flèches verticales sont induites par les inclusions:

$$\begin{array}{ccccccccccc}
\to H_2(F) & \to & H_1(F_\infty \cap F_a) & \to & H_1(F_\infty) \oplus H_1(F_a) & \to & H_1(F) & \to & H_0(F_\infty \cap F_a) & \to \\
\downarrow & & \downarrow & & \downarrow & & \downarrow & & \downarrow & \\
\to H_2(V) & \to & H_1(V_\infty \cap V_a) & \to & H_1(V_\infty) \oplus H_1(V_a) & \to & H_1(V) & \to & H_0(V_\infty \cap V_a) & \to
\end{array}$$

Lemme 12. *Les inclusions $F_a \hookrightarrow V_a$ et $F_a \cap F_\infty \hookrightarrow V_a \cap V_\infty$ sont des équivalences d'homotopie. Par conséquent, les flèches verticales induites par les inclusions*

$$H_1(F_a) \to H_1(V_a) \quad et \quad H_1(F_a \cap F_\infty) \to H_1(V_a \cap V_\infty)$$

sont des isomorphismes.

Preuve. Puisque $V_a \cap V_\infty$ est un voisinage tubulaire dans S^3 de $F_a \cap F_\infty$, l'assertion est évidente pour la deuxième inclusion. Considérons l'application $\tilde{f} : V_a \to \overline{D_\delta}$ induite par f. Nous avons:

- $\tilde{f}$ est propre;
- $\tilde{f}$ est une fibration localement triviale $V_a \setminus F_a \to D_\delta^*$;
- $\tilde{f}$ est une fibration triviale au bord $V_a \cap V_\infty$ compatible avec la précédente.

Dans cette situation, $F_a \hookrightarrow V_a$ est une équivalence d'homotopie. $\qquad \square$

Etudions la suite de Mayer-Viétoris de F; il est immédiate que $H_2(F) = 0$ (il s'agit d'une surface de Riemann ouverte). D'autre part, dans les hypothèses du lemme 7, F est connexe. Ceci implique que

$$0 \to H_0(F_\infty \cap F_a) \to H_0(F_\infty) \oplus H_0(F_a) \to H_0(F) \to 0$$

est une suite exacte courte. Par conséquent, la flèche $H_1(F) \to H_0(F_\infty \cap F_a)$ est nulle. Nous pouvons remplacer la suite supérieure du diagramme par la suite exacte courte:

$$0 \to H_1(F_\infty \cap F_a) \to H_1(F_\infty) \oplus H_1(F_a) \to H_1(F) \to 0.$$

Nous avons vu dans le Lemme 1 que $H_2(T_\delta; \mathbb{C}) = 0$. Puisque T_δ et V ont le même type d'homotopie, nous en déduisons que $H_2(V; \mathbb{Z})$ est de torsion. Puisque $H_1(F_a \cap F_\infty; \mathbb{Z})$ est libre, la flèche $H_2(V; \mathbb{Z}) \to H_1(F_a \cap F_\infty; \mathbb{Z})$ est nulle. Le diagramme précédent devient:

$$
\begin{array}{ccccccccc}
0 & \to & H_1(F_\infty \cap F_a) & \to & H_1(F_\infty) \oplus H_1(F_a) & \to & H_1(F) & \to & 0 \\
 & & \cong \downarrow & & \downarrow \quad\quad \downarrow \cong & & \downarrow & & \downarrow \\
0 & \to & H_1(V_\infty \cap V_a) & \to & H_1(V_\infty) \oplus H_1(V_a) & \to & H_1(V) & \to & H_0(V_\infty \cap V_a)
\end{array}
$$

Démonstration du Lemme 7. Nous supposons que F est connexe et irrégulière à l'infini. Remarquons que si F est régulière à l'infini, $n(P) = 1$, $\forall P \in \mathrm{Dic}_0$. Nous distinguons deux cas:

Cas A. *Il existe $P \in \mathrm{Dic}_0$ tel que $n(P) > 1$.*

On vérifie aisément dans ce cas que la flèche $H_0(V_a \cap V_\infty) \to H_0(V_a) \oplus H_0(V_\infty)$, de la suite exacte de Mayer-Viétoris pour V, n'est pas injective. Par conséquent, $H_1(V) \to H_0(V_a \cap V_\infty)$ n'est pas l'application nulle, ce qui implique que le morphisme $H_1(V_a) \oplus H_1(V_\infty) \to H_1(V)$ n'est pas surjectif. Or, il est évident que ceci implique que $j_* : H_1(F) \to H_1(V)$ ne peut pas être surjective. Nous venons de résoudre le cas A.

Cas B. *$\forall P \in \mathrm{Dic}_0$, nous avons $n(P) = 1$.*

A cause d'une version faible du lemme des cinq la surjectivité de l'application $j_* : H_1(F) \to H_1(V)$ est équivalente à celle de $j_* : H_1(F_\infty) \to H_1(V_\infty)$. Or, les deux espaces se décomposent en somme directe de sorte que pour achever la preuve du Lemme 7 il suffit de montrer:

Assertion. *Il existe $P \in \mathrm{Dic}_0$ tel que $j_* : H_1(F_P) \to H_1(V_P)$ n'est pas surjective.*

Nous appliquons le critère d'irrégularité à l'infini de [LW]. Il y a deux cas possibles:

Cas B1. *Il existe $P \in \mathrm{Dic}_0$ avec $P \in \bar{F}$ tel que si D est le dicritique qui contient P, alors $\tilde{\phi}_{|D}$ a un point critique en P.*

L'hypothèse du cas B1 implique que $m_P = (\bar{F}_P \cdot D)_P > 1$. Considérons $j_* :$ $H_1(F_P) \to H_1(V_P)$. Il est clair que $H_1(F_P) \cong H_1(V_P) \cong \mathbb{Z}$, avec des générateurs $[K_P] \in H_1(F_P)$ et $[M_P] \in H_1(V_P)$. Nous avons vu que $j_*([K_P]) = m_P[M_P]$. Dans ce cas-ci, l'application n'est pas surjective. Nous avons résolu le cas B1.

Cas B2. *Il existe $P \in Dic_0$ tel que $P \in C$ où C est un bambou associé a la valeur $t = 0$.*

Nous reprenons la notation de l'étude de ce type de points. Nous avons $C = D_1 \cup \cdots \cup D_{r(P)}$, avec $r(P) \geq 1$. Puisqu'il n'y a qu'une branche, nous n'avons qu'une suite $m_P = (m_P(1), \ldots, m_P(r(P)))$.

Comme dans le cas précédent, soit $j_* : H_1(F_P) \to H_1(V_P)$ avec les isomorphismes $H_1(F_P) \cong H_1(V_P) \cong \mathbb{Z}$; ils sont engendrés par $[K_P] \in H_1(F_P)$ et $[M_P] \in H_1(V_P)$. Nous avons vu que

$$j_*([K_P]) = \Big(\sum_{k=1}^{r(P)} m_P(k) l_k \Big) [M_P].$$

D'après le Lemme 10, nous voyons que j_* est surjective si et seulement si

$$m_P = (0, \ldots, 0, 1).$$

Nous aurons fini la preuve du Lemme 7 si l'on démontre que ce cas est impossible. Géometriquement, ce cas implique que il y a exactement une branche $\bar{F}$ qui s'attache au bambou C; cette branche est lisse et intersecte transversalement C en un point lisse de C dans $D_{r(P)}$ (i.e., la dernière composante irréductible du bambou). Si l'on regarde la suite d'éclatements nécessaire pour obtenir $\tilde{\phi}$, cette composante est la première à apparaître parmi celles du bambou.

Nous considérons l'éclatement qui produit $D_{r(P)}$. Soit Q le point à éclater. Nous allons construire une famille analytique de germes de singularités isolées de courbes en Q paramétrée par un disque D_η, avec $\eta > 0$ suffisamment petit.

Le disque D_η est construit comme suit. Il s'agit de l'image d'un voisinage U de P dans le dicritique D, de façon à ce que $\tilde{\phi}_{|U} : U \to D_\eta$ soit propre et il n'y a pas de points critiques (sauf éventuellement P). Nous pouvons supposer que $U \subset B_P$; on considère la famille analytique de courbes $\{\tilde{\phi}^{-1}(t) \cap B_P\}_{t \in D_\eta}$. En prenant les images par la suite de contractions au-dessus de Q, elle définit une famille analytique $\{A_t\}_{t \in D_\eta}$ de germes de singularités isolées en Q dont le seul élément où la topologie change est A_0. Or, A_0 vérifie qu'après un éclatement, sa transformée stricte est une branche lisse transverse au lieu exceptionnel, ce qui implique que A_0 lui-même est lisse; par conséquent, son nombre de Milnor est $\mu_0 = 0$. Notons μ le nombre de Milnor générique de ce pinceau; si la topologie change, il est bien connu que $\mu < \mu_0 = 0$, ce qui est impossible.

Exemple 4.(suite) Nous allons étudier l'application $j_* : H_1(F_0) \to H_1(V)$ pour les fibres irrégulières à l'infini de l'exemple de Briançon. Nous rappelons que dans ce cas il y a deux dicritiques E_1 et E_2 tels que:

1. – E_1 est un *bon* dicritique, c'est-à-dire, il est de multiplicité un et il n'y a pas de bambou qui s'y accroche;

2. – E_2 est un dicritique de multiplicité deux et il y a un bambou qui s'y accroche. Ce bambou correspond à la fibre $t = 0$ et il consiste en une composante irréductible de self-intersection -2 et telle que le point d'intersection du dicritique avec le bambou n'est pas critique pour $\tilde{\phi}_{|E_2}$.

Remarquons que la fibre pour $t = 0$ est irrégulière à cause du bambou. Notons P_0 le point de Dic_0 correspondant au bambou. Il y a deux branches dans $\bar{F}_{P_0}$. Par conséquent, $j_* : H_1(F_0) \to H_1(V)$ n'est ni injective ni surjective.

Considérons maintenant la fibre $t = -16/9$. Dans ce cas-ci, il y a deux points dans Dic_0. Soit P_1 celui qui se trouve dans E_1; il n'y qu'une branche de la fibre en ce point, et il est facile de voir que $j_* : H_1(F_{P_1}) \to H_1(V_{P_1})$ est un isomorphisme de groupes infinis cycliques.

Soit P_2 celui qui se trouve dans E_2. La résolution donnée en section 3 ci-dessus montre que $\bar{F}_{P_2}$ est un point double irréductible de type $\mathbb{A}_2$ transverse au dicritique. Alors $j_* : H_1(F_{P_2}) \to H_1(V_{P_2})$ est un monomorphisme de groupes infinis cycliques dont l'image est d'indice 2. Par conséquent, $j_* : H_1(F_0) \to H_1(V)$ est une application de groupes abeliens libres de rang 2, dont l'image est d'indice 2.

Exemple 6. Nous finissons avec deux exemples qui montrent que, en général, si la fibre n'est pas réduite, nous pouvons avoir l'équivalence d'homotopie de l'énoncé du Théorème 3.

Le premier exemple est $f(x, y) = x^n y$, $n \geq 2$. L'équivalence d'homotopie est évidente à cause de la quasihomogénéité. Dans cet exemple, si l'on fait la résolution $\tilde{\phi}$, on trouve un bambou (de longueur $n - 1$) et l'adhérence de la transformée stricte de $x = 0$ rencontre transversalement en un point la dernière composante du bambou (ce cas a été exclu dans le cas réduit).

Le deuxième exemple est $f(x, y) = (x+y^{10})(y(xy-1)^2+x^5)^2$. Ce polynôme est primitif, la fibre en $t = 0$ est non-réduite mais l'injection du Théorème 3 est une équivalence d'homotopie. Il n'y a pas de bambou dans la résolution $\tilde{\phi}$ de ce polynôme.

References

[AM] S. Abhyankar; T. Moh: *Embeddings of the line in the plane*. J. Reine Angew. Math. **267**, 148–166, (1975).

[A'C] N. A'Campo: *La fonction Zêta d'une monodromie*. Comment. Math. Helvetici **50**, 233–248, (1975).

[A] E. Artal Bartolo: *Une démonstration géométrique du théorème d' Abyankar-Moh*. J. Reine angew. Math **464**, 97–108, (1995).

[ACL] E. Artal Bartolo; P. Cassou-Noguès; I. Luengo Velasco: *On polynomials whose fibers are irreducible with no critical points.* Math. Ann. **299**, 477–490, (1994).

[BK] E. Brieskorn; H. Knörrer: *Plane Algebraic Curves.* Birkhäuser, Boston, (1986).

[B] S.A. Broughton: *Milnor number and the topology of polynomial hypersurfaces.* Invent. Math. **92**, 217–241, (1988).

[C] C.H. Clemens: *Degeneration of Kähler manifolds.* Duke Math. J. **44**, 215–290, (1977).

[CD] A.D.R. Choudary; A. Dimca: *Complex hypersurfaces diffeomorphic to affine spaces.* Kodai Math. J. **17**, 171–178, (1994).

[De] P. Deligne: *Théorie de Hodge II.* Publ. Math. I.H.E.S. **40**, 5–58, (1971).

[D1] A. Dimca: *On the connectivity of complex affine hypersurfaces.* Topology **29**, 511–514, (1990).

[D2] A. Dimca: *Singularities and topology of hypersurfaces.* Universitext, Springer, (1992).

[GN] R. García López; A. Nemethi: *On the monodromy at infinity of a polynomial map, I.* Compositio Math. **100**, 205–231, (1996).

[HZ] Ha Huy Vui; A. Zaharia: *Families of polynomials with total Milnor number constant.* Math. Ann. **304**, 481–488, (1996).

[F] L. Fourrier: *Topologie d'un polynôme de deux variables complexes au voisinage de l'infini.* Thèse Université de Toulouse, (1993).

[Fu] W. Fulton: *Introduction to toric varieties* Annals of Math. Studies **131**. Princeton Univ. Press, (1993).

[H] H. Hamm: *Zum Homotopietyp Steinscher Raeume.* J. reine angew. Math. **338**, 121–135, (1983).

[JS] A.J. de Jong; J.H.M. Steenbrink: *Proof of a conjecture of W. Veys.* Indag. Math. N. S. **6**, 99–104, (1995).

[K] S. Kaliman: *Two remarks on polynomials in two variables.* Pacific J. Math. **154**, 285–295, (1992).

[LW] Le Dung Trang; C. Weber: *Polynômes à fibres rationnelles et conjecture jacobienne à 2 variables* C. R. Acad. Sci. Paris, **t. 320**, 581–584, (1995).

[L] S. Lojasiewicz: *Triangulation of semi-analytic sets.* Ann. Sc. Norm. Sup. Pisa **18**, 449–474, (1964).

[Mi] J. Milnor: *Singular Points of Complex Hypersurfaces*. Annals of Math. Studies **61**, Princeton, (1968).

[MS] M. Miyanishi; T. Sugie: *Generically rational polynomials*. Osaka J. Math. **17**, 339–362, (1980).

[Mu] D. Mumford: *The topology of normal singularities of an algebraic surface and a criterion for simplicity*. Publ. Math. I.H.E.S. **9**, 5–22, (1961).

[NZ] A. Nemethi; A. Zaharia: *Milnor fibration at infinity*. Indag. Math. **3**, 323–335, (1992).

[P] F. Pham: *Vanishing homologies and the n variable saddlepoint method*. Proc. Symp. Pure Math. **40**, 319–333, (1983).

[Pa] A. Parusinski: *A note on singularities at infinity of complex polynomials*. Banach Center Math. Publ.

[Se] J.P. Serre: *Corps locaux*. Hermann, Paris, (1968).

[ST] D. Siersma; M. Tibar: *Singularities at infinity and their vanishing cycles*. Duke Math. J. **80**, 771–783, (1995).

[S] E.H. Spanier: *Algebraic Topology*. Mc Graw Hill, (1966).

[Z] O. Zariski: *Algebraic Surfaces*. Chelsea Publishing Press, (1948).

Progress in Mathematics, Vol. 162, © 1998 Birkhäuser Verlag Basel/Switzerland

Five Definitions of Critical Point at Infinity

Alan H. Durfee
Department of Mathematics
Mount Holyoke College
South Hadley, MA 01075
U.S.A.

Abstract

This survey paper discusses five equivalent ways of defining a "critical point at infinity" for a complex polynomial of two variables.

1 Introduction

A proper smooth map without critical points from one manifold to another is a locally trivial fibration by a well-known theorem of Ehresmann. On the other hand, a nonproper map without critical points may not be a fibration. This phenomenon occurs for complex polynomials. A simple example is provided by $f : \mathbb{C}^2 \to \mathbb{C}$ defined by the polynomial $f(x,y) = y(xy - 1)$. This map has no critical points, but the fiber over the origin is different from the other fibers. (In fact, the fiber over the origin is two rational curves, one punctured at two points and the other at one point, whereas the general fiber is a cubic curve, punctured at two points.) One would like to identify these "critical values" where the topology changes and their corresponding "critical points at infinity". We first review the history of this subject.

Let $f : \mathbb{C}^n \to \mathbb{C}$ be a complex polynomial. There is a finite set $\Sigma \in \mathbb{C}$ such that

$$f : \mathbb{C}^n - f^{-1}(\Sigma) \to \mathbb{C} - \Sigma$$

is a fibration. This is a form of Sard's theorem for polynomials; the set Σ is finite because it is algebraic. For a proof, see [Bro83, Proposition 1] (based on work of Verdier), [Pha83, Appendix A1], [HL84, Theorem 1] or [Ha89]. We let

$$\Sigma = \Sigma_{fin} \cup \Sigma_{\infty}$$

where Σ_{fin} is the set of critical values coming from critical points in $\mathbb{C}^n$, and Σ_{∞} is the set of critical values "coming from infinity". Of course these two sets may have nonempty intersection.

Broughton in [Bro83, Bro88] calls the polynomial f *tame* if there is a $\delta > 0$ such that the set $\{x : |grad\, f(x)| \leq \delta\}$ is compact. He proved that if f is tame, then Σ_∞ is empty.

Thus if the gradient of a polynomial goes to zero along some path going to infinity, then something bad may happen. Topics surrounding the gradient of the polynomial are treated in Section 4 of this paper. The speed at which the gradient of f goes to zero is measured by the Lojasiewicz number at infinity; see [Ha90, Ha91a, Ha94, CNH94, CNH95].

There followed many efforts in the case $n = 2$ to identify the set Σ_∞ more precisely. Suzuki [Suz74, Corollary 1] provides an estimate on the number of points in Σ. In [HL84] it is shown that $c \in \Sigma$ if and only if $\chi(f^{-1}(c)) \neq \chi(f^{-1}(t))$, where $f^{-1}(t)$ is a generic fiber of f and χ denotes Euler characteristic. Further work on identifying Σ_∞ ca be found in [HN89, Ha89, NZ90, NZ92, LO95].

The homology and homotopy of the fibers of the polynomial f were also computed by many authors, leading to various numerical invariants which will be discussed in the Section 2 of this paper. The earliest result is probably due to Suzuki [Suz74, Proposition 2], who shows that

$$rank\, H_1(f^{-1}(t)) = \mu + \lambda$$

where $f^{-1}(t)$ is a generic fiber, μ is the sum of the Milnor numbers at the critical points of f in $\mathbb{C}^2$, and λ is the sum of all the "jumps" in the Milnor numbers at infinity. (In the terminology of Section 2, $\lambda = \sum \nu_{p,c}$, where the sum is over $c \in \mathbb{C}$ and $p \in \mathbb{L}_\infty$, and $\nu_{p,c}$ is the jump in the Milnor number at the point $p \in \mathbb{L}_\infty$ and value $c \in \mathbb{C}$.)

The polynomial f extends to a function on projective space $\mathbb{P}^2$ which is well defined except at a finite number of points. The points of indeterminacy can be easily resolved, and the structure of the resolution contains information about these points [LW95, LW96]. These topics are discussed in Section 3.

Other topics investigated (but not discussed in this paper) include Newton diagrams [Bro88, Proposition 3.4], [NZ90, CN96], knots [Neu89, Ha91b], and the Jacobian conjecture [LW95].

Papers in higher dimensions (that is, $n \geq 2$) include [Lib93, Par95, ST95, Tib96]. Broughton [Bro88, Proposition 3.2] shows that the tame polynomials form a dense constructible set in the set of polynomials of a given degree; Cassou-Nogues [CN96, Example V] gives an example to show that this set is not open in dimension $n = 3$. Connections with number theory are given in [LS95].

Although the scene in higher dimensions is not yet settled, the situation in dimension two is now clear. The purpose of this paper is to collect together five definitions of "critical point at infinity" in this low-dimensional case and prove that they are equivalent. These definitions have appeared in the literature in some form or other, usually in a global affine context; the purpose of this paper is to give these definitions and prove their equivalence in a purely local setting near a point on the line at infinity. Many examples are also given. It should be noted that this material can be tricky, despite its apparent simplicity, and one should take care to make precise statements and proofs as well as to check examples.

If $f(x, y)$ is a polynomial, $p \in \mathbb{L}_\infty$ is a point through which the level curves of f pass, and $c \in \mathbb{C}$, we say that the pair (p, c) is a *regular point at infinity* for $f(x, y)$ if it satisfies any one of the following equivalent conditions. (Otherwise it is a *critical point at infinity*.)

- Condition M (2.15): There is no jump in the Milnor number (2.1): $\nu_{p,c} = 0$.

- Condition E (2.16): The family of germs $f(x, y) = tz^d$ at p is equisingular at $t = c$.

- Condition F (2.17): The map f is a smooth fiber bundle near p and the value c.

- Condition R (3.1): There is a resolution $\tilde{f} : M \to \mathbb{P}^1$ with $\pi : M \to \mathbb{P}^2$ and a neighborhood U of $p \in \mathbb{P}^2$ such that $\{\tilde{f} = c\} \cap \pi^{-1}(U)$ is smooth and intersects the exceptional set $\pi^{-1}(p)$ transversally.

- Condition G (4.1): There does not exist a sequence of points $\{p_k\} \in \mathbb{C}^2$ with $p_k \to p$, $\operatorname{grad} f(p_k) \to 0$ and $f(p_k) \to c$ as $k \to \infty$.

Most of these equivalences are well known; we give either proofs or references for proofs in the pages that follow.

There are several new results in this paper. First, we define (2.6) an invariant $\nu_{p,\infty}$ which measures the number of vanishing cycles at a point p on the line at infinity for the critical value infinity, and show that this invariant has many of the same properties that $\nu_{p,c}$ does for $c \in \mathbb{C}$. Secondly, we define $g_{p,c}$ to be the number of isotopy classes of paths $\alpha : \mathbb{R} \to \mathbb{C}^2$ such that $\alpha(t) \to p$, $\operatorname{grad} f(\alpha(t)) \to 0$ and $f(\alpha(t)) \to c$ as $t \to +\infty$. We use this to give a new proof that Condition M implies Condition G. In fact, we will show (Proposition 4.12) that $\nu_{p,c} \geq g_{p,c}$.

The work described in this paper started in 1989 when the author supervised a group of undergraduates in the Mount Holyoke Summer Research Institute in Mathematics who were working on corresponding problems for real polynomials. These results are described in [DKM$^+$93], with further results in [Dur]. The work for this paper was carried out at the Tata Institute, Bombay, Martin-Luther University, Halle (with support from IREX, the International Research and Exchanges Board), the University of Nijmegen, Warwick University, the Massachusetts Institute of Technology and the University of Bordeaux. The author would like to thank all of them for their hospitality.

Earlier versions of this paper included results on deformations of critical points at infinity; these will appear elsewhere.

2   Numerical Invariants

We will use coordinates (x, y) for the complex plane $\mathbb{C}^2$, and coordinates $[x, y, z]$ for the projective plane $\mathbb{P}^2$. We let

$$\mathbb{L}_\infty = \{[x, y, z] \in \mathbb{P}^2 : z = 0\}$$

be the line at infinity. We let d be the degree of the polynomial $f(x, y)$. We let f_d denote the homogeneous term of degree d in f. If $p = [a, b, 0] \in \mathbb{L}_\infty$, we let d_p be the multiplicity of the factor $(bx - ay)$ in f_d.

Suppose that the level sets of f intersect $\mathbb{L}_\infty$ at p. Let

$$F_t(x, y, z) = z^d f(x/z, y/z) - tz^d$$

be the homogenization of the polynomial $f(x, y) - t$, where $t \in \mathbb{C}$, and let $g_{p,t}$ be the local equation of F_t at p. If $p = [1, 0, 0]$, then $g_{p.t}$ is given in local coordinates

$$(u, v) = (y/x, 1/x)$$

by

$$g_{p,t}(u, v) = F_t(1, u, v) = v^d f(1/v, u/v) - tv^d .$$

Note that the multiplicity of $g_{p,t}$ at $(0, 0)$ is at most d_p.

Definition 2.1 The *Milnor number* $\mu_{p,t}$ of $f(x, y)$ at $(p, t) \in \mathbb{L}_\infty \times \mathbb{C}$ is the Milnor number of the germ $g_{p,t}$ at $(0, 0)$ in the usual sense. The *generic Milnor number* $\mu_{p,gen}$ is the Milnor number $\mu_{p,t}$ for generic t. The number of *vanishing cycles* at (p, t) is

$$\nu_{p,t} = \mu_{p,t} - \mu_{p,gen} .$$

Example 2.2 Let $f(x, y) = y(xy - 1)$ and $p = [1, 0, 0]$. Then $g_{p,t}(u, v) = u^2 - uv^2 - tv^3$. We have $\mu_{p,gen} = 2$, $\nu_{p,0} = 1$, and all other $\nu_{p,t} = 0$. In fact, for $t \neq 0$, the singularity is of type A_2, and for $t = 0$, the singularity is of type A_3. This well-known example is the simplest "critical point at infinity". More generally, if $f(x, y) = y(x^a y - 1)$ then for $t \neq 0$, $\mu_{p,t} = a + 1$ and there is a singularity of type A_{a+1}. For $t = 0$, $\mu_{p,0} = 2a + 1$ and there is a singularity of type A_{2a+1}.

Example 2.3 Let $f(x, y) = x(y^2 - 1)$ and $p = [1, 0, 0]$. Then $g_{p,t}(u, v) = u^2 - v^2 - tv^3$. For all t, $\mu_{p,t} = 1$; the family is equisingular, and there is no "critical point at infinity". This is another basic example.

Example 2.4 Here is a more complicated example (see [DKM$^+$93, Dur]): Let $f(x, y) = (xy^2 - y - 1)^2 + (y^2 - 1)^2$. At $p = [1, 0, 0]$ we have $\mu_{gen} = 15$, $\nu_{p,1} = 2$, $\nu_{p,2} = 1$ and $\nu_{p,c} = 0$ for all other c.

Next we relate $\nu_{p,t}$ to homological vanishing cycles. Fix $p \in \mathbb{L}_\infty$ and $c \in \mathbb{C} \cup \{\infty\}$. Let $U \subset \mathbb{C}^2$ be an open set such that the closure in projective space of the set

$$\{(x,y) \in \mathbb{C}^2 : (x,y) \in U \text{ and } f(x,y) = t\}$$

is p for t near c. Choose $C > 0$ large. We define the *Milnor fiber* of f at (p,c) to be

$$\tilde{F}_{p,c} = \overline{\{(x,y) \in \mathbb{C}^2 : (x,y) \in U \text{ and } |(x,y)| \geq C \text{ and } f(x,y) = t\}}$$

where the overbar indicates closure in projective space, and, if $c \in \mathbb{C}$, then t is near, but not equal to, c, and if $c = \infty$, then t is large.

Proposition 2.5 *For $p \in \mathbb{L}_\infty$ and $c \in \mathbb{C}$,*

$$\nu_{p,c} = rank\, H_1(\tilde{F}_{p,c})\,.$$

Proof. Without loss of generality, we may assume that $p = [1,0,0]$. The number $\nu_{p,c}$ is the difference of the Milnor number $\mu_{p,c}$ and the generic Milnor number $\mu_{p,gen}$. The number $\mu_{p,gen}$ is the Milnor number of $g_{p,t}$ for t near, but not equal to, c. By the usual argument, this difference is $rank\, H_1(\{g_{p,c} = 0\} \cap B_0)$ where B_0 is the small ball for the Milnor number of $g_{p,t}$. We may replace $\{g_{p,c} = 0\} \cap B_0$ by

$$F'_{p,c} = \{(u,v) \in \mathbb{C}^2 : |v| \leq \epsilon' \text{ and } g_{p,t}(u,v) = 0\}\,.$$

We may replace $\tilde{F}_{p,c}$ by

$$\tilde{F}'_{p,c} = \overline{\{(x,y) \in \mathbb{C}^2 : (x,y) \in U \text{ and } |x| \geq C \text{ and } f(x,y) = t\}}\,.$$

The change of coordinates $x = 1/v$ and $y = u/v$ takes $\tilde{F}_{p,c}$ to $F'_{p,c}$. $\qquad\square$

To define "vanishing cycles" for the critical value $c = \infty$, we take the above proposition to be a definition:

Definition 2.6 For $p \in \mathbb{L}_\infty$ we let

$$\nu_{p,\infty} = rank\, H_1(\tilde{F}_{p,\infty})\,.$$

Remark 2.7 Here is a topological interpretation of the number of vanishing cycles at infinity: Suppose the level curves of the polynomial f of degree d intersect $\mathbb{L}_\infty$ at k points (counted without multiplicities). Then $\nu_{p,\infty} = 0$ for all $p \in \mathbb{L}_\infty$ if and only if $\overline{\{f(x,y) = t\}}$ for t large is homeomorphic to a d-fold cover of $\mathbb{L}_\infty$ branched at k points. For example, $y(xy - 1) = t$ (where $\nu_{p,\infty} = 0$ for all p) is a three-fold cover of $\mathbb{P}^1$ branched at two points, but $y^2 - x = t$ (where $\nu_{[1,0,0],\infty} = 1$) is not a two-fold cover of $\mathbb{P}^1$ branched at one point.

Next we describe three ways of computing the number of vanishing cycles $\nu_{p,c}$. The first is to compute (perhaps with a computer algebra progam) $\mu_{p,c}$ and $\mu_{p,gen}$ and subtract. The second is by counting nondegerate critical points, as described in the proposition below. This is similar to computing the usual Milnor number by counting the number of nondegerate critical points in a Morsification (see [AGZV85, vol II, p. 31]), and the proof is similar.

Proposition 2.8 *Let $p \in \mathbb{L}_\infty$ and $c \in \mathbb{C} \cup \{\infty\}$. The number $\nu_{p,c}$ is equal to the number of critical points (assumed nondegenerate) $q \neq (0,0)$ of the function $g_{p,t}$ such that $q \to (0,0)$ as $t \to c$.*

Example 2.9 Let $f(x,y) = y(xy-1)$ and $p = [1,0,0]$. Then $g_{p,t}(u,v) = u^2 - uv^2 - tv^3$. For $t \neq 0$ the function $g_{p,t}$ has a (degenerate) critical point at $(0,0)$ with critical value 0, and a nondegenerate critical point at $((9/2)t^2, -3t)$ with critical value $(27/4)t^4$. As $t \to 0$ the second critical point approaches $(0,0)$. Thus $\nu_{p,0} = 1$.

Example 2.10 Let $f(x,y) = x - y^2$ and $p = [1,0,0]$. Then $g_{p,t}(u,v) = v - u^2 - tu^2$. The function $g_{p,t}$ has a single nondegenerate critical point at $(0, 1/(2t))$ with critical value $1/(4t)$. As $t \to \infty$ this critical point approaches $(0,0)$, so $\nu_{p,\infty} = 1$.

The next proposition describes the result of computing $\nu_{p,\infty}$ by similar methods.

Proposition 2.11 *For $p \in \mathbb{L}_\infty$,*

$$\nu_{p,\infty} = (d_p - 1)(d - 1) - \mu_{p,gen} .$$

Proof. Without loss of generality $p = [1,0,0]$. The intersection multiplicity of the curves $(g_{p,t})_u$ and $(g_{p,t})_v$ at $(0,0)$ for $t = \infty$, where (u,v) are local coordinates at $(0,0)$, can be computed using the algorithm in [Ful69], and is found to be $(d_p - 1)(d - 1)$. (To compute the intersection multiplicity at $t = \infty$, we let $s = 1/t$ and compute it at $s = 0$.) For large $t \neq \infty$, the intersections split into those at $(0,0)$, the number of which is $\mu_{p,gen}$, and those not at $(0,0)$, the number of which is $\nu_{p,\infty}$. $\square$

Example 2.12 The polynomial $f(x,y) = y^a + x^{a-2}y + x$ has $\nu_{p,\infty} = a^2 - 2a$ at the point $p = [1,0,0]$, and all other $\nu_{p,c} = 0$.

Finally, $\nu_{p,t}$ can computed a third way by using polar curves, as described below. (See [HN89, 1.6, 1.8].) This method also shows that some vanishing cycles are easy to "see" from a contour plot, since they are where the level curves of the polynomial have a vertical tangent.

Proposition 2.13 *Suppose $p = [1,0,0]$, $c \in \mathbb{C} \cup \{\infty\}$ and the level sets of f pass through p. Then $\nu_{p,c}$ is the number of points of intersection $q \in \mathbb{C}^2$ (assumed transverse) of the curves $f = t$ and $f_y = 0$ in $\mathbb{C}^2$ such that $q \to p$ as $t \to c$.*

Proof. The set $F'_{p,c}$ from the proof of Proposition 2.5 is a connected branched cover of the disk $|v| \le \epsilon'$ in the uv-plane. Two sheets come together at each branch point, and all the sheets come together over p. The result follows from Hurwitz's formula. $\qquad\square$

Example 2.14 If $f(x,y) = x(y^2 - 1)$, the curves $f = t$ and $f_y = 0$ intersect at $(-t,0)$. As $t \to \infty$, the intersection point $(-t,0) \to [1,0,0]$ and $f(t,0) \to \infty$. Thus $\nu_{[1,0,0],\infty} = 1$. All other $\nu_{[1,0,0],c} = 0$.

Next we give three definitions of "critical point at infinity".

Definition 2.15 The polynomial $f(x,y)$ satisfies Condition M at the point $p \in \mathbb{L}_\infty$ and $c \in \mathbb{C} \cup \{\infty\}$ if $\nu_{p,c} = 0$.

Definition 2.16 The polynomial $f(x,y)$ satisfies Condition E at the point $(p,c) \in \mathbb{L}_\infty \times \mathbb{C}$ if the family of germs $g_{p,t}$ at $(0,0)$ is equisingular at $t = c$.

A proof that Condition M for $c \in \mathbb{C}$ is equivalent to Condition E may be found at the end of [LR76]).

There are various equivalent ways of specifying equisingularity; see for instance the papers by Zariski in volume IV of [Zar79]. One that will be useful for us is the following: The family of germs $g_{p,t}$ is equisingular if the germs $g_{p,t} = 0$ at $(0,0)$ form a fiber bundle near $t = c$.

Definition 2.17 The polynomial $f(x,y)$ satisfies Condition F at a point $(p,c) \in \mathbb{L}_\infty \times \mathbb{C}$ if the map f is a smooth fiber bundle near p and the value c. (More precisely, a polynomial satisfies Condition F if there is a $U \subset \mathbb{C}^2$ with p in the closure of U in projective space and $C > 0$ and $\beta > 0$ such that, letting

$$B = \{t \in \mathbb{C} : |t - c| \le \beta\}$$

and

$$N = \{(x,y) \in \mathbb{C}^2 : (x,y) \in U \text{ and } |(x,y)| \ge C \text{ and } f(x,y) \in B\}$$

then

$$f : N \to B$$

is a smooth fiber bundle.)

Proposition 2.18 *A polynomial $f(x,y)$ satisfies Condition E at a point $(p,c) \in \mathbb{L}_\infty \times \mathbb{C}$ if and only if it satisfies Condition F at that point.*

Proof. The proof is straight-forward, and just involves replacing the "spherical" Milnor fiber by one in a "box": Without loss of generality, we may assume that $p = [1, 0, 0]$. We may replace Condition E by the following: There is an $\epsilon' > 0$ and a $\delta' > 0$ such that, letting

$$D' = \{t \in \mathbb{C} : |t - c| < \delta'\}$$

and

$$M' = \{(u, v, t) \in \mathbb{C}^2 \times \mathbb{C} : |v| \leq \epsilon' \text{ and } t \in D' \text{ and } g_{p,t}(u, v) = 0\}$$

then the restriction of the projection to the third coordinate

$$\pi : M' \to D'$$

is a fiber bundle. We may do this since the germs $g_{p,t}(u, v) = 0$ never have $v = 0$ as a component.

We may also replace Condition F by the following: There is a $U' \subset \mathbb{C}^2$ with p in the closure of U' in projective space and $C' > 0$ and $\beta' > 0$ such that, letting

$$B' = \{t \in \mathbb{C} : |t - c| \leq \beta'\}$$

and

$$N' = \{(x, y) \in \mathbb{C}^2 : (x, y) \in U' \text{ and } |x| \geq C' \text{ and } f(x, y) \in B'\}$$

then

$$f : N' \to B'$$

is a smooth fiber bundle.

The change of coordinates $x = 1/v$ and $y = u/v$ takes $f(x, y) = t$ to $g_{p,t}(u, v) = 0$ and N' to M'. $\square$

3   Resolutions

The polynomial $f : \mathbb{C}^2 \to \mathbb{C}$ extends to a map

$$\hat{f} : \mathbb{P}^2 \to \mathbb{P}$$

which is undefined at a finite number of points on the line at infinity $\mathbb{L}_\infty$. By blowing up these points one gets a manifold M and a map $\pi : M \to \mathbb{P}^2$ such that the map

$$\tilde{f} : M \to \mathbb{P}$$

which is the lift of $\hat{f}$ is everywhere defined. We call the map $\tilde{f}$ a *resolution of f*. Some interesting results on the structure of resolutions are announced in [LW95, Theorems 2, 3, 4].

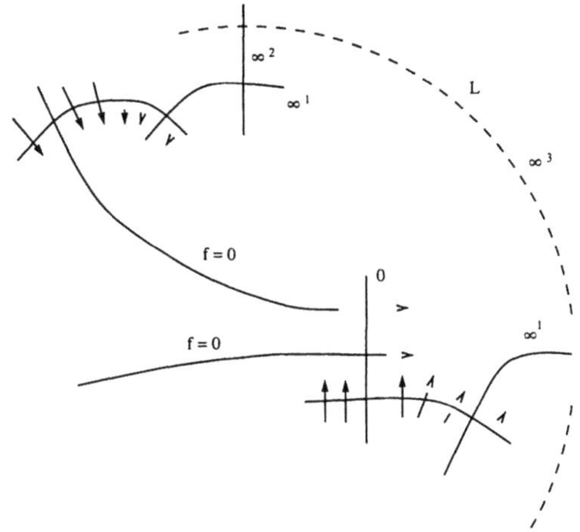

Figure 1: Resolution of $y(xy - 1)$

For example, a resolution (the minimal resolution) of $y(xy - 1)$ is given in Figure 1. The symbol c^m next to a divisor means that at each smooth point of the divisor there are local coordinates (z, w) in a neighborhood of the point such that the divisor is $z = 0$ and $\tilde{f}(z, w) = (z - c)^m$. The proper transform of level curves of f have arrowheads on them; the exceptional sets do not.

Resolution are easy compute. For example, starting with $f(x, y) = y(xy - 1)$ which we wish to resolve near $[1, 0, 0]$, the function in local coordinates at $[1, 0, 0]$ is $u(u - v^2)/v^3$, and we blow up in the standard fashion until it is everywhere defined. More examples are shown in Figures 2 and 3.

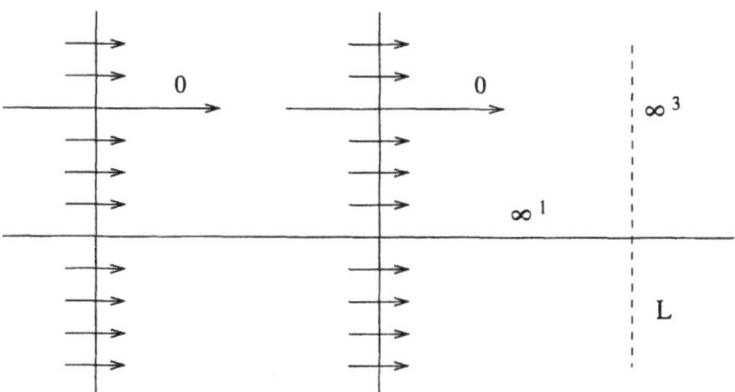

Figure 2: Resolution of $x(y^2 - 1)$ at $[1, 0, 0]$

Next we give a condition for "regular point at infinity" in terms of a resolution. (See also [LW95, Theorem 5], [Lib93].)

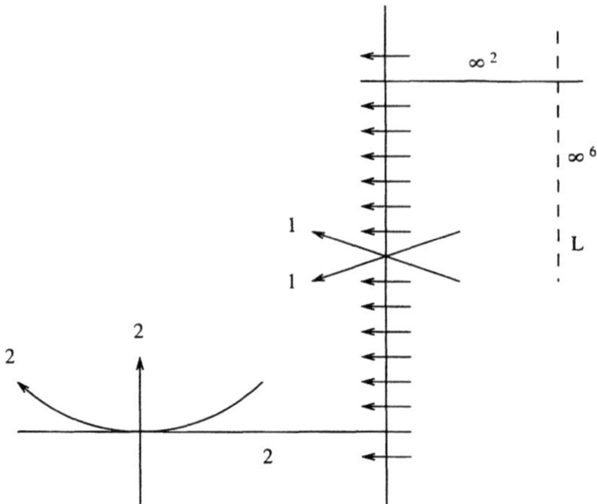

Figure 3: Resolution of $(xy^2 - y - 1)^2 + (y^2 - 1)^2$ at $[1, 0, 0]$

Definition 3.1 The polynomial $f(x, y)$ satisfies Condition R at a point $(p, c) \in \mathbb{L}_\infty \times \mathbb{C}$ if there is a resolution $\tilde{f} : M \to \mathbb{P}^1$ with $\pi : M \to \mathbb{P}^2$ and a neighborhood U of $p \in \mathbb{P}^2$ such that $\{\tilde{f} = c\} \cap \pi^{-1}(U)$ is smooth and intersects the exceptional set $\pi^{-1}(p)$ transversally.

Example 3.2 Let $f(x, y) = y - (xy - 1)^2$ near $[1, 0, 0]$ (See [Kra91]). In this example the level curve of the function $\tilde{f} = 0$ is smooth, but it does not intersect the exceptional divisor transversally; see Figure 4. Hence $(p, c) = ([1, 0, 0], 0)$ does not satisfy Condition R. (Here $\nu_{[1,0,0],0} = 1$.)

Proposition 3.3 *A point $p \in \mathbb{L}_\infty$ and a value $c \in \mathbb{C}$ for a polynomial $f(x, y)$ satisfies Condition E if and only if it satisfies Condition R.*

Proof. Suppose (p, c) satisfies Condition E. Let U be a neighborhood of p in $\mathbb{P}^2$ containing no critical points of f in $\mathbb{C}^2$ or points on $\mathbb{L}_\infty$ though which the level curves of f pass. Find a resolution $\tilde{f}$ of f. By further blowing up (if necessary), we may assume that $\tilde{f}^{-1}(c)$ is a divisor with normal crossings transversally intersecting the exceptional set where $\tilde{f}$ is not constant. Equisingularity in the form of Zariski's (b)-equivalence [Zar65, p. 513] implies that the functions $g_{p,t}$ for t near c have the same resolution as the function $g_{p,c}$. This can only happen if $\{\tilde{f}^{-1}(t) \cap \pi^{-1}(U)\}$ is smooth and transversally intersects the exceptional set $\pi^{-1}(p)$. Thus p and c satisfy Condition R.

     Conversely, if p and c satisfy Condition R, then the resolutions of $\{g_{p,t} = 0\}$ for t near c are (b)-equivalent and hence equisingular. $\qquad\qquad\square$

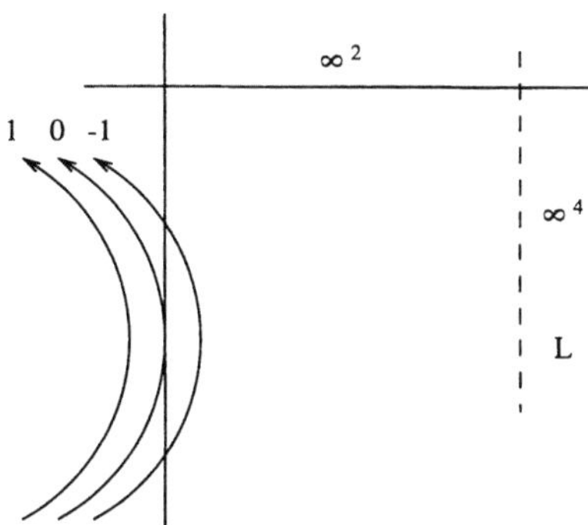

Figure 4: Resolution of $y - (xy - 1)^2$ at $[1, 0, 0]$

4 The Gradient

If f is a complex polynomial, we define $grad\, f$ as in [Mil68] to be the complex conjugate of the vector of partial derivatives.

Of course $p \in \mathbb{C}^2$ is a regular point for a function f with regular value $c \in \mathbb{C}$ if $f(p) = c$ and $grad\, f(p) \neq 0$. An equivalent definition would be to say that there is no sequence of points $\{p_k\}$ with $p_k \to p$, $grad\, f(p_k) \to 0$ and $f(p_k) \to c$ as $k \to \infty$. We can now imitate this definition for $p \in \mathbb{L}_\infty$ as follows:

Definition 4.1 The polynomial $f(x, y)$ satisfies Condition G at a point $p \in \mathbb{L}_\infty$ and $c \in \mathbb{C} \cup \{\infty\}$ if there does not exist a sequence of points $\{p_k\} \in \mathbb{C}^2$ with $p_k \to p$, $grad\, f(p_k) \to 0$ and $f(p_k) \to c$ as $k \to \infty$.

If (p, c) does not satisfy Condition G, then a version of Milnor's curve selection lemma (see for instance [Ha91a, Lemma 3.1] or [NZ92, Lemma 2]) implies that the sequence of points can be replaced by a curve:

Lemma 4.2 *If (p, c) does not satisfy Condition G, then there is a smooth real algebraic curve $\alpha : \mathbb{R}^+ \to \mathbb{C}^2$ such that $\alpha(t) \to p$, $grad\, f(\alpha(t)) \to 0$ and $f(\alpha(t)) \to c$ as $t \to +\infty$.*

By "real algebraic curve" we mean that the image of α in $\mathbb{C}^2$ is contained in an irreducible component of the zero locus of a real polynomial.

Example 4.3 Let $f(x, y) = y(xy - 1)$. Let $\alpha(t) = (t, 1/(2t))$. As $t \to +\infty$, $\alpha(t) \to [1, 0, 0]$, the gradient of f goes to 0 and the value of the function approaches 0.

Example 4.4 (c.f. Example 2.4.) Let $f(x, y) = (xy^2 - y - 1)^2 + (y^2 - 1)^2$. Let $\alpha(t) = (t + t^2, \pm 1/t)$. As $t \to +\infty$, $\alpha(t) \to [1, 0, 0]$, the gradient of f goes to 0 and the function approaches the value 1. If $\beta(t) = (t/2, 1/t)$, then as $t \to +\infty$, $\beta(t) \to [1, 0, 0]$, the gradient of f goes to 0 and the function approaches the value 2. (These paths were found by Ian Robertson in the Mount Holyoke REU program in the summer of 1992.)

Example 4.5 If $f(x, y) = x^2 y + xy^2 + x^5 y^3 + x^3 y^5$ and $q \to [1, 0, 0]$ along the curve $y^2 x^3 = -1/3$, then $grad\, f(q) \to 0$ and $f(q) \to \infty$. Here $v_{p,\infty} = 1$. This polynomial is "quasi-tame" but not "tame" (see [NZ92] and the references therein). It would be interesting to find more examples like this.

The following proposition is well-known. It was first proved in the global case by Broughton [Bro88]; see also [NZ90], proof of Theorem 1, and [ST95], proof of Proposition 5.5. The idea of the proof is to use integral curves of the vector field $grad\, f / |grad\, f|^2$ to identify the fibers.

Proposition 4.6 *If a polynomial $f(x, y)$ satisfies Condition G at $(p, c) \in \mathbb{L}_\infty \times \mathbb{C}$, then it satisfies Condition F at this point.*

Next we will show that Condition M implies Condition G; this has been shown in [Ha90, Ha91a, ST95, Par95]. Here we will prove a stronger result by different methods.

Definition 4.7 For $p \in \mathbb{L}_\infty$ and $c \in \mathbb{C} \cup \{\infty\}$, let $g_{p,c}$ be the number of isotopy classes of smooth real algebraic curves $\alpha : \mathbb{R} \to \mathbb{C}^2$ such that $\alpha(t) \to p$, $grad\, f(\alpha(t)) \to 0$ and $f(\alpha(t)) \to c$ as $t \to +\infty$.

Example 4.8 If $f(x, y) = y^5 + x^2 y^3 - y$ and $p = [1, 0, 0]$, then $\nu_{p,0} = 2$. There are two isotopy classes of curves approaching p along which $grad\, f$ goes to zero, namely the ones containing the two branches of the curve $f_y = 0$ at p. Hence $g_{p,0} = 2$. (This example is from [DKM$^+$93].)

Clearly $(p, c) \in \mathbb{L}_\infty \times (\mathbb{C} \cup \{\infty\})$ satisfies Condition G if and only if $g_{p,c} = 0$. Now let $\pi : M \to \mathbb{P}^2$ be a resolution of f, f_x, and f_y (so that $\tilde{f}$, $\widetilde{(f_x)}$ and $\widetilde{(f_y)}$ are defined on M), and let

$$G_{p,c} = \{q \in M : \pi(q) = p,\ \tilde{f}(q) = c,\ \widetilde{(f_x)}(q) = 0\ \text{ and }\ \widetilde{(f_y)}(q) = 0\}.$$

Definition 4.9 For $p \in \mathbb{L}_\infty$ and $c \in \mathbb{C} \cup \{\infty\}$, let $\tilde{g}_{p,c}$ be the number of connected components of $G_{p,c}$.

The number $\tilde{g}_{p,c}$ is independent of the resolution by the usual argument.

Example 4.10 In the minimal resolution of $f(x, y) = y(xy - 1)$ at $p = [1, 0, 0]$ (Figure 1), the functions f_x and f_y are defined. The zero locus of the lift of f_x contains the exceptional set where the lift of f is zero, and the zero locus of the lift of f_y intersects this set transversally. Thus $G_{p,0}$ consists of a single point, and $\tilde{g}_{p,0} = 1$. If $f(x,y) = y^5 + x^2 y^3 - y$ and $p = [1,0,0]$, one finds similarly that $G_{p,0}$ consists of two points.

The two definitions are equivalent by the following proposition.

Proposition 4.11 *For $p \in \mathbb{L}_\infty$ and $c \in \mathbb{C} \cup \{\infty\}$, $g_{p,c} = \tilde{g}_{p,c}$.*

Proof. Let $\pi : M \to \mathbb{P}^2$ be a resolution of f, f_x and f_y. We will show that there is a one-one correspondence between isotopy classes of curves satisfying (4.7) and connected components of $G_{p,c}$. Suppose that $\alpha : \mathbb{R}^+ \to \mathbb{C}^2$ is a smooth real algebraic curve satisfying the conditions of (4.7). Since α is real algebraic, it lifts to to a map $\tilde{\alpha} : \mathbb{R}^+ \cup \{\infty\} \to M$ with $\tilde{\alpha}(\infty) \in \pi^{-1}(\mathbb{L}_\infty)$. Let $q = \tilde{\alpha}(\infty)$. Then $\tilde{f}(q) = c$ and $\widetilde{(f_x)}(q) = 0$ and $\widetilde{(f_y)}(q) = 0$. Thus $q \in G_{p,c}$.

If α_0 is isotopic to α_1 through curves α_t satisfying (4.7), then the curves α_t lift to M and are isotopic. In particular, $\tilde{\alpha}_0(\infty)$ and $\tilde{\alpha}_1(\infty)$ are in the same connected component of $G_{p,c}$.

For each $q \in G_{p,c}$ there is an algebraic curve $\tilde{\alpha} : \mathbb{R}^+ \cup \{\infty\} \to M$ with $\tilde{\alpha}(\infty) = q$ and $\tilde{\alpha}(R^+) \subset \pi^{-1}(\mathbb{C}^2)$. Let $\alpha = \pi \circ \tilde{\alpha} : \mathbb{R}^+ \to \mathbb{C}^2$. Then α satisfies the conditions of (4.7). If $q_0, q_1 \in G_{p,c}$, then there are two such curves $\tilde{\alpha}_0, \tilde{\alpha}_1$. If q_0 and q_1 are in the same connected component of $G_{p,c}$, then $\tilde{\alpha}_0$ is isotopic to $\tilde{\alpha}_1$ through a family of such curves $\tilde{\alpha}_t$. Hence α_0 is isotopic to α_1 through curves satisfying (4.7). If q_0 and q_1 are in different connected components, then α_0 is not isotopic to α_1 through curves satisfying (4.7). Thus $g_{p,c} = \tilde{g}_{p,c}$. $\square$

Proposition 4.12 *For $p \in \mathbb{L}_\infty$ and $c \in \mathbb{C} \cup \{\infty\}$,*

$$\nu_{p,c} \geq g_{p,c} .$$

Proof. We will show that $\nu_{p,c} \geq \tilde{g}_{p,c}$ and will use Proposition 2.13 to compute $\nu_{p,c}$. We may assume without loss of generality that $p = [1,0,0]$. Pick a connected component G' of $G_{p,c}$. Let t be near c. We will show that $f = t$ intersects $f_y = 0$ in $\mathbb{C}^2$ near G'.

There is a $q \in G'$ and a component C of $f_y = 0$ in $\mathbb{C}^2$ such that q is in the closure of C in M: We have that $\tilde{f}_y = 0$ on G'. Blow down G' to a point q', and let E' be the image of $\pi^{-1}(p)$. Then the lift of f_y is not constant on E' near q', so there is a component of $f_y = 0$ passing through q'. Lift this component back to M.

Next, f is not constant on C: If it were, then the gradient vector of f would be horizontal, so C would be of the form $x = const$, and p would not be in the closure of C.

Thus $f = t$ intersects C near q for small $\epsilon \neq 0$, and the intersection points are in $\mathbb{C}^2$. $\square$

Remark 4.13 The inequality of the proposition can be strict, as it is for the polynomial $y(x^2 y - 1)$ at $p = [1,0,0]$ and $c = 0$, where $\nu_{p,c} = 2$ and $g_{p,c} = 1$.

Corollary 4.14 *If $p \in \mathbb{L}_\infty$ and $c \in \mathbb{C} \cup \{\infty\}$ satisfy Condition M, then they satisfy Condition G.*

Remark 4.15 The converse to this corollary is not true for $c = \infty$: For example, the polynomial $x(y^2 - 1)$ has a gradient whose magnitude is bounded below for large x and hence satisfies Condition G at $p = [1, 0, 0]$, yet $\nu_{p,\infty} = 1$.

References

[AGZV85] V.I. Arnold, S.M. Gusein-Zade, and A.N. Varchenko. *Singularities of Differentiable Maps*. Birkhäuser, Boston, 1985.

[Bro83] S.A. Broughton. On the topology of polynomial hypersurfaces. In P. Orlik, editor, *Proceedings of Symposia in Pure Mathematics, Vol. 40*, pages 167–178, Providence RI, 1983. American Mathematical Society.

[Bro88] S.A. Broughton. Milnor numbers and the topology of polynomial hypersurfaces. *Inventiones Math.*, **92**, 217–241, 1988.

[CN96] Pierrette Cassou-Noguès. Sur la généralisation d'un théorème de Kouchnirenko. *Compositio Math*, **103**, 95–121, 1996.

[CNH94] Pierrette Cassou-Noguès and Huy Vui Ha. Théorème de Kuiper-Kuo-Bochnak-Lojasiewicz à l'infini. Preprint, Université Bordeaux, 1994.

[CNH95] Pierrette Cassou-Noguès and Huy Vui Ha. Sur le nombre de Lojasiewicz à l'infini d'un polynôme. *Annales Polonici Math.*, **62**, 23–44, 1995.

[DKM$^+$93] A. Durfee, N. Kronenfeld, H. Munson, J. Roy, and I. Westby. Counting critical points of a real polynomial in two variables. *American Math Monthly*, **100**, 255–271, 1993.

[Dur] Alan H. Durfee. The index of $grad\, f(x, y)$. Duke alg-geom preprint 9506002. To appear (in revised form), *Topology*.

[Ful69] William Fulton. *Algebraic Curves*. Benjamin, New York, 1969.

[Ha89] Huy Vui Ha. Sur la fibration globale des polynômes de deux variables complexes. *C. R. Acad. Sci Paris*, **309**, 231–234, 1989.

[Ha90] Huy Vui Ha. Nombres de Lojasiewicz et singularités a l'infini des polynômes de deux variables complexes. *C. R. Acad. Sci Paris*, **311**, 429–432, 1990.

[Ha91a]   Huy Vui Ha. On the irregular at infinity algebraic plane curves. Preprint, Mathematical Institute, Hanoi, 1991.

[Ha91b]   Huy Vui Ha. Sur l'irrégularité du diagramme splice pour l'entrelacement á l'infini des courbes planes. *C. R. Acad. Sci Paris*, **313**, 277–280, 1991.

[Ha94]   Huy Vui Ha. A version at infinity of the Kuiper-Kuo theorem. *Acta Math Vietnamica*, **19**, 3–12, 1994.

[HL84]   Huy Vui Ha and Dung Trang Le. Sur la topologie des polynômes complexes. *Acta Math Vietnamica*, **9**, 21–32, 1984.

[HN89]   Huy Vui Ha and Le Anh Nguyen. Le comportement géometrique à l'infini des polynômes de deux variables complexes. *C. R. Acad. Sci Paris*, **309**, 183–186, 1989.

[Kra91]   T. Krasinski. On branches at infinity of a pencil of polynomials in two complex variables. *Ann. Polon. Math.*, **55**, 213–220, 1991.

[Lib93]   Anatoly Libgober. Topological invariants of affine hypersurfaces: connectivity, ends and signature. *Duke Math. J.*, **70**, 207–227, 1993.

[LO95]   Van Thanh Le and Mutsuo Oka. Estimation of the number of the critical values at infinity of a polynomial function $f : C^2 \to c$. *Publ. Res. Inst. Math. Sci.*, **31**, 577–598, 1995.

[LR76]   Dung Trang Le and C.P. Ramanujam. The invariance of Milnor number implies the invariance of the topological type. *Amer. J. Math*, **98**, 67–78, 1976.

[LS95]   A. Libgober and S. Sperber. On the zeta function of monodromy of a polynomial map. *Compositio Math*, **95**, 287–307, 1995.

[LW95]   Dung Trang Le and Claude Weber. Polynômes à fibres rationnelles et conjecture jacobienne à 2 variables. *C. R. Acad. Sci Paris*, **320**, 581–584, 1995.

[LW96]   Dung Trang Le and Claude Weber. Equisingularité dans les pinceaux de germes de courbes planes et c^0-suffisance. Preprint, University of Geneva, 1996.

[Mil68]   John Milnor. *Singular points of complex hypersurfaces*. Princeton University Press, Princeton, 1968.

[Neu89]   Walter D. Neumann. Complex algebraic plane curves via their links at infinity. *Invent. math.*, **98**, 445–489, 1989.

[NZ90]   A. Némethi and A. Zaharia. On the bifurcation set of a polynomial function and Newton boundary. *Publ. Math RIMS Kyoto*, **26**, 681–689, 1990.

[NZ92]   A. Némethi and A. Zaharia. Milnor fibration at infinity. *Indag Math*, **3**, 323–335, 1992.

[Par95]   Adam Parusinski. A note on singularities at infinity of complex polynomials. Preprint no. 426, University of Nice-Sophia-Antipolis, 1995.

[Pha83]   Frederic Pham. Vanishing homologies and the n variable saddle-point method. In P. Orlik, editor, *Proceedings of Symposia in Pure Mathematics, Vol. 40*, pages 319–333, Providence RI, 1983. American Mathematical Society.

[ST95]   Dirk Siersma and Mihai Tibar. Singularities at infinity and their vanishing cycles. *Duke Math. J.*, **80**, 771–783, 1995.

[Suz74]   Masakazu Suzuki. Propriétés topologiques des polynômes de deux variables complexes, et automorphismes algébriques de l'espace c^2. *J. Math.Soc. Japan*, **26**, 141–157, 1974.

[Tib96]   Mihai Tibar. Topology at infinity of polynomial mappings and Thom regularity condition. Preprint, University of Angers, 1996.

[Zar65]   O. Zariski. Studies in equisingularity i. *Amer. J. Math.*, **87**, 507–536, 1965.

[Zar79]   Oscar Zariski. *Collected Papers*. The MIT Press, Cambridge, 1979.

Progress in Mathematics, Vol. 162, © 1998 Birkhäuser Verlag Basel/Switzerland

Evaluation of Fermion Loops
by Iterated Residues

Joel Feldman *
Department of Mathematics
University of British Columbia
Vancouver, B.C.
CANADA V6T 1Z2

Horst Knörrer [†]
Mathematik
ETH-Zentrum
CH-8092 Zürich
SWITZERLAND

Robert Sinclair
Mathematik
ETH-Zentrum
CH-8092 Zürich
SWITZERLAND

Eugene Trubowitz
Mathematik
ETH-Zentrum
CH-8092 Zürich
SWITZERLAND

1 Introduction

Let

$$e(\mathbf{k}) \;=\; \frac{1}{2m}\,|\mathbf{k}|^2 - \mu\,, \quad \mathbf{k} \in \mathbb{R}^2\,,$$

be the dispersion relation for a two dimensional electron gas with chemical potential $\mu > 0$. By definition, the amplitude of the $(n+1)$ Fermion loop with external momenta $q_i = (q_{i\,0}, \mathbf{q}_i) \in \mathbb{R} \times \mathbb{R}^2$, $i = 1, 2, \ldots, n+1$, is

$$I(q_1, \ldots, q_{n+1}) \;=\; \frac{1}{(2\pi)^3} \int_{\mathbb{R}^2} dk_1\, dk_2 \int_{-\infty}^{+\infty} \frac{dk_0}{\prod_{i=1}^{n+1} \big(\imath(k - q_i)_0 - e(\mathbf{k} - \mathbf{q}_i)\big)}\,.$$

Recall that the amplitude of an arbitrary diagram contributing to the formal power series expansion of the corresponding many fermion Green's functions in powers of the coupling constant is obtained by integrating products of Fermion loops and interactions.

---

*Research supported in part by the Natural Sciences and Engineering Research Council of Canada, the Schweizerischer Nationalfonds zur Förderung der wissenschaftlichen Forschung and the Forschungsinstitut für Mathematik, ETH Zürich.

[†]Research supported in part by the IHES and SFB288 Differentialgeometrie und Quantenphysik (Berlin).

It is not hard to evaluate the 2-loop (the "polarization bubble") $I(q_1, q_2)$ when $(q_1 - q_2)_0 \neq 0$ and $\mathbf{q}_1 - \mathbf{q}_2 \neq 0$. One finds (see, for example, [2], Proposition 2.1)

$$I(q_1, q_2) = -\frac{m}{2\pi} + \frac{mk_F}{2\pi\,|\mathbf{q}_1 - \mathbf{q}_2|}\, \Re\left(\alpha - \frac{1}{\alpha}\right)$$

where α is the root outside the unit circle of

$$z^2 - \frac{1}{k_F}\left(|\mathbf{q}_1 - \mathbf{q}_2| - 2m\imath\,\frac{(q_1 - q_2)_0}{|\mathbf{q}_1 - \mathbf{q}_2|}\right) z + 1 = 0$$

and $k_F = \sqrt{2m\mu}$. The zero frequency limit has a particularly interesting behaviour. Namely,

$$I\big((0, \mathbf{q}_1), (0, \mathbf{q}_2)\big) = \begin{cases} -\dfrac{m}{2\pi}\,, & \text{if } 0 \leq |\mathbf{q}_1 - \mathbf{q}_2| \leq 2k_F\,, \\[2ex] -\dfrac{m}{2\pi}\left(1 - \dfrac{\sqrt{|\mathbf{q}_1 - \mathbf{q}_2|^2 - 4k_F^2}}{|\mathbf{q}_1 - \mathbf{q}_2|}\right)\,, & \text{if } |\mathbf{q}_1 - \mathbf{q}_2| \geq 2k_F\,. \end{cases}$$

In particular, the value of the 2-loop is $-\frac{m}{2\pi}$ whenever the disks of radius k_F centered at $\mathbf{q}_1$ and $\mathbf{q}_2$ overlap.

In this paper, we evaluate the $(n+1)$ Fermion loop explicitly for all $n \geq 2$. Our result (see, Theorem 3.1 and the Remarks after it) is

$$I(q_1, \ldots, q_{n+1}) = \frac{m^n}{2\pi\imath} \sum_{\substack{i,j=1 \\ i \neq j}}^{n+1} \int_{w_{ij}} \varphi_{ij}(z)\, dz,$$

where φ_{ij} is an explicitly computable rational function of one complex variable and w_{ij} is an explicitly given curve in $\mathbb{C}$. For $n = 2$, that is the 3-loop, we have written out the function φ_{ij} and the path w_{ij} in all detail.

Again, the zero frequency limit

$$J(\mathbf{p}_1, \ldots, \mathbf{p}_{n+1}) = \lim_{\substack{q_i \to (0, \mathbf{p}_i) \\ q_{i0} \neq q_{j0}}} I(q_1, \ldots, q_{n+1})$$

is interesting. The third author (see, [7]) evaluated $J(\mathbf{p}_1, \mathbf{p}_2, \mathbf{p}_3)$ numerically and observed that

$$J(\mathbf{p}_1, \mathbf{p}_2, \mathbf{p}_3) = 0,$$

whenever the disks of radius k_F centered at $\mathbf{p}_1$, $\mathbf{p}_2$ and $\mathbf{p}_3$ have a point in common. We have evaluated (Theorem 3.2) $J(\mathbf{p}_1, \ldots, \mathbf{p}_{n+1})$. In particular (Corollary 3.3 (ii)), for all $n \geq 2$,

$$J(\mathbf{p}_1, \ldots, \mathbf{p}_{n+1}) = 0,$$

if all $n+1$ disks with radius k_F around the points $\mathbf{p}_1, \ldots, \mathbf{p}_{n+1}$ have at least one point in common.[1] Also (Corollary 3.3 (iii)),

$$J(\mathbf{p}_1, \mathbf{p}_2, \mathbf{p}_3) = \frac{m^2}{2 \times \text{area of the triangle with vertices } \mathbf{p}_1, \mathbf{p}_2, \mathbf{p}_3}\,,$$

---

[1] The Appendix contains an elementary proof of this fact for the special case that $n = 2$ and the triangle with vertices $\mathbf{p}_1, \mathbf{p}_2, \mathbf{p}_3$ is acute.

if the three disks with radius k_F around the points $\mathbf{p}_1, \mathbf{p}_2, \mathbf{p}_3$ have no point in common but any two of the disks intersect.

To evaluate $I(q_1, \ldots, q_{n+1})$, we perform (see, the beginning of Section 3) the integral over k_0 to obtain

$$I(q_1, \ldots, q_{n+1}) \;=\; \frac{m^n}{(2\pi)^2} \sum_{i=1}^{n+1} \int_{|\mathbf{k}-\mathbf{q}_i|<k_F} \frac{dk_1\, dk_2}{\displaystyle\prod_{\substack{j=1 \\ j\neq i}}^{n+1} f_{ij}(\mathbf{k})} \,,$$

where

$$f_{ij}(\mathbf{k}) \;=\; (\mathbf{q}_j - \mathbf{q}_i)\cdot\mathbf{k} + \tfrac{1}{2}\left(\mathbf{q}_i^2 - \mathbf{q}_j^2\right) + \imath\, m\, (q_{i0} - q_{j0}).$$

Next, it is observed that each summand

$$\int_{|\mathbf{k}-\mathbf{q}_i|<k_F} \frac{dk_1\, dk_2}{\displaystyle\prod_{\substack{j=1 \\ j\neq i}}^{n+1} f_{ij}(\mathbf{k})}$$

is of the generic form

$$\int_{D_\rho(\mathbf{q})} \frac{dx_1 \wedge dx_2}{\prod_{j=1}^{n}(a_{j1}x_1 + a_{j2}x_2 - b_j)} \,,$$

where the coefficients a_{j1}, a_{j2}, $j = 1, \ldots, n$, are real and b_j is complex. Here, $D_\rho(\mathbf{q})$ is the disk of radius $\rho > 0$ centered at $\mathbf{q}$. In Section 2, we replace the disk by a homologous cycle in the complex projective plane $\mathbb{P}^2$ where iterated residues can be applied.

It is a great pleasure to thank Andrea Cavalli and Peter Wagner for many helpful suggestions. Peter Wagner also found a different proof of Theorem 3.2, using methods similar to those of [4]. He has generously allowed us to incorporate some of his improvements in the statement of Theorem 3.2 itself.

2   The integral of a particular rational function over a disk

Let $n \geq 2$ and fix a real matrix

$$A \;=\; \begin{pmatrix} a_{11} & a_{12} \\ a_{21} & a_{22} \\ \vdots & \vdots \\ a_{n1} & a_{n2} \end{pmatrix},$$

such that no two by two subdeterminant vanishes. Fix $b \in \mathbb{C}^n$ such that $\Im b_j \neq 0$, for all $j = 1, \ldots, n$. For each $\mathbf{q} \in \mathbb{R}^2$ and $\rho \geq 0$ let

$$D_\rho(\mathbf{q}) \;=\; \left\{ (x_1, x_2) \in \mathbb{R}^2 \,\middle|\, (x_1 - \mathbf{q}_1)^2 + (x_2 - \mathbf{q}_2)^2 \leq \rho^2 \right\}$$

be the closed disk of radius ρ with the standard orientation, centered at $\mathbf{q}$. We shall compute

$$\int_{D_\rho(\mathbf{q})} \frac{dx_1 \wedge dx_2}{\prod_{j=1}^n (a_{j1}x_1 + a_{j2}x_2 - b_j)}.$$

To formulate the result, let L_j, $j = 1\ldots,n$, be the line in the complex projective plane $\mathbb{P}^2$ given by

$$L_j = \Big\{ [z_0, z_1, z_2] \in \mathbb{P}^2 \ \Big|\ a_{j1}z_1 + a_{j2}z_2 = b_j z_0 \Big\},$$

and let

$$L_\infty = \Big\{ [z_0, z_1, z_2] \in \mathbb{P}^2 \ \Big|\ z_0 = 0 \Big\}.$$

We assume that the intersections

$$L_\alpha \cap L_\beta = \big\{ d^{\alpha,\beta} \big\}$$

for $\alpha, \beta = 1, \ldots, n, \infty$ with $\alpha \neq \beta$ are pairwise different and do not lie on the projective quadric

$$Q_s = \Big\{ [z_0, z_1, z_2] \in \mathbb{P}^2 \ \Big|\ (z_1 - \mathbf{q}_1 z_0)^2 + (z_2 - \mathbf{q}_2 z_0)^2 = s^2 z_0^2 \Big\}$$

for either $s = 0$ or $s = \rho$.

If $n \geq 3$, let η_j, $j = 1, \ldots, n$, be the unique rational one-form on the projective line L_j that has a simple pole at the point $d^{j,k}$ for all $k = 1, \ldots, n$, $k \neq j$, with residue

$$\frac{1}{\det \begin{pmatrix} a_{j1} & a_{j2} \\ a_{k1} & a_{k2} \end{pmatrix}} \prod_{\ell \neq j,k} \Big(\frac{1}{d_0^{j,k}} \big(a_{\ell 1} d_1^{j,k} + a_{\ell 2} d_2^{j,k} - b_\ell d_0^{j,k} \big) \Big)^{-1},$$

no other poles , and a zero of order $n - 3$ at $d^{j,\infty}$. We will see below that the sum of these residues is zero and for this reason the form η_j does in fact exist. If $n = 2$, let η_j, $j = 1, 2$, be the unique rational one form on the projective line L_j that has a simple pole at the point $d^{j,k}$ for $k \neq j$, with residue

$$\frac{1}{\det \begin{pmatrix} a_{j1} & a_{j2} \\ a_{k1} & a_{k2} \end{pmatrix}},$$

a simple pole at $d^{j,\infty}$, and no other poles. For all $s \in \mathbb{C}$, set

$$Q_s^+ = \Big\{ [z_0, z_1, z_2] \in Q_s \ \Big|\ \Im \tfrac{z_2 - \mathbf{q}_2}{z_1 - \mathbf{q}_1} > 0 \Big\},$$

$$Q_s^- = \Big\{ [z_0, z_1, z_2] \in Q_s \ \Big|\ \Im \tfrac{z_2 - \mathbf{q}_2}{z_1 - \mathbf{q}_1} < 0 \Big\}.$$

We will show, see Lemma 2.2, that for all $s \geq 0$ and all $j = 1, \ldots, n$, the intersection $Q_s^+ \cap L_j$ consists of exactly one point, say, $c_j^+(s)$ and the intersection $Q_s^- \cap L_j$ consists of exactly one point, say, $c_j^-(s)$.

THEOREM 2.1 *Suppose that for each $j = 1, \ldots, n$, the paths $c_j^\pm(s)$, $0 \le s \le \rho$, do not pass through any of the double points $L_j \cap L_k$, $j, k = 1, \ldots, n$, $j \ne k$. Then*

$$\int_{\mathrm{D}_\rho(\mathbf{q})} \frac{dx_1 \wedge dx_2}{\prod_{j=1}^n (a_{j1}x_1 + a_{j2}x_2 - b_j)} \;=\; -2\pi\imath \sum_{j=1}^n \int_{\left\{ c_j^+(s) \mid 0 \le s \le \rho \right\}} \eta_j$$

$$\;=\; +2\pi\imath \sum_{j=1}^n \int_{\left\{ c_j^-(s) \mid 0 \le s \le \rho \right\}} \eta_j \,.$$

Remark. Observe that for any path $s(\sigma)$, $0 \le \sigma \le 1$, in $\mathbb{C}$ joining 0 to ρ and lying sufficiently close to the interval $[0, \rho]$, the intersection $Q^\pm_{s(\sigma)} \cap L_j$ consists of exactly one point $c_j^\pm(s(\sigma))$. Our hypotheses imply that $c_j^\pm(0)$ and $c_j^\pm(\rho)$ are not double points $L_j \cap L_k$. It is therefore possible to choose a path $s(\sigma)$, $0 \le \sigma \le 1$, arbitrarily close to $[0, \rho]$ such that $c_j^\pm(s(\sigma))$ is not a double point for all $j = 1, \ldots, n$ and $0 \le \sigma \le 1$. With this choice

$$\int_{\mathrm{D}_\rho(\mathbf{q})} \frac{dx_1 \wedge dx_2}{\prod_{j=1}^n (a_{j1}x_1 + a_{j2}x_2 - b_j)} \;=\; 2\pi\imath \sum_{j=1}^n \int_{\left\{ c_j^+(s(\sigma)) \mid 0 \le \sigma \le 1 \right\}} \eta_j$$

$$\;=\; -2\pi\imath \sum_{j=1}^n \int_{\left\{ c_j^-(s(\sigma)) \mid 0 \le \sigma \le 1 \right\}} \eta_j \,.$$

Two lemmas are required for the proof of Theorem 2.1. Observe that by translation invariance, we may assume that $\mathbf{q} = 0$.

LEMMA 2.2 *For all $s \ge 0$ and all $\alpha = 1, \ldots, n, \infty$, the intersection $Q_s^+ \cap L_\alpha$ consists of exactly one point, say, $c_\alpha^+(s)$ and the intersection $Q_s^- \cap L_\alpha$ consists of exactly one point, say, $c_\alpha^-(s)$. Furthermore, $c_\infty^+(s)$ and $c_\infty^-(s)$ do not depend on s.*

Proof. First, fix $1 \le j \le n$. We can rotate (z_1, z_2) around zero to make $a_{j2} = 0$. The intersection $Q_s \cap L_j$ is determined by the equation

$$z_1^2 + z_2^2 \;=\; s^2 \left(\frac{a_{j1}}{b_j} \right)^2 z_1^2 \,.$$

Thus,

$$\left(\frac{z_2}{z_1} \right)^2 \;=\; s^2 \left(\frac{a_{j1}}{b_j} \right)^2 - 1 \,.$$

It follows that there is exactly one root with positive imaginary part and one with negative imaginary part since a_{j1} is real and $\Im b_j \ne 0$. By inspection,

$$Q_s \cap L_\infty \;=\; \left\{ \, [0, 1, \imath] \,,\, [0, 1, -\imath] \, \right\} \,. \qquad \square$$

Embed $\mathbb{C}^2$ in $\mathbb{P}^2$ by the standard map

$$(z_1, z_2) \in \mathbb{C}^2 \; \longrightarrow \; [z_0, z_1, z_2] \in \mathbb{P}^2.$$

Let ω be the meromorphic two form on $\mathbb{P}^2$ given by

$$\omega \; = \; \frac{dz_1 \wedge dz_2}{\prod_{j=1}^{n}(a_{j1}z_1 + a_{j2}z_2 - b_j)}.$$

If $n \geq 3$, then ω has simple poles on the lines $L_1, \ldots, L_n$ and is holomorphic on the difference $\mathbb{P}^2 \setminus (L_1 \cup L_2 \cup \cdots \cup L_n)$. If $n = 2$, then ω has simple poles on $L_1 L_2 L_\infty$, and is holomorphic on $\mathbb{P}^2 \setminus (L_1 \cup L_2 \cup L_\infty)$.

Regard the union of lines

$$C \; = \; L_1 \cup L_2 \cup \cdots \cup L_n \cup L_\infty$$

as a singular algebraic curve of degree $n+1$ in $\mathbb{P}^2$. Then, the Poincaré residue of ω along the curve C (see, for example [3], p.147),

$$\eta \; = \; \mathrm{res}_C \, \omega$$

is a Rosenlicht differential on C (see [6], Ch. IV.9).

Fix, $j = 1, \ldots, n$. For each $k \neq j$,

$$\omega \; = \; \frac{\dfrac{d(a_{j1}z_1 + a_{j2}z_2 - b_j)}{a_{j1}z_1 + a_{j2}z_2 - b_j} \wedge \dfrac{d(a_{k1}z_1 + a_{k2}z_2 - b_k)}{a_{k1}z_1 + a_{k2}z_2 - b_k}}{\det \begin{pmatrix} a_{j1} & a_{j2} \\ a_{k1} & a_{k2} \end{pmatrix} \displaystyle\prod_{\ell \neq j, k}^{n} (a_{\ell 1}z_1 + a_{\ell 2}z_2 - b_\ell)}.$$

It follows from this representation that the restriction of η to the projective line L_j has a simple pole at the point $d^{j,k}$ for all $k = 1, \ldots, n$, $k \neq j$, with residue

$$\frac{1}{\det \begin{pmatrix} a_{j1} & a_{j2} \\ a_{k1} & a_{k2} \end{pmatrix}} \prod_{\ell \neq j, k} \left(\frac{1}{d_0^{j,k}} \left(a_{\ell 1} d_1^{j,k} + a_{\ell 2} d_2^{j,k} - b_\ell d_0^{j,k} \right) \right)^{-1}.$$

Furthermore, it is holomorphic and does not vanish on $L_j \setminus \{ d^{j,\alpha} \,|\, \alpha = 1, \ldots, n, \infty, j \neq \alpha \}$. Consequently, $\eta|_{L_j}$ has a zero of order $n - 3$ at $d^{j,\infty}$, when $n \geq 3$, and a simple pole when $n = 2$. It follows from the preceding remarks that the restriction $\eta|_{L_j}$, has all the properties of the rational one form η_j introduced above. In particular, η_j exists and

$$\eta_j \; = \; \eta\big|_{L_j}.$$

Set

$$\eta_\infty \; = \; \eta\big|_{L_\infty}.$$

For $n \geq 3$, the restriction $\eta_\infty = 0$, since ω is regular along L_∞. If $n = 2$, then η_∞ has simple poles at the points $d^{1,\infty}$ and $d^{2,\infty}$, and no other poles.

We need the following residue formula.

LEMMA 2.3 *Fix* $\alpha = 1, \ldots, n, \infty$. *Let* p_1 *and* p_2 *be any two points on*

$$L_\alpha \setminus \{ d^{\alpha,\beta} \,|\, \beta = 1, \ldots, n, \infty, \, \beta \neq \alpha \}.$$

Suppose S_1 *and* S_2 *are algebraic curves in* $\mathbb{P}^2$ *such that* L_α *meets* S_1 *transversally at* p_1 *and* S_2 *transversally at* p_2. *Let*

$$\pi \,:\, T_\alpha \,\longrightarrow\, L_\alpha \setminus \{ d^{\alpha,\beta} \,|\, \beta = 1, \ldots, n, \infty, \, \beta \neq \alpha \}$$

be a tubular neighborhood of $L_\alpha \setminus \{ d^{\alpha,\beta} \,|\, \beta = 1, \ldots, n, \infty, \, \beta \neq \alpha \}$ *such that*

$$S_i \cap T_\alpha \,=\, \pi^{-1}(p_i)$$

for $i = 1, 2$. *Finally, let* γ *be a path on* $L_\alpha \setminus \{ d^{\alpha,\beta} \,|\, \beta = 1, \ldots, n, \infty, \, \beta \neq \alpha \}$ *joining* p_1 *to* p_2 *and set*

$$\Gamma \,=\, \pi^{-1}(\gamma) \cap \partial T.$$

The "cylinder" Γ *is oriented so that*

$$\partial \Gamma \,=\, \partial(S_2 \cap T_\alpha) \,-\, \partial(S_1 \cap T_\alpha).$$

Here, $S_i \cap T_\alpha$, $i = 1, 2$, *is given the standard orientation of an open subset of the Riemann surface* S_i. *Then,*

$$\int_\Gamma \omega \,=\, -2\pi\imath \int_\gamma \eta_\alpha\,.$$

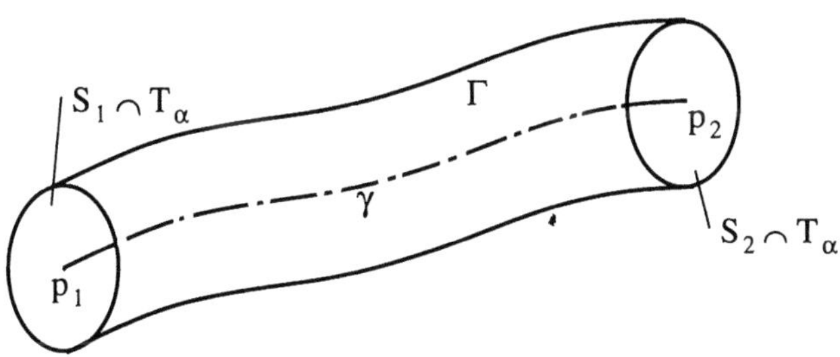

Figure 1

Proof. The proof is similar to that of [5]. $\square$

Remark. By convention, the boundary of an oriented surface is oriented such that the ordered pair of the outward pointing normal and an oriented tangent vector gives the orientation of the surface.

Proof of Theorem 2.1. By construction, for all $\alpha = 1, \ldots, n, \infty$, the projective line L_α intersects the quadric $\mathcal{Q}_s$ transversally at the points $c_\alpha^\pm(s)$. For $0 \le s \le \rho$ let

$$\pi_{\alpha,s},: T_{\alpha,s} \longrightarrow L_\alpha \setminus \left\{ d^{\alpha,\beta} \,\middle|\, \beta = 1, \ldots, n, \infty, \, \beta \ne \alpha \right\}$$

be tubular neighborhoods of $L_\alpha \setminus \left\{ d^{\alpha,\beta} \,\middle|\, \beta = 1, \ldots, n, \infty, \, \beta \ne \alpha \right\}$ such that

$$\mathcal{Q}_t^\pm \cap T_{\alpha,s} \;=\; \pi_{\alpha,s}^{-1}\bigl(c_\alpha^\pm(t)\bigr)$$

for $t = 0, s$, and

$$\Delta_{\alpha,0}^\pm \;=\; \mathcal{Q}_0^\pm \cap T_{\alpha,s}$$

is independent of s. Set

$$\Delta_{\alpha,s}^\pm \;=\; \mathcal{Q}_s^\pm \cap T_{\alpha,s} \,.$$

By convention, the disks $\Delta_{\alpha,t}^\pm$, $t = 0, s$, are oriented as subsets of the Riemann surface $\mathcal{Q}_t^\pm$. Also, set

$$\Gamma_{\alpha,s}^\pm \;=\; \pi_{\alpha,s}^{-1}\left(\left\{ c_\alpha^\pm(t) \mid 0 \le t \le s \right\}\right) \cap \partial T_{\alpha,s} \,.$$

The "cylinder" $\Gamma_{\alpha,s}^\pm$ is oriented so that

$$\partial \Gamma_{\alpha,s}^\pm \;=\; \partial \Delta_{\alpha,s}^\pm - \partial \Delta_{\alpha,0}^\pm \,.$$

By Lemma 2.3,

$$\int_{\Gamma_{\alpha,s}^\pm} \omega \;=\; -2\pi\imath \int_{\left\{ c_\alpha^\pm(t) \mid 0 \le t \le s \right\}} \eta_\alpha \,.$$

Now, for each $0 \le s \le \rho$, set

$$Z_s^\pm \;=\; \pm D_s(0) - \left(\mathcal{P}_s^\pm - \mathcal{P}_0^\pm + \sum_{j=1}^{n} \Gamma_{j,s}^\pm + \Gamma_{\infty,s}^\pm \right),$$

where, for $t = 0, s$

$$\mathcal{P}_t^\pm \;=\; \mathcal{Q}_t^\pm \setminus \left(\bigcup_{j=1}^{n} \Delta_{j,t}^\pm \cup \Delta_{\infty,t}^\pm \right).$$

We have for $t = 0, s$

$$\partial \mathcal{P}_t^\pm \;=\; \partial \mathcal{Q}_t^\pm - \sum_{j=1}^{n} \partial \Delta_{j,t}^\pm - \partial \Delta_{\infty,t}^\pm$$

and

$$\partial \mathcal{Q}_s^\pm \;=\; \pm \partial D_s(0)\,, \qquad \partial \overline{\mathcal{Q}_0^\mp} \;=\; \emptyset.$$

Therefore,

$$
\begin{aligned}
\partial Z_s^\pm \;=\;& \pm \partial D_s(0) - \left(\partial \mathcal{Q}_s^\pm - \sum_{j=1}^{n} \partial \Delta_{j,s}^\pm - \partial \Delta_{\infty,s}^\pm \right) \\
& + \left(\partial \mathcal{Q}_0^\pm - \sum_{j=1}^{n} \partial \Delta_{j,0}^\pm - \partial \Delta_{\infty,0}^\pm \right) \\
& - \sum_{j=1}^{n} \left(\partial \Delta_{j,s}^\pm - \partial \Delta_{j,0}^\pm \right) - \left(\partial \Delta_{\infty,s}^\pm - \partial \Delta_{\infty,0}^\pm \right) \\
\;=\;& 0.
\end{aligned}
$$

In other words, $Z_s^\pm$ is a 2-cycle in $\mathbb{P}^2 \setminus C$.

Observe that the homology class $[Z_s^\pm] \in H_2(\mathbb{P}^2 \setminus C, \mathbb{Z})$ represented by $Z_s^\pm$ is 0 since, by construction, $Z_s^\pm$ depends continuously on s and $[Z_0^\pm] = 0$. The integral of the holomorphic two form ω over the one dimensional complex manifolds $\mathcal{P}_t^\pm$, $t = 0, s$ vanishes. It follows that

$$
\begin{aligned}
0 \;=\;& \int_{Z_s^\pm} \omega = \pm \int_{D_s(0)} \omega - \sum_{j=1}^{n} \int_{\Gamma_{j,s}^\pm} \omega - \int_{\Gamma_{\infty,s}^\pm} \\
\;=\;& \pm \int_{D_s(0)} \omega + 2\pi\imath \sum_{j=1}^{n} \int_{\left\{ c_j^\pm(s) \,\middle|\, 0 \le s \le \rho \right\}} \eta_j + 2\pi\imath \int_{\left\{ c_\infty^\pm(s) \,\middle|\, 0 \le s \le \rho \right\}} \eta_\infty\,.
\end{aligned}
$$

Since, $c_\infty^+(s)$ and $c_\infty^-(s)$ do not depend on s,

$$\int_{\left\{ c_\infty^\pm(s) \,\middle|\, 0 \le s \le \rho \right\}} \eta_\infty \;=\; 0.$$

Therefore

$$\int_{D_\rho(\mathbf{q})} \omega \;=\; \mp\, 2\pi\imath \sum_{j=1}^{n} \int_{\left\{ c_j^\pm(s) \,\middle|\, 0 \le s \le \rho \right\}} \eta_j\,. \qquad \square$$

Remark. For all $0 \le s \le \rho$, the divisor of the rational function

$$\frac{z_1^2 + z_2^2 - s^2 z_0^2}{z_1^2 + z_2^2}$$

on the curve C is

$$\sum_{j=1}^{n} \left(c_j^+(s) - c_j^+(0) \right) + \left(c_j^-(s) - c_j^-(0) \right).$$

It therefore also follows from Abel's theorem for the singular curve C and the Rosenlicht differential η that

$$\sum_{j=1}^{n} \int_{\{\, c_j^+(s)\,|\, 0\leq s\leq\rho \,\}} \eta_j \;=\; -\sum_{j=1}^{n} \int_{\{\, c_j^-(s)\,|\, 0\leq s\leq\rho \,\}} \eta_j.$$

3　The evaluation of Fermion loops

Let $q_i = (q_{i0}, \mathbf{q}_i)$, $i = 1, 2, \ldots, n+1$, be vectors in $\mathbb{R} \times \mathbb{R}^2$. We assume that no three of the points $\mathbf{q}_1, \ldots, \mathbf{q}_{n+1}$ lie on a line. The amplitude of the $(n+1)$-loop with momenta $q_1, q_2, \ldots, q_{n+1}$ is

$$I(q_1, \ldots, q_{n+1}) \;=\; \frac{1}{(2\pi)^3} \int_{\mathbb{R}^2} dk_1 \, dk_2 \int_{-\infty}^{+\infty} \frac{dk_0}{\displaystyle\prod_{i=1}^{n+1} \big(\imath(k - q_i)_0 - e(\mathbf{k} - \mathbf{q}_i)\big)},$$

where the dispersion relation $e(\mathbf{k})$ is given by

$$e(\mathbf{k}) \;=\; \frac{1}{2m}\big(\mathbf{k}^2 - k_F^2\big).$$

Here, $k_F = \sqrt{2\mathbf{m}\mu}$ is the Fermi momentum.

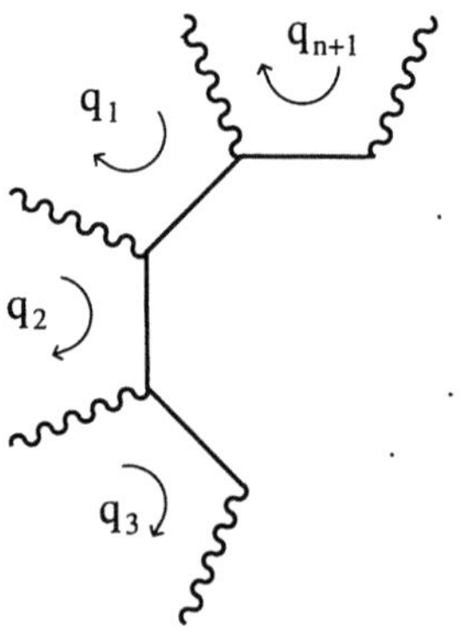

Figure 2

If $e(\mathbf{k} - \mathbf{q}_i) \neq 0$ for $i = 1, \ldots, n+1$, and $n \geq 1$ then, by the residue theorem (closing the contour in the upper half plane),

$$\int_{-\infty}^{+\infty} \frac{dk_0}{\displaystyle\prod_{i=1}^{n+1} \big(i(k_0 - q_{i0}) - e(\mathbf{k} - \mathbf{q}_i)\big)} \;=\; 2\pi \, m^n \sum_{i=1}^{n+1} \frac{\chi\big(e(\mathbf{k} - \mathbf{q}_i) < 0\big)}{\displaystyle\prod_{\substack{j=1 \\ j\neq i}}^{n+1} f_{ij}(\mathbf{k})},$$

where

$$
\begin{aligned}
f_{ij}(\mathbf{k}) &= m\left(e(\mathbf{k}-\mathbf{q}_i) - e(\mathbf{k}-\mathbf{q}_j) + \imath(q_{i0}-q_{j0})\right) \\
&= (\mathbf{q}_j - \mathbf{q}_i)\cdot\mathbf{k} + \tfrac{1}{2}\left(\mathbf{q}_i^2 - \mathbf{q}_j^2\right) + \imath m\,(q_{i0}-q_{j0}).
\end{aligned}
$$

Observe that $f_{ij}(\mathbf{k})$, $i,j = 1,\ldots n+1$, is an affine linear function of $\mathbf{k}$. Substituting, we obtain

$$
I(q_1,\ldots,q_{n+1}) = \frac{m^n}{(2\pi)^2}\sum_{i=1}^{n+1}\int_{|\mathbf{k}-\mathbf{q}_i|<k_F}\frac{dk_1\,dk_2}{\displaystyle\prod_{\substack{j=1\\ j\neq i}}^{n+1} f_{ij}(\mathbf{k})}\,.
$$

It follows from this representation that $I(q_1,\ldots,q_{n+1})$ is a continuous function on the set

$$
\left\{\,((q_{10},\mathbf{q}_1),\ldots,(q_{n+10},\mathbf{q}_{n+1})) \in \left(\mathbb{R}\times\mathbb{R}^2\right)^{n+1} \;\middle|\; q_{i0}\neq q_{j0}\ \text{for}\ i\neq j\,\right\}.
$$

Observe that Theorem 2.1 can be used to compute each of the summands

$$
\int_{|\mathbf{k}-\mathbf{q}_i|<k_F}\frac{dk_1\,dk_2}{\displaystyle\prod_{\substack{j=1\\ j\neq i}}^{n+1} f_{ij}(\mathbf{k})}\,.
$$

For each pair $1\leq i\neq j\leq n+1$,

$$
\ell_{ij} = \left\{z\in\mathbb{C}^2 \mid f_{ij}(z)=0\right\}
$$

is a line in $\mathbb{C}^2$. We have

$$
\ell_{ij} = \ell_{ji}\,,
$$

since $f_{ij} = -f_{ji}$. If $q_{i0} = q_{j0}$ the line $\ell_{ij}\cap\mathbb{R}^2$ is the perpendicular bisector of the points $\mathbf{q}_i$ and $\mathbf{q}_j$. Observe that for any three different indices $1\leq i,j,k\leq n+1$ the lines ℓ_{ij} and ℓ_{jk} intersect in exactly one point $\mathbf{d}^{ijk}$ because $\mathbf{q}_i$, $\mathbf{q}_j$ and $\mathbf{q}_k$ do not lie on a line. Furthermore, $\mathbf{d}^{ijk}$ is the common intersection point of the three lines $\ell_{ij}=\ell_{ji}$, $\ell_{jk}=\ell_{kj}$, $\ell_{ki}=\ell_{ik}$, since

$$
f_{ij} + f_{jk} + f_{ki} = 0.
$$

Observe that $\Re\,\mathbf{d}^{ijk}$ is the center of the circle circumscribing the triangle with vertices $\mathbf{q}_i$, $\mathbf{q}_j$, $\mathbf{q}_k$. If $q_{i0}=q_{i0}=q_{i0}$, then the point $\mathbf{d}^{ijk}$ is real. For any three different indices $1\leq i,j,k\leq n+1$, put

$$
r_{ijk} = \frac{1}{\det\left(\mathbf{q}_j - \mathbf{q}_i,\,\mathbf{q}_k - \mathbf{q}_i\right)}\,\frac{1}{\displaystyle\prod_{\substack{\nu=1\\ \nu\neq i,j,k}}^{n+1} f_{i\nu}(\mathbf{d}^{ijk})}\,,
$$

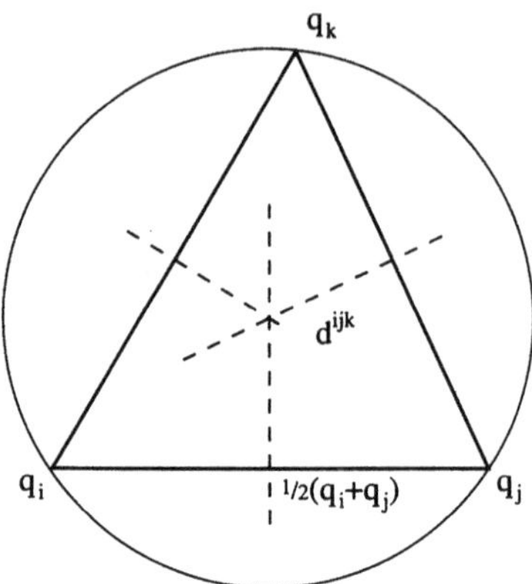

Figure 3

where $\det\left(\mathbf{q}_j - \mathbf{q}_i, \mathbf{q}_k - \mathbf{q}_i\right)$ is the determinant of the matrix with rows $\mathbf{q}_j - \mathbf{q}_i$ and $\mathbf{q}_k - \mathbf{q}_i$. Observe that $\left|\det\left(\mathbf{q}_j - \mathbf{q}_i, \mathbf{q}_k - \mathbf{q}_i\right)\right|$ is twice the area of the triangle with vertices $\mathbf{q}_i$, $\mathbf{q}_j$, $\mathbf{q}_k$. By construction,

$$0 = f_{ij}(\mathbf{d}^{ijk}) = -f_{j\nu}(\mathbf{d}^{ijk}) - f_{\nu i}(\mathbf{d}^{ijk}) = -f_{j\nu}(\mathbf{d}^{ijk}) + f_{i\nu}(\mathbf{d}^{ijk}),$$

so that

$$r_{ijk} = -r_{jik}.$$

Let $\bar{\ell}_{ij}$, $i,j = 1,\ldots,n+1$, be the projective closure of ℓ_{ij} in the complex projective plane $\mathbb{P}^2$ and, as in the last section, let L_∞ be the line at infinity. For each pair of indices $1 \leq i \neq j \leq n+1$, let η_{ij} be the unique meromorphic differential form on $\bar{\ell}_{ij}$ that is holomorphic outside the points $\mathbf{d}^{ijk}$, $k = 1,\ldots,n+1$, $k \neq i,j$, and $\bar{\ell}_{ij} \cap L_\infty$, that has a simple pole with residue r_{ijk} at the point $\mathbf{d}^{ijk}$, and that has a zero of order $n-3$ (respectively, a simple pole for $n = 2$) at $\bar{\ell}_{ij} \cap L_\infty$. By construction,

$$\eta_{ij} = -\eta_{ji}.$$

If $1 \leq i \neq j \leq n+1$, $q_{i0} \neq q_{j0}$ and $0 < s < k_F$, let $c^{\pm}_{ij}(s)$ be the point where the line ℓ_{ij} meets the quadric

$$Q^{\pm}_i(s) = \left\{ z \in \mathbb{C}^2 \,\big|\, (z_1 - \mathbf{q}_{i1})^2 + (z_2 - \mathbf{q}_{i2})^2 = s^2 \text{ and } \pm \Im \tfrac{z_2 - \mathbf{q}_{i2}}{z_1 - \mathbf{q}_{i1}} > 0 \right\}.$$

In general $c^{\pm}_{ij}(s) \neq c^{\pm}_{ji}(s)$.

THEOREM 3.1 Let $n \geq 2$. Suppose that no three of the points $\mathbf{q}_1,\ldots,\mathbf{q}_{n+1}$ in $\mathbb{R}^2$ lie on a line, that $q_{i0} \neq q_{j0}$ for $i \neq j$, and that $\mathbf{d}^{ijk} \neq \mathbf{d}^{ijk'}$ for $k \neq k'$.

Further, suppose that $\mathbf{d}^{ijk} \neq c_{ij}^{\pm}(s)$, $0 < s < k_F$, *for all* $i \neq j$ *and all* $k \neq i, j$. *Then,*

$$
\begin{aligned}
I(q_1, \ldots, q_{n+1}) &= +\frac{m^n}{2\pi\imath} \sum_{\substack{i,j=1 \\ i \neq j}}^{n+1} \int_{\{c_{ij}^{+}(s) \mid 0 < s < k_F\}} \eta_{ij} \\
&= -\frac{m^n}{2\pi\imath} \sum_{\substack{i,j=1 \\ i \neq j}}^{n+1} \int_{\{c_{ij}^{-}(s) \mid 0 < s < k_F\}} \eta_{ij} .
\end{aligned}
$$

Proof. By Theorem 2.1,

$$
\begin{aligned}
\int_{|\mathbf{k}-\mathbf{q}_i| < k_F} \frac{dk_1 \, dk_2}{\prod_{\substack{j=1 \\ j \neq i}}^{n+1} f_{ij}(\mathbf{k})} &= -2\pi\imath \sum_{j \neq i} \int_{\{c_{ij}^{+}(s) \mid 0 < s < k_F\}} \eta_{ij} \\
&= +2\pi\imath \sum_{j \neq i} \int_{\{c_{ij}^{-}(s) \mid 0 < s < k_F\}} \eta_{ij}
\end{aligned}
$$

for $i = 1, \ldots n+1$. $\qquad\square$

Remark. For each pair $1 \leq i, j \leq n+1$, $i \neq j$, choose $k \neq i, j$. We parameterize the line ℓ_{ij}, by

$$
\pi_{ij}(z) = z\left((\mathbf{q}_j - \mathbf{q}_i)_2, -(\mathbf{q}_j - \mathbf{q}_i)_1\right) + \mathbf{d}^{ijk} , \quad z \in \mathbb{C} .
$$

The pull back $\pi_{ij}^{*} \eta_{ij}$ of η_{ij} is a rational form on $\mathbb{C}$ with simple poles. Consequently, there is a rational function $\varphi_{ij}(z)$ such that

$$
\varphi_{ij} \, dz = \pi_{ij}^{*} \eta_{ij} .
$$

We can therefore reformulate the statement of Theorem 3.1 as

$$
I(q_1, \ldots, q_{n+1}) = \frac{m^n}{2\pi\imath} \sum_{\substack{i,j=1 \\ i \neq j}}^{n+1} \int_{w_{ij}} \varphi_{ij}(z) \, dz ,
$$

where

$$
w_{ij} = \pi_{ij}^{-1}\left(\{c_{ij}^{+}(s) \mid 0 < s < k_F\}\right) .
$$

Remark. For $n = 2$, that is, the 3-loop, we parameterize the line ℓ_{ij}, $i, j = 1, 2, 3$, $i \neq j$, by

$$
\pi_{ij}(z) = z\left((\mathbf{q}_j - \mathbf{q}_i)_2, -(\mathbf{q}_j - \mathbf{q}_i)_1\right) + \mathbf{d} , \quad z \in \mathbb{C} ,
$$

where $\mathbf{d} = \mathbf{d}^{123}$ is the solution to the pair of equations

$$(\mathbf{q}_2 - \mathbf{q}_1) \cdot \mathbf{k} + \tfrac{1}{2}\left(\mathbf{q}_1^2 - \mathbf{q}_2^2\right) + \imath m(q_{10} - q_{20}) \;=\; 0\,,$$
$$(\mathbf{q}_3 - \mathbf{q}_1) \cdot \mathbf{k} + \tfrac{1}{2}\left(\mathbf{q}_1^2 - \mathbf{q}_3^2\right) + \imath m(q_{10} - q_{30}) \;=\; 0\,.$$

Observe that $\pi_{ij}^* \, \eta_{ij}$ is a rational form on $\mathbb{C}$ that has simple poles at 0 and ∞ and no others. By construction,

$$\pi_{ij}^* \, \eta_{ij} \;=\; \pm\, \frac{1}{\det(\mathbf{q}_2 - \mathbf{q}_1, \mathbf{q}_3 - \mathbf{q}_1)}\, \frac{dz}{z} \;=\; \pm\, \frac{1}{\det(\mathbf{q}_2 - \mathbf{q}_1, \mathbf{q}_3 - \mathbf{q}_1)}\, d\log z$$

with $+$ for $(i,j) = (1,2),(2,3),(3,1)$ and $-$ for $(i,j) = (2,1),(3,2),(1,3)$. Let $w_{ij}(s)$ be the unique root of

$$\left(z(\mathbf{q}_j - \mathbf{q}_i)_2 + \mathbf{d}_1 - \mathbf{q}_{i1}\right)^2 + \left(-z(\mathbf{q}_j - \mathbf{q}_i)_1 + \mathbf{d}_2 - \mathbf{q}_{i2}\right)^2 \;=\; s^2$$

with

$$\Im\, \frac{-w_{ij}(s)(\mathbf{q}_j - \mathbf{q}_i)_1 + \mathbf{d}_2 - \mathbf{q}_{i2}}{w_{ij}(s)(\mathbf{q}_j - \mathbf{q}_i)_2 + \mathbf{d}_1 - \mathbf{q}_{i1}} \;>\; 0\,.$$

We have

$$I(q_1, q_2, q_3) \;=\; \frac{1}{2\pi\imath}\, \frac{m^2}{\det(\mathbf{q}_2 - \mathbf{q}_1, \mathbf{q}_3 - \mathbf{q}_1)}\, \sum_{\substack{i,j=1 \\ i \neq j}}^{3} \int_{\{w_{ij}(s)\,|\,0 \leq s \leq k_F\}} \pm\, d\log z\,.$$

Clearly, each of the six integrals on the right-hand side can be performed explicitly.

Remark. For each pair $1 \leq i,j \leq n+1$, $i \neq j$, and $k \neq i,j$ let η_{ijk} be the unique meromorphic differential form on $\bar{\ell}_{ij}$ that is holomorphic outside the points $\mathbf{d}^{ijk}$, and $\bar{\ell}_{ij} \cap L_\infty$, that has a simple pole with residue r_{ijk} at the point $\mathbf{d}^{ijk}$, and that has a simple pole at $\bar{\ell}_{ij} \cap L_\infty$. Then

$$\eta_{ij} \;=\; \sum_{k \neq i,j} \eta_{ijk}\,.$$

Consequently

$$I(q_1, \ldots, q_{n+1}) \;=\; \frac{m^n}{2\pi\imath}\, \sum_{\substack{i,j,k=1 \\ \text{pairwise different}}}^{n+1} \int_{\{c_{ij}^+(s)\,|\,0 < s < k_F\}} \eta_{ijk}\,.$$

So $I(q_1, \ldots, q_{n+1})$ is decomposed into a sum of terms each of which only depends on a triple $\{q_i, q_j, q_k\}$ of momenta. The fact that such a decomposition is possible follows directly from the identity

$$\frac{1}{\prod\limits_{i=1}^{n+1}\left(\imath(k-q_i)_0 - e(\mathbf{k}-\mathbf{q}_i)\right)}$$

$$= \sum_{1\le i<j<k\le n+1} \frac{1}{\prod\limits_{\substack{\nu=1\\ \nu\ne i,j,k}}^{n+1} f_{i\nu}(\mathbf{d}^{ijk})} \frac{1}{\prod\limits_{\nu=i,j,k}\left(\imath(k-q_\nu)_0 - e(\mathbf{k}-\mathbf{q}_\nu)\right)}.$$

This identity was pointed out to us by A. Cavalli and P. Wagner.

Fix $\mathbf{p}_1,\ldots,\mathbf{p}_{n+1}$ in $\mathbb{R}^2$. We assume that no three of these points lie on a line and no four of them lie on a circle. The main purpose of this section is to show that the limit

$$J(\mathbf{p}_1,\ldots,\mathbf{p}_{n+1}) = \lim_{\substack{q_i\to(0,\mathbf{p}_i)\\ q_{i0}\ne q_{j0}}} I(q_1,\ldots,q_{n+1})$$

exists and to evaluate it. To do this, first set

$$F_{ij}(\mathbf{k}) = m\left(e(\mathbf{k}-\mathbf{p}_i) - e(\mathbf{k}-\mathbf{p}_j)\right) = (\mathbf{p}_j-\mathbf{p}_i)\cdot\mathbf{k} + \tfrac{1}{2}\left(\mathbf{p}_i^2-\mathbf{p}_j^2\right)$$

for all $i,j=1,\ldots,n+1$, and observe that

$$\lim_{q_i\to(0,\mathbf{p}_i)} f_{ij}(\mathbf{k}) = F_{ij}(\mathbf{k}).$$

Furthermore, for each $1\le i\ne j\le n+1$ the line ℓ_{ij} converges to the perpendicular bisector

$$L_{ij} = \left\{z\in\mathbb{C}^2 \mid F_{ij}(z) = 0\right\}$$

of $\mathbf{p}_i$ and $\mathbf{p}_j$. Also, let $\mathbf{D}^{ijk}$ be the center of the circle circumscribing the triangle with vertices $\mathbf{p}_i$, $\mathbf{p}_j$ and $\mathbf{p}_k$. Then,

$$\lim_{q_i\to(0,\mathbf{p}_i)} \mathbf{d}^{ijk} = \mathbf{D}^{ijk}.$$

We have $\mathbf{D}^{ijk}\ne\mathbf{D}^{ijk'}$ for all $k\ne k'$, since no four of the points $\mathbf{p}_1,\ldots,\mathbf{p}_{n+1}$ lie on a circle.

For any three different indices $1\le i,j,k\le n+1$, let

$$R_{ijk} = \frac{1}{\det\left(\mathbf{p}_j-\mathbf{p}_i, \mathbf{p}_k-\mathbf{p}_i\right)} \frac{1}{\prod\limits_{\substack{\nu=1\\ \nu\ne i,j,k}}^{n+1} F_{i\nu}(\mathbf{D}^{ijk})}.$$

Then,

$$\lim_{q_i\to(0,\mathbf{p}_i)} r_{ijk} = R_{ijk}.$$

Observe that

$$R_{ijk} \;=\; R_{jki} \;=\; R_{kij} \;=\; -R_{jik}\,.$$

Let

$$\triangle_{(\mathbf{p}_i,\mathbf{p}_j,\mathbf{p}_k)} \;=\; \big\{\mathbf{p}_i,\mathbf{p}_j,\mathbf{p}_k\big\}$$

be the triangle with vertices $\mathbf{p}_i,\mathbf{p}_j,\mathbf{p}_k$. We can, using the antisymmetry of R_{ijk}, define

$$R_\triangle \;=\; \begin{cases} R_{ijk}, & \text{if } \mathbf{p}_i,\mathbf{p}_j,\mathbf{p}_k \text{ are oriented counter clockwise} \\ -R_{ijk}, & \text{if } \mathbf{p}_i,\mathbf{p}_j,\mathbf{p}_k \text{ are oriented clockwise} \end{cases}$$

for every triangle $\triangle = \triangle_{(\mathbf{p}_i,\mathbf{p}_j,\mathbf{p}_k)}$. Let

$$\rho_\triangle \;=\; \big|\mathbf{D}^{ijk} - \mathbf{p}_i\big| \;=\; \big|\mathbf{D}^{ijk} - \mathbf{p}_j\big| \;=\; \big|\mathbf{D}^{ijk} - \mathbf{p}_k\big|$$

be the radius of the circle circumscribing $\triangle$. Since

$$\begin{aligned} F_{i\nu}(\mathbf{D}^{ijk}) \;&=\; m\left(e(\mathbf{D}^{ijk} - \mathbf{p}_i) - e(\mathbf{D}^{ijk} - \mathbf{p}_\nu)\right) \\ &=\; \tfrac{1}{2}\left(\rho_\triangle^2 - |\mathbf{D}^{ijk} - \mathbf{p}_\nu|^2\right), \end{aligned}$$

we have the geometric description of R_{ijk}

$$R_{ijk} \;=\; \frac{1}{\det\left(\mathbf{p}_j - \mathbf{p}_i,\, \mathbf{p}_k - \mathbf{p}_i\right)}\; \frac{2^{n-2}}{\displaystyle\prod_{\substack{\nu=1 \\ \nu \neq i,j,k}}^{n+1} \left(\rho_\triangle^2 - |\mathbf{D}^{ijk} - \mathbf{p}_\nu|^2\right)}\,.$$

THEOREM 3.2 *Let $n \geq 2$ and let $\mathbf{p}_1,\ldots.\mathbf{p}_{n+1}$ be points in $\mathbb{R}^2$ such that no three lie on a line and no four lie on a circle. Furthermore, assume that for every triple $1 \leq i,j,k \leq n+1$ of pairwise distinct indices the triangle $\triangle = \triangle_{(\mathbf{p}_i,\mathbf{p}_j,\mathbf{p}_k)}$ is not a right triangle and $\rho_\triangle \neq k_F$. Let $\mathcal{T}$ be the set of all triangles $\triangle$ with vertices in the set $\{\mathbf{p}_1,\ldots,\mathbf{p}_{n+1}\}$, and let $\mathcal{T}_{\mathrm{ac}}$ be the set of those triangles in $\mathcal{T}$ that have no angle bigger than $90°$, i.e. acute triangles. Then,*

$$J(\mathbf{p}_1,\ldots,\mathbf{p}_{n+1}) \;=\; m^n \sum_{\substack{\triangle \in \mathcal{T}_{\mathrm{ac}} \\ \rho_\triangle > k_F}} R_\triangle \;-\; 2m^n \sum_{\substack{\triangle \in \mathcal{T} \\ s \text{ edge of } \triangle \\ |s| > 2k_F}} \epsilon_{\triangle,s}\, R_\triangle \operatorname{arccot}\sqrt{\frac{4\rho_\triangle^2 - |s|^2}{|s|^2 - 4k_F^2}}\,.$$

Here, for an edge s of a triangle $\triangle$ we denote by $|s|$ the length of s and set

$$\epsilon_{\triangle,s} \;=\; \begin{cases} +1 & \text{if the angle of } \triangle \text{ at the vertex opposite to } s \text{ is acute,} \\ -1 & \text{if the angle of } \triangle \text{ at the vertex opposite to } s \text{ is obtuse.} \end{cases}$$

COROLLARY 3.3   *(i) If $|\mathbf{p}_i - \mathbf{p}_j| < 2k_F$ for all $i, j = 1, \ldots, n+1$, then*

$$J(\mathbf{p}_1, \ldots, \mathbf{p}_{n+1}) \;=\; m^n \sum_{\substack{\triangle \in \mathcal{T}_{\mathrm{ac}} \\ \rho_\triangle > k_F}} R_\triangle \,.$$

If, in particular, $k_F < k_F'$ such that $k_F > \frac{1}{2} \min\limits_{k=1,\ldots,n+1} |\mathbf{p}_i - \mathbf{p}_j|$ and such that there is no triangle $\triangle \in \mathcal{T}_{\mathrm{ac}}$ with $k_F < \rho_\triangle < k_F'$, then

$$J(\mathbf{p}_1, \ldots, \mathbf{p}_{n+1}; k_F) \;=\; J(\mathbf{p}_1, \ldots, \mathbf{p}_{n+1}; k_F') \,.$$

(ii) If all $n+1$ disks with radius k_F around the points $\mathbf{p}_1, \ldots, \mathbf{p}_{n+1}$ have at least one point in common, then

$$J(\mathbf{p}_1, \ldots, \mathbf{p}_{n+1}) \;=\; 0 \,.$$

(iii) Suppose $n = 2$ and $|\mathbf{p}_i - \mathbf{p}_j| < 2k_F$ for all $i, j = 1, 2, 3$. Let $\triangle = \triangle_{(\mathbf{p}_1, \mathbf{p}_2, \mathbf{p}_3)}$. Then,

$$J(\mathbf{p}_1, \mathbf{p}_2, \mathbf{p}_3) \;=\; \begin{cases} 0 \,, & \text{if the disks of radius } k_F \text{ around } \mathbf{p}_1, \mathbf{p}_2, \mathbf{p}_3 \\ & \text{have at least one point in common,} \\[4pt] \dfrac{m^2}{2 \times \text{area of } \triangle} \,, & \text{otherwise.} \end{cases}$$

Proof of the Corollary. Part (i) follows immediately from Theorem 3.2. If $\triangle = \triangle_{(\mathbf{p}_1, \mathbf{p}_2, \mathbf{p}_3)}$ is a triangle with only acute angles and the disks around the points $\mathbf{p}_1, \mathbf{p}_2, \mathbf{p}_3$ with radius k_F have at least one point in common, then $\rho_\triangle \leq k_F$. So, part (ii) follows from part (i). Finally, part (iii) is a special case of (i) and (ii). $\qquad\square$

The rest of this section is devoted to the proof of Theorem 3.2. For each pair of indices $1 \leq i \neq j \leq n+1$, let ω_{ij} be the unique meromorphic differential form on $\overline{L}_{ij}$ ($=$ projective closure of L_{ij}) that is holomorphic outside the points $\mathbf{D}^{ijk}$, $k = 1, \ldots, n+1$, $k \neq i, j$, and $\overline{L}_{ij} \cap L_\infty$, that has a simple pole with residue R_{ijk} at the point $\mathbf{D}^{ijk}$, and that has a zero of order $n-3$ (respectively, a simple pole for $n = 2$) at $\overline{L}_{ij} \cap L_\infty$. By construction,

$$\omega_{ij} = -\omega_{ji} \,.$$

If $0 \leq k_F \leq \frac{1}{2}|\mathbf{p}_i - \mathbf{p}_j|$, let $C_{ij}^{\pm}$ be the intersection of the line L_{ij} with

$$\mathcal{Q}_i^{\pm}(k_F) = \left\{ z \in \mathbb{C}^2 \,\middle|\, (z_1 - \mathbf{p}_{i1})^2 + (z_2 - \mathbf{p}_{i2})^2 = k_F^2 \text{ and } \pm \Im \frac{z_2 - \mathbf{p}_{i2}}{z_1 - \mathbf{p}_{i1}} > 0 \right\} \,.$$

Observe that $C_{ij}^{+} = C_{ji}^{-}$. It is useful to introduce the notation

$$p = \big((0, \mathbf{p}_1), \ldots, (0, \mathbf{p}_{n+1}) \big) \in \big(\mathbb{R} \times \mathbb{R}^2 \big)^{n+1}$$

and

$$\mathcal{M} = \left\{ q = ((q_{10}, \mathbf{q}_1), \ldots, (q_{n+10}, \mathbf{q}_{n+1})) \in \left(\mathbb{R} \times \mathbb{R}^2\right)^{n+1} \,\middle|\, q_{i0} \neq q_{j0} \text{ for } i \neq j \right\}.$$

Now, by Theorem 3.1,

$$J(\mathbf{p}_1, \ldots, \mathbf{p}_{n+1}) = \lim_{\substack{q \to p \\ q \in \mathcal{M}}} \frac{m^n}{2\pi\imath} \sum_{\substack{i,j=1 \\ i \neq j}}^{n+1} \int_{\left\{ c_{ij}^+(s) \,\mid\, 0 < s < k_F \right\}} \eta_{ij}.$$

It is easy to see that

$$\lim_{\substack{q \to p \\ q \in \mathcal{M}}} \eta_{ij} = \omega_{ij},$$

pointwise. However, it takes some work to determine the limiting behavior

$$\lim_{\substack{q \to p \\ q \in \mathcal{M}}} \left\{ c_{ij}^+(s) \,\mid\, 0 < s < k_F \right\}$$

of the path $\left\{ c_{ij}^+(s) \,\mid\, 0 < s < k_F \right\}$. There are two cases, $0 \leq s < \frac{1}{2}|\mathbf{p}_i - \mathbf{p}_j|$ and $k_F \geq s \geq \frac{1}{2}|\mathbf{p}_i - \mathbf{p}_j|$. Suppose $0 \leq s < \frac{1}{2}|\mathbf{p}_i - \mathbf{p}_j|$ for some $1 \leq i, j \leq n+1$. Then, the line L_{ij} intersects

$$\mathcal{Q}_i^{\pm}(s) = \left\{ z \in \mathbb{C}^2 \,\middle|\, (z_1 - \mathbf{p}_{i1})^2 + (z_2 - \mathbf{p}_{i2})^2 = s^2 \text{ and } \pm \Im \tfrac{z_2 - \mathbf{p}_{i2}}{z_1 - \mathbf{p}_{i1}} > 0 \right\}$$

in exactly one point which is denoted by $C_{ij}^{\pm}(s)$. Again, $C_{ij}^+(s) = C_{ji}^-(s)$ and

$$C_{ij}^{\pm}(s) = \lim_{\substack{q \to p \\ q \in \mathcal{M}}} c_{ij}^{\pm}(s).$$

Also,

$$\Re\, C_{ij}^{\pm}(s) = \tfrac{1}{2}(\mathbf{p}_i + \mathbf{p}_j).$$

Suppose that $s > \frac{1}{2}|\mathbf{p}_i - \mathbf{p}_j|$. Then,

$$L_{ij} \cap \left\{ \mathbf{x} \in \mathbb{R}^2 \,\mid\, |\mathbf{x} - \mathbf{p}_i| = s \right\} = L_{ij} \cap \left\{ \mathbf{x} \in \mathbb{R}^2 \,\mid\, |\mathbf{x} - \mathbf{p}_j| = s \right\}$$

consists of two points which we again denote by $C_{ij}^{\pm}(s)$. By convention, $C_{ij}^+(s)$ is the point that makes the triangle with vertices $\mathbf{p}_i$, $\mathbf{p}_j$ and $C_{ij}^+(s)$ counter clockwise oriented, and $C_{ij}^-(s)$ is the point that makes the triangle with vertices $\mathbf{p}_i$, $\mathbf{p}_j$ and $C_{ij}^-(s)$ clockwise oriented.

LEMMA 3.4   *Let $1 \leq i, j \leq n+1$.*

(i) *Suppose that $0 \leq s < \frac{1}{2}|\mathbf{p}_i - \mathbf{p}_j|$. Let $\mathbf{x} \neq \frac{1}{2}(\mathbf{p}_i + \mathbf{p}_j)$ in $L_{jk} \cap \mathbb{R}^2$ be a real point on the perpendicular bisector of $\mathbf{p}_i$ and $\mathbf{p}_j$. Then, the triangle in the complex line L_{ij} with the vertices $C_{ij}^+(s)$, $C_{ij}^-(s)$ and $\mathbf{x}$ has the same orientation as the triangle in $\mathbb{R}^2$ with the vertices $\mathbf{p}_i$, $\mathbf{p}_j$ and $\mathbf{x}$.*

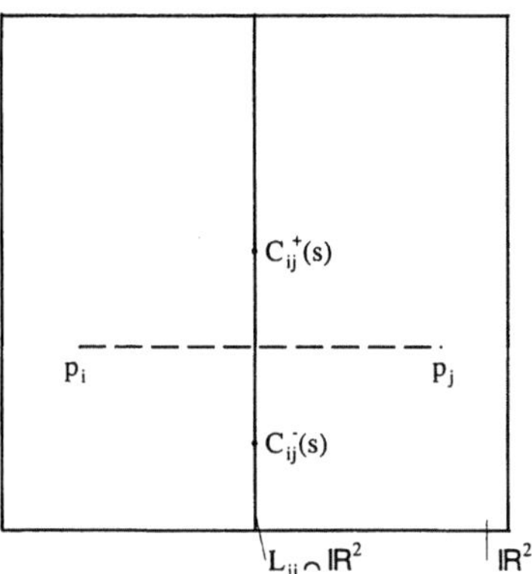

Figure 4

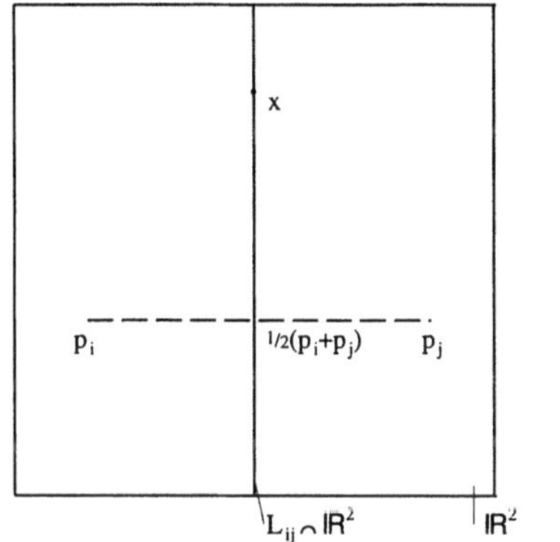

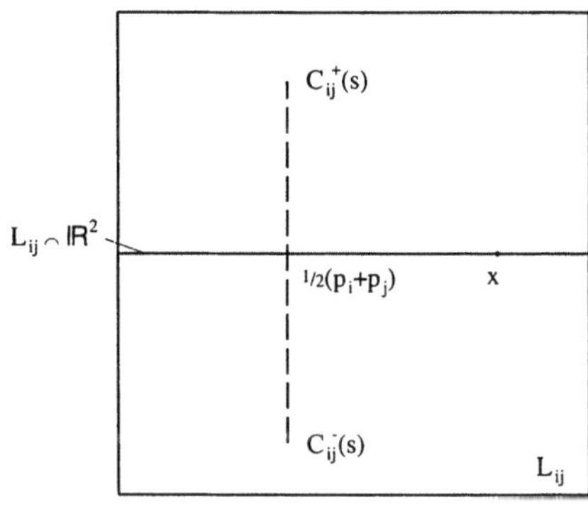

Figure 5

(ii) Suppose that $s > \frac{1}{2}|\mathbf{p}_i - \mathbf{p}_j|$. Then, $C_{ij}^{\pm}(s) = C_{ji}^{\mp}(s)$ and

$$
\begin{aligned}
C_{ij}^{\pm}(s) &= \lim_{\substack{q \to p \\ q \in \mathcal{M} \\ q_{i0}-q_{j0}>0}} c_{ij}^{\pm}(s) = \lim_{\substack{q \to p \\ q \in \mathcal{M} \\ q_{i0}-q_{j0}>0}} c_{ji}^{\pm}(s) \\
&= \lim_{\substack{q \to p \\ q \in \mathcal{M} \\ q_{i0}-q_{j0}<0}} c_{ji}^{\mp}(s) = \lim_{\substack{q \to p \\ q \in \mathcal{M} \\ q_{i0}-q_{j0}<0}} c_{ij}^{\mp}(s).
\end{aligned}
$$

Proof. For convenience, we set $m = 1$. We may assume, by applying an orientation preserving isometry, that

$$
\mathbf{p}_i = (-a, 0) \quad \text{and} \quad \mathbf{p}_j = (a, 0)
$$

with $a > 0$. Now,

$$L_{ij} = \{ (0, z_2) \mid z_2 \in \mathbb{C} \}$$

and

$$\mathcal{Q}_i^{\pm}(s) = \left\{ z \in \mathbb{C}^2 \mid (z_1 + a)^2 + z_2^2 = s^2 \text{ and } \pm \Im \tfrac{z_2}{z_1 + a} > 0 \right\}.$$

If $s < a$, then the first statement of Lemma 3.4 follows from

$$C_{ij}^{\pm}(s) = \left(0, \pm i \sqrt{a^2 - s^2} \right).$$

If, on the other hand, $s > a$,

$$C_{ij}^{\pm}(s) = \left(0, \pm \sqrt{s^2 - a^2} \right) = C_{ji}^{\mp}(s).$$

Suppose $q \in \mathcal{M}$ is close to p. Then, by a q-dependent orientation preserving isometry close to the identity, we may assume that

$$\mathbf{q}_i = (-a', 0) \quad \text{and} \quad \mathbf{q}_j = (a', 0)$$

with $a' > 0$ close to a. Now,

$$\ell_{ij} = \left\{ z \in \mathbb{C}^2 \mid z_1 = -i \tfrac{q_{i0} - q_{j0}}{2a'} \right\},$$

and the second statement of Lemma 3.4 follows directly from

$$c_{ij}^{\pm}(s) = \left(-i \tfrac{q_{i0} - q_{j0}}{2a'} , \; \pm \operatorname{sgn}(q_{i0} - q_{j0}) \sqrt{s^2 - a'^2 + \tfrac{(q_{i0} - q_{j0})^2}{4a'^2}} + i(q_{i0} - q_{j0}) \right),$$

$$c_{ji}^{\pm}(s) = \left(-i \tfrac{q_{i0} - q_{j0}}{2a'} , \; \pm \operatorname{sgn}(q_{i0} - q_{j0}) \sqrt{s^2 - a'^2 + \tfrac{(q_{i0} - q_{j0})^2}{4a'^2}} - i(q_{i0} - q_{j0}) \right),$$

where we make a branch cut for the square root along the negative real axis. $\qquad\square$

If $0 \le k_F < \tfrac{1}{2} |\mathbf{p}^{(j)} - \mathbf{p}^{(k)}|$, then, by construction,

$$\lim_{\substack{q \in \mathcal{M} \\ q \to p}} \left\{ c_{ij}^{\pm}(s) \mid 0 \le s \le k_F \right\} = \left\{ C_{ij}^{\pm}(s) \mid 0 \le s \le k_F \right\}$$

is the straight line segment $[C_{ij}^{\pm}(0), C_{ij}^{\pm}]$ joining the point $C_{ij}^{\pm}(0)$ to the point $C_{ij}^{\pm} = C_{ij}^{\pm}(k_F)$ in L_{ij}.

If $k_F > \tfrac{1}{2} |\mathbf{p}^{(j)} - \mathbf{p}^{(k)}|$, then by Lemma 3.4 (ii),

$$\lim_{\substack{q \to p \\ q \in \mathcal{M} \\ \pm(q_{i0} - q_{j0}) > 0}} \left\{ c_{ij}^{+}(s) \mid 0 \le s \le k_F \right\} \cup \left\{ c_{ji}^{+}(s) \mid k_F \ge s \ge 0 \right\}$$

is the path obtained by composing the three paths:

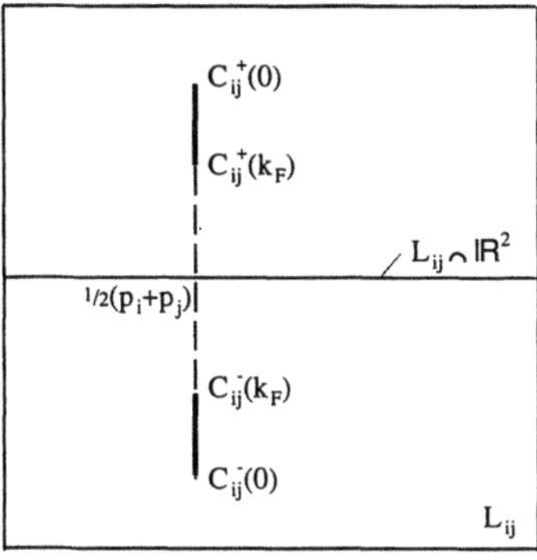

Figure 6

(1) the straight segment in L_{ij} joining $C_{ij}^+(0)$ to $\frac{1}{2}(\mathbf{p}_i + \mathbf{p}_j)$

(2) the segment on $L_{ij} \cap \mathbb{R}^2$ joining $\frac{1}{2}(\mathbf{p}_i + \mathbf{p}_j)$ to $C_{ij}^{\pm}(k_F)$, followed by the inverse of this segment

(3) the straight segment in L_{ij} joining $\frac{1}{2}(\mathbf{p}_i + \mathbf{p}_j)$ to $C_{ij}^-(0)$.

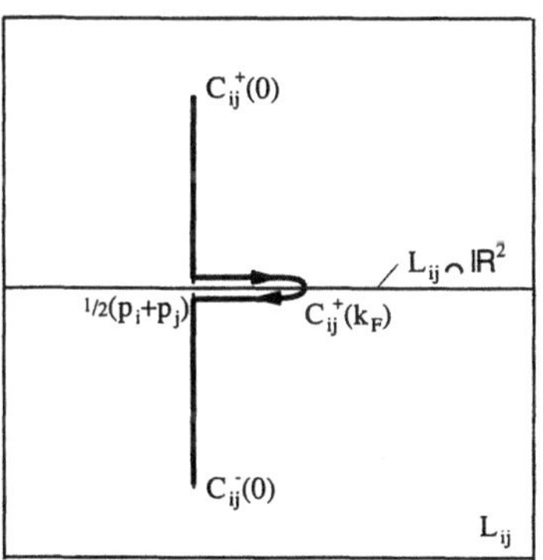

Figure 7

We encounter two problems in calculating

$$\lim_{\substack{q \to p \\ q \in \mathcal{M} \\ \pm(q_{i0} - q_{j0}) > 0}} \left(\int_{\{c_{ij}^+(s)\,|\,0<s<k_F\}} \eta_{ij} + \int_{\{c_{ji}^+(s)\,|\,0<s<k_F\}} \eta_{ji} \right)$$

$$= \lim_{\substack{q \to p \\ q \in \mathcal{M} \\ \pm(q_{i0} - q_{j0}) > 0}} \int_{\{c_{ij}^+(s)\,|\,0\leq s\leq k_F\} \cup \{c_{ji}^+(s)\,|\,k_F\geq s\geq 0\}} \eta_{ij}\,.$$

First, the path (2) depends on the sign of $q_{i0} - q_{j0}$. Second, the path (2) may contain one or more of the points $\mathbf{D}^{ijk}$, that are the poles of the limiting meromorphic form ω_{ij}.

Let $q \in \mathcal{M}$ be close to p. For all indices $1 \leq j \neq k \leq n+1$ with $k_F > \frac{1}{2}|\mathbf{p}_i - \mathbf{p}_j|$, let $[c_{ij}^+(k_F), c_{ji}^+(k_F)]$ be the straight line segment in ℓ_{ij} joining $c_{ij}^+(k_F)$ to $c_{ji}^+(k_F)$. The composition of the paths

$$\left\{ c_{ij}^+(s) \mid 0 \leq s \leq k_F \right\}, \quad [c_{ij}^+(k_F), c_{ji}^+(k_F)], \quad \left\{ c_{ji}^+(s) \mid k_F \geq s \geq 0 \right\}$$

is a path joining $c_{ij}^+(0)$ to $c_{ji}^+(0)$.

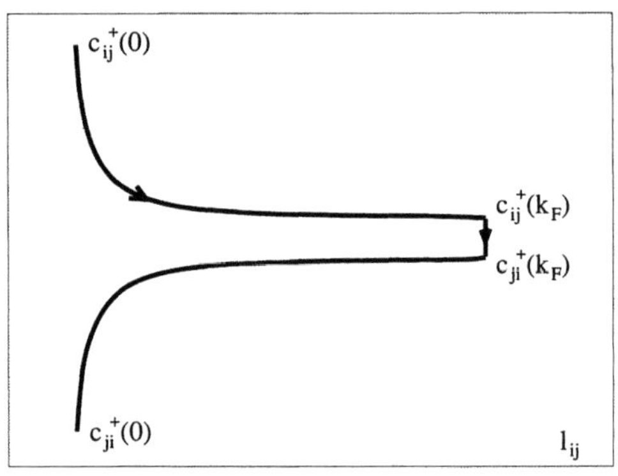

Figure 8

For any three different indices $1 \leq i, j, k \leq n+1$, recall that $\Re\, \mathbf{d}^{ijk}$ is the center of the circle circumscribing the triangle with the vertices $\mathbf{q}_i, \mathbf{q}_j, \mathbf{q}_k$. Let

$$\alpha_{ijk} = \tfrac{1}{2}\det(\mathbf{q}_i - \Re\, \mathbf{d}^{ijk}, \mathbf{q}_j - \Re\, \mathbf{d}^{ijk})$$

be the oriented area of the triangle with vertices $\mathbf{q}_i, \mathbf{q}_j, \Re\, \mathbf{d}^{ijk}$, and

$$A_{ijk} = \tfrac{1}{2}\det(\mathbf{q}_i - \mathbf{q}_k, \mathbf{q}_j - \mathbf{q}_k)$$

the oriented area of the full triangle $\mathbf{q}_i, \mathbf{q}_j, \mathbf{q}_k$.

For all $1 \leq i \neq j \leq n+1$, set

$$\begin{aligned}
\mathcal{I}_{ij} &= \left\{ 1 \leq k \leq n+1 \mid |\mathbf{D}^{ijk} - \mathbf{p}_i| < k_F \right\} \\
&= \left\{ 1 \leq k \leq n+1 \mid \rho_{\triangle(\mathbf{p}_i, \mathbf{p}_j, \mathbf{p}_k)} < k_F \right\}
\end{aligned}$$

$$\mathcal{I}_{ij}^+(q) = \left\{ k \in \mathcal{I}_{ij} \;\middle|\; \begin{array}{l} q_{i0} - q_{j0} > 0 \text{ and the triangle with vertices} \\ \mathbf{p}_i, \mathbf{p}_j, \mathbf{D}^{ijk} \text{ is counterclockwise oriented} \end{array} \right\}$$

$$\mathcal{I}_{ij}^-(q) = \left\{ k \in \mathcal{I}_{ij} \;\middle|\; \begin{array}{l} q_{i0} - q_{j0} < 0 \text{ and the triangle with vertices} \\ \mathbf{p}_i, \mathbf{p}_j, \mathbf{D}^{ijk} \text{ is clockwise oriented} \end{array} \right\}$$

Finally, recall that for any three different indices $1 \leq i,j,k \leq n+1$, r_{ijk} is the residue of η_{ij} at $\mathbf{d}^{ijk}$, and if $\triangle = \triangle_{(\mathbf{q}_i, \mathbf{q}_j, \mathbf{q}_k)}$ is the triangle with vertices $\mathbf{q}_i, \mathbf{q}_j, \mathbf{q}_k$, set

$$r_{\triangle} = \begin{cases} r_{ijk}, & \text{if } \mathbf{q}_i, \mathbf{q}_j, \mathbf{q}_k \text{ are oriented counter clockwise,} \\ -r_{ijk}, & \text{if } \mathbf{q}_i, \mathbf{q}_j, \mathbf{q}_k \text{ are oriented clockwise.} \end{cases}$$

LEMMA 3.5 *Let $q \in \mathcal{M}$ be close to p and suppose that for all pairwise distinct indices $1 \leq i,j,k \leq n+1$,*

$$\frac{\alpha_{ijk}}{A_{ijk}} \frac{q_{k0} - q_{i0}}{q_{i0} - q_{j0}} \neq \frac{\alpha_{kij}}{A_{kij}}.$$

(i) Fix $1 \leq i \neq j \leq n+1$ with $k_F > \frac{1}{2}|\mathbf{p}_i - \mathbf{p}_j|$. Then,

$$\int_{\{c_{ij}^+(s)|0 \leq s \leq k_F\}} \eta_{ij} + \int_{[c_{ij}^+(k_F), c_{ji}^+(k_F)]} \eta_{ij} + \int_{\{c_{ji}^+(s)|0 \leq s \leq k_F\}} \eta_{ji}$$

$$= \int_{c_{ij}^+(0)}^{c_{ji}^+(0)} \eta_{ij} - 2\pi i \sum_{\substack{k=1 \\ k \neq i,j}}^{n+1} \chi_{ijk}\, r_{ijk}.$$

Here, $\displaystyle\int_{c_{ij}^+(0)}^{c_{ji}^+(0)} \eta_{ij}$ is the integral along the straight line segment $[c_{ij}^+(0),$ $c_{ji}^+(0)]$ in ℓ_{ij}, and

$$\chi_{ijk} = \begin{cases} 0, & \text{unless } k \in \mathcal{I}_{ij}^+(q) \cup \mathcal{I}_{ij}^-(q) \text{ and} \\ & \frac{\alpha_{ijk}}{A_{ijk}} \frac{q_{k0}-q_{i0}}{q_{i0}-q_{j0}} < \frac{\alpha_{kij}}{A_{kij}} \text{ and } \frac{\alpha_{ijk}}{A_{ijk}} \frac{q_{i0}-q_{k0}}{q_{i0}-q_{j0}} < \frac{\alpha_{jki}}{A_{jki}}, \\ \pm 1, & \text{if } k \in \mathcal{I}_{ij}^{\pm}(q) \text{ and} \\ & \frac{\alpha_{ijk}}{A_{ijk}} \frac{q_{k0}-q_{i0}}{q_{i0}-q_{j0}} < \frac{\alpha_{kij}}{A_{kij}} \text{ and } \frac{\alpha_{ijk}}{A_{ijk}} \frac{q_{j0}-q_{k0}}{q_{i0}-q_{j0}} < \frac{\alpha_{jki}}{A_{jki}}. \end{cases}$$

(ii) Let $1 \leq i,j,k \leq n+1$ be any three different indices and let $\triangle = \triangle_{(\mathbf{q}_i, \mathbf{q}_j, \mathbf{q}_k)}$. Then,

$$\sum_{\{i',j',k'\}=\{i,j,k\}} \chi_{i'j'k'}\, r_{i'j'k'} = \begin{cases} 2r_{\triangle}, & \text{if } \rho_{\triangle} < k_F \text{ and all angles in the} \\ & \text{triangle } \triangle \text{ are smaller than } 90^{\circ}, \\ 0, & \text{otherwise.} \end{cases}$$

Lemma 3.5 is now used to obtain an intermediate expression for the function $J(\mathbf{p}_1, \ldots, \mathbf{p}_{n+1})$ appearing in Theorem 3.2. Recall that $\mathbf{p}_1, \ldots, \mathbf{p}_{n+1}$ are points in $\mathbb{R}^2$ such that no three of them lie on a line and no four of them lie on a circle, and furthermore, that for every triple $1 \leq i,j,k \leq n+1$ of pairwise distinct indices the triangle $\triangle = \triangle_{(\mathbf{p}_i, \mathbf{p}_j, \mathbf{p}_k)}$ is not a right triangle and $\rho_{\triangle} \neq k_F$. Here, as above, $\rho_{\triangle}$ is the radius of the circle circumscribing the triangle $\triangle$.

PROPOSITION 3.6

$$J(\mathbf{p}_1,\ldots,\mathbf{p}_{n+1}) = \frac{m^n}{4\pi\imath} \sum_{1\le i\ne j\le n+1} \int_{C_{ij}^+(0)}^{C_{ij}^-(0)} \omega_{ij}$$

$$+ \frac{m^n}{4\pi\imath} \sum_{\substack{1\le i\ne j\le n+1 \\ |\mathbf{p}_i-\mathbf{p}_j|>2k_F}} \int_{C_{ij}^-}^{C_{ij}^+} \omega_{ij} \;-\; m^n \sum_{\substack{\triangle\in T_{ac} \\ \rho_\triangle<k_F}} R_\triangle\,.$$

Proof. By Theorem 3.1 and Lemma 3.5 (i),

$$\frac{2}{m^n} I(q_1,\ldots,q_{n+1})$$

$$= \frac{1}{2\pi\imath} \sum_{1\le i\ne j\le n+1} \left(\int_{\{c_{ij}^+(s)\,|\,0<s<k_F\}} \eta_{ij} + \int_{\{c_{ji}^+(s)\,|\,0<s<k_F\}} \eta_{ji} \right)$$

$$= \frac{1}{2\pi\imath} \sum_{\substack{1\le i\ne j\le n+1 \\ |\mathbf{p}_i-\mathbf{p}_j|>2k_F}} \left(\int_{c_{ij}^+(0)}^{c_{ij}^+(k_F)} \eta_{ij} + \int_{c_{ji}^+(0)}^{c_{ji}^+(k_F)} \eta_{ji} \right)$$

$$+ \frac{1}{2\pi\imath} \sum_{\substack{1\le i\ne j\le n+1 \\ |\mathbf{p}_i-\mathbf{p}_j|<2k_F}} \left(\int_{c_{ij}^+(0)}^{c_{ji}^+(0)} \eta_{ij} - \int_{[c_{ij}^+(k_F),\,c_{ji}^+(k_F)]} \eta_{ij} \right)$$

$$- \sum_{1\le i\ne j\le n+1} \sum_{\substack{k=1 \\ k\ne i,j}}^{n+1} \chi_{ijk}\, r_{ijk}$$

on an open, dense subset of $\mathcal{M}$. By Lemma 3.5 (ii),

$$\frac{2}{m^n} I(q_1,\ldots,q_{n+1})$$

$$= \frac{1}{2\pi\imath} \sum_{\substack{1\le i\ne j\le n+1 \\ |\mathbf{p}_i-\mathbf{p}_j|>2k_F}} \left(\int_{c_{ij}^+(0)}^{c_{ij}^+(k_F)} \eta_{ij} + \int_{c_{ji}^+(0)}^{c_{ji}^+(k_F)} \eta_{ji} \right)$$

$$+ \frac{1}{2\pi\imath} \sum_{\substack{1\le i\ne j\le n+1 \\ |\mathbf{p}_i-\mathbf{p}_j|<2k_F}} \left(\int_{c_{ij}^+(0)}^{c_{ji}^+(0)} \eta_{ij} - \int_{[c_{ij}^+(k_F),\,c_{ji}^+(k_F)]} \eta_{ij} \right)$$

$$- 2 \sum_{\substack{\triangle\in T_{ac} \\ \rho_\triangle<k_F}} r_\triangle$$

on an open, dense subset of $\mathcal{M}$. In fact the last identity holds for all q in $\mathcal{M}$ near p, since both sides are continuous functions on $\mathcal{M}$ near p. Finally, by

Lemma 3.4,

$$\frac{2}{m^n} J(\mathbf{p}_1,\ldots,\mathbf{p}_{n+1}) = 2 \lim_{\substack{q \in \mathcal{M} \\ q \to p}} I(q_1,\ldots,q_{n+1})$$

$$= \frac{1}{2\pi i} \sum_{\substack{1 \le i \ne j \le n+1 \\ |\mathbf{p}_i - \mathbf{p}_j| > 2k_F}} \left(\int_{C_{ij}^+(0)}^{C_{ij}^+(k_F)} \omega_{ij} - \int_{C_{ji}^+(0)}^{C_{ji}^+(k_F)} \omega_{ij} \right)$$

$$+ \frac{1}{2\pi i} \sum_{\substack{1 \le i \ne j \le n+1 \\ |\mathbf{p}_i - \mathbf{p}_j| < 2k_F}} \int_{C_{ij}^+(0)}^{C_{ji}^+(0)} \omega_{ij} - 2 \sum_{\substack{\triangle \in \mathcal{T}_{\mathrm{ac}} \\ \rho_\triangle < k_F}} R_\triangle$$

$$= \frac{1}{2\pi i} \sum_{\substack{1 \le i \ne j \le n+1 \\ |\mathbf{p}_i - \mathbf{p}_j| > 2k_F}} \left(\int_{C_{ij}^+(0)}^{C_{ji}^+(0)} \omega_{ij} - \int_{C_{ij}^+(k_F)}^{C_{ji}^+(k_F)} \omega_{ij} \right)$$

$$+ \frac{1}{2\pi i} \sum_{\substack{1 \le i \ne j \le n+1 \\ |\mathbf{p}_i - \mathbf{p}_j| < 2k_F}} \int_{C_{ij}^+(0)}^{C_{ji}^+(0)} \omega_{ij} - 2 \sum_{\substack{\triangle \in \mathcal{T}_{\mathrm{ac}} \\ \rho_\triangle < k_F}} R_\triangle \qquad \square$$

Proof of Lemma 3.5. For convenience, we again set $m = 1$. To prove part (i), let $\gamma_{ij}(q)$ be the closed path obtained by composing $\{c_{ij}^+(s) \mid 0 \le s \le k_F\}$, $[c_{ij}^+(k_F), c_{ji}^+(k_F)]$, $\{c_{ji}^+(s) \mid k_F \ge s \ge 0\}$ and $[c_{ji}^+(0), c_{ij}^+(0)]$. We have

$$\int_{\{c_{ij}^+(s) \mid 0 \le s \le k_F\}} \eta_{ij} + \int_{[c_{ij}^+(k_F), c_{ji}^+(k_F)]} \eta_{ij} + \int_{\{c_{ji}^+(s) \mid 0 \le s \le k_F\}} \eta_{ji}$$

$$= \int_{c_{ij}^+(0)}^{c_{ji}^+(0)} \eta_{ij} + \int_{\gamma_{ij}(q)} \eta_{ij} ,$$

since, $\eta_{ij} = -\eta_{ji}$. By the residue theorem

$$\int_{\gamma_{ij}(q)} \eta_{ij} = 2\pi i \sum_{\substack{k=1 \\ k \ne i,j}}^{n+1} (\text{winding number of } \gamma_{ij} \text{ around } \mathbf{d}^{ijk}) \times \mathrm{res}_{\eta_{ij}}(\mathbf{d}^{ijk})$$

$$= 2\pi i \sum_{\substack{k=1 \\ k \ne i,j}}^{n+1} (\text{winding number of } \gamma_{ij} \text{ around } \mathbf{d}^{ijk}) \, r_{ijk} .$$

We shall show that the winding number of $\gamma_{ij}(q)$ around $\mathbf{d}^{ijk}$ is $-\chi_{ijk}$.

Fix $1 \le k \le n+1$, $k \ne i,j$. A necessary condition that the winding number of $\gamma_{ij}(q)$ around $\mathbf{d}^{ijk}$ is not zero is that $\mathbf{D}^{ijk}$ be a point on the limiting path (see Figure 8)

$$\lim_{\substack{q' \to p \\ q' \in \mathcal{M} \\ \mathrm{sgn}(q'_{i0} - q'_{j0}) = \mathrm{sgn}(q_{i0} - q_{j0})}} \gamma_{ij}(q') .$$

By Lemma 3.4 (ii), our criterion becomes

$$\mathbf{D}^{ijk} \in \begin{cases} \left[\tfrac{1}{2}(\mathbf{p}_i + \mathbf{p}_j),\, C_{ij}^+(k_F)\right], & q_{i0} - q_{j0} > 0, \\ \left[\tfrac{1}{2}(\mathbf{p}_i + \mathbf{p}_j),\, C_{ij}^-(k_F)\right], & q_{i0} - q_{j0} < 0, \end{cases}$$

since $\mathbf{D}^{ijk}$ is real. Observe that, by construction, the distance between $\mathbf{D}^{ijk}$ and the midpoint $\tfrac{1}{2}(\mathbf{p}_i + \mathbf{p}_j)$ is smaller than the distance between $C_{ij}^\pm(k_F)$ and $\tfrac{1}{2}(\mathbf{p}_i + \mathbf{p}_j)$ if and only if $|\mathbf{D}^{ijk} - \mathbf{p}_i| < k_F$. Also, observe that the point $\mathbf{D}^{ijk}$ lies on the same side of $\tfrac{1}{2}(\mathbf{p}_i + \mathbf{p}_j)$ as $C_{ij}^+(k_F)$ if and only if the triangle with vertices $\mathbf{p}_i$, $\mathbf{p}_j$, $\mathbf{D}^{ijk}$ is counter clockwise oriented, and on the same side of $\tfrac{1}{2}(\mathbf{p}_i + \mathbf{p}_j)$ as $C_{ij}^-(k_F)$ if and only if the triangle with vertices $\mathbf{p}_i$, $\mathbf{p}_j$, $\mathbf{D}^{ijk}$ is clockwise oriented. Thus, the winding number of $\gamma_{ij}(q)$ around $\mathbf{d}^{ijk}$ is zero, when $k \notin \mathcal{I}_{ij}^+(q) \cup \mathcal{I}_{ij}^-(q)$.

Now, suppose that $k \in \mathcal{I}_{ij}^+(q) \cup \mathcal{I}_{ij}^-(q)$. As in the proof of Lemma 3.4 we may assume that

$$\mathbf{p}_i = (-a, 0) \quad \text{and} \quad \mathbf{p}_j = (a, 0)$$

with $a > 0$ and

$$\mathbf{q}_i = (-a', 0) \quad \text{and} \quad \mathbf{q}_j = (a', 0)$$

with a' close to a. We have

$$\mathbf{d}^{ijk} = \left(-i\, \frac{q_{i0} - q_{j0}}{2a'}\, ,\; \frac{\left(|\mathbf{q}_k|^2 - a'^2\right)}{2\mathbf{q}_{k2}} + i\, \frac{\left(q_{k0} + \frac{\mathbf{q}_{k1} - a'}{2a'} q_{i0} - \frac{\mathbf{q}_{k1} + a'}{2a'} q_{j0}\right)}{\mathbf{q}_{k2}} \right).$$

The function

$$g(t) = \left(-i\, \frac{q_{i0} - q_{j0}}{2a'}\, ,\; \frac{\left(|\mathbf{q}_k|^2 - a'^2\right)}{2\mathbf{q}_{k2}} + i\,t \right),$$

defined for $t \in \mathbb{R}$, parameterizes the real line in ℓ_{ij} passing through $\mathbf{d}^{ijk}$ parallel to the imaginary z_2-axis. By construction $g(t_0) = \mathbf{d}^{ijk}$, where

$$t_0 = \frac{1}{\mathbf{q}_{k2}} \left(q_{k0} + \frac{\mathbf{q}_{k1} - a'}{2a'} q_{i0} - \frac{\mathbf{q}_{k1} + a'}{2a'} q_{j0} \right).$$

It is easy to see that

$$t_i = \frac{\mathbf{q}_{k2}}{|\mathbf{q}_k|^2 - a'^2} \left(q_{i0} - q_{j0} \right)$$

is the only real number for which there exists an $s > 0$ such that $g(t) \in Q_i^+(s)$. Similarly,

$$t_j = -\frac{\mathbf{q}_{k2}}{|\mathbf{q}_k|^2 - a'^2} \left(q_{i0} - q_{j0} \right)$$

is the only real number for which there exists an $s > 0$ such that $g(t) \in Q_j^+(s)$. Our hypothesis (see, the statement of Theorem 3.2) implies that $\mathbf{q}_i$, $\mathbf{q}_j$ and $\mathbf{q}_k$ are not the vertices of a right triangle. By Thales' theorem (see, Euclid, Book III, §31), the denominator $|\mathbf{q}_k|^2 - a'^2$ does not vanish and

$$\frac{\mathbf{q}_{k\,2}}{|\mathbf{q}_k|^2 - a'^2} > 0,$$

when $k \in \mathcal{I}_{ij}^+(q)$, and

$$\frac{\mathbf{q}_{k\,2}}{|\mathbf{q}_k|^2 - a'^2} < 0,$$

when $k \in \mathcal{I}_{ij}^-(q)$. Consequently,

$$t_j < 0 < t_i,$$

since $k \in \mathcal{I}_{ij}^+(q) \cup \mathcal{I}_{ij}^-(q)$. The line $g(t)$, $-\infty < t < \infty$, meets the loop γ_{ij} in exactly the two points $g(t_i)$ and $g(t_j)$. Furthermore, the winding number of γ_{ij} around $\mathbf{d}^{ijk} = g(t_0)$ is

$$\begin{cases} 0, & \text{if } t_i < t_0 \text{ or } t_0 < t_j, \\ \mp 1, & \text{if } t_j < t_0 < t_i \text{ and } k \in \mathcal{I}_{ij}^\pm(q). \end{cases}$$

The first case is immediate from Figure 9

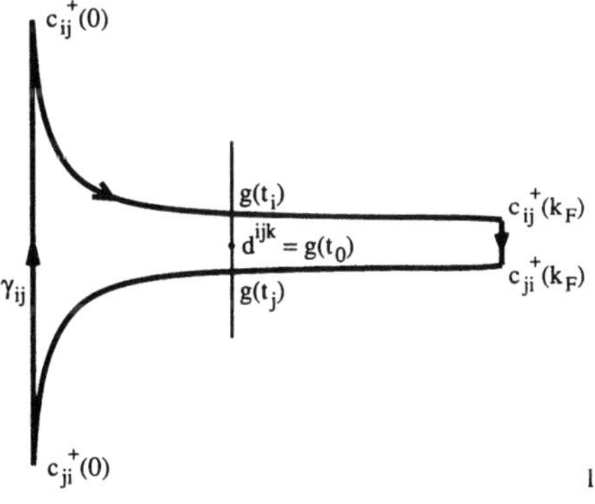

Figure 9

If, on the other hand, $t_j < t_0 < t_i$, then the winding number is -1 when the triangle with vertices $c_{ij}^+(0)$, $c_{ji}^+(0)$, $\mathbf{d}^{ijk}$ is counter clockwise oriented. This occurs, if and only if the limiting triangle with vertices $C_{ij}^+(0)$, $C_{ij}^-(0) = C_{ji}^+(0)$, $\mathbf{D}^{ijk}$ is counter clockwise oriented. By Lemma 3.4 (i), this happens if and only if the triangle with vertices $\mathbf{p}_i$, $\mathbf{p}_j$, $\mathbf{D}^{ijk}$ is counter clockwise oriented. That is, $k \in \mathcal{I}_{ij}^+(q)$. The case of winding number $+1$ is similar.

The last step in the proof of part (i) is to express the conditions on t_i, t_0 and t_j that determine the winding number of γ_{ij} around $\mathbf{d}^{ijk}$ in terms of the quantities α_{ijk} and A_{ijk}. To start with,

$$t_0 = \frac{1}{\mathbf{q}_{k2}}\left(q_{k0} + \frac{\mathbf{q}_{k1} - a'}{2a'}q_{i0} - \frac{\mathbf{q}_{k1} + a'}{2a'}q_{j0}\right) < \frac{\mathbf{q}_{k2}}{|\mathbf{q}_k|^2 - a'^2}(q_{i0} - q_{j0}) = t_i$$

is equivalent to

$$\frac{\alpha_{ijk}}{A_{ijk}}\frac{1}{q_{i0} - q_{j0}}\left(q_{k0} + \frac{\mathbf{q}_{k1} - a'}{2a'}q_{i0} - \frac{\mathbf{q}_{k1} + a'}{2a'}q_{j0}\right) < \frac{1}{2},$$

since

$$\alpha_{ijk} = \frac{a'}{2\mathbf{q}_{k2}}\left(|\mathbf{q}_k|^2 - a'^2\right), \quad A_{ijk} = a'\,\mathbf{q}_{k2},$$

and $t_i > 0$. This inequality is in turn equivalent to

$$\frac{\alpha_{ijk}}{A_{ijk}}\frac{q_{k0} - q_{i0}}{q_{i0} - q_{j0}} < \frac{1}{2} - \frac{\alpha_{ijk}}{A_{ijk}}\frac{1}{q_{i0} - q_{j0}}\left(q_{i0} + \frac{\mathbf{q}_{k1} - a'}{2a'}q_{i0} - \frac{\mathbf{q}_{k1} + a'}{2a'}q_{j0}\right)$$

$$= \frac{1}{2}\left(1 - \frac{\alpha_{ijk}}{A_{ijk}}\frac{\mathbf{q}_{k1} + a'}{a'}\right).$$

We will show that

$$\frac{1}{2}\left(1 - \frac{\alpha_{ijk}}{A_{ijk}}\frac{\mathbf{q}_{k1} + a'}{a'}\right) = \frac{\alpha_{ijk}}{A_{ijk}}.$$

In other words,

$$t_0 < t_i \Leftrightarrow \frac{\alpha_{ijk}}{A_{ijk}}\frac{q_{k0} - q_{i0}}{q_{i0} - q_{j0}} < \frac{\alpha_{kij}}{A_{kij}}.$$

Let $2e_\ell > 0$, $\ell = i, j, k$, be the length of the edge opposite to the vertex $\mathbf{q}_\ell$ in the triangle with vertices $\mathbf{q}_i$, $\mathbf{q}_j$, $\mathbf{q}_k$, and let h_ℓ be the (signed) height of the triangle over the edge opposite to the vertex $\mathbf{q}_\ell$. Clearly,

$$A_{ijk} = e_\ell \cdot h_\ell$$

for $\ell = i, j, k$. We also let s_ℓ be the (signed) height of $\Re\,\mathbf{d}^{ijk}$ over the edge opposite to the vertex $\mathbf{q}_\ell$. Clearly,

$$\alpha_{ijk} = e_k \cdot s_k = a' \cdot s_k,$$
$$\alpha_{jki} = e_i \cdot s_i,$$
$$\alpha_{kij} = e_j \cdot s_j.$$

Finally, let $\mathbf{b}$ be the point where the perpendicular bisector of $\mathbf{q}_i, \mathbf{q}_j$ meets the line through $\mathbf{q}_i$ and $\mathbf{q}_k$.

The two triangles in Figure 10 with vertex $\mathbf{b}$ and solid edges are similar.

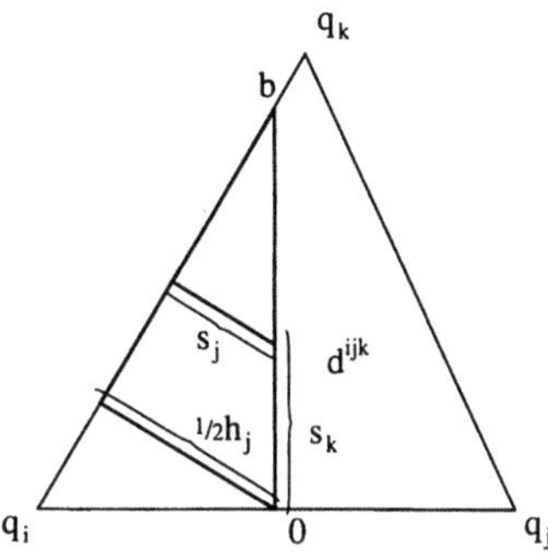

Figure 10

Therefore,

$$\frac{s_j}{\frac{1}{2}h_j} = \frac{\mathbf{b}_2 - s_k}{\mathbf{b}_2} = 1 - \frac{s_k}{\mathbf{b}_2}.$$

Also, the triangle with vertices $\mathbf{q}_i = (-a', 0), 0, \mathbf{b}$ is similar to the triangle with vertices $\mathbf{q}_i = (-a', 0), (\mathbf{q}_{k1}, 0), \mathbf{q}_k$. Therefore,

$$\frac{\mathbf{q}_{k2}}{\mathbf{b}_2} = \frac{a' + \mathbf{q}_{k1}}{a'}.$$

It follows that

$$\begin{aligned}
\frac{\alpha_{kij}}{A_{kij}} &= \frac{s_j}{h_j} = \frac{1}{2}\left(1 - \frac{s_k}{\mathbf{b}_2}\right) = \frac{1}{2}\left(1 - \frac{s_k}{\mathbf{q}_{k2}}\frac{a' + \mathbf{q}_{k1}}{a'}\right) \\
&= \frac{1}{2}\left(1 - \frac{\alpha_{ijk}}{A_{ijk}}\frac{a' + \mathbf{q}_{k1}}{a'}\right),
\end{aligned}$$

as claimed above.

By the same sort of argument,

$$t_j < t_0 \quad \Leftrightarrow \quad \frac{\alpha_{ijk}}{A_{ijk}}\frac{q_{j0} - q_{k0}}{q_{i0} - q_{j0}} < \frac{\alpha_{jki}}{A_{jki}}.$$

The proof of the first part of Lemma 3.5 is now complete.

(ii) If $k_F < \rho_\triangle$, then, by definition, $\chi_{i'j'k'} = 0$ for all permutations i', j', k' of i, j, k. We may therefore assume that $\rho_\triangle < k_F$. We may further assume that the triangle with vertices $\mathbf{q}_i, \mathbf{q}_j, \mathbf{q}_k$ is counter clockwise oriented since, by direct inspection, $\chi_{ijk} = -\chi_{jik}$ and, as observed before, $r_{ijk} = -r_{jik}$. It thus suffices to show that

$$\chi_{ijk} + \chi_{jki} + \chi_{kij} \;=\; \begin{cases} 1\,, & \text{if } \triangle \text{ is an acute triangle,} \\ 0\,, & \text{if } \triangle \text{ is an obtuse triangle.} \end{cases}$$

Clearly, one or two of the differences

$$q_{i0} - q_{j0}\,,\; q_{j0} - q_{k0}\,,\; q_{k0} - q_{i0}$$

is positive, since $\left(q_{i0} - q_{j0}\right) + \left(q_{j0} - q_{k0}\right) + \left(q_{k0} - q_{i0}\right) \;=\; 0$. We first verify the statement of the last paragraph in the case that $\triangle_{(\mathbf{q}_i, \mathbf{q}_j, \mathbf{q}_k)}$ is an acute triangle whose vertices $\mathbf{q}_i, \mathbf{q}_j, \mathbf{q}_k$ are counter clockwise oriented and exactly one of the differences is positive. Observe that $0 < A_{ijk} = A_{jki} = A_{kij}$ since $\mathbf{q}_i, \mathbf{q}_j, \mathbf{q}_k$ are counter clockwise oriented and that $\alpha_{ijk}, \alpha_{jki}, \alpha_{kij} > 0$ since $\triangle_{(\mathbf{q}_i, \mathbf{q}_j, \mathbf{q}_k)}$ is acute.

We may assume, without loss of generality, that

$$q_{i0} - q_{j0} > 0\,,\; q_{j0} - q_{k0} < 0\,,\; q_{k0} - q_{i0} < 0\,.$$

By construction, $i \notin \mathcal{I}_{jk}^{+}(q) \cup \mathcal{I}_{jk}^{-}(q)$ and $j \notin \mathcal{I}_{ki}^{+}(q) \cup \mathcal{I}_{ki}^{-}(q)$ and therefore, by definition,

$$\chi_{jki} \;=\; \chi_{kij} \;=\; 0\,.$$

On the other hand, $k \in \mathcal{I}_{ij}^{+}(q)$ and

$$\frac{\alpha_{ijk}}{A_{ijk}}\frac{q_{k0} - q_{i0}}{q_{i0} - q_{j0}} \;<\; 0 \;<\; \frac{\alpha_{kij}}{A_{kij}}\,,$$

$$\frac{\alpha_{ijk}}{A_{ijk}}\frac{q_{j0} - q_{k0}}{q_{i0} - q_{j0}} \;<\; 0 \;<\; \frac{\alpha_{jki}}{A_{jki}}\,,$$

so that, by definition,

$$\chi_{ijk} \;=\; 1\,.$$

We next treat the case in which $\triangle_{(\mathbf{q}_i, \mathbf{q}_j, \mathbf{q}_k)}$ is an acute triangle whose vertices $\mathbf{q}_i, \mathbf{q}_j, \mathbf{q}_k$ are counter clockwise oriented and exactly two of the differences are positive. We assume, without loss of generality, that

$$q_{i0} - q_{j0} > 0\,,\; q_{j0} - q_{k0} > 0\,,\; q_{k0} - q_{i0} < 0\,.$$

As above, the third inequality implies $\chi_{kij} = 0$. On the other hand, the first two inequalities imply $k \in \mathcal{I}_{ij}^{+}(q)$, $i \in \mathcal{I}_{jk}^{+}(q)$ and consequently,

$$\chi_{ijk} \;=\; \begin{cases} 1\,, & \text{if } \frac{\alpha_{ijk}}{A_{ijk}}\frac{q_{j0} - q_{k0}}{q_{i0} - q_{j0}} < \frac{\alpha_{jki}}{A_{jki}}\,, \\ 0\,, & \text{otherwise.} \end{cases}$$

$$\chi_{jki} \;=\; \begin{cases} 1 & \text{if } \frac{\alpha_{jki}}{A_{jki}}\frac{q_{i0} - q_{j0}}{q_{j0} - q_{k0}} < \frac{\alpha_{ijk}}{A_{ijk}}\,, \\ 0 & \text{otherwise.} \end{cases}$$

This shows that

$$\chi_{ijk} + \chi_{jki} = 1.$$

The proof that

$$\chi_{ijk} + \chi_{jki} + \chi_{kij} = 1,$$

when $\triangle$ is an accute, counter clockwise oriented triangle is now complete.

The case of an obtuse triangle is similar. $\qquad\square$

The proof of Theorem 3.2 requires two ingredients in addition to Proposition 3.6.

PROPOSITION 3.7 *For each pair* $1 \le i,j \le n+1$, $i \ne j$, *and* $k \ne i,j$ *let* ω_{ijk} *be the unique meromorphic differential form on* $\overline{L}_{ij}$ *that is holomorphic outside the points* $\mathbf{D}^{ijk}$, *and* $\overline{L}_{ij} \cap L_\infty$, *that has a simple pole with residue 1 at the point* $\mathbf{D}^{ijk}$, *and that has a simple pole at* $\overline{L}_{ij} \cap L_\infty$. *If* $|\mathbf{p}_i - \mathbf{p}_j| > 2k_F$ *then*

$$\int_{C_{ij}^-}^{C_{ij}^+} \omega_{ijk} = -4\pi\imath\,\epsilon_{\triangle,(\mathbf{p}_i,\mathbf{p}_j)}\,R_\triangle\,\mathrm{arccot}\sqrt{\frac{4\rho_\triangle^2 - |\mathbf{p}_i-\mathbf{p}_j|^2}{|\mathbf{p}_i-\mathbf{p}_j|^2 - 4k_F^2}},$$

where $\triangle = \triangle(\mathbf{p}_i,\mathbf{p}_j,\mathbf{p}_k)$ *and*

$$\epsilon_{\triangle,(\mathbf{p}_i,\mathbf{p}_j)} = \begin{cases} +1 & \text{if the angle of } \triangle \text{ at } \mathbf{p}_k \text{ is acute,} \\ -1 & \text{if the angle of } \triangle \text{ at } \mathbf{p}_k \text{ is obtuse.} \end{cases}$$

PROPOSITION 3.8

$$\sum_{1 \le i \ne j \le n+1} \int_{C_{ij}^+(0)}^{C_{ij}^-(0)} \omega_{ij} = 4\pi i \sum_{\triangle \in \mathcal{T}_{\mathrm{ac}}} R_\triangle.$$

Before proving Proposition 3.7 and Proposition 3.8, we give the

Proof of Theorem 3.2. By Proposition 3.6 and Proposition 3.8,

$$\frac{1}{m^n} J(\mathbf{p}_1, \ldots, \mathbf{p}_{n+1})$$

$$= \frac{1}{4\pi\imath} \sum_{1 \le i \ne j \le n+1} \int_{C_{ij}^+(0)}^{C_{ij}^-(0)} \omega_{ij} + \frac{1}{4\pi\imath} \sum_{\substack{1 \le i \ne j \le n+1 \\ |\mathbf{p}_i - \mathbf{p}_j| > 2k_F}} \int_{C_{ij}^-}^{C_{ij}^+} \omega_{ij} - \sum_{\substack{\triangle \in \mathcal{T}_{\mathrm{ac}} \\ \rho_\triangle < k_F}} R_\triangle$$

$$= \sum_{\triangle \in \mathcal{T}_{\mathrm{ac}}} R_\triangle + \frac{1}{4\pi\imath} \sum_{\substack{1 \le i \ne j \le n+1 \\ |\mathbf{p}_i - \mathbf{p}_j| > 2k_F}} \int_{C_{ij}^-}^{C_{ij}^+} \omega_{ij} - \sum_{\substack{\triangle \in \mathcal{T}_{\mathrm{ac}} \\ \rho_\triangle < k_F}} R_\triangle$$

$$= \frac{1}{4\pi\imath} \sum_{\substack{1 \le i \ne j \le n+1 \\ |\mathbf{p}_i - \mathbf{p}_j| > 2k_F}} \int_{C_{ij}^-}^{C_{ij}^+} \omega_{ij} + \sum_{\substack{\triangle \in \mathcal{T}_{\mathrm{ac}} \\ \rho_\triangle > k_F}} R_\triangle.$$

By construction for $1 \leq i,j \leq n+1$, $i \neq j$

$$\omega_{ij} \;=\; \sum_{k \neq i,j} R_{ijk}\, \omega_{ijk}\,.$$

Therefore, by Proposition 3.7

$$\frac{1}{4\pi\imath} \sum_{\substack{1 \leq i \neq j \leq n+1 \\ |\mathbf{p}_i - \mathbf{p}_j| > 2k_F}} \int_{C_{ij}^-}^{C_{ij}^+} \omega_{ij}$$

$$= \; - \sum_{\substack{1 \leq i \neq j \leq n+1 \\ |\mathbf{p}_i - \mathbf{p}_j| > 2k_F}} \sum_{\substack{1 \leq k \leq n+1 \\ k \neq i,j}} \epsilon_{\triangle(\mathbf{p}_i,\mathbf{p}_j,\mathbf{p}_k),(\mathbf{p}_i,\mathbf{p}_j)}\, R_{\triangle(\mathbf{p}_i,\mathbf{p}_j,\mathbf{p}_k)}$$

$$\operatorname{arccot}\sqrt{\frac{4\rho^2_{\triangle(\mathbf{p}_i,\mathbf{p}_j,\mathbf{p}_k)} - |\mathbf{p}_i - \mathbf{p}_j|^2}{|\mathbf{p}_i - \mathbf{p}_j|^2 - 4k_F^2}}$$

$$= \; -2 \sum_{\substack{\triangle \in \mathcal{T} \\ s \text{ edge of } \triangle \\ |s| > 2k_F}} \epsilon_{\triangle,s}\, R_{\triangle} \operatorname{arccot}\sqrt{\frac{4\rho^2_{\triangle} - |s|^2}{|s|^2 - 4k_F^2}}\,. \qquad \square$$

Proof of Proposition 3.7. By possibly interchanging i and j we may assume that the triangle with vertices $\mathbf{p}_i, \mathbf{p}_j, \mathbf{p}_k$ is counterclockwise oriented. As in the proof of Lemma 3.4 we may assume that $\mathbf{p}_i = (-a, 0)$ and $\mathbf{p}_j = (a, 0)$ with $a > 0$. Then there is $b \neq 0$ such that $\mathbf{D}^{ijk} = (0, b)$. Then

$$a^2 + b^2 \;=\; \rho^2_{\triangle}, \qquad\qquad\qquad b \;=\; \epsilon_{\triangle,(\mathbf{p}_i,\mathbf{p}_j)}\,|b|\,,$$

$$L_{ij} \;=\; \left\{ (z_1, z_2) \in \mathbb{C}^2 \mid z_1 = 0 \right\}, \qquad C_{ij}^{\pm} \;=\; (0, \pm\imath\sqrt{a^2 - k_F^2})\,,$$

$$\omega_{ijk} \;=\; \frac{dz_2}{z_2 - b}\,.$$

Consequently,

$$\int_{C_{ij}^-}^{C_{ij}^+} \omega_{ijk} \;=\; \int_{-\imath\sqrt{a^2 - k_F^2}}^{+\imath\sqrt{a^2 - k_F^2}} \frac{dz}{z - b} \;=\; \int_{-b-\imath\sqrt{a^2 - k_F^2}}^{-b+\imath\sqrt{a^2 - k_F^2}} \frac{dz}{z}$$

$$= \; -4\pi\imath \operatorname{arccot}\frac{b}{\sqrt{a^2 - k_F^2}} \;=\; -4\pi\imath\,\epsilon_{\triangle,(\mathbf{p}_i,\mathbf{p}_j)} \operatorname{arccot}\sqrt{\frac{\rho^2_{\triangle} - a^2}{a^2 - k_F^2}}\,. \qquad \square$$

The proof of Proposition 3.8 requires a lemma. To prepare for Lemma 3.9 pick a generic vector $(v_1, v_2) \in \mathbb{R}^2$ and, for each $i = 1, \ldots, n+1$, let

$$g_i \;=\; \left\{ z \in \mathbb{C}^2 \mid v_1 z_1 + v_2 z_2 = v_1 \mathbf{p}_{i1} + v_2 \mathbf{p}_{i2} \right\}$$

be the line through $\mathbf{p}_i$ whose slope is determined by (v_1, v_2). For all pairs $i \neq j$, let E_{ij} be the point where the lines g_i and L_{ij} meet. Furthermore, let $\omega_{i\infty}$ be the rational differential form on the line L_∞ at infinity that is holomorphic outside the points of $L_\infty \cap \overline{L}_{ij}$ and has a pole of order one at $L_\infty \cap \overline{L}_{ij}$ with residue $-\sum_{k \neq i,j} R_{ijk}$. Finally, let γ_∞ be the union of the paths

$$\{[0, -(1-t)i - tv_2, 1 - t + tv_1] \mid 0 \le t \le 1\},$$
$$\{[0, +(1-t)i - tv_2, 1 - t + tv_1] \mid 1 \ge t \ge 0\},$$

lying in L_∞.

LEMMA 3.9 *For all* $1 \le i \le n+1$,

$$\sum_{j \neq i} \int_{C_{ij}^+(0)}^{C_{ij}^-(0)} \omega_{ij} = \int_{\gamma_\infty} \omega_{i\infty} + 2\pi\imath \sum_{\substack{j,k \neq i \\ j \neq k}} \beta_{ijk} R_{ijk},$$

where

$$\beta_{ijk} = \begin{cases} +1, & \textit{if } \mathbf{D}^{ijk} \textit{ lies between } \tfrac{1}{2}(\mathbf{p}_i + \mathbf{p}_j) \textit{ and } E_{ij}, \\ & \textit{and } \mathbf{p}_i, \mathbf{p}_j, E_{ij} \textit{ are counterclockwise oriented,} \\ -1, & \textit{if } \mathbf{D}^{ijk} \textit{ lies between } \tfrac{1}{2}(\mathbf{p}_i + \mathbf{p}_j) \textit{ and } E_{ij}, \\ & \textit{and } \mathbf{p}_i, \mathbf{p}_j, E_{ij} \textit{ are clockwise oriented,} \\ 0, & \textit{otherwise.} \end{cases}$$

Proof. Put

$$F_t^\pm(z_1, z_2) = (1-t)\big((z_1 - \mathbf{p}_{i1}) \pm i(z_2 - \mathbf{p}_{i2})\big) + t\big(v_1 z_1 + v_2 z_2 - v_1 \mathbf{p}_{i1} - v_2 \mathbf{p}_{i2}\big),$$

and let $E_{ij}^\pm(t)$ be the root of $F_t^\pm$ on L_{ij}. Clearly, $E_{ij}^\pm(1) = E_{ij}$, and

$$\begin{aligned} g_i &= \{(z_1, z_2) \in \mathbb{C}^2 \mid F_1^\pm(z_1, z_2) = 0\}, \\ \overline{Q_i^\pm(0)} &= \{(z_1, z_2) \in \mathbb{C}^2 \mid F_0^\pm(z_1, z_2) = 0\}, \end{aligned}$$

and consequently,

$$\delta_{ij} = [C_{ij}^+(0), C_{ij}^-(0)] \cup \{E_{ij}^-(t) \mid 0 \le t \le 1\} \cup \{E_{ij}^+(t) \mid 1 \ge t \ge 0\}$$

is a loop that meets $L_{ij} \cap \mathbb{R}^2$ only in the points $\tfrac{1}{2}(\mathbf{p}_i + \mathbf{p}_j)$ and E_{ij}. Observe that the projective closure of $\{(z_1, z_2) \in \mathbb{C}^2 \mid F_t^\pm(z_1, z_2) = 0\}$ meets the line L_∞ at the point

$$E_{i\infty}(t) = [0, \mp(1-t)\imath - tv_2, 1 - t + tv_1].$$

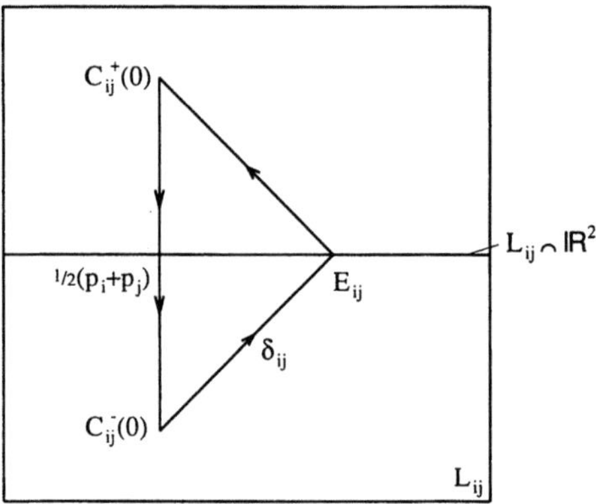

Figure 11

We have

$$\gamma_\infty = \left\{ E_{i\infty}^+(t) \mid 0 \le t \le 1 \right\} \cup \left\{ E_{i\infty}^-(t) \mid 1 \ge t \ge 1 \right\}$$

for all $i = 1, \ldots, n+1$. The union of lines

$$X_i = L_\infty \cup \bigcup_{j \ne i} \overline{L}_{ij}$$

is a singular curve in $\mathbb{P}^2$. The forms $\omega_{ij}, j = 1, \ldots, n+1, \infty, j \ne i$, define a Rosenlicht differential on X_i. For each $t \in [0,1]$ the divisor of the meromorphic function

$$\frac{F_t^- \cdot F_0^+}{F_0^- \cdot F_t^+}$$

on the curve X_i is

$$\sum_{\substack{j \ne i \\ j=1,\ldots,n+1,\,\infty}} \left(E_{ij}^-(t) - E_{ij}^-(0) \right) + \left(E_{ij}^+(0) - E_{ij}^+(t) \right).$$

By Abel's Theorem ([6], Ch. V.10, Proposition 5),

$$\sum_{\substack{j \ne i \\ j=1,\ldots,n+1,\,\infty}} \left(\int_{\left\{ E_{ij}^-(t)\,\mid\,0 \le t \le s \right\}} \omega_{ij} + \int_{\left\{ E_{ij}^+(t)\,\mid\,s \ge t \ge 0 \right\}} \omega_{ij} \right)$$

is independent of s. For $s = 0$ this quantity is zero. Therefore,

$$\sum_{\substack{j \ne i \\ j=1,\ldots,n+1,\,\infty}} \int_{\left\{ E_{ij}^-(t)\,\mid\,0 \le t \le 1 \right\} \cup \left\{ E_{ij}^+(t)\,\mid\,1 \ge t \ge 0 \right\}} \omega_{ij} = 0.$$

By the definition of the loop δ_{ij} and the conclusion of the last paragraph,

$$\sum_{\substack{j=1 \\ j \neq i}}^{n+1} \int_{C_{ij}^+(0)}^{C_{ij}^-(0)} \omega_{ij} \;=\; \sum_{\substack{j=1 \\ j \neq i}}^{n+1} \int_{\delta_{ij}} \omega_{ij} \;+\; \int_{\gamma_\infty} \omega_{i\infty}$$

$$- \sum_{\substack{j \neq i \\ j=1,\ldots,n+1,\,\infty}} \int_{\{E_{ij}^-(t)\,|\,0 \leq t \leq 1\} \cup \{E_{ij}^+(t)\,|\,1 \geq t \geq 0\}} \omega_{ij}$$

$$=\; 2\pi\imath \sum_{\substack{j,k \neq i \\ j \neq k}} \beta_{ijk}\, R_{ijk} \;+\; \int_{\gamma_\infty} \omega_{i\infty}\,,$$

since, by Lemma 3.4 (i), the winding number of δ_{ij} around $\mathbf{D}^{ijk}$ is equal to β_{ijk}. $\qquad\square$

Proof of Proposition 3.8. Observe that

$$\sum_{i=1}^{n+1} \omega_{i\infty} \;=\; 0\,,$$

since, by construction, the residues at each pole add up to zero. Now, by Lemma 3.9,

$$\sum_{1 \leq j \neq i \leq n+1} \int_{C_{ij}^+(0)}^{C_{ij}^-(0)} \omega_{ij} \;=\; \int_{\gamma_\infty} \sum_{i=1}^{n+1} \omega_{i\infty} \;+\; 2\pi\imath \sum_{\substack{\text{pairwise different} \\ 1 \leq i,j,k \leq n+1}} \beta_{ijk}\, R_{ijk}$$

$$=\; 2\pi\imath \sum_{\substack{\text{pairwise different} \\ 1 \leq i,j,k \leq n+1}} \beta_{ijk}\, R_{ijk}\,.$$

Fix three pairwise different indices i, j, k and let $\triangle = \triangle_{(\mathbf{p}_i, \mathbf{p}_j, \mathbf{p}_k)}$. Let

$$h \;=\; \left\{ (x_1, x_2) \in \mathbb{R}^2 \;\middle|\; v_1 x_1 + v_2 x_2 = v_1 \mathbf{D}_1^{ijk} + v_2 \mathbf{D}_2^{ijk} \right\}$$

be the real line through the center $\mathbf{D}^{ijk}$ of the circle circumscribing $\triangle$ whose slope is determined by (v_1, v_2). Then,

$$\beta_{ijk} \;=\; \begin{cases} 0\,, & \text{unless } h \text{ meets the segment between } \mathbf{p}_i \text{ and } \tfrac{1}{2}(\mathbf{p}_i + \mathbf{p}_j), \\ \pm 1\,, & \text{if } h \text{ meets the segment between } \mathbf{p}_i \text{ and } \tfrac{1}{2}(\mathbf{p}_i + \mathbf{p}_j) \text{ and the} \\ & \text{triangle } \mathbf{p}_i, \mathbf{p}_j, \mathbf{D}^{ijk} \text{ is counterclockwise (clockwise) oriented.} \end{cases}$$

In particular, if $\triangle$ is acute, there are exactly two permutations i', k', j' of i, j, k for which $\beta_{i'k'j'} \neq 0$.

Furthermore, if $\beta_{i'k'j'} \neq 0$, then

$$\beta_{i'k'j'} = \pm 1 \quad\Leftrightarrow\quad \mathbf{p}_{i'}, \mathbf{p}_{j'}, \mathbf{p}_{k'} \text{ counterclockwise (clockwise) oriented}$$

$$\Leftrightarrow\quad R_{i'k'j'} = \pm R_\triangle\,.$$

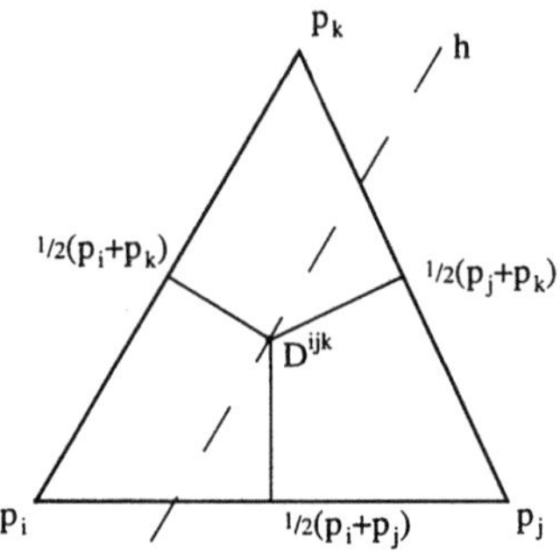

Figure 12

Consequently,

$$\sum_{\{i',j',k'\}=\{i,j,k\}} \beta_{i'j'k'}\, R_{i'j'k'} \;=\; 2R_\Delta\,.$$

Similarly, if Δ is obtuse,

$$\sum_{\{i',j',k'\}=\{i,j,k\}} \beta_{i'j'k'}\, R_{i'j'k'} \;=\; 0\,. \qquad\qquad \square$$

Appendix

In this appendix we assume that $J(\mathbf{p}_1, \mathbf{p}_2, \mathbf{p}_3)$ exists and give a direct proof of

$$J(\mathbf{p}_1, \mathbf{p}_2, \mathbf{p}_3) \;=\; 0\,,$$

when the circle circumscribing the triangle with vertices $\mathbf{p}_1, \mathbf{p}_2, \mathbf{p}_3$ has radius less than k_F. If the triangle is acute, the previous condition is equivalent to the statement that the three open disks K_1, K_2, K_3 with radius k_F around the points $\mathbf{p}_1, \mathbf{p}_2, \mathbf{p}_3$ have a point in common. By translation invariance, the center of the circle circumscribing the triangle can be placed at the origin.

By definition,

$$J(\mathbf{p}_1, \mathbf{p}_2, \mathbf{p}_3) \;=\; \frac{1}{(2\pi)^3} \int_{\mathbb{R}^2} dk_1\, dk_2 \int_{-\infty}^{+\infty} \frac{dk_0}{\prod_{i=1}^{3}\big(\imath k_0 - e(\mathbf{k} - \mathbf{p}_i)\big)}\,.$$

The integral

$$\int_{-\infty}^{+\infty} \frac{dk_0}{\prod_{i=1}^{3}\big(\imath k_0 - e(\mathbf{k} - \mathbf{p}_i)\big)}$$

is evaluated by closing the contour in the upper half plane when the point $\mathbf{k} \in \mathbb{R}^2$ belongs to at most one of the disks K_1, K_2, K_3 and closing in the lower half plane when it belongs to at least two of the disks. In particular, the integral vanishes when $\mathbf{k} \notin K_1 \cup K_2 \cup K_3$ or $\mathbf{k} \in K_1 \cap K_2 \cap K_3$. We obtain

$$
\frac{(2\pi)^2}{m^2} \, J(\mathbf{p}_1, \mathbf{p}_2, \mathbf{p}_3)
$$

$$
= \int_{K_1 \setminus (K_2 \cup K_3)} \frac{dk_1 \, dk_2}{F_{12}(\mathbf{k}) \, F_{13}(\mathbf{k})} - \int_{(K_2 \cup K_3) \setminus K_1} \frac{dk_1 \, dk_2}{F_{12}(\mathbf{k}) \, F_{13}(\mathbf{k})}
$$

$$
+ \int_{K_3 \setminus (K_1 \cup K_2)} \frac{dk_1 \, dk_2}{F_{31}(\mathbf{k}) \, F_{32}(\mathbf{k})} - \int_{(K_1 \cup K_2) \setminus K_3} \frac{dk_1 \, dk_2}{F_{31}(\mathbf{k}) \, F_{32}(\mathbf{k})}
$$

$$
+ \int_{K_2 \setminus (K_3 \cup K_1)} \frac{dk_1 \, dk_2}{F_{23}(\mathbf{k}) \, F_{21}(\mathbf{k})} - \int_{(K_3 \cup K_1) \setminus K_2} \frac{dk_1 \, dk_2}{F_{23}(\mathbf{k}) \, F_{21}(\mathbf{k})} \, ,
$$

where, for all $i, j = 1, 2, 3$,

$$
F_{ij}(\mathbf{k}) \; = \; m \left(e(\mathbf{k} - \mathbf{p}_i) - e(\mathbf{k} - \mathbf{p}_j) \right) = \; (\mathbf{p}_j - \mathbf{p}_i) \cdot \mathbf{k},
$$

since $\mathbf{p}_1^2 = \mathbf{p}_2^2 = \mathbf{p}_3^2$.

Let $2r$ be the length of the secant to the circle ∂K_1 through 0 that is perpendicular to the line through 0 and $\mathbf{p}_1$. Then the map

$$
\phi : \; \mathbb{R}^2 \setminus \{0\} \; \longrightarrow \; \mathbb{R}^2 \setminus \{0\}
$$

$$
\mathbf{k} \; \longmapsto \; -\frac{r^2}{|\mathbf{k}|^2} \, \mathbf{k}
$$

maps each of the circles ∂K_1, ∂K_2, ∂K_3 to itself ([Be], 10.8). By hypothesis, $0 \in K_i$, $i = 1, 2, 3$, and therefore, ϕ maps K_i to the exterior of ∂K_i and conversely. In particular,

$$
\phi \left(K_1 \setminus \left(\overline{K}_2 \cup \overline{K}_3 \right) \right) \; = \; (K_2 \cap K_3) \setminus \overline{K}_1.
$$

Substituting,

$$
F_{12}\big(\phi(\mathbf{k})\big) \, F_{13}\big(\phi(\mathbf{k})\big) \; = \; \frac{r^4}{|\mathbf{k}|^4} \, F_{12}(\mathbf{k}) \, F_{13}(\mathbf{k}).
$$

Also,

$$
\phi^* \left(dk_1 \wedge dk_2 \right) \; = \; \frac{r^4}{|\mathbf{k}|^4} \, dk_1 \wedge dk_2.
$$

The last three equations imply that

$$
\int_{K_1 \setminus (K_2 \cup K_3)} \frac{dk_1 \, dk_2}{F_{12}(\mathbf{k}) \, F_{13}(\mathbf{k})} - \int_{(K_2 \cap K_3) \setminus K_1} \frac{dk_1 \, dk_2}{F_{12}(\mathbf{k}) \, F_{13}(\mathbf{k})} \; = \; 0.
$$

The other pairs also cancel.

        J. *Feldman, H. Knörrer, R. Sinclair and E. Trubowitz*

References

[1] M. Berger: Geometry. Springer-Verlag (1987).

[2] J. Feldman, H. Knörrer, R. Sinclair, E. Trubowitz: Superconductivity in a Repulsive Model. Helvetica Physica Acta **70**, 154–191 (1997)

[3] P. Griffiths, J. Harris: Principles of Algebraic Geometry. Wiley-Interscience (1978).

[4] N. Ortner, P. Wagner: On the evaluation of one–loop Feynman amplitudes in Euclidean quantum field theory. *Ann. Inst. Henri Poincaré (Physique théorique)* **63**, 81–110 (1995).

[5] M. Sebastiani: Un Exposé de la Formule de Leray-Norguet. *Boletim da Sociedade Brasileira de Matematica*, vol. 1, No. 2, 47–57, (1970).

[6] J.P. Serre: Groupes Algébriques et Corps de Classes. Hermann, Paris (1954).

[7] R. Sinclair: The evaluation of Feyman graphs for a (2+1)-dimensional nonrelativistic electron gas. *Computer Physics Communications* **81** , 173–184 (1994).

Progress in Mathematics, Vol. 162, © 1998 Birkhäuser Verlag Basel/Switzerland

Möbius and Odd Real Trigonometric M-Functions

Victor Goryunov
Department of Mathematical Sciences
Division of Pure Mathematics
The University of Liverpool
Liverpool L69 3BX
UK

Clare Baines
Department of Mathematical Sciences
Division of Pure Mathematics
The University of Liverpool
Liverpool L69 3BX
UK

Abstract

We study two series of spaces of special real trigonometric polynomials of fixed degree having the maximal possible number of distinct critical values. Those are functions such that either $g(\varphi+\pi) \equiv -g(\varphi)$ or $g(-\varphi) \equiv -g(\varphi)$. For each of the spaces, we calculate the number of its connected components and identify, within the mirror arrangement of the Weyl group of series B, a convex polyhedral model for its closure.

Introduction

A real trigonometric *M-polynomial* of degree n is one with the maximal number $2n$ of real critical points. In his recent paper [4], V. I. Arnold constructed a polyhedral model for the manifold of such polynomials and calculated the number of topologically different M-polynomials with all their critical values distinct. The model was provided by a convex cone in the space equipped with the mirror arrangement of the reflection group A_{2n-1}. The enumeration was done in terms of updown sequences (also called *A-snakes*) of [1, 2].

In the present note we establish a similar result for *Möbius* trigonometric M-polynomials, that is those with the property $g(\varphi + \pi) \equiv -g(\varphi)$. The polyhedral model is now a simplex within the mirror arrangement of the reflection group of series B. Analogous constructions are carried out for odd trigonometric functions.

Also we enumerate topological types of our M-functions with non-coinciding critical values. The enumeration is given by so-called β- and γ-snakes (see [2]), which have been lacking a direct singularity theory interpretation.

All our results describe properties of the real Lyashko-Looijenga mapping which associates to an M-function the ordered set of its critical values. They are parallel to the results on the sets of real M-functions in the other natural families (see [2, 6, 7] and papers cited there).

1   Möbius polynomials

1.1   Functions of degree 3

We start with an example that illustrates general properties of Möbius M-polynomials which we are going to establish.

Consider the 2-parameter family

$$f(\varphi, a, b) = \tfrac{1}{3}\cos 3\varphi + a\cos\varphi + b\sin\varphi$$

of all possible real Möbius trigonometric polynomials of degree 3 with the fixed leading term $\tfrac{1}{3}\cos 3\varphi$.

The *bifurcation diagram* Σ of the family is the set of all the values of the parameters (a, b) for which the function $f(\cdot, a, b)$ is non-generic. It consists of three curves:

Σ_c, caustic, — functions $f(\cdot, a, b)$ have non-Morse critical points;

Σ_m, Maxwell stratum, — functions $f(\cdot, a, b)$ have coinciding non-zero critical values;

Σ_0, *special* Maxwell stratum, — functions $f(\cdot, a, b)$ have critical points on their zero-levels.

To find Σ_c one can proceed as follows. Rewrite the family as

$$F(\varphi, A, \delta) = \tfrac{1}{3}\cos 3\varphi + A\cos(\varphi - \delta), \qquad a = A\cos\delta, \quad b = A\sin\delta.$$

Let prime denote the derivation with respect to φ. Then the equation for Σ_c is

$$0 = -F'' - iF' = 2e^{3i\varphi} + e^{-3i\varphi} + Ae^{i(\varphi - \delta)}.$$

Thus, $Ae^{-i\delta} = -2e^{2i\varphi} - e^{-4i\varphi}$. This is a hypocycloid with 3 cusps, the bigger one of Fig. 1.

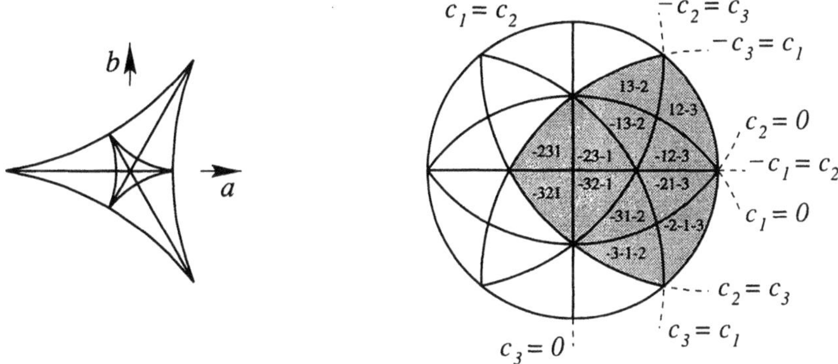

Figure 1: *Stratification of the space of Möbius functions of degree 3 with a fixed leading term and the mirror arrangement of the reflection group B_3. The spherical triangles are marked with the corresponding β_3-snakes (see Sect.1.2).*

The smaller hypocycloid of Fig. 1 is Σ_0 defined by the equation $F - iF' = 0$, that is $Ae^{-i\delta} = -\frac{2}{3}e^{2i\varphi} + \frac{1}{3}e^{-4i\varphi}$.

The Maxwell stratum Σ_m consists of the three intervals. The calculations for it are, as usual, a bit more complicated. We do not show them here.

The set of M-functions $f(\cdot, a, b)$, with all 6 critical points real, is that inside Σ_c. The subdivision of its closure by the strata Σ_m and Σ_0 is seen to be homeomorphic to that of the spherical domain in the cone $c_1 \leq c_2 \geq c_3 \leq -c_1$ in the coordinate 3-space by the mirrors of the group B_3. The elementary triangles of the domain are in one-to-one correspondence with permutations described in the next section.

1.2 β-snakes

Consider a Möbius trigonometric M-polynomial of order n (n is odd). Walk counterclockwise along the source circle starting from some initial point so that the first critical point to meet is a local minimum. This provides us with a sequence $\{c_i\}$ of the critical values satisfying the relations

$$c_1 < c_2 > c_3 < \cdots > c_{2n+1} = c_1 \quad \text{and} \quad c_{i+n} = -c_i, \ i = 1, 2, \ldots, n.$$

The space of Möbius M-functions will be shown to be closely related to the space of such sequences. So let us consider the latter in some detail.

An arbitrary sequence of $2n$ real numbers subject to the above relations in which all the inequalities are allowed to be non-strict will be called a *Möbius snake of order $2n$*. A snake all of whose elements are distinct will be called *generic*.

Since a Möbius snake is completely defined by its initial half, the set S_n of all Möbius snakes of order $2n$ is the cone

$$c_1 \le c_2 \ge c_3 \le \ldots \ge c_n \le -c_1$$

in $\mathbb{R}^n$. The set S_n^Σ of non-generic snakes is its intersection with the set of mirrors

$$c_i = \pm c_j, \quad c_k = 0$$

of the reflection group B_n.

THEOREM 1.1 *The number $\mathcal{M}_r$ of connected components of the set of generic Möbius snakes of order $2(2r+1)$ is equal to $(2r+1)\mathcal{S}_r$, where the numbers $\mathcal{S}_r$ are given by the exponential generating function*

$$\sum_{r=0}^{\infty} \mathcal{S}_r \frac{t^{2r}}{(2r)!} = \sec 2t \; .$$

Thus $\mathcal{M}_r = (2r+1)2^{2r}E_r$, where the sequence $\{E_r\}$ is that of the Euler numbers: 1, 1, 5, 61, 1385, $\ldots$. Fig. 1 is an illustration to $\mathcal{M}_1 = 12$.

Proof. Let $n = 2r+1$. Each of the connected components under consideration is contractible and contains one and only one n-sequence which is the initial half of a *normalised* Möbius snake, that is a Möbius snake which is a permutation of the numbers $\pm 1, \pm 2, \ldots, \pm n$. So we need to show that the number of the normalised snakes is that claimed in the theorem.

Let us delete the absolute maximum $c_\ell = n$ and minimum $c_{\ell+n} = -n$ of a normalised Möbius snake and consider the $(n-1)$-sequence $s_i = c_{\ell+i}$, $i = 1, \ldots, n-1$ (the indexation in the original snake is taken modulo $2n$). The obtained integer sequence satisfies the conditions:

$$s_1 < s_2 > s_3 < \ldots s_{n-1} \quad \text{and} \quad \{|s_i|\} = \{1, 2, \ldots, n-1\} \, .$$

Following [2], we call such a sequence a β_{n-1}-*snake* (we do not call it normalised since we will not need any others).

Denote by $\mathcal{B}_{n-1}$ the number of all possible β_{n-1}-snakes. The following lemma, allowing n to be of any parity (so that the last inequality in the above chain is ">" for even n), identifies the generating function $\mathcal{B}(t) = \sum_{k \ge 0} \mathcal{B}_k \frac{t^k}{k!}$.

LEMMA 1.2 *(part of Theorem 24 of [2])*

$$\mathcal{B}(b) = \sec 2t + \tan 2t \, .$$

The lemma implies our theorem. Indeed, $\mathcal{S}_r = \mathcal{B}_{2r}$ and the maximal-minimal pair $(c_\ell, c_{\ell+n}) = (n, -n)$ of a normalised Möbius snake of order $2n = 2(2r+1)$ can stay in any of the $2r+1$ different positions. $\square$

1.3 Homeomorphism of the configurations

Consider the space $\mathbb{R}^{2r}$ of all Möbius trigonometric polynomials

$$\cos(2r+1)\varphi + \sum_{k=0}^{r-1} a_k \cos(2k+1)\varphi + b_k \sin(2k+1)\varphi$$

of degree $2r+1$ with fixed leading term. Let $M_r \subset \mathbb{R}^{2r}$ be the closure of the set of all M-functions, and $M_r^\Sigma \subset M_r$ its intersection with the bifurcation diagram of the family. The following theorem relates these two sets to the set $S_{2r+1} \subset \mathbb{R}^{2r+1}$ of all Möbius snakes of order $2(2r+1)$ and its subset S_{2r+1}^Σ of all non-generic snakes.

THEOREM 1.3 *The pair (M_r, M_r^Σ) is homeomorphic to the intersection of the pair $(S_{2r+1}, S_{2r+1}^\Sigma)$ with a sphere in $\mathbb{R}^{2r+1}$ centered at the origin. The homeomorphism is a diffeomorphism at any of the internal points of M_r.*

Proof. Our statement will follow from the main result of [4] which asserts that the real analog of the Lyashko-Looijenga mapping [5, 3] provides a similar homeomorphism for trigonometric M-functions which are not necessarily Möbius. We recall the definiton of the mapping and the result.

Consider the space $\mathbb{R}^{2n-2}$ of *all* trigonometric polynomials of degree n with the fixed leading term $\cos(n\varphi)$ and no free term. Take one of M-functions in it and order its $2n$ critical points as in Section 1.2 starting from a local minimum. Since the closure $N_n \subset \mathbb{R}^{2n-2}$ of the set of all M-functions is contractible [4], this induces the ordering of the critical points of any function in the interior of N_n. Now to each of the functions in the interior we associate the ordered set $c_1 < c_2 > c_3 < \cdots < c_{2n} > c_1$ of its critical values. Shift all the values by their arithmetic mean. This gives a mapping into the hyperplane $c_1 + c_2 + \cdots + c_{2n} = 0$ in the $2n$-dimensional coordinate c-space equipped with the diagonals $c_i = c_j$. Its composition with the radial projection onto a sphere in it centred at the origin is a diffeomorphism between the interior of N_n and the spherical domain defined by the above chain of inequalities. The diffeomorphism maps the Maxwell stratum to the diagonals, and extends to a homeomorphism $\mathcal{L}_n$ of the closures [4].

Returning to the Möbius M-functions, consider M_r as a subset of N_{2r+1}. The composition of the restriction of $\mathcal{L}_{2r+1}$ to M_r with the further projection forgetting the last half of the coordinates in the c-space $\mathbb{R}^{4r+2}$ is the homeomorphism required in the claim of the theorem. $\qquad\square$

In Fig. 1, the ordering of the critical values is induced by that for the function $\frac{1}{3}\cos 3\varphi$ starting from its local minimum at $\varphi = \pi/3$.

1.4 Monodromy

Now consider trigonometric polynomials of order n with the leading term varying and no free term:

$$\sum_{k=1}^{n} a_k \cos k\varphi + b_k \sin k\varphi = \sum_{k=1}^{n} A_k \cos\left(k\varphi - \delta_k\right), \quad A_n \neq 0.$$

The set of all such polynomials is $S^1 \times \mathbb{R}^{2n-1}$.

For any real τ, the shift

$$\varphi \mapsto \varphi + \tau, \qquad \delta_k \mapsto \delta_k + k\tau, \quad k = 1,\ldots,n,$$

does not change the value of a function. Thus, when δ_n changes from 0 to 2π, the bifurcation diagram of a subfamily with the fixed leading term maps onto itself with the twist by $2k\pi/n$ in the coordinate (a_k, b_k)-plane. For the diagram of Fig. 1 this is the counterclockwise rotation by $2\pi/3$.

In terms of the critical values, the above monodromy cancels the difference between every n snakes obtained from each other by rotation of the source circle. Thus the number of different topological types of generic Möbius trigonometric M-functions is given by

COROLLARY 1.4 *In the set $S^1 \times \mathbb{R}^{2r+1}$ of all Möbius trigonometric polynomials of degree $2r+1$ with an arbitrary leading term, the subset of M-functions with all their critical values distinct has $2^{2r} E_r$ connected components each contractible onto a circle.*

2 Odd trigonometric polynomials

2.1 Degree 3 family

Again we start with an illustration to our other general statement. Consider the family

$$\sin\varphi \left(\cos^2\varphi + a\cos\varphi + b\right)$$

of degree 3 odd trigonometric polynomials.

The set of non-generic functions in the family (Fig. 2) consists of:

Σ_c, caustic, — functions with non-Morse critical points not at multiples of π;

Σ_b, *boundary* caustic, — functions with non-Morse critical points at multiples of π, thus on their zero-levels;

Σ_m, Maxwell stratum, — functions with coinciding critical values on the open interval $(0, \pi)$;

Σ_m^-, *anti*-Maxwell stratum, — functions with anti-coinciding critical values $c_i = -c_j$ on the open interval $(0, \pi)$;

Σ_0, *special* Maxwell stratum, — functions with critical points on their zero-levels not at multiples of π.

In Fig. 2, the asymptote for the anti-Maxwell strata is $b = -\frac{1}{2}$. The strata Σ_m^- and Σ_m intersect at $b = -\frac{1}{4}$, that is at $\frac{1}{4}\sin 3\varphi$. The strata Σ_m and Σ_0 meet at the origin. The intersections of Σ_b and Σ_m^- are at $(\pm 3(1 - 2^{-1/3}), 2 - 3 \cdot 2^{-1/3})$.

The shaded region in the parameter space is that of M-functions. Its configuration is isomorphic to the spherical domain $0 < c_1 > c_2 < c_3 > 0$ in the 3-space containing the B_3 arrangement.

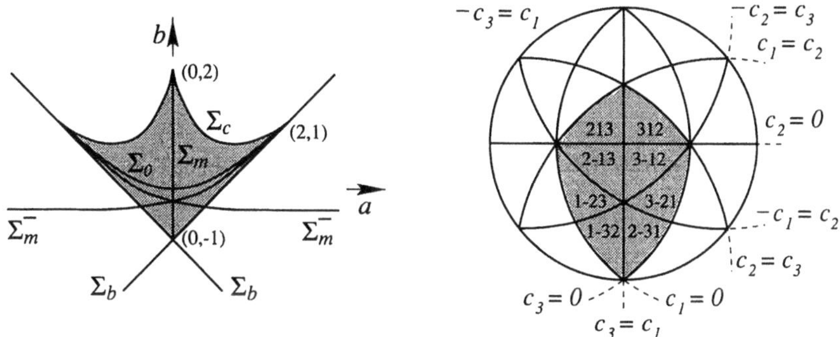

Figure 2: *Stratification of the space of odd functions of degree 3 compared with the B_3 mirror arrangement. The spherical triangles are marked with the β_3-snakes of critical values of the functions in the open interval $(0, \pi)$.*

2.2 The polyhedral model

Up to the choice of the sign of a function, we may assume that any real odd trigonometric polynomial of degree n enters the family

$$\Phi_n = \sin\varphi \left(\cos^{n-1}\varphi + a_1\cos^{n-2}\varphi + \cdots + a_{n-2}\cos\varphi + a_{n-1}\right).$$

Consider the restriction of the real Lyashko-Looijenga mapping $\mathcal{L}_n$ from the space of all trigonometric M-polynomials of degree n with the fixed leading term $\frac{1}{n}\sin n\varphi$ to M-functions of the family Φ_n. Let the ordering of the critical values for $\mathcal{L}_n$ be induced by that of the function $\frac{1}{n}\sin n\varphi$ with respect to the increase of the coordinate $\varphi \in (0, 2\pi)$ of its critical points. According to [2], our restriction of $\mathcal{L}_n$ maps the set of all M-functions of the family Φ_n diffeomorphically onto the spherical domain in $\mathbb{R}^{2n}$ lying in the cone

$$c_1 > c_2 < \cdots > c_{2n} < c_1, \qquad c_i = -c_{2n+1-i}, \quad i = 1,\ldots,n.$$

Thus, the set of all M-functions in Φ_n is connected. Denote by O_n its closure.

The subset of functions in O_n for which $c_1 = 0$ (and thus $c_{2n} = 0$) lies on the boundary of O_n. Indeed this special critical value in general corresponds to a cubical critical point at $\varphi = 0$.

So, for any M-function from Φ_n, the above ordering of the critical values is that given by the increase of the coordinate $\varphi \in (0, 2\pi)$ of its own critical points. In particular, for such a function the first critical point is a local maximum and $c_1 > 0$. Similarly, $(-1)^{n-1} c_n > 0$.

Now let $O_n^\Sigma \subset O_n$ be the subset of non-generic odd functions in the parameter space $\mathbb{R}^{n-1}$ of the family Φ_n. Denote by P_n the intersection of the cone

$$0 \le c_1 \ge c_2 \le c_3 \ge \ldots c_n \,, \qquad (-1)^{n-1} c_n \ge 0$$

with the sphere in the coordinate space $\mathbb{R}^n$ centred at the origin. Let P_n^Σ be the intersection of P_n with the mirror arrangement of the group B_n.

The above discussion proves

THEOREM 2.1 *The pairs (O_n, O_n^Σ) and (P_n, P_n^Σ) are homeomorphic. The homeomorphism between them is a diffeomorphism at the internal points.*

The homeomorphism is provided by the composition of the restriction of the mapping $\mathcal{L}_n$ with the projection forgetting the critical values $c_{n+1}, \ldots, c_{2n}$.

2.3 Enumeration of the topological types

THEOREM 2.2 *The numbers $\mathcal{O}_n$ of topologically distinct real odd trigonometric M-functions of degree n fit in the exponential generating function*

$$\sum_{n=0}^{\infty} \mathcal{O}_n \frac{t^n}{n!} = \sec 2t + \tan 2t \,.$$

For $n = 0$ this makes a reasonable sense: there is only one odd function, $g \equiv 0$, of degree 0 with the maximal possible number 1 of distinct critical values.

The topological equivalence of the theorem is that via odd diffeomorphisms of the source circle preserving its orientation and fixing $\varphi = 0$, and orientation-preserving odd diffeomorphisms of the target real axis. Thus the family Φ_n contains only half of all the types. For example, in Fig. 2 we see $8 = \mathcal{O}_3/2$ of them.

Proof of the theorem. Consider for the moment only the topological types represented in Φ_n. From the previous section we see that their number is that of β_n-snakes (more precisely, of the negatives of them) which might be continued to the left and right by zeros so that the inequality chain stays alternating:

$$0 < p_1 > p_2 < p_3 > \ldots p_n \,, \qquad (-1)^{n-1} p_n > 0 \,, \qquad \{|p_i|\} = \{1, 2, \ldots, n\} \,.$$

An odd diffeomorphism of the real axis adding 1 to each natural number establishes a one-to-one correspondence between the above snakes and sequences

$$1 = \bar{p}_0 < \bar{p}_1 > \bar{p}_2 < \bar{p}_3 > \dots \bar{p}_n, \qquad \{|\bar{p}_i|\} = \{1, 2, \dots, n+1\}.$$

extendible by 0 to the right in the similar way. Giving up the restriction $\bar{p}_0 = 1$ we get exactly the definition of Arnold's γ_{n+1}-snakes [2].

Let $\mathcal{G}^k_{n+1}$ be the number of γ_{n+1}-snakes with the fixed beginning $\bar{p}_0 = k$, and $\mathcal{B}^k_{n+1}$ the number of β_{n+1}-snakes starting with $k = \bar{p}_0 < \bar{p}_1 > \dots$. We have to identify $\mathcal{G}^1_{n+1} = \mathcal{O}_n/2$.

LEMMA 2.3 *(part of Theorem 15 of [2])* For $k > 0$,

$$\mathcal{B}^{-k}_{n+1} = \mathcal{G}^{k-(n+2)}_{n+1} + \mathcal{G}^{(n+2)-k}_{n+1}.$$

Setting $k = n + 1$ we obtain

COROLLARY 2.4 $\qquad 2\mathcal{G}^1_{n+1} = \mathcal{B}_n$.

Indeed, according to the lemma, $\mathcal{B}^{-(n+1)}_{n+1} = \mathcal{G}^{-1}_{n+1} + \mathcal{G}^1_{n+1}$. On the other hand, $\mathcal{B}^{-(n+1)}_{n+1} = \mathcal{B}_n$ and $\mathcal{G}^{-1}_{n+1} = \mathcal{G}^1_{n+1}$.

According to Lemma 1.2, the exponential generating function for the numbers $\mathcal{B}_n$ is $\sec 2t + \tan 2t$. $\qquad\qquad\qquad\qquad\qquad\qquad\square$

REMARK 2.5 The exponential generating function $\sum_{n \geq 0} \mathcal{E}_{n-1} \frac{t^n}{n!}$ for the numbers $\mathcal{E}_n$ of topological types of real even trigonometric M-functions of degree n is easily seen to be $2(\sec t + \tan t)$, with a very formal setting $\mathcal{E}_{-1} = 2 = \mathcal{E}_0$.

References

[1] V.I. Arnold, *Bernoulli updown numbers associated with function singularities, their combinatorics and arithmetic*, Duke Math. J. **63** (1991), 537–555.

[2] V.I. Arnold, *The calculus of snakes and the combinatorics of Bernoulli, Euler and Springer numbers of Coxeter groups*, Russian Math. Surveys **47** (1992), no.1, 1–51.

[3] V.I. Arnold, *Topological classification of complex trigonometric polynomials and combinatorics of graphs with equal number of vertices and egdes*, Functional Analysis and its Applications **30** (1996), no.1, 1–18.

[4] V.I. Arnold, *Topological classification of real trigonometric polynomials and cyclic serpents polyhedron*, preprint, 1996.

[5] V.V. Goryunov, *Geometry of bifurcation diagrams of simple projections onto the line*, Functional Analysis and its Applications **15** (1981), no.2, 77–82.

[6] V.V. Goryunov, *Subprincipal Springer cones and morsifications of Laurent polynomials and D_μ singularities*, in: 'Singularities and Bifurcations' (V. I. Arnold, ed.), Advances in Soviet Mathematics **21** (1994), AMS, Providence, RI, 163–188.

[7] V.V. Goryunov, *Morsifications of rational functions*, in: 'Topology of Real Algebraic Varieties and Related Topics' (V.M. Kharlamov, O.Ya. Viro e.a., eds.), American Mathematical Society Translations (2) **173** (1996), AMS, Providence, RI, 85–96.

Progress in Mathematics, Vol. 162, © 1998 Birkhäuser Verlag Basel/Switzerland

Moduli Space of Smooth Affine Curves of a Given Genus with one Place at Infinity

Mutsuo Oka
Department of Mathematics
Tokyo Metropolitan University
Minami-Ohsawa, Hachioji-shi
Tokyo 192-03, JAPAN

Abstract

We consider the space of smooth affine curves with one place at infinity and a fixed genus. We will show that the quotient space by the algebraic automorphism group of $\mathbf{C}^2$ has a structure of an algebraic variety which has finite connected components and each component is isomorphic to a cyclic quotient of a rational variety.

1 Introduction

Let $f(x,y) \in \mathbf{C}[x,y]$ be a polynomial. We consider an affine curve $C^a(f) := \{(x,y) \in \mathbf{C}^2; f(x,y) = 0\}$ and the projective curve $C(f)$ is defined by the closure of $C^a(f)$ in $\mathbf{P}^2$. We say that f (or $C^a(f)$) has one place at infinity if the intersection with the line at infinity L_∞ is one point and $C(f)$ is locally irreducible at that point. Let $\mathcal{POL}$ be the set of polynomials with one place at infinity, let $\mathcal{POL}^{(r)}$ be the set of polynomials $f \in \mathcal{POL}$ which defines a smooth affine curve $C^a(f)$ and let $\mathcal{POL}^{(r)}(g)$ be the set of polynomials $f \in \mathcal{POL}^{(r)}$ for which the genus of $C(f)$ is g.

Let $G = \mathrm{Aut}(\mathbf{C}^2)$ be the group of algebraic automorphisms of $\mathbf{C}^2$. The action of G on $\mathcal{POL}$ is defined by $f^\varphi(x,y) := f(\varphi^{-1}(x,y))$ for $f \in \mathcal{POL}$ and $\varphi \in G$. The subspaces $\mathcal{POL}^{(r)}$ and $\mathcal{POL}^{(r)}(g)$ are invariant under this action. The main result of this paper is:

MAIN THEOREM 5.4. *The quotient set $\mathcal{POL}^{(r)}(g)/G$ has a structure of algebraic variety which has a finite number of connected components*

$$\mathcal{POL}^{(r)}(\mathcal{W}^j)/Stab(\mathcal{POL}^{(r)}(\mathcal{W}^j)), \quad j = 1, \ldots, m$$

and each component is isomorphic to a finite cyclic quotient of a rational variety.

2 Moduli of polynomials with a fixed irreducible singularity

2.1 Tschirnhausen-good resolution tower

We recall results of [A-O] and [O4]. Let $\mathcal{P}$ be the set of polynomial $h(u,v)$ which are locally irreducible at the origin and monic in v and $h(0,v) = v^{\deg_v h}$. For any $h(u,v) \in \mathcal{P}$, we can write $h(u,v)$ as

$$h(u,v) = (v^{a_1} + \xi_1 u^{b_1})^{A_2} + \text{(higher terms)},$$
$$n = a_1 A_2, \ \xi_1 \in \mathbf{C}^*, \ a_1 > 1 \tag{2.1.1}$$

where $\gcd(a_1, b_1) = 1$ and (higher terms) is a linear combination of monomials $u^a v^b$ satisfying $a_1 a + b_1 b > a_1 b_1 A_2$ and $b < n$. Let

$$C := \{(u,v) \in \mathbf{C}^2; h(u,v) = 0\}$$

and assume that C is irreducible at the origin. Let $\mathcal{T} = \{p_j : X_j \to X_{j-1}; j = 1, \ldots, k\}$ be a toric resolution tower of C at the origin with $X_0 = \mathbf{C}^2$. Let $\Psi_i := p_1 \circ \cdots \circ p_i : X_i \to X_0$ be the composition. Then $\Psi_k : X_k \to X_0$ is a good resolution of C. For each $i \leq k - 1$, let $\Xi_i \in X_i$ be the singular point of the strict transform $C^{(i)}$ of C to X_i. This is the center of the next toric modification $p_{i+1} : X_{i+1} \to X_i$. Let E_i be the irreducible component of the exceptional divisor of the toric modification $p_i : X_i \to X_{i-1}$ so that $\Xi_i \in E_i$. The divisor E_i corresponds to the weight vector P_i. We use a coordinate system (u_i, v_i) with the center Ξ_i to construct a toric modification $p_{i+1} : X_{i+1} \to X_i$ so that $\{u_i = 0\}$ is the defining equation of the exceptional divisor E_i. Here $(u_0, v_0) = (u, v)$.

A toric resolution tower $\mathcal{T}$ is called *Tschirnhausen-good* if $a_i \geq 2$, $i = 1, \ldots, k$. The first weight vector $P_1 = {}^t(a_1, b_1)$ is given by the expression (2.1.1). Put $A_{i,j} = a_i a_{i+1} \cdots a_j$, $i \leq j \leq k$. As $n = a_1 \cdots a_k$, $A_{2,k} = A_2$. Let $h_i(u,v)$ be the $A_{i+1,k}$-th Tschirnhausen approximate polynomial of $h(u,v)$ and let $C_i = \{(u,v) \in \mathbf{C}^2; h_i(u,v) = 0\}$ for $i = 1, \ldots, k$. Here $h_k = h$ and $C_k = C$. Recall that *the Tschirnhausen approximate polynomial* $h_i(u,v)$ is the monic polynomial in v of degree $A_{1,i}$ characterized by the property:

$$\deg_v(h(u,v) - h_i(u,v)^{A_{i+1,k}}) < n - A_{1,i}.$$

Let μ be the Milnor number of $h(u,v)$ at the origin and let $I(C_{k-1}, C; O)$ be the intersection number at the origin. The following properties are satisfied for a given Tschirnhausen-good resolution tower $\mathcal{T}$ ([A-O], Theorem 4.5, Theorem 4.7, Theorem 5.1).

THEOREM 2.1 *Assume that $\mathcal{T}$ is a Tschirnhausen-good resolution tower.*
1. The pull backs of polynomials h_i, $i = -1, 0, \ldots, k$ are written as
(a)

$$\Psi_i^* h_\ell(u_i, v_i) = \begin{cases} u_i^{m_{i,\ell}}\left((v_i^{a_{i+1}} + \xi_{i+1} u_i^{b_{i+1}})^{A_{i+2,\ell}} + (\text{higher terms})\right), & i < \ell \\ u_i^{m_{i,i}} V_i, & i = \ell \\ u_i^{m_{i,i}} U_{i,\ell}, & U_{i,\ell}: \text{ a unit, } i > \ell \end{cases}$$

where V_i is written as $V_i = v_i + R_i(u_i, v_i)$ and R_i has no linear term and thus $C_i^{(i)}$ is smooth and defined by $\{V_i = 0\}$. The multiplicities $m_{0,\ell}, \ldots, m_{k,\ell}$ for $\ell = -1, \ldots, k$ are determined inductively by the equalities:

(b) $\quad m_{0,-1} = 1, \ m_{0,\ell} = 0 \ (\ell \geq 0), \ m_{i,\ell} = \begin{cases} a_i m_{i-1,\ell} + a_i b_i A_{i+1,\ell}, & i \leq \ell \\ a_{\ell+1} m_{\ell,\ell} + b_{\ell+1}, & i = \ell + 1 \\ a_i m_{i-1,\ell}, & i > \ell + 1 \end{cases}$

2. (Intersection number) $I(C_{k-1}, C; O) = \sum_{i=1}^k a_i b_i A_{i+1,k}^2 / a_k$.
3. (Milnor number) $\mu = 1 - A_{1,k} + \sum_{i=1}^k (A_{i,k} - 1) b_i A_{i+1,k}$.
4. The sequence of the weight vectors $\mathcal{W} = \{P_1, \ldots, P_k\}$ and the multiplicity sequence $\{m_{i,j}\}$ does not depend on the choice of a Tschirnhausen-good toric resolution tower.

We call $\mathcal{W}$ the sequence of the weight vectors of C at the origin and denote it by $\mathcal{W}(C; O)$ hereafter.

2.2 Moduli of polynomials with a given tower of weight vectors

Let $Q = {}^t(p, q)$ be a weight vector and let $d(Q; u^a v^b) = pa + qb$. For a polynomial $g(u, v) = \sum_{(a,b)} c_{a,b} u^a v^b$, we define $d(Q; g) = \min\{d(Q; u^a v^b); c_{a,b} \neq 0\}$ and $D(Q; g) = \max\{d(Q; u^a v^b); c_{a,b} \neq 0\}$ and let $g_Q(u, v)$ and $g^Q(u, v)$ be the sum of $c_{a,b} u^a v^b$ where they satisfy $d(Q; u^a v^b) = d(Q; g)$ and $d(Q; u^a v^b) = D(Q; g)$ respectively.

Let $\mathcal{W} = \{P_1, \ldots, P_k\}$ be a given sequence of primitive weight vectors and put $\mathcal{W}_\ell = \{P_1, \ldots, P_\ell\}$ for $\ell \leq k$. We consider subsets of polynomials $\mathcal{P}(\mathcal{W}_\ell; Q) \subset \mathcal{P}(\mathcal{W}_\ell) \subset \mathcal{P}$, which are defined by

$$\begin{aligned} \mathcal{P}(\mathcal{W}_\ell) &:= \{g \in \mathcal{P}; \mathcal{W}(C(g); O) = \mathcal{W}_\ell\}, \\ \mathcal{P}(\mathcal{W}_\ell; Q) &:= \{g \in \mathcal{P}(\mathcal{W}_\ell); D(Q; g) = q A_{1,\ell}\} \end{aligned}$$

They give the moduli spaces in which we are interested. Suppose that $h \in \mathcal{P}(\mathcal{W})$ (respectively $h \in \mathcal{P}(\mathcal{W}; Q)$) and let h_i be the corresponding $A_{i+1,k}$-th Tschirnhausen approximate polynomial of h for $i = 0, 1, \ldots, k - 1$. Using a Tschirnhausen toric resolution tower, it is easy to see that $h_j \in \mathcal{P}(\mathcal{W}_j)$ (resp. $h_j \in \mathcal{P}(\mathcal{W}_j; Q)$) for $j \leq k$ ([O4]).

A polynomial $h_{-1}^{\nu_{-1}} h_0^{\nu_0} \cdots h_\ell^{\nu_\ell}$ is called a *Tschirnhausen monomial* and it is said to be *admissible* if $\nu_i < a_{i+1}$ for any $1 \le i \le \ell$. Here $h_{-1} = u$. Note that h_0 is written as $h_0(u,v) = v + \sum_{j=1}^{s} c_j u^j$, $s = [q/p]$ and $c_1, \ldots, c_s \in \mathbf{C}$. (Recall that $[q/p]$ is the maximal integer m such that $m \le q/p$.) Let $Q_0 = {}^t(1,1)$. In this paper, we only consider the case $Q = Q_0$. So we can write as $h_0(u,v) = v + c_1 u$. For a non-negative multi-integer $M = (\nu_{-1}, \nu_0, \ldots, \nu_\ell)$, we denote the corresponding Tschirnhausen monomial $h_{-1}^{\nu_{-1}} \cdots h_\ell^{\nu_\ell}$ by $\mathbf{h}^M$. Any polynomial $g(u,v)$ with $\deg_v g(u,v) < A_{1,\ell+1}$ can be uniquely expanded as a linear combination of admissible Tschirnhausen monomials of $h_{-1}, \ldots, h_\ell$. In particular, by the definition of the Tschirnhausen polynomial, we have the expression:

$$h_{\ell+1} = h_\ell^{a_{\ell+1}} + R_{\ell+1}, \quad R_{\ell+1} = \sum_{M = (\nu_{-1}, \nu_0, \ldots, \nu_\ell) \in \mathbf{N}^{\ell+2}} C_M \mathbf{h}^M \tag{2.2.1}$$

where the Tschirnhausen monomials $h_{-1}^{\nu_{-1}} \cdots h_\ell^{\nu_\ell}$ satisfy $\nu_\ell < a_\ell - 1$ and $\nu_j < a_{j+1}$ for $0 \le j \le \ell - 1$ ([A-M2], [A-O]).

2.3 Algebraic construction of the moduli space

Let $\mathcal{W} = \{P_i = {}^t(a_i, b_i); i = 1, \ldots, k\}$ be a given weight vector sequence and let $\mathcal{N}_{\ell+1}(\mathcal{W})$ be the set of non-negative multi-integers $M = (\nu_{-1}, \nu_0, \ldots, \nu_\ell) \in \mathbf{N}^{\ell+2}$ such that

$$a_{\ell+1} \sum_{i=-1}^{\ell} \nu_i m_{\ell,i} + b_{\ell+1} \nu_\ell \ge a_{\ell+1}^2 m_{\ell,\ell} + a_{\ell+1} b_{\ell+1} \tag{2.3.1}$$

$$0 \le \nu_\ell < a_{\ell+1} - 1, \quad 0 \le \nu_j < a_{j+1}, \ j = 0, \ldots, \ell - 1. \tag{2.3.2}$$

Here the positive integers $m_{\ell,i}$, $0 \le \ell \le k$, $-1 \le i \le k$, are inductively defined by (b) in Theorem 2.1. Note that $\mathcal{N}_{\ell+1}(\mathcal{W})$ is an infinite set, as ν_{-1} can be arbitrarily large. The inequality (2.3.1) corresponds to the multiplicity condition of the pull-back $\Psi_{\ell+1}^* \mathbf{h}^M$ along the divisor $E_{\ell+1}$: $m_{\ell+1}(\mathbf{h}^M) \ge m_{\ell+1}(h_\ell^{a_{\ell+1}})$. Let $\mathcal{N}_{\ell+1}(\mathcal{W}; Q)$ be the subset of $\mathcal{N}_{\ell+1}(\mathcal{W})$ which consists of multi-integers $M = (\nu_{-1}, \nu_0, \ldots, \nu_\ell)$ satisfying (2.3.1), (2.3.2) and

$$p\nu_{-1} + q \sum_{j=0}^{\ell} \nu_j A_{1,j} \le q A_{1,\ell+1}. \tag{2.3.3}$$

Now $\mathcal{N}_{\ell+1}(\mathcal{W}; Q)$ is a finite set as ν_{-1} is also bounded. The inequality (2.3.3) says that $D(Q; \mathbf{h}^M) \le \mathbf{D}(\mathbf{Q}; \mathbf{v}^{\mathbf{A}_{1,\ell+1}}) = \mathbf{q}\mathbf{\Lambda}_{1,\ell+1}$ for any $M \in \mathcal{N}_{\ell+1}(\mathcal{W}; Q)$.

THEOREM 2.2 *Let $\mathcal{W} = \{P_i = {}^t(a_i, b_i); i = 1, \ldots, k\}$ be a given sequence of weight vectors with $a_i \ge 2$ for $i = 1, \ldots, k$.*

1. *There exists a unique multi-integer*

$$M_{\ell+1} = (\nu^{\#1}_{-1}\ell + 1, \nu^{\#1}_0\ell + 1, \ldots, \nu^{\#1}_\ell\ell + 1) \in \mathcal{N}_{\ell+1}(\mathcal{W})$$

which satisfies the equality in (2.3.1) and $M_{\ell+1}$ depend only on $\mathcal{W}_{\ell+1}$ and $\nu^{\#1}_\ell\ell + 1 = 0$. The set of polynomials $\mathcal{P}(\mathcal{W};Q)$ is non-empty if and only if $M_{\ell+1} \in \mathcal{N}_{\ell+1}(\mathcal{W};Q)$ for $\ell = 0, \ldots, k-1$.

2. *Assume that $h = h_k \in P(\mathcal{W};Q)$. Then $h_{\ell+1} \in P(\mathcal{W}_{\ell+1};Q)$ and it is written as*

$$(a) \qquad h_{\ell+1} = h_\ell^{a_{\ell+1}} + \sum_{M \in \mathcal{N}_{\ell+1}(\mathcal{W};Q)} C_M \mathbf{h}^M$$

for $\ell = 0, \ldots, k-1$. The corresponding coefficient $C_{M_{\ell+1}}$ is non-zero.

3. *Conversely, assume that $\tilde{h}_\ell \in P(\mathcal{W}_\ell;Q)$ and let $\tilde{h}_i$ be the $A_{i+1,\ell}$-th Tschirnhausen approximate polynomials of $\tilde{h}_\ell$, $i \leq \ell$. Assume further that $M_{\ell+1} \in \mathcal{N}_{\ell+1}(\mathcal{W};Q)$. For any coefficients $\{\widetilde{C}_M; M \in \mathcal{N}_{\ell+1}(\mathcal{W};Q)\}$ with $\widetilde{C}_{M_{\ell+1}} \neq 0$, the polynomial $\tilde{h}_{\ell+1}$ defined by*

$$\tilde{h}_{\ell+1} := \tilde{h}_\ell^{a_{\ell+1}} + \sum_{M \in \mathcal{N}_{\ell+1}(\mathcal{W};Q)} \widetilde{C}_M \tilde{\mathbf{h}}^M$$

is contained in $P(\mathcal{W}_{\ell+1};Q)$.

COROLLARY 2.3 *Assume that $M_\ell \in \mathcal{N}_\ell(\mathcal{W};Q)$ for $\ell = 1, \ldots, k$ and let r_ℓ be the cardinality of $\mathcal{N}_\ell(\mathcal{W};Q)$. Then the projection $\pi_{\ell+1} : P(\mathcal{W}_{\ell+1};Q) \to P(\mathcal{W}_\ell;Q)$ is a trivial fibration with fiber $\mathbf{C}^* \times \mathbf{C}^{r_{\ell+1}-1}$, where $\pi_{\ell+1}(h_{\ell+1})$ is defined by the $a_{\ell+1}$-th Tschirnhausen approximate polynomial of $h_{\ell+1}$. In particular, $P(\mathcal{W};Q)$ is isomorphic to $\mathbf{C}^{r_0} \times (\mathbf{C}^* \times \mathbf{C}^{r_1-1}) \times \cdots \times (\mathbf{C}^* \times \mathbf{C}^{r_k-1}) \cong \mathbf{C}^{*k} \times \mathbf{C}^{r-k}$ where $r = \sum_{i=0}^k r_i$ and $r_0 = [q/p]$.*

3 Correspondence to affine curves with one place at infinity

3.1 Space of polynomials with a given singularity at infinity

Let $\mathcal{POL}$ be the set of polynomials introduced in Section 1. Take $f(x,y) \in \mathcal{POL}$ and let $C^a(f) = \{(x,y) \in \mathbf{C}^2; f(x,y) = 0\}$ be the corresponding affine curve with one place at infinity. Taking a linear change of coordinates if necessary, we may assume that $C(f) \cap L_\infty = \{[1;0;0]\}$. Put $\rho_\infty := [1;0;0]$. As $C(f) \cap L_\infty =$

$\{\rho_\infty\}$ and $C(f)$ is locally irreducible at ρ_∞, we can write $f(x,y)$, up to a multiplication of a constant, that

$$f(x,y) = (y^{a_1} + \xi_1 x^{c_1})^{A_2} + \text{(lower terms)},$$
$$a_1 > c_1 \geq 1, \quad \gcd(a_1, c_1) = 1 \tag{3.1.1}$$

and let $F(X,Y,Z) = f(X/Z, Y/Z)Z^n$ and $h(u,v) = F(1, v, u)$ where $n = a_1 A_2$. See for example [L-O]. Here (lower term) is a linear combination of monomials $x^\nu y^\mu$ such that $a_1\nu + c_1\mu < a_1 c_1 A_2$. Note that $F(X,Y,Z)$ (respectively $h(u,v)$) is a monic polynomial in Y (resp. in v) and $\deg_Y F(X,Y,Z) = \deg F(X,Y,Z) = n$, $\deg_v h(u,v) = \deg h(u,v) = n$. Then $C := \{h(u,v) = 0\}$ defines an irreducible singularity at the origin and

$$h(u,v) = (v^{a_1} + \xi_1 u^{b_1})^{A_2} + \text{(higher terms)}, \quad b_1 = a_1 - c_1. \tag{3.1.2}$$

Let $\mathcal{W} = \{P_i = {}^t(a_i, b_i), \ i = 1, \ldots, k\}$ be the sequence of primitive weight vectors of (C, O) with $a_i \geq 2$, and $b_i \geq 1$ and $a_1 > b_1 \geq 1$. We call $\{P_1, \ldots, P_k\}$ *the sequence of the weight vectors of f at infinity* and we denote it by $\mathcal{W}(f; \rho_\infty)$. Let $F_i(X,Y,Z)$ be the $A_{i+1,k}$-th Tschirnhausen approximate polynomial of $F(X,Y,Z)$ and put $f_i(x,y) = F_i(x,y,1)$ and $h_i(1, v, u)$. Then f_i and h_i are the $A_{i+1,k}$-th Tschirnhausen approximate polynomials of $f(x,y)$ and $h(u,v)$ respectively and they are related by the equality: $f_i(x,y) = F_i(x,y,1) = h_i(1/x, y/x)x^{A_{i+1,k}}$. We define a subset $\mathcal{POL}(\mathcal{W}) \subset \mathcal{POL}$ by

$$\mathcal{POL}(\mathcal{W}) := \{g(x,y); h(u,v) := g(1/u, v/u)u^{A_{1,k}} \in \mathcal{P}(\mathcal{W}; Q_0)\}$$

where $Q_0 = {}^t(1,1)$. Thus if $g(x,y) \in \mathcal{POL}(\mathcal{W})$, $g(x,y)$ is a monic polynomial of degree $A_{1,k}$ and the affine curve $C(g) = \{g(x,y) = 0\}$ has one place at infinity which is $\rho_\infty := [1; 0; 0]$ in the homogeneous coordinates. Note that the condition (2.3.3) for Q_0 is simply equivalent to $\deg h_{\ell+1} \leq A_{1,\ell+1}$. We consider the subset of polynomials $f \in \mathcal{POL}(\mathcal{W})$, which defines a smooth affine curve $C^a(f)$ and we denote this subset by $\mathcal{POL}^{(r)}(\mathcal{W})$. Let $\mathcal{N}_{\ell+1}(\mathcal{W}; Q_0)$ be as in Section 2.

3.2 Algebraic description of $\mathcal{N}_{\ell+1}(\mathcal{W}; Q_0)$

Let us define $\lambda_{\ell+1} : \mathbf{N}^{\ell+2} \to \mathbf{N}^{\ell+2}$ by $\lambda_{\ell+1}(\nu_{-1}, \ldots, \nu_\ell) = (\mu_{-1}, \ldots, \mu_\ell)$, where

$$\mu_{-1} = A_{1,\ell+1} - \left(\sum_{j=0}^{\ell} \nu_j A_{1,j} + \nu_{-1}\right), \quad \mu_i = \nu_i, \ i \geq 0$$

and let $\mathcal{N}'_{\ell+1}(\mathcal{W}) := \lambda_{\ell+1}(\mathcal{N}_{\ell+1}(\mathcal{W}))$, $\mathcal{N}'_{\ell+1}(\mathcal{W}; Q_0) := \lambda_{\ell+1}(\mathcal{N}_{\ell+1}(\mathcal{W}; Q_0))$ and $M'_{\ell+1} := \lambda_{\ell+1}(M_{\ell+1}) \in \mathcal{N}'_{\ell+1}(\mathcal{W})$. For given multi-integers

$$M = (\nu_{-1}, \ldots, \nu_\ell) \in \mathcal{N}_{\ell+1}(\mathcal{W}; Q_0) \quad \text{and} \quad M' = (\mu_{-1}, \ldots, \mu_\ell) \in \mathcal{N}'_{\ell+1}(\mathcal{W}; Q_0),$$

we associate the respective Tschirnhausen monomials

$$\mathbf{h}^M := h_{-1}^{\nu_{-1}} \cdots h_{\ell}^{\nu_{\ell}}, \quad \mathbf{f}^{M'} := f_{-1}^{\mu_{-1}} \cdots f_{\ell}^{\mu_{\ell}} \tag{3.2.1}$$

where $h_{-1} = u$, $f_{-1} = x$ and $h_0 = v + cu$, $f_0 = y + c$. We also use the notation:

$$\lambda_{\ell+1}(\mathbf{h}^M) := \mathbf{f}^{\lambda_{\ell+1}(M)}, \quad \lambda_{\ell+1}^{-1}(\mathbf{f}^{M'}) := \mathbf{h}^{\lambda_{\ell+1}^{-1}(M')}. \tag{3.2.2}$$

Assume that $g(x,y)$ is a polynomial of degree $\deg g(x,y) \le A_{1,\ell+1}$ and let $g = \sum_{M'} \gamma_{M'} \mathbf{f}^{M'}$ be the Tschirnhausen expansion. We also define

$$\lambda_{\ell+1}^{-1}(g)(u,v) := g(1/u, v/u)u^{A_{1,\ell+1}} = \sum_{M'} \gamma_{M'} \lambda_{\ell+1}^{-1}(\mathbf{f}^{M'}). \tag{3.2.3}$$

Note that $\lambda_{\ell+1}^{-1}(\mathbf{f}^{M'})(u,v) = \mathbf{f}^{M'}(1/u, v/u)u^{A_{1,\ell+1}}$. Now Corollary 2.3 gives the following:

THEOREM 3.1

1. *The set of polynomials* $\mathcal{POL}(\mathcal{W})$ *is non-empty if and only if* $M_{\ell+1} \in \mathcal{N}_{\ell+1}(\mathcal{W}; Q_0)$ *for* $\ell = 0, \ldots, k-1$.

2. *Assume that* $f(x,y) \in \mathcal{POL}(\mathcal{W})$ *and* $f(x,y)$ *is expressed as (3.1.1). Then* $f_{\ell+1} \in \mathcal{POL}(\mathcal{W}_{k+1})$ *and* $f_{\ell+1}(x,y)$ *can be written as* $f_{\ell+1} = f_\ell^{a_{\ell+1}} + \sum_M C_M \mathbf{f}^M$ *and* $C_{M'_{\ell+1}} \ne 0$ *for* $\ell = 0, \ldots, k-1$ *where the summation is taken for* $M \in \mathcal{N}'_{\ell+1}(\mathcal{W}; Q_0)$.

3. *Assume that* $\mathcal{POL}(\mathcal{W}_{\ell+1}) \ne \emptyset$ *and* $\tilde{f}_\ell \in \mathcal{POL}(\mathcal{W}_\ell)$ *and let let* $\tilde{f}_1, \ldots, \tilde{f}_{\ell-1}$ *be Tschirnhausen approximate polynomials of* $\tilde{f}_\ell$. *For any coefficients* $\{\widetilde{C}_M\}$, $M \in \mathcal{N}'_{\ell+1}(\mathcal{W}; Q_0)$, *with* $\widetilde{C}_{M'_{\ell+1}} \ne 0$, *the polynomial* $\tilde{f}_{\ell+1} := f_\ell^{a_{\ell+1}} + \sum_M \widetilde{C}_M \tilde{\mathbf{f}}^M$ *is contained in* $\mathcal{POL}(\mathcal{W}_{\ell+1})$.

4. *If* $\mathcal{POL}(\mathcal{W}) \ne \emptyset$, $\mathcal{POL}(\mathcal{W})$ *is isomorphic to* $\mathbf{C}^{r_0} \times (\mathbf{C}^* \times \mathbf{C}^{r_1-1}) \times \cdots \times (\mathbf{C}^* \times \mathbf{C}^{r_k-1}) \cong \mathbf{C}^{*k} \times \mathbf{C}^{r-k}$ *where* $r = \sum_{i=0}^k r_i$ *and* $r_0 = 1$.

PROPOSITION 3.2 *The space* $\mathcal{POL}^{(r)}(\mathcal{W}_{\ell+1})$ *is a Zariski open dense subset of* $\mathcal{POL}(\mathcal{W}_{\ell+1})$. *In fact, we have*

$$M_c := (A_{1,\ell+1}, 0, \ldots, 0) \in \mathcal{N}_{\ell+1}(\mathcal{W}; Q_0), \quad (0, \ldots, 0) \in \mathcal{N}'_{\ell+1}(\mathcal{W}; Q_0). \tag{a}$$

Proof. For the proof of the assertion (a), see Appendix, §3, [N-O] or Theorem 6.2, [A-O]. The corresponding Tschirnhausen monomial gives the constant term of $f_{\ell+1}$. Therefore for a given $f_{\ell+1} \in \mathcal{POL}(\mathcal{W}_{\ell+1})$, $f_{\ell+1} + t$ is also in $\mathcal{POL}(\mathcal{W}_{\ell+1})$ and by Bertini's theorem, we can choose a generic t so that $\{f_{\ell+1}(x,y) + t = 0\}$ is smooth. So $f_{\ell+1} + t \in \mathcal{POL}^{(r)}(\mathcal{W}_{\ell+1})$. This implies the Zariski-openness. $\qquad\square$

DEFINITION 3.3 Let $f(x,y) \in \mathcal{POL}$. We define

$$\deg^+(f) = \max\{\deg_x(f), \deg_y(f)\}, \quad \deg^-(f) = \min\{\deg_x(f), \deg_y(f)\}.$$

The pair of integers $(\deg^+(f), \deg^-(f)) \in \mathbf{N}^2$ is called the *degree-type* of $f(x,y)$. We put the lexicographic order in $\mathbf{N}^2$. Assume that $f \in \mathcal{POL}$ is written as

$$f(x,y) = (\xi y^{a_1} + \lambda x^{c_1})^{A_2} + \text{(lower terms)}, \quad \lambda, \ \xi \neq 0, \ \gcd(a_1, c_1) = 1.$$

We say that $f(x,y)$ is *minimal* if $\min(a_1, c_1) > 1$. Assuming $a_1 \geq c_1$, f is minimal if and only if the ratio $\deg^+(f)/\deg^-(f)$ is not an integer, as $\gcd(a_1, c_1) = 1$.

We say that f and g are *w-equivalent* if their weight sequences at infinity coincide. When f and g are w-equivalent, we denote it by $f \overset{w}{\sim} g$. By definition, two polynomials $f, g \in \mathcal{POL}(\mathcal{W})$ are w-equivalent. If $f \overset{w}{\sim} g$, their degree-types are equal and the germs of plane curves $(C(f), \rho_\infty)$ and $(C(g), \rho_\infty)$ are topologically equivalent.

4 $\mathrm{Aut}(\mathbf{C}^2)$ and its action on the space of polynomials

Let $G := \mathrm{Aut}(\mathbf{C}^2)$ be the group of algebraic automorphisms of $\mathbf{C}^2$. First we recall basic properties about G. Let (x, y) be a fixed system of affine coordinates of $\mathbf{C}^2$. The following 3 types of automorphisms are called *primitive*.

$$\alpha_r^{a,b} : (x,y) \mapsto (ax + r(y), by) \qquad\qquad \text{(Type I)}$$
$$\beta_r^{a,b} : (x,y) \mapsto (bx, ay + r(x)) \qquad\qquad \text{(Type II)}$$
$$\tau : (x,y) \mapsto (y, x) \qquad \text{(permutation)} \qquad \text{(Type III)}$$

where $r(T) \in \mathbf{C}[T]$ and $a, b \neq 0$. As special cases of primitive automorphisms, we have a change of scale $\iota_{a,b} : (x,y) \mapsto (ax, by)$, and parallel translations $\theta_a : (x,y) \mapsto (x + a, y)$ and $\theta_a' : (x,y) \mapsto (x, y + a)$. Put $\theta_{a,b} := \theta_a \theta_b'$. We say that $g \in G$ is *reduced* if it fixes the origin, i.e., $g(O) = O$ where $O = (0,0)$. We denote the subgroup of the reduced automorphisms by $\widetilde{G}$. Note that $\tau \in \widetilde{G}$ and $\alpha_r^{a,b}, \beta_r^{a,b}, \in \widetilde{G}$ if and only if $r(0) = 0$. We define $\deg \alpha_r^{a,b} = \deg r$ and $\deg \beta_r^{a,b} = \deg r$. Let Θ be the subgroup of G which consists of the parallel translations $\{\theta_{a,b}; a, b \in \mathbf{C}\}$ and let $\mathcal{I}$ be the subgroup generated by the changes of scales $\{\iota_{a,b}; a, b \in \mathbf{C}^*\}$. Then $\Theta \cong \mathbf{C}^2$ and $\mathcal{I} \cong \mathbf{C}^* \times \mathbf{C}^*$. Let $\varphi, \psi \in G$. The multiplication $\varphi\psi$ is defined by the composition of automorphisms,

$$\varphi\psi = \psi \circ \varphi : \mathbf{C}^2 \to \mathbf{C}^2.$$

For any given $g \in G$, putting $g(O) = (a, b)$ and $g^{-1}(O) = (c, d)$, we can write g as a product $g = g_1 \theta_{a,b} = \theta_{c,d}^{-1} g_2$, where $g_1 := g\theta_{a,b}^{-1}$, $g_2 := \theta_{c,d} g$. Observe that $g_1, g_2 \in \widetilde{G}$. Therefore $G = \Theta\widetilde{G} = \widetilde{G}\Theta$.

LEMMA 4.1 ([Ju ,[A])] 1. *The group of automorphisms G (respectively $\widetilde{G}$) is generated by primitive automorphisms (resp. reduced primitive automorphisms).*
2. *Assume that $\varphi = (f, g) \in \widetilde{G}$. If neither φ nor $\varphi\tau$ are primitive, f and g are written as*

$$f(x, y) = (\xi y^{a_1} + \lambda x^{c_1})^m + \text{(lower terms)},$$
$$g(x, y) = (\xi y^{a_1} + \lambda x^{c_1})^n + \text{(lower terms)}$$

where $\xi, \lambda \neq 0$, $\gcd(a_1, c_1) = 1$ and $\min(a_1, c_1) = 1$.
3. *For any $\widetilde{\varphi} = (f, g) \in \widetilde{G}$, either $\deg_x(f)$ divides $\deg_x(g)$ or $\deg_x(g)$ divides $\deg_x(f)$. The same assertion holds for $\deg_y(f)$ and $\deg_y(g)$.*

Proof. Assertion 1 is proved by [Ju]. Assume that $\varphi \in G$ and let $(c, d) = \varphi^{-1}(O)$ and put $\widetilde{\varphi} = \theta_{c,d}\varphi$. Then $\widetilde{\varphi} \in \widetilde{G}$ and $\varphi = \theta_{a,b}^{-1}\widetilde{\varphi}$. Let $\widetilde{\varphi}(x, y) = (f(x, y), g(x, y))$. If f or g is a monomial, it must be either ax or ay by the Jacobian condition: $J(f, g) = f_x g_y - f_y g_x \equiv c \neq 0$. If $f(x, y) = ax$ or ay for example, the Jacobian condition $J(f, g) = f_x g_y - f_y g_x \equiv c \neq 0$ implies that $g(x, y) = by + p(x)$ or $bx + p(y)$ respectively for some $b \neq 0$ and $p(x) \in \mathbf{C}[x]$. Thus either φ or $\widetilde{\varphi}\tau$ is primitive. Thus Assertions $1 \sim 3$ are obvious in this case. Now we assume that neither $f(x, y)$ nor $g(x, y)$ are monomials. As φ has an algebraic inverse, it is easy to see that $f, g \in \mathcal{POL}^{(r)}(0)$. So they have one place at infinity and they are written as

$$f(x, y) = (\xi y^{a_1} + \lambda x^{c_1})^m + \text{(lower terms)},$$
$$g(x, y) = (\xi y^{a_1} + \lambda x^{c_1})^n + \text{(lower terms)}.$$

As the Jacobian $J(f, g)$ is a non-zero constant, we need to have $\min(a_1, c_1) = 1$ ([A], [O1]). This proves Assertion 2. Suppose that (f, g) is written as above and $c_1 = 1$ for example. Let $\psi \in \widetilde{G}$ be defined by $\psi(x, y) = (\xi y^{a_1} + \lambda x, y)$. (In the case of $a_1 = 1$, we take $\psi \in \widetilde{G}$ defined by $\psi(x, y) = (x, \xi y + \lambda x^{c_1})$.) Note that $\deg_x f^\psi = m$ and $\deg_x g^\psi = n$. We assume for brevity that $m \leq n$. If either $\psi^{-1}\widetilde{\varphi} = (f^\psi, g^\psi)$ or $\psi^{-1}\widetilde{\varphi}\tau$ is primitive, we need to have $m = 1$ and $f(x, y) = \xi y^{a_1} + \lambda x$ from the beginning. Assertions 1, 2, 3 are obvious in this case. If $\psi^{-1}\widetilde{\varphi}$ is not still primitive, we can write $\psi^{-1}\widetilde{\varphi} = (f^\psi, g^\psi)$ again as

$$f^\psi(x, y) = (\xi' y^{a_1'} + \lambda' x^{c_1'})^{m'} + \text{(lower terms)},$$
$$g^\psi(x, y) = (\xi' y^{a_1'} + \lambda' x^{c_1'})^{n'} + \text{(lower terms)}$$

where $c_1' m' = m$, $c_1' n' = n$, $\lambda' \neq 0$ and $\min(a_1', c_1') = 1$. By the definition of ψ, it is easy to see that $(\deg^+(f^\psi), \deg^-(f^\psi)) < (\deg^+(f), \deg^-(f))$. Thus Assertion 1 can be easily proved by the induction on the degree-type $(\deg^+(f), \deg^-(f))$. The last assertion about the divisibility follows also by induction as $m/n = m'/n'$. $\qquad\square$

DEFINITION 4.2 We consider the following subsets of the group of automorphisms G

$$G_I := \{\alpha_r^{a,b}\}, \quad G_I(s) := \{\alpha_r^{a,b}; \deg r \leq s\}, \quad G_I(s)' := \{\alpha_r; \deg r \leq s\}$$
$$\widetilde{G}_I := G_I \cap \widetilde{G}, \quad \widetilde{G}_I(s) := G_I(s) \cap \widetilde{G}, \quad \widetilde{G}_I(s)' := G_I(s)' \cap \widetilde{G}.$$

Here $r \in \mathbf{C}[T]$ and $\alpha_r := \alpha_r^{1,1} : (x,y) \mapsto (x + r(y), y)$. The equality $\alpha_r^{a,b}\alpha_{r'}^{a',b'} = \alpha_{r''}^{aa',bb'}$ with $r''(y) := a'r(y) + r'(by)$, implies that these subsets are, in fact, subgroups of G. In particular, $G_I(s)', \widetilde{G}_I(s)'$ are isomorphic to the additive groups $\mathbf{C}^{s+1}$ and $\mathbf{C}^s$ respectively. By the relations

$$\iota_{a^{-1},b^{-1}}\alpha_r\iota_{a,b} = \alpha_{r'}, \qquad r'(y) = ar(y/b),$$
$$\alpha_r^{a,b} = \iota_{a,b}\alpha_\ell, \quad \ell(y) = r(y/b), \tag{4.2.1}$$

$G_I(s)'$ is a normal subgroup of $G_I(s)$ and the quotient is isomorphic to $\mathcal{I}$. Similarly, we can consider the primitive automorphisms of type II, $\beta_r, \beta_r^{a,b} \in G$ and we define subgroups $G_{II}, \widetilde{G}_{II}, G_{II}(s), \widetilde{G}_{II}(s), G_{II}(s)', \widetilde{G}_{II}(s)'$ similarly. Note that

$$\tau\alpha_r^{a,b}\tau = \beta_r^{a,b}, \quad \tau\beta_r^{a,b}\tau = \alpha_r^{a,b}. \tag{4.2.2}$$

We study the canonical action of G from the right side on the space of polynomials $\mathcal{POL}, \mathcal{POL}^{(r)}, \mathcal{POL}^{(r)}(g)$ which are introduced in Section 1. Recall that $f \in \mathcal{POL}^{(r)}(g)$ if and only if $f \in \mathcal{POL}$, the affine curve $C^a(f)$ is smooth and the normalization of $C(f)$ is a Riemann surface of genus g. The action is defined by $\mathcal{POL} \times G \to \mathcal{POL}, \quad (f, \varphi) \mapsto f^\varphi$ where $f^\varphi(x,y) := f(\varphi^{-1}(x,y))$. The following is immediate from the definition.

PROPOSITION 4.3 *For any $f \in \mathcal{POL}$ and $\varphi \in G_I$, $\deg_x f^\varphi = \deg_x f$. For any $\psi \in G_{II}$, $\deg_y f^\psi = \deg_y f$. We have also*

$$(\deg^+(f^\tau), \deg^-(f^\tau)) = (\deg^+(f), \deg^-(f))$$

where τ is the permutation automorphism.

Recall that the *stabilizer* of a subset $S \subset \mathcal{POL}$ is defined by

$$\mathrm{Stab}(S) := \{\varphi \in G; f^\varphi \in S, \forall f \in S\}.$$

Now we consider the stabilizer of $\mathcal{POL}(\mathcal{W}) \subset \mathcal{POL}$ where $\mathcal{W} = \{P_i = {}^t(a_i, b_i); i = 1, \ldots, k\}$ are a given sequence of weight vectors with $a_i \geq 2$, $i = 1, \ldots, k$. We put $n = A_{1,k}$ as before.

LEMMA 4.4 *Assume that $\mathcal{POL}(\mathcal{W})$ is not empty. Then the two stabilizers $\mathrm{Stab}(\mathcal{POL}(\mathcal{W}))$ and $\mathrm{Stab}(\mathcal{POL}^{(r)}(\mathcal{W}))$ coincide.*

Proof. In fact, the inclusion $\mathrm{Stab}(\mathcal{POL}(\mathcal{W})) \subset \mathrm{Stab}(\mathcal{POL}^{(r)}(\mathcal{W}))$ is obvious. If $f \in \mathcal{POL}(\mathcal{W})$, we can find a constant $c \in \mathbf{C}$ so that $f + c \in \mathcal{POL}^{(r)}(\mathcal{W})$ by Proposition 3.2. Thus for any $\varphi \in \mathrm{Stab}(\mathcal{POL}^{(r)}(\mathcal{W}))$,

$$f^\varphi = (f + c)^\varphi - c \in \mathcal{POL}(\mathcal{W}).$$

Hence we have the opposite inclusion: $\mathrm{Stab}(\mathcal{POL}(\mathcal{W})) \supset \mathrm{Stab}(\mathcal{POL}^{(r)}(\mathcal{W}))$. $\square$

Let us consider subgroups $\Theta_{II} \subset \Theta$ and $G_I(s;n) \subset G_I(s)$ defined by

$$\Theta_{II} = \langle \theta'_a ; a \in \mathbf{C} \rangle, \quad G_I(s;n) = \langle \alpha_r^{a,b} \in G_I(s); b^n = 1 \rangle.$$

The product $\Theta_{II} G_I(s;n) = \{\theta'_c \alpha_r^{a,b}; \theta_c \in \Theta_{II}, \alpha_r^{a,b} \in G_I(s;n)\}$ is a subgroup of G by the following relation: $\alpha_r^{a,b} \theta'_c = \theta'_{c/b} \alpha_{r'}^{a,b}$ where $r'(y) = r(y - c/b)$.

THEOREM 4.5 *Assume that $a_1 > b_1 \geq 1$. Put $c_1 = a_1 - b_1$ and let $s =$ be the maximal integer such that $s c_1 < a_1$.*

1. *Then we have the inclusion: $\Theta_{II} G_I(s;n) \subset \mathrm{Stab}(\mathcal{POL}(\mathcal{W}))$.*

2. *Assume that $a_1 > c_1 > 1$ and $f \in \mathcal{POL}(\mathcal{W})$. For any $\varphi \in \widetilde{G}$ which satisfies $\deg_y f^\varphi \geq \deg_x f^\varphi$, we have $(\deg_y f^\varphi, \deg_x f^\varphi) \geq (\deg_y f, \deg_x f)$ and the equality is taken if and only if $\varphi \in \widetilde{G}_I(s)$.*

3. *Assume that $a_1 > c_1 > 1$. Then $\mathrm{Stab}(\mathcal{POL}(\mathcal{W})) = \Theta_{II} G_I(s;n)$.*

First we prepare a key lemma. Let $\varphi = \alpha_r \in G_I(s)'$ and $f(x,y) \in \mathcal{POL}(\mathcal{W})$ where $r \in \mathbf{C}[T]$ and $\deg r \leq s$. Let $f_i(x,y)$ and $h_i(u,v)$ be $A_{i+1,k}$-th Tschirnhausen approximate polynomials of $f(x,y)$ and $h(x,y) := f(1/u, v/u)u^{A_{1,k}}$ respectively. As we have the expression, $f_i(x,y) = (y^{a_1} + \xi_1 x^{c_1})^{A_{2,i}} + $ (lower terms), and $\deg r < a_1/c_1$ by the assumption, it is clear that $\deg f_i^\varphi(x,y) = A_{1,i}$ and the leading degree term $(f_i^\varphi)^P$ of f_i^φ with respect to $P = {}^t(a_1, c_1)$ is given by $(f_i^\varphi)^P = (y^{a_1} + \xi_1 x^{c_1})^{A_{2,i}}$. We define $h_i^\varphi(u,v) := \lambda_i^{-1}(f_i^\varphi) = f_i^\varphi(1/u, v/u)u^{A_{1,i}}$ and let $C_i^\varphi = \{h_i^\varphi(u,v) = 0\}$. Note that C_i^φ is irreducible at ρ_∞ because C_i has one place at infinity. Under this situation, we assert:

LEMMA 4.6 *There exists a Tschirnhausen-good toric resolution tower of (C^φ, ρ_∞): $\mathcal{T} = \{\psi_j : X_j \to X_{j-1}; j = 1, \ldots, k\}$, $X_0 = \mathbf{C}^2$ which has $\mathcal{W}$ as the sequence of weight vectors, and $\mathcal{T}$ satisfies the following property. Let $\Psi_i : X_i \to X_0$ be the composition $\psi_1 \circ \cdots \circ \psi_i$ and let E_i be the component of the exceptional divisors of ψ_i which intersects with the strict transform $(C^\varphi)^{(i)}$ of C^φ to X_i and let $\Xi_i = E_i \cap (C^\varphi)^{(i)}$. Then $\Psi_i : X_i \to X_0$ is a good resolution of C_i^φ and with respect to the chosen coordinate (u_i, v_i) centered at Ξ_i and*

$E_i = \{u_i = 0\}$ and for $j \geq 0$,

$$\Psi_i^* h_j^\varphi(u_i, v_i) = \begin{cases} u_i^{m_{i,j}} \times (\text{a unit}), & j \leq i-1 \\ u_i^{m_{i,i}} v_i, & j = i \\ u_i^{m_{i,j}}\left((v_i^{a_{i+1}} + \xi_{i+1} u_i^{b_{i+1}})^{A_{i+2,k}} + (\text{higher terms})\right), & j > i \end{cases}$$

and $\Psi_i^* u(u_i, v_i) = u_i^{m_{i,-1}} \times (\text{a unit})$ where the integers $m_{i,j}, -1 \leq j \leq k$ are inductively defined by (2.3) and $\xi_{i+1} \neq 0$. In particular, $\mathcal{W}(f^\varphi; \rho_\infty) = \mathcal{W}$ and therefore $f^\varphi \overset{w}{\sim} f$ and $f_j^\varphi \overset{w}{\sim} f_j$ for $j = 1, \ldots, k-1$.

We remark that f_i^φ is not necessarily the $A_{i+1,k}$-th Tschirnhausen approximate polynomial of f^φ. Therefore the resolution tower constructed in Lemma 4.6 is not necessarily a Tschirnhausen resolution tower. The proof is essentially parallel to the proof of Theorem 4.5 of [A-O] and is given in §7.

Proof of Theorem 4.5. Firstly, we show Assertion 1. Take $f \in \mathcal{POL}(\mathcal{W})$ and $\varphi = \theta_c' \alpha_r^{a,b} \in \Theta_{II} G_I(s; n)$. As $\theta_c' \in \text{Stab}(\mathcal{POL}(\mathcal{W}))$, we may assume that $\varphi = \alpha_r^{a,b}$.

Moreover, as $\alpha_r^{a,b} = \iota_{a,b} \alpha_{r'}$, $r'(y) = r(y/b)$ and $\iota_{a,b} \in \text{Stab}(\mathcal{POL}(\mathcal{W}))$, the essential point is to show that $\alpha_{r'} \in \text{Stab}(\mathcal{POL}(\mathcal{W}))$. This results from Lemma 4.6 and Theorem 4.7 of [A-O]. This proves Assertion 1.

Now we prove Assertion 2 of Theorem 4.5. Assume that $f \in \mathcal{POL}(\mathcal{W})$ and $f(x, y)$ is written as

$$f(x, y) = (y^{a_1} + \xi_1 x^{c_1})^{A_{2,k}} + \sum_{M=(\nu,\mu)} c_M x^\nu y^\mu, \quad a_1 > c_1 > 1, \ n = a_1 A_{2,k}.$$

For any weight vector $P = {}^t(p, q)$, we have $D(P; g_1(x,y)^\alpha g_2(x,y)^\beta) = \alpha D(P; g_1) + \beta D(P; g_2)$. We define a weight vector $P^\varphi := {}^t(D(P; g_1), D(P; g_2))$. Then we have $D(P^\varphi; x^\alpha y^\beta) = D(P; g_1(x,y)^\alpha g_2(x,y)^\beta)$. Assume that $\varphi \in \widetilde{G}$ and let $\varphi^{-1} = (g_1, g_2)$. Note that

$$f^\varphi(x, y) = (g_2(x,y)^{a_1} + \xi_1 g_1(x,y)^{c_1})^{A_{2,k}} + \sum_{M=(\nu,\mu)} c_M g_1(x,y)^\nu g_2(x,y)^\mu.$$

Therefore we have

$$D(P; f^\varphi(x,y)) \leq \max\{D(P; g_1^{nc_1/a_1}), D(P; g_2^n)\} = D(P^\varphi; f).$$

The equality holds if $D(P; g_1^{nc_1/a_1}) \neq D(P; g_2^n)$. Note that we have

$$D(E_1; g(x,y)) = \deg_x g(x,y) \ \text{ and } \ D(E_2; g(x,y)) = \deg_y g(x,y)$$

where $E_1 = {}^t(1, 0)$, $E_2 = {}^t(0, 1)$. Taking $P = E_i$, we assert:

ASSERTION 4.7 *Assume that $a_1 > c_1 > 1$. Then $\deg_x g_1^{nc_1/a_1} \neq \deg_x g_2^n$ and $\deg_y g_1^{nc_1/a_1} \neq \deg_y g_2^n$. In particular, we have*

$$\deg_y f^\varphi = \max\{\deg_y g_1^{nc_1/a_1}, \deg_y g_2^n\},$$
$$\deg_x f^\varphi = \max\{\deg_x g_1^{nc_1/a_1}, \deg_x g_2^n\}.$$

Proof. Assume that $\deg_x g_1^{nc_1/a_1} = \deg_x g_2^n$, for example. Then we get the equality:

$$nc_1/a_1 \times \deg_x g_1 = n \deg_x g_2$$

and this implies that $\deg_x g_1, \deg_x g_2 \neq 0$ and $a_1/c_1 = \deg_x g_1/\deg_x g_2$. As $a_1/c_1 > 1$, the right side is a positive integer by Lemma 4.1. This is a contradiction as $a_1/c_1 \notin \mathbf{Z}$. The same discussion works for $\deg_y g_i$. $\qquad\square$

Now we are ready to finish the *Proof of Assertion 2 in Theorem 4.5*. Assume that $\deg_y f^\varphi \geq \deg_x f^\varphi$. It is clear that φ is not the permutation τ. Assume first that $\deg_x g_2 \neq 0$.

Then $\deg_x f^\varphi = \max\{nc_1/a_1 \deg_x g_1, n \deg_x g_2\} \geq n \deg_x g_2 \geq n$ and therefore $\deg_y f^\varphi \geq \deg_x f^\varphi \geq n$. Thus we have the strict inequality:

$$(\deg_y f^\varphi, \deg_x f^\varphi) > (\deg_y f, \deg_x f).$$

Next, assume that $\deg_x g_2 = 0$. This implies that $g_2(x, y) = by$ for some $b \in \mathbf{C}^*$ and therefore $\varphi^{-1} = \alpha_r^{a,b}$ for some $a \in \mathbf{C}^*$ and $r(T) \in C[T]$ with $r(0) = 0$ and we may assume $\deg r > 0$. Then we get $\deg_y f^\varphi = \max\{nc_1/a_1 \times \deg r, n\} \geq n$ and $\deg_x f^\varphi = nc_1/a_1$. Thus we see that $(\deg_y f^\varphi, \deg_x f^\varphi) \geq (\deg_y f, \deg_x f)$ and the equality holds if and only if $\deg r \leq a_1/c_1$, namely if $\varphi \in \widetilde{G}_I(s)$ with $s = [a_1/c_1]$. This proves Assertion 2.

Now we assume that $\varphi \in \mathrm{Stab}(\mathcal{POL}(\mathcal{W}))$. Let $(a, b) = \varphi^{-1}(O)$. Then we can write $\varphi = \theta_{a,b}^{-1}\widetilde{\varphi}$ with $\widetilde{\varphi} \in \widetilde{G}$. As $\theta_{a,b} \in \mathrm{Stab}(\mathcal{POL}(\mathcal{W}))$, we have $\widetilde{\varphi} \in \mathrm{Stab}(\mathcal{POL}(\mathcal{W}))$. Thus by Assertion 2, $\widetilde{\varphi} \in \widetilde{G}_I(s)$. Put $\widetilde{\varphi} = \alpha_r^{c,d}$ with $\deg r \leq s$. As $\alpha_r^{c,d} \in \mathrm{Stab}(\mathcal{POL}(\mathcal{W}))$, it is necessary that $d^n = 1$. Thus $\alpha_r^{c,d} \in G_I(s; n)$. Thus we conclude $\varphi = \theta_{a,b}\alpha_r^{c,d} = \theta_b'\alpha_{r'}^{c,d} \in \Theta_{II}G_I(s; n)$ with $r'(y) = r(y) + ca$. This completes the proof of Assertion 3. $\qquad\square$

As a corollary of Theorem 4.5, we get the uniqueness theorem for a minimal model:

THEOREM 4.8 *For a given $f \in \mathcal{POL}$, there exists an automorphism $\varphi \in G$ so that $f^\varphi(x, y)$ is minimal. The sequence of weight vectors at infinity $\mathcal{W}(f^\varphi; \rho_\infty)$ of f^φ does not depend on the choice of φ.*

Hereafter we refer $\mathcal{W}(f^\varphi; \rho_\infty)$ *the minimal weight sequence of f* and we denote it by $\mathcal{MW}(f)$.

COROLLARY 4.9 *A polynomial $f \in \mathcal{POL}$ is minimal if and only if the degree-type of f is minimal among the orbit $G \cdot f$.*

Proof of Theorem 4.8. Assume that there exists two automorphisms $\varphi, \psi \in G$ so that f^φ and f^ψ are minimal. Taking the action of τ if necessary, we may assume that $\deg^+ f^\varphi = \deg_y f^\varphi$ and $\deg^+ f^\psi = \deg_y f^\psi$. Put $\xi := \varphi^{-1}\psi \in G$. Then we have the relation $(f^\varphi)^\xi = f^\psi$ among two minimal polynomials f^φ and f^ψ. Write $\xi = \tilde{\xi}\theta_{a,b}$ with $\tilde{\xi} \in \tilde{G}$ and put $\psi' = \psi\theta_{-a,-b}$.

Then $f^{\psi'} = (f^\varphi)^{\tilde{\xi}}$. Note that f^ψ and $f^{\psi'}$ have the same degree type. By Theorem 4.5, we see that $(\deg_y f^\psi, \deg_x f^\psi) \geq (\deg_y f^\varphi, \deg_x f^\varphi)$. By the same discussion for $(f^\psi)^{\psi^{-1}\varphi} = f^\varphi$, we get the opposite inequality. Thus f^φ and f^ψ have also the same degree type.

We show the existence. Assume that f is not minimal. We assume for simplicity that $\deg_y f \geq \deg_x f$. Then we can write $f(x,y)$ as

$$f(x,y) = (\lambda y^{a_1} + \xi x)^{A_2} + \text{(lower terms)}, \quad \lambda, \xi \in \mathbf{C}^*.$$

Take $g \in G$ defined by $g(x,y) = (\lambda y^{a_1} + \xi x, y)$. Then we see that $\deg_x f^g(x,y) = A_2$ and $\deg_y f^g(x,y) < a_1 A_2$. Thus $(\deg^+ f^g, \deg^- f^g) < (\deg^+ f, \deg^- f)$ and the existence of an automorphism which minimalizes f follows from an inductive argument on the degree type of f. $\square$

5   Structure of the quotient $\mathcal{POL}^{(r)}(g)/G$

5.1   Geometric minimal w-solution

First we recall the following fact. Let f be a polynomial of degree n in $\mathcal{POL}$ and let $C(f) \subset \mathbf{P}^2$ be the corresponding projective curve. Let $q_0, q_1, \ldots, q_s$ be the singular points of $C(f)$ where $q_0 = \rho_\infty$ and $q_1, \ldots, q_s \in C^a(f)$ and let $\mu_i := \mu(f; q_i)$ be the Milnor number at q_i. Then by Plücker's formula and by Mayer-Vietoris argument, the topological Euler number of the projective curve $\chi(C(f))$ is given by

$$\text{(Modified Plücker formula)}: \quad \chi(C(f)) = 3n - n^2 + \sum_{i=0}^{s} \mu_i. \tag{5.1.1}$$

We consider a subspace $\mathcal{POL}(\{\mu_1, \ldots, \mu_s\}; \nu) \subset \mathcal{POL}$ which is defined as follows. A polynomials $f \in \mathcal{POL}$ is contained in $\mathcal{POL}(\{\mu_1, \ldots, \mu_s\}; \nu)$ if and only if an affine curve $C^a(f)$ has exactly s singular points with given Milnor numbers $\mu_1, \ldots, \mu_s$ respectively and the topological Euler number $\chi(C(f))$ is equal to ν. In the case of $s = 0$, we denote it by $\mathcal{POL}(\{\emptyset\}; \nu)$. Thus the space $\mathcal{POL}^{(r)}(g)$ is nothing but $\mathcal{POL}(\{\emptyset\}; 2 - 2g)$ in this notation. It is obvious that $\mathcal{POL}(\{\mu_1, \ldots, \mu_s\}; \nu)$ is stable by the action of G. Let

$$\mathcal{POL}(\mathcal{W}; \{\nu_1, \ldots, \mu_s\}; \nu) := \mathcal{POL}(\mathcal{W}) \cap \mathcal{POL}(\{\mu_1, \ldots, \mu_s\}; \nu).$$

Assume that $f \in \mathcal{POL}(\mathcal{W}; \{\nu_1, \ldots, \mu_s\}; \nu)$ where $\mathcal{W} = \{P_1, \ldots, P_k\}$ and $P_i = {}^t(a_i, b_i)$. Using (5.1.1) and Theorem 2.1, we have the equality:

$$\sum_{i=1}^{k}(A_{i,k} - 1)b_i A_{i+1,k} = \nu - 2 + (n-1)^2 - \sum_{i=1}^{s}\mu_i, \quad n = A_{1,k} \qquad (5.1.2)$$

where $\nu = \chi(C(f))$. On the other hand, Bezout Theorem gives the inequality:

$$C(f) \cdot C(f_{k-1}) = n^2/a_k \geq I(C(f), C(f_{k-1}); \rho_\infty)$$

where $C(f) \cdot C(f_{k-1})$ is the intersection number in $\mathbf{P}^2$. Put $c_1 = a_1 - b_1$ as before. Using Theorem 2.3 again, we obtain the inequality:

$$\sum_{i=1}^{k} a_i b_i A_{i+1,k}^2 \leq A_{1,k}^2, \quad \text{or equivalently,} \quad \sum_{i=2}^{k} a_i b_i A_{i+1,k}^2 \leq c_1 a_1 A_{2,k}^2. \qquad (5.1.3)$$

Assume that $k \geq 2$. Then taking the sum : $(5.1.2) \times A_{2,k} + (5.1.3) \times (1 - A_{2,k})$, we obtain

$$0 \geq \sum_{i=2}^{k} b_i A_{i+1,k}(A_{i,k} - A_{2,k})$$

$$\geq A_{2,k}^2((a_1 - 1)(c_1 - 1) - 1) - (1 - \nu)A_{2,k} - A_{2,k}\sum_{i=1}^{s}\mu_i.$$

We say that the sequence of weight vectors $\mathcal{W}$ is *minimal* if the first weight vector P_1 satisfies the inequality: $a_1 > c_1 > 1$. Assuming that $\mathcal{W}$ is minimal and $k \geq 2$, we have

$$A_{2,k} \leq A_{2,k}((a_1 - 1)(c_1 - 1) - 1) \leq (1 - \nu) + \sum_{i=1}^{s}\mu_i. \qquad (5.1.4)$$

A minimal weight sequence $\mathcal{W}$ which satisfies (5.1.2) and (5.1.3) is called a *numerical minimal w-solution* of $\mathcal{POL}(\{\mu_1, \ldots, \mu_s\}; \nu)$ ([N-O]). A numerical minimal w-solution $\mathcal{W}$ is called a *geometric minimal w-solution* of $\mathcal{POL}(\{\mu_1, \ldots, \mu_s\}; \nu)$ if $\mathcal{POL}(\mathcal{W}; \{\mu_1, \ldots, \mu_s\}; \nu) \neq \emptyset$.

THEOREM 5.1

1. *The number of numerical minimal w-solutions of $\mathcal{POL}(\{\mu_1, \ldots, \mu_s\}; \nu)$ is finite. Thus the number of geometric minimal w-solutions is also finite.*

2. *([N-O]). In particular, $\mathcal{POL}^{(r)}(g)$ has a finite number of geometric minimal w-solutions. They satisfy (5.1.2) and (5.1.3), where $\nu = 2 - 2g$.*

3. $POL^{(r)}(\mathcal{W})$ is open dense in $POL(\mathcal{W})$ and the boundary

$$\partial POL^{(r)}(\mathcal{W}) := POL(\mathcal{W}) - POL^{(r)}(\mathcal{W})$$

is given by the finite union $\bigcup POL(\mathcal{W}; \{\mu_1, \ldots, \mu_s\}; \nu)$ where the union is taken for $\{s; \mu_1, \ldots, \mu_s; \nu\}$, $s \geq 1$ and $\nu \leq 2$ satisfing (5.1.2).

Proof. Assume that $f \in POL(\mathcal{W}; \{\mu_1, \ldots, \mu_s\}; \nu)$ and $a_1 > c_1 > 1$. In the case $k = 1$, there are obviously finite possible (a_1, c_1)'s by (5.1.2): $(a_1 - 1)(c_1 - 1) = 2 - \nu + \sum_{i=1}^{s} \mu_i$. For $k \geq 2$, we have the inequality by (5.1.4): $A_{2,k} \leq (1 - \nu) + \sum_{i=1}^{s} \mu_i$. As $A_{2,k} \geq 2^{k-1}$, we get the inequality: $k - 1 \leq \log_2((1 - \nu) + \sum_{i=1}^{s} \mu_i)$. Thus the number of possible k's is finite and for any fixed k, there exist a finite possible $\{a_2, \ldots, a_k\}$, $a_i \geq 2$ which satisfies (5.1.4) and for each of them, a finite possibility for $\{a_1, c_1\}$ by the second inequality of (5.1.4). For a fixed choice of $\{k, a_1, \ldots, a_k\}$, we consider the inequality (5.1.3). The left side is a monotone increasing function of $b_1, \ldots b_k$ and the right side is fixed. Thus there exist only finite possible k-ples of positive integers $\{(b_1, \ldots, b_k)\}$ satisfying (5.1.2) and this inequality with $\gcd(a_i, b_i) = 1$. This proves Assertion 1. Assertion 2 is immediate from Assertion 1 and (5.1.2) as $s = 0$ and $\nu = 2 - 2g$. Now we consider the boundary of $POL^{(r)}(\mathcal{W})$. Assume that $f \in POL(\mathcal{W}; \{\mu_1, \ldots, \mu_s\}; \nu)$ and let $\nu = \chi(C(f))$. Then $\{\nu; \mu_1, \ldots, \mu_s\}$ satisfies (5.1.2):

$$2 - \nu + \sum_{i=1}^{s} \mu_i = (n-1)^2 - \sum_{i=1}^{k}(A_{i,k} - 1)b_i A_{i+1,k}\,.$$

As $2 - \nu$ and μ_i are non-negative, there are only finite possible $\{s; \mu_1, \ldots, \mu_s\}$ which satisfies this equality. The denseness of $POL^{(r)}(\mathcal{W})$ is already observed in Proposition 3.2. The openness and the other assertion are obvious. $\square$

We remark here that $POL(\mathcal{W}; \{\mu_1, \ldots, \mu_s\}; \nu)$ may be empty and it may not be connected. By Theorem 5.1, we obtain:

COROLLARY 5.2 *Let $\mathcal{W}^1, \ldots, \mathcal{W}^m$ be the geometric minimal w-solutions of $POL^{(r)}(g)$, i.e., let $\{\mathcal{W}^1, \ldots, \mathcal{W}^m\} = \{\mathcal{MW}(f); f \in POL^{(r)}(g)\}$. Then*

$$G \cdot POL^{(r)}(\mathcal{W}^1) \cup \cdots \cup G \cdot POL^{(r)}(\mathcal{W}^m)$$

gives a disjoint partition of $POL^{(r)}(g)$ where $G \cdot POL^{(r)}(\mathcal{W}^j) := \{f^\varphi; f \in POL^{(r)}(\mathcal{W}^j), \varphi \in G\}$. In particular, the quotient set $POL^{(r)}(g)/G$ is the disjoint union of the quotients $POL^{(r)}(\mathcal{W}^j)/Stab(POL^{(r)}(\mathcal{W}^j))$ for $j = 1, \ldots, m$.

The quotient $POL(\mathcal{W})/Stab(POL(\mathcal{W}))$ has a canonical structure of an algebraic variety which is described by:

THEOREM 5.3 *Let $\mathcal{W} = \{P_j = {}^t(a_j, b_j); j = 1, \ldots, k\}$ be a sequence of weight vectors and put $n = a_1 \cdots a_k$, $c_1 = a_1 - b_1$, $A_{1,\ell} = a_1 \cdots a_\ell$ and $s = [a_1/c_1]$. We assume that $\mathcal{W}$ is minimal and $POL(\mathcal{W}) \neq \emptyset$. Let r_ℓ be the cardinality of $\mathcal{N}_\ell(\mathcal{W}; A_{1,\ell})$ and $r = \sum_{j=0}^k r_j$, where $r_0 = 1$.*

Then $POL(\mathcal{W})/Stab(POL(\mathcal{W}))$ is isomorphic to a quotient of $\mathbf{C}^{*(k-1)} \times \mathbf{C}^{r-k-s-2}$ by the cyclic group $\mathbf{Z}/nc_1\mathbf{Z}$. Thus $\dim POL(\mathcal{W})/Stab(POL(\mathcal{W})) = r - s - 3$. The quotient space $POL^{(r)}(\mathcal{W})/Stab(POL(\mathcal{W}))$ is Zariski open in $POL(\mathcal{W})/Stab(POL(\mathcal{W}))$.

By Theorem 5.3 and Corollary 5.2, we get the following Main Theorem.

MAIN THEOREM 5.4 *The quotient set $POL^{(r)}(g)/G$ has a structure of algebraic variety which has a finite number of connected components*

$$POL^{(r)}(\mathcal{W}^j)/Stab(POL^{(r)}(\mathcal{W}^j)), \quad j = 1, \ldots, m$$

and each component is isomorphic to a finite cyclic quotient of a rational variety.

REMARK 5.5 It is possible to consider the canonical topology in $POL^{(r)}(g)$ and the quotient topology in $POL(g)/G$. However we do not know if the partition given by Corollary 5.2 is the one by the connected components. Assume that $f_t(x, y)$, $t \in \mathbf{C}$ be a algebraic family of polynomials such that $f_t(x, y) \in POL^{(r)}(g)$. Is the minimal weight sequence $\mathcal{MW}(f_t)$ independent of t?

5.2 Strategy

For the proof of Theorem 5.3, we will find a rational subvariety $POL(\mathcal{W})''$ of codimension $s+3$ in $POL(\mathcal{W})$ so that every orbit Gf intersects with $POL(\mathcal{W})''$ and the stabilizer $Stab(POL(\mathcal{W})'')$ is a finite cyclic group. First we normalize the coefficient of $x^{c_1 A_{2,k}}$ to be 1 and we consider the following subspaces:

$$POL(\mathcal{W})' := \{f \in POL(\mathcal{W}); f(x, y) = (y^{a_1} + x^{c_1})^{A_{2,k}} + (\text{lower terms})\}$$
$$P(\mathcal{W}; Q_0)' := \{h \in P(\mathcal{W}; Q_0); h(u, v) = (v^{a_1} + u^{b_1})^{A_{2,k}} + (\text{higher terms})\}.$$

It is easy to see that for each $f \in POL(\mathcal{W})$, we can find $\varphi \in G$ so that $f^\varphi \in POL(\mathcal{W})'$. Let us define a subgroup $G_I(s; n)'$ of $G_I(s; n)$ as $G_I(s; n)' = \{\alpha_r^{a,b} \in G_I(s); a^{c_1} = b^{a_1}, b^n = 1\}$.

PROPOSITION 5.6 *The stabilizer $Stab(POL(\mathcal{W})')$ is equal to $\Theta_{II}G_I(s; n)'$.*

Proof. The inclusion $\Theta_{II}G_I(s; n)' \subset Stab(POL(\mathcal{W})')$ is obvious. Assume that $\varphi \in Stab(POL(\mathcal{W})')$. Write $\varphi = \theta_{d,e}\widetilde{\varphi}$ with $\varphi \in \widetilde{G}$. As $\theta_{d,e} \in Stab(POL(\mathcal{W})')$,

$\widetilde{\varphi} \in \mathrm{Stab}(\mathcal{POL}(\mathcal{W})')$. Take $f \in \mathcal{POL}(\mathcal{W})'$. Then we have $f^{\widetilde{\varphi}} \in \mathcal{POL}(\mathcal{W})'$. Thus by 2 of Theorem 4.5, $\widetilde{\varphi} \in \widetilde{G}_I(s)$ and therefore we can write $\widetilde{\varphi} = \alpha_r^{a,b} \in G_I(s)$. As $\widetilde{\varphi} \in \mathrm{Stab}(\mathcal{POL}(\mathcal{W})')$, it is necessary that $\alpha_r^{a,b} \in G_I(s;n)'$. Now we have $\varphi = \theta_{d,e}\alpha_r^{a,b} = \theta'_e \alpha_{r'}^{a,b} \in \Theta_{II} G_I(s;n)'$ where $r'(y) = r(y) + ad$. $\qquad\square$

Take $f \in \mathcal{POL}(\mathcal{W})'$ and write $f(x,y)$ and $h(u,v) := f(1/u, v/u)u^n$ as

$$f(x,y) = \sum_{M=(\mu_1,\mu_2)} \gamma_M x^{\mu_1} y^{\mu_2}, \quad h(u,v) = \sum_{N=(\nu_1,\nu_2)} \delta_N u^{\nu_1} v^{\nu_2}. \qquad (5.2.1)$$

There are several coefficients of f which we can eliminate by the action of the stabilizer $\mathrm{Stab}(\mathcal{POL}(\mathcal{W})')$. Let

$$\begin{cases} R = (0, n-1), & R_j = (c_1 - 1, n - a_1 + j) \\ S = (1, n-1), & S_j = (b_1 - j + 1, n - a_1 + j) \end{cases}, \quad j = 0, \ldots, s. \qquad (5.2.2)$$

We consider the subspaces $\mathcal{POL}(\mathcal{W})'' \subset \mathcal{POL}(\mathcal{W})'$ and $\mathcal{P}(\mathcal{W}; Q_0)'' \subset \mathcal{P}(\mathcal{W}; Q_0)'$ which are defined by

$$\mathcal{POL}(\mathcal{W})'' = \{f \in \mathcal{POL}(\mathcal{W})'; \gamma_M = 0, M = R, R_j, \ j = 0, \ldots, s\}$$
$$\mathcal{P}(\mathcal{W}; Q_0)'' = \{h \in \mathcal{P}(\mathcal{W}; Q_0)'; \delta_N = 0, \ N = S, S_j, \ j = 0, \ldots, s\}.$$

The correspondence $\lambda_k : \mathcal{P}(\mathcal{W}; Q_0)'' \to \mathcal{POL}(\mathcal{W})''$, which is introduced in §3, gives a birational mapping.

Let us consider a subgroup: $\mathcal{I}(a_1, c_1; n) := \{\iota_{a,b}; a^{c_1} = b^{a_1}, \ b^n = 1\}$. It is easy to see that $\mathcal{I}(a_1, c_1; n)$ is a cyclic group of order $c_1 n$ with the generator $\iota := \iota_{\omega(n'), \omega(n)} \in \mathcal{I}(a_1, c_1; n)$ where $n' := c_1 n / a_1$ and $\omega(m) := \exp(2\pi i / m)$. We first claim that

LEMMA 5.7 1. For each $f \in \mathcal{POL}(\mathcal{W})'$, there exists $\varphi \in \mathrm{Stab}(\mathcal{POL}(\mathcal{W})')$ so that $f^\varphi \in \mathcal{POL}(\mathcal{W})''$, i.e. $Gf \cap \mathcal{POL}(\mathcal{W})'' \neq \emptyset$.
2. The stabilizer $\mathrm{Stab}(\mathcal{POL}(\mathcal{W})'')$ is equal to $\mathcal{I}(a_1, c_1; n)$, which is isomorphic to the cyclic group $\mathbf{Z}/nc_1\mathbf{Z}$. Therefore the cardinality $|\mathcal{POL}(\mathcal{W})'' \cap G \cdot f|$ is bounded by nc_1.

Proof. First, take a polynomial $f \in \mathcal{POL}(\mathcal{W})'$ and $\varphi \in \mathrm{Stab}(\mathcal{POL}(\mathcal{W})')$. Let us write $f^\varphi(x,y) = \sum \gamma_M^\varphi x^{\mu_1} y^{\mu_2}$. Assume that there exists a non-zero coefficient γ_{R_j} of f and let $j_0 := \max\{j; \gamma_{R_j} \neq 0\}$. We consider an automorphism $\varphi_{j_0} = \alpha_{ty^{j_0}}$. We look at the action of φ on the leading term of $f(x,y)$:

$$(y^{a_1} + x^{c_1})^{A_{2,k}} = y^n + A_{2,k} y^{a_1(A_{2,k}-1)} + \cdots + x^{c_1 A_{2,k}}.$$

Under the action of φ_{j_0} on f, the monomial $A_{2,k} y^{n-a_1} x^{c_1}$ changes into

$$A_{2,k} y^{n-a_1} (x - ty^{j_0})^{c_1} = A_{2,k} y^{n-a_1} x^{c_1} - c_1 A_{2,k} t y^{n-a_1+j_0} x^{c_1-1} + \cdots.$$

By a simple consideration, those monomials which can produce $x^{c_1-1}y^{n-a_1+j_0}$ under the substitution $x \mapsto x - ty^{j_0}$ are $x^{c_1-1+\ell}y^{n-a_1-(\ell-1)j_0}$ for $\ell = 0, 1, \ldots$, but the multi-integer $(c_1 - 1 + \ell, n - a_1 - (\ell - 1)j_0)$ is outside of the Newton diagram $\Delta(f)$ for $\ell \geq 2$, as

$$a_1(c_1 - 1 + \ell) + c_1(n - a_1 - (\ell - 1)j_0) = c_1 n + (\ell - 1)(a_1 - c_1 j_0) > c_1 n.$$

Therefore no other monomial of $f(x, y)$ except $y^{n-a_1+j_0}x^{c_1-1}$ can produce the monomial $y^{n-a_1+j_0}x^{c_1-1}$. There exists a unique $t \in \mathbf{C}$ so that the coefficient of $y^{n-a_1+j_0}x^{c_1-1}$ in $f^{\varphi_{j_0}}(x, y)$ vanishes. (Take $t = -\gamma_{R_{j_0}}/c_1 A_{2,k}$.) Note also that the action of φ_{j_0} does not change γ_{R_j}, $j > j_0$ and therefore $\gamma_{R_j}^{\varphi_{j_0}} = 0$, $j \geq j_0$. If $f^{\varphi_{j_0}}$ still has some non-zero coefficient $\gamma_{R_j}^{\varphi_{j_0}}$, $j < j_0$, we take a $\varphi_j = \alpha_{t_j y^j}$ for a suitable $t_j \in \mathbf{C}$ to kill such coefficients. By induction, we get $\psi \in G_I(s)'$ so that $\gamma_{R_j}^{\psi} = 0$ for $j = 0, \ldots, s$. Finally to kill γ_R^{ψ}, the coefficient of y^{n-1} in $f^{\psi}(x, y)$, we consider the automorphism $\theta_a' : (x, y) \mapsto (x, y + a)$ with $a = -\gamma_R/n$. So y^n produces $(y + a)^n = y^n + nay^{n-1} + \cdots$ and we can kill the coefficient of y^{n-1}. Then Assertion 1 follows by taking the product $\varphi = \psi\theta_a'$. (It is easy to observe that $\gamma_{R_j}^{\varphi} = 0$.) This proves Assertion 1. Now we prove Assertion 2. Assume that $\varphi \in \mathrm{Stab}(\mathcal{POL}(\mathcal{W})'')$ and $f(x, y) \in \mathcal{POL}(\mathcal{W})''$. Write it first as $\varphi = \theta_{c,d}\tilde{\varphi}$ with $\tilde{\varphi} \in \tilde{G}$. By Assertion 2 of Theorem 4.5, we can write $\tilde{\varphi} = \alpha_r^{a,b} \in \tilde{G}_I(s)$. Thus we can write $\varphi = \theta_d'\alpha_{r'}^{a,b}$ with $r'(y) = y(y) + ac$. As the coefficients γ_{R_j} of f, $j = 0, \ldots, s$, are zero by the assumption, we also have $\gamma_{R_j}^{\theta_d'} = 0$ for $j = 0, \ldots, s$. Therefore it is necessary that $r(y) = 0$ as $(f^{\theta_d'})^{\alpha_{r'}^{a,b}} \in \mathcal{POL}(\mathcal{W})''$ and the coefficient $\gamma_{R_j}^{\varphi}$ vanishes for $j = 0, \ldots, s$. Thus $\varphi = \theta_d'\iota_{a,b}$. By looking at the coefficient γ_R, we see that $d = 0$. Thus as $\varphi = \iota_{a,b}$. Now looking at the coefficients of $((y^{a_1} + x^{c_1})^{A_{2,k}})^{\varphi}$, we conclude that $a^{c_1} = b^{a_1}$ and $b^n = 1$. Thus $\mathrm{Stab}(\mathcal{POL}(\mathcal{W})'') \subset \mathcal{I}(a_1, c_1; n)$. The opposite inclusion $\supset$ is immediate by a simple computation. $\qquad\square$

Proof of Theorem 5.3. Now we are ready to complete the proof of Theorem 5.3. Let r_ℓ be the cardinality of $\mathcal{N}_\ell(\mathcal{W}; Q_0)$ and put $r = \sum_{j=0}^k r_j$. We have seen in Corollary 2.3 that $\mathcal{P}(\mathcal{W}; Q_0)$ is isomorphic to $\mathbf{C} \times (\mathbf{C}^* \times \mathbf{C}^{r_1-1}) \times \cdots \times (\mathbf{C}^* \times \mathbf{C}^{r_k-1}) \cong \mathbf{C}^{*k} \times \mathbf{C}^{r-k}$. Let $h(u, v) = f(1/u, v/u)u^n$ and let h_j be the $A_{j+1,k}$-th Tschirnhausen approximate polynomial of $h(u, v)$ for $j = 0, 1, \ldots, k$. We compare the expressions

$$h_1 = h_0^{a_1} + \sum_{M=(\nu,\mu)\in\mathcal{N}_1(\mathcal{W};Q_0)} C_M h_{-1}^{\nu} h_0^{\mu} = v^{a_1} + \sum_{M=(\nu,\mu)} C_M' u^{\nu} v^{\mu}. \tag{5.2.3}$$

where the first equality is the Tschirnhausen expansion of h_1 with respect to $h_0 = v + c$ and $h_{-1} = u$, and the second expression is the expansion as a polynomial of u and v. However the condition $\gamma_R = 0$ implies $f_0(x, y) = y$ and $h_0(u, v) = v$, and therefore $C_{(1,a_1-1)}' = 0$ and $C_M = C_M'$ for any M, if

$f \in POL(\mathcal{W})''$. The normalized condition for f is given by $C_{M_1} = 1$ where $M_1 = (b_1, 0)$. Let us write

$$h_1(u,v)^{A_{2,k}} = \sum_{N=(\nu,\mu)} \delta'_N u^\nu v^\mu. \tag{5.2.4}$$

We define a subvariety $\mathcal{P}_1(\mathcal{W}; Q_0)''$ of codimension $s+3$ of $\mathcal{P}_1(\mathcal{W}; Q_0)$ by

$$\mathcal{P}_1(\mathcal{W}; Q_0)'' := \{h_1 \in \mathcal{P}_1(\mathcal{W}; Q_0); h_0 = v,\ C_{M_1} = 1,\ \delta'_S = \delta'_{S_j} = 0,\ j = 0, ..., s\}$$

where S and S_j are defined as in (5.2.2). Then the proof of Theorem 5.3 follows from Lemma 5.7 and the following. $\qquad\square$

LEMMA 5.8 *Let* $f \in POL(\mathcal{W})'$. *Then*

1. *The space* $\mathcal{P}_1(\mathcal{W}; Q_0)''$ *is a rational subvariety of* $\mathcal{P}_1(\mathcal{W}; Q_0)$ *with codimension* $s+3$.

2. *A polynomial* $f \in POL(\mathcal{W})'$ *is in* $POL(\mathcal{W})''$ *if and only if* $h_1 \in \mathcal{P}_1(\mathcal{W}; Q_0)''$, *where* h_1 *is the* $A_{2,k}$-*th Tschirnhausen approximate polynomial of* $h(u,v) := f(1/u, v/u)u^n$.

3. *The subset* $POL(\mathcal{W})''$ *is a rational subvariety of* $POL(\mathcal{W})$ *of codimension* $s+3$.

Proof. First we will show Assertion 1. Let $S'_j = (b_1 - j + 1, j)$ for $j = 0, \ldots, s$. As $d(P_1; S'_j) = b_1 j + a_1(b_1 - j + 1) > a_1 b_1$ and $D(Q_0; S'_j) = b_1 + 1 < a_1$, we see that $S'_j \in \mathcal{N}_1(\mathcal{W}; Q_0)$. Note that $S_j = S'_j + (A_{2,k} - 1) \times (0, a_1)$. Expanding

$$h_1(u,v)^{A_{2,k}} = \left(v^{a_1} + \sum_{M=(\nu,\mu) \in \mathcal{N}_1(\mathcal{W};Q_0)} C_M u^\nu v^\mu \right)^{A_{2,k}}$$

in a polynomial of variables u, v, we have the polynomial expression of δ'_{S_j} in $\{C_M; M \in \mathcal{N}_1(W_1; Q_0)\}$:

$$\delta'_{S_j} = \sum_{N_1 + \cdots + N_{A_{2,k}} = S_j} C_{N_1} \cdots C_{N_{A_{2,k}}}, \quad \text{where} \quad C_{(0,a_1)} := 1. \tag{5.2.5}$$

Put $N_i = (\alpha_i, \beta_i)$. Then $\beta_i \leq a_i$ and therefore by considering the second component of $N_1 + \cdots + N_{A_{2,k}}$, we have $n - a_1 + j = \sum_{i=1}^{A_{2,k}} \beta_i \leq n$. Put

$$\mathcal{N}^j := \{N = (\alpha, \beta) \in \mathcal{N}_1(\mathcal{W}_1; Q_0); \beta > j\}.$$

Then we have the inclusion relations: $\mathcal{N}^0 \supset \mathcal{N}^1 \supset \cdots \supset \mathcal{N}^s \supset \mathcal{N}^{s+1} = \emptyset$. We order the multi-integers of $\mathcal{N}_1(\mathcal{W}_1; Q_0)$ so that $\mathcal{N}_1(\mathcal{W}_1; Q_0) = \{L(1), \ldots, L(r_1)\}$ and

$$\mathcal{N}^j = \{L(1), \ldots, L(\ell_j)\}, \qquad j = s, \ldots, 0, \quad \ell_s \leq \ell_{s-1} \leq \cdots \leq \ell_0 \leq r_1.$$

Assume that $N_1 = S'_j = (b_1 - j + 1, j)$ and $N_1 + \cdots + N_{A_{2,k}} = S_j$. Then the only possibility of $N_2, \ldots, N_{A_{2,k}}$ is the case $N_2 = \cdots = N_{A_{2,k}} = (0, a_1)$. By the same reason, N_i which appears in the summation of (5.2.5) must satisfy $\beta_i \geq j$. Therefore by (5.2.5), we can write δ'_{S_j} as

$$\delta'_{S_j} = A_{2,k} C_{S'_j} + T_j(C_{L(1)}, \ldots, C_{L(\ell_j)})$$

where $T_j(C_{L(1)}, \ldots, C_{L(\ell_j)})$ is a polynomial of $C_{L(1)}, \ldots, C_{L(\ell_j)}$. In particular, we can solve the equality $\delta'_{S_j} = 0$ in $C_{S'_j}$, for $j = 0, \ldots, s$ as

$$C_{S'_j}(C_{L(1)}, \ldots, C_{L(\ell_j)}) := -T_j(C_{L(1)}, \ldots, C_{L(\ell_j)})/A_{2,k}.$$

Let us consider the subspace of the coefficients $X'' \cong \mathbf{C}^{r_1 - (s+2)}$ defined by $X'' = \{(C_M); M \in \mathcal{N}_1(\mathcal{W}; Q_0)''\}$ where

$$\mathcal{N}_1(\mathcal{W}; Q_0)'' = \mathcal{N}_1(\mathcal{W}; Q_0) - \{M_1, S'_0, \ldots, S'_s\}.$$

Then the projection $\pi : \mathcal{POL}(\mathcal{W}_1) \to X''$, $h_1 \mapsto (C_M)_{M \in \mathcal{N}_1(\mathcal{W};Q_0)''}$ is an isomorphism. The inverse of π is given by $(C_M) \mapsto h_1$ where the other coefficients of h_1 is given by $h_0 = v$, $C_{M_1} = 1$ and $C_{S'_j} = -T_j(C_{L(1)}, \ldots, C_{L(\ell_j)})/A_{2,k}$ for $j = 0, \ldots, s$. They are solved first for $j = s$ and then $j = s-1, \ldots, 0$, substituting the already solved C_{S_ℓ} with $\ell > j$. As $\mathcal{P}_1(\mathcal{W}; Q_0)'' \cong \mathbf{C}^{r_1 - (s+2)} \cong \mathbf{C}^{r_1 + r_0 - (s+3)}$, Assertion 1 follows immediately.

Now we prove Assertion 2. We compare the following expansions of $h(u,v)$

$$h(u,v) = \sum_{N=(\nu_1,\nu_2)} \delta_N u^{\nu_1} v^{\nu_2} = h_{k-1}^{a_k} + \sum_{M \in \mathcal{N}_k(\mathcal{W};Q_0)} C_M \mathbf{h}^M$$

Let $h_{-1}^{\nu_{-1}} \cdots h_\ell^{\nu_\ell}$ be an admissible Tschirnhausen monomial in the expansion of $h_{\ell+1}$. We consider $h_{-1}^{\nu_{-1}} \cdots h_\ell^{\nu_\ell}$ as a polynomial in u, v, given by the obvious substitutions $h_i \mapsto h_i(u,v) \in \mathbf{C}[u,v]$, $i = -1, \ldots, \ell$, where $h_{-1} = u$ and $h_0 = v$. First we assert

$$\deg_v h_{-1}^{\nu_{-1}} \cdots h_\ell^{\nu_\ell}(u,v) < A_{1,\ell+1} - A_{1,\ell}, \quad \text{for any } (\nu_{-1}, \ldots, \nu_\ell) \in \mathcal{N}_{\ell+1}(\mathcal{W}; Q_0).$$

In fact, we have

$$\deg_v h_\ell^{\nu_\ell} \cdots h_{-1}^{\nu_{-1}}(u,v) \leq (a_{\ell+1} - 2)A_{1,\ell} + (a_\ell - 1)A_{1,\ell-1} + \cdots + (a_1 - 1)A_{1,0}$$
$$\leq A_{1,\ell+1} - A_{1,\ell} - 1.$$

Thus the monomial $u^\nu v^\mu$ with $\mu \geq n - a_1$ in $h(u,v)$ only comes from $h_{k-1}^{a_k}(u,v)$. By the downward induction on ℓ using the expression:

$$h_{\ell-1}^{A_{\ell,k}} = \left(h_{\ell-2}^{a_{\ell-1}} + \sum_{N \in \mathcal{N}_{\ell-1}(\mathcal{W};Q_0)} \mathbf{h}^N \right)^{A_{\ell,k}}$$

we can show that the monomial $u^\nu v^\mu$ with $\mu \geq n - a_1$ of $h_{\ell-1}^{A_{\ell,k}}$ can only come from $h_{\ell-2}^{A_{\ell-1,k}}$. Thus the monomial $u^\nu v^\mu$ of $h(u,v)$ with $\mu \geq n - a_1$ only comes from $h_1(u,v)^{A_{2,k}}$, and this shows that $\delta_N = \delta'_N$ for $N = (\nu,\mu)$, $\mu \geq n - a_1$. This implies Assertion 2. Assertion 3 is immediate from Assertions 1, 2 and Corollary 2.3. $\qquad\qquad\square$

REMARK 5.9 It is also possible to consider the scalar action on the target space:
$$\mathbf{C}^* \times \mathcal{POL} \ni (c, f) \mapsto cf \in \mathcal{POL}.$$

Thus the quotient space of $\mathcal{POL}$ by the action of $G \times \mathbf{C}^*$ is one dimension smaller than the one we have considered.

6 Examples

EXAMPLE 1 ($g = 0$). Consider $\mathcal{POL}^{(r)}(0)$. It is known that for any $f \in \mathcal{POL}^{(r)}(0)$, there is an automorphism $\varphi \in G$ so that $f^\varphi = x$ ([A-M3], [S], [Ne], [A-O]). This means that $\mathcal{POL}^{(0)}/G$ is a single point.

EXAMPLE 2 ($g = 1$). In [A-O], we have shown that $\mathcal{POL}^{(r)}(1)$ has only one minimal class: $\mathcal{POL}^{(r)}(\mathcal{W})$ where $\mathcal{W} = \{P_1 = {}^t(3,1)\}$. Using the notation $\omega(j) := \exp(2\pi i/j)$, we have $r_1 = 5, s = 1$ and

$$\mathcal{POL}(\mathcal{W}) = \{f(x,y) = y^3 + t_{0,2}y^2 + t_{0,1}y + t_{0,0} + t_{1,1}xy + t_{1,0}x + \xi_1 x^2\}$$
$$\mathcal{POL}(\mathcal{W})'' = \{f(x,y) = y^3 + x^2 + t_{0,1}y + t_{0,0}\}$$

and $\mathrm{Stab}(\mathcal{POL}(\mathcal{W})'')$ is generated by $\iota_{\omega(2),\omega(3)}$. This induces the action on the parameter space by $(t_{0,1}, t_{0,0}) \mapsto (t_{0,1}\omega(3), t_{0,0})$. The quotient space is isomorphic to $\mathbf{C}^2$. By an easy computation, $\mathcal{POL}(\mathcal{W})'' - \mathcal{POL}^{(r)}(\mathcal{W})''$ is defined by $27t_{0,0}^2 + 4t_{0,1}^3 = 0$.

EXAMPLE 3 ($g = 2$). We have shown [A-O] that $\mathcal{POL}^{(r)}(2)$ have also a unique minimal geometric w-solution $\mathcal{W} = \{P_1 = {}^t(5,3)\}$. Thus we have

$$\mathcal{POL}(\mathcal{W})'' = \{y^5 + x^2 + t_{0,0} + t_{0,1}y + t_{0,2}y^2 + t_{0,3}y^3\}$$

and $\mathrm{Stab}(\mathcal{POL}(\mathcal{W})'') = \langle \iota_{\omega(2),\omega(5)} \rangle$, where $\omega(j) := \exp(2\pi i/j)$. The moduli space is given by $\mathbf{C} \times (\mathbf{C}^3/\mathbf{Z}/10\mathbf{Z}) \cong \mathbf{C} \times (\mathbf{C}^3/\mathbf{Z}/5\mathbf{Z})$ where the action of $\iota_{\omega(2),\omega(5)}$ is given by

$$(t_{0,0}, t_{0,1}, t_{0,2}, t_{0,3}) \mapsto (t_{0,0}, t_{0,1}\omega(5), t_{0,2}\omega(5)^2, t_{0,3}\omega(5)^3).$$

More generally we consider $\mathcal{POL}(\mathcal{W})$ where $\mathcal{W} = \{P_1 = {}^t(2g+1, 2g-1)\}$. Then we have $\mathcal{POL}(\mathcal{W})'' = \{f(x,y) = y^{2g+1} + x^2 + t_{0,0} + t_{0,1}y + \cdots + t_{0,2g-1}y^{2g-1}\}$

and $\mathrm{Stab}(\mathcal{POL}(\mathcal{W})'') = \langle \iota_{\omega(2),\omega(2g+1)} \rangle$. Observe that $f(x,y)$ is the unfolding of $y^{2g+1} + x^2$. The moduli space is given by $\mathbf{C} \times (\mathbf{C}^{g-1}/\mathbf{Z}/(2g+1)\mathbf{Z})$. The generic polynomial f defines a smooth curve with genus g. Thus this quotient gives a component of $\mathcal{POL}^{(r)}(g)/G$. The boundary $\partial\mathcal{POL}^{(r)}(\mathcal{W})$ is defined by the discriminant locus $\Sigma(p)$ of the polynomial

$$p(y) = y^{2g+1} + t_{0,2g-1}y^{2g-1} + \cdots + t_{0,1}y + t_{0,0}$$

and it has a canonical regular stratification

$$\mathcal{S} = \{\Delta(\alpha_1,\ldots,\alpha_\ell); \alpha_1 + \cdots + \alpha_\ell = 2g+1, \ \alpha_i \geq 1, \ i = 1,\ldots,\ell\}.$$

The stratification $\mathcal{S}$ is the image of the canonical stratification $\tilde{\mathcal{S}}$ of $\mathbf{C}^{2g+1}$ which is induced by the A_{2g+1}-arrangement in $\mathbf{C}^{2g+1}$. See [O2] for detail. For any point $(t_{0,0},\ldots,t_{0,2g-1}) \in \Delta(\alpha_1,\ldots,\alpha_\ell)$, the corresponding affine curve $C^a(f)$ has A_{α_i-1}-singularity for each $\alpha_j \geq 2$.

EXAMPLE 4 $(g \geq 3)$. The moduli space $\mathcal{POL}^{(r)}(g)/G$ has several components for $g \geq 3$. For example, $\mathcal{POL}^{(r)}(3)$ (thus $\mathcal{POL}^{(r)}(3)/G$ also) has 3 connected components: $\mathcal{POL}(\mathcal{W}^1)$ with $\mathcal{W}^{(1)} = \{P_1 = {}^t(7,5)\}$ (see Example 3), $\mathcal{POL}(\mathcal{W}^2)$ with $\mathcal{W}^2 = \{P_1 = {}^t(4,1)\}$ and $\mathcal{POL}(\mathcal{W}^3)$ with $\mathcal{W}^3 = \{P_1 = {}^t(3,1), \ P_2 = {}^t(2,9)\}$ (see [A-O], [N-O]).

Let us study $\mathcal{POL}(\mathcal{W}^3)''/\mathrm{Stab}(\mathcal{POL}(\mathcal{W}^3)'')$. First we have

$$\begin{aligned} f_1(x,y) &= y^3 + x^2 + t_{0,1}y + t_{0,0} \\ f_2(x,y) &= f_1(x,y)^2 + \xi_2 x + s_0 + s_1 y \\ &= (y^3 + x^2 + t_{0,1}y + t_{0,0})^2 + \xi_2 x + s_0 + s_1 y \end{aligned}$$

and the dimension of the moduli space is 5. The stabilizer $\mathrm{Stab}(\mathcal{POL}(\mathcal{W}^3)'')$ is given by $\langle \iota_{\omega(4),\omega(6)} \rangle$ and the action is given by

$$(t_{0,0},\ldots,s_1) \mapsto (-t_{0,0}, -t_{0,1}\omega(3), \xi_2\omega(4), s_0, s_1\omega(6)).$$

7 Appendix. Proof of Lemma 4.6

We prove Lemma 4.6. The proof is completely parallel to the proof of Theorem 4.5 of [A-O]. See also the proof of Theorem 4.7 of [A-O]. Let $\varphi = \alpha_r$ and write $r(y) = \sum_{i=0}^{s} \gamma_j y^j$. Put $\varphi_i := \alpha_{\gamma_i y^i}$. As $\alpha_r = \varphi_s \cdots \varphi_0$, we may assume that $\varphi = \alpha_{\gamma_\alpha y^\alpha}$ with $\alpha c_1 < a_1$ for simplicity. As $\varphi_\alpha : \mathbf{C}^2 \to \mathbf{C}^2$ extends to an automorphism of $\mathbf{P}^2$ if $s = 0$ or 1, we may also assume that $\alpha \geq 2$. Recall that $f_0 = y + c$ and $h_0 = v + cu$. We will show the assertion by the induction on k. Assume that we have constructed a tower of toric modifications satisfying the assertion for h_ℓ^φ for some ℓ, $(0 \leq \ell < k)$: $\psi_j : X_j \to X_{j-1}$, $j = 1,\ldots,\ell$ where $X_0 = \mathbf{C}^2$ and let (u_ℓ, v_ℓ) be the chosen coordinates centered at Ξ_ℓ. We study

$\Psi_\ell^*(h_{\ell+1}^\varphi)$, where $\Psi_\ell : X_\ell \to X_0$ is the composition $\psi_1 \circ \cdots \circ \psi_\ell$. We start from the Tschirnhausen expansion

$$h_{\ell+1} = h_\ell^{a_{\ell+1}} + \sum_{M \in \mathcal{N}_{\ell+1}(\mathcal{W};Q_0)} C_M \mathbf{h}^M. \tag{7.1}$$

We consider the pull back $\Psi_\ell^* h_{\ell+1}(u_\ell, v_\ell)$. First we have

$$\Psi_\ell^*(h_\ell^\varphi)^{a_{\ell+1}}(u_\ell, v_\ell) = u_\ell^{a_{\ell+1} m_{\ell,\ell}} v_\ell^{a_{\ell+1}}.$$

Take $M = (\nu_{-1}, \ldots, \nu_\ell) \in \mathcal{N}_{\ell+1}(\mathcal{W};Q_0)$ and put $\lambda_{\ell+1}(M) = (\mu_{-1}, \nu_0, \ldots, \nu_\ell)$ where $\mu_{-1} = A_{1,\ell+1} - \sum_{j=0}^\ell A_{1,j}\nu_j - \nu_{-1}$. The inequalities $D(Q_0; \mathbf{h}^M) \leq D(Q_0; h_\ell^{a_{\ell+1}})$ and $d(P_1; \mathbf{h}^M) \geq d(P_1; h_\ell^{a_{\ell+1}})$ give:

$$\sum_{j=0}^\ell A_{1,j}\nu_j + \nu_{-1} \leq A_{1,\ell+1}, \quad b_1 \sum_{j=0}^\ell A_{1,j}\nu_j + a_1\nu_{-1} \geq a_1 b_1 A_{2,\ell+1}. \tag{7.2}$$

Under the action of φ, we get

$$h_{\ell+1}^\varphi = (h_\ell^\varphi)^{a_{\ell+1}} + \sum_{M \in \mathcal{N}_{\ell+1}(\mathcal{W};Q_0)} C_M(\mathbf{h}^M)^\varphi,$$

$$\begin{aligned}(\mathbf{h}^M)^\varphi &= \lambda_{\ell+1}^{-1}((x + \gamma_\alpha y^\alpha)^{\mu-1}(f_0^\varphi)^{\nu_0} \cdots (f_\ell^\varphi)^{\nu_\ell}) \\ &= (u^{\alpha-1} + \gamma_\alpha v^\alpha)^{\mu-1} u^\delta (h_0^\varphi)^{\nu_0} \cdots (h_\ell^\varphi)^{\nu_\ell}\end{aligned} \tag{7.0.1}$$

where $\delta = A_{1,\ell+1} - \sum_{j=0}^\ell \nu_j A_{1,j} - \mu_{-1}\alpha$. The positivity of δ follows easily from (7.2). By the induction's assumption, we have $\Psi_\ell^*(u)(u_\ell, v_\ell) = u_\ell^{m_{\ell,-1}} U_{\ell,-1}$, $m_{\ell,-1} = A_{1,\ell}$ and $\Psi_\ell^*(h_0)(u_\ell, v_\ell) = u_\ell^{m_{\ell,0}} U_{\ell,0}$, $m_{\ell,0} = b_1 A_{2,\ell}$ ($U_{\ell,-1}, U_{\ell,0}$: units). Thus $\Psi_\ell^*(v) = \Psi_\ell^*(h_0 - cu) = u_\ell^{m_{\ell,0}} \times$ (a unit). As $\alpha b_1 A_{2,\ell} - (\alpha-1)A_{1,\ell} = (a_1 - \alpha c_1)A_{2,\ell} > 0$, $\Psi_\ell^*(u^{\alpha-1} + \gamma_\alpha v^\alpha) = u_\ell^{(\alpha-1)A_{1,\ell}} U$ with a unit U. Therefore we have

$$\Psi_\ell^*((u^{\alpha-1} + \gamma_\alpha v^\alpha)^{\mu-1} u^\delta)(u_\ell, v_\ell) = u_\ell^{\nu_{-1}A_{1,\ell}} \times \text{(a unit)}.$$

Thus by induction's hypothesis, we have that

$$\Psi_\ell^*\left((u^{\alpha-1} + \gamma_\alpha h_0^\alpha)^{\mu-1} u^\delta (h_0^\varphi)^{\nu_0} \cdots (h_\ell^\varphi)^{\nu_\ell}\right)(u_\ell, v_\ell) = u_\ell^{m_\ell(\nu_\ell, \ldots, \nu_{-1})} v_\ell^{\nu_\ell} U_M$$

where $m_\ell(\nu_\ell, \ldots, \nu_{-1}) = \sum_{j=-1}^\ell \nu_j m_{\ell,j}$. In particular, we have

$$\begin{aligned}d(P_{\ell+1}; \Psi_\ell^*((\mathbf{h}^M)^\varphi)) &= d(P_{\ell+1}; \Psi_\ell^* \mathbf{h}^M) \\ &= a_{\ell+1} m_\ell(\nu_{-1}, \ldots, \nu_\ell) + b_{\ell+1}\nu_\ell \geq a_{\ell+1}^2 m_{\ell,\ell} + a_{\ell+1} b_{\ell+1}\end{aligned}$$

by (2.3.1) and the equality is taken only for $M = M_{\ell+1}$ and $\Psi_\ell^*((\mathbf{h}^\varphi)^{M_{\ell+1}}) = u_\ell^{a_{\ell+1} m_{\ell,\ell} + b_{\ell+1}} U_\ell$ with some unit U_ℓ. Put $\xi_{\ell+1} = U(0)$. Then by the above consideration, we can write

$$\Psi_\ell^* h_{\ell+1}^*(u_\ell, v_\ell) = u^{a_{\ell+1} m_{\ell,\ell}} \left((v_\ell^{a_{\ell+1}} + \xi_{\ell+1} u_\ell^{b_{\ell+1}}) + (\text{higher terms}) \right)$$

which show that $\Psi_\ell^* h_{\ell+1}^*(u_\ell, v_\ell)$ is non-degenerate in this coordinate. By the exact same argument as in the proof of Theorem 4.5, [A-O], we also get

$$\Psi_\ell^* h_j^*(u_\ell, v_\ell) = u^{a_{\ell+1} m_{\ell,\ell}} \left((v_\ell^{a_{\ell+1}} + \xi_{\ell+1} u_\ell^{b_{\ell+1}})^{A_{\ell+1,j}} + (\text{higher terms}) \right).$$

Take an admissible toric modification $\psi_{\ell+1} : X_{\ell+1} \to X_\ell$ and take a change of coordinates $(u_{\ell+1}, v_{\ell+1})$ centered at $\Xi_{\ell+1}$ so that $E_{\ell+1} = \{u_{\ell+1} = 0\}$ and

$$\Psi_{\ell+1}^* h_{\ell+1}^\varphi(u_{\ell+1}, v_{\ell+1}) = u_{\ell+1}^{a_{\ell+1} m_{\ell,\ell} + a_{\ell+1} b_{\ell+1}} v_{\ell+1} = u_{\ell+1}^{m_{\ell+1,\ell+1}} v_{\ell+1}.$$

Then we have $\Psi_{\ell+1}^* h_j(u_{\ell+1}, v_{\ell+1}) = u_{\ell+1}^{m_{\ell+1,j}} \times (\text{a unit})$ for $j \leq \ell + 1$.

The assertion for $\Psi_{\ell+1}^* h_j(u_{\ell+1}, v_{\ell+1})$, $j > \ell + 1$, follows by the same argument as above. $\qquad\square$

REMARK 7.1 Consider the one-parameter family of automorphisms

$$\varphi_t := \alpha_{t\gamma_\alpha y^\alpha}, \ t \in \mathbf{C}.$$

Note that $\deg f^{\varphi_t} = n$, $\varphi_0 = \text{id}$, $\varphi_1 = \varphi$ and $\varphi_t : C^a(f^{\varphi_t}) \to C^a(f)$ is an isomorphism. Thus by Modified Plücker formula, we see that the Milnor number at infinity ρ_∞ of $C(f^{\varphi_t})$ is constant. Therefore using the topological invariance of Puiseux pairs under μ constant family and Lemma 4.6 follows from the argument in the proof of Theorem 4.7 of [A-O].

References

[A]   S.S. Abhyankar:*Expansion techniques in algebraic geometry*. Tata Inst. Fundamental Research, Bombay, (1977)

[A-M1] S.S. Abhyankar; T. Moh: *Newton-Puiseux expansion and generalized Tschirnhausen transformation I*, J. Reine Angew. Math., **260**, 47–83, (1973).

[A-M2] S.S. Abhyankar; T. Moh: *Newton-Puiseux expansion and generalized Tschirnhausen transformation II*, J. Reine Angew. Math., **261**, 29–54, (1973).

[A-M3] S.S. Abhyankar; T. Moh: *Embeddings of line in the plane*, J. Reine Angew. Math., **276**, 148–166, (1975).

[A-O]   N. A'Campo; M. Oka: *Geometry of plane curves via Tschirnhausen resolution tower*, toappear in Osaka J. Math., (1996)

[Ja]   P. Jaworski: *Normal forms and bases of local rings of irreducible germs of functions of two variables*, J. Soviet Math., **50**, 1350–1364, (1990).

[Ju]   H.W.E. Jung: *Über ganze birationale Transformationen der Ebene*, J. Reine Angew. Math., **184**, 1–15, (1942).

[L-O]   V.T. Le; M. Oka: *Estimation of the number of the critical values at infinity of a polynomial function* $f : \mathbf{C}^2 \to \mathbf{C}$, Publ. R.I.M.S., **31, No. 4**, 577–598, (1995).

[Mi]   J. Milnor: *Singular Points of Complex Hypersurface*. Annals Math. Studies **No. 61**, Princeton Univ. Press, Princeton, (1968).

[Mo]   T.T. Moh: *On analytic irreducibility at* ∞ *of a pencil of curves*. Proc. Amer. Soc., **44**, 22–24, (1974).

[N-O]   Y. Nakazawa; M. Oka: *Smooth plane curves with one place at infinity*. toappear in J. Math. Soc. Japan, (1996)

[Ne]   W. Neumann: *Complex algebraic curves via their links at infinity*. Invent. Math., **98**, 445–489, (1989).

[O1]   M. Oka: *On the boundary obstructions to the Jacobian problem*. Kodai Math. J., **6, No. 3**, 419–433, (1983).

[O2]   M. Oka: *On the stratification of the discriminant varieties*. Kodai Math. J., **12, No. 2**, 210–227, (1989).

[O3]   M. Oka: *Geometry of plane curves via toroidal resolution*. Algebraic Geometry and Singularities, Edited by A. Campillo and etc., Progress in Math. 134, Birkhäuser, Basel, 95–118, (1996).

[O4]   M. Oka: *Polynomial normal form of a plane curve with a given weight sequence* Chinese Quarterly Journal of Math., **10, No. 4**, 53–61, (1995).

[S]   M. Suzuki: *Propriétés topologiques des polynômes de deux variables complexes, et automorphismes algébriques de l'espace* $\mathbf{C}^2$. Journal of Math. Soc. Japan, **26**, 241–257, (1974).

[T]   E.W.v. Tschirnhausen, Acta eruditorum, Leipzig, (1683)

Progress in Mathematics, Vol. 162, © 1998 Birkhäuser Verlag Basel/Switzerland

Shadows of Legendrian Links and J^+-Theory of Curves

Michael Polyak

IHES

Le Bois-Marie, 35

Route de Chartres

F91440 Bures-sur-Yvette

FRANCE

and

Dept. of Mathematics

Hebrew University of Jerusalem

Givat Ram, Jerusalem

91904 ISRAEL

Abstract

We introduce invariants of 2-component fronts similar to Arnold's [1] invariants $J^\pm$ following approach of Viro [22] and generalize Viro's formulae to invariants of 1 and 2-component 0-homologous fronts on surfaces of non-zero Euler characteristic. We modify Turaev's construction [19] of link shadows and define shadows of Legendrian links in ST^*S^2. This enables us to relate integral formulae for J^+-type invariants of fronts to Turaev's [19] shadow formulae for linking and self-linking numbers applied to Legendrian shadows. Other applications of Legendrian shadows, e.g. quantum J^+-type invariants of fronts are discussed.

1 Introduction

Generic circle immersions into the plane have only transversal double points. Immersions with points of self-tangency create a singular discriminant in the space of all immersions. The consideration of this discriminant enabled Arnold [1] to introduce recently new basic invariants J^+ and J^- of generic immersions and to generalize them to invariants of fronts (i.e., loosely speaking, cooriented curves with cusp singularities). A renormalized version of these invariants are conformally invariant ([2]), i.e. are actually invariants of generic fronts on the sphere.

Fronts can be canonically lifted to Legendrian knots in the manifold $ST^*S^2 \cong RP^3$ of cooriented contact elements of the sphere. Arnold [1] noticed that the invariant J^+ is closely related to the Bennequin invariant– the self-linking number– of Legendrian knots. In a similar way, J^- is related to the linking number of Legendrian link obtained by lifting the same front with two opposite coorientations. The relation of J^+ to the self-linking number suggests that a natural object for study should be similar invariants of 2-component curves related (in the same way) to the linking number of Legendrian link. Strangely, though, such a definition seems to be missing (exept for an invariant of [22], which does not distinguish between the same and different components). We introduce the corresponding invariants of 2-component curves and fronts by explicit formula in the spirit of Viro [22]. We then relate these invariants to Arnold's invariants $J^\pm$.

Various formulae for $J^\pm$ are known; see e.g. Tabachnikov [17], Viro [22], Polyak [12], [13], Chmutov-Goryunov [5]. Formulas of [17] and [5] are fairly well-understood at the moment, as they are directly derived from the relation of J^+ to Legendrian knots. On the other hand, formulae of Viro [22] and the author [13] remained rather mysterious from this point of view, since their original motivation comes from quite different subjects: Rokhlin's formula of complex orientation for algebraic curves for [22] and the splitting operation, generating the Poisson bracket on the space of loops, for [13]. This lack of understanding may, in particular, be the reason why a natural generalization of Viro's formula to the case of fronts was missing. This formula is very similar to an interesting formula (e.g. [7], [11]) for Whitney index of planar curves, which also had no direct topological interpretation.

In this note we fill these gaps by relating the formulae of [22] to self-linking of Legendrian knots. The main tool in establishing this missing relation is Turaev's theory of shadows. Shadows of knots were introduced in [19] as a substitute of knot diagrams for knots in manifolds fibered over surfaces with a circle fiber. A shadow consists of a knot projection equipped with a set of half-integers– the gleams– assigned to the regions (instead of signs in double points for usual knot diagrams). We modify the original Turaev's construction to define shadows of Legendrian links. These shadows consist of a front equipped with Legendrian gleams, which can be reconstructed directly from the front. This construction allows us to explain various formulae for invariants of curves. In particular, the expression for index of a curve via indices of its regions turns out to be the shadow formula for homology class of Legendrian knot in a solid torus. In a similar way, Viro's formula of [22] turns out to be Turaev's shadow formula [19] for self-linking of shadow knots. As we learned recently, another construction of Legendrian shadows was introduced independently by Tchernov [18]. Although the idea to use shadows to obtain invariants of Legendrian links is mentioned in [18], no applications are considered there (also, there is no treatment of the framed case, which leads, in particular, to self-linking numbers).

Our approach leads us, also, to a natural generalization of Viro's formula to invariants of planar fronts and, even more generally, to invariants of 0-homologous fronts on any surface of non-zero Euler characteristic. To incorporate cusps in our approach, we use a generalized version of Euler characteristic, which has an elementary geometric meaning in our setting. We should mention, that formulae of [13] may be derived from the same Turaev's shadow formula for the linking number (rewritten via "gleams of double points"), as we plan to explain elsewhere.

The J-theory initiated by Arnold reminds the construction of Vassiliev knot invariants. However, J-theory of quantum invariants of curves and fronts, similar in spirit to quantum invariants of links, is not yet fully developed: the only results in this direction we are aware of at present are skein formulae for the invariants corresponding to Jones and HOMFLY polynomials [6]. Our approach to shadows of Legendrian links via geometry of fronts has numerous consequences. One of them is a simple general construction of invariants of (colored) curves and fronts from any quantum group, based on the quantum link invariants (see Kirillov-Reshetikhin [10], Reshetikhin-Turaev [15]) modified for shadow links by Turaev [19].

The paper is arranged in the following way. In Section 2 we briefly recall the definitions of Arnold's $J^{\pm}$ invariants of planar and spherical curves in fronts. Viro's [22] formula for J^- is discussed, modified and generalized to fronts in Section 3. In Section 4 we introduce J-type invariants of 2-component fronts (via explicit formulae in the spirit of Section 3) and establish their relation with $J^{\pm}$ invariants of 1-component fronts. Sections 5, 6.2 are devoted to the interpretation of these formulae in terms of linking numbers of shadow links [19] and to the construction of shadows for Legendrian links. Finally, the construction of quantum J-type invariants of fronts is the subject of Section 7.

2 Arnold's invariants $J^{\pm}$ of curves and fronts

2.1 Curves and fronts

Let Σ be an oriented compact surface. An n-component *curve* C on Σ is an immersion C of n oriented circles into Σ. A smooth mapping C of n circles to Σ equipped with a coorienting normal direction on the image, so that C is immersion except for a finite set of cusp points, is called a *front* on Σ. Further we will usually address planar fronts or curves just as fronts or curves, without mentioning the underlying surface $\Sigma = \mathbb{R}^2$ unless a confusion may occur. Denote by $\mathcal{C}_{\Sigma}^{\backslash}$ and $\mathcal{F}_{\Sigma}^{\backslash}$ the spaces of 0-homologous n-component curves and fronts on Σ respectively. Denote also by $\mathcal{C}_{\Sigma}$, $\mathcal{F}_{\Sigma}$ the graded spaces $\oplus \mathcal{C}_{\Sigma}^{\backslash}$, $\oplus \mathcal{F}_{\Sigma}^{\backslash}$ of all 0-homologous curves and fronts. The orientation of Σ and an orientation of a curve C induce a coorientation of C (we choose the one for which the frame of coorienting and orienting vectors orient Σ positively). Thus one

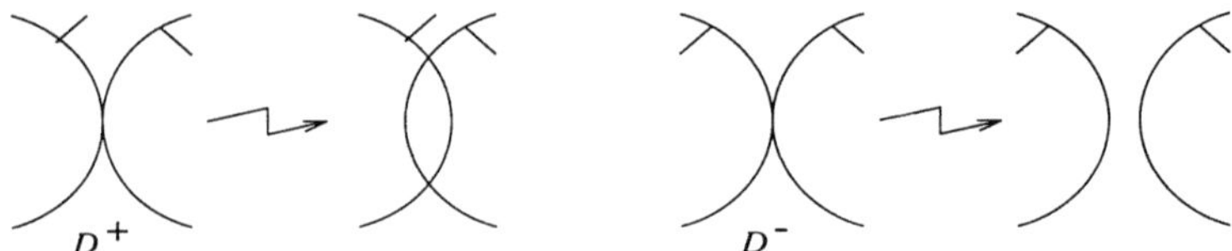

Figure 1: Strata D^+, D^- and their positive resolution

can identify $\mathcal{C}_\Sigma$ with a subspace of $\mathcal{F}_\Sigma$, which consists of fronts without cusps. A front is *generic*, if all of its singular points are, i.e. are transversal double points for curves and transversal double points and semi-cubical cusps for fronts. Images of non-generic maps form a *discriminant hypersurface* in the infinite-dimensional space $\mathcal{F}_\Sigma$. The discriminant consists of 5 main open strata of codimension 1 in $\mathcal{F}_\Sigma$, corresponding to 5 types of singularities: 2 types of tangencies, triple points, cusp crossings and cusp births. Only 3 of them intersect $\mathcal{C}_\Sigma$, so the structure of the discriminant is considerably simpler for curves.

Further on we will be interested only in the strata of tangencies. These strata D^+ and D^- correspond to maps with all the singularities generic except exactly one dangerous or safe tangency point. Here the tangency is *dangerous*, if the coorienting normal vectors to both branches coincide in this point and *safe* otherwise, see Fig 1. Call a tangency *direct*, if the directions of both tangent vectors coincide there and *opposite* otherwise. Since for curves the orientation and coorientation are equivalent, a dangerous tangency of curves is always direct and a safe one is opposite. For $n > 1$ strata $D^\pm$ can be subdivided further (depending on whether the tangency belong to the same or different components).

2.2 Invariants of plane and spherical curves

In [1] the structure of discriminant for $\Sigma = \mathbb{R}^2$ was studied and it was shown that D^+, D^- admit a natural coorientation. In other words, for any tangency point of a front there's a consistent choice of one of its resolutions; such a resolution is called *positive*, the other one *negative*. Such a choice is actually easy to make: the resolution resulting in a front with a larger number of double points is positive for D^+ and negative for D^-. Positive resolution of tangencies is illustrated in Fig 1. We call a 1-parameter local deformation of a front crossing the discriminant $D^\pm$ in the positive direction a (dangerous or safe) tangency move.

Recently, Arnold [1] introduced the basic invariants J^+ and J^- of regular homotopy of generic 1-component curves on $\Sigma = \mathbb{R}^2$ and generalized them later to the case of planar fronts. These invariants are uniquely defined [1] by the following properties. $J^\pm$ are additive under the operation of connected summation and independent on the orientation and coorientation of fronts. J^+ changes by $+2$ under a dangerous tangency move. Under a safe tangency move J^- changes by $+2$. Under the other moves J^+ and J^- do not change.

It was immediately noticed in [1] that the invariants $J^{\pm}$ of a curve C with n crossings differ only by n, i.e. $J^+(C) - J^-(C) = n$. Similar relation for invariants of fronts was established in [12]. Let C be a front with c cusps and let n_+, n_- be the number of crossings, such that the frames of orienting and coorienting vectors of intersecting branches define the same (respectively opposite) orientations of the plane. Then $J^+(C) - J^-(C) = n_+ - n_- - \frac{c}{2}$. It allows one to concentrate on the study of just one of these invariants, the other being immediately recovered from the simple relation above.

REMARK 2.1 After a change of normalization $J^{\pm}$-invariants of curves become conformal, i.e. give invariants $SJ^{\pm}$ on the space $\mathcal{C}_{S^2}^{\infty}$ of spherical curves, see [2] and Section 3.6. Also, the results of Viro [22] imply that such invariants can be defined on the space $\mathcal{C}_{\Sigma}^{\infty}$ for any surface Σ. In Section 3.6 we extend $J^{\pm}$ in a similar way to invariants on the space $\mathcal{F}_{\Sigma}^{\infty}$ of 0-homologous fronts on any surface Σ and then construct related invariants of 2-component fronts in Section 4.4.

2.3 Fronts and Legendrian links

Arnold [1] noticed a deep relation of $J^{\pm}$-theory to invariants of Legendrian knots. To explain this point we need some preliminaries.

A *contact element* on S^2 is a line in the tangent space TS^2. The manifold $ST^*S^2 \cong \mathbb{RP}^3$ of cooriented contact elements in S^2 is the spherized cotangent bundle $\pi : ST^*S^2 \to S^2$ (with the circle fiber over a point in S^2 being all the contact elements in this point). This manifold has a *tautological contact structure*, i.e. a completely non-integrable field of tangent planes defined as follows. A point in $x \in ST^*S^2$ is a contact element on S^2. The contact plane in x is defined as the inverse image of this contact element under π_*. A *Legendrian link* in ST^*S^2 is an embedded closed 1-manifold, tangent to a contact plane in each point; its canonical framing is given by the field of normals to contact planes. Self-linking number $l(L)$ of a Legendrian link L in $\mathbb{RP}^3$ (see Remark 6.1) is called the Bennequin invariant of L.

To any Legendrian link there corresponds a front on S^2, given by the bundle projection $C = \pi(L)$ to S^2. Vice versa, any front $C \subset S^2$ uniquely lifts to a Legendrian link $L = \pi^{-1}(C)$ by taking all contact elements tangent to C. A Legendrian isotopy of L can be visualized via its front C. Out of all singularities of C only dangerous tangencies correspond to singularities (i.e. self-crossings) of L. Thus J^+-theory is directly related to the theory of Legendrian imbeddings. The properties of $J^+(C)$ imply that up to normalization $J^+(C)$ coincides with the Bennequin invariant $l(L)$. Involution $C \to -C$ of coorientation reversal induces the involution $L \to -L$ of Legendrian links. In a similar way, safe self-tangencies correspond to crossings of L with $-L$ and $1 - J^-(C)$ counts the linking number of L with $-L$.

3 $J^\pm$ invariants of 1-component fronts

The approach of Viro [22] provides a convenient tool for study and generalization of $J^\pm$-type invariants. We start with some preliminary definitions.

3.1 Indices of points

Let $C \in \mathcal{F}_\Sigma$ be a (possibly multi-component) generic 0-homologous front on an oriented surface Σ. Denote by D the set of all crossings of C. Define the $\mathrm{Ind}_C : \Sigma \times \Sigma \to \frac{1}{2}\mathbb{Z}$ as follows. Denote by $[x] - [y]$ 0-chain consisting of the point $x \in \Sigma$ taken with the positive orientation and a point $y \in \Sigma$ taken with the negative orientation. For $x, y \in \Sigma - C$ define $\mathrm{Ind}(x,y) = \mathrm{lk}([x] - [y], C)$ as a linking number in Σ of 0-chain $[x] - [y]$ with (0-homologous) 1-chain C. For points with $x \in C$ (and/or $y \in C$) define $\mathrm{Ind}(x) \in \frac{1}{2}\mathbb{Z}$ by averaging the values of Ind_C of all the adjacent regions of $\Sigma - C$ (both regions for a generic or cusp point x and all four regions for a double point).

For a fixed point $a \in \Sigma - C$ a function $ind_C^a : \Sigma \to \frac{1}{2}\mathbb{Z}$ is defined by $ind_C^a(x) = \mathrm{Ind}(x,a)$. On $\Sigma = \mathbb{R}^2$ define also a closely related function $ind_C : \mathbb{R}^2 \to \frac{1}{2}\mathbb{Z}$ by $ind_C(x) = ind_C^\infty = \mathrm{Ind}_C(x, \infty)$, where ∞ is any point near infinity.

3.2 Integration with respect to Euler characteristic

In what follows we use the integration with respect to Euler characteristic, introduced in [21]. For a detailed treatment see [21], [14].

Let M be a stratified manifold $M = \cup_{i=1}^m \tau_i$ and f be a function constant on each stratum τ_i. Define

$$\int_M f d\chi = \sum_{i=1}^m f(\tau_i)\chi(\tau_i),$$

where $\chi(\tau_i)$ is the Euler characteristic of τ_i. The integral defined in this way is independent on the stratification $\{\tau_i\}_1^m$ of M (see [21], [14]).

3.3 Whitney index and J^-

Let C be a 1-component curve with n crossings. Smoothing all the crossings of C respecting the orientation, as illustrated in Fig 3a, we obtain a multi-component curve $\widetilde{C}$. Define $\widetilde{ind} = ind_{\widetilde{C}}$ as above. The curve $\widetilde{C}$ defines a stratification of $\mathbb{R}^2$ with strata being connected components of $\widetilde{C}$ and $\mathbb{R}^2 - \widetilde{C}$. The function $\widetilde{ind}$ is constant on each stratum, so its integral over $\mathbb{R}^2$ or $\mathbb{R}^2 - \widetilde{C}$ with respect to Euler characteristic is well-defined.

An important function of a curve C is its *Whitney index* (rotation number). It is the degree of Gauss map, mapping a point on C to the direction of tangent vector in this point. The Whitney index can be expressed via ind in the following, somehow mysterious, way (e.g. [7], [11]; a new interpretation in terms of shadows is proposed in Section 6):

$$\text{index}(C) = \int_{\mathbb{R}^2 - \widetilde{C}} \widetilde{\text{ind}} \, d\chi. \tag{1}$$

Integrating in a similar way the square of index, one obtains Viro's formula [22] for computation of Arnold's invariant $J^-(C)$:

THEOREM 3.1 (VIRO [22])

$$J^-(C) = 1 - \int_{\mathbb{R}^2 - \widetilde{C}} \widetilde{\text{ind}}^2 \, d\chi. \tag{2}$$

REMARK 3.2 Note, that in computation of the right hand side (and its behavior under different moves) the fact that C has only 1 component is not used. Thus, as pointed out in [22], this formula can be also used for multi-component curves.

3.4 Modifying Viro's formula

Let us modify Viro's formula using the index ind_C with respect to non-smoothening curve C. The curve C also defines a stratification of $\mathbb{R}^2$ with strata being *vertices*, *edges* and *regions* of C. Here we consider C as a planar graph and define its vertices, edges and regions as the crossings $d \in D$ of C, connected components of $C - D$ and connected components of $\mathbb{R}^2 - C$ respectively. The function ind_C is constant on each stratum, so its integral over $\mathbb{R}^2$ (or $\mathbb{R}^2 - C$) with respect to Euler characteristic is well-defined.

PROPOSITION 3.3 (CF. [22])

$$\begin{aligned} J^{\pm}(C) &= 1 - \int_{\mathbb{R}^2} \text{ind}_C^2 \, d\chi \pm \frac{n}{2} \\ &= 1 - \int_{\mathbb{R}^2 - C} \text{ind}_C^2 \, d\chi + \sum_{d \in D} \text{ind}_C(d)^2 + \frac{n}{2}(1 \pm 1). \end{aligned} \tag{3}$$

Proof. To prove the first equality, note that the result of integration over $\mathbb{R}^2 - \widetilde{C}$ is the same as over $\mathbb{R}^2$, since the Euler characteristic of (each component of) a 1-dimensional manifold $\widetilde{C}$ is 0. Now we can use the fact that the integral with respect to Euler characteristic is independent of the stratification to calculate the terms resulting from smoothing. We change the stratification in a neighborhood of a crossing as shown in Fig. 2 and compare values of ind_C and $\widetilde{ind}$ for this stratification.

Easy to check that the smoothing of a crossing d changes the contribution only of 0-cells by $(i - \frac{1}{2})^2 + (i + \frac{1}{2})^2 - 2i^2 = \frac{1}{2}$, where $i = ind_C(d)$. Comparing

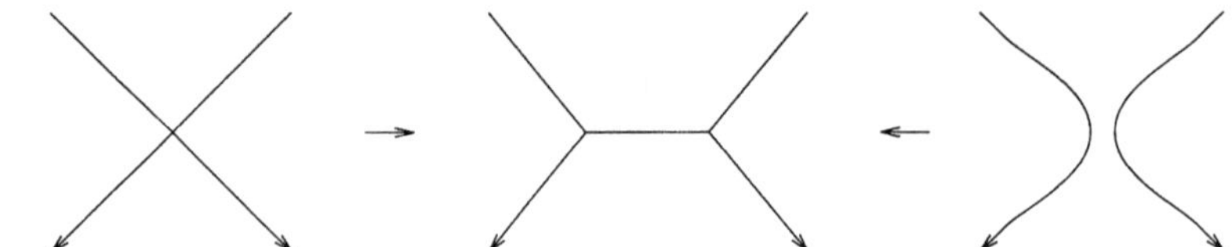

Figure 2: Changing the stratification near the crossing

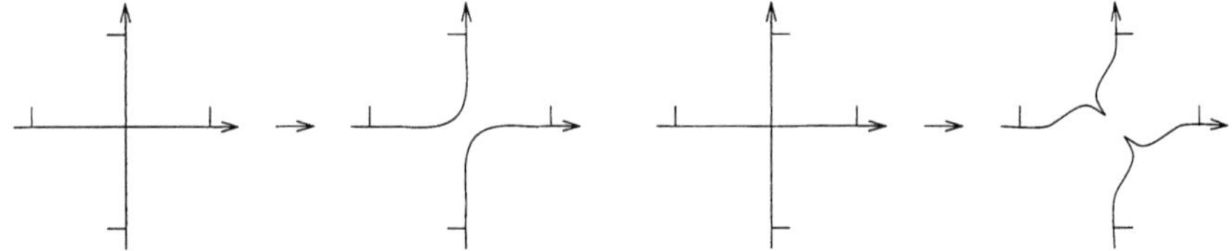

Figure 3: Smoothening of double points for fronts

now the contribution to the integral of a vertex d (with $\chi = 1$) with half of the contribution for its incident edges (with $\chi = -1$), we obtain the last equality (see also [22]). $\qquad\qquad\qquad\qquad\qquad\qquad\qquad\qquad\qquad\qquad\qquad\qquad\qquad\quad\square$

REMARK 3.4 Formula (1) may be rewritten in the same way:

$$\text{index}(C) = \int_{\mathbb{R}^2 - C} \text{ind}_C \, d\chi - \sum_{d \in D} \text{ind}_C(d) \,. \tag{4}$$

We will interpret this equation as a shadow formula for computation of a homology class of Legendrian link in Section 6.

3.5 Setting the scene for fronts

The formulae for $J^\pm$ above has a natural generalization for fronts. We start from some preliminaries.

Let C be a generic n-component plane front with c cusps. "Smoothing" C in each double point respecting orientation and coorientation, as depicted in Fig 3, we obtain a front $\widetilde{C}$ with cusps, but no double points. The function $\widetilde{\text{ind}} = \text{ind}_{\widetilde{C}}$ is defined as in Section 3.1.

Finally, to incorporate cusps on the boundary of regions in the computation of Euler characteristic, we consider the following modification χ' of Euler characteristics for fronts. For a component f of $\mathbb{R}^2 - C$ we add $+\frac{1}{2}$ to $\chi(f)$ for each cusp on $\partial \tau$ turned inwards f and $-\frac{1}{2}$ for each cusp turned outwards. For 1- and 0-dimensional strata $\chi' = \chi$. The geometrical meaning of this modification will be cleared in Section 6.2. We are ready now to formulate the following generalization of Viro's formula (2):

THEOREM 3.5 *For the invariant J^- of a generic front C we have*

$$J^-(C) = 1 - \int_{\mathbb{R}^2 - \widetilde{C}} \widetilde{ind}^2 \, d\chi'.$$

Proof. Straightforward calculation assures that this expression has prescribed jumps under all the moves, does not depend on the orientation and coorientation of C and is additive under connected sum. $\qquad\square$

Its modification via the function ind_C repeats the considerations of Proposition 3.3, so we omit the proof:

PROPOSITION 3.6 *For a generic front C with c cusps and n crossings we have*

$$J^{\pm}(C) = 1 - \int_{\mathbb{R}^2} \mathrm{ind}_C^2 \, d\chi' \pm \frac{n_+ - n_-}{2} - \frac{c}{4}(1 \pm 1), \qquad (5)$$

where n_+, n_- are the numbers of crossings such that the frames of orienting and coorienting vectors of intersecting branches define the same (respectively opposite) orientations of the plane.

3.6 Invariants of fronts on surfaces

The function ind_C is defined in a more general case of 0-homologous fronts on an open surface of arbitrary genus (see [22] for the case of curves). Thus the invariants $J^{\pm}$ generalize to this situation. Moreover, $J^{\pm}$ can be defined (in the spirit of [22] and [2]) even for fronts on a closed surface Σ with Euler characteristic $\chi(\Sigma) \neq 0$.

Indeed, for a front C on Σ fix $a \in \Sigma - \widetilde{C}$ and define $\widetilde{ind^a}(x) = \mathrm{Ind}_{\widetilde{C}}(x, a)$ as in Section 3.1. Define also the "Whitney index" $\mathrm{index}^a(C)$ similarly to (1) by $\mathrm{index}^a(C) = \int_{\Sigma} \widetilde{ind^a} d\chi'$. The dependence of $\int_{\Sigma} (\widetilde{ind^a})^2 d\chi'$ and $\mathrm{index}^a(C)$ on a as we move a to adjacent region is easily computable and leads us to the following

THEOREM 3.7 *Let C be a generic front on a compact surface Σ with $\chi(\Sigma) \neq 0$. Pick a point $a \in \Sigma - \widetilde{C}$ and define $J_{\Sigma}^-(C)$ by*

$$J_{\Sigma}^-(C) = 1 - \int_{\Sigma - \widetilde{C}} (\widetilde{ind^a})^2 d\chi' + \frac{1}{\chi(\Sigma)} (\mathrm{index}^a(C))^2. \qquad (6)$$

The invariant J_{Σ}^- does not depend on the choice of the point $a \in \Sigma - C$ and on the orientation and the coorientation of C. Under a positive direct (respectively opposite) safe tangency move J_{Σ}^- increases (respectively decreases) by 2 and is preserved under the other moves.

In the case of curves a similar formula was obtained in [22]. For curves on $\Sigma = S^2$ this relates the invariants $SJ^- = J^-_{S^2}$ and J^- of spherical and plane curves and reproduces the definition $SJ^-(C) = J^-(C) + \frac{1}{2}\,\mathrm{index}(C)^2$ of the conformal invariant SJ^- in [2]. The formulae above involve a choice of $a \in \Sigma - \widetilde{C}$, which hides the symmetric nature of the invariants. More insight into this symmetry is given by using the relative index $\widetilde{\mathrm{Ind}} = \mathrm{Ind}_{\widetilde{C}}$ of pairs of points instead of ind^a (see Section 3.1):

THEOREM 3.8 *For a generic front on a surface Σ with $\chi(\Sigma) \neq 0$ we have*

$$J^-_{\Sigma}(C) = 1 - \frac{1}{2\chi(\Sigma)} \int_{(\Sigma - \widetilde{C})^2} (\widetilde{\mathrm{Ind}})^2 (d\chi')^2.$$

Proof. Fix a point a in a region f of $\Sigma - \widetilde{C}$ and use the obvious equalities

$$\chi(f) = \chi(\Sigma) - \int_{\Sigma - \widetilde{C} - f} d\chi', \quad \widetilde{Ind}(x, y) = \widetilde{\mathrm{ind}^a}(x) - \widetilde{\mathrm{ind}^a}(y)$$

to deduce (6) from the expression above. $\qquad\qquad\qquad\qquad\qquad\qquad$ $\square$

REMARK 3.9 A similar construction was used by Shumakovich in [16] for symmetrization of Turaev's shadow expression [19] for the linking number of links in S^3. Relation of these subjects is discussed in Section 6.2.

4 J-type invariants of multi-component fronts

Arnold's $J^\pm$ invariants were defined only for 1-component curves, though the structure of the appropriate strata of discriminant and their coorientation are similar in the case of multi-component curves. Consideration of 2-component curves leads to a natural counterpart of $J^\pm$.

4.1   Invariants of 2-component curves

Consider a 2-component curve $C_1 \cup C_2 \subset \mathbb{R}^2$. Formula (3) for $J^\pm$ suggests the following definition. Define $J^\pm(C_1, C_2)$ by

$$J^\pm(C_1, C_2) = 1 - \int_{\mathbb{R}^2} \mathrm{ind}_{C_1}\,\mathrm{ind}_{C_2}\,d\chi \pm \frac{n_{12}}{4}, \qquad (7)$$

where n_{12} is the number of crossings of C_1 with C_2.

Obviously, $J^\pm(C_1, C_2)$ do not depend on the ordering of the components and are invariants of regular homotopy. Their properties under various moves are described by

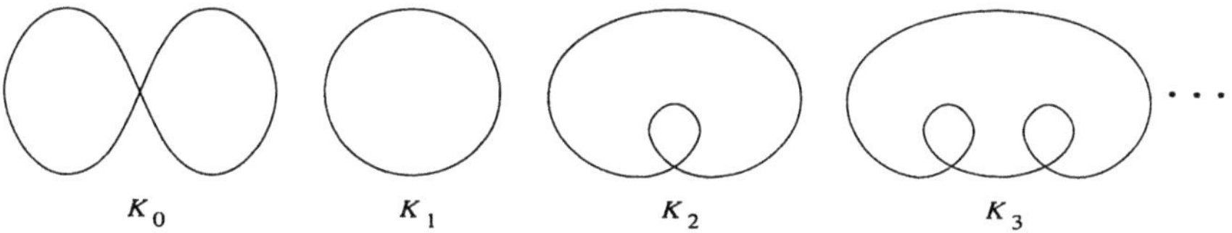

Figure 4: Standard curves of indices $0, \pm 1, \pm 2, \ldots$

THEOREM 4.1 *Let $C = C_1 \cup C_2$ be a generic 2-component plane curve. Under a direct (resp. opposite) dangerous tangency of C_1 with C_2 $J^+(C_1, C_2)$ increases (resp. decreases) by 1. Under a direct (resp. opposite) safe tangency of C_1 with C_2 $J^-(C_1, C_2)$ decreases (resp. increases) by 1. $J^\pm(C_1, C_2)$ remain invariant under the other moves of C.*

Proof. Lengthy, but a straightforward check assures that the invariants $J^\pm(C_1, C_2)$ have prescribed jumps under different moves of the curve C. $\quad \square$

Comparing the contribution to (7) of vertices and their incident edges as in the proof of Proposition 3.3, we obtain

$$- \int_C \operatorname{ind}_{C_1} \operatorname{ind}_{C_2} d\chi = \sum_d \operatorname{ind}_{C_1}(d) \operatorname{ind}_{C_2}(d).$$

COROLLARY 1

$$J^\pm(C_1, C_2) = 1 - \int_{\mathbb{R}^2 - C} \operatorname{ind}_{C_1} \operatorname{ind}_{C_2} d\chi + \sum_{d \in D} \operatorname{ind}_{C_1}(d) \operatorname{ind}_{C_2}(d) \pm \frac{n_{12}}{4}. \quad (8)$$

4.2 Expressing $J^\pm(C)$ via $J^\pm(C_1, C_2)$

Arnold's invariants $J^\pm(C)$ can be expressed via $J^\pm(C_1, C_2)$ in the following natural way.

THEOREM 4.2 *Let C be a (cooriented) 1-component curve. Denote by C' its push-off along the coorientation. Then $J^\pm(C, C') = J^\pm(C)$.*

Proof. An invariant of 1-component curves is completely determined by its jumps under self-tangency and triple point moves and the normalization, e.g. its values on the standard curves K_i of Whitney indexes $\operatorname{index}(K_i) = \pm i$ depicted in Fig. 4.

From the behavior of $J^\pm(C, C')$ under the moves we conclude that $J^\pm(C, C')$ coincides with $J^\pm(C)$ up to normalization. It remains to compare the values of $J^\pm(C, C')$ and $J^\pm(C)$ on the standard curves K_i. Clearly, $J^\pm(C, C')$ does not depend on the orientation of C. Choosing the positive orientation of K_i and computing the values of $J^\pm(K_i, K_i')$, we obtain that $J^\pm(K_0, K_0') = J^\pm(K_0) = \frac{1}{2}(-1 \pm 1)$ and $J^+(K_{j+1}, K_{j+1}') = J^+(K_{j+1}) = -2j$, $J^-(K_{j+1}, K_{j+1}') = J^-(K_{j+1}) = -3j$, $j = 0, 1, 2, \ldots$ $\qquad\square$

REMARK 4.3 Recall that $J^+(C)$ coincides, up to normalization, with the Bennequin invariant counting the self-linking number of the corresponding Legendrian knot (see e.g. [1], [13] and Sections 2.3, 6). In a similar way, $J^+(C_1, C_2)$ counts the linking number of the corresponding Legendrian link (see Section 6).

4.3 Expressing $J^\pm(C_1, C_2)$ via $J^\pm(C)$

As we have seen above, Arnold's invariants $J^\pm(C)$ can be expressed via the invariants $J^\pm(C_1, C_2)$. One can as well express the invariants $J^\pm(C_1, C_2)$ of 2-component curves via Arnold's invariants $J^\pm(C)$. This requires, actually, no more than the equality $2ab = (a+b)^2 - a^2 - b^2$ (with $a = \mathrm{ind}_{C_1}$, $b = \mathrm{ind}_{C_2}$).

View for the moment $C_1 \cup C_2$ as one curve C, i.e. consider it without distinguishing the components. Viro's formula (2) for computation of J^- may be applied in this situation (see remark 3.2). Alternatively, formula (3) may be used with $\mathrm{ind}_C = \mathrm{ind}_{C_1} + \mathrm{ind}_{C_2}$. Compute also $J^\pm(C_1)$, $J^\pm(C_2)$ for each component separately. Comparing (3) with (7) we immediately obtain the following theorem.

THEOREM 4.4 *Let $C = C_1 \cup C_2$ be a 2-component curve. Then*

$$2J^\pm(C_1, C_2) = J^\pm(C) - J^\pm(C_1) - J^\pm(C_2) + 3$$

4.4 Invariants of 2-component fronts

The definition of $J^\pm(C_1, C_2)$ from the previous section readily extends to the case of 2-component fronts $C = C_1 \cup C_2$ by merging (7) and Proposition 3.6. Define the invariants $J^\pm(C_1, C_2)$ of 2-component fronts by

$$J^\pm(C_1, C_2) = 1 - \int_{\mathbb{R}^2} \mathrm{ind}_{C_1}\, \mathrm{ind}_{C_2}\, d\chi' \pm \frac{1}{4}(n_{12+} - n_{12-}), \qquad (9)$$

where n_{12+} (resp. n_{12-}) are the numbers of crossings $d \in C_1 \cap C_2$, such that the frames of orienting and coorienting vectors to C_1 and C_2 in d determine the same (resp. opposite) orientation of $\mathbb{R}^2$. The following statement generalizes Theorem 4.1 to fronts.

THEOREM 4.5 *Let $C = C_1 \cup C_2$ be a generic 2-component plane front. Under a direct (resp. opposite) dangerous tangency of C_1 with C_2 $J^+(C_1, C_2)$ increases (resp. decreases) by 1. Under a direct (resp. opposite) safe tangency of C_1 with C_2 $J^-(C_1, C_2)$ decreases (resp. increases) by 1. $J^\pm(C_1, C_2)$ remain invariant under the other moves of C.*

Since the functions Ind_{C_1}, Ind_{C_2} are defined for 0-homologous fronts on a surface Σ of arbitrary genus, the invariant $J(C_1, C_2)$ extend to an invariant on $\mathcal{F}_\Sigma^\in$ for $\chi(\Sigma) \neq 0$ exactly as in Section 3.6. Let us formulate the analog of Theorem 3.8 (the analog of Theorem 3.7 is completely similar):

THEOREM 4.6 *Let $C = C1 \cup C_2$ be a generic 2-component front on a surface Σ with $\chi(\Sigma) \neq 0$. Define*

$$SJ_\Sigma^-(C) = 1 - \frac{1}{2\chi(\Sigma)} \int_{\Sigma \times \Sigma} \mathrm{Ind}_{C_1} \mathrm{Ind}_{C_2} (d\chi')^2 - \frac{1}{4}(n_{12+} - n_{12-}).$$

Under a direct (resp. opposite) safe tangency move SJ_Σ^- increases (resp. decreases) by 1 and is preserved under the other moves.

As in Section 3.6, conformal invariants $SJ^\pm = SJ_{S^2}^\pm$ of spherical curves are related to the invariants of planar curves by

$$SJ^\pm(C_1, C_2) = J^\pm(C_1, C_2) + \frac{1}{2}\mathrm{index}(C_1)\,\mathrm{index}(C_2).$$

5   Shadows and Legendrian shadows

5.1   Diagrams and shadows

A usual way to present a link L in $\mathbb{R}^3 \subset S^3$ is by means of a diagram, i.e. its image under orthogonal projection to $\mathbb{R}^2$ equipped with a sign (the local writhe) in each crossing point d. The sign indicates which of the two preimages of d is higher in the fiber $\mathbb{R}^1$. Another important projection is the Hopf map $\pi : S^3 \to S^2$. In this case one can not define sign of the crossing point, since the fiber is S^1 and one can not say which preimage is "higher". The solution, suggested by Turaev [19] for links in circle bundles over surfaces, is to consider *shadows* of links by assigning some additional data to the regions of the bundle projection instead of the crossings. Similarly to link diagrams, a link in S^3 may be reconstructed from its shadow in S^2 up to isotopy. Recall briefly Turaev's construction of shadows in the particular case of links in circle bundles over S^2. For details and more general construction of shadows for links in arbitrary 3-manifolds see [20]. Note also, that opposite sign conventions were used in [19] and [20]; we choose the latest one ([20]).

5.2   Shadows of links

Consider an oriented circle bundle $\pi : E \to S^2$ over S^2. Let link L in E be generic, i.e. such that its projection $C = \pi(L)$ is a generic immersed curve. Denote the set of all regions of C (i.e. connected components of $S^2 - C$) by F. A *shadow* (C, gl) of L is the curve C equipped with a set of half-integer

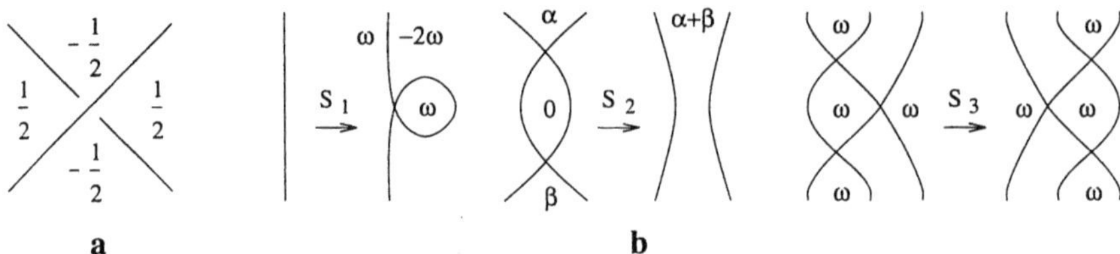

Figure 5: Standard gleams and shadow moves

gleams $gl(f)$, $f \in F$. The gleam $gl(f)$ (or "modified gleam" in the language of [19]) of a region f is, roughly speaking, a number of turns in the fiber S^1 made by the preimage in E of the boundary ∂f of f. Unfortunately, the preimage $p^{-1}(\partial f) \subset E$ consists of some non-connected pieces of L, so an additional step is needed to make this definition rigorous. Namely, we should apply to L a *vertical isotopy*, i.e. an isotopy along the fibers (so that the projection $p(L)$ remains the same). By a vertical isotopy we deform L to a position where the preimages of any double point of $p(L)$ become a pair of opposite points in the fiber S^1. Consider our fibration as a composition of a double cover projection $\pi_2 : E \to E/-1$ identifying the opposite points in each fiber and the fibration $\tilde{\pi} : E/-1 \to S^2$ (with fiber S^1). Now the inverse image $\gamma \subset \pi_2(L)$ of ∂f under $\tilde{\pi}$ is a collection of closed curves. An orientation of S^2 induces an orientation of the boundary ∂f of f which, in turn, induces an orientation of γ. Denote by $p : \tilde{\pi}^{-1}(f) \to S^1$ the natural projection to the fiber of the (trivial) S^1 bundle $\tilde{\pi}^{-1}(f)$ over f. We take $2gl(f)$ to be the homology class $p_*([\gamma]) \in H_1(S^1) \cong \mathbb{Z}$ of the cycle $p(\gamma)$. The sum of gleams over all the regions of C equals $\chi(\pi)$, where $\chi(\pi)$ is the Euler number of π. In particular, it equals ± 1 for $E = S^3$ and ± 2 for $E = \mathbb{RP}^3$.

5.3 Shadow moves

Link isotopy in E may be visualized on the level of shadows via the *shadow moves* [19], similar to Reidemeister moves for link diagrams.

These moves can be easily deduced from the usual Reidemeister moves in the following way: any link in $\mathbb{R}^3$ may be considered as lying in a slice between two meridional discs of $\mathbb{R}^2 \times S^1$. Completing $\mathbb{R}^2$ to S^2 we can view our link as being in S^3. The resulting shadow, called a *standard shadow* of a link in R^3, can be obtained by the following simple rule. Assign local gleams to each crossing as shown in Fig 5a and define the global gleam of a region as the sum of local gleams over the crossings on its boundary. Translating the Reidemeister moves Ω_1 Ω_3 to this language, we get shadow moves S_1–S_3. These moves are depicted in Fig 5b; there $\omega = \pm \frac{1}{2}$ and only the local changes of gleams are shown. Note that unlike in the case of Reidemeister moves, the inverse S_2^{-1} is not unique, as the gleam of central region may be split in different ways.

If two links in E are isotopic, their shadows may be connected by a (finite) sequence of moves $S_1^{\pm 1}$, $S_2^{\pm 1}$ and S_3. Link invariants are, in this setting, functions on the equivalence classes of shadows, i.e. functions of shadows which are invariant under the shadow moves. Equivalence classes of shadows are called, following Turaev, *shadow links*. As further we will be interested only in shadow links, we will sometimes address these also as shadows, if a mistake is unlikely to occur.

Each knot may be equipped with the blackboard framing (the one tangent to the fibers of π). Moves S_2 and S_3 preserve this framing, while S_1 changes it by 2ω. Similarly to the case of usual Reidemeister moves, one may introduce a set of shadow moves which preserve the framing by taking, instead of S_1, a new move S_1'. This move is obtained by inserting one under another a pair of small kinks given by S_1 moves with opposite ω (so that the changes of framing caused by each one cancel out). We consider further each knot with its blackboard framing, unless the opposite is explicitly stated.

5.4   Setting the scene for Legendrian shadows

We would like to apply the technique of shadows to Legendrian links in the manifold $ST^*S^2 \cong \mathbb{RP}^3$ of cooriented contact elements of the sphere. One can, of course, first deform any Legendrian link to a generic link close to it, and then apply the usual shadow technique; this was the approach chosen in [18]. Unfortunately, in this way we loose all the specifically Legendrian information (including the framing) encoded in the link. Thus, having in mind further developements and applications, we prefer to extend the shadow technique to incorporate Legendrian links in our setting. Some minor generalization of Turaev's original construction arc needed for this purpose, as we face now three new obstacles. First, we would like to perform all the isotopies in the class of Legendrian links rather than in the class of generic links in $\mathbb{RP}^3$. Second, generic Legendrian links are not in the general position with respect to the bundle projection π. Finally, the standard framing of Legendrian knots is different from the blackboard one. All three problems can be easily solved; let us address these questions one by one.

First of all, recall that the basic tool of Turaev's construction was vertical isotopy, used to deform a link to a special position where the preimages of crossings on its projection become pairs of opposite points in the S^1 fiber. As such an isotopy can not be performed in the class of Legendrian links, we would like to consider a more general type of isotopy. Fortunately, any isotopy of L in E in course of which the regular homotopy class of the curve $C = \pi(L)$ does not change, preserves the gleams and, thus, results in the same shadow. In the same way, some non-generic singularities of C may be allowed, in particular two branches intersecting in the double point may be tangent there (see Fig. 7). This, as we will see below, settles the first problem.

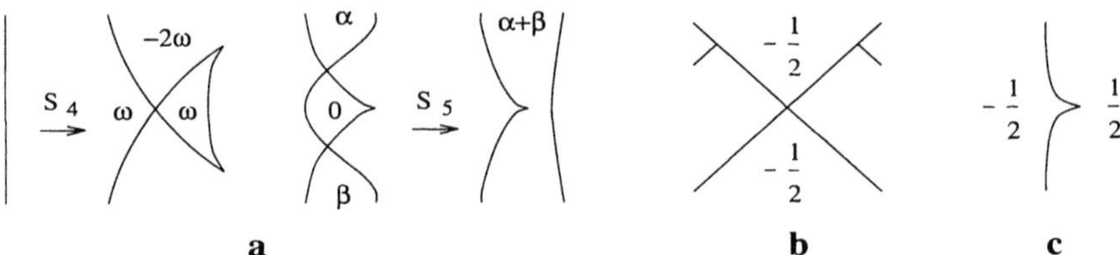

Figure 6: New shadow moves and Legendrian gleams

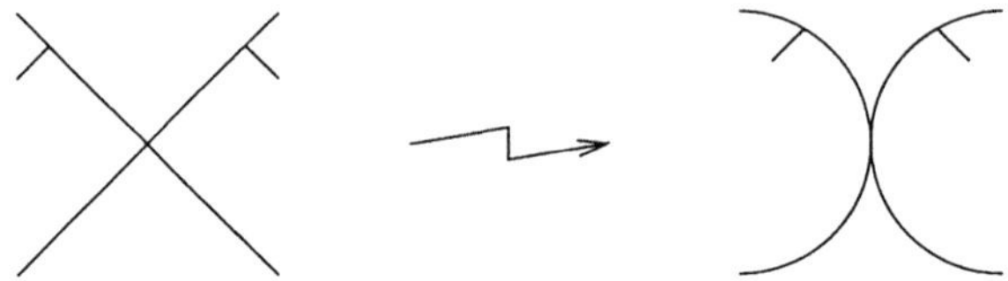

Figure 7: Deforming the front by Legendrian isotopy

Now, note that generic projections of Legendrian knots have cusps (which, unlike before, can not be removed by a small deformation in the class of Legendrian knots). To tackle this problem, we generalize the original setting. In addition to all generic knots, we include knots, which are tangent to the fiber in a discrete set of points and project to fronts with cusps in these points. The definition of gleams remains as before, so in this generalized setting a shadow is a pair (C, gl) where C is a front, rather than a curve, equipped with some gleams. The only change is that two new moves shown in Fig 6a should be added to the set of shadow moves; here again $\omega = \pm\frac{1}{2}$. This solves the second question. Note also, that if we restrict ourselves only to the case of Legendrian links, the situation simplifies further: the move S_1 does not exist and for S_4 only $\omega = \frac{1}{2}$ is possible.

However, we should be careful about framings: the blackboard framing tangent to S^1 fibers is not defined over the cusp points. Generally speaking, this can be settled via different conventions about framings, e.g. one can require the blackboard framing to extend over the preimages of cusps by continuity. However, having in mind our specific case, i.e. Legendrian knots, we choose a different convention. Namely, we assume that in a small neighborhood of each point on L, which projects to a casp, the framing makes *positive half-twist*, turning by 180 degrees, while outside this neighborhood it remains tangent to the fibers. Indeed, this exactly reproduces the behavior of the Legendrian framing for a Legendrian knot, as compared to the blackboard framing of a generic (in fact, transversal) knot close to it, see e.g. [3].

Now we are in a position to define shadows of Legendrian knots.

5.5 Shadows of Legendrian links

The shadow of a Legendrian link L in $E = ST^*S^2$ consists, as discussed above, of its front C equipped with some set of gleams. These gleams may be computed by the following rule. Define the local gleams of crossings and cusps of C as shown in Fig 6b and Fig 6c respectively.

Set the global gleam $gl_L(f)$ of a region f as $\chi(f)$ plus the sum of local gleams over the crossings and cusps on its boundary (see also [18]). We call these gleams and the resulting shadow (C, gl_L) *Legendrian*. Note, that the sum of Legendrian gleams over all the regions equals 2 in full agreement with $\chi(\pi) = 2$ for $\pi : ST^*S^2 \to S^2$.

THEOREM 5.1 *The shadow of a Legendrian link L in $E = ST^*S^2$ consists of its front C equipped with the Legendrian gleams gl_L.*

Proof. Recall that a front C can be uniquely lifted to the corresponding link L in the manifold $E = ST^*S^2$ of contact elements by taking the direction of coorienting vector in a point of C as a contact element in this point. We deform C so, that the intersecting branches in each crossing become tangent with the opposite tangent vectors, as illustrated in Fig. 7.

The resulting front C' can be lifted to a Legendrian link L', Legendrian isotopic to L. Note that by its construction the link L' is in a special position, namely the preimages of crossings of its projection C' consist of pairs of opposite points in the S^1 fiber. Therefore we may apply the usual technique (see section 5.2) to define the gleam $gl(f')$ for each region f' of C'. As C' was obtained from C by a regularly homotopy of generic fronts, the regions of C' are in one-to-one correspondence with the regions of C and the resulting set of gleams $gl(f) = gl(f')$ give the shadow (C, gl) of C.

The gleam $gl(f')$ is given by the Whitney index of the boundary $\partial f'$ of f, i.e. by the number of turns made by the coorienting vector as we pass along $\partial f'$ in the positive direction. This rotation number clearly coincides with the (modified) Euler characteristic $\chi'(f')$ of the region f', introduced in section 3.5, so $gl(f') = \chi'(f')$. However, the Euler characteristic chi' for f' and for the corresponding region f of C are different. Indeed, it is easy to see on Fig. 7, that some crossings on ∂f result in (outside-looking) cusps on $\partial f'$. Thus $gl(f') = chi'(f')$ equals $chi'(f)$ minus half the number of the crossings of C, where the coorienting vector of exactly one of the intersecting branches points inside f. Comparing now $\chi'(f)$ with $\chi(f)$, we recover the assignment used for Legendrian gleams. $\qquad\square$

REMARK 5.2 The construction of Legendrian shadows above remains the same for fronts on arbitrary oriented surface Σ and gives a shadow description of Legendrian links in the manifold of contact elements $ST * \Sigma$.

6 J-invariants and shadow links

6.1   Invariants of shadow links

Isotopy class of a link is completely characterized by its shadow ([19]). Thus, in particular, link invariants may, in theory, be computed via shadows. Shadow version of quantum link invariants derived from quantum groups was established by Turaev [19] via a clever modification of Kirillov-Reshetikhin construction [10]. There was less progress with finite-degree invariants. Results of Goussarov [8] and Burri [4] imply that Vassiliev invariants depend on gleams polynomially (with the degree at most twice the degree of the invariant). Explicit expressions for these polynomials were obtained only for the Vassiliev knot invariant v_2 ([16], see also [4]) and the linking number of a 2-component link ([19], see also [16]).

Let $(C_1 \cup C_2, gl)$ be a shadow of 2-component link $L = L_1 \cup L_2$ in a circle bundle $E \to S^2$. Pick an "infinite" point ∞ on S^2 and identify $S^2 - \infty$ wih R^2. Turaev [19] gives the following formula for the linking number lk of L:

$$lk = \sum_{f \in F} \operatorname{ind}_{C_1}(f) \operatorname{ind}_{C_2}(f) gl(f) - \frac{1}{\chi(\pi)} \operatorname{rot}(C_1) \operatorname{rot}(C_2), \qquad (10)$$

where F is the set of all the finite regions of C in $\mathbb{R}^2 = S^2 - \infty$ and $\operatorname{rot}(C_i) = [L_i] \in \mathbb{Z} \cong H_1(R^2 \times S^1)$ is the homology class of L_i in the space $\pi^{-1}(\mathbb{R}^2) \cong R^2 \times S^1$ over $R^2 = S^2 - \infty$. This homology class may be easily computed [19] in terms of the shadow as

$$\operatorname{rot}(C_i) = \sum_{f \in F} \operatorname{ind}_{C_i}(f) gl(f). \qquad (11)$$

The linking number defined in this way does not depend on the choice of $\infty \in S^2$. The self-linking number l of a shadow is defined similarly by

$$l = \sum_{f \in F} \operatorname{ind}_C(f)^2 gl(f) - \frac{1}{\chi(\pi)} \operatorname{rot}(C)^2. \qquad (12)$$

Further we will apply these formulae for $\chi(\pi) = 2$, corresponding to the bundle $E = ST^*S^2 \cong \mathbb{RP}^3$ of cooriented contact elements. Recall also that the Euler number $\chi(\pi)$ may be recovered from the shadow as the sum of gleams of all (including infinite) regions.

REMARK 6.1 Linking numbers are defined for links in any lens spaces. The elementary way to visualize the linking number lk for a 2-component link L in ST^*S^2 is the following (e.g. [2]). Note that the bundle E is double covered by the Hopf bundle. The inverse image $L' = \pi_2^{-1}(L)$ of L under the covering map $\pi_2 : S^3 \to E$ is a link in S^3. Each component L_i, $i = 1, 2$ of L is covered by either one or two component sublink $L'_i = \pi_2^{-1}(L_i)$ of L'. One defines $lk(L_1, L_2)$ as half of the linking number $lk(L'_1, L'_2)$ of L'_1 with L'_2. Here if L'_1 or L'_2 has 2 components, by $lk(L'_1, L'_2)$ we mean the sum of the corresponding linking numbers with each of the components.

6.2 Invariants of fronts and Legendrian shadows

All the formulae for invariants of shadows extend to shadows of Legendrian links. In particular, formula (10) for the linking number remains the same. As for the self-linking number, recall that the Legendrian framing makes an additional positive half-twist over each cusp point. Thus an additional term $\frac{c}{2}$, where c is the number of cusps, should be added to formula (12):

$$l = \sum_{f \in F} \mathrm{ind}_C(f)^2 gl(f) + \frac{c}{2} - \frac{1}{\chi(\pi)} \mathrm{rot}(C)^2. \tag{13}$$

Let us show that formulae of Sections 3–4 for the invariants $\mathrm{index}(C)$, $J^{\pm}$ and $J(C_1, C_2)$ of fronts are (up to normalization) just shadow formulae for $\mathrm{rot}(C)$, $l(C)$ and $lk(C_1, C_2)$ applied to Legendrian shadows.

We start from a simple proposition showing that formula (1) (or (4)) for $\mathrm{index}(C)$ is a restriction of (11) for $\mathrm{rot}(C)$ of the Legendrian shadow (C, gl_L). Indeed, $\mathrm{index}(C)$ has the same meaning as $\mathrm{rot}(C)$ for Legendrian knots, namely it is clearly the homology class $[L] \in \mathbb{Z} \cong H_1(\mathbb{R}^2 \times S^1)$ of the corresponding Legendrian knot $L = \pi^{-1}(C)$ in the space $\pi^{-1}(\mathbb{R}^2) \cong \mathbb{R}^2 \times S^1$ over $\mathbb{R}^2$.

PROPOSITION 6.2 *Let C be a generic curve. Then formula (11) for* $\mathrm{rot}(C)$ *of the Legendrian shadow (C, gl_L) coincide with equation (4) for* $\mathrm{index}(C)$.

Proof. We substitute the Legendrian gleams $gl_L(f)$ into (11) and regroup the gleams, leaving $\chi(f) = 1$ in the region and moving the local gleams back to the crossings $d \in D$. The contribution of local gleams to $\mathrm{rot}(C)$ equals $-\sum_{d \in D} \mathrm{ind}_C(d)$, while the unit gleams left in the regions contribute $\sum_{f \in F} \mathrm{ind}_C(f) \chi(f)$, thus resulting in (4). $\qquad \square$

Considering in the same way 2-component fronts and products of indices, we obtain that the conformal invariant $SJ^+(C_1, C_2)$ coincides (up to normalization) with the linking number of Legendrian shadows:

THEOREM 6.3 *Let $C = C_1 \cup C_2$ be a generic 2-component front and let $SJ^+(C_1, C_2)$ be the conformal invariant of spherical fronts. For the linking number lk of 2-component Legendrian shadow (C, gl_L) we have $SJ^+(C_1, C_2) = 1 - lk$.*

Proof. The proof repeats the one of Proposition 6.2: we substitute Legendrian gleams $gl_L(f)$ into (10) (with $\chi(\pi) = 2$) and regroup the gleams as above. The contribution of the local gleams to lk equals $-\operatorname{ind}_{C_1}(d)\operatorname{ind}_{C_2}(d)$ for each self-crossing d of C_1 or C_2 and $-\operatorname{ind}_{C_1}(d)\operatorname{ind}_{C_2}(d)-\frac{1}{4}$ for each crossing $d \in C_1 \cap C_2$. Therefore

$$lk = \sum_{f \in F}\operatorname{ind}_{C_1}(f)\operatorname{ind}_{C_2}(f)\chi(f) - \sum_{d \in D}\operatorname{ind}_{C_1}(d)\operatorname{ind}_{C_2}(d)$$
$$- \frac{n_{12}}{4} - \frac{1}{2}\operatorname{rot}(C_1)\operatorname{rot}(C_2).$$

Recall that the conformal invariant SJ^+ is related to J^+ by $SJ^+(C_1, C_2) = J^+(C_1, C_2) + \frac{1}{2}\operatorname{index}(C_1)\operatorname{index}(C_2)$; thus the theorem follows from (8) and Proposition 6.2. $\qquad\square$

COROLLARY 2 *Let C be a generic curve and let $SJ^+(C)$ be the conformal invariant of spherical curves. For the self-linking number l of the Legendrian shadow (C, gl_L) we have $SJ^+(C) = 1 - l$.*

7 Quantum J^+-type invariants of fronts

Quantum invariants of shadows were introduced in [19] via a modification of quantum link invariants [10], [15] derived from the quantum group $U_q(\mathrm{sl}_2)$; more general and formal construction including, in particular, invariants derived from any quantum group $U_q(\mathcal{G})$, was established later in [20]. An application of this construction to Legendrian shadows produces quantum $J^{\pm}$-type invariants of curves and fronts on surfaces. For the sake of simplicity we restrict ourselves, following [19], to the data used for quantum $U_q(\mathrm{sl}_2)$ invariants of shadows; see [20] for more general setting.

7.1 Invariants of shadows

Let us review (with minor modifications) parts of Turaev's construction [19] needed for our purposes; see [19] for details.

Let I be a (possibly infinite) set (so-called set of "colors") with a distinguished element 0 and a fixed set Adm of unordered triples $i, j, k \in I$ called *admissible triples*. Denote by $\delta : I^3 \to 0, 1$ the characteristic function of Adm. We require that $\delta(0, i, j) = \delta_{i,j}$ and for fixed $i, j \in I$ there is only a finite number of $k \in I$ s.t. $\delta(i, j, k) \neq 0$. Assume that with each 6-tuple (i, j, k, l, m, n) is associated a *symbol* of this tuple $\begin{vmatrix} i & j & k \\ l & m & n \end{vmatrix} \in \mathbb{C}$, which satisfies the symmetry

$\begin{vmatrix} i & j & k \\ l & m & n \end{vmatrix} = \begin{vmatrix} l & n & j \\ i & k & m \end{vmatrix}$ and equals 0 unless $\delta(i, j, k) = 1$. We say that this algebraic data satisfies the *invariance conditions*, if there exist $v_i, u_i \in \mathbb{C}$, $i \in I$ s.t.

for any i, j, k, l, m, n, n', p, r, s, t the following holds:

$$\sum_k v_k \begin{vmatrix} i & j & k \\ l & m & n \end{vmatrix} \cdot \begin{vmatrix} l & j & n' \\ i & m & k \end{vmatrix} = \delta_{n,n'}\delta(i,j,k) \tag{14}$$

$$\sum_k v_k \exp(u_k + u_n + u_r + u_t) \cdot \begin{vmatrix} i & j & k \\ l & m & n \end{vmatrix} \cdot \begin{vmatrix} p & j & r \\ j & s & k \end{vmatrix} \cdot \begin{vmatrix} p & k & s \\ l & t & m \end{vmatrix} =$$

$$\sum_k v_k \exp(u_k + u_j + u_m + u_s) \cdot \begin{vmatrix} p & n & k \\ i & t & m \end{vmatrix} \cdot \begin{vmatrix} p & j & r \\ l & k & n \end{vmatrix} \cdot \begin{vmatrix} i & r & s \\ l & t & k \end{vmatrix} \tag{15}$$

$$\sum_{n,n'} v_n v_{n'} \exp(u_n - u_{n'}) \cdot \begin{vmatrix} i & j & k \\ i & n & k \end{vmatrix} \cdot \begin{vmatrix} i & j & k \\ i & n' & k \end{vmatrix} = \delta(i,j,k) \tag{16}$$

There are different examples of algebraic data satisfying conditions (14)–(16). The most usual example (e.g. [10], [19]) is given by quantum $6j$-symbols derived from the representation theory of the quantum group $U_q(\mathrm{sl}_2)$ (see [10], [19] for the explicit formulae). For generic q the set $I = 0, \frac{1}{2}, 1, \frac{3}{2}, \ldots$ is infinite; for $q = e^{\frac{\pi i h}{2r}}$ being a primitive $4r$-th root of unity the set $I = 0, \frac{1}{2}, 1, \ldots \frac{r-2}{2}$ is finite. A generalization of the algebraic setting and conditions above (allowing, in particular, to incorporate in the same way $6j$-symbols derived from other quantum groups) can be found in [20]. Another, more recent, simple example was given in [9] by studying the $6j$-symbols derived from quantum dilogarithm.

A *coloring* of some set J is a function $J \to I$; in particular, a *colored curve* is a curve C in Σ with an element of I assigned to each component C_i of C and to each boundary region of Σ if $\partial\Sigma \neq \emptyset$. In a similar way, an *area-coloring* is a function $\eta : F - F_\partial \to I$, where F is the set of all regions and F_∂ is the set of boundary regions of Σ. Let (C, gl) be a colored shadow (i.e. shadow with colored curve C) and let η be an area coloring of C. Pick a crossing d of, say, i and l colored components. The *weight* $w_\eta(d)$ of d is defined as the symbol corresponding to the 6-tuple (i, j, k, l, m, n), where j, k, m, n are the colors of four adjacent regions, see Fig 8a. Put

$$|(C, gl)|_\eta = \prod_d w_\eta(d) \times \prod_f v_{\eta(f)}^{\chi(f)} \exp(2u_{\eta(f)}gl(f)).$$

THEOREM 7.1 ([19], [20]; CF. [10]) *Let Σ be an oriented surface and (C, gl) be a colored shadow on Σ. Let either I be a finite set or $\partial\Sigma \neq \emptyset$. The state sum $|(C, gl)| = \sum_\eta |(C, gl)|_\eta$ is invariant under the framed shadow moves S_1'–S_3.*

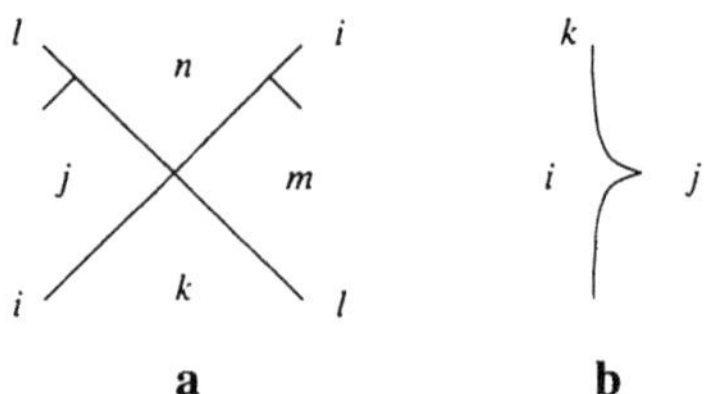

Figure 8: Colored crossing and cusp

7.2 Quantum J^+-type invariants of fronts

Modifying this construction to Legendrian shadows (C, gl_L) of fronts (and re-grouping Legendrian gleams back to the crossings and cusps) we get, in the notations introduced above, the following assignment of weights. Let C be a front with colored components on a surface Σ (see Remark 5.2) and η be an area-coloring. The weight of a crossing $d \in C$ of i and l colored components, as depicted in Fig 8a, is defined by

$$W_\eta(d) = \exp(-u_k - u_n)w_\eta(d).$$

The weight of a cusp c looking from, say, i to j colored region, as depicted in Fig 8b, is defined by

$$W_\eta(c) = v_i^{-\frac{1}{2}} v_j^{\frac{1}{2}} \exp(-u_i + u_j).$$

Finally, define

$$|C|_\eta = \prod_d W_\eta(d) \times \prod_c W_\eta(c) \times \prod_f v_{\eta(f)}^{\chi(f)} \exp(2u_{\eta(f)}).$$

From the results of Sections 5.4, 5.5 and Theorem 7.1 we conclude that the state sum constructed from the Legendrian shadow (C, gl_L) remains invariant under the shadow moves S_1'–S_5, thus

THEOREM 7.2 *Let Σ be an oriented surface and C be a colored front on Σ. Let either I be finite or $\partial \Sigma \neq \emptyset$. The state sum $|C| = \sum_\eta |C|_\eta$ is invariant under the safe tangency, triple point, cusp crossing and birth moves (hence gives an invariant of the corresponding Legendrian link in the space $ST^*\Sigma$ of contact elements).*

REMARK 7.3 One can treat the case of $\Sigma = S^2$ in a similar way. Indeed, cut out a point in $S^2 - C$ and apply the previous theorem. Easy to show that the resulting state sum does not depend on the choice of coloring for the boundary region (and on its choice).

REMARK 7.4 For a curve on Σ one may construct an abstract shadow (i.e. not corresponding to any link), taking all gleams of all finite regions to be 0. As the construction of the state sum $|(C, gl)|$ above can be applied in this situation, this gives a quantum invariant of a curve. It is interesting to note, that the orthogonality condition (14) implies that this invariant is of St-type, i.e. is invariant under both types of tangencies. Condition (15) is not necessary in this case (as it was needed only to provide invariance under S_3), so any data satisfying (14) and (16) gives a quantum invariant of St-type. We plan to explain this subject in more details elsewhere.

ACKNOWLEDGMENT   I benefited much from numerous discussions with O. Viro. I am grateful to A. Shumakovitch and S. Tabachnikov for stimulating remarks.

References

[1] V.I. Arnold, *Topological invariants of plane curves and caustics*, University lecture series (Providence RI) **5** (1994); *Plane curves, their invariants, perestroikas and classifications*, Singularities and bifurcations (ed. V.I. Arnold), Adv. Sov. Math. bf 21 (1994) 33–91.

[2] V.I. Arnold, *Geometry of spherical curves and the algebra of quaternions*, Rus. Math. Surv. **50**, 1(301), (1995), 3–68.

[3] D. Bennequin, *Entrelacements et equations de Pfaff*, Astérique bf 107–108 (1983), 83–161.

[4] U. Burri, *For a fixed Turaev shadow Jones' Vassiliev invariants depend polynomially on the gleams*, Preprint Mat. Inst. Univ. Basel (1995).

[5] S. Chmutov, V. Goryunov, *Kauffman bracket of plane curves*, Comm. Math. Phys. (to appear).

[6] S. Chmutov, V. Goryunov, *Regular Legendrian knots and the HOMFLY polynomial of immersed plane curves*, Preprint Univ. of Liverpool **4-96** (1996); *Polynomial invariants of Legendrian links and wave fronts*, Proc. Conf. on Knot Theory, Waseda Univ. (to appear).

[7] G. Cairns, M. McIntyre, *A new formula for winding number*, Geom. Dedicata, **46** (1993), 149–160.

[8] M. Goussarov, *Interdependent modifications of links and invariants of finite degree*, Preprint Uppsala Univ. UUDM-1995:26 (1995)

[9] R.M. Kashaev, *Quantum dilogarithm as a 6j-symbol*, Modern Phys. Let. A bf 9, no. 40 (1994) 3757–3768.

[10] A.N. Kirillov, N. Reshetikhin, *Representations of the algebra $U_q(sl_2)$, q-orthogonal polynomials and invariants of links*, In: Infinite dimensional Lie algebras and groups (ed. V.G. Kac), Adv.Ser. in Math. Phys. **7** (1988) 285–339.

[11] G. Mikhalkin, M. Polyak, *Whitney formula in higher dimensions*, J. Diff. Geom., to appear.

[12] M. Polyak, *Invariants of plane curves and fronts via Gauss diagrams*, Preprint MPI 1994-116 (1994).

[13] M. Polyak, *On the Bennequin invariant of Legendrian curves and its quantization*, Comp. Rend. Ac. Sci. Paris **322**, Série **I** (1996), 77–82.

[14] A.V. Pukhlikov, A.G. Khovanskii, *Finitely additive measures of virtual polytopes*, St. Petersburg Math. J., **4** (1993), 337–356.

[15] N. Reshetikhin, V. Turaev, *Ribbon graphs and their invariants derived from quantum groups*, Comm. Math. Phys. **127** (1990), 1–26.

[16] A. Shumakovitch, *Shadow formula for the Vassiliev invariant of degree two*, Topology, to appear.

[17] S. Tabachnikov, *Computation of the Bennequin invariant of a Legendrian curve from the geometry of its front*, Func. Anal. Appl. **22** (1988) no. 3, 89–90.

[18] V. Tchernov, *First degree Vassiliev invariants of knots in $\mathbb{R}^1$- and S^1-fibrations*, preprint Uppsala Univ. (1996).

[19] V. Turaev, *Quantum invariants of 3-manifolds and a glimpse of shadow topology*, Comp. Rend. Ac. Sci. Paris **313**, Ser. I (1991), 395–398; *Shadow links and face models of statistical mechanics*, J. Diff. Geom., **36** (1992), 35–74.

[20] V. Turaev, *Quantum invariants of knots and 3-manifolds*, de Gruyter (1994).

[21] O. Viro, *Some integral calculus based on Euler characteristic*, Lect. Notes Math., **1346** (1988), 127–138.

[22] O. Viro, *Generic immersions of the circle to surfaces and the complex topology of real algebraic curves*, AMS Transl. (2), **173** (1996), 231–252.

Progress in Mathematics

Edited by:

H. Bass
Columbia University
New York
10027
U.S.A.

J. Oesterlé
Dépt. de Mathématiques
Université de Paris VI
4, Place Jussieu
75230 Paris Cedex 05, France

A. Weinstein
Dept. of Mathematics
University of CaliforniaNY
Berkeley, CA 94720
U.S.A.

Progress in Mathematics is a series of books intended for professional mathematicians and scientists, encompassing all areas of pure mathematics. This distinguished series, which began in 1979, includes research level monographs, polished notes arising from seminars or lecture series, graduate level textbooks, and proceedings of focused and refereed conferences. It is designed as a vehicle for reporting ongoing research as well as expositions of particular subject areas.

If you have any concerns about our products,
you can contact us on
ProductSafety@springernature.com

In case Publisher is established outside the EU,
the EU authorized representative is:
Springer Nature Customer Service Center GmbH
Europaplatz 3, 69115 Heidelberg, Germany

Printed by Libri Plureos GmbH
in Hamburg, Germany